Dominique Noguet · Klaus Moessner
Jacques Palicot (Eds.)

Cognitive Radio Oriented Wireless Networks

11th International Conference, CROWNCOM 2016
Grenoble, France, May 30 – June 1, 2016
Proceedings

 Springer

Editors
Dominique Noguet
CEA-LETI
Grenoble
France

Klaus Moessner
University of Surrey
Guildford
UK

Jacques Palicot
CentraleSupélec – Campus de Rennes
Institut d'Electronique et de
   Télécommunications de Rennes,
   UMR CNRS 6164
Cesson-Sévigné
France

ISSN 1867-8211   ISSN 1867-822X   (electronic)
Lecture Notes of the Institute for Computer Sciences, Social Informatics
and Telecommunications Engineering
ISBN 978-3-319-40351-9   ISBN 978-3-319-40352-6   (eBook)
DOI 10.1007/978-3-319-40352-6

Library of Congress Control Number: 2016940880

Printed on acid-free paper

This Springer imprint is published by Springer Nature
The registered company is Springer International Publishing AG Switzerland

CROWNCOM 2016

Preface

The 11th EAI International Conference on Cognitive Radio Oriented Wireless Networks (CROWNCOM 2016) was hosted by CEA-LETI and was held in Grenoble, France, the capital of the Alps. 2016 was the 30th anniversary of wireless activities at CEA-LETI, but Grenoble has a more ancient prestigious scientific history, still very fresh in the signal processing area as Joseph Fourier worked there on his fundamental research, and established the Imperial Faculty of Grenoble in 1810, now the Joseph Fourier University.

This year, the main themes of the conference centered on the application of cognitive radio to 5G and to the Internet of Things (IoT). According to the current trend, the requirements of 5G and the IoT will increase demands on the wireless spectrum, accelerating the spectrum scarcity problem further. Both academic and regulatory bodies have focused on dynamic spectrum access and or dynamic spectrum usage to optimize scarce spectrum resources. Cognitive radio, with the capability to flexibly adapt its parameters, has been proposed as the enabling technology for unlicensed secondary users to dynamically access the licensed spectrum owned by legacy primary users on a negotiated or an opportunistic basis. It is now perceived in a much broader paradigm that will contribute toward solving the resource allocation problem that 5G requirements raise.

The program of the conference was structured to address these issues from the perspectives of industry, regulation bodies, and academia. In this transition period, where several visions of 5G and spectrum usage coexist, the CROWNCOM Committee decided to have a very strong presence of keynote speeches from key stakeholders. We had the pleasure of welcoming keynotes from industry 5G leaders such as Huawei and Qualcomm, the IoT network operator Sigfox, and the European Commission with perspectives on policy and regulation. We were also honored to welcome prestigious academic views from the Carnegie Mellon University and Zhejiang University. Along with the keynote speeches, we had the opportunity to debate the impact of massive IoT deployments in future spectrum use with a panel gathering high-profile experts in the field, thanks to the help of the conference panel chair.

The committee also wanted to emphasize how cognitive radio techniques could be applied in different areas of wireless communication in a pragmatic way, in a context where cognitive radio is often perceived as a theoretical approach. With this in mind, a significant number of demonstrations and exhibits were presented at CROWNCOM. Seven demonstrations and two exhibition booths were present during the conference, showcasing the maturity level of the technology. In this regard, we would like to express our gratitude to the conference demonstration and exhibit chair for his outstanding role in the organization of the demonstration area in conjunction with the conference local chairs. In addition, we decided to integrate a series of workshops that can be seen as special sessions where specific aspects of cognitive radio regarding applications, regulatory frameworks, and research were discussed. Four such workshops were organized as part of the conference embracing topics such as modern spectrum management, 5G technology enabler, software-defined networks and virtualization, and cloud technologies. This year, delegates could also attend three tutorials organized as part of the conference. We are thankful to the tutorial chair for organizing these tutorials.

Of course, the core of the conference was composed of regular paper presentations covering the seven tracks of the conference topics. We received a fair number of submissions, out of which 62 high-quality papers were selected. The paper selection was the result of a rigorous and high-standard review process involving more than 150 Technical Program Committee (TPC) members with a rejection rate of 25 %. We are grateful to all TPC members and the 14 track co-chairs for providing high-quality reviews. This year, the committee decided to reward the best papers of the conference in two ways. First, the highest-ranked presented papers were invited to submit an extended version of their work to a special issue of the *EURASIP Journal on Wireless Communications and Networking* (Springer) on "Dynamic Spectrum Access and Cognitive Techniques for 5G." This special issue is planned for publication at the end of 2016. Then, a smaller number of papers with the highest review scores competed for the conference "Best Paper Award," sponsored by Orange. The committee would like to thank the authors for submitting high-quality papers to the conference, thereby maintaining COWNCOM as a key conference in the field.

Of course the organization of this event would have been impossible without the constant guidance and support of the European Alliance for Innovation (EAI). We would like to warmly thank the organization team at EAI and in particular Anna Horvathova, the conference manager of CROWNCOM 2016. We would also like to express our thanks to the team at CEA: our conference secretary and the conference webchair for their steady support and our local chairs for setting up all the logistics.

Finally, the committee would like express its gratitude to the conference sponsors. In 2016, CROWNCOM was very pleased by the support of a large number of prestigious sponsors, stressing the importance of the conference for the wireless community. Namely, we express our warmest thanks to Keysight Technologies, Huawei, Nokia, National Instruments, Qualcomm, and the European ForeMont project.

We hope you will enjoy the proceedings of CROWNCOM 2016.

April 2016

Dominique Noguet
Klaus Moessner
Jacques Palicot

Lecture Notes of the Institute for Computer Sciences, Social Informatics and Telecommunications Engineering 172

More information about this series at http://www.springer.com/series/8197

Organization

Steering Committee

Imrich Chlamtac	Create-Net, EAI, Italy
Thomas Hou	Virginia Tech, USA
Abdur Rahim Biswas	Create-Net, Italy
Tao Chen	VTT – Technical Research Centre of Finland, Finland
Tinku Rasheed	CREATE-NET, Italy
Athanasios Vasilakos	Kuwait University, Kuwait

Organizing Committee

General Chair

Dominique Noguet   CEA-LETI, France

Conference Secretary

Sandrine Bertola   CEA-LETI, France

Technical Program Chair

Klaus Moessner   University of Surrey, UK

Publicity and Social Media Chairs

Emilio Calvanese-Strinati	CEA-LETI, France
Masayuki Ariyoshi	NEC, Japan
Apurva Mody	BAE systems, USA

Publication Chair

Jacques Palicot   Centrale Supélec, France

Keynote and Panel Chair

Paulo Marques   Instituto de Telecomunicacoes, Portugal

Workshop Chair

Jordi Pérez-Romero   UPC, Spain

Tutorial Chair

Sofie Polin   KU Leuven, Belgium

Exhibition and Demonstration Chair

Benoit Miscopein   CEA-LETI, France

Local Chairs

Pascal Conche   CEA, France
Audrey Scaringella   CEA, France

Web Chair

Yoann Roth   CEA-LETI, France

Technical Program Committee Chairs

Track 1: Dynamic Spectrum Access/Management and Database

Co-chairs

Oliver Holland   King's College, London, UK
Kaushik Chowdhury   Northeastern University, USA

Track 2: Networking Protocols for Cognitive Radio

Co-chairs

Luca De Nardis   Sapienza University of Rome, Italy
Christophe Le Martret   Thales Communications & Security, France

Track 3: PHY and Sensing

Co-chairs

Friedrich Jondral   Karlsruhe Institute of Technology, Germany
Stanislav Filin   NICT, Japan

Track 4: Modelling and Theory

Co-chairs

Panagiotis Demestichas   University of Piraeus, Greece
Danijela Cabric   UCLA, USA

Track 5: Hardware Architecture and Implementation

Co-chairs

Seungwon Choi   Hanyang University, Korea
Olivier Sentieys   Inria, France

Track 6: Next Generation of Cognitive Networks

Co-chairs

Bernd Bochow   FhG Focus, Germany
Zaheer Khan   University of Oulu, Finland

Track 7: Standards, Policies, and Business Models

Co-chairs

Martin Weiss	University of Pittsburgh, USA
Baykas Tuncer	Istanbul Medipol University, Turkey

Technical Program Committee Members

Hamed Ahmadi	University College Dublin, Ireland
Adnan Aijaz	Toshiba Research European Labs, UK
Ozgur Barış Akan	Koc University, Turkey
Anwer Al-Dulaimi	University of Toronto, Canada
Mohammed Altamimi	Communications and IT Commission, Saudi Arabia
Onur Altintas	Toyota, Japan
Osama Amin	KAUST, Saudi Arabia
Xueli An	European Research Centre, Huawei Technologies, Germany
Peter Anker	Netherlands Ministry of Economic Affairs, The Netherlands
Angelos Antonopoulos	Telecommunications Technological Centre of Catalonia (CTTC), Spain
Ludovic Apvrille	Telecom-ParisTech, France
Masayuki Ariyoshi	NEC, Japan
Stefan Aust	NEC, Japan
Faouzi Bader	Centrale Supélec Rennes, France
Arturo Basaure	Aalto University, Finland
Bernd Bochow	Fraunhofer Focus, Germany
Tadilo Bogale	Institut National de recherche Scientifique, Canada
Doug Brake	Information Technology and Innovation Foundation, USA
Omer Bulakci	Huawei, Germany
Carlos E. Caicedo	Syracuse University, USA
Emilio Calvanese-Strinati	CEA-LETI, France
Chan-Byoung Chae	Yonsei University, Korea
Ranveer Chandra	Microsoft, USA
Periklis Chatzimisios	Alexander TEI of Thessaloniki, Greece
Pravir Chawdhry	Joint Research Center EC, Italy
Luiz Da Silva	Virginia Tech, USA
Antonio De Domenico	CEA-LETI, France
Luca De Nardis	Sapienza University of Rome, Italy
Thierry Defaix	DGA-MI, France
Simon Delaere	iMinds, Belgium
Panagiotis Demestichas	University of Piraeus, Greece
Marco Di Felice	University of Bologna, Italy
Rogério Dionisio	Instituto de Telecomunicações, Portugal
Jean-Baptiste Doré	CEA-LETI, France

Marc Emmelmann	Fraunhofer FOKUS, Germany
Ozgur Ergul	Koc University, Turkey
Serhat Erkucuk	Kadir Has University, Turkey
Takeo Fujii	University of Electro-Communications, Japan
Piotr Gajewski	Military University of Technology, Poland
Yue Gao	Queen Mary University of London, UK
Matthieu Gautier	IRISA, France
Liljana Gavrilovska	Ss Cyril and Methodius University of Skopje, Republic of Macedonia
Andrea Giorgetti	WiLAB, University of Bologna, Italy
Michael Gundlach	Nokia, Germany
Hiroshi Harada	Kyoto University, Japan
Ali Hassa	NUST School of Electrical Engineering and Computer Science (SEECS), Pakistan
Stefano Iellamo	Crete (at ICS-FORTH), Greece
Florian Kaltenberger	Eurecom, France
Du Ho Kang	Ericsson, Sweden
Pawel Kaniewski	Military Communication Institute, Poland
Mika Kasslin	Nokia, Finland
Shuzo Kato	Tohoku University, Japan
Seong-Lyun Kim	Yonsei University, Korea
Adrian Kliks	Poznan University of Technology, Poland
Heikki Kokkinen	Fairspectrum, Finland
Kimon Kontovasilis	NCSR Demokritos, Greece
Pawel Kryszkiewicz	Poznan University of Technology, Poland
Samson Lasaulce	L2S-CNRS, France
Vincent Le Nir	Royal Military Academia, Belgium
Didier Le Ruyet	CNAM, France
Per H. Lehne	Telenor, Norway
William Lehr	MIT, USA
Janne Lehtomäki	University of Oulu, Finland
Zexian Li	Nokia Networks, Finland
Dan Lubar	RelayServices, USA
Irene MacAluso	Trinity College Dublin, Ireland
Allen, Brantley MacKenzie	Virginia Tech, USA
Milind Madhav Buddhikot	Alcatel Lucent Bell Labs, USA
Petri Mähönen	RWTH Aachen University, Germany
Vuk Marojevic	Virginia Tech, USA
Paulo Marques	Instituto de Telecomunicacoes, Portugal
Torleiv Maseng	FFI, Norway
Daniel Massicotte	UQTR, Canada
Raphaël Massin	Thales Communications & Security, France
Marja Matinmikko	VTT, Finland
Arturas Medeisis	ITU Representative, Saudi Arabia
Albena Mihovska	Aalborg University, Denmark
Christophe Moy	Centrale Supélec Rennes, France

Contents

Networking Protocols for Cognitive Radio

PHY and Sensing

Modelling and Theory

Hardware Architecture and Implementation

Next Generation of Cognitive Networks

Standards, Policies and Business Models

Workshop Papers

Dynamic Spectrum Access/Management and Database

A New Evaluation Criteria for Learning Capability in OSA Context

Navikkumar Modi[1(✉)], Christophe Moy[1], Philippe Mary[2],
and Jacques Palicot[1]

[1] CentraleSupelec/IETR, Avenue de la Boulaie, 35576 Cesson Sevigne, France
{navikkumar.modi,christophe.moy,jacques.palicot}@centralesupelec.fr
[2] INSA de Rennes, IETR, UMR CNRS 6164, 35043 Rennes, France
philippe.mary@insa-rennes.fr

Abstract. The activity pattern of different primary users (PUs) in the spectrum bands has a severe effect on the ability of the multi-armed bandit (MAB) policies to exploit spectrum opportunities. In order to apply MAB paradigm to opportunistic spectrum access (OSA), we must find out first whether the target channel set contains sufficient structure, over an appropriate time scale, to be identified by MAB policies. In this paper, we propose a criteria for analyzing suitability of MAB learning policies for OSA scenario. We propose a new criteria to evaluate the structure of random samples measured over time and referred as Optimal Arm Identification (OI) factor. OI factor refers to the difficulty associated with the identification of the optimal channel for opportunistic access. We found in particular that the ability of a secondary user to learn the activity of PUs spectrum is highly correlated to the OI factor but not really to the well known LZ complexity measure. Moreover, in case of very high OI factor, MAB policies achieve very little percentage of improvement compared to random channel selection (RCS) approach.

Keywords: Cognitive radio · Opportunistic spectrum access · Reinforcement learning · Multi-armed bandit · Lempel-Ziv (LZ) complexity · Optimal Arm Identification (OI) factor

1 Introduction

Spectrum learning and decision making is a core part of the cognitive radio (CR) to get access to the underutilized spectrum when not occupied by a licensed or primary users (PUs). In particular, we deal with multi-armed bandit (MAB) paradigm, which allows unlicensed or secondary user (SU) to make action to select a free channel to transmit when no PUs are using it, and finally learns about the *optimal channel*, i.e. channel with the highest probability of being vacant, in the long run [1–3].

The PU activity pattern, i.e. presence or absence of PU signal in the spectrum band, can be modeled as a 2-state Markov Process [3–5]. In case of SU trying to learn about the probability for a channel to be vacant, the success of

© ICST Institute for Computer Sciences, Social Informatics and Telecommunications Engineering 2016
D. Noguet et al. (Eds.): CROWNCOM 2016, LNICST 172, pp. 3–14, 2016.
DOI: 10.1007/978-3-319-40352-6_1

MAB policy is affected by the amount of structure obtained in the PUs activity pattern in these channels [6]. There exist several pieces of work on opportunistic spectrum access (OSA), by means of learning and predicting an opportunity, which deals with CR to find out the PUs activity pattern. In this paper, we address the following fundamental questions, affecting MAB performance: (i) when is it advantageous to apply MAB learning framework to address the problem of opportunistic access for SU? (ii) Is the use of MAB policies for OSA is justified over a simple non-intelligent approach?

In almost all cases, the performance of MAB learning policies has been studied with respect to PUs activity level, i.e. probability that PUs occupy radio channels [2,3]. In some cases, spectrum utilization is modeled as an independent and identically distributed (i.i.d) process [1], and it does not take into account the likely sequential activity patterns of PUs in the spectrum. To address the first question raised above, the Lempel-Ziv (LZ) complexity was introduced in [6] to characterize the PU activity pattern for general reinforcement learning (RL) problem. However, MAB paradigm is a special kind of RL game where SU maximizes its long term reward by making action to learn about the optimal channel, opposed to general RL framework where SU is interacting with a system by making actions and learns about the underlying structure of the system. Moreover, in this paper, we propose the *Optimal Arm Identification (OI) factor* to identify the difficulty associated with prediction of an optimal channel having highest probability of being vacant from the set of channels. Finally, the last question raised above is answered by comparing the performance of MAB policies against the random channel selection (RCS) approach (a non-intelligent approach).

We found out that, for several spectrum utilization patterns, MAB policies can be beneficial compared to non-intelligent approach, but the percentage of improvement is highly correlated with the level of OI factor and very little affected by the level of LZ complexity. This result does not just emphasize aphorism that performance of MAB policies in OSA framework depends extremely on the OI factor associated with the selected channel set. The work presented in this paper can be the answer to the question raised in several papers about the effectiveness of MAB framework for OSA scenario. The remainder of this paper is organized as follows. In Sect. 2, we introduce MAB framework and RL policies which are used to verify the effect of PUs activity pattern on spectrum learning performance. Section 3 and 4 contain our main contributions where in Sect. 3, LZ complexity is revisited and a new criteria measuring the structure of spectrum utilization pattern, named OI, is introduced and in Sect. 4, numerical results giving the efficiency of MAB policies w.r.t. the output of LZ and OI criteria are presented. Finally, Sect. 5 concludes the paper.

2   System Model and RL Policies

We consider a network with a single[1] secondary transceiver pair (Tx-Rx) and a set of channel $\mathbb{K} = \{1, \cdots, K\}$. SU can access one of the K channels if it is not

---

[1] The presented analysis can also be justified for multiple SUs scenario, where each SU tries to find optimal channel following underlying activity pattern of PUs.

occupied by PUs. The i-th channel is modeled by an irreducible and aperiodic discrete time Markov chain with finite state space S^i. $P^i = \{p^i_{kl}, (k, l \in \{0, 1\})\}$ denotes the state transition probability matrix of the i-th channel, where 0 and 1 are the Markov states, i.e. occupied and free respectively. Let, $\boldsymbol{\pi}^i$ be the stationary distribution of the Markov chain defined as:

$$\boldsymbol{\pi}^i = [\pi^i_0, \pi^i_1] = \left[\frac{p^i_{10}}{p^i_{10} + p^i_{01}}, \frac{p^i_{01}}{p^i_{10} + p^i_{01}}\right]. \tag{1}$$

$S^i(t)$ being the state of the channel i at time t and $r^i(t) \in \mathbb{R}$ is the reward associated to the band i. Without loss of generality we can assume, that $r^i(t) = S^i(t)$, i.e. $S^i(t) = 1$ if sensed free and $S^i(t) = 0$ if sensed occupied. The stationary mean reward μ^i of the i-th channel under stationary distribution π^i is given by: $\mu^i = \pi^i$. A channel is said optimal when it has the highest mean reward μ^{i^*}, such that $\mu^{i^*} > \mu^i$ and $i^* \neq i, i \in \{1, \cdots, K\}$, i.e. a channel with the highest probability to be vacant. The mean reward optimality gap is defined as $\Delta_i = \mu^{i^*} - \mu^i$. The regret $R(t)$ of a MAB policy up to time t, is defined as the reward loss due to selecting sub-optimal channel μ^i:

$$R(t) = t\mu^{i^*} - \sum_{m=0}^{t} r(m), \tag{2}$$

2.1 Reinforcement Learning (RL) Approaches

We consider two different reinforcement learning (RL) strategies, i.e. UCB1 and Thomson-Sampling (TS), in order to evaluate the learning efficiency of MAB policies on channel set containing different PUs activity pattern. These policies are based on RL algorithms introduced in [7–9] as an approach to solve MAB problem and they attempt to identify the most vacant channel in order to maximize their long term reward. Figure 1 illustrates a realization of the random process: 'occupancy of spectrum bands by PUs'. In this figure, all channels do not have the same occupancy ratio and it seems intuitively clear that the more different the channel occupations are, the easier the learning.

Upper Confidence Bound (UCB) Policy. It has been shown previously in [1] that UCB1 allows spectrum learning and decision making in OSA context in order to maximize the transmission opportunities. UCB1 is a RL based policy, learning about the optimal channel from previously observed rewards starting from scratch, i.e. without any *a priori* knowledge on the activity within the set of channels. For each time t, UCB1 policy updates indices named as $B_{t,i,T_i(t)}$, where $T_i(t)$ is the number of times the i-th channel has been sensed up to time t, and returns the channel index $a_t = i$ of the maximum UCB1 index. UCB1 is detailed in Algorithm 1 where α is the exploration-exploitation coefficient. If α increases, the bias $A_{t,i,T_i(t)}$ dominates and UCB1 policy explores new channels. Otherwise, if α decreases, the index computation is governed by $\bar{X}_{i,T_i(t)}$ and the policy tends to exploit the previously observed optimal channel.

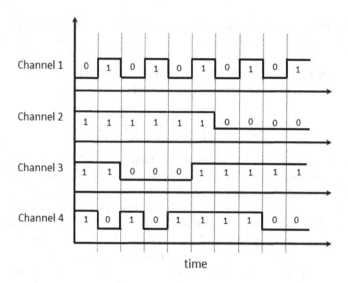

Fig. 1. PUs activity pattern

Algorithm 1. UCB1 policy

Input: K, α
Output: a_t
 1: **for** $t = 1$ to n **do**
 2:    **if** $t \leq K$ **then**
 3:       $a_t = t + 1$
 4:    **else**
 5:       $T_i(t) = \sum\limits_{m=0}^{t-1} \mathbb{1}_{a_m=i}, \forall i$
 6:       $\bar{X}_{i,T_i(t)} = \dfrac{\sum\limits_{m=0}^{t-1} S^i(m)\mathbb{1}_{a_m=i}}{T_i(t)}, \quad A_{t,i,T_i(t)} = \sqrt{\dfrac{\alpha \ln t}{T_i(t)}}, \forall i$
 7:       $B_{t,i,T_i(t)} = \bar{X}_{i,T_i(t)} + A_{t,i,T_i(t)}, \forall i$
 8:       $a_t = \arg\max_i(B_{t,i,T_i(t)})$
 9:    **end if**
10: **end for**

Thomson-Sampling (TS) Policy. Introduced in [8,9] and detailed in Algorithm 2, TS selects a channel having the highest $J_{t,i,T_i(t)}$ index, computed with the β function w.r.t. two arguments, i.e. $G_{i,T^i(t)} = \sum\limits_{m=0}^{t-1} S^i(m)\mathbb{1}_{a_m=i}$ and $F_{i,T^i(t)} = T^i(t) - G_{i,T^i(t)}$, where $T^i(t)$ has the same meaning than previously. The former argument is the total number of free state observed up to time t for channel i and the second is the total number of occupied state. For start, no prior knowledge on the mean reward of each channel is assumed i.e. uniform distribution and hence the index for all channels is set to $\beta(1,1)$. TS policy updates the distribution on mean reward μ^i as $\beta\left(G_{i,T^i(t)} + 1, F_{i,T^i(t)} + 1\right)$.

Algorithm 2. Thomson-Sampling (TS) policy

Input: K, $G_{i,1} = 0$, $F_{i,1} = 0$
Output: a_t
1: **for** $t = 1$ **to** n **do**
2:     $J_{t,i,T_i(t)} = \beta(G_{i,T^i(t)} + 1, F_{i,T^i(t)} + 1)$
3:     Sense channel $a_t = \arg\max_i \left(J_{t,i,T_i(t)}\right)$
4:     Observe state $S^i(t)$
5:     $T_i(t) = \sum_{m=0}^{t-1} \mathbb{1}_{a_m=i}, \forall i,$ $G_{i,T_i(t)} = \sum_{m=0}^{t-1} S^i(m)\mathbb{1}_{a_m=i}, \forall i$
6:     $F_{i,T_i(t)} = T_i(t) - G_{i,T_i(t)}, \forall i$
7: **end for**

---

3 PUs Activity Pattern vs Difficulties of Prediction

In general, multi-armed bandit (MAB) algorithms are evaluated with the PUs traffic load which characterizes the occupancy of the spectrum band. Intuitively, the higher occupancy of the channel by PUs, the more difficult the opportunistic access for SU will be. However, traffic load of the PUs is not sufficient for evaluating the efficiency of MAB policies. In fact, performance of MAB policies leverages on the structure of the PUs activity pattern and also on the difficulties associated with identification of the optimal channel, i.e. channel with optimal mean reward distribution μ^{i^*}. The ON/OFF PUs activity model approximates the spectrum usage pattern as depicted in Fig. 1. Moreover, if the separation between the mean reward distribution of the optimal and a sub-optimal channel is large, SU should be able to converge to the optimal channel faster, and thus achieves a higher number of opportunistic accesses. Therefore, estimating the amount of structure present in the PUs activity pattern is of essential interest for applying machine learning strategies to OSA.

3.1 Lempel-Ziv (LZ) Complexity

Lempel-Ziv (LZ) complexity was proposed in [11] as a measure for characterizing randomness of sequences. It has been widely adopted in several research areas such as biomedical signal analysis, data compression and pattern recognition. Lempel and Ziv, in [11], have associated to every sequence a complexity c which is estimated by looking at the sequence and incrementing c every time a new substring of consecutive symbols is available. Then c is normalized via the asymptotic limit $n/\log_2(n)$, where n is the length of the sequence. LZ complexity is a property of individual sequences and it can be estimated regardless of any assumptions about the underlying process that generated the data. In [6], the authors have applied the LZ definition to the production rate of new patterns in Markovian processes. This is of particular interest when PUs activity is modeled as Markov process to evaluate the efficiency of MAB policies. For an ergodic source, LZ complexity equals the entropy rate of the source, which for a Markov chain S is given by [6]:

$$h(S) = -\sum_{k,l} \pi_k p_{k,l} \log p_{k,l}, \qquad k,l \in \{0,1\}, \tag{3}$$

where $p_{k,l}$ is the transition probability between state k and l. System with LZ complexity equal to 1 implies very high rate of new patterns production and thus it could make difficult for the learning policy to predict the next sequence. For example in Fig. 1, channels 1 to 4 have different PUs activity pattern characterized by normalized LZ complexity of 0.05, 0.30, 0.60 and 0.66, respectively. It is clear that prediction of next vacancy is an easy task in case of channel 1 which has lower LZ complexity, whereas it becomes more and more difficult to predict next vacancy in channel 4 which has higher LZ complexity.

3.2 Optimal Arm Identification (OI) Factor

As stated before, performance of MAB policy applied to OSA context also leverages on the separation between optimal and sub-optimal channels mean reward distribution. Here, we define another criteria to characterize the difficulty for a MAB to learn the PUs spectrum occupancy. The MAB policy learns, based on past observations, which channel is optimal in term of mean reward distribution in the long run. The optimal arm identification for MAB framework has been studied since the 1950s under the name 'ranking and identification problems' [12,13].

In recent advances in MAB context, an important focus was set on a different perspective, in which each observation is considered as a reward: the user tries to maximize his cumulative reward. Equivalently, its goal is to minimize the expected regret $R(t)$, as defined in (2), up to time $t > 1$. As stated in [7,14], regret $R(t)$, defined as the reward loss due to the selection of sub-optimal channels, up to time t is upper bounded uniformly by a logarithmic function:

$$R(t) \leq a \sum_{i:\mu^i < \mu^{i^*}} \frac{\ln t}{(\mu^{i^*} - \mu^i)} + b \sum_{i:\mu^i < \mu^{i^*}} (\mu^{i^*} - \mu^i), \tag{4}$$

where a and b are constants independent from channel parameters and time t. As stated in (4), upper bound on regret of MAB policy is scaled by the change in mean reward optimality gap $\Delta_i = (\mu^{i^*} - \mu^i)$. Intuitively, decreasing Δ_i makes the upper bound looser and thus increases the uncertainty on MAB policies performance. In this paper, we propose the OI factor H_1 as a measure of difficulty associated with finding an optimal channel among several other channels:

$$H_1 = 1 - \sum_{i=1}^{K} \frac{(\mu^{i^*} - \mu^i)}{K}, \tag{5}$$

where, μ^i and μ^{i^*} are the mean reward distribution of sub-optimal and optimal channels respectively, K is the number of channels and i^* is the index of optimal channel. H_1 measures how close the mean reward of all sub-optimal channels are from the mean reward of the optimal channel. If H_1 is close to 1 then all channels have very closely distributed mean reward, thus it becomes almost impossible for learning policy to identify the optimal channel from the set of channels.

4   Impact of LZ Complexity and OI Factor on Prediction Accuracy

In this section, two MAB policies, i.e. UCB1 and Thomson-Sampling (TS), are investigated and the performance they achieve are put in correlation with the information given by LZ complexity and OI factor H_1. Markov chains with several levels of stationary distribution $\pi = [\pi_0, \pi_1]$, LZ complexity and H_1 factor are generated for further numerical analysis. For simulation convenience, some parameters need to be set. Indeed, (1) is an undetermined system with two unknowns p_{01} and p_{10}. Therefore, as a side step, we considered 9 different levels of π_1, i.e. probability of being vacant, as $0.1, 0.2, \cdots, 0.9$. For these values of π_1, we obtained 45 different transition probability matrices P, each corresponding to different LZ complexity. A total of $\binom{45}{5}$ combinations are obtained by considering 45 different transition probability matrices and $K = 5$ channels, and those correspond to various H_1 factor. Finally, MAB policies are applied to the randomly selected 2000 combinations from a total of $\binom{45}{5}$ combinations. Every point in each figure corresponds to one realization of MAB policies. For each realization, policy is executed over 10^2 iterations of 10^4 time slots each. Moreover, the exploitation-exploration coefficient in UCB1 is set to $\alpha = 0.5$ which is proved to be efficient for maintaining a good tradeoff between exploration and exploitation [10].

4.1   Probability of Success

Probability of success P_{Succ} is computed by considering the number of times vacant channel is explored over the number of iterations. The success probability depends on the probability P_f that there exists at least one free channel from the set of channels K. Considering that the channel occupation is independent from one channel to another, we have [6]:

$$P_f = 1 - \prod_{i=1}^{K} \pi_0^i, \tag{6}$$

where π_0^i is the probability that the i-th channel is occupied.

Figure 2(a) and (b) depict the probability of success P_{Succ} of UCB1 and TS policies, i.e. the probability that these policies access to a free channel, according to the probability that at least one channel is free, i.e. P_f and LZ complexity. In both figures, success probability increases with P_f for a given level of LZ complexity. However, in Fig. 2(a) and (b), for a given P_f, several values of LZ complexity lead to the same level of performance for UCB1 and TS algorithms. This reveals that LZ complexity is not really related to the ability of UCB1 and TS policies to learn the scenario. For instance in Fig. 2(a) and (b), SU is able to achieve more than 90 % of probability of success on a channel set with LZ complexity of 0.2 and $P_f = 0.98$, whereas it only achieve 75 % of probability of success on a channel set with LZ complexity of 0.6 and $P_f = 0.98$. In that case,

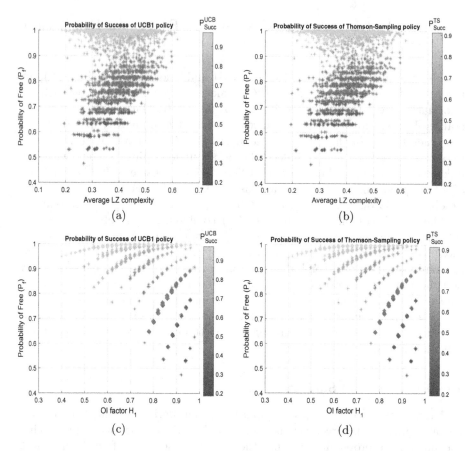

Fig. 2. (a), (b) Probability of success P_{Succ} of MAB policies, i.e. UCB1 and TS, with respect to the average LZ complexity and the probability of free P_f. Each point denotes a particular realization of MAB policies applied to $K = 5$ channels. The number of random combinations which we analyzed is 2000. (c), (d) Probability of success P_{Succ} of MAB policies, i.e. UCB1 and TS, with respect to the OI factor H_1 and the probability of free P_f applied to same set of channels.

the variation in the probability of success is up to 15 % however the variation of P_{Succ} along the x-axis can be even less important for lower values of P_f.

On the other hand, Fig. 2(c) and (d) show the probability of success of UCB1 and TS policies according to H_1 and the probability of free P_f. As we can see that H_1 is highly correlated to UCB1 and TS policies performance on a given scenario. In order to achieve very high level of P_{Succ}, H_1 required to be low. For instance in Fig. 2(c) and (d), SU is able to achieve more than 90 % of P_{Succ} on a channel set when H_1 is 0.4 and $P_f = 0.95$, whereas it only achieves 50 % of P_{Succ} on a channel set when $H_1 = 0.95$ and $P_f = 0.95$. Thus, we can state that P_{Succ}^{UCB} varies up to 40 % according to the changes in H_1, along x-axis, for certain values of P_f.

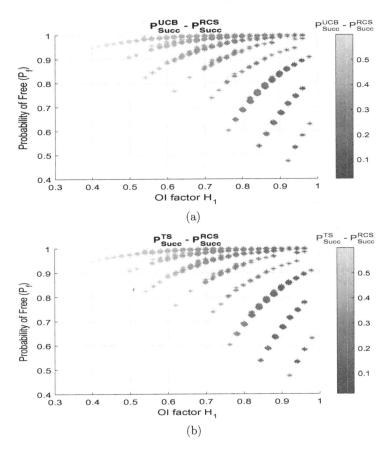

Fig. 3. (a), (b) Each point denotes the difference between the probability of success of MAB policies, i.e. UCB1 and TS, and the probability of success of the random channel selection (RCS) approach applied to $K = 5$ channels, respectively. 2000 random combinations have been analyzed.

4.2 Comparison with Random Channel Selection Policy

Figure 3(a) and (b) show difference between probability of success of MAB policies, i.e. UCB1 and TS, and random channel selection (RCS) approach. As expected, MAB policies outperform RCS approach in general, but difference becomes negligible for very high H_1 regime, i.e. the mean rewards of sub-optimal and optimal channels become equivalent. For a given P_f, performance of MAB policies decreases when H_1 increases. For instance in Fig. 3(a) and (b), we can notice that $P_{Succ}^{UCB} - P_{Succ}^{RCS}$ and $P_{Succ}^{TS} - P_{Succ}^{RCS}$ vary up to 50 % along the x-axis, i.e. H_1.

Figure 4 shows the average percentage of improvement in the probability of success achieved by MAB policies, i.e. UCB1 and TS, with respect to RCS approach under various PUs activity pattern. As we stated before, percentage

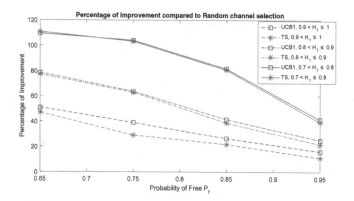

Fig. 4. Average percentage of improvement in the probability of success of MAB policies, i.e. UCB1 and TS, with respect to RCS policy as a function of the OI factor H_1 and the probability of free P_f. Each point denotes an average percentage of improvement achieved by MAB policies applied to several combinations of H_1 and P_f.

of improvement of MAB policies compared to RCS approach decreases when P_f increases, because RCS approach is able to find more opportunities in high P_f regime. On the contrary, average percentage of improvement of MAB policies also decreases when P_f decreases after certain limit. It is due to the fact that there are not many opportunities available to exploit for MAB policies in low P_f regime. As stated in Fig. 4, combinations with low H_1, i.e. $0.7 < H_1 \leq 0.8$, increases the percentage of improvement of MAB policies compared to RCS approach. Even for high H_1, i.e. $0.9 < H_1 \leq 1$, the relative improvement of learning policies is still noticeable, i.e. more than 15 %. It also reveals that all MAB policies, i.e. UCB1 and TS, achieve nearly same level of percentage of improvement for low H_1, i.e. $0.7 < H_1 \leq 0.8$, whereas in case of high H_1, i.e. $0.9 < H_1 \leq 1$, UCB1 policy significantly outperforms TS policy. Figures 3 and 4 prove that OI factor H_1 is rather well suitable compared to LZ complexity to analyze learning capability of MAB policies in OSA context.

5   Conclusions

While MAB policies, e.g. UCB1 and TS, are often assumed to be beneficial for OSA context, the problem of characterizing the scenarios where they are effective is barely studied. In this paper, we propose a new criteria, named OI factor, to characterize the situations where MAB policies will be good a priori. We evaluate the performance of UCB1 and TS on various scenarios, and correlate this to the output of OI factor and LZ complexity. Our findings show that LZ complexity does not give sufficient insights on how MAB policies behave on learning scenarios. On the other hand, OI factor is well connected to the percentage of success of MAB policies. Hence, we suggest to use OI factor in order to know if learning compared to random channel selection is beneficial for a given scenario or not. Moreover, MAB learning can achieve more than 50 % of improvement in the probability of success compared to the non-intelligent approach in scenarios presenting low OI factors.

Acknowledgments. This work has received a French government support granted to the CominLabs excellence laboratory and managed by the National Research Agency in the "Investing for the Future" program under reference No. ANR-10-LABX-07-01. The authors would also like to thank the Region Bretagne, France, for its support of this work. Authors would like to thank Hamed Ahmadi, from University College of Dublin, Ireland and CONNECT research center, for introducing Lempel-Ziv complexity to us. Authors would also like to thank Sumit J. Darak, from IIIT-Delhi, India, for introducing Thomson-Sampling policy to us.

References

1. Jouini, W., Ernst, D., Moy, C., Palicot, J.: Upper confidence bound based decision making strategies and dynamic spectrum access. In: proceedings of IEEE International Conference on Communications (ICC), pp. 1–5 (2010)
2. Liu, H., Liu, K., Zhao, Q.: Learning in a changing world: restless multiarmed bandit with unknown dynamics. IEEE Trans. Inf. Theor. **59**(3), 1902–1916 (2013)
3. Modi, N., Mary, P., Moy, C.: QoS driven channel selection algorithm for opportunistic spectrum access. In: Proceedings of IEEE GC 2015 Workshop on Advances in Software Defined Radio Access Networks and Context-aware Cognitive Networks, San Diego, USA (2015)
4. Rehmani, M.H., Viana, A.C., Khalife, H., Fdida, S.: Activity pattern impact of primary radio nodes on channel selection strategies. In: Proceedings of the 4th International Conference on Cognitive Radio and Advanced Spectrum Management (CogART 2011), Barcelona, Spain (2011)
5. Yuan, G., Grammenos, R.C., Yang, Y., Wang, W.: Performance analysis of selective opportunistic spectrum access with traffic prediction. IEEE Trans. Veh. Technol. **59**(4), 1949–1959 (2010)
6. Macaluso, I., Finn, D., Ozgul, B., DaSilva, L.A.: Complexity of spectrum activity and benefits of reinforcement learning for dynamic channel selection. IEEE J. Sel. Areas Commun. **31**(11), 2237–2248 (2013)
7. Auer, P., Cesa-Bianchi, N., Paul, F.: Finite-time analysis of the multiarmed bandit problem. J. Mach. Learn. **47**(2–3), 235–256 (2002)
8. Russo, D., Van Roy, B.: An information-theoretic analysis of Thompson sampling. Computing Research Repositor (2014)
9. Agrawal, S., Goyal, N.: Analysis of Thompson sampling for the multi-armed bandit problem. In: Proceedings of the 25th Annual Conference on Learning Theory (COLT) (2012)
10. Melián-Gutiérrez, L., Modi, N., Moy, C., Pérez-lvarez, I., Bader, F., Zazo, S.: Upper confidence bound learning approach for real HF measurements. In: proceedings of IEEE ICC 2015-Workshop on Advances in Software Defined and Context Aware Cognitive Networks, London, UK, pp. 387–392 (2015)
11. Lempel, A., Ziv, J.: On the complexity of finite sequences. IEEE Trans. Inf. Theor. **22**(1), 75–81 (1976)

12. Audibert, J., Bubeck, S., Munos, R.: Best arm identification in multi-armed bandits. In: Proceedings of the 23th Annual Conference on Learning Theory (COLT), Haifa, Israel, pp. 41–53 (2010)
13. Kalyanakrishnan, S., Tewari, A., Auer, P., Stone, P.: PAC subset selection in stochastic multi-armed bandits. In: Proceedings of the 29th International Conference on Machine Learning ICML 2012, Edinburgh, Scotland, UK (2012)
14. Tekin, C., Liu, M.: Online algorithms for the multi-armed bandit problem with Markovian rewards. In: Proceedings of the 48th Annual Allerton Conference on Communication, Control, and Computing (Allerton) (2010)

A Two-Stage Precoding Algorithm for Spectrum Access Systems with Different Priorities of Spectrum Utilization

Yiteng Wang[1], Youping Zhao[1(✉)], Xin Guo[2], and Chen Sun[2]

[1] School of Electronic and Information Engineering,
Beijing Jiaotong University, Beijing, China
yozhao@bjtu.edu.cn
[2] Sony China Research Laboratory, Beijing, China
{Xin.Guo,Chen.Sun}@sony.com.cn

Abstract. In this paper, a two-stage precoding algorithm, termed as subspace-projection prioritized signal-to-leakage-and-noise ratio (SP-PSLNR) algorithm, is proposed for dynamic spectrum access systems with different priorities of spectrum utilization. In the first stage precoding, the interference from the secondary users (SUs) to the primary users (PUs) is canceled; while in the second stage precoding, we differentiate the interference protection and quality of service (QoS) among SUs with different priorities of spectrum access. For this purpose, we newly introduce a parameter called as "interference leakage weight (ILW)" to be used in the optimization of signal to leakage and noise ratio (SLNR). The simulation results show that the proposed method can increase SUs' maximum allowed transmit power while maintaining protection to the PUs. Moreover, this method can jointly optimize the transmit power of SUs and minimize the interference among SUs. Furthermore, the QoS of SUs can be differentiated by adjusting the ILWs.

Keywords: 5G · Interference leakage weight · Incumbent user protection · Prioritized dynamic spectrum access · Spectrum access system

1 Introduction

To meet the requirements of the emerging fifth generation (5G) wireless communication systems, such as even higher system capacity and spectrum utilization, cognitive radio (CR) technology has been widely investigated as an important enabling technology. In CR-enabled dynamic spectrum access systems, secondary users (SUs) are allowed to access the spectrum of licensed primary users (PUs) on a non-interference basis. For future wireless networks, a large variety of macrocells, microcells, and femtocells will coexist together with numerous device-to-device (D2D) or machine type communications. Thus, multi-tier or hierarchical wireless systems, in which each tier has different

---

This work is supported in part by Sony China Research Laboratory.

© ICST Institute for Computer Sciences, Social Informatics and Telecommunications Engineering 2016
D. Noguet et al. (Eds.): CROWNCOM 2016, LNICST 172, pp. 15–28, 2016.
DOI: 10.1007/978-3-319-40352-6_2

quality of service (QoS) requirements, are envisioned [1, 2]. Notably, the U.S. President's Council of Advisors on Science and Technology (PCAST) recommends a three-tier hierarchy (i.e., Federal primary access, priority secondary access, and general authorized access) for access to Federal spectrum [7]. In this three-tier architecture, the first tier users would be entitled to interference protection to a level such that their communication performance requirements are satisfied. The second tier users would receive short-term priority authorizations. The third tier users would be entitled to use the spectrum on an opportunistic basis and would not be entitled to interference protection. In this study, we consider a two-tier system in which the first tier is the primary user system (PS) and the second tier is the secondary user system (SS). Further, we consider that the SUs in the second tier may have *different* priority levels of spectrum utilization. The high-priority SUs will have better QoS than the low-priority SUs. A new problem which needs to be addressed is how to support *different* QoS requirements of those prioritized SUs while maintaining protection to the PUs.

For most study on cognitive multiple-input multiple-output (MIMO) systems, the SUs have been treated with the same priority. Even though there are many algorithms (e.g., block diagonalization [3], minimum mean square error (MMSE) [3, 4], interference alignment [5]) to mitigate the co-channel interference between the SUs, these algorithms cannot support *different* interference protection and QoS requirements of the prioritized SUs. Ekram Hossain also summaries the challenges of traditional interference management methods (e.g., power control, cell association, etc.) and argues that the existing methods will not be able to address the interference management problem in 5G multi-tier networks because of the more complex interference dynamics (e.g., disparate QoS requirements and priorities at different tiers, huge traffic load imbalance, etc.) [1]. To support the different priorities of interference protection and QoS for spectrum access systems, new interference management algorithms need to be developed.

In this paper, a two-stage precoding algorithm, termed as subspace-projection prioritized signal-to-leakage-and-noise ratio (SP-PSLNR) algorithm, is proposed for spectrum access systems with different priority levels of spectrum utilization. The first stage precoding is based on subspace projection (SP), which mitigates the PU's interference (PUI) caused by SUs. The second stage precoding is based on maximizing the prioritized signal-to-leakage-and-noise ratio (PSLNR), which suppresses the interference between the SUs and supports different priorities of interference protection. A new parameter called as "*interference leakage weight* (ILW)" is introduced at the second stage precoding to account for the resulting interference leakage from one SU to the other SUs. Simulations are conducted to verify the effectiveness of the proposed algorithm.

The rest of this paper is organized as follows. In Sect. 2, the system model and system parameters are discussed. In Sect. 3, the proposed two-stage precoding algorithm is analyzed in more details. How to assign the appropriate ILWs to SUs with different priority levels is also explained. Simulation results are presented in Sect. 4 followed by the conclusion of this paper.

Notations: we use A^H, $E\{A\}$ to denote the conjugate transpose and the statistical expectation of matrix A, respectively. $Tr\{A\}$ denotes the trace of matrix A. I_N denotes an $N \times N$ identity matrix.

2 System Model

Figure 1 shows the system model of spectrum access systems with different priority levels of spectrum utilization. For the system model discussed in this paper, one PS coexists with k SSs. These SSs have different priority levels of spectrum utilization. All SUs are assumed to operate in the same spectrum used by PU while the interference to the PU should be kept below the predefined threshold. In this system model, a pair of active transmitter and receiver equipped with multiple antennas is considered in the PS or SS. As shown in Fig. 1, N_{Tp} and N_{Ts} represents the number of transmitting antennas at PU and SU, respectively. N_{Rp}, N_{Rs} represents the number of receiving antennas at PU and SU, respectively. In Fig. 1, the solid arrow lines represent the desired signals, while the dashed arrow lines stand for the interference. A database is employed to store/retrieve the priority information, geolocation information as well as the channel state information of the PUs and the SUs.

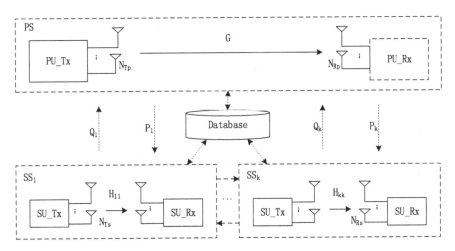

Fig. 1. System model of spectrum access systems with different priority levels of spectrum utilization

The received signal vectors y_p and y_{si} at the PU receiver and the i-th SU receiver are expressed as follows:

$$y_p = Gs + \sum_{i=1}^{k} Q_i F_i x_i + n_p , \qquad (1)$$

$$y_{si} = H_{ii} F_i x_i + \sum_{r=1, r \neq i}^{k} H_{ir} F_r x_r + P_i s + n_{si} , \qquad (2)$$

where s is the transmitted symbol vector from the PU transmitter; x_i is the transmitted symbol vector from the i-th SU transmitter; G *is* the channel matrix between the PU transmitter and receiver; Q_i is the channel matrix between the i-th SU transmitter and PU receiver; the term F_i is the precoding matrix of the i-th SU; H_{ii} is the channel matrix between the i-th SU transmitter and receiver; H_{ir} is the channel matrix between the r-th SU transmitter and the i-th SU receiver; P_i is the channel matrix between the PU transmitter and i-th SU receiver; n_p and n_{si} are the additive white Gaussian noise (AWGN) vectors with zero mean and unit variance.

The first term in (1) and (2) represents the desired signal at the receiver of PU and i-th SU, respectively. The second term in (1) represents the aggregated PUI caused by SUs. The second and third term in (2) represents the SU's interference (SUI) caused by the other SUs and the PU, respectively. The last term in (1) and (2) is the AWGN noise. In this paper, unless stated otherwise, it is assumed that perfect channel state information (CSI) is known at the transmitters and receivers of SUs.

3   Two-Stage Precoding Algorithm (SP-PSLNR)

In this section, the proposed two-stage precoding algorithm "SP-PSLNR" is discussed in a stage-by-stage approach in the first two subsections. Then the combination of two stages, i.e., the SP-PSLNR algorithm, is presented in the third subsection. In the last subsection, how to assign the appropriate ILWs to SUs of different priority levels is further discussed.

3.1   First Stage Precoding: SP Algorithm

The first stage precoding is based on SP, which eliminates the PUI caused by the SUs operating in the same spectrum. In a CR-enabled dynamic spectrum access system, the SUs are allowed to access the same spectrum used by the PU only if the resulting interference to the PU remains below the predefined PUI threshold. Therefore, the SUs' maximum allowed transmit power has to be limited by the predefined interference threshold at the PU. Consequently, the SUs' transmit power might be too low to meet the communication quality requirements. To increase the allowed transmit power of SUs, more effective interference suppression algorithm is quite needed. The SP algorithm can help SUs to transmit signals in the null space of the interference channel (i.e., the channel between SUs transmitter and PU receiver), thus eliminating the PUI caused by SUs. In this way, the maximum allowed transmit power of SUs with the SP-based precoding can be significantly higher than that when using the traditional power control method.

Based on the geolocation database approach such as the advanced geolocation engine (AGE) database (please refer to [6] for more details about AGE database), the i-th SU first finds out the PU within its interference range, and then the channel matrix Q_i is retrieved when evaluating the PUI caused by the i-th SU transmitter. By applying the singular value decomposition (SVD) of Q_i, the null of Q_i, i.e., $V^{(0)}$, which is the first stage precoding vector $F_i^{(1)}$ for the i-th SU transmitter, can be obtained as follows.

$$Q_i = U \sum [V^{(1)} V^{(0)}]^H ,\tag{3}$$

$$F_i^{(1)} = V^{(0)} ,\tag{4}$$

where $V^{(1)}$ is the matrix composed of the right singular vectors corresponding to non-zero singular values of Q_i, and $V^{(0)}$ is the matrix composed of the zero singular values of Q_i. To obtain the null space of non-zero elements, the number of the SU's transmitting antennas should be no less than the number of the PU's receiving antennas.

3.2 Second Stage Precoding: PSLNR Algorithm

To support different priority levels of QoS requirements for SUs, the PSLNR algorithm is proposed by introducing a new parameter called as "ILW" into the traditional SLNR algorithm. The PSLNR measured by the i-th SU receiver is expressed as follows:

$$PSLNR_i = \frac{E\left\{|H_{ii}F_i x_i|^2\right\}}{E\left\{|\alpha_i \sum_{r=1,r\neq i}^{k} H_{ri} F_i x_i + n_{si}|^2\right\}}$$

$$= \frac{Tr(F_i^H H_{ii}^H H_{ii} F_i)}{Tr(\alpha_i^2 \sum_{r=1,r\neq i}^{k} F_i^H H_{ri}^H H_{ri} F_i) + I_{N_{Rs}}} ,\tag{5}$$

where α_i represents the i-th SU's ILW. The higher the α_i value of i-th SU, the stronger constraint is forced to the investigated SU on its interference leakage to the other SUs. Therefore, a smaller ILW will be assigned to the high-priority SU. That is to say, the high-priority SU is endowed with looser constraint on its interference leakage to other lower-priority SUs.

The second stage precoding is based on maximizing the PSLNR, which suppresses the interference among the SUs and supports different priorities of interference protection and QoS.

The optimization problem is formulated as follows:

$$\begin{cases} \max_{F_i^{(2)}} PSLNR_i \\ s.t. ||F_i||^2 \leq p_{Si} \\ F_i = F_i^{(1)} F_i^{(2)} , \end{cases}\tag{6}$$

where $F_i^{(2)}$ represents the second stage precoding matrix of the i-th SU and p_{Si} represents the transmit power of i-th SU.

By solving the above optimization problem according to the solution of traditional SLNR algorithm, the second stage precoding matrix for the i-th SU can be expressed as follows:

$$F_i^{(2)} = \left[f_i^{(2)} \cdots f_i^{(2)} \right]_{N_{Ts} \times N_{Rs}}, \tag{7}$$

where $f_i^{(2)} = \Phi \left[\left(\alpha_i^2 \sum\limits_{r=1, r \neq i}^{k} F_i^{(1)H} H_{ri}^H H_{ri} F_i^{(1)} + I_{N_{Rs}} \right)^{-1} F_i^{(1)H} H_{ii}^H H_{ii} F_i^{(1)} \right]$, and $\Phi[A]$ represents the eigenvector corresponding to the largest eigenvalue of A.

Supposing there are two SSs of different priorities, the ratio of two SUs' signal to interference plus noise ratio *(SINR)* can be written as:

$$\frac{SINR_1}{SINR_2} = \frac{Tr(F_1^H H_{11}^H H_{11} F_1)}{Tr(F_2^H H_{22}^H H_{22} F_2)} \times \frac{Tr(F_1^H H_{21}^H H_{21} F_1) + Tr(P_2^H P_2) + N_0}{Tr(F_2^H H_{12}^H H_{12} F_2) + Tr(P_1^H P_1) + N_0}, \tag{8}$$

where $F_1 = F_1^{(1)} F_1^{(2)}$, $F_2 = F_2^{(1)} F_2^{(2)}$, and N_0 represents the noise power. As a special case, when these two pairs of SU transmitters and receivers are symmetrically distributed with regarding to the PU transmitter (as shown in Fig. 3 in the next Section), the ratio of two SUs' *SINR* is only affected by the second precoding matrix. As a result, the ratio of two SUs' *SINR* will be mainly determined by the ILWs assigned to these two SUs.

3.3   Two-Stage Precoding Algorithm (SP-PSLNR)

The proposed scheme, termed as SP-PSLNR, is the combination of SP precoding and PSLNR precoding. By using the SP-PSLNR algorithm, the different priorities of interference protection and QoS requirements can be supported for spectrum access systems with different priority levels of spectrum utilization. The two-stage precoding scheme at the *i*-th SU transmitter is carried out by the following 4 steps:

(1) Identify the PU within the SU's interference range. We get the channel matrix Q_i which is the interference channel matrix between the *i*-th SU transmitter and the PU receiver.
(2) Obtain the null of Q_i, (i.e., the first stage precoding matrix) by applying the SVD of Q_i.
(3) Assign the appropriate ILW to the *i*-th SU transmitter according to the procedures detailed in the next subsection.
(4) Obtain the second stage precoding matrix by using the PSLNR criterion.

The proposed two-stage precoding algorithm is employed at the SU transmitter. The flow chart of the desired signal from the *i*-th SU transmitter to its intended receiver is depicted in Fig. 2.

Fig. 2. Flow chart of the desired signal from the i-th SU transmitter to the i-th SU receiver

3.4 ILW Assignment

As mentioned in the Subsect. 3.2, ILW is introduced to account for the interference leakage power in the objective function of SU. In this way, the high-priority SU has a looser constraint on its interference leakage to the other low-priority SUs, contrarily the low-priority SU has to obey a stronger constraint on its interference leakage to the other high-priority SUs. Therefore, the high-priority SU can obtain better interference protection and QoS guarantee.

When adopting the proposed SP-PSLNR algorithm, how to assign the appropriate ILWs to SUs of different priority levels is an important issue. According to the PSLNR expressed by (5) and the SINR ratio expressed by (8), the priority level, required QoS (e.g., SINR) and the transmit power of SUs are the key factors to be considered when making the ILW assignment. Based on simulations or field tests, the ratio of ILWs can be pre-determined according to the required QoS (say, SINR) difference and the transmit power at different SUs.

As shown from the simulation results presented in the next section, the ratio of ILWs has significant impact on the differentiation of the SINR at different prioritized SUs. Therefore, a proportional ratio method is proposed to adjust ILWs for SUs with different priorities. The sum of ILWs assigned to all active SUs is normalized to 1.

The procedures of the ILW assignment are detailed as follows:

(1) Estimate the interference range of the SU based on the maximal transmit power, the height of transmitting antennas and other related information;
(2) Find all the active SUs (i.e., the SUs in active communications) in this interference range, and then sort their priorities;
(3) Determine the ratio of ILWs for the SUs in a proportional manner according to the required SINR difference and the transmit power at different SUs;
(4) Assign an appropriate ILW to each active SU such that the sum of ILWs assigned to all active SUs is equal to 1.

4 Performance Simulation

Simulations are conducted to further evaluate the performance of the proposed two-stage precoding algorithm. This section presents the simulation results, which demonstrate the effectiveness of the proposed SP-PSLNR algorithm.

The simulation model is depicted in Fig. 3, in which one PS and two prioritized SSs share the same spectrum. Assuming that at a given time instance, there is only one pair of active communication users in the PS and SS. SS_1 (serving the SU_1) has higher priority than SS_2 (serving the SU_2). The cell radius of the PS and SS is set as 30 m. The transmitter is located at the cell center. As shown in Fig. 3, the propagation distances of the desired signal are designated as d_{11} and d_{22}; the propagation distances of the interference from the neighboring SS are designated as d_{12} and d_{21}; and the propagation distance of interference from PS is designated as d_{01} and d_{02}. Rayleigh fading channel model and free-space path loss model are assumed in the simulation. The modulation type is binary phase shift keying (BPSK). Moreover, it is assumed that the complete CSI is known at the transmitters of SS$_1$ and SS$_2$. Some other system parameters are listed in Table 1.

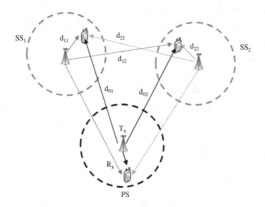

Fig. 3. Simulation model of spectrum access of SUs with different priority levels

Table 1. System parameters used in the simulation (unless stated otherwise)

Parameters	Value
Operating frequency	2 GHz
Channel bandwidth	10 MHz
Noise figure at the PU/SU receiver	5 dB
PU transmit power	5 dBm
SU transmit power	5 dBm
Number of transmitting antennas at PU	2
Number of receiving antennas at PU	2
Number of transmitting antennas at SU	2
Number of receiving antennas at SU	2

Without loss of generality, the SU receiver is randomly distributed in the small cell of SS, and thus the desired signal distance or the interference distance may not be equal. The simulation results are shown in Figs. 4, 5, 6 and 7. Then, we introduce a special case to show the differentiated SUs' performance due to different ILW assignments, in which we set the two pairs of SU transmitters and receivers symmetrically distributed with regarding to the PU transmitter. For this special case, both the desired signal distances and the interference distances are equal for the receivers of two SUs, i.e., $d_{11} = d_{22}$ and $d_{01} = d_{02}$, the corresponding simulation results are shown in Figs. 8, 9 and 10.

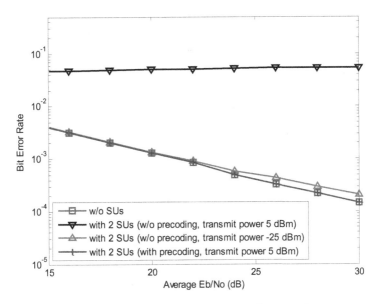

Fig. 4. BER performance of PU under different settings

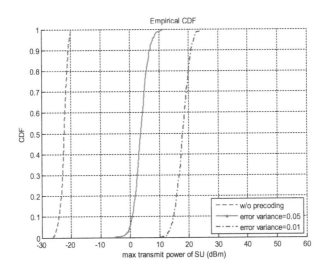

Fig. 5. CDF of SUs' maximum allowed transmit power

Figure 4 shows the BER performance of PU under different settings. Especially, the performance of SP-based precoding algorithm is compared against that of the traditional power control method (i.e., without precoding). The PU receiver is located at the cell edge of PS. The simulation results show that if all the SUs employ the SP-based precoding with the perfect CSI, the PUI can be completely eliminated. More importantly, the maximum allowed transmit power at SUs can be increased significantly by

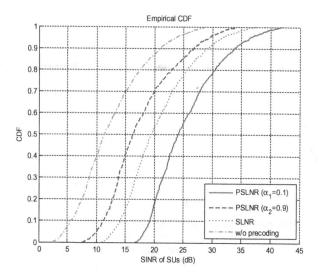

Fig. 6. CDF of SUs' SINR

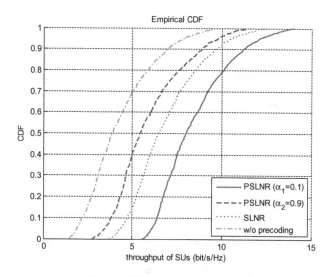

Fig. 7. CDF of SUs' throughput

about 30 dB as compared to the traditional power control method (i.e., without precoding). The reason is that SP precoding enables SUs to transmit signal in the null space of interference channel to the PU. Therefore, the proposed SP-PSLNR algorithm can protect the PU's QoS while significantly relaxing the limitation on SUs' maximum allowed transmit power.

As mentioned above, theoretically, the SP-based precoding can completely eliminate the PUI with perfect CSI. However, in practice, the estimated channel matrix

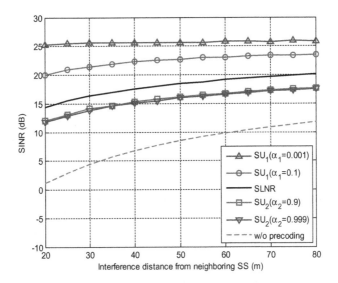

Fig. 8. SINR of SUs with different algorithms or ILW values

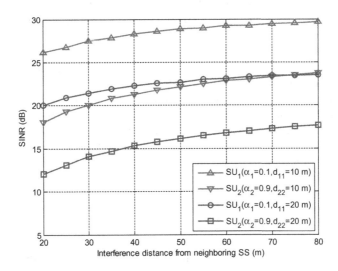

Fig. 9. SINR of SUs with different desired signal distances

always contains some errors. In Fig. 5, the element of error matrix is considered to be a random variable, e_{ij}, following the complex Gaussian distribution with zero mean and σ^2 variance, i.e., CN $(0, \sigma^2)$. In this simulation, the interference to noise ratio (INR) threshold at PU is set to 0 dB. Note that, for the simulation shown in Fig. 5, the number of SUs' transmitting antennas is set to 8. The maximum allowed transmit power of SU refers to the transmit power of SU when the PUI caused by this SU is equal to the interference tolerance threshold of PU. And the maximum allowed transmit

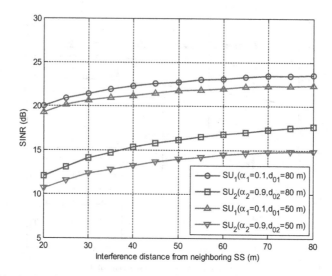

Fig. 10. SINR of SUs under different interference distance from the PS transmitter

power is related with the INR threshold of PU, the distance between SU and PU, and the adopted interference suppression algorithm of SU. As can be seen in Fig. 5, when the error variance is set to 0.05 and 0.01, the maximum allowed transmit power can be increased by about $23 \sim 30$ dB, $38 \sim 45$ dB, respectively. Larger error variance results in larger PUI caused by SUs. As expected, the smaller the error variance, the higher the maximum allowed transmit power.

Figures 6 and 7 present the cumulative distribution function (CDF) of the SUs' SINR and throughput, respectively. It shows that when using the proposed two-stage precoding algorithm, the high-priority SU always has better SINR and higher throughout as compared to that using the traditional SLNR algorithm. This is because the high-priority SU (with lower ILW) has looser constraint on its interference leakage than the SU with traditional SLNR algorithm. However, the low-priority SU always has poorer SINR and lower throughout as compared to that using the traditional SLNR algorithm. This observation implies that the performance improvement of the high-priority SU is obtained at the cost of performance degradation of the low-priority SU.

Figure 8 shows the SINR of different prioritized SUs. In this simulation, $d_{11} = d_{22} = 20$ m and $d_{01} = d_{02} = 80$ m. The ILWs assigned to the SUs with different priority levels of spectrum utilization are shown in the legend.

Figure 8 shows that, firstly, PSLNR can reduce the interference between SSs, and significant benefits can be obtained as compared to that without precoding; secondly, according to the comparison of PSLNR and SLNR, the higher prioritized user has obtained better SINR as compared to that with SLNR algorithm, whereas the lower prioritized user has even lower SINR than that with SLNR algorithm; thirdly, as expected, the SU with smaller ILW results in higher SINR than the SU with larger ILW. When the ratio of the ILWs between the high-priority SU and the low-priority SU gets smaller, the SINR difference of between these SUs becomes even larger. For example, when the ratio of the SUs' ILWs (i.e., α_1/α_2 in this simulation) is reduced to

1/999 from 1/9, the SINR difference of these SUs is increased to 13.5 dB from about 8 dB (when the interference distance from the neighboring SS transmitter is 20 m). In sum, the smaller the SUs' ILW ratio, the larger the SUs' SINR difference. Fourthly, for a given SUs' ILW ratio, the difference of SUs' SINR decreases as the distance from the neighboring SS transmitter increases. As the distance from the neighboring SS transmitter increases, the interference between prioritized SUs becomes even smaller. As a result, the SUs' SINR difference narrows down.

Figure 9 shows the impact of the desired signal distance on the SUs' SINR difference when the ILWs (i.e., 0.1 and 0.9) to the two SUs remain the same. It shows that when the desired signal distance is reduced from 20 m to 10 m, the SINR of SUs increases about 6 dB on average, while the SUs' SINR difference changes little. This is because the SUI is independent with the desired signal distance.

Figure 10 shows when the interference distance from the PS transmitter decreases from 80 m to 50 m, the SINR of SUs decreases in general and the SINR difference of two SUs may increase about 2 dB. When the distance from the PS transmitter decreases, the SUI becomes more serious. In general, the stronger the SUI, the larger the SUs' SINR difference. Hence, the SINR difference increases when the interference distance from the PS transmitter decreases.

As we expected, simulation results shown in Figs. 8, 9 and 10 demonstrate that the SINR difference of SUs mainly depends on the ratio of SUs' ILWs (i.e., α_1/α_2 in this paper).

5 Conclusion

To support the coexistence of PU and prioritized SUs in spectrum access systems with different priority levels of spectrum utilization, a two-stage precoding algorithm (termed as "SP-PSLNR algorithm") is proposed in this paper. With the help of the first stage precoding based on SP, the SUs can access the spectrum without degrading the QoS of PU. In the meanwhile, SUs' maximum allowed transmit power can be significantly increased as compared to that when using the traditional power control method. By maximizing the PSLNR with the newly introduced parameter "ILW", the second stage precoding can differentiate the priorities of interference protection and QoS (e.g., the received SINR) for prioritized SUs. In this way, the high-priority SUs can obtain better SINR than the low-priority SUs, and the SUs' SINR difference can be adjusted by the ratio of ILWs. The stronger the SUI, the larger SUs' SINR difference can be obtained. Performances of the proposed algorithm are verified through simulations. For the future work, in the second stage precoding, the SU's interference caused by the PU will also be taken into account. The proposed algorithm could also be further investigated and exploited to deal with the challenging interference issues in 5G mobile communication systems.

Acknowledgments. Special thanks to Jinxing Li, Qixin Tai, Songpeng Li, Wei Ding and Zhichao Hou for their great help in debugging the simulation code for this work.

References

1. Hossain, E., Rasti, M., Tabassum, H., Abdelnasser, A.: Evolution toward 5G multi-tier cellular wireless networks: an interference management perspective. IEEE Wirel. Commun. **21**(3), 118–127 (2014)
2. ElSawy, H., Hossain, E., Haenggi, M.: Stochastic geometry for modeling, analysis, and design of multi-tier and cognitive cellular wireless networks: a survey. IEEE Commun. Surv. Tutorials Third Q. **15**(3), 996–1019 (2013)
3. Lee, K.-J., Lee, I.: MMSE based block diagonalization for cognitive radio MIMO broadcast channels. IEEE Trans. Wireless Commun. **10**(10), 3139–3144 (2011)
4. Yiftach, R., Bergel, I.: MMSE-SLNR Precoding for Multi-Antenna Cognitive Radio. IEEE Trans. Signal Process. **62**(10), 2719–2729 (2014)
5. Xu, Y., Mao, S.: Distributed interference alignment in cognitive radio networks. In: 2013 IEEE 22nd International Conference on Computer Communications and Networks (ICCCN), Nassau, Bahamas, pp. 1–7 (2013)
6. Li, S., Zhao, Y., Sun, C., Guo, X.: Development of an advanced geolocation engine-based cognitive radio testbed. In: 2014 IEEE/CIC International Conference on Communications in China (ICCC), Shanghai, China, pp. 528–533 (2014)
7. Executive Office of the President and President's Council of Advisors on Science and Technology, Report to the President - Realizing the Full Potential of Government-Held Spectrum to Spur Economic Growth, July 2012. www.whitehouse.gov/ostp/pcast

Closed Form Expression of the Saddle Point in Cognitive Radio and Jammer Power Allocation Game

Feten Slimeni[1(✉)], Bart Scheers[2], Vincent Le Nir[2], Zied Chtourou[1], and Rabah Attia[3]

[1] VRIT Lab, Military Academy of Tunisia, 8000 Nabeul, Tunisia
feten.slimeni@gmail.com, ziedchtourou@gmail.com
[2] CISS Department, Royal Military Academy (RMA), 1000 Brussels, Belgium
{bart.scheers,vincent.lenir}@rma.ac.be
[3] SERCOM Lab, EPT University of Carthage, 2078 Marsa, Tunisia
rabah_attia@yahoo.fr

Abstract. In this paper, we study the power allocation problem for a cognitive radio in the presence of a smart jammer over parallel Gaussian channels. The objective of the jammer is to minimize the total capacity achievable by the cognitive radio. We model the interaction between the two players as a zero-sum game, for which we derive the saddle point closed form expression. First, we start by solving each player's unilateral game to find its optimal power allocation. These games will be played iteratively until reaching the Nash equilibrium. It turns out that it is possible to develop analytical expressions for the optimal strategies characterizing the saddle point of this minimax problem, under certain condition. The analytic expressions will be compared to the simulation results of the Nash equilibrium.

Keywords: Cognitive radio · Jammer · Power allocation · Saddle point · Nash equilibrium

1 Introduction

Cognitive radio (CR) technology is presented in [1] as a promising solution to the spectrum scarcity problem due to its dynamic spectrum access capability. However, a challenging problem for a CR is the presence of malicious users such as smart jammers. A jammer equipped with a cognitive technology may always prevent CR users from efficiently exploiting the free frequency bands through a real time adaptation of its transmission parameters such as the jamming power. Since game theory is a process of modeling the strategic interaction between players, it is a suitable tool to understand and analyze this adversarial system.

The optimal power allocation problem in presence of smart jammer has been investigated from game theoretical point of view in both wireless [2–6] and cognitive [7–9] networks with diverse utility functions such as the SINR, transmission

© ICST Institute for Computer Sciences, Social Informatics and Telecommunications Engineering 2016
D. Noguet et al. (Eds.): CROWNCOM 2016, LNICST 172, pp. 29–40, 2016.
DOI: 10.1007/978-3-319-40352-6_3

capacity and number of successful channel access. Most of related papers proved the existence and uniqueness of the pure strategy Nash equilibrium (NE), but only some papers have dealt with analytic computation of the optimal strategies. In [2,3], Altman proved the existence and uniqueness of NE considering the transmission capacity as the utility function. To develop the closed form analytic expressions of the optimal power allocations, in the first paper [2] he proposed an algorithm based on the bisection method. In the second paper [3], he converted the problem to a minimax problem since the NE strategy of a zero-sum game is equal to the optimal minimax strategy [10], and he considered the particular case of proportional channel fading coefficients faced by both the jammer and the transmitter. Considering finite strategy sets for both the transmitter and the jammer, the authors in [6] prove the existence of NE in pure (deterministic) strategies and characterize the optimal power allocations in asymptotic regimes over independent parallel Gaussian wiretap channels.

In the context of cognitive radio networks, the interaction between a jammer and a CR is presented in [7] as Colonel Blotto game where the two opponents distribute limited resources over a number of battlefields with the payoff equal to SINR, and the equilibrium is derived in terms of mixed (probabilistic) strategy via power randomization. Likewise, the authors in [9] adopt a Bayesian approach in studying the power allocation game between the CR and the jammer, and provide the Cumulative Distribution Functions (CDFs) of the transmission powers that should be adopted by the CR and the jammer at NE to optimize the utility function equal to the number of successful transmissions.

In this paper, we model the interaction between a CR and a smart jammer as a two-person zero-sum game, considering the transmission capacity as the utility function, and the power allocation over multiple channels as both players strategy sets. We consider that the game is continuous since the players choose from an uncountably infinite strategy sets (the allocated power to each channel can take any decimal value). The proof of existence of a saddle point is given by Nikaido in [11] who generalizes Von Neumann minimax theorem for infinite strategy sets. The saddle point of this game corresponds to the optimal power allocations for both the jammer and the CR. Computing its closed form through exhaustive search over all the possible power allocations of the two players turns out to be hard to do in terms of resource and time consumption. We start by solving the unilateral games; in each one, only one player has to make the decision about how to distribute his power between the available channels considering that the other player has a fixed power allocation. The unilateral games will be played iteratively until reaching the NE. Then, we analytically determine the closed form expressions of the optimal power allocations characterizing the saddle point, under the assumption that all channels are used by both the CR and the jammer. The explicit solution to this game allows the CR to study the jamming strategy and to proactively use the corresponding optimal anti-jamming power allocation. Finally, we compare the analytical saddle point to the NE simulation result.

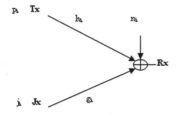

Fig. 1. Scenario of CR jamming attack

2   System Model

We consider that the CR has the capacity of accessing multiple frequency bands at the same time with a limited power budget P. The jammer is also assumed to be able to inject interference to all channels with a limited power budget J, which is known as barrage jamming. Whether the attackers can successfully jam communication in a particular channel will depend on how much power the CR and the jammer allocate on that channel. The system is described in Fig. 1. The CR adopt the 'listen-before-talk' rule, that is, sensing for spectrum opportunities at the beginning of each timeslot. On finding M available channels, the CR allocates power $p_k \geq 0$ to the channel $k \in [1, M]$ such that:

$$\sum_{k=1}^{M} p_k = P \tag{1}$$

An action of the CR is designed by the vector $\mathbf{p} = (p_1, \cdots, p_k, \cdots, p_M)$ and the goal is to maximize its transmission capacity subject to (1). At the same time, the jammer injects power $j_k \geq 0$ to the channel k such that:

$$\sum_{k=1}^{M} j_k = J \tag{2}$$

An action of the jammer is designed by the vector $\mathbf{j} = (j_1, \cdots, j_k, \cdots, j_M)$ and the goal is to minimize the transmission capacity of the CR, subject to (2).

In this paper, the CR is trying to maximize its total transmission capacity over the available channels and the jammer is trying to minimize this capacity, so this interaction can be seen as a two person zero-sum game. In each iteration of this game, each player updates its power allocation over the available channels to maximize its payoff. Because each element of the vectors $\mathbf{p}$ and $\mathbf{j}$ can take as value any element in $[0, P]$ and $[0, J]$ respectively, we have a continuous set of actions for both the CR and the jammer. The CR's capacity is proportional to

$$f(\mathbf{p}, \mathbf{j}) = \sum_{k=1}^{M} log_2(1 + \frac{|h_k|^2 p_k}{|g_k|^2 j_k + n_k}) \tag{3}$$

n_k is the noise variance of channel k, h_k and g_k are the gains of channel k for the CR and the jammer respectively. In this paper, we assume that all channel gains are common knowledge to both players, and we consider that the M channels are parallel Gaussian channels. We consider $f(\mathbf{p}, \mathbf{j})$ and $-f(\mathbf{p}, \mathbf{j})$ the utility functions of the CR and the jammer, respectively.

This game can be seen as a succession of the two following unilateral games, in which one player is trying to update its power allocation after observing the other player's power allocation.

3 Unilateral Games

We start by considering the extreme cases, where only one player has to decide for a one-time how to allocate his total power, the other player has fixed strategy.

3.1   CR Unilateral Game

In each iteration of the game, the CR can consider that the jammer's power allocation is momentarily fixed, and the game degenerates to a classical power allocation problem, where the CR assigns its power into the current noise plus jamming space to maximize the capacity. Mathematically, it can be formulated as the following nonlinear optimization problem:

$$\begin{aligned} \underset{\mathbf{p}}{\text{maximize}} \ &\sum_{k=1}^{M} log_2(1 + \frac{|h_k|^2 p_k}{|g_k|^2 j_k + n_k}) \\ \text{subject to} \ &\sum_{k=1}^{M} p_k \leq P \end{aligned} \tag{4}$$

Karush-Kuhn-Tucker (KKT) equations give first order necessary conditions for a solution in nonlinear programming to be optimal, provided that some regularity conditions are satisfied. Allowing inequality constraints, the KKT approach to nonlinear programming generalizes the method of Lagrange multipliers, which allows only equality constraints. The Lagrangian is then

$$L(\mathbf{p}, \mathbf{j}, \lambda) = \sum_{k=1}^{M} log_2(1 + \frac{|h_k|^2 p_k}{|g_k|^2 j_k + n_k}) - \lambda(\sum_{k=1}^{M} p_k - P) \tag{5}$$

Since L is separable in p_k, we can separately optimize each term.

$$\frac{\partial L}{\partial p_k} = \frac{|h_k|^2}{|h_k|^2 p_k + |g_k|^2 j_k + n_k} - \lambda \tag{6}$$

The optimal solution of this optimization problem yields a waterfilling strategy

$$p_k = (\frac{1}{\lambda} - N_k)^+ \tag{7}$$

where $\frac{1}{\lambda}$ is called the waterlevel and the KKT multiplier $\lambda > 0$, that can be found by bisection, is chosen to satisfy (1), N_k is the effective noise power on each channel,

$$N_k = \frac{|g_k|^2 j_k + n_k}{|h_k|^2} \tag{8}$$

and $(x)^+ = \max(0, x)$.

3.2 Jammer Unilateral Game

On the other hand, in each iteration the jammer can consider that the CR has a fixed power allocation for the moment of making its decision, and the game degenerates to a jamming unilateral optimization. In such a circumstance, the jammer will allocate its jamming power to minimize the total capacity. Mathematically, this is expressed as the following minimizing problem

$$\begin{aligned} \underset{\mathbf{j}}{\text{minimize}} \quad & f(\mathbf{p}, \mathbf{j}) \\ \text{subject to} \quad & \sum_{k=1}^{M} j_k \leq J \end{aligned} \tag{9}$$

We can write the Lagrangian

$$L(\mathbf{j}, \mu) = -f(\mathbf{p}, \mathbf{j}) - \mu \left(\sum_{k=1}^{M} j_k - J \right) \tag{10}$$

Since L is separable in j_k, we can separately minimize each term.

$$\frac{\partial L}{\partial j_k} = \frac{|g_k|^2 |h_k|^2 p_k}{(|h_k|^2 p_k + |g_k|^2 j_k + n_k)(|g_k|^2 j_k + n_k)} - \mu \tag{11}$$

After solving the resulting second order equation in j_k, we get

$$j_k = \left(\frac{1}{2} \sqrt{ \left(\frac{|h_k|^2 p_k}{|g_k|^2} \right)^2 + 4 \frac{|h_k|^2 p_k}{|g_k|^2 \mu} } - \frac{|h_k|^2 p_k}{2|g_k|^2} - \frac{n_k}{|g_k|^2} \right)^+ \tag{12}$$

where the KKT multiplier μ should satisfy (2) and can be found by bisection. Note that unlike the CR who uses the waterfilling strategy, the jammer applies a different strategy as given by (12) to dynamically allocate its power.

After solving the optimization problems independently for the CR and the jammer, we can consider Nash game scenario in which both the CR and the jammer make decisions but sequentially. In game theory, a sequential game is a game where one player chooses his action before the others choose theirs. The later players must have some information of the first's choice. The implementation of this game consists in implementing the two unilateral games between the CR and the jammer in an iterative way until convergence to almost fixed power allocation per channel. We consider that the duration of the iterative process until convergence is inferior to the channel coherence time. For the CR, we will use the expression (7) and we proceed by bisection until reaching the value of λ corresponding to the allocation of the total CR power. For the jammer we exploit another strategy with respect the expression (12) and we proceed by bisection until reaching the value of μ corresponding to the allocation of the total jamming power.

In Sect. 5, the NE resulting from the sequential game will be compared to the closed form expression of the saddle point.

4   The Closed Form Expression of the Saddle Point

The saddle point is so called because if we represent the payoff values as a matrix, the equilibrium value is the minimum in its row and the maximum in its column, this value is the value of the game, and the players' actions are the row and column that intersect at that point. This description of the saddle-point refers to a saddle sits on a horse's back at the lowest point on its head-to-tail axis and highest point on its flank-to flank axis [12]. As example, we determine in Fig. 2 of Subsect. 5.1 the saddle point of this game over two flat fading channels.

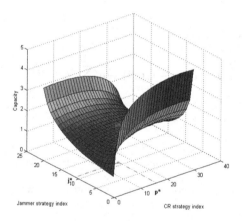

Fig. 2. The saddle point for two channels

The proof of existence of a saddle point is given by Nikaido in [11] who generalizes Von Neumann minimax theorem for infinite strategy sets. The saddle point of this game corresponds to the optimal power allocations for both the jammer and the CR. Since it is hard to derive it through exhaustive search over the continuous strategy sets, we develop its closed form analytic expression.

4.1   General Case

Based on the equations (6) and (11) given to solve each player's decision problem, a vector of powers $(\mathbf{p}, \mathbf{j})$ constitutes the saddle point if and only if there are KKT multipliers λ and μ such that [13]:

$$\frac{\partial f}{\partial p_k} = \frac{|h_k|^2}{|h_k|^2 p_k + |g_k|^2 j_k + n_k} = \lambda \tag{13}$$

and

$$\frac{\partial(-f)}{\partial j_k} = \frac{|g_k|^2|h_k|^2 p_k}{(|h_k|^2 p_k + |g_k|^2 j_k + n_k)(|g_k|^2 j_k + n_k)} = \mu \tag{14}$$

Equation (13) gives the expression of p_k as

$$p_k = \frac{1}{\lambda} - \frac{n_k + |g_k|^2 j_k}{|h_k|^2} \tag{15}$$

Now we replace p_k in (14) by the expression (15) to find j_k

$$j_k = \frac{|h_k|^2}{\lambda|g_k|^2 + \mu|h_k|^2} - \frac{n_k}{|g_k|^2} \tag{16}$$

If $j_k \geq 0$, we can replace j_k in (15) to get the expression of p_k

$$p_k = \frac{\mu}{\lambda} \frac{|h_k|^2}{\lambda|g_k|^2 + \mu|h_k|^2} \tag{17}$$

So, we can give the equilibrium strategies closed forms for $k \in [1, M]$

$$p_k = \begin{cases} \frac{\mu}{\lambda} \frac{|h_k|^2}{\lambda|g_k|^2 + \mu|h_k|^2} & \text{if } \frac{n_k}{|h_k|^2} < \frac{|g_k|^2}{\lambda|g_k|^2 + \mu|h_k|^2} \\ \frac{1}{\lambda} - \frac{n_k}{|h_k|^2} & \text{if } \frac{|g_k|^2}{\lambda|g_k|^2 + \mu|h_k|^2} \leq \frac{n_k}{|h_k|^2} < \frac{1}{\lambda} \\ 0 & \text{if } \frac{n_k}{|h_k|^2} > \frac{1}{\lambda} \end{cases} \tag{18}$$

and

$$j_k = \begin{cases} \frac{|h_k|^2}{\lambda|g_k|^2 + \mu|h_k|^2} - \frac{n_k}{|g_k|^2} & \text{if } \frac{n_k}{|h|^2} < \frac{|g_k|^2}{\lambda|g_k|^2 + \mu|h_k|^2} \\ 0 & \text{if } \frac{n_k}{|h_k|^2} \geq \frac{|g_k|^2}{\lambda|g_k|^2 + \mu|h_k|^2} \end{cases} \tag{19}$$

To simplify and explain these power allocation expressions, we define a new parameter $\tau_k = \lambda + \mu\frac{|h_k|^2}{|g_k|^2}$. We get $\forall k \in [1, M]$

$$p_k = \begin{cases} \frac{\mu}{\lambda} \frac{|h_k|^2}{\lambda|g_k|^2 + \mu|h_k|^2} & \text{if } \frac{n_k}{|h_k|^2} < \frac{1}{\tau_k} \\ \frac{1}{\lambda} - \frac{n_k}{|h_k|^2} & \text{if } \frac{1}{\tau_k} \leq \frac{n_k}{|h_k|^2} < \frac{1}{\lambda} \\ 0 & \text{if } \frac{n_k}{|h_k|^2} > \frac{1}{\lambda} \end{cases} \tag{20}$$

and

$$j_k = \begin{cases} \frac{|h_k|^2}{|g_k|^2}\left(\frac{1}{\tau_k} - \frac{n_k}{|h_k|^2}\right) & \text{if } \frac{n_k}{|h|^2} < \frac{1}{\tau_k} \\ 0 & \text{if } \frac{n_k}{|h_k|^2} \geq \frac{1}{\tau_k} \end{cases} \tag{21}$$

We can draw the following three cases controlled by the three power levels $\frac{1}{\lambda}$ related to the CR, $\frac{1}{\tau_k}$ related to the jammer and $\frac{n_k}{|h_k|^2}$ related to the noise:

- (a) Since $\frac{1}{\tau_k} < \frac{1}{\lambda}$, $\forall k \in [1, M]$, a bad channel for the CR ($\frac{n_k}{|h_k|^2} > \frac{1}{\lambda}$) is also a bad channel for the jammer ($\frac{n_k}{|h_k|^2} > \frac{1}{\tau_k}$). The jammer does not attack a channel which is not occupied by the CR, i.e. if $p_k = 0$ then $j_k = 0$

- (b) In channels verifying $\frac{1}{\tau_k} \leq \frac{n_k}{|h_k|^2} < \frac{1}{\lambda}$, the CR occupies these channels without being jammed; i.e. $p_k > 0$ and $j_k = 0$, these channels are considered unfavorable for the jammer. It avoids these channels may be because of low g_k values which may force it to send with very high power to achieve the CR attack. A solution for the jammer to minimize the number of channels verifying this condition (since it can be considered as favorable opportunity for the CR), is to be close to the receiver node in order to get high g_k values and so $\frac{1}{\tau_k} \approx \frac{1}{\lambda}$.
- (c) If $\frac{n_k}{|h_k|^2} < \frac{1}{\tau_k}$, the channel is considered good for the two players and so occupied by both the CR and the jammer.

We provide in Subsect. 5.2 an example covering these three situations, see Fig. 3 .

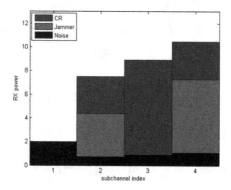

Fig. 3. The strategies at the NE in general case (Color figure online)

4.2   Case All Channels are used by both the CR and the Jammer

Under the assumption that the jammer and the CR use all the channels (p_k, j_k $> 0, \forall\, k \in [1, M]$), which means $\frac{|g_k|^2}{\lambda|g_k|^2 + \mu|h_k|^2} \geq \frac{n_k}{|h_k|^2}$, then we can give the power allocation closed forms at the NE for $k \in [1, M]$

$$\begin{cases} p_k &= \frac{\mu}{\lambda} \frac{|h_k|^2}{\lambda|g_k|^2 + \mu|h_k|^2} \\ j_k &= \frac{|h_k|^2}{\lambda|g_k|^2 + \mu|h_k|^2} - \frac{n_k}{|g_k|^2} \end{cases} \qquad (22)$$

The power allocations should respect the conditions (1) and (2) which give

$$\begin{cases} \frac{\mu}{\lambda} \sum_{k=1}^{M} \frac{|h_k|^2}{\lambda|g_k|^2 + \mu|h_k|^2} &= P \\ \sum_{k=1}^{M} \frac{|h_k|^2}{\lambda|g_k|^2 + \mu|h_k|^2} - \sum_{k=1}^{M} \frac{n_k}{|g_k|^2} &= J \end{cases} \qquad (23)$$

it gives the following relation between λ and μ

$$\frac{\lambda}{\mu} = \frac{J + \sum_{k=1}^{M} \frac{n_k}{|g_k|^2}}{P} \qquad (24)$$

so, we can replace μ in p_k, and λ in j_k to get

$$
\begin{cases}
p_k & = 1/\lambda \left(1 + \frac{|g_k|^2}{|h_k|^2} \frac{J + \sum \frac{n_k}{|g_k|^2}}{P}\right) \\
j_k = 1/\mu \left(1 + \frac{|g_k|^2}{|h_k|^2} \frac{\left(J + \sum \frac{n_k}{|g_k|^2}\right)}{P}\right) - \frac{n_k}{|g_k|^2}
\end{cases}
\tag{25}
$$

Using the conditions (1) and (2), we get the closed form expressions of λ and μ

$$
\begin{cases}
\lambda & = \sum_{k=1}^{M} 1/\left(P + \frac{|g_k|^2}{|h_k|^2}\left(J + \sum \frac{n_k}{|g_k|^2}\right)\right) \\
\mu = \frac{1}{J + \sum \frac{n_k}{|g_k|^2}} \sum_{k=1}^{M} 1/\left(1 + \frac{|g_k|^2}{|h_k|^2} \frac{\left(J + \sum \frac{n_k}{|g_k|^2}\right)}{P}\right)
\end{cases}
\tag{26}
$$

Finally, replacing λ and μ in (25) gives the closed form expressions of the power allocations at the NE, and the following relation

$$
j_k = \frac{J + \sum \frac{n_k}{|g_k|^2}}{P} p_k - \frac{n_k}{|g_k|^2}
\tag{27}
$$

This analytical result will be compared in Subsect. 5.3 with the NE found by simulation through playing iteratively the unilateral games.

4.3   Case of Proportional Fading Channels

Now, let's consider the particular case studied by Altman in [3] of proportional fading coefficients,

$$
g_k = \gamma h_k, \forall k \in [1, M]
\tag{28}
$$

and we define

$$
\tau = \lambda \gamma + \mu
\tag{29}
$$

So, the expression of λ in (26) becomes

$$
\lambda = \frac{M}{P + \gamma(J + \sum_{k=1}^{M} \frac{n_k}{|g_k|^2})}
\tag{30}
$$

Replacing λ in (25) results in

$$
\begin{cases}
p_k & = \frac{P}{M} \\
j_k = \frac{J + \sum_{k=1}^{M} \frac{n_k}{|g_k|^2}}{M} - \frac{n_k}{|g_k|^2}
\end{cases}
\tag{31}
$$

which brings us to the same conclusion as [3] about uniform power allocation; i.e. if the jammer tries to jam all the channels, then the optimal anti-jamming strategy for the CR is to allocate its power equally over the channels, under the assumption of proportional fading coefficients.

5   Simulation and Comparison with the Closed Form Expressions

5.1   Saddle Point Example

Just to illustrate the concept of saddle-point, we have considered $M = 2$ flat fading channels with gain coefficients $h_k = g_k = 1, \forall k \in [1, M]$. We used P=30 and J=20 as the total power for the CR and the jammer respectively. We considered only finite sets of power allocations with steps of 1, so $p_k \in \{0, 1, 2, \cdots, P\}$ and $j_k \in \{0, 1, 2, \cdots, J\}$. We have implemented this scenario in Matlab using the exhaustive search over the finite set of possible power allocations. We calculated a matrix of capacity values, its rows are the possible jammer's power allocations and its columns are the CR's power allocations. We found the optimal maxmin CR's power allocation: $\mathbf{p}^* = (15, 15)$ corresponding to the column number 16, and optimal minimax jammer's power allocation $\mathbf{j}^* = (10, 10)$ corresponding to the row 11. Figure 2 illustrates this saddle point given by the indexes of $\mathbf{p}^*$ and $\mathbf{j}^*$.

5.2   Nash Equilibrium in the General Case

In this simulation we consider the Nash game between the CR and the jammer, which consists in playing iteratively the two unilateral games presented in Sect. 3. In each iteration of this game, the CR applies waterfilling technique according to equation (7), and the jammer uses bisection and equation (12) to update its power allocation. To cover the general case detailed in Subsect. 4.1, we consider the system model described in Fig. 1 with $M = 4$ parallel Gaussian channels, $P = 10$ and $J = 10$ as the total CR's and jammer's powers in watts, the background noise over the four channels $\mathbf{n} = (2, 0.75, 0.9, 1.1)$ and the channel gain coefficients $\mathbf{h} = (0.1, 1.1, 1.2, 1.3)$, $\mathbf{g} = (0.7, 0.8, 0.1, 1.2)$.

After convergence of the iterative game to almost fixed power allocations with tolerance $\epsilon = 1e - 10$, we get $\mathbf{j} = (0, 5.704, 0, 4.296)$ and $\mathbf{p} = (0, 2.5543, 5.5661, 1.8797)$. Which results in a payoff value of $C = 4.5978$ with $\frac{1}{\lambda} = 6.1911$ and $\mu = 0.06$. Figure 3 gives the received power due to the noise, jammer and CR's powers in each channel at the NE.

We can see that in channel 1, $p_k = j_k = 0$ since $\frac{n_1}{|h_1|^2} > \frac{1}{\lambda}$ which corresponds to the case (a) in paragraph Subsect. 4.1. Channel 3 receives $p_k > 0$ but $j_k = 0$, since $\frac{1}{3} < \frac{n_3}{|h_3|^2} < \frac{1}{\lambda}$ which corresponds to case (b). Channels 2 and 4 corresponds to case (c) since $\frac{n_k}{|h_k|^2} < \frac{1}{\tau_k}$ which results in $p_k > 0$ and $j_k > 0$.

5.3   Comparison of NE and Closed Form of the Saddle Point

In this subsection, we consider the case of all channels used by the CR and the jammer, studied in Subsect. 4.2. To compare the NE of the iterative power allocation game, with the closed form expressions of the power allocations at the saddle point, we consider the system model described in Fig. 1 with the following parameters: $M = 4$ parallel Gaussian channels, $P = 10$ and $J = 10$ as the total CR's and jammer's powers in watts, the background noise over the four channels $\mathbf{n} = (0.25, 0.75, 0.9, 1.1)$ and the channel gain coefficients $\mathbf{h} = (0.9, 1.1, 1.2, 1.3)$ and $\mathbf{g} = (0.7, 0.8, 1, 1.2)$.

Analytical Saddle Point. Let's start by replacing the parameters $(|\mathbf{h}|^2, |\mathbf{g}|^2, \mathbf{n}, P, J, M)$ in the closed form expressions given by analytical calculation in Subsect. 4.2. According to the optimal power allocations given by the expressions (25) and (26), we get $\mathbf{j} = (2.9625, 2.5073, 2.3574, 2.1729)$ and $\mathbf{p} = (2.602, 2.7568, 2.4407, 2.2005)$. Which results in a payoff value of $C = 4.4017$. Let's compare this analytical result with the simulation result found at the convergence of the game considering complete information.

Simulation NE. Under the same conditions, after convergence of the iterative game to almost fixed power allocations with tolerance $\epsilon = 1e - 10$, we get the same power allocation vectors and the same payoff value as given by closed form expressions, which validate our analytical calculation of the NE. Figure 4 gives the received power due to the noise, jammer and CR's powers in each channel at the NE.

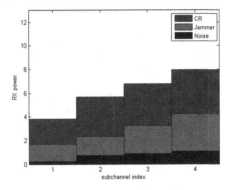

Fig. 4. The strategies at the NE (Color figure online)

6 Conclusion

In this paper, we considered a continuous power allocation zero-sum game between a jammer and a CR over parallel Gaussian channels. We provided the optimal strategy for each player depending on the other player's power allocation and on one parameter that usually found using the bisection method. Then, we provided analytical expressions for the optimal power allocations characterizing the saddle point of this game, under the assumption that both the CR and the jammer are using all the channels (i.e. $p_k, j_k > 0, \forall k \in [1, M]$). Finally, by means of numerical example we found that the analytical expressions are equal to the NE simulation result found by playing iteratively the unilateral games using bisection in each iteration.

References

1. Mitola, J.: Cognitive radio for flexible mobile multimedia communications. In: 1999 IEEE International Workshop on Mobile Multimedia Communications (MoMuC 1999), pp. 3–10 (1999)
2. Altman, E., Avrachenkov, K., Garnaev, A.: A jamming game in wireless networks with transmission cost. In: Chahed, T., Tuffin, B. (eds.) NET-COOP 2007. LNCS, vol. 4465, pp. 1–12. Springer, Heidelberg (2007)
3. Altman, E., Avrachenkov, K., Garnaev, A.: Fair resource allocation in wireless networks in the presence of a jammer. In: 3rd International ICST Conference on Performance Evaluation Methodologies and Tools, VALUETOOLS 2008, Athens, Greece, 20–24 October, 2008, p. 33 (2008)
4. Yang, D., Xue, G., Zhang, J., Richa, A.W., Fang, X.: Coping with a smart jammer in wireless networks: a stackelberg game approach. IEEE Trans. Wireless Commun. **12**(8), 4038–4047 (2013)
5. Gohary, R.H., Huang, Y., Luo, Z.-Q., Pang, J.-S.: A generalized iterative water-filling algorithm for distributed power control in the presence of a jammer. In: ICASSP 2009, pp. 2373–2376 (2009)
6. Ara, M., Reboredo, H., Ghanem, S., Rodrigues, M.: A zero-sum power allocation game in the parallel gaussian wiretap channel with an unfriendly jammer. In: IEEE International Conference on Communication Systems (ICCS), pp. 60–64, Nov 2012
7. Wu, Y., Wang, B., Liu, K.J.R., Clancy, T.C.: Anti-jamming games in multi-channel cognitive radio networks. IEEE J. Sel. Areas Commun. **30**(1), 4–15 (2012)
8. Dabcevic, K., Betancourt, A., Marcenaro, L., Regazzoni, C.: Intelligent cognitive radio jamming-a game-theoretical approach. EURASIP J. Adv. Sign. Process. **2014**, 1–18 (2014)
9. El-Bardan, R., Brahma, S., Varshney, P.: Power control with jammer location uncertainty: a game theoretic perspective. In: 2014 48th Annual Conference on Information Sciences and Systems (CISS), pp. 1–6, March 2014
10. Yin, Z., Korzhyk, D., Kiekintveld, C., Conitzer, V., Tambe, M.: Stackelberg vs. nash in security games: interchangeability, equivalence, and uniqueness. In: AAMAS. IFAAMAS, pp. 1139–1146 (2010)
11. Nikaid, H.: On von Neumann's minimax theorem. Pac. J. Math. **4**(1), 65–72 (1954)
12. Colman, A.: Game theory, its applications in the social, biological sciences. International Series in Social Psychology, 2nd edn. Butterworth-Heinemann, Routledge, Oxford, London (1995)
13. Ca, J.: Optimisation: thorie et algorithmes. Paris (1971)

Code-Aware Power Allocation for Irregular LDPC Codes

Zeina Mheich and Valentin Savin$^{(\boxtimes)}$

CEA-LETI, MINATEC Campus, 38054 Grenoble, France
{Zeina.Mheich,Valentin.Savin}@cea.fr

Abstract. In this paper, we investigate a code-dependent unequal power allocation method for Gaussian channels using irregular LDPC codes. This method allocates the power for each set of coded bits depending on the degree of their equivalent variable nodes. We propose a new algorithm to optimize the power allocation vector using density evolution algorithm under the Gaussian approximation. We show that unequal power allocation can bring noticeable gains on the threshold of some irregular LDPC codes with respect to the classical equal power allocation method depending on the code and the maximum number of decoding iterations.

Keywords: Irregular LDPC codes · Unequal power allocation · Density evolution · Gaussian approximation

1 Introduction

The Shannon capacity of the power constrained point-to-point Gaussian channel is achieved using independent and identically distributed (i.i.d.) symbols. Therefore, unequal power allocation (UPA) does not increase capacity. However, Shannon did not provide any practical coding/decoding scheme to achieve this capacity. Nowadays, there exists powerful codes approaching this capacity as LDPC codes. It was shown in [1] that irregular LDPC codes perform better than regular ones in terms of threshold. In irregular LDPC codes, the variable nodes do not have the same degree and thus are not equally protected. Thus we expect that the performance of the code will be affected if the power allocated for some set of symbols associated to variable nodes of a certain degree is different from that allocated for a set of symbols associated to variable nodes of an another degree. Therefore, it is not known if equal power allocation (EPA) is also optimal for practical irregular LDPC codes. This will be the subject of investigation in our work. In reference [2], the authors investigate the UPA problem for irregular LDPC codes for point-to-point Gaussian channels. They obtained up to 0.25 dB of gain with respect to EPA method. However, the authors use in [2] Monte-Carlo simulations in order to optimize the power allocation vector, which

---

This work was carried out within the framework of Celtic-Plus SHARING project.

D. Noguet et al. (Eds.): CROWNCOM 2016, LNICST 172, pp. 41–52, 2016.
DOI: 10.1007/978-3-319-40352-6_4

is time-consuming. Usually, the LDPC code is chosen to have a good threshold assuming that the decoder can perform unlimited number of iterations. In this work, we consider that a given irregular LDPC code is used at the transmitter and we optimize the power allocation for a target bit error rate and using a decoder with limited number of iterations. We propose a new algorithm to optimize the power allocation vector such that the noise threshold of the existing irregular LDPC code family is maximized. Hence, we propose a modified density evolution algorithm using Gaussian approximation [3] when the UPA method is used at the transmitter. We apply this algorithm for the Gaussian point-to-point channel and the Gaussian relay channel.

2   LDPC Codes and Density Evolution

This section recalls some of the basics about LDPC codes. We consider the point-to-point Gaussian channel case. At the source side, a message of k information bits is encoded by a LDPC encoder to a n-bit codeword. When the transmitter uses an equal power allocation strategy, the coded bits are modulated using a BPSK constellation, such that the bits 0 and 1 are mapped into $+1$ and -1 respectively. The resulting sequence after modulation is denoted by $x^n = [x_1, \cdots, x_n]$. The symbols x_i, $i \in \{1, \cdots, n\}$ are transmitted over a discrete-time memoryless additive white Gaussian noise channel. The channel output corresponding to the input x is $y = x + z$, where z is a noise following a normal distribution of zero mean and of variance σ^2. Consider an LDPC code characterized by a Tanner graph $\mathcal{H}$, with n variable nodes and m check nodes ($m = n - k$). An irregular LDPC code is characterized by bit nodes and check nodes with varying degrees. The fraction of edges which are connected to degree-i variable nodes is denoted λ_i, and the fraction of edges which are connected to degree-i check nodes, is denoted ρ_i. The functions $\lambda(x) = \sum_{i=2}^{d_v} \lambda_i x^{i-1}$ and $\rho(x) = \sum_{i=2}^{d_c} \rho_i x^{i-1}$ are defined to describe the degree distributions from the perspective of Tanner graph edges. By definition $\lambda(1) = 1$ and $\rho(1) = 1$. An alternative characterization of the degree distribution for the variable nodes $\Lambda(x) = \sum_{i=2}^{d_v} \Lambda_i x^i$, from the perspective of Tanner graph nodes, will be used also in this paper. Hence, Λ_i designates the fraction of degree-i variable nodes. The decoding algorithms used to decode LDPC codes are collectively called message-passing algorithms since they operate by the passing of messages along the edges of a Tanner graph. Under a message-passing algorithm, variable nodes and check nodes exchanges messages iteratively. Each node processes the received messages on the edges connected to it and sends messages back to its neighbors such that the output message is a function of all incoming messages to the node except the incoming message on the edge where the output message will be sent. The sum-product decoding is a message-passing algorithm in which the messages are log likelihood ratios (LLR) and the calculations at the variable and check nodes are performed using sum and product operations. Hence, a message can be written as the LLR of the equally probable random variable $x \in \{+1, -1\}$:

$$\text{LLR} = \log \frac{p(x = +1|w)}{p(x = -1|w)}, \qquad (1)$$

where w is a random variable describing all the information incorporated into this message. The sum-product algorithm iteratively computes an approximation of the maximum a posteriori (MAP value) for each code bit. However, the a posteriori probabilities returned by the sum-product decoder are only exact MAP probabilities if the Tanner graph is cycle free. Under the "cycle free assumption", the analysis of the decoding algorithm is straightforward because the incoming messages to every node are independent. In sum-product decoding the extrinsic message from a check node to a variable node, u, at the ℓth iteration, is

$$u^{(\ell)} = 2 \cdot \tanh^{-1}\left(\prod_{j=1}^{d_c-1} \tanh \frac{v_j^{(\ell)}}{2} \right), \tag{2}$$

where d_c is the check node degree and $v_j, j = 1, \cdots, d_c - 1$ are the received messages from all neighbors of the check node except the variable node that gets the message u. The message from a variable node to a check node, v at the ℓth iteration is equal to

$$v^{(\ell)} = \sum_{i=1}^{d_v-1} u_i^{(\ell-1)} + u_0, \tag{3}$$

where d_v is the variable node degree, $u_i, i = 1, \cdots, d_v - 1$ are the received messages from all neighbors of the variable node except the check node that gets the message v with $u_i^{(0)} = 0$ and u_0 is the input a priori LLR of the output bit associated with the variable node. For an AWGN channel the a priori LLRs are given by $u_0 = \frac{2}{\sigma^2}y$. The total LLR of the i-th bit is $\text{LLR}_i^{(\ell)} = \sum_{i=1}^{d_v} u_i^{(\ell)} + u_0$. The i-th bit is decided to be a 0 if $\text{LLR}_i > 0$, and 1 otherwise. The decoding process ends when the decoded sequence is a codeword or until the maximum number of iterations is reached.

Density evolution is an algorithm where the evolution of probability density functions of the exchanged messages are tracked through the message-passing algorithm. It determines the behavior of an ensemble of Tanner graphs if the channel is memoryless and under the assumption that the Tanner graphs are all cycle free. Due to the symmetry of the channel and the decoder, the density evolution equations can be derived without loss of generality by assuming that all-zero codeword is sent through the channel $(x_i = +1, \forall i \in \{1, \cdots, n\})$. Thus, negative messages indicates errors. In particular, the evolution of the error probability can be determined, via density evolution, as a function of the iteration number of the message-passing decoding algorithm. The density evolution algorithm enables to compute the *noise threshold* of a family of LDPC codes which is the maximum level (e.g. variance) of channel noise such that the probability of error converges to zero as the number of iterations tends to infinity.

Chung *et al.* investigated in [3] the sum-product decoding of LDPC codes using a Gaussian (for regular LDPC codes), or a Gaussian mixtures (for irregular LDPC codes), approximation for message densities (of u and v) under density evolution to simplify the analysis of the decoding algorithm and the design of irregular LDPC codes for AWGN channels. The authors show that the mean of a Gaussian density, which is a one-dimensional quantity, can act as faithful surrogate for the message density, which is an infinite-dimensional vector.

In order to present some results of [3] on the density evolution using Gaussian approximation, assume that the all-zero codeword is sent through the channel. Thus, the LLR message $u_0 = \frac{2}{\sigma^2}y$ from the channel is Gaussian with mean $\frac{2}{\sigma^2}$ and variance $\frac{4}{\sigma^2}$. The *symmetry condition* for a Gaussian variable with mean m and variance σ^2 reduces to $\sigma^2 = 2m$, thus we need only to keep the mean during the density evolution process. We denote the means of the messages u and v by m_u and m_v respectively. The Gaussian approximation method in [3] for irregular LDPC codes assumes that the individual output of a variable or a check node is Gaussian. Thus, the mean of the output of a variable node of degree i at the ℓth iteration, $m_{v,i}^{(\ell)}$, is given by

$$m_{v,i}^{(\ell)} = m_{u_0} + (i-1)m_u^{(\ell-1)}, \tag{4}$$

where m_{u_0} and $m_u^{(\ell-1)}$ are the means of u_0 and $u^{(\ell-1)}$ respectively. Therefore, a message v sent by the variable node to its neighbors check nodes at the ℓth iteration has a density function $f_v^{(\ell)}$ following a Gaussian mixture:

$$f_v^{(\ell)} = \sum_{i=2}^{d_v} \lambda_i \mathcal{N}(m_{v,i}^{(\ell)}, 2m_{v,i}^{(\ell)}) \tag{5}$$

Using (2), the authors demonstrate also in [3] that the mean of the Gaussian output message $u_j^{(\ell)}$ of a degree-j check node at the ℓth iteration, $m_{u,j}^{(\ell)}$ can be written as:

$$m_{u,j}^{(\ell)} = \phi^{-1}\left(1 - \left[1 - \sum_{i=2}^{d_v} \lambda_i \phi(m_{v,i}^{(\ell)})\right]^{j-1}\right), \tag{6}$$

where

$$\phi(x) = \begin{cases} 1 - \frac{1}{\sqrt{4\pi x}} \int_{\mathbb{R}} \tanh \frac{u}{2} e^{-\frac{(u-x)^2}{4x}} \, du & \text{if } x > 0 \\ 1, & \text{if } x = 0. \end{cases} \tag{7}$$

Hence, the mean of $u^{(\ell)}$ at the ℓth iteration, $m_u^{(\ell)}$, is obtained by linearly combining $m_{u,j}^{(\ell)}$ with weights $\rho_j, 2 \leq j \leq d_c$:

$$m_u^{(\ell)} = \sum_{j=2}^{d_c} \rho_j \phi^{-1}\left(1 - \left[1 - \sum_{i=2}^{d_v} \lambda_i \phi(m_{v,i}^{(\ell)})\right]^{j-1}\right). \tag{8}$$

The noise threshold σ^* is the supremum of all $\sigma \in \mathbb{R}^+$ such that $m_u^{(\ell)} \to \infty$ as $\ell \to \infty$. In [3], the functions $h_i(s,r)$ and $h(s,r)$ are defined as:

$$h_i(s,r) = \phi\left(s + (i-1)\sum_{j=2}^{d_c} \rho_j \phi^{-1}(1 - (1-r)^{j-1})\right), \tag{9}$$

$$h(s,r) = \sum_{i=2}^{d_v} \lambda_i h_i(s,r). \tag{10}$$

The Eq. (8) is written equivalently in [3] as

$$r_\ell = h(s, r_{\ell-1}), \tag{11}$$

where $s = m_{u_0} = \frac{2}{\sigma^2}$, and $r_0 = \phi(s)$. It is demonstrated in [3] that the convergence condition is equivalent to $r_\ell \underset{\ell \to \infty}{\to} 0$ and is satisfied iff $r > h(s, r), \forall r \in (0, \phi(s))$.

2.1 Unequal Power Allocation

Problem Formulation. Given an irregular LDPC code with a variable node degree distribution $\Lambda(x) = \sum_{i=2}^{d_v} \Lambda_i x^i$, we denote by P_i the power allocated at the transmitter to a symbol associated to a variable node of degree i. Thus $P_i > 0$ if $\Lambda_i > 0$ (Λ_i is the portion of variable nodes of degree i). The bits 0 and 1 are mapped into $+\sqrt{P_i}$ and $-\sqrt{P_i}$ respectively. We assume that the destination is aware of the power allocation strategy at the source. Without loss of generality, we assume a total power constraint $P = 1$. The power constraint at the transmitter can be written as $\sum_{i=2}^{d_v} \Lambda_i P_i = 1$. In this work, we propose to choose $P_i, i \in \{2, \cdots, d_v\}$ in order to optimize the threshold of the irregular LDPC code family under consideration via density evolution. This is because density evolution, using Gaussian approximation, is a simple tool to evaluate the asymptotic performance of a family of LDPC codes. For convenience, we denote by $\mathbf{P}$ the vector whose elements are $P_i, i \in \{2, \cdots, d_v\}$. Thus, the optimization problem under consideration is the following:

$$\sigma_{th} = \max_{\mathbf{P}} \sigma^*(\mathbf{P})$$

$$\text{subject to} \quad \sum_{i=2}^{d_v} \Lambda_i P_i = 1, \tag{12}$$

where $\sigma^*(\mathbf{P})$ is the noise threshold for a given $\mathbf{P}$. When a 0-bit is transmitted with a power P_i, the mean of the message from the channel, $m_{u_{0i}} = \frac{2}{\sigma^2} P_i$ is varying with the node degree unlike the case with equal power allocation. We can easily extend the equations of the evolution of message means using Gaussian approximation in (4–11) to the case with unequal power allocation at the transmitter. Therefore, we can demonstrate that these equations become:

$$m_{v,i}^{(\ell)} = m_{u_{0i}} + (i-1)m_u^{(\ell-1)}, \tag{13}$$

$$m_v^{(\ell)} = \sum_{i=2}^{d_v} \lambda_i m_{v,i}^{(\ell)}, \tag{14}$$

$$m_{u,j}^{(\ell)} = \phi^{-1}\left(1 - \left[1 - \sum_{i=2}^{d_v} \lambda_i \phi(m_{v,i}^{(\ell)})\right]^{j-1}\right), \tag{15}$$

$$m_u^{(\ell)} = \sum_{j=2}^{d_c} \rho_j \phi^{-1}\left(1 - \left[1 - \sum_{i=2}^{d_v} \lambda_i \phi(m_{v,i}^{(\ell)})\right]^{j-1}\right). \tag{16}$$

$$h_i(s_i, r) = \phi\left(s_i + (i-1)\sum_{j=2}^{d_c} \rho_j \phi^{-1}(1 - (1-r)^{j-1})\right), \tag{17}$$

$$h(\mathbf{s}, r) = \sum_{i=2}^{d_v} \lambda_i h_i(s_i, r). \tag{18}$$

$$r_\ell = h(\mathbf{s}, r_{\ell-1}), \tag{19}$$

where $s_i = m_{u_{0i}} = \frac{2}{\sigma^2} P_i$, $\mathbf{s} = \{s_i\}$, and $r_0 = \sum_{i=2}^{d_v} \lambda_i \phi(s_i)$. The convergence condition to the threshold is the same of [3]. Therefore, for a given vector $\mathbf{P}$, the threshold $\sigma^*(\mathbf{P})$ is the supremum of all $\sigma \in \mathbb{R}^+$ such that $r_\ell \to 0$ as $\ell \to \infty$.

Proposed Solution. In order to solve the optimization problem in (12), we propose an algorithm with less complexity comparing to exhaustive search, in which the number of optimization variables is independent of the LDPC code. The proposed solution is inspired from the expression of r_ℓ in (19). Indeed, we rewrite r_ℓ as

$$r_\ell(\mathbf{P}, \sigma) = \sum_{i=2}^{d_v} \lambda_i \phi\left(\frac{2}{\sigma^2} P_i + (i-1)k_{\ell-1}\right), \tag{20}$$

where $k_{\ell-1} = \sum_{j=2}^{d_c} \rho_j \phi^{-1}(1 - (1 - r_{\ell-1})^{j-1}) \in (0, r_0)$. We recall that $\phi(x)$ is continuous and monotonically decreasing on $[0, +\infty)$, with $\phi(0) = 1$ and $\phi(\infty) = 0$ [3]. Hence $r_\ell \geq 0$. Since the convergence condition to the threshold requires that $r_\ell \underset{\ell \to \infty}{\to} 0$, we consider the parametric family of functions $\{f_k, k \geq 0\}$ with parameter k, where f_k is defined by

$$f_k(\sigma, \mathbf{P}) = \sum_{i=2}^{d_v} \lambda_i \phi\left(\frac{2}{\sigma^2} P_i + (i-1)k\right), \tag{21}$$

and for a fixed σ, we look only for the vectors $\mathbf{P}$ which are the minimas of $\{f_k\}_{k \geq 0}$. Therefore, for a fixed σ and k, we consider the following optimization problem

$$\mathbf{P}^*(\sigma, k) = \arg\min_{\mathbf{P}} f_k(\sigma, \mathbf{P})$$

$$\text{subject to } \sum_{i=2}^{d_v} \Lambda_i P_i = 1. \tag{22}$$

Then, we define $r(k, \sigma) = r_\infty(\mathbf{P}^*(\sigma, k), \sigma)$ which can be calculated using (20). For a fixed k, the threshold $\sigma^*(k)$ is defined as the maximal value of σ such that $r(k, \sigma) \to 0$. Finally, k is chosen to maximize the threshold, thus we denote

$$k^* = \arg\max_k \sigma^*(k). \tag{23}$$

For a given k and σ, we should solve the optimization problem in (22). The expression of $\phi(x)$ in (7) makes very difficult to have a closed form expression of $\mathbf{P}^*(\sigma, k)$. In [3] the following approximation of $\phi(x)$ is used $\phi(x) \sim e^{-\alpha x^\gamma + \beta}$, where $\alpha = 0.4527$, $\beta = 0.0218$ and $\gamma = 0.86$. Even with this approximation, it is difficult to obtain an analytic solution for the optimization problem (22). Therefore, we propose to approximate $\phi(x)$ by a convex function of the form $\phi(x) \sim e^{-ax}$ with $a > 0$. Since the value for which the function ϕ should be evaluated in (21) depends on i and since e^{-ax} cannot approximate exactly $e^{-\alpha x^\gamma + \beta}$ for all values of x, a, we define a function $\phi_i(x) = e^{-a_i x}$ for each $i \in \{2, \cdots, d_v\}$ with $a_i > 0$. Thus, using the latest approximation, the optimization algorithm in (22) becomes:

$$\mathbf{P}^*(\sigma, k) = \arg\min_{\mathbf{P}} \sum_{i=2}^{d_v} \lambda_i \phi_i \left(\frac{2}{\sigma^2} P_i + (i-1)k \right)$$

$$\text{subject to} \quad \sum_{i=2}^{d_v} \Lambda_i P_i = 1. \tag{24}$$

Proposition 1. *The solution of the optimization algorithm in (24) is*

$$P_i^*(\sigma, k) = -\frac{(i-1)k}{k_0} - \frac{1}{a_i k_0} \log\left(\frac{\Lambda_i}{a_i \lambda_i k_0} \right)$$
$$- \left(\frac{1}{a_i \sum_i \frac{\Lambda_i}{a_i}} \right) \left(-1 - k \sum_{i=2}^{d_v} (i-1)\frac{\Lambda_i}{k_0} - \sum_{i=2}^{d_v} \frac{\Lambda_i}{a_i k_0} \log\left(\frac{\Lambda_i}{a_i \lambda_i k_0} \right) \right), \tag{25}$$

where $k_0 \triangleq \frac{2}{\sigma^2}$ and $i \in \{2, \cdots, d_v\}$.

Proof. First, we form the Lagrangian of problem (24)

$$L(\mathbf{P}, \theta; \sigma, k) = \sum_{i=2}^{d_v} \lambda_i \phi_i \left(\frac{2}{\sigma^2} P_i + (i-1)k \right) + \theta \left(\sum_{i=2}^{d_v} \Lambda_i P_i - 1 \right). \tag{26}$$

where θ is the Lagrange multiplier. Since the objective function in problem (24) is convex and the constraint is affine, the KKT conditions guarantee the global optimality of the solution. Thus, the solution $(\mathbf{P}^*, \theta^*)$ of problem (24) verifies the following equations

$$\begin{cases} \frac{\partial L}{\partial P_i}(P_i^*) = -a_i k_0 \lambda_i e^{-a_i(k_0 P_i^* + (i-1)k)} + \theta^* \Lambda_i = 0, & \forall i \in \{2, \cdots, d_v\} \\ \sum_{i=2}^{d_v} \Lambda_i P_i^* - 1 = 0. \end{cases} \tag{27}$$

For simplicity of notation, we drop the dependence of P_i^* on σ and k. From the first equation in (27), we get

$$P_i^* = -\frac{k(i-1)}{k_0} - \frac{\log\left(\frac{\theta^* \Lambda_i}{a_i \lambda_i k_0} \right)}{a_i k_0}. \tag{28}$$

After substituting (28) in the equation $\sum_{i=2}^{d_v} \Lambda_i P_i^* - 1 = 0$, we get

$$\theta^* = e^{\frac{-k\sum_i (i-1)\Lambda_i - k_0 - \sum_i \frac{\Lambda_i}{a_i} \log\left(\frac{\Lambda_i}{a_i \Lambda_i k_0}\right)}{\sum_i \frac{\Lambda_i}{a_i}}}. \tag{29}$$

Finally, after substituting (29) in (28), we get (25).

In order to determine the set $\{a_i\}_{i>1}$, we set $e^{-a_i x} = e^{-\alpha x^\gamma + \beta}$ where x is the value for which the function ϕ should be evaluated. Thus $a_i = \frac{-\alpha x^\gamma + \beta}{-x}$. Since we should evaluate $\phi_i(x)$ in (24) where $x = \frac{2}{\sigma^2} P_i + (i-1)k$ and that P_i is unknown, we set in our experiments $a_i = \frac{-\alpha \hat{x}^\gamma + \beta}{-\hat{x}}$ where $\hat{x} = \frac{2}{\sigma^2} + (i-1)k$.

The proposed solution is summarized in Algorithm 1.

Remark: In the optimization problem (24), we discarded the constraint $P_i \geq 0$. The solution obtained in Proposition 1 can be written as $P_i^* = A_i + \frac{1}{k_0} B_i$ where $A_i > 0$ and $k_0 = 2/\sigma^2$. Thus if for some $i \in \{2, \cdots, d_v\}$ and k, the obtained solution P_i^* is non-positive, we should decrease the value of σ while searching for the threshold, as stated in Algorithm 1.

Algorithm 1. Algorithm to solve (12)

1: $\epsilon \leftarrow 10^{-10}$, MaxIter $\leftarrow 10^4$.
2: **for** $k = 0 : k_{\max}$ **do**
3:      $\sigma_{\min} \leftarrow 0$, $\sigma_{\max} \leftarrow 10$
4:      $\sigma \leftarrow \frac{\sigma_{\max} + \sigma_{\min}}{2}$
5:      **while** $\sigma_{\max} - \sigma_{\min} > \epsilon$ **do**
6:          Compute $\mathbf{P}^*(\sigma, k)$ using Proposition 1
7:          **if** there exists at least one $P_i^* < 0$ **then** decrease $\sigma_{\max}$
8:          **else**
9:              $m_{u_{0i}} \leftarrow \frac{2}{\sigma^2} P_i^*(\sigma, k)$, for $i \in \{2, \cdots, d_v\}$
10:             **for** $\ell = 1 : $ MaxIter **do**
11:                 Calculate $r_\ell = h(\mathbf{s}, r_{\ell-1})$ using (19)
12:                 **if** $r_\ell < \epsilon$ **then** break
13:             **if** $r_\ell < \epsilon$ **then** $\sigma_{\min} \leftarrow \sigma$
14:             **else** $\sigma_{\max} \leftarrow \sigma$
15:             $\sigma \leftarrow \frac{\sigma_{\max} + \sigma_{\min}}{2}$
16:     Save $\sigma^*(k) = \sigma$
17: $\sigma_{\text{th}} \leftarrow \max_k \sigma^*(k)$

In Algorithm 1, "MaxIter" represents the maximum number of iterations that can be performed by the decoder. In our work, we will study also the unequal power allocation for decoders with limited number of iterations L. In this case, we set MaxIter$= L$ in Algorithm 1. Finally, we should note that the power allocation vector maximizing the noise threshold in Algorithm 1 to get zero error probability could achieve worst performance than equal power allocation for practical (low) bit error rate values (BER). Hence we propose to optimize the noise threshold

$\sigma_{\text{th}}(\eta)$ for a target error probability η. Using the density evolution algorithm with Gaussian approximation, the error probability at the Lth iteration of the decoder, $P_e^{(L)}$, can be calculated. Therefore, for a target BER η, the convergence condition to the threshold $\sigma_{\text{th}}(\eta)$ becomes $P_e^{(L)} < \eta$, instead of $r_L \to 0$.

3   Unequal Power Allocation for Relay Channels

In this section, we consider the Gaussian relay channel case. The indexes s, r and d will refer to the source, the relay and the destination respectively. The relay uses the "amplify and forward" strategy. Without loss of generality, we consider an average power constraint at the source $P_s = 1$ and an average power constraint at the relay $P_r = 1$.

In the relay channel case, the source and the relay can use simultaneously different power allocation strategies but this makes the problem difficult to solve due to the large number of variables. Therefore, we will study in the following, a strategy where the source only uses an unequal power allocation strategy. The transmitted signal x_s by the source is given by $x_s = \sqrt{P_{si}}x$ where $x \in \{+1, -1\}$ and P_{si} is the power allocated at the source for a bit associated to variable node of degree i. We denote by $\mathbf{P_s}$ the vector whose elements are P_{si}, $i \in \{2, \cdots, d_v\}$. The received signal by the relay is $y_{sr} = \sqrt{P_{si}}x + z_{sr}$, $z_{sr} \sim \mathcal{N}(0, \sigma_{sr}^2)$. The transmitted signal by the relay is given by $x_r = f y_{sr}$, where $f = \frac{1}{\sqrt{P_s + \sigma_{sr}^2}} = \frac{1}{\sqrt{1 + \sigma_{sr}^2}}$. The destination receives $y_{sd} = x_s + z_{sd}$ from the source ($z_{sd} \sim \mathcal{N}(0, \sigma_{sd}^2)$) and $y_{rd} = x_r + z_{rd} = f\sqrt{P_{si}}x + f z_{sr} + z_{rd}$ from the relay ($z_{rd} \sim \mathcal{N}(0, \sigma_{rd}^2)$). The destination, aware of the power allocation strategy, can determine y'_{sd} and y'_{rd} as

$$y'_{sd} = \frac{y_{sd}}{\sqrt{P_{si}}} = x + z'_{sd}, \text{ where } z'_{sd} = \frac{z_{sd}}{\sqrt{P_{si}}} \sim \mathcal{N}(0, \frac{\sigma_{sd}^2}{P_{si}}), \text{ and } y'_{rd} = \frac{y_{rd}}{f\sqrt{P_{si}}} = x + z'_{rd},$$

where $z'_{rd} = \left(\frac{z_{sr}}{\sqrt{P_{si}}} + \frac{z_{rd}}{f\sqrt{P_{si}}} \right) \sim \mathcal{N}(0, \frac{\sigma_{sr}^2}{P_{si}} + \frac{\sigma_{rd}^2}{f^2 P_{si}})$. At the destination, the LLR message at a variable node of degree i from the relay channel is given by

$$u_{0i} = \log \frac{p(x = +1 | y'_{sd}, y'_{rd})}{p(x = -1 | y'_{sd}, y'_{rd})} \tag{30}$$

$$= \log \frac{p(y'_{sd} | x = +1)}{p(y'_{sd} | x = -1)} + \log \frac{p(y'_{rd} | x = +1)}{p(y'_{rd} | x = -1)} \tag{31}$$

$$= \qquad u_{0i}^s \qquad + \qquad u_{0i}^r \tag{32}$$

where $u_{0i}^s = \frac{2y'_{sd} P_{si}}{\sigma_{sd}^2}$ and $u_{0i}^r = \frac{2y'_{rd} P_{si}}{\sigma_{sr}^2 + \frac{\sigma_{rd}^2}{f^2}}$. Hence, when the all-zero codeword is assumed to be sent by the source, the mean of the LLR message u_{0i} is given by $m_{u_{0i}} = m_{u_{0i}^s} + m_{u_{0i}^r}$, where $m_{u_{0i}^s} = \frac{2P_{si}}{\sigma_{sd}^2}$ and $m_{u_{0i}^r} = \frac{2P_{si}}{\sigma_{sr}^2 + \frac{\sigma_{rd}^2}{f^2}}$. In the relay channel case, we have three independent channels. Therefore, we should fix the SNR of two channels and find the power allocation vector which maximizes the noise threshold for the remaining channel. In this work, we consider the setting where the source and the relay know both the SNRs of the channels source-relay and source-destination. The power allocation vector at the source is optimized in order to maximize the threshold of the relay-destination channel. This optimization can be solved in the same manner as in the point-to-point case. Indeed, the Algorithm 1 can be extended for the relay channel case after replacing the expression of $m_{u_{0i}}$ in Algorithm 1 by its expression in the relay case depending on the strategy used. The output σ_{th} of Algorithm 1 will refer to $\sigma_{rd_{\text{th}}}$ in the relay case. Moreover, the Proposition 1 should be updated for the relay channel case in Algorithm 1. Hence, it is easy to demonstrate that the expression of P_{si}^* as function of k and the channel noise variances is obtained from Proposition 1 by replacing k_0 with $k_0 = \frac{2}{\sigma_{sd}^2} + \frac{2}{\sigma_{sr}^2 + \frac{\sigma_{rd}^2}{f^2}}$.

4   Simulation Results and Discussion

This section presents simulation results on unequal power allocation for irregular LDPC codes. Table 1 shows the SNR threshold of some irregular LDPC codes using Gaussian approximation with both equal power allocation and unequal power allocation. It gives also the power allocation function $P(x)$ obtained by Algorithm 1, for each code in Table 1, where $P(x) = \sum_i P_i x^{i-1}$ and P_i is the power allocated for the degree-i variable node. We observed in our simulations that our proposed algorithm gives the same threshold values as the exhaustive search method. We observe in Table 1 that a gain up to 0.22 dB can be obtained on the threshold with respect to equal power allocation strategy. However, for some codes EPA is optimal. We recall that the proposed power allocation method in Algorithm 1 relies on computing the threshold SNR$^*(k)$ for each parameter $k \geq 0$ and then to choose k^* which gives the better threshold (cf. (22) and 23). Figure 1 shows SNR$^*(k)$ as a function of k for the LDPC code of rate 1/2 and $\Lambda(x) = 0.5x^2 + 0.5x^4$. We observe that the SNR threshold decreases with k until k^*. Intuitively, k is small means that the density evolution algorithm is far from the convergence (cf. 20). Thus, the power allocation vector optimized for small values of k could be not the optimal power allocation vector when the density evolution algorithm is near convergence. Moreover, when k becomes large, the objective function to minimize in problem (24) decreases (since $\phi(x)$ is decreasing with x), and the objective function becomes less dependent on the power allocation vector (since its value is close to zero when k is large). Thus there is a trade-off in the optimal value of k which minimizes the SNR threshold.

Remark: for k and σ fixed, the objective function to minimize in problem (24) can be written as $f(\mathbf{P}) = \sum_{i=2}^{d_v} \lambda_i f_i(P_i)$, where $f_i(P_i) = \phi_i\left(\frac{2}{\sigma^2}P_i + (i-1)k\right)$. When the irregular LDPC code involves a degree i which is too large, $f_i(P_i)$ becomes too small and the value of P_i will not affect too much the value of the objective function. Thus, in our simulations, we define a "saturation" parameter K_{sat} and we set $f_i(P_i) = \phi_i\left(\frac{2}{\sigma^2}P_i + K_{\text{sat}}\right), \forall(i,k)$ such that $(i-1)k \geq K_{\text{sat}}$. This introduces an additional optimization variable in Algorithm 1 (K_{sat}), however, this value is optimized in our simulations one time for all the simulated codes.

Table 1. The threshold in dB of some LDPC code families of rate 1/2 with EPA and UPA for the BIAWGN point-to-point channel.

$\Lambda(x)$	SNR$_{th}$ EPA	SNR$_{th}$ UPA	$P(x)$
$0.5x^2 + 0.5x^4$	0.8733	0.8725	$0.9775x^2 + 1.0225x^4$
$0.5x^2 + 0.5x^6$	1.0854	0.9492	$0.8183x^2 + 1.1817x^6$
$0.5x^2 + 0.5x^8$	1.4520	1.2301	$0.8509x^2 + 1.1491x^8$

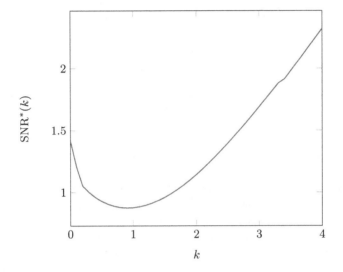

Fig. 1. SNR$^*(k)$ as a function of k, for the code of rate 1/2 and $\Lambda(x) = 0.5x^2 + 0.5x^4$.

Figure 2 shows the SNR threshold as a function of the decoder maximum number of iterations L for a target BER less than 10^{-8}. We observe that a gain of 0.5 dB on the SNR threshold can be obtained by optimizing the power allocation with respect to equal power allocation. We should note that in this work, we optimize the power allocation for a given code but rather it is also possible to optimize the code degree distributions for a decoder of fixed number of iterations L and using EPA. Figure 3 shows the SNR threshold for the relay-destination channel as a function of the decoder maximum number of iterations L for a target BER less than 10^{-8}, when $\sigma_{sd} = 1.1$ and $\sigma_{sr} = 0.7$. The gain of the unequal power allocation method with respect to the equal power allocation seems to be more important for the relay case (up to 2.4 dB). This is because we have more degrees of freedom in the relay case where we can fix the SNR of two channels and determine the threshold for the remaining channel. When L increases, the gain brought by UPA decreases in Fig. 2 and Fig. 3.

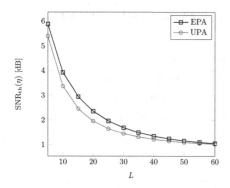
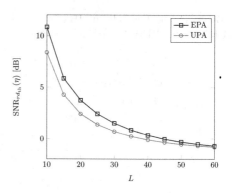

Fig. 2. $SNR_{th}(\eta = 10^{-8})$ as a function of L, for the code of rate $1/2$ and $\Lambda(x) = 0.5x^2 + 0.5x^4$, in the point-to-point channel case.

Fig. 3. $SNR_{rd_{th}}(\eta = 10^{-8})$ as a function of L, for the code of rate $1/2$ and $\Lambda(x) = 0.5x^2 + 0.5x^4$, in the relay channel case with $\sigma_{sd} = 1.1$ and $\sigma_{sr} = 0.7$.

5 Conclusion

In this paper, we investigated an unequal power allocation method for irregular LDPC codes where the power allocated of a coded bit depends on the degree of its associated variable node in the Tanner graph. We proposed an algorithm to optimize the power allocation vector using a modified density evolution algorithm under the Gaussian approximation. Simulation results show that in some cases, the unequal power allocation leads to a gain on the threshold of the LDPC code comparing to equal power allocation.

References

1. Richardson, T.J., Shokrollahi, M.A., Urbanke, R.L.: Design of capacity-approaching irregular low-density parity-check codes. IEEE Trans. Inf. Theor. **47**(2), 619–637 (2001)
2. Qi, H., Malone, D., Subramanian, V.: Does every bit need the same power? An investigation on unequal power allocation for irregular LDPC codes. In: WCSP, Nanjing, pp. 1–5 (2009)
3. Chung, S.Y., Richardson, T., Urbanke, R.L.: Analysis of sum-product decoding of low-density parity-check codes using a Gaussian approximation. IEEE Trans. Inf. Theor. **47**(2), 657–670 (2001)

Cooperative Game and Relay Pairing in Cognitive Radio Networks

Lifeng Hao, Sixing Yin[✉], and Zhaowei Qu

Beijing University of Posts and Telecommunications, Beijing 100876, China
{hlf415,yinsixing,zwqu}@bupt.edu.cn

Abstract. In this paper, we focus on cooperative spectrum access in a cognitive radio networks (CRN), where secondary users (SUs) serve as relays for primary users (PUs) to improve their throughput, and in return SUs can gain transmission opportunities. To optimize the overall utility of a cooperative CRN, we first investigate the cooperation between a single pair of PU and SU with Stackelberg game model, where PU determines access time allocation while SU determines relaying power for the PU. Based on the analytical results, cooperation pairing between multiple PUs and SUs is modeled as a bipartite matching problem and solved using *Gale-Shapley* algorithm. Numerical results demonstrate that, with the proposed schemes, overall utility for PUs and SUs can be balanced with low computational complexity.

Keywords: Cognitive radio · Cooperative communications · Stable matching

1 Introduction

With the rapid development in wireless applications and services, the demand for radio spectrum resource has significantly increased. However, the radio spectrum is limited and much of it has already been licensed exclusively to existing services. What's more, it is widely recognized that the licensed spectrum is in fact underutilized since licensed users typically do not fully utilize their allocated spectrum at most of the time. On the contrary, unlicensed users are starved for spectrum availability [1].

To cope with such a dilemma, a great number of solutions have been discussed and cognitive radio (CR) turns out to be the one with most potential by allowing secondary users (SUs) opportunistically to utilize the spectrum resource, which is found temporarily unused by primary (licensed) users (PUs) via spectrum sensing. However, due to PU's dynamics and unreliability of spectrum sensing resulted from channel fading or shadowing, SUs are forced to terminate the ongoing transmission once it detects that the spectrum band is reoccupied by a PU, which making SU's transmission highly unstable.

This work is supported by the grant from the National Natural Science Foundation of China (No. 61372109 and No. 61401034).

Moreover, cooperative spectrum access has emerged as a powerful technique. In CR systems, instead of keeping silent when PUs are busy, SUs can actively relay PUs' data and in return gain opportunities for its own transmission [2,3]. In [4], a scenario in which the SU acts as a relay for the packets that the SU can receive from the primary source but the primary destination can't, is considered and the stable throughput of the SU under this model is derived. The authors in [5] propose that the PU has the possibility to lease the owned spectrum to an *ad hoc* network of secondary nodes in exchange for cooperation in the form of distributed space-time coding. In [6], the authors studied the optimal cooperation strategy with Energy Harvesting by discuss the cooperation and none-cooperation modes. In [7], the authors investigate optimization for the cooperative spectrum sensing with an improved energy detector to minimize the total error rate (sum of the probability of false alarm and miss detection). However, most of the existing works involve only one pair of PU and SU, which may not be fully applied to the whole network.

In this paper, we consider a CR system with multiple pairs of primary and secondary transceivers, which operate in time-slotted mode. Each PU operates on a unique channel and can choose only one SU as relay using *Decode-and-Forward* (DF) mode. This work will use the stable mating algorithm to determine cooperation pairing for PUs and SUs. Moreover, we investigate the optimal cooperation strategy in this CR system, i.e., PUs decides the optimal allocation of channel resource to maximize their summarize transmission rate and SUs decide their optimal cooperative transmitting power to maximize their summarize transmission rate without spending too much power for relaying. The contribution of this work can be summarized as follows. First, we study the cooperation between multiple PUs and SUs and between single PU and single SU. Second, we study the cooperation by using DF mode but most of others are AF mode. Finally, cooperation between PUs and SUs is studied to maximize the primary network utility and secondary network utility.

The remainder of the paper is organized as follows. The detailed description of the system model is given in Sect. 2. Cooperation between one PU and one SU is studied in Sect. 3 and cooperation between multiple PUs and SUs is studied in Sect. 4. Simulation is provided in Sect. 5. Concluding remarks are provided in Sect. 6.

1.1 System Model

As shown in Fig. 1, we consider a CRN that consists of M PUs and N SUs, in which each PU transmits data on an unique channel. Instead of keeping silent when the PU is busy, an SU can alternatively act as a cooperative relay to improve the PU's throughput, which makes SU benefit a fraction of time slot for secondary transmission as reward. In this paper, we consider the DF mode for the cooperation, i.e., the SU firstly decodes the received signal from the PU transmitter, re-encodes it, and then transmits it to the corresponding PU receiver. Notations used in this paper are summarized in Table 1.

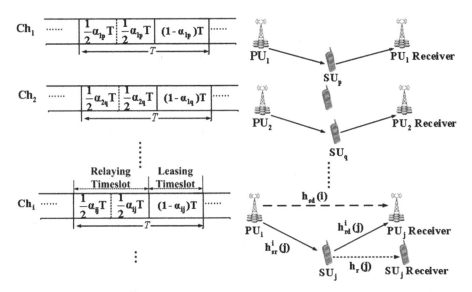

Fig. 1. Cooperative cognitive radio network with multiple channels.

Table 1. Notations

T	Timeslot duration
N_0	The white Gaussian noise
W	The spectrum bandwidth
α_{ij}	The fraction of timeslot that PU_i and PU_j cooperate
P_s^i	The PU_i's transmitting power
$P_r^i(j)$	The SU_j's transmitting power when cooperated with PU_i
h_{sd}^i	The channel gains in the PU_i's direct transmission
$h_{sr}^i(j)$	The channel gains from PU_i transmitter to SU_j
$h_{rd}^i(j)$	The channel gains from SU_j to PU_i's corresponding receiver
$h_r(j)$	The channel gains from SU_j to its corresponding receiver

Both of PUs and SUs operate in time-slotted mode as shown in Fig. 1. A fraction α_{ij} $(0 < \alpha_{ij} < 1)$ of the time slot duration T is used for cooperation between PU_i and SU_j. In the first duration of $\frac{\alpha_{ij}}{2}T$, PU_i transmits its data to SU_j. In the next duration of $\frac{\alpha_{ij}}{2}T$, SU_j relays the received data to the PU_i receiver. In the last period of $(1 - \alpha_{ij})T$, the cooperating SU_j is rewarded to transmit its own data while PU_i is silent. A common control channel is assumed for exchanging information on cooperation decision among PUs and SUs.

In such a cooperation model, we study the optimal strategy for the overall utility of the CRN through two steps. Firstly, by analyzing the cooperation between single pair of PU and SU, the optimal decision of cooperation between a single pair of PU and SU, i.e., PU's cooperation fraction and SU's cooperation power is determined. Secondly, based on the analytical result of cooperation between single pair of PU and SU, the overall utility of such a CRN is further investigated by properly pairing PUs and SUs for cooperation.

2   Cooperation Between Single PU and SU

In this section, we will discuss the cooperation between a single pair PU and SU. For ease of presentation, the PU's and SU's index are omitted, e.g., α_{ij} turns to α and $h^i_{sr}(j)$ turns to h_{sr} and so on. In this section, the SU can increase the PU's throughput by relaying PU's data, and in return gains a fraction of time to transmit its own data. The cooperation between PU and SU is modeled as a Stackelberg game. In such a game, utilities of both the PU and the SU are presented and analyzed and close-form solutions are derived.

2.1   Stackelberg Game Between PU and SU

Since both PU and SU are selfish and rational and they just wish to maximize its own utility, i.e., the PU wishes to maximize its throughput while the SU wishes to consume less energy for relaying PU's data in addition to improving throughput. Therefore, the cooperation between PU and SU can be modeled as a Stackelberg game, where the PU acts as the leader and the SU acts as the follower. As the leader, the PU can choose the best strategy, awaring of the effect of its decision on the strategy of the follower (the SU); the SU can just choose its own strategy based on the PU's strategy. The utility functions for both PU and SU are respectively defined in the following. By analyzing the game, the optimal cooperation strategy of both PU and SU can be determined.

PU Utility. Given the fixed time duration T, increasing the throughput is equivalent to increasing the average transmission rate. Suppose the cooperated SU's relay power is known, the PU decides the slot allocation parameter α to maximize the potential profit.

Without cooperation, the transmission rate of the direct communication can be given by

$$R_d = W\log_2(1 + \frac{P_s|h_{sd}|^2}{N_0})$$

With cooperation, the transmission rate R_p through DF cooperative communication between the PU and SU, which serves as the utility function of a single PU, is given as follows:

$$R_p = \min\{\frac{\alpha}{2}W\log(1 + \frac{P_s|h_{sr}|^2}{N_0}), \frac{\alpha}{2}W\log(1 + \frac{P_r|h_{rd}|^2}{N_0})\} \tag{1}$$

The factor $\frac{\alpha}{2}$ accounts for the fact that αT is used for cooperative relaying, which is further split into two phases. Here we specify that both PUs and SUs transmit at a constant power, which are denoted by P_s for PUs and P_r for SUs, respectively. We assume that the PU chooses to cooperate only when the cooperative transmission rate is greater than that achieved by direct transmission. The objective of the PU is to maximize its own utility function by properly choosing the cooperation α and its utility function $U_p = R_p$.

SU Utility. The SU can gain transmission opportunities through cooperation with PU. In particular, the SU relays PU's data in the second phase and transmits its own data in the last phase. When cooperating, the SU decides its transmission power, pertaining to the given α. The target of the SU is to maximize throughput (equivalent to the transmission rate) without expending too much energy. Following the cooperation agreement, the SU spends the same power P_r for both cooperation and its own transmissions. In particular, the transmission rate R_s for secondary transmission from SU to its corresponding receiver is given by

$$R_s = (1 - \alpha)W \log(1 + \frac{P_r|h_r|^2}{N_0}) \tag{2}$$

With energy consumption $(1 - \frac{\alpha}{2})P_rT$, the utility function of SU can be represented by $R_sT - cP_r(1 - \frac{\alpha}{2})T$, where $c(0 < c < 1)$ is the weight of energy consumption in the overall utility. Over the period of T, the utility function of SU is given by

$$U_s = (1 - \alpha)W \log(1 + \frac{P_r|h_r|^2}{N_0}) - c(1 - \frac{\alpha}{2})P_r \tag{3}$$

The objective of SU in the game is to maximize its utility by choosing the optional transmission power P_r.

2.2 Game Analysis

As a sequential game, the Stackelberg game can be analyzed by the backward induction method. First, assuming the strategy of the PU (the leader) is fixed, the optimal strategy of the SU (the follower) is analyzed. Second, the PU decides the optimal strategy, knowing the results of the first step. By doing so, the best response functions of both the PU and the SU are derived such that the corresponding utilities can be maximized. Then, the Stackelberg equilibrium of the proposed game can be achieved based on the best response functions.

SU's Best Response Function. Assuming that the PU uses α for cooperation, SU selects the optimal transmission power to maximize its utility, which can be formulated as the following optimization problem:

$$\max_{P_r} U_s(\alpha) = (1 - \alpha)W \log(1 + \frac{P_r|h_r|^2}{N_0}) - c(1 - \frac{\alpha}{2})P_r$$

Solving the above problem, the optimal transmission power can be determined.

Definition 1. *Let $P_r^*(\alpha)$ be the best response function of the secondary user if the utility of SU can achieve the maximum value. When $P_r^*(\alpha)$ is selected, for any given α, i.e., $\forall 0 < \alpha < 1$, $U_s(P_r^*(\alpha), \alpha) \geq U_s(P_r(\alpha), \alpha)$.*

Theorem 1. *The best response function of the secondary user is given by*

$$P_r^*(\alpha) = \frac{(1-\alpha)W}{c(1-\frac{\alpha}{2})\ln 2} - \frac{N_0}{|h_r|^2} \qquad (4)$$

Proof. Given the time allocation coefficient α, the utility function of SU is given as (3). From the Eq. (3), it is easy to prove that the utility function first increases and then decreases with the increase of P_r without considering the power constraint. Therefore, there exists an optimal power such that U_s can reach the maximum value at that transmission power. Taking the first order partial derivative of the utility function with respect to P_r yields

$$\frac{\partial U_s}{\partial P_r} = \frac{(1-\alpha)W|h_r|^2}{(1 + \frac{P_r|h_r|^2}{N_0})N_0 \ln 2} - c(1 - \frac{\alpha}{2})$$

Setting $\frac{\partial U_s}{\partial P_r} = 0$ yields the optimal transmission power. The best $P_r^*(\alpha)$ response function will be

$$P_r^*(\alpha) = \frac{(1-\alpha)W}{c(1-\frac{\alpha}{2})\ln 2} - \frac{N_0}{|h_r|^2}$$

This completes the proof.

PU's Best Response Function. Awaring of the best response function of the SU, the PU decides its own best strategy for utility maximization.

Definition 2. *Let α^* be associated with the best response function of the primary user if the utility of the PU can achieve the maximum value when this strategy is selected.*

Theorem 2. *The best response function of the primary user α^* can be given as follows:*

$$\alpha^* = \begin{cases} \alpha_1^*, & if\ U_p(\alpha_1^*) \geq U_p(\alpha_2^*) \\ \alpha_2^*, & otherwise \end{cases} \qquad (5)$$

Where α_1^* and α_2^* are respectively the optimal function of the first item and the second item of the two processes from (1). α_1^* and α_2^* are given as follows:

$$\alpha_1^* = \begin{cases} \alpha_1', & if\ \ 0 < \alpha_1' < 1 \\ 0, & otherwise \end{cases} \qquad (6)$$

and

$$\alpha_2^* = \begin{cases} \alpha_2', & if\ \ 0 < \alpha_2' < 1\ \ and\ \ \ \ \ R_p(\alpha_2') \geq R_d \\ 0, & otherwise \end{cases} \qquad (7)$$

where

$$\alpha'_2 = \arg\max(U_p)$$

$$\alpha'_1 = \max(\Psi_1, \Psi_2)$$

and

$$\Psi_1 = 2[1 - \frac{W}{c\ln 2(\frac{2W}{c\ln 2} - \frac{N_0}{|h_r|^2} - \frac{P_s|h_{sr}|^2}{|h_{rd}|^2})}]$$

$$\Psi_2 = 2\frac{\log(1 + \frac{P_s|h_{sd}|^2}{N_0})}{\log(1 + \frac{P_s|h_{sr}|^2}{N_0})}$$

Proof. We can see the cooperative rate of the primary user is determined by smaller item of the two processes from (1). Now we will solve this problem in two ways. (a) When

$$\frac{\alpha}{2}W\log(1 + \frac{P_s|h_{sr}|^2}{N_0}) \leq \frac{\alpha}{2}W\log(1 + \frac{P_r|h_{rd}|^2}{N_0}) \tag{8}$$

then $U_p = \frac{\alpha}{2}W\log(1 + \frac{P_s|h_{sr}|^2}{N_0})$. From (8) we have

$$P_r \geq \frac{P_s|h_{sr}|^2}{|h_{rd}|^2} \tag{9}$$

Substituting (4) into function (9), the inequality can be expressed as

$$\frac{(1-\alpha)W}{c(1 - \frac{\alpha}{2})\ln 2} - \frac{N_0}{|h_r|^2} \geq \frac{P_s|h_{sr}|^2}{.|h_{rd}|^2} \tag{10}$$

which is a function of α. It can be derived from (10) that

$$\alpha \leq 2[1 - \frac{W}{c\ln 2(\frac{2W}{c\ln 2} - \frac{N_0}{|h_r|^2} - \frac{P_s|h_{sr}|^2}{|h_{rd}|^2})}] \tag{11}$$

The PU chooses cooperation only when the transmission rate via cooperation is greater than that of the direct communication. It can be expressed by

$$\frac{\alpha}{2}W\log(1 + \frac{P_s|h_{sr}|^2}{N_0}) \geq W\log(1 + \frac{P_s|h_{sd}|^2}{N_0}) \tag{12}$$

It can be derived from (12) that

$$\alpha \geq 2\frac{\log(1 + \frac{P_s|h_{sd}|^2}{N_0})}{\log(1 + \frac{P_s|h_{sr}|^2}{N_0})} \tag{13}$$

It's easy to see that U_p increased while α increased from $U_p = \frac{\alpha}{2}W\log(1 + \frac{P_s|h_{sr}|^2}{N_0})$. Now we can conclude from (11) and (13) that

$$\alpha'_1 = \max(\Psi_1, \Psi_2)$$

where

$$\Psi_1 = 2[1 - \frac{W}{c \ln 2(\frac{2W}{c \ln 2} - \frac{N_0}{|h_r|^2} - \frac{P_s|h_{sr}|^2}{|h_{rd}|^2})}]$$

$$\Psi_2 = 2\frac{\log(1 + \frac{P_s|h_{sd}|^2}{N_0})}{\log(1 + \frac{P_s|h_{sr}|^2}{N_0})}$$

Since $0 < \alpha < 1$, we have

$$\alpha_1^* = \begin{cases} \alpha_1', & \text{if } 0 < \alpha_1' < 1 \\ 0, & \text{otherwise} \end{cases} \tag{14}$$

(b) When

$$\frac{\alpha}{2}W \log(1 + \frac{P_s|h_{sr}|^2}{N_0}) > \frac{\alpha}{2}W \log(1 + \frac{P_r|h_{rd}|^2}{N_0}) \tag{15}$$

then $U_p = \frac{\alpha}{2}W \log(1 + \frac{P_r|h_{rd}|^2}{N_0})$. From (15) we have

$$P_r < \frac{P_s|h_{sr}|^2}{|h_{rd}|^2} \tag{16}$$

Similar (10) and (11), by substituting (4) into function (16), we have

$$\alpha > 2[1 - \frac{W}{c \ln 2(\frac{2W}{c \ln 2} - \frac{N_0}{|h_r|^2} - \frac{P_s|h_{sr}|^2}{|h_{rd}|^2})}] \tag{17}$$

Substituting (4) into utility function $U_p = \frac{\alpha}{2}W \log(1 + \frac{P_r|h_{rd}|^2}{N_0})$, the utility can be expressed by

$$U_p = \frac{\alpha}{2}W \log[1 + (\frac{(1-\alpha)W}{c(1 - \frac{\alpha}{2})\ln 2} - \frac{N_0}{|h_r|^2})\frac{|h_{rd}|^2}{N_0})]$$

which is a function of α.

U_p can be maximized because it is easy to proof $\frac{\partial U^2}{\partial \alpha^2} < 0$. The optimal α_2' is given by

$$\alpha_2' = \arg\max(U_p) \tag{18}$$

The PU chooses cooperation only when the transmission rate via cooperation is greater than that of the direct communication. It can be expressed by

$$\frac{\alpha}{2}W \log(1 + \frac{P_r|h_{rd}|^2}{N_0}) \geq W \log(1 + \frac{P_s|h_{sd}|^2}{N_0}) \tag{19}$$

We can get from (19) that

$$P_r \geq [(1 + \frac{P_s|h_{sd}|^2}{N_0})^{2/\alpha} - 1]\frac{N_0}{|h_{rd}|^2} \tag{20}$$

Substituting (4) into function (20), the inequality can be expressed by

$$\frac{(1-\alpha)W}{c(1-\frac{\alpha}{2})\ln 2} - \frac{N_0}{|h_r|^2} \geq [(1 + \frac{P_s|h_{sd}|^2}{N_0})^{2/\alpha} - 1]\frac{N_0}{|h_{rd}|^2} \tag{21}$$

Also, the time slot α_2^* must satisfy $0 < \alpha_2^* < 1$.

Now we can conclude from (17), (18) and (21) that

$$\alpha_2^* = \begin{cases} \alpha_2', & \text{if } 0 < \alpha_2' < 1 \quad \text{and} \quad R_p(\alpha_2') \geq R_d \\ 0, & \text{otherwise} \end{cases} \tag{22}$$

where

$$\alpha_2' = \arg\max(U_p)$$

3 Cooperation Between PUs and SUs

Although bring benefit to single PU and single SU, the approach aforementioned for the single cooperation can not bring the maximum benefit to the whole network because it only optimizes the interest of individual users. Therefore, it is necessary to consider the cooperation over whole network, which involves multiple PUs and SUs, to exploit the cooperation benefit. A common control channel is assumed for exchanging information among PUs and SUs, it can guide PUs and SUs select the suitable cooperator.

There are M PUs and N SUs in the network. Denote by x_{ij} the indicator which indicates whether PU_i cooperates with SU_j or not. Then, we have

$$x_{ij} = \begin{cases} 1, & \text{if } PU_i \text{ and } SU_j \text{ is paired for cooperation} \\ 0, & \text{otherwise} \end{cases} \tag{23}$$

From primary network perspective and from secondary network perspective, there are two utilities.

Primary Network Utility: The objective of pimary network is maximize total utility of the primary network. Note that, when a certain PU selects a certain SU, the throughput of this PU is obtained using Stackelberg Equilibrium strategy. Then, the utility function of primary network is given by

$$U_p^o = \max \sum_{i=1}^{M} \sum_{j=1}^{N} U_p^{ij} x_{ij} + U_p'$$

$$s.t. \quad x_{ij} \in \{0,1\} \,\forall i \in \{1,2,...,M\}, j \in \{1,2,...,N\} \tag{24}$$

$$\sum_{i=1}^{M} x_{ij} \leq 1 \qquad\qquad \forall j = 1,2,...,N$$

$$\sum_{j=1}^{N} x_{ij} \leq 1 \qquad\qquad \forall i = 1,2,...,M$$

where

$$U_p' = \sum_i R_d^i \ , i \in \{i | \sum_{j=1}^N x_{ij} = 0, i = 1, 2, ..., M\} \tag{25}$$

refers to the sum rate of direct transmission (since PUs choose to cooperate only when cooperative rate is greater than direct transmission rate) and U_p^{ij} is the utility function of PU_i while cooperating with SU_j. Because one PU can at most cooperate with only one SU and one SU can at most cooperate with one PU, we have $\sum_{i=1}^M x_{ij} \leq 1$ and $\sum_{j=1}^N x_{ij} \leq 1$.

Secondary Network Utility: Same as the primary network utility, the secondary network utility is given by

$$U_s^o = \max \sum_{j=1}^N \sum_{i=1}^M U_s^{ji} x_{ij}$$

$$s.t. \quad x_{ij} \in \{0, 1\} \ \forall i \in \{1, 2, ..., M\}, j \in \{1, 2, ..., N\} \tag{26}$$
$$\sum_{i=1}^M x_{ij} \leq 1 \qquad\qquad \forall j = 1, 2, ..., N$$
$$\sum_{j=1}^N x_{ij} \leq 1 \qquad\qquad \forall i = 1, 2, ..., M$$

Note that U_s^{ji} is the utility of SU_j when cooperated with PU_i, which is given by (3).

Using Stackelberg Equilibrium strategy calculate U_p^{ij} and U_s^{ji} ($i \in \{1, 2, ..., M\}$, $j \in \{1, 2, ..., N\}$), the above problem can be transformed into the bipartite matching problem(deciding x_{ij}). The KM algorithm is known as the most suitable algorithm for maximum matching. However, two utilities, U_p^o and U_s^o, exist in in this model. Therefore, it is not suitable for this model. This will be verified in the section simulation.

The Gale-Shapley Stable Marriage Theorem [8] is very general and states that finding a stable matching between two equally sized sets of elements given an ordering of preferences for each element. It is suitable for this model because each element has its own utility regarding the opposite element like this model. So we use G-S Stable Matching to solve the matching problem. Since number of PUs is not necessarily equal to that of SUs, there are inevitably PUs or SUs left uncooperated. The spectrum allocation algorithm based on stable matching model is summarized in Algorithm 1.

4   Numerical Results

In order to evaluate the performance of the proposed algorithm, we use Matlab to simulate our algorithm. We set up the first simulation as a scenario with a single PU and SU. The PU's transmission power P_s is 5. We set $W = 1$ and

---

Algorithm 1. Stable Matching

Step1: Initialization

1. Caculate U_p^{ij} and U_s^{ji} where $i \in \{1, 2, ..., M\}$ and $j \in \{1, 2, ..., N\}$.
2. PU_i ranks a preference list of SUs by U_p^{ij}.
3. SU_j ranks a preference list of PUs by U_s^{ji}.
4. construct $x_{ij} = 0$ where $i \in \{1, 2, ..., M\}$ and $j \in \{1, 2, ..., N\}$.
5. construct a list of all PUs which not matched and has SUs not asked.
   for cooperation, denoted by MATCHLIST $= \{PU_1, PU_2,...,PU_N\}$.

Step2: Matching process

6. **while** MATCHLIST is not empty **do**
7.   select $PU_k \in$ MATCHLIST.
8.   find SU_j which is in highest preference and $x_{kj}==0$.
9.   **if** $\sum_{i=1}^{M} x_{ij} == 0$ which means SU_j not matched **do**
10. $x_{kj} = 1$ and remove PU_k from MATCHLIST
11.   **else**
12.   find PU_h that SU_j is already matched(denoted by $x_{hj} = 1$).
13.   **if** SU_j has a higher preference of PU_k **do**
14. $x_{hj} = 0$.
15.   put PU_h into MATCHLIST if PU_h still has SU not asked.
16. $x_{kj} = 1$ and remove PU_k from MATCHLIST.
17.   **end if**
18.   **end if**
19.**end while**

---

set $N_0 = 1$ for simplicity. The power gains between PU transmitter and PU receiver, and between SU transmitter and SU receiver, are $h_{sd} = 0.3$ and $h_r = 5$, respectively. h_{sd} is setted relatively small to encourage PU to cooperates with SU. The average power gains between the PU transmitter and SU, and between the SU and PU receiver, are $h_{sr} = 5$ and $h_{rd} = 5$, respectively. The weight c ranges from 0.1 to 1 by step adding 0.02. Figure 2 and the left one of Fig. 3 show that time slot of α, the SU's transmission power P_r and the throughput of PU are decreasing in overall trend with the increase of weight c until $c = 0.66$ where PU transmits data directly. The right one of Fig. 3 shows the SU utility is increased with the increase of weight until where PU transmit data directly. $c = 0.28$ is a mutation point because in left of the point the U_p is determined by the rate from the PU transmitter to the SU and in right of the point the U_p is determined by the rate from the SU to the PU receiver.

Another simulation scenario is similar to the one above, impact of the power gains between PU transmitter and PU receiver (h_{sd}) is investigated, which ranges from 0.3 to 1.3 with step adding 0.1. Different from above, the weight c is static and $c = 0.1$. The PU number is setted 100 and the SU number becomes 100 too. Other variable are the same. Figures 4 and 5 show cooperation can bring great benefit to PUs and SUs by versus none cooperation with stable mathing and other two KM algorithems. The benefit is specially greater when the PU's

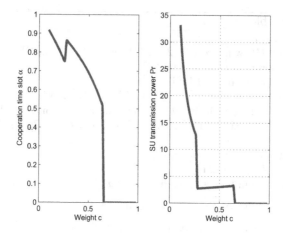

Fig. 2. Cooperation time slot a versus the weight c(left); SU transmission power versus the weight c(right).

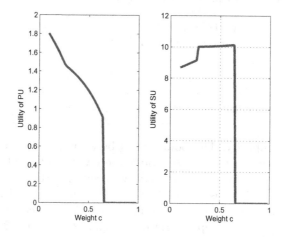

Fig. 3. Utility of PU versus the weight c(left); Utility of SU versus the weight c(right).

direct power gain (h_{sd}) is fairly weak, this is because the transmitting environment is so bad that more cooperation happened. Figures 4 and 5 show that the cooperation matching algorithm of the stable matching algorithm can get less primary network utility but more secondary network utility than the matching algorithm of the KM algorithm by using U_p^{ij} as weight. It also show that the cooperation matching algorithm of the stable matching algorithm can get more secondary utility but less primary network utility than the matching algorithm of the KM algorithm by using U_s^{ji} as weight. So, Figs. 4 and 5 show that the cooperation matching algorithm of the stable matching algorithm can make primary network and secondary both approach optimal. What's more, it's known that stable matching has the lower computational complexity.

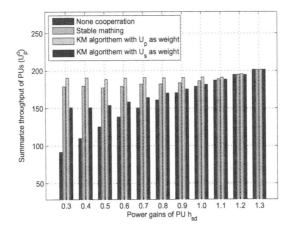

Fig. 4. Summarize throughput of PUs versus the power gains of PU h_{sd}.

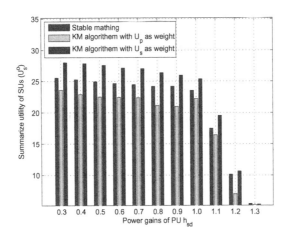

Fig. 5. Summarize utility of SUs versus the power gains of PU h_{sd}.

5   Conclusion

In this paper, we focus on cooperative spectrum access in a cognitive radio networks (CRN), where secondary users (SUs) serve as relays for primary users (PUs) to improve their throughput, and in return SUs can gain transmission opportunities. We first model the cooperation with single PU and single SU as Stackelberg game, through which PU's cooperation fraction and SU's cooperation power are derived. Then based on the above results, we using stable matching algorithm to study the cooperation between multiple PUs and multiple SUs. Through simulations, we show that with the proposed schemes utilities of both PUs and SUs can be balanced. PUs can achieve higher throughput and SUs can obtain more access opportunities. Numerical results also show that the stable matching algorithm is weak Pareto optimal with low complexity.

References

1. Docket, E.T.: Spectrum policy task force. FCC **2**(3), 135 (2002)
2. Zhang, J., Zhang, Q.: Stackelberg game for utility-based cooperative cognitiveradio networks. In: Proceedings of the Tenth ACM International Symposium on Mobile Ad Hoc Networking, Computing, vol. 2(3), p. 135, November 2009
3. Laneman, J.N., Tse, D.N.C., Wornell, G.W.: Cooperative diversity in wireless networks: Efficient protocols and outage behavior. IEEE Trans. Inf. Theory **50**(12), 3062–3080 (2004)
4. Simeone, O., Bar-Ness, Y., Spagnolini, U.: Stable throughput of cognitive radios with and without relaying capability. IEEE Trans. Commun. **55**(12), 2351–2360 (2007)
5. Simeone, O., Stanojev, I., Savazzi, S., Bar-Ness, Y., Spagnolini, U., Pickholtz, R.: Spectrum leasing to cooperating secondary ad hoc networks. IEEE J. Sel. Areas Commun. **26**(1), 203–213 (2008)
6. Yin, S., Zhang, E., Zhaowei, Q., Yin, L., Li, S.: Optimal cooperation strategy in cognitive radio systems with energy harvesting. IEEE Trans. Wireless Commun. **13**(9), 4693–4707 (2014)
7. Singh, A., Bhatnagar, M.R., Mallik, R.K.: Cooperative spectrum sensing with an improved energy detector in cognitive radio network. In: 2011 National Conference on Communications (NCC), pp. 1–5, January 2011
8. Gale, D., Shapley, L.S.: College admissions and the stability of marriage. Am. Math. Monthly **69**(1), 9–15 (1962)

Effect of Primary User Traffic on Largest Eigenvalue Based Spectrum Sensing Technique

Pawan Dhakal[1(✉)], Shree K. Sharma[2], Symeon Chatzinotas[2], Björn Ottersten[2], and Daniel Riviello[1]

[1] Politecnico di Torino, Turin, Italy
{pawan.dhakal,daniel.riviello}@polito.it
[2] University of Luxembourg, SnT, Kirchberg, Luxembourg
{shree.sharma,symeon.chatzinotas,bjorn.ottersten}@uni.lu

Abstract. In this paper, the effect of primary user (PU) traffic on the performance of largest eigenvalue based spectrum sensing technique (Roy's Largest Root Test (RLRT)) is investigated. A simple and realistic discrete time modeling of PU traffic is considered which is only based on the discrete time distribution of PU free and busy periods. Furthermore, in order to analyze the effect of PU traffic on the detection performance, analytical expressions for the probability density functions of the decision statistic are derived and validated by Monte-Carlo simulations. Numerical results demonstrate that the sensing performance of RLRT is no more monotonically increasing with the length of the sensing duration and also with SNR which contrasts with the common property of the spectrum sensing techniques under known PU traffic scenario. Furthermore, it is shown that the performance gain due to the multiple antennas in the sensing unit is significantly suppressed by the effect of the PU traffic when the frequency of the PU traffic transitions is higher.

Keywords: Eigenvalue based detection · Cognitive radio · Spectrum sensing · RLRT · Primary user traffic

1 Introduction

By accessing the idle spectrum band of Primary User (PU) network (licensed user), Cognitive Radio (CR) based dynamic spectrum sharing is initially intended to alleviate one of the most challenging problems of future wireless communications, namely, spectrum scarcity. With the real-time perception of surroundings and bandwidth availability using spectrum sensing functionality of a CR, secondary users (unlicensed users) may dynamically use the vacant spectrum and perform opportunistic transmissions [1]. Thus, the domain of spectrum sensing techniques has long been investigated by many researchers: a detailed bibliography of the contributions in this area can be found in [2,3]. Despite the significant volume of available literature on spectrum sensing under ideal scenarios, investigation under practical constraints and imperfections are still lacking [3]. Thus, recent research efforts are devoted to improve the accuracy and efficiency of sensing techniques under practical constraints and imperfections.

© ICST Institute for Computer Sciences, Social Informatics and Telecommunications Engineering 2016
D. Noguet et al. (Eds.): CROWNCOM 2016, LNICST 172, pp. 67–78, 2016.
DOI: 10.1007/978-3-319-40352-6_6

Among many practical imperfections and constraints for spectrum sensing in CR scenarios mentioned in the literature, the unknown PU traffic is one of the important constraints which significantly limits the sensing performance of the secondary user. In the existing literature on spectrum sensing, the SUs are assumed to have a perfect knowledge of the exact time slot structure of PU transmissions providing a solid basis for guaranteeing that PU traffic transitions occur only at the beginning of the SU sensing slots. However, in practice, the SU may not have the knowledge of exact time slot structure of PU transmissions or it is also possible that the communications among PUs are not based on synchronous schemes at all [4,5]. Thus, it is necessary to analyze the sensing performance of existing spectrum sensing techniques under unknown PU traffic.

A first attempt to analyze the performance of a detector in unknown PU traffic was made in [6]. The author analyzed the sensing performance of the well known semi-blind spectrum sensing techniques including Energy Detection (ED) and Roy's Largest Root Test (RLRT) under bursty PU traffic. The PU traffic model used in [6] is limited to a constant burst length of the PU data whose length is assumed to be always shorter than the SU sensing duration. However, the burst length of the PU may vary with time following some stochastic models [7,8]. A more general scenario in which the PU traffic transitions are completely random has been considered in [9–13]. By modeling PU traffic as a two state Markov process, authors in [9–12] analyzed the effect of PU traffic on the sensing performance and the sensing-throughput trade-off considering an ED technique under the half duplex scenario. Moreover, the effect of multiple PUs traffic on the sensing-throughput trade-off of the secondary system has been studied in [13]. Although all the aforementioned contributions recognized the fact that the PU traffic affects the sensing performance including sensing-throughput trade-off, none of them analyzed the sensing performance of other spectrum sensing techniques including Eigenvalue Based Detection (EBD) techniques under unknown primary user traffic.

In this paper, the effect of PU traffic on the performance of RLRT is evaluated. First, a realistic discrete time modeling of PU traffic is considered which is only based on the discrete time distribution of PU free and busy periods. Next, the analytical expressions for the probability density functions (pdfs) of the decision statistic are derived and validated by Monte-Carlo simulations. Finally, an analytical performance evaluation of the decision statistic in terms of receiver operating characteristics (ROC) under the considered scenario is carried out.

2   System Model

We consider a single source scenario (single primary transmitter) whereas multiple antennas are employed by an SU. Suppose the SU has K antennas and each antenna receives N samples in each sensing slot. In a given sensing frame, the detector calculates its decision statistic T_D by collecting N samples from each one of the K antennas. Subsequently, the received samples are collected by the detector in the form of a $K \times N$ matrix $\mathbf{Y}$. As described in Sect. 1, when the primary transmissions are not based on some synchronous schemes or the sensing unit at the SU does not have any information about the primary traffic pattern, the received vector at the sensing unit may consist of partly the samples from one PU state and the remaining from alternate PU state. To simplify the scenario, we begin with the following classification of the sensing slots based on the PU traffic status, which is also illustrated in Fig. 1.

1. Steady State (SS) sensing slot: In such type of sensing slot, all the received samples in one sensing slot are obtained from the same PU state.
2. Transient State (TS) sensing slot: In such type of sensing slot, a part of the received samples within the sensing slot are obtained from one PU state and the remaining from the another PU state.

In general, the probabilities of receiving SS and TS sensing slots are dependent on the PUs traffic model. In contrast to the commonly used hypothesis definition in spectrum sensing literature, we define two hypotheses in the following way:

$$\mathcal{H}_0 : \text{the channel is going to be free,}$$
$$\mathcal{H}_1 : \text{the channel is going to be busy.}$$

This hypothesis formulation implies that the decision is based on the PU status at the end of the sensing interval. Thus, in a TS sensing slot, a transition from the PU busy state to the PU free state is considered $\mathcal{H}_0$, while a transition from the PU free state to the PU busy state is considered $\mathcal{H}_1$. In the considered scenario, in an SS sensing interval, the generic received signal matrix under each hypothesis can be written as,

$$\boldsymbol{Y}_{SS} = \begin{cases} \boldsymbol{V}_{[K,N]} & (\mathcal{H}_0), \\ \boldsymbol{S}_{[K,N]} & (\mathcal{H}_1), \end{cases} \tag{1}$$

where $\boldsymbol{V}_{[K,N]} \triangleq [\boldsymbol{v}(1) \cdots \boldsymbol{v}(n) \cdots \boldsymbol{v}(N)]$ is the $K \times N$ noise matrix, $\boldsymbol{S}_{[K,N]} = \boldsymbol{h}_{[K,1]}\boldsymbol{s}_{[1,N]} + \boldsymbol{V}_{[K,N]}$ is the $K \times N$ received noisy signal matrix when PU signal is present. $\boldsymbol{h}_{[K,1]} = [h_1 \cdots h_K]^T$ is the channel vector and $\boldsymbol{s}_{[1,N]} \triangleq [s(1) \cdots s(n) \cdots s(N)]$ is a $1 \times N$ PU signal vector. And in the TS sensing interval, the generic received signal matrix under each hypothesis can be written as,

$$\boldsymbol{Y}_{TS} = \begin{cases} [\mathbf{S}_{[K,N-D_0]}|\mathbf{V}_{[K,D_0]}] \ (\mathcal{H}_0), \\ [\mathbf{V}_{[K,N-D_1]}|\mathbf{S}_{[K,D_1]}] \ (\mathcal{H}_1), \end{cases} \tag{2}$$

where D_0 represents the number of pure noise samples in TS sensing slot under $\mathcal{H}_0$, D_1 represents the number of noise plus PU signal samples in TS sensing slot under $\mathcal{H}_1$, $\mathbf{S}_{[K,N-D_0]} = \mathbf{h}_{[K,1]}\mathbf{s}_{[1,N-D_0]} + \mathbf{V}_{[K,N-D_0]}$ is the $(K \times N - D_0)$ received noisy signal matrix when PU signal is present only for $(N - D_0)$ sample periods. Similarly, $\mathbf{S}_{[K,D_1]} = \mathbf{h}_{[K,1]}\mathbf{s}_{[1,D_1]} + \mathbf{V}_{[K,D_1]}$ is the $K \times D_1$ received noisy signal matrix when PU signal is present only for D_1 sample periods. In each of these, the unknown primary transmitted signal $s(n)$ at time instant n is modeled as independent and identically distributed (i.i.d.) complex Gaussian with zero mean and variance $\sigma_s^2 : s(n) \sim \mathcal{N}_{\mathbb{C}}(0, \sigma_s^2)$. The noise sample $v_k(n)$ at the kth antenna of the SU at the time instant n is also modeled as complex Gaussian

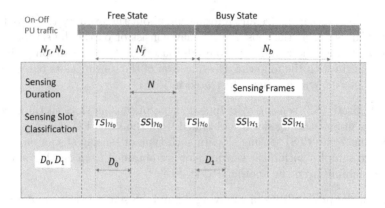

Fig. 1. Primary user traffic scenario and sensing slot classification

with mean zero and variance $\sigma_v^2 : v_k(n) \sim \mathcal{N}_{\mathbb{C}}(0, \sigma_v^2)$. The channel coefficient h_k of the kth antenna is assumed to be constant and memory-less during the sensing interval. The average SNR at the receiver is defined as, $\rho = \frac{\sigma_s^2 \|\boldsymbol{h}\|^2}{K\sigma_v^2}$, where $\|.\|$ denotes the Euclidean norm.

3   Characterization of Primary User Traffic

In this section, we characterize the mathematical model of PU traffic. Based on the proposed stochastic PU traffic model, we construct the PU's probability transition matrix, which lead to analytical formulation of the SU's probability of receiving SS sensing frame and TS sensing frame under each null and alternate hypothesis.

In this paper, we model the PU traffic as a two state Markovs process (On-Off process: PU 'On' representing busy state and PU 'Off' representing free state). The length of free as well as busy period are independent geometrically distributed random variables with parameters α and β, respectively. Essentially, the parameters α and β represent the state transition probabilities in single sample duration. The mean length of free period M_f and busy period M_b of PU traffic can be related to parameters α and β as, $M_f = \frac{1}{\alpha}$ and $M_b = \frac{1}{\beta}$, respectively.

At any time instant, the PU is in free state with probability $P_f = \frac{M_f}{M_b + M_f}$ and similarly, in the busy state with probability, $P_b = \frac{M_b}{M_b + M_f}$. We further assume that the parameters (α and β) of geometrically distributed length of PU free and busy periods are constant over time. Thus, the corresponding two-state Markovs process can be considered homogeneous in nature. Using this homogeneity property and the Chapman-Kolmogorov equation gives the PU n-step transition probability matrix as,

$$\mathbf{P}^n = \begin{bmatrix} p_{00}^n & p_{01}^n \\ p_{10}^n & p_{11}^n \end{bmatrix}$$

$$= \frac{1}{\alpha + \beta} \begin{bmatrix} \beta + \alpha(1-\alpha-\beta)^n & \alpha - \alpha(1-\alpha-\beta)^n \\ \beta - \beta(1-\alpha-\beta)^n & \alpha + \beta(1-\alpha-\beta)^n \end{bmatrix}, \tag{3}$$

which reduces to (4) for single step transition matrix as,

$$\mathbf{P} = \begin{bmatrix} p_{00} & p_{01} \\ p_{10} & p_{11} \end{bmatrix} = \begin{bmatrix} 1-\alpha & \alpha \\ \beta & 1-\beta \end{bmatrix} \tag{4}$$

As already mentioned earlier in Sect. 2, the stochastic nature of the PU state transition gives a mixed nature of received signals in a TS sensing slot resulting in random variables (RVs) D_0 and D_1. Thus, in each PU state transition from *Busy* to *Free State*, the sensing unit has to decide based on D_0 pure noise samples and $(N - D_0)$ noise plus primary signal samples which actually affects the overall sensing performance. Thus, with the support of above analysis and also keeping (1) and (2) in reference, it is clear that, to find the distribution of the decision statistic under different hypotheses, the prior deduction of the chances of occurrence of SS sensing slot, TS sensing slot, probability mass function (pmf) of D_0 and the pmf of D_1 are inevitable. It can be shown that[1], under the assumption that the lengths of busy period and free period of PU have comparable mean parameters M_f and M_b, the pmf of D_0 which represents the probability of having D_0 noise only (PU signal free) samples in a TS sensing slot under $\mathcal{H}_0$ reduces to,

$$P_{D_0}(D_0 = d_0) = \begin{cases} \frac{1}{N-1} & 1 \leq d_0 < N, \\ 0 & \text{otherwise.} \end{cases} \tag{5}$$

Similarly, the pmf of D_1 which represents the probability of having D_1 PU-signal-plus-noise samples in a TS sensing slot under $\mathcal{H}_1$ reduces to,

$$P_{D_1}(D_1 = d_1) = \begin{cases} \frac{1}{N-1} & 1 \leq d_1 < N, \\ 0 & \text{otherwise.} \end{cases} \tag{6}$$

Also, the probability of receiving SS sensing slot under $\mathcal{H}_0$ is given by,

$$P_{SS|\mathcal{H}_0} = \frac{1}{1 + \frac{N}{M_f}}. \tag{7}$$

and the probability of receiving SS sensing slot under $\mathcal{H}_1$ is given by,

$$P_{SS|\mathcal{H}_1} = \frac{1}{1 + \frac{N}{M_b}}. \tag{8}$$

---

[1] Due to space limitation, we omit the proofs of following equations and will include in the journal version of the paper.

4   Sensing Performance Analysis

In this section, we consider an another important class of detection techniques, designed for multi-sensor detectors, based on the eigenvalues of the received signal covariance matrix. Receive diversity can be achieved either by multiple users (cooperative detection) or by multiple antennas. Given a $K \times N$ received signal matrix $\mathbf{Y}$, the sample covariance matrix is defined as, $\mathbf{R} \triangleq \frac{1}{N}\mathbf{Y}\mathbf{Y}^H$ and $\lambda_1 \geq \cdots \geq \lambda_k$ its eigenvalues sorted in the decreasing order.

Eigenvalue based detection techniques infer the presence of signal from eigenvalues λ_i. In particular, the detection technique which considers the largest one (λ_1) and compare against the noise variance is known in statistics as Roy's Largest Root Test (RLRT) [15] and its test statistics is,

$$T_{RLRT} \triangleq \frac{\lambda_1}{\sigma_v^2}. \tag{9}$$

RLRT is "semi-blind" as it requires the exact knowledge of noise variance and is considered to be asymptotically optimum test in this setting [17]. Other related tests have been proposed in the literature for example λ_1 against smallest eigenvalue [16], λ_1 against trace of covariance matrix [18]. These are considered "blind" as they do not require the prior knowledge of the noise variance.

Here, we analyze in detail the RLRT method. However, the results can be extended to the other methods as well. To analyze the RLRT performance, it is necessary to express the pdf of the test statistics for the case of unknown PU traffic. The following theorem computes the pdf of the RLRT decision statistic under both the hypotheses using the PU traffic characterization presented in Sect. 3.

Theorem 1. *Given a multi-antenna sensing unit with K receive antennas, N received samples in each slot and the random PU traffic with geometrically distributed free and busy state duration, let $c = K/N$, N_s a independent parameter and define:*

$$\mu_1(N_s) = \left(\frac{N_s}{N}K\rho + 1\right)\left(1 + \frac{K-1}{N_s K\rho}\right), \quad \sigma_1^2(N_s) = \frac{N_s}{N^2}(K\rho + 1)\left(1 - \frac{K-1}{N_s K^2 \rho^2}\right) \tag{10}$$

$$\mu_{N,K} = \left[1 + \sqrt{c}\right]^2, \quad \sigma_{N,K} = N^{-2/3}\left[1 + \sqrt{c}\right]\left[1 + \frac{1}{\sqrt{c}}\right]^{1/3}. \tag{11}$$

Then, the pdfs of RLRT decision statistic under $\mathcal{H}_0$ and $\mathcal{H}_1$ are given by (12) and (13) respectively,

$$f_{T_{RLRT}|H_0}(x) = P_{SS|H_0} f_{TW2}\left(\frac{x - \mu_{N,K}}{\sigma_{N,K}}\right) + P_{TS|H_0}\sum_{d_0=1}^{N-1} P_{D_0}(d_0) f_D(x, N - d_0), \tag{12}$$

$$f_{T_{RLRT}|H_1}(x) = P_{SS|\mathcal{H}_1} f_D(x, N) + P_{TS|\mathcal{H}_1}\sum_{d_1=1}^{N-1} P_{D_1}(d_1) f_D(x, d_1). \tag{13}$$

where,

$$f_D(x,d) = \begin{cases} f_\mathcal{N}(\mu_1(d), \sigma_1^2(d)) & \text{if, } d > \frac{K-1}{K^2\rho^2}, \\ f_{TW2}(\mu_{N,K}, \sigma_{N,K}) & \text{otherwise,} \end{cases} \tag{14}$$

In (14), $f_\mathcal{N}(\mu_1(d), \sigma_1^2(d))$ denote a Gaussian pdf with mean $\mu_1(N_s)$ and variance $\sigma_1^2(N_s)$ provided in (10) at $N_s = d$. Next, $f_{TW2}(\mu_{N,K}, \sigma_{N,K})$ is a pdf of Tracy-Widom distribution of order 2 with parameters $\mu_{N,K}$ and $\sigma_{N,K}$ provided in (11).

Proof. As noted from Sect. 2, the nature of the received signal matrix is different for the SS sensing slot and TS sensing slot. Under null hypothesis $\mathcal{H}_0$, received sample covariance matrix $\mathbf{R}|_{\mathcal{H}_0}$ can be decomposed as a probabilistic sum of $\mathbf{R}_{SS}|_{\mathcal{H}_0}$ and $\mathbf{R}_{TS}|_{\mathcal{H}_0}$ in the following way,

$$\mathbf{R}|_{\mathcal{H}_0} = P_{SS}|_{\mathcal{H}_0}\mathbf{R}_{SS}|_{\mathcal{H}_0} + P_{TS}|_{\mathcal{H}_0}\mathbf{R}_{TS}|_{\mathcal{H}_0}. \tag{15}$$

Since sensing slots are independent from each other, we treat each covariance matrix in (15) independently. Given an SS sensing slot under null hypothesis, all the received samples $y_k(n)$ are homogeneous in nature comprising the i.i.d. Gaussian noise samples with mean zero and variance σ_v^2. Thus, the sample covariance matrix $\mathbf{R}_{SS}|_{\mathcal{H}_0}$ follows a Wishart distribution whose largest eigenvalue normalized by noise variance can be expressed by a Tracy-Widom distribution of second order [17,20].

$$\frac{\lambda_1^{SS}|_{\mathcal{H}_0}}{\sigma_v^2} = f_{TW2}\left(\frac{x - \mu_{N,K}}{\sigma_{N,K}}\right), \tag{16}$$

where $\mu_{N,K}$ and $\sigma_{N,K}$ are given in (11).

Next, given a TS sensing slot under null hypothesis, all the received samples $y_k(n)$ are not homogeneous in nature. To provide a better understanding, we express the covariance matrix in a TS sensing slot under $\mathcal{H}_0$ as,

$$\mathbf{R}_{TS}|_{\mathcal{H}_0} = \mathbf{R}_S(N - D_0) + \mathbf{R}_N(D_0), \tag{17}$$

where,

$$\mathbf{R}_S(N - D_0) \triangleq \frac{1}{N - D_0}\mathbf{S}_{[K,N-D_0]}\mathbf{S}_{[K,N-D_0]}^H, \tag{18}$$

$$\mathbf{R}_N(D_0) \triangleq \frac{1}{D_0}\mathbf{V}_{[K,D_0]}\mathbf{V}_{[K,D_0]}^H, \tag{19}$$

are the partial covariance matrices constructed respectively from signal-plus-noise and only-noise samples. $\mathbf{R}_S(N - D_0)$ is a standard spiked population covariance matrix of rank-1 and $\mathbf{R}_N(D_0)$ is Wishart matrix. The largest eigenvalue of $\mathbf{R}_N(D_0)$ is negligible compared to the largest eigenvalue of $\mathbf{R}_S(N - D_0)$ given a signal identifiability condition is met [6]. It is known that the fluctuation of the largest eigenvalue of a rank-1 spiked population matrix normalized by the noise variance are asymptotically Gaussian [17,19] if the signal identifiability condition is met otherwise its distribution is again a Tracy-Widom of order 2.

$$\frac{\lambda_1^{TS}|_{\mathcal{H}_0}}{\sigma_v^2} = f_D(x, (N - D_0)) \tag{20}$$

Using the results from (16) and (20), the RLRT decision statistic under null hypothesis can be written as,

$$T_{RLRT}|_{\mathcal{H}_0} = \frac{\lambda_1|_{\mathcal{H}_0}}{\sigma_v^2} \tag{21}$$

$$= p_{SS}|_{\mathcal{H}_0} \frac{\lambda_1^{SS}|_{\mathcal{H}_0}}{\sigma_v^2} + p_{TS}|_{\mathcal{H}_0} \frac{\lambda_1^{TS}|_{\mathcal{H}_0}}{\sigma_v^2} \tag{22}$$

$$= p_{SS}|_{\mathcal{H}_0} f_{TW}\left(\frac{t - \mu_{N,K}}{\sigma_{N,K}}\right) + p_{TS}|_{\mathcal{H}_0} f_D(x, N - D_0). \tag{23}$$

Using the fact that D_0 is a random variable distributed as in (5), we obtain the final distribution of the decision statistic of RLRT test under null hypothesis as in (12).

We consider now the case when the PU signal is present (hypothesis $\mathcal{H}_1$). In this case, an error is made if the presence of PU signal is not detected. Under alternate hypothesis $\mathcal{H}_1$, the received sample covariance matrix $\mathbf{R}|_{\mathcal{H}_1}$ can be decomposed as a probabilistic sum of $\mathbf{R}_{SS}|_{\mathcal{H}_1}$ and $\mathbf{R}_{TS}|_{\mathcal{H}_1}$.

$$\mathbf{R}|_{\mathcal{H}_1} = p_{SS}|_{\mathcal{H}_1}\mathbf{R}_{SS}|_{\mathcal{H}_1} + p_{TS}|_{\mathcal{H}_1}\mathbf{R}_{TS}|_{\mathcal{H}_1}. \tag{24}$$

Since $\mathbf{R}_{SS}|_{\mathcal{H}_1}$ is a standard spiked population covariance matrix of rank-1, the distribution of the largest eigenvalue normalized by the noise variance in a SS sensing slot under $\mathcal{H}_1$ can be approximated as [17, 19],

$$\frac{\lambda_1^{SS}|_{\mathcal{H}_1}}{\sigma_v^2} = f_D(x, N). \tag{25}$$

Using the same line of reasoning as in $\mathcal{H}_0$, we get,

$$\frac{\lambda_1^{TS}|_{\mathcal{H}_1}}{\sigma_v^2} = f_D(x, D_1). \tag{26}$$

Using (25) and (26), the distribution of the RLRT decision statistic under alternate hypothesis can be written as,

$$T_{RLRT}|_{\mathcal{H}_1} = \frac{\lambda_1|_{\mathcal{H}_1}}{\sigma_v^2} \tag{27}$$

$$= p_{SS}|_{\mathcal{H}_1} \frac{\lambda_1^{SS}|_{\mathcal{H}_1}}{\sigma_v^2} + p_{TS}|_{\mathcal{H}_1} \frac{\lambda_1^{TS}|_{\mathcal{H}_1}}{\sigma_v^2} \tag{28}$$

$$= p_{SS}|_{\mathcal{H}_1} f_D(x, N) + p_{TS}|_{\mathcal{H}_1} f_D(x, D_1). \tag{29}$$

Incorporating the pmf of D_1 (derived in (5)) in (29) yields (13). $\square$

A. *Probability of False Alarm:* Given the pdf of the decision statistic in (12), we can compute the false-alarm probability. Under $\mathcal{H}_0$, the PU is in free state at the end of the sensing interval, but the decision statistic is erroneously above the threshold τ and the PU signal is declared present. For defining the probability of false-alarm P_F in our case, the following Corollary of Theorem 1 holds.

Corollary 1. *The false-alarm probability of the RLRT test under unknown PU traffic and complex signal space scenario is:*

$$P_F = P(T_{RLRT}|\mathcal{H}_0 \geq \tau) \equiv \int_{\tau}^{+\infty} f_{T_{RLRT}|H_0}(x)dx. \tag{30}$$

B. Probability of Detection: Given the pdf of the decision statistic in (13), we can compute the detection probability. Under $\mathcal{H}_1$, i.e., the PU is in busy state at the end of the sensing interval. Under this scenario, if the decision statistic is above the threshold, the PU signal is declared present. The following Corollary of Theorem 1 holds for defining the probability of detection P_D.

Corollary 2. *The detection probability of the RLRT test under unknown PU traffic and the complex signal space scenario is:*

$$P_D = P(T_{RLRT}|\mathcal{H}_1 \geq \tau) \equiv \int_{\tau}^{+\infty} f_{T_{RLRT}|H_1}(x)dx. \tag{31}$$

5   Numerical Results and Discussion

In this section, the effect of PU traffic on the RLRT detection method is analyzed based on the traffic model developed in Sect. 3. The length of the free and busy periods of the PU traffic are measured in terms of the discrete number of samples where each of them has Geometric distribution with mean parameters M_f and M_b, respectively as described in Sect. 3. Under multiple antenna sensing scenario, the average SNR at the receiver is defined as, $\rho = \frac{\sigma_s^2\|\boldsymbol{h}\|^2}{K\sigma_v^2}$, where $\|.\|$ denotes the Euclidean norm. The analytical expressions derived in Sect. 4 are validated via numerical simulations.

In Fig. 2, the sensing performance of RLRT under unknown PU traffic is compared with the ideal RLRT performance. It can be well understood that the conventional model with perfect synchronization of the PU-SU sensing slots performs better than the one with unknown PU traffic. In addition, the accuracy of the derived analytical expressions of P_F and P_D are confirmed where the theoretical formulas are compared against the numerical results obtained by Monte-Carlo simulations. The perfect match of the theoretical and numerical curves validates the derived analytical expressions. The Receiver Operating Characteristics (ROC) performance of RLRT in the considered PU traffic model for different PU traffic parameters is presented in Fig. 2(a). The sensing performance degrades significantly when the mean lengths of busy and free periods are comparable with the length of the sensing interval or in a few multiples of it. However, an improvement in the sensing performance can be seen if the length of the mean parameters M_f and M_b is increased. We present in Fig. 2(b), the missed detection probability (P_{Md}) as a function of SNR. From this figure, it is seen that, for a given PU traffic parameters, increasing the SNR improves the sensing performance for certain lower range of SNR. However, in contrast to RLRT sensing performance under known PU traffic, the RLRT sensing performance under unknown PU traffic levels to some point $(1 > P_{Md} >> 0)$ above

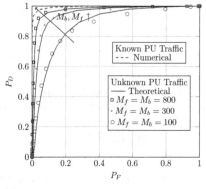

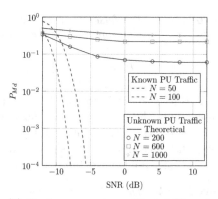

(a) ROC performance, Parameters: $N = 50$, $K = 4$ and SNR $= -6$ dB.

(b) Performance curve (Probability of Missed detection vs. SNR), Parameters: $K = 8$, $P_F = 0.01$ and $M_f = M_b = 3000$.

Fig. 2. Sensing performance of RLRT under unknown PU traffic.

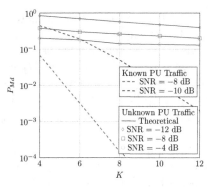

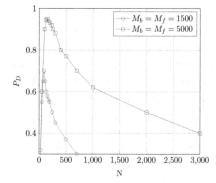

(a) Prob. of Missed Detection vs. the number of receiving antennas (K), Parameters: $N = 100$, $P_F = 0.01$ and $M_f = M_b = 1500$.

(b) Variation of Detection Probability with increasing N, Parameters: SNR $= -10$ dB, $K = 8$ and $P_F = 0.01$

Fig. 3. RLRT sensing performance comparison for different sensing parameters (N and K).

certain SNR. This is due to the effect of the mixing of the PU signal-plus-noise and only-noise samples in the TS sensing slot.

In Fig. 3, the RLRT sensing performance is plotted as a function of sensing parameters N and K. The variation of the sensing performance of RLRT detector for different number of receiving antennas (K) is plotted in Fig. 3(a). It can be observed that, unlike the rapid increase in sensing performance (decrease in missed-detection probability) with the increasing number of receiving antennas under synchronized PU-SU sensing slot scenario, the RLRT sensing performance under unknown PU traffic is almost constant even if we increase the number of antennas. During a TS sensing slot, from each receiving antenna, the received signal samples are the mixture of pure noise samples and the samples with both noise and PU signal. Thus, even if we use multiple antennas, the nature of the received signal doesn't change much which is the reason the sensing performance improvement is suppressed by the unknown PU traffic (more specifically, the TS sensing performance) when the length of the free and busy periods of PU traffic are quite small (a few multiples of the length of the sensing window). Furthermore, we present in Fig. 3(b), the numerical simulation of detection probability (P_D) as a function of sensing window (N). Note that, unlike RLRT detection probability under known PU traffic which monotonically increases indefinitely until '$P_D = 1$' with increasing length of sensing window, the detection probability of RLRT under unknown PU traffic do not have a monotonic property as a function of the length of the sensing window.

6   Conclusion

In this paper, the effect of PU traffic on the performance of largest eigenvalue based detection technique (RLRT) has been studied under the complex domain of PU signal, noise and channel. A realistic and simple PU traffic model has been considered which is based only on the discrete time distribution of PU free and busy periods. Moreover, an analytical evaluation of the spectrum sensing performance under the considered scenario has been carried out. It has been observed that the sensing performance of RLRT is no more monotonically increasing with the increase in the length of the sensing duration and also with SNR which contrasts with the common property of the spectrum sensing techniques under known PU traffic scenario. Finally, it has been observed that the performance gain due to multiple antennas in the sensing unit is significantly suppressed by the effect of the PU traffic when the frequencies of the PU traffic transitions are more frequent.

Acknowledgment. This work was partially supported by FNR, Luxembourg under the CORE projects "SeMIGod" and "SATSENT" and also by the EC under FP7 NEWCOM# (Grant agreement no. 318306).

References

1. Haykin, S.: Cognitive radio: Brain-empowered wireless communications. IEEE J. Sel. Areas Commun. **23**(2), 201–220 (2005)
2. Axell, E., Leus, G., Larsson, E.G., Poor, H.V.: Spectrum sensing for cognitive radio: State-of-the-art and recent advances. IEEE Signal Process. Mag. **29**(3), 101–116 (2012)
3. Sharma, S.K., Bogale, T.E., Chatzinotas, S., Ottersten, B., Le, L.B., Wang, X.: Cognitive radio techniques under practical imperfections: a survey. IEEE Commun. Surv. Tutorials **17**, 1858–1884 (2015)

4. Consultative Committee for Space Data Systems, "TC synchronization and channel coding", CCSDS 231.0-B-2 Blue Book, September 2010

5. Zorzi, M., Gluhak, A., Lange, S., Bassi, A.: From today's INTRAnet of things to a future INTERnet of things: a wireless- and mobility-related view. IEEE Wireless Commun. Mag. 17(6), 44–51 (2010)

6. Penna, F., Garello, R.: Detection of discontinuous signals for cognitive radio applications. IET Commun. 5(10), 1453–1461 (2011)

7. Palit, R., Naik, K., Singh, A.: Anatomy of WiFi access traffic of smart-phones and implications for energy saving techniques. Intl. J. Energy, Inform. Commun. 3, February 2012

8. Ghosh, A., Jana, R., Ramaswami, V., Rowland, J., Shankaranarayanan, N.K.: Modeling and characterization of large-scale Wi-Fi traffic in public hot-spots. In: IEEE Proceedings of the INFOCOM, pp. 2921–2929, April 2011

9. MacDonald, S.L., Popescu, D.C.: Impact of primary user activity on the performance of energy-based spectrum sensing in cognitive radio systems. In: IEEE Proceedings of the GLOBECOM, pp. 3224–3228, December 2013

10. Wang, T., Chen, Y., Hines, E., Zhao, B.: Analysis of effect of primary user traffic on spectrum sensing performance. In: Proceedings of the International Conference Communications and Networking in China, pp. 1–5, August 2005

11. Wu, J.Y., Huang, P.H., Wang, T.Y., Wong, V.W.S.: Energy detection based spectrum sensing with random arrival and departure of primary user's signal. In: IEEE Proceedings of the GLOBECOM, pp. 380–384, December 2013

12. Tang, L., Chen, Y., Hines, E.L., Alouini, M.S.: Effect of primary user traffic on sensing-throughput tradeoff for cognitive radios. IEEE Trans. Wireless Commun. 10(4), 1063–1068 (2011)

13. Pradhan, H., Kalamkar, A.S., Banerjee, A.: Sensing-throughput tradeoff in cognitive radio with random arrivals and departures of multiple primary users. IEEE Commun. Letters 19(3), 415–418 (2015)

14. Hsu, H.P.: Theory and Problems of Probability, Random Variables, and Random Processes. Schaum's outline series. McGraw-hill (1996)

15. Roy, S.N.: On a heuristic method of test construction and its use in multivariate analysis. Ann. Math. Stat. 24, 220–238 (1953)

16. Zeng, Y.H., Liang, Y.C.: Eigenvalue based spectrum sensing algorithms for cognitive radio. IEEE Trans. Commun. 57(6), 1784–1793 (2009)

17. Dhakal, P., Riviello, D., Penna, F., Garello, R.: Impact of noise estimation on energy detection and eigenvalue based spectrum sensing algorithms. In: IEEE Proceedings of the ICC, pp. 1367–1372, June 2014

18. Bianchi, P., Najim, J., Alfano, G., Debbah, M.: Asymptotics of eigenbased collaborative sensing. In: Proceedings of the IEEE Information Theory Workshop (ITW 2009), October 2009

19. Penna, F., Garello, R., Spirito, M.A.: Probability of missed detection in eigenvalue ratio spectrum sensing. In: Fifth IEEE International Conference on WiMob, October 2009

20. Bai, Z.D.: Methodologies in spectral analysis of large-dimensional random matrices, a review. Stat. Sin. 9, 611–677 (1999)

Energy Efficient Information Sharing in Social Cognitive Radio Networks

Anna Vizziello[✉] and Riccardo Amadeo

Department of Electrical, Computer, and Biomedical Engineering,
University of Pavia, 27100 Pavia, Italy
anna.vizziello@unipv.it, riccardo.amadeo01@ateneopv.it

Abstract. The growing usage of the devices and the popular high data rate services is increasing the demand for radio resources and the associated energy consumption. Cognitive radio (CR) technology has been proposed as a solution to improve spectrum utilization, enabling the devices to identify alternate frequency bands and opportunistically use them. We propose a novel social-cognitive radio (S-CR) architecture able to improve the network energy efficiency. The key aspect of the such solution is the resulting indirect form of cooperation. Indeed, the devices only share information about channel availability and device profile information, without any predefined cooperation to perform a common task, such as cooperative spectrum sensing. In this way, it is possible to reduce some energy demanding operations of the cognitive cycle, such as the sensing procedure, since the social information sharing requires less energy. A novel S-CR protocol is developed to share environmental information and allocate spectrum resources to the CRs. Simulation results compare the S-CR protocol with its non-social version and reveal the effectiveness of the proposed solution in terms of energy consumption reduction.

Keywords: Cognitive radio networks · Social networks · Sharing · Energy efficiency

1 Introduction

Over the last years, the data volume through the network is increasing by a factor of ten each five years, contributing to the increment of radio resources demand and the grow of the Information and Communication Technologies (ICT) energy consumption in the range of 15–20 % [1].

The development of new radio and network solutions have become essential to improve energy efficiency and to bring commercial benefits to both end-users and operators [2].

Moreover, given the limited available spectrum, user devices are called to more complex tasks to exploit the best spectrum opportunities, as specified in the concept of cognitive radio (CR). Indeed, the CR framework allows CR users to detect, use and share the available spectrum, in a way that the licensed or primary users (PUs) are not affected [1,3], at the cost of more complex devices.

© ICST Institute for Computer Sciences, Social Informatics and Telecommunications Engineering 2016
D. Noguet et al. (Eds.): CROWNCOM 2016, LNICST 172, pp. 79–90, 2016.
DOI: 10.1007/978-3-319-40352-6_7

Several works have deeply studied the cognitive cycle within each radio device. However, none of them exploits the social network paradigm to reduce the considerable burden at device level.

Social networks have already demonstrated the power of information sharing among users, a characteristic that can be helpful also in this context. We propose to use the social cognitive network concept to shift the work load from individual devices to a network level cooperation and decision making, through cognitive and social networking.

Some studies, such as [3], have already evaluated the energy consumed by a CR to perform some power hungry functionalities like spectrum sensing. The social network paradigm would be able to avoid such energy demanding operations, since spectrum information can be obtained and shared through a social network, improving the overall system energy efficiency.

The main contributions of the paper are:

- a social cognitive radio architecture, with a detailed model and the main definitions, such as the type of information to be shared, the community, and the concept of indirect cooperation;
- a social cognitive sharing protocol to exchange environmental information and allocate radio resources in a fully distributed manner.

The paper is organized as follows: Sect. 2 describes the definition of social network and its application to cognitive radio systems, Sect. 3 shows the proposed social cognitive network architectures and details the model, Sect. 4 focuses on the proposed social sharing protocol, Sect. 5 shows some interesting simulation results, and, finally, some concluding remarks wrap up and close the paper in Sect. 6.

2   Social Structure and Application

Social network concept is attracting the attention of researchers in both academic and industry fields, intrigued by their peculiar features [4].

Here we first present the general definition of social network, useful to fully understand the potential of this type of network, and then describe its application to cognitive radio networks (CRNs).

2.1   Social Network Definition

The authors in [4] describe the characteristics of social network sites (SNSs) and propose a comprehensive definition of it as web-based services that allow individuals to (i) construct a public or semi-public profile in a limited system, (ii) create a list of other users with whom sharing information/connection, (iii) view their list of connections and the ones made by others inside the system.

The main feature of social network sites is not allowing people to meet strangers, but rather enabling users to create and make visible their social connections/networks. Indeed, since friends list contains links to each friend's profile, public connections allow users to sweep the network graph by exploring the friend lists.

2.2 Social Behavior in Cognitive Radio Networks

Some recent research in CRNs is exploiting environmental and relations among the actors to develop more efficient protocols. Two main directions have been developed:

Social Collaboration for Sensing in CRNs. Generally, cooperative sensing assumes that all CRs cooperate for the procedure. In [5] the authors describe a more realistic cooperation willingness according to the social relation between the CRs. Indeed, a CR would sense for a CR that is considered a "friend", while it may not for one considered a "stranger".

After denoting a CR by some parameters, such as community, cooperation tendency, friend list, sympathy list, and cooperation score list, a cooperation set selection procedure is developed to define the CRs involved in the sensing operation [5]. Finally, specific scores are updated for each CR according to its behavior in the cooperative sensing procedure.

Social Recommendation System for CRNs. The collaboration among CRs in terms of channel recommendation has been investigated in [6]. The recommendation procedure consists in sharing channel preferences of CR users, which can change over time, and results in a behavior propagation in a social network. The dynamics of such social behavior in CRNs has been studied as a stochastic dynamical system [6].

3   Social CRN Model

Unlike some recent solutions in CRNs that exploit only some features of social concept [5,6], the proposed social CR (S-CR) scheme has been developed starting from the general definition of a social network to properly exploit all its capabilities. As shown in Fig. 1, we consider a distributed CR network without any centralized base station (BS) to coordinate the resource allocation for CRs. In this scenario, CRs send sensing information each other in accordance to a novel social framework.

Specifically, we assume N primary users (PUs) and M CRs in a certain location. The CRs are allowed to access the unused spectrum resources of PUs, but, as soon as a PU is detected in the same channel, a CR has to vacate the spectrum band to avoid interference towards PUs.

We consider that CRs may transmit different traffic types and organize them in communities according to the traffic type they use mostly.

In the proposed social framework, a CR that needs to transmit, first requests the information about the available channels to the CRs belonging to its communities, and, only if there is non-reliable information, the CR will start the sensing operation.

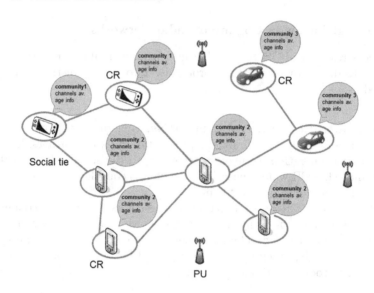

Fig. 1. Social CR architecture

In particular, the main features of the proposed social CRN framework are:

- a device-centric social framework, where the social ties are considered among the devices, not humans.
- As stated in [4], the main concept of SNSs it that the actors of the network share their profile. Thus, differently from classical CRNs, besides environmental information, like the available channels, we propose to share CR device information, such as the required bandwidth.
- A fully distributed system is considered, where each CR has its own repository about channels availability and a possible different age of information for each channel.
- Finally, in line with human social networks we develop an indirect type of cooperation. Specifically, CRs will not directly be involved in a common operation, such as cooperative spectrum sensing, but rather they will just share their own information about channels, as users in human social network have the option to share information without direct cooperation.

3.1 Social CRN Definitions

In the following we outline the main definitions of the proposed social CR framework.

Types of Sharing Information: we consider both environmental and CR device profile data, which are then exploited in the S-CR protocol to improve the overall performance of the network (see Sect. 4). We here list some possible types of data, although in the S-CR protocol we use only some of them:

– environmental information:
  – available channels,
  – age of information,
– profile information:
  – power battery level,
  – traffic type commonly used by the user (video, call, text),
  – required bandwidth,
  – will to share.

In particular, given a CR repository, the ages of information about the channels may be different, as shown in Fig. 2. Indeed, as an example, as soon as a CR finds an available channel through sensing, the CR stops the spectrum sensing on the other channels. Thus, the information of the spanned channels is younger and more reliable than the channels not involved in the sensing process.

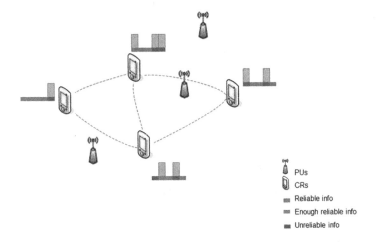

Fig. 2. Ages of shared information

Community: the friends list is used to define the community a CR belong to. We define three communities depending on the CR traffic type, among video, call and text. For the time being, we do not consider location restriction, assuming that all CRs belonging to the same community can interact with each other.

We assume that the CRs share their information, about the environment and their profile, only with the users belonging to their same community. In this way we limit the channels to sense and the information sharing is performed only among the interested CRs. As an example, the CRs interested in video traffic will not need information about all the available channels, but only about the ones with larger bandwidth.

Fully Distributed Social CRN: following the definition of SNs, we consider a fully distributed network where each CR has its own repository. A CR may obtain information about available channels in two ways: (i) through social sharing, (ii) through its spectrum sensing operation.

Indirect Social Cooperation: the social framework differentiates from cooperation since CRs do not perform together a common operation, such as cooperative spectrum sensing. Also, if a CR requests available channels, the other CRs do not perform any sensing operation on its behalf. In this sense, there is no direct cooperation among the CRs. A CR just share information already stored in its repository if another CR requests them.

3.2 Spectrum Sensing Procedure

A CR that needs to transmit, first asks for information about the available channels among the CRs in its community, then, if there are no reliable information, starts sensing the channels. As soon as the CR finds a channel free from PU transmissions, it stops the sensing operation and uses that channel to transmit. As consequence, each CR will have on its repository a different age of information per each channel.

The CRs may employ any type of detector for spectrum sensing, such as an energy-based detector, to detect PU signals. The energy detector measures the energy of the received signal, i.e., the output signal of bandpass filter with bandwidth B is squared and integrated over the observation interval [7]. The output of the integrator Z_y is then compared with a threshold λ, to decide if a PU is present or not [7]:

$$
\begin{cases}
Z_y > \lambda & decide\ H_1 \\
\\
otherwise & decide\ H_0
\end{cases}
\tag{1}
$$

where H_0 and H_1 represent respectively the hypothesis that the PU is inactive and active.

According to recent findings, we assume PU activities with death rate α and birth rate β. We can estimate the a posteriori probabilities as follows [7]:

$$
P_{on} = \frac{\beta}{\alpha + \beta}
$$
$$
P_{off} = \frac{\alpha}{\alpha + \beta}
\tag{2}
$$

where P_{on} is the probability of the period used by PUs, and P_{off} is the probability of the PU idle period.

From the definition of maximum a posteriori detection [7], PU detection probability P_d and false-alarm probability P_f are given by

$$
P_d = Pr[Z_y > \lambda|H_1] \cdot P_{on} = \bar{P}_d \cdot P_{on}
$$
$$
P_f = Pr[Z_y > \lambda|H_0] \cdot P_{off} = \bar{P}_f \cdot P_{off}
\tag{3}
$$

where $\bar{P}_d$ and $\bar{P}_f$ are the detection and false-alarm probabilities of a CR using an energy-based detector, respectively [7].

4   Social CRN Information Sharing Protocol

The proposed solution allows to reduce the number of burden sensing operations performed by each CR, by using the information about available channels shared in the social architecture, which is less energy demanding (see Table 1).

4.1   Description of CR States

At each round, a CR can be in *busy* or *idle* mode. A CR is in *busy* state if:

– the CR was transmitting its data in the previous round and the transmission has not finished yet. If so, at the current round, the CR continues the transmission on the channel it is already assigned, on the condition that no PU starts transmitting on that channel.
– the CR was not transmitting in the previous round but:
  – its data were queued in the previous round because, after sensing all the channels, no free channel was available;
  – a PU starts transmitting in the channel on which the CR was transmitting, so its data are queued until a new free channel becomes available.

A CR is in *idle* state if it finished/did not need to transmit in the previous round, meaning that it is waiting for new data to transmit.

4.2   S-CR Protocol

Figure 3 shows an overview of the developed S-CR procedure to find the available channels and to share available spectrum resources. At the beginning, a CR checks the reliability of its own information and the one received from the social sharing. If the information are not reliable, the CR performs spectrum sensing to obtain the available channels. Then, the CR posts such information on its repository, along with its own device profile information and share them on the social network. We assume that only the CRs belonging to the same community are allowed to see that information.

Going in more details, the CR k that needs to access a channel first checks its own repository about available channels information. If none of the available channels has information age below a certain threshold T_{th}, the CR considers its data as unreliable and asks for information in the social system. The acceptable age threshold T_{th} defines how old an information can be before becoming unreliable.

Only the CRs belonging to its community will be involved in the social process. For the time being, we consider only the will of sharing information and the community a CR belongs to as profile information.

In particular, the CR k sends a request to the other CRs at the energy cost $E_{tx_k} = P_{tx_k} \cdot T_{tx_k}$, which is significantly less than the energy $E_{s_k} = P_{s_k} \cdot T_{s_k}$ needed to scan a channel. This is mainly due to the different time necessary to send a request packet and the one to sense a channel (see Table 1). The time to

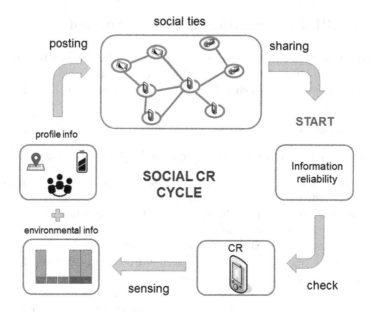

Fig. 3. S-CR protocol

sense a channel T_s has been set in order to achieve good detection probability P_d and low false alarm probability P_f [3].

All the CRs belonging to the community will spend $E_{rx_m} = P_{rx_m} \cdot T_{rx_m}$ to decode the request. Among them, only the CRs with free channels in their repositories and whose age of information is below the threshold could answer to the request. The CR k will pick randomly a CR m among those CRs.

CR m will consume $E_{tx_m} = P_{tx_m} \cdot T_{tx_m}$ to send its repository information to CR k. On the other side, CR k will spend $E_{rx_k} = P_{rx_k} \cdot T_{rx_k}$ to decode such information.

The information about the available channels sent by CR m to CR k will have different ages of information since each channel could have been updated at a different stage. Figure 2 illustrates the repository of each CR, showing a possible different age of information for each channel.

Among the available channels sent by CR m, the CR k will choose the free channel with the most reliable information for its transmission. Moreover, CR k will update its repository according to the received information by CR m.

However, if there is not a CR in the community that replies to CR k, it will start the sensing operation a the cost of $E_{s_k} = P_{s_k} \cdot T_{s_k}$ for each channel sensed. As soon as CR k find a free channel, it starts transmitting on that channel.

Note that only the information and the age of the channels scanned during the sensing operation will be update in the repository of CR k. Thus, at the next round, the information about those channels will be more reliable than the one of the other channels.

5 Simulation Results

5.1 Simulation Environment

We simulate as much as possible a real environment with heterogeneous ad time varying conditions, where the number of transmitting CRs, the number of PUs occupying the channels, the length of data to transmit, the available channels, etc., may vary at each event, while it is time constant during a single event. The number of channels is set equal to 250, while the number of CRs ranges from 10 to 220.

The probability P_{on} that a PU transmits and the probability P_{off} that a PU finishes its transmission are set according to the case of high opportunity for CR transmissions, as defined in [7]. Specifically, each channel shows $P_{on} = 0.3$ and $P_{off} = 0.6$, which results, on average, in one third of the channels occupied by PUs at each event. In the current simulation set we assume perfect spectrum sensing estimation, i.e. $P_d = 1$ and $P_f = 0$, so that the unreliable information may occur only through social network in case old information is shared.

The CR probability to start a new transmission is set equal 0.8. This applies only when the CR already finished its previous data transmission. Thus, the CR transmission probability at each event is even higher than this value, since it accounts for both the already ongoing transmission and the new ones. Given the high CR transmission probabilities, the average number of free channels per event is inversely proportional to the number of CRs in the system. Simulations show that it varies from an average value of 160, when there are only 10 CRs, to zero when the total number of users (PUs and CRs) exceeds the total number of channels.

As explained in Sect. 3.1, we consider three communities, one for a different CR traffic type. The traffic type is distinguished by the required bandwidth. Moreover, we consider that different traffic types need a different amount of events to be completely transmitted, which we set equal to 3, 10 and 25 events.

The following Table 1 shows the main variables definition and value settings. The value of energy and time parameters to transmit/receive reports (shared on the social network) and to sense a channel are chosen according to [3]. Note that the time and power to receive a report refers to the information sharing through the social network, i.e. for control messages about the available channels, it does not concern CR traffic transmissions. We compare the proposed S-CR protocol with its version without social interaction. In order not to cause interference towards PUs, in the proposed S-CR protocol we made conservative assumptions by setting the acceptable age threshold T_{th} equal to 1. In this way only the information calculated at the previous event is considered reliable. As an example, if a CR performed sensing in the previous event and shares its own information about available channels in the current event, such information is considered reliable. On the contrary, the information is unreliable if calculated at previous events, thus the social interaction is not reliable and sensing operation would then be necessary.

Table 1. Simulation parameters

Parameter	Symbol	Value
Time for sensing one channel	T_s	50 ms
Time to transmit a report	T_{tx}	0.08 ms
Time to receive a report	T_{rx}	0.08 ms
Power for sensing one channel	P_s	700 mW
Power to transmit a report	P_{tx}	750 mW
Power to receive a report	P_{rx}	750 mW
Energy for sensing one channel	$E_s = P_s \cdot T_s$	350 mJ
Energy to transmit a report	$E_{tx} = P_{tx} \cdot T_{tx}$	0.06 mJ
Energy to receive a report	$E_{rx} = P_{rx} \cdot T_{rx}$	0.06 mJ
Number of CRs	M	[10 − 220]
Number of channels	C	250

The non-social protocol has T_{th} equal to 0. Indeed, non-social information will fulfill the reliable requirement if T_{th} is set to 0, and only sensing will be performed to obtain information about the available channels.

5.2   Performance Evaluation

We evaluate the proposed S-CR protocol in terms of energy consumed by the whole system and compare it with its non-social version (see Fig. 4).

Figure 4 shows that the energy saving of the S-CR protocol is negligible when the social system includes just few members (number of CRs M = 10), while, when the number of CRs increases, the S-CR protocol reveals relevant energy savings of the system. However, when the total number of users, including both CRs and PUs, exceeds the total number of channels, the system saturates and tends to re-balance itself to a situation where no relevant energy saving is possible.

This imitates the behavior of common social networks [8]: when the number of users is low, interactions among members are unlikely and the perceived system value is scarce. On the contrary, when the number of users increases and exceeds a certain threshold (usually defined as critical mass), the exchange of useful information increases too, causing an increment of the perceived value of the system by the users.

In the considered scenario, the number of CRs representing the critical mass is in the range of [10–40].

Simulation results show that the average energy saving of the proposed S-CR protocol over the classic non-social procedure is around 13.3 % (see Table 2). This average value corresponds to save 131.43 J for the whole system and 1.30 J for a single CR (see Table 2).

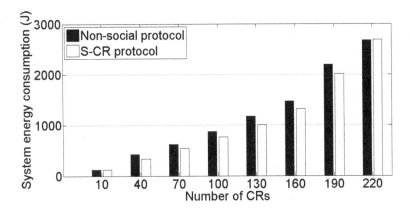

Fig. 4. System energy consumption varying the number of CRs

Table 2. Energy consumption

# CRs	Non-social consumption (J)	S-CR consumption (J)	System saving (J)	System saving (%)	Single CR saving (J)
10	125.00	121.13	3.88	3.10	0.39
40	427.96	337.50	90.47	21.14	2.26
70	625.17	543.46	81.72	13.07	1.16
100	875.92	769.60	106.31	12.14	1.06
130	1175.50	1005.05	170.44	14.50	1.31
160	1475.97	1320.29	155.67	10.55	0.97
190	2195.95	2012.01	183.94	8.38	0.97
220	2677.04	2689.00	−11.96	−0.45	−0.05

Such average result accounts for several simulation settings, given a number of CRs between 40 and 190. However, as soon as the number of CRs reaches 190, the system saturates and it is not possible to achieve any energy saving. The number of CRs that saturates the system is related to the number of channels and the probability that PUs occupy the channels.

6   Conclusion

In this paper we have proposed a new social-cognitive radio architecture to improve the energy efficiency of the overall network. By sharing some information about device profile and available channels, the CRs are able to reduce some energy demanding operation of the cognitive cycle, such as spectrum sensing.

A novel S-CR protocol have been developed to share environmental information and to allocate spectrum resources to the CRs in the network. Simulation results confirmed that social information sharing combined with spectrum sensing is less energy demanding than the pure spectrum sensing operation. Moreover, results reveal the effectiveness of the proposed S-CR solution in terms of energy consumption.

References

1. CrownCom 2009, 4th International Conference on Cognitive Radio Oriented Wireless Networks and Communications, Hannover, 22-24 June 2009. http://www.crowncom2009.org/p/Conference/Panel-Discussion/
2. Ericsson, Press Release, June 2008
3. Syeda, R., Namboodiri, V.: Energy-Efficiency of Cooperative Sensing Schemes in Ad Hoc WLAN Cognitive Radios (2012)
4. Boyd, D.M.: Social network sites: definition, history, and scholarship. J. Comput.-Mediated Commun. **13**, 210–230 (2008)
5. Guven, C., Bayhan, S., Alagoz, F.: Effect of social relations on cooperative sensing in cognitive radio networks. In: 2013 First International Black Sea Conference on Communications and Networking (BlackSeaCom), pp. 247–251, 3-5 July 2013
6. Li, H., Song, J.B., Chen, C., Lai, L., Qiu, R.C.: Behavior propagation in cognitive radio networks: a social network approach. IEEE Trans. Wirel. Commun. **13**(2), 646–657 (2014)
7. Lee, W., Akyildiz, I.F.: Optimal spectrum sensing framework for cognitive radio networks. IEEE Wirel. Commun. **7**(10), 3845–3857 (2008)
8. Katona, Z., Zubcsek, P.P., Sarvary, M.: Network effects and personal influences: the diffusion of an online social network. J. Mark. Res. **48**(3), 425–443 (2011)

Fair Channel Sharing by Wi-Fi and LTE-U Networks with Equal Priority

Andrey Garnaev[1,2]($\boxtimes$), Shweta Sagari[2], and Wade Trappe[2]

[1] Saint Petersburg State University, St Petersburg, Russia
garnaev@yahoo.com
[2] WINLAB, Rutgers University, North Brunswick, USA
{shsagari,trappe}@winlab.rutgers.edu

Abstract. The paper is concerned with the problem Wi-Fi and LTE-U networks sharing access to a band of communication channels, while also considering the issue of fairness in how the channel is being shared. As a criteria of fairness for such joint access, α-fairness and maxmin fairness with regards to expected throughput are explored as fairness metrics. Optimal solutions are found in closed form, and it is shown that these solutions can be either: (a) a channel on/off strategy in which access to the channels is performed sequentially, or (b) a channel sharing strategy, i.e., where simultaneous joint access to the channels is applied. A criteria for switching between these two type of optimal strategies is found, and its robustness on the fairness coefficient is established, as well as the effectiveness of the fairness coefficient to control the underlying protocol of the joint access to the shared resource is managed. Finally, we note that the approach that is explored is general, and it might be adapted to different problems for accessing a sharing resource, like joint sharing of voice and data traffic by cellular carriers.

Keywords: Fairness · Maxmin fairness · LTE-U · Wi-Fi

1 Introduction

With the emergence of new wireless applications and devices, the demand being placed on limited radio spectrum has been dramatically increasing over the last decade. Developing methods by which different wireless technologies can effectively share under-utilized spectrum bands is an important step towards meeting this growing demand. Consequently, maintaining fair coexistence in unlicensed spectrum, e.g. technologies like LTE-U and Wi-Fi, with respect to throughput and latency has become a widely discussed topic by the wireless community [1], even though the specification for LTE-U has not yet been finalized.

Currently, there are only a few works dealing with LTE-U and Wi-Fi coexistence. In [2], based on simulation results it was showed that LTE system performance might be slightly affected by coexistence, whereas Wi-Fi is significantly impacted by LTE transmissions. In [3], it was pointed out that the coexistence of

© ICST Institute for Computer Sciences, Social Informatics and Telecommunications Engineering 2016
D. Noguet et al. (Eds.): CROWNCOM 2016, LNICST 172, pp. 91–103, 2016.
DOI: 10.1007/978-3-319-40352-6_8

LTE and Wi-Fi needs to be carefully investigated since, as it was illustrated, Wi-Fi might be severely impacted by LTE transmissions. The performance of coexisting femtocell and Wi-Fi networks operating over a fully-utilized unlicensed band were analytically modeled in [4]. The effects of Wi-Fi channel access parameters on the performance of Wi-Fi and femtocell networks were investigated in [5]. A fair and QoS-based unlicensed spectrum splitting strategy between Wi-Fi and femtocell networks was studied in [6], and experimental results for the coexistence of Wi-Fi and LAA-LTE were presented in [7]. Modeling the coexistence of LTE and Wi-Fi heterogeneous networks was performed in [9], and a proportional fair allocation scheme for them was develop in [8].

One critical challenge facing coexistence of such technologies is having an architecture that can support dynamic spectrum management of LTE-U and Wi-Fi networks [10,11]. In [10], a system for coordinating between multiple heterogeneous networks to improve spectrum utilization and facilitate co-existence, which is built on the principles of Software Defined Networking to support logically centralized dynamic spectrum management involving multiple autonomous networks, was presented. Based on this architecture, an optimization model to maximize the aggregated Wi-Fi+LTE throughput was designed and tested in [11]. This optimization problem was divided into two steps: in the first step, based on information about networks' infrastructure and agents exploiting networks' facilities, power control optimization problems were solved to get optimal throughput for Wi-Fi only access and joint Wi-Fi+LTE access to the channels. In the second step, a throughput maxmin problem dealing with joint time division channel access was solved. Evaluation of such joint coordination showed that such a dual optimization approach might increase the aggregated Wi-Fi+LTE throughput. Our work builds upon these prior efforts, and in particular, in this paper we provide strong analytical evidence supporting this result, and consider joint coordination of Wi-Fi+LTE under a unified fairness criteria.

The organization of this paper is as follows: in Sect. 2, we first give formulation of α-fairness problem. Then, in Sect. 3, we solve it in closed form. In Sect. 4, the problem of maxmin fairness is formulated and solved. Finally, in Sect. 5, discussions are supplied.

2 Formulation of α-Fairness Problem

As a system for coordinating between multiple heterogeneous networks for the improvement spectrum utilization and to facilitate co-existence, we consider the model suggested in [11]. The core element of this system is a Global Controller (GC), which employs information about the network "ecology" and the associated access points to calculate throughput under separate or joint access between Wi-Fi and LTE networks. In this paper we add to the system an additional component, the Fairness Decision Maker (FDM) that, based on this information, finds the fair *joint* time division channel access and returns it to the GC to optimize throughput (Fig. 1).

To deal with the problem of fairly allocating the fraction of time each system can access the channel, it is necessary to employ an appropriate fairness metric.

Fig. 1. Coordination between GC and FDM involves passing parameters and solutions to the optimization

A survey of different fairness concepts as used in wireless communication is given in [12]. Generally, in the formulation of fairness, there are n agents, each of which has an utility depending on its share of a common resource. The fair allocation of a common resource depends directly on the criteria for fairness being used, and maxmin is one possible criteria that is popular in the literature. We focus, however, on α-fairness, which provides a unified framework for considering a wide array of fairness concepts, such as bargaining (for $\alpha = 1$) and maxmin (for α tending to infinity). We note that α-fairness has been applied previously in the literature, in [13] it was applied to a throughput assignment problem, while in [14], it was applied to fair power control for femtocell networks. In [15], generalized α-fairness concept explored and applied to optimizing resource allocation in downlink cellular networks. In [16,17], a problem of fair resource allocation under a malicious attack was investigated for SINR (signal to interference plus noise ratio) and throughput as the user utilities.

We note that although we deal only with two agents (LTE-U and Wi-Fi networks with throughput as their utilities), the problem generalizes to a classical fairness problem since an agent's utility depends on the amount of the resource it uses (individual access to the channels) as well as on the joint resource (joint access to the channels). In this more general situation, we can also observe that increasing the coefficient of fairness to ∞ yields the maxmin criteria.

Let us formulate the problem of fairly allocating the fraction of time that Wi-Fi and LTE-U access a channel. In order to get insight into the problem, similar to what was studied in [11], we assume that the total throughput of each network is proportional to the fraction of time a technology access the channel and on whether the channel access by Wi-Fi and LTE-U is simultaneous or not. In this paper, we consider a model where there is equal right to access the channels for both Wi-Fi and LTE-U networks. To describe the problem let us introduce the following notations:

 (i) q^W is the fraction of time the channel is accessed by Wi-Fi network only (Wi-Fi access mode).
 (ii) q^L is the fraction of time the channel is accessed by LTE-U network only (LTE-U access mode).
 (iii) q is the fraction of time the channel is accessed by both the networks simultaneously (Joint Wi-Fi and LTE-U access mode).
 (iv) Without loss of generality we can assume that total time slot for access to the channel is denoted $[0,1]$. Thus, $q^W + q^L + q = 1$, and the vector of time fractions is $\boldsymbol{q} = (q^L, q, q^W)$.

(v) P^W is the throughput of the Wi-Fi network per time unit, when the network is in Wi-Fi access mode.

(vi) P^L is the throughput of the LTE-U network per time unit, when the network is in LTE-U access mode.

(vii) P_W^L and P_L^W are the throughputs of LTE-U and Wi-Fi networks per time unit, when the system is in joint Wi-Fi and LTE-U access mode, where both networks access the channel simultaneously. It is natural to assume that the extra interference in the network reduces its throughput, i.e., $P_W^L \leq P^L$ and $P_L^W \leq P^W$.

(viii) $\overline{P}^W$ is the total throughput of the Wi-Fi network, i.e., $\overline{P}^W = q^W P^W + q P_L^W$.

(ix) $\overline{P}^L$ is the total throughput of the LTE-U network, i.e., $\overline{P}^L = q^L P^L + q P_W^L$.

If $q = 0$, we call such strategy $\boldsymbol{q}$ as a *channel on/off strategy*, i.e., the networks do not access the channel simultaneously, but rather one by one. If $q > 0$, we call such strategy $\boldsymbol{q}$ as a *channel sharing strategy*, i.e., in which the networks might access the channel simultaneously. Note that, different resource sharing strategies have arisen in different network optimization problems, c.f. for channel sharing [18–21], for bandwidth scanning [22–24], for time sharing [25], and for node protection [26]. To deal with the problem of the joint access to a shared channel for LTE-U and Wi-Fi networks we apply α-fair approach, which also incorporate the maxmin approach. The considered α-fairness problem can be formulated as follows:

$$v_\alpha = \max_q v_\alpha(\boldsymbol{q}), \tag{1}$$

with

$$v_\alpha(\boldsymbol{q}) = \begin{cases} \dfrac{(q^W P^W + q P_L^W)^{1-\alpha}}{1-\alpha} + \dfrac{(q P_W^L + q^L P^L)^{1-\alpha}}{1-\alpha}, & \alpha \neq 1, \\ \ln(q^W P^W + q P_L^W) + \ln(q P_W^L + q^L P^L), & \alpha = 1. \end{cases}$$

Let $\boldsymbol{q}_\alpha$ be the optimal α-fair strategy, i.e., $\boldsymbol{q}_\alpha := (q_\alpha^L, q_\alpha, q_\alpha^W) = \arg\max_q v_\alpha(\boldsymbol{q})$.

3   Optimal α-Fair Strategies

In this section, Theorem 1 gives the optimal α-fair strategy $\boldsymbol{q}_\alpha$ in closed form as well as the condition for it to be either a channel on/off strategy or a channel sharing strategy.

Theorem 1. *(a) Let*

$$P_L^W / P^W + P_W^L / P^L < 1. \tag{2}$$

Then the optimal α-fair strategy $\boldsymbol{q}_\alpha = (q_\alpha^W, q_\alpha, q_\alpha^L)$ is channel on/off strategy, and it is given as follows:

$$(q_\alpha^W, q_\alpha, q_\alpha^L) = \left((P^W)^{(1-\alpha)/\alpha} / \left((P^L)^{(1-\alpha)/\alpha} + (P^W)^{(1-\alpha)/\alpha} \right), 0, 1 - q_\alpha^W \right). \tag{3}$$

(b) Let (2) do not hold and

$$(P_W^L)^{1-\alpha}(P_L^W)^\alpha \le P^W - P_L^W \text{ and } (P_W^L)^\alpha(P_L^W)^{1-\alpha} > P^L - P_W^L. \qquad (4)$$

Then

$$(q_\alpha^W, q_\alpha, q_\alpha^L) = \left(1 - q_\alpha, \frac{(P_W^L)^{(1-\alpha)/\alpha}/(1 - P_L^W/P^W)^{1/\alpha}}{(P^W)^{(1-\alpha)/\alpha} + \left(P_W^L/(1 - P_L^W/P^W)\right)^{(1-\alpha)/\alpha}}, 0\right). \qquad (5)$$

(c) Let (2) do not hold and

$$(P_W^L)^\alpha(P_L^W)^{1-\alpha} \le P^L - P_W^L \text{ and } (P_W^L)^{1-\alpha}(P_L^W)^\alpha > P^W - P_L^W. \qquad (6)$$

Then

$$(q_\alpha^W, q_\alpha, q_\alpha^L) = \left(0, \frac{(P_L^W)^{(1-\alpha)/\alpha}/(1 - P_W^L/P^L)^{1/\alpha}}{(P^L)^{(1-\alpha)/\alpha} + \left(P_L^W/(1 - P_W^L/P^L)\right)^{(1-\alpha)/\alpha}}, 1 - q_\alpha\right). \qquad (7)$$

(d) Let (2) do not hold and

$$(P_W^L)^{1-\alpha}(P_L^W)^\alpha \ge P^W - P_L^W \text{ and } (P_W^L)^\alpha(P_L^W)^{1-\alpha} \ge P^L - P_W^L. \qquad (8)$$

Then $(q_\alpha^W, q_\alpha, q_\alpha^L) = (0, 1, 0)$.
(e) Let (2) do not hold and

$$(P_W^L)^{1-\alpha}(P_L^W)^\alpha \le P^W - P_L^W \text{ and } (P_W^L)^\alpha(P_L^W)^{1-\alpha} \le P^L - P_W^L. \qquad (9)$$

Then $\boldsymbol{q}_\alpha$ is given by (5) for

$$(P^W)^{(1-\alpha)/\alpha} + \left(\frac{P_W^L}{1 - P_L^W/P^W}\right)^{(1-\alpha)/\alpha} > (P^L)^{(1-\alpha)/\alpha} + \left(\frac{P_L^W}{1 - P_W^L/P^L}\right)^{(1-\alpha)/\alpha}, \qquad (10)$$

and $\boldsymbol{q}_\alpha$ is given by (7) if (9) does not hold.

The case $\alpha = 0$ is a limiting case for this theorem in which we examine α tending to zero. Then, $\boldsymbol{q}_0 = (1, 0, 0)$ if $P^W \ge \max\{P_L^W + P_W^L, P^L\}$, $\boldsymbol{q}_0 = (0, 0, 1)$ if $P^L \ge \max\{P_L^W + P_W^L, P^W\}$ and $\boldsymbol{q}_0 = (0, 1, 0)$ if $P_L^W + P_W^L \ge \max\{P^W, P^L\}$.

Figures 2 and 3 illustrate the optimal α-fair fraction of time for applying the access mode for $P_L^W \in [0.1, P^W]$, $P_W^L \in [0.1, P^L]$ and $P^W = P^L = 3$ and $\alpha = 0.1, 2$. These figures illustrate that increasing the α-coefficient causes a reduction in the zone of permanent joint access to the channel, with it diminishing while α is increasing. Also, the figures illustrate the robustness of the zone for applying channel on/off strategies on the fairness coefficient.

4   Maxmin Fairness

The maxmin strategy for joint access between Wi-Fi and LTE-U networks to a shared channel in terms of maxmin throughput $\overline{P}$ can be formulated as follows: $\overline{P} = \max_q \min\{q^W P^W + q P_L^W, q^L P^L + q P_W^L\}$. The following theorem gives the optimal maxmin strategy q in closed form, as well as the condition for existence of optimal channel on/off and channel sharing strategies.

Theorem 2. *(a) The maxmin solution q is a channel sharing one, i.e., $q > 0$, if and only if (2) does not hold.*
  (a_1) If $P_L^W \geq P_W^L$ then $(q^W, q, q^L) = (0, 1/(1 + (P_L^W - P_W^L)/P^L), 1 - q)$;
  (a_2) If $P_L^W \leq P_W^L$ then $(q^W, q, q^L) = (1 - q, 1/(1 + (P_W^L - P_L^W)/P^W), 0)$.
(b) If (2) holds then $q = 0$, i.e., maxmin solution is an channel on/off strategy, and $(q^W, q, q^L) = (P^L/(P^L + P^W), 0, P^W/(P^L + P^W))$.

This theorem shows the difference between the maxmin and the α-fair solution. Namely, permanent joint access to the channel (i.e., when $q_\alpha = 1$) cannot be a maxmin solution. Meanwhile, the more general α-fairness accepts a permanent joint access as an optimal solution.

Figure 4 illustrates zones (P_W^L, P_L^W) for applying optimal maxmin channel sharing and on/off strategies, and the switching lines between them. The dotted domain depicts the zone where the optimal solutions for both models coincide

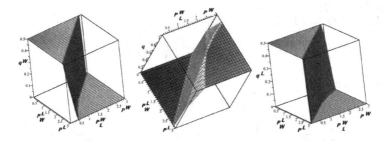

Fig. 2. The optimal α-fair time fractions q^W (left), q (center) and q^L (right) for $P_L^W \in [0.1, P^W]$, $P_W^L \in [0.1, P^L]$ and $P^W = P^L = 3$ and $\alpha = 0.1$

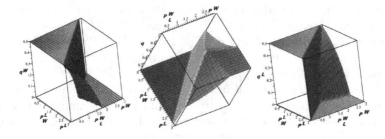

Fig. 3. The optimal α-fair time fractions q^W (left), q (center) and q^L (right) for $P_L^W \in [0.1, P^W]$, $P_W^L \in [0.1, P^L]$ and $P^W = P^L = 3$ and $\alpha = 2$.

with each other. Also, Fig. 4 illustrates the optimal time fractions q^L, q and q^W as a function of throughput in Wi-Fi and LTE-U access mode $P_L^W \in [0.1, P^W]$ and $P_W^L \in [0.1, P^L]$ for $P^W = P^L = 3$. In the zone of using such a mode we have that

(a) if $P_L^W > P_W^L$ then the time fraction q to use such mode is increasing in P_W^L and decreasing in P_L^W. Wi-Fi access mode is not used at all, while the LTE-U access mode is decreasing in P_W^L and increasing in P_L^W;

(b) if $P_L^W < P_W^L$ then the frequency q to use such a mode is increasing in P_L^W and decreasing in P_W^L. LTE-U access mode is not used at all, meanwhile Wi-Fi access mode is decreasing in P_L^W and increasing in P_W^L.

Thus, in the zone of joint access mode, the network with higher throughput yields longer access to the channel to the network with lower throughput. If either P_W^L or P_L^W becomes too large, the optimal maxmin access to the channel switches to a channel on/off strategy.

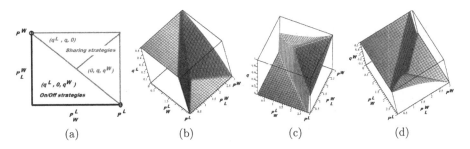

Fig. 4. (a) Zones of applying maxmin sharing and on/off strategies, and the optimal time fractions q^W (b), q (c) and q^L (d) for $P_L^W \in [0.1, P^W]$, $P_W^L \in [0.1, P^L]$ and $P^W = P^L = 3$.

5   Discussion

This paper examined the problem of Wi-Fi and LTE-U networks sharing access to a channel in support of each network's throughput needs. As a criteria for how the two networks jointly access the channel, we formulated the sharing using the α-fairness over expected throughput criteria. Maxmin access to the channel results as a limit case for the considered criteria as the fairness coefficient tends to infinity. It was shown that the optimal solution can be either (a) channel on/off strategies, i.e., in which access to the channel is performed sequentially, or (b) channel sharing strategies, i.e., in which simultaneous joint access to the channels is possible. A criteria for switching between these two types of optimal strategies was found, and the robustness of this criteria to the fairness coefficient was established. It was shown that the α-fair solution tends to a maxmin solution as α tends to infinity, and that a strategy supporting permanent joint access cannot be optimal for the maxmin problem, yet it can be optimal for the α-fairness problem. In particular, we showed that the fairness coefficient can be an efficient tool to control the protocol that determines what fraction of time Wi-Fi and LTE-U networks are dedicated to use the channel by themselves or the fraction of time they simultaneously use a channel. Finally, we note that the suggested scheme for access to a joint resource is general, and its realization depends on the implementation of the global controller, who is in charge for processing all the data on networks' facilities and users's demands associated with a specific problem. For Wi-Fi and LTE-U networks, a model of the global controller was developed in [11] and the model-based results were partially validated via experimental evaluations using USRP based SDR platforms on the ORBIT testbed. Numerical modelling in [11] showed significant gains in both Wi-Fi and LTE performance under the global controller's moderation. A goal of our future work is to adapt the formalism presented to the problem of multiple cellular providers sharing access to a communication medium in support of different classes of users during a disaster scenario.

A   Appendix I: Proof of Theorem 1

To find the optimal $\boldsymbol{q} = bq_\alpha$ define the Lagrangian $L_\omega(\boldsymbol{q}) = v_\alpha(\boldsymbol{q}) + \omega(1 - q^W - q - q^L)$. Thus, $\boldsymbol{q}$ is the optimal probability vector then the following conditions has to hold:

$$\frac{P^W}{(q^W P^W + q P_L^W)^\alpha} \begin{cases} = \omega, & q^W > 0, \\ \leq \omega, & q^W = 0, \end{cases} \tag{11}$$

$$\frac{P^L}{(q^L P^L + q P_W^L)^\alpha} \begin{cases} = \omega, & q^L > 0, \\ \leq \omega, & q^L = 0, \end{cases} \tag{12}$$

and

$$\frac{P_L^W}{(q^W P^W + q P_L^W)^\alpha} + \frac{P_W^L}{(q^L P^L + q P_W^L)^\alpha} \begin{cases} = \omega, & q > 0, \\ \leq \omega, & q = 0. \end{cases} \tag{13}$$

Thus, the boundary strategies $\boldsymbol{q} = (1, 0, 0)$ and $\boldsymbol{q} = (0, 0, 1)$ cannot be optimal.

First we find the condition when the rest boundary strategy $\boldsymbol{q} = (0, 1, 0)$ can be optimal. Substituting it into (11)–(13) implies that the following condition has to hold: $\omega = (P_L^W)^{1-\alpha} + (P_W^L)^{1-\alpha} \geq \max\{P^L/(P_W^L)^\alpha, P^W/(P_L^W)^\alpha\}$. This condition is equivalent to (8), and (d) follows.

Let us pass to finding channel sharing optimal strategy, i.e., with $q > 0$. Then, either $q^W = 0$ or $q^L = 0$.

Let $q^L = 0$. Then, (11)–(13) turn into the following conditions

$$P^W/(q^W P^W + q P_L^W)^\alpha = \omega, \tag{14}$$

$$P^L/(q P_W^L)^\alpha \leq \omega \tag{15}$$

and
$$P_L^W/(q^W P^W + qP_L^W)^\alpha + P_W^L/(qP_W^L)^\alpha = \omega. \tag{16}$$

By (14),
$$q^W P^W + qP_L^W = (P^W/\omega)^{1/\alpha}. \tag{17}$$

Since $q^W + q = 1$ then $(1-q)P^W + qP_L^W = (P^W/\omega)^{1/\alpha}$. So,

$$\cdot \qquad q = \left(1 - (P^W)^{(1-\alpha)/\alpha}/\omega^{1/\alpha}\right) / \left(1 - P_L^W/P^W\right). \tag{18}$$

By (14) and (16),
$$\frac{P_L^W}{P^W}\omega + \frac{P_W^L}{(qP_W^L)^\alpha} = \omega. \tag{19}$$

Thus,
$$q = \frac{(P_W^L)^{(1-\alpha)/\alpha}}{(1 - P_L^W/P^W)^{1/\alpha}\omega^{1/\alpha}}. \tag{20}$$

By (18) and (20),

$$\frac{1 - (P^W)^{(1-\alpha)/\alpha}/\omega^{1/\alpha}}{1 - P_L^W/P^W} = \frac{(P_W^L)^{(1-\alpha)/\alpha}}{(1 - P_L^W/P^W)^{1/\alpha}\omega^{1/\alpha}}. \tag{21}$$

Thus,

$$\omega^{1/\alpha} = (P^W)^{(1-\alpha)/\alpha} + \left(\frac{P_W^L}{1 - P_L^W/P^W}\right)^{(1-\alpha)/\alpha}. \tag{22}$$

Thus,

$$q = \frac{\dfrac{(P_W^L)^{(1-\alpha)/\alpha}}{(1 - P_L^W/P^W)^{1/\alpha}}}{(P^W)^{(1-\alpha)/\alpha} + \left(\dfrac{P_W^L}{1 - P_L^W/P^W}\right)^{(1-\alpha)/\alpha}}. \tag{23}$$

It is clear that $q > 0$. Since, $\boldsymbol{q}$ is the probability vector we have to find only the condition for q being less or equal to 1. By (23), it is equivalent to

$$\frac{(P_W^L)^{(1-\alpha)/\alpha}}{(1 - P_L^W/P^W)^{1/\alpha}} \le (P^W)^{(1-\alpha)/\alpha} + \left(\frac{P_W^L}{1 - P_L^W/P^W}\right)^{(1-\alpha)/\alpha}. \tag{24}$$

The last inequality is equivalent to

$$(P_W^L)^{1-\alpha}(P_L^W)^\alpha \le P^W - P_L^W. \tag{25}$$

Finally we have to find the condition that (15) holds. Substituting q from (20) into (15) implies (b).

The case $q^W = 0$ as well as the case $q = 0$, $q^W > 0$ and $q^L > 0$ can be considered similarly, and (a) and (c) follow. To deal with (e) denote by $\boldsymbol{q}_b$ and

q_c the optimal strategies given by (b) and (b). The previous analyze yields that in (d) the optimal strategy is q_b if $v_\alpha(q_b) > v_\alpha(q_c)$, and it is q_c if $v_\alpha(q_b) < v_\alpha(q_c)$.

Note that, by (17) and (19),

$$v_\alpha(q_b) = \frac{1}{1-\alpha}\left(\frac{P^W}{\omega}\right)^{(1-\alpha)/\alpha} + \frac{1}{1-\alpha}\left(\frac{P_W^L}{(1-P_L^W/P^W)\omega}\right)^{(1-\alpha)/\alpha}. \tag{26}$$

Substituting (22) into (26) yields

$$(1-\alpha)v_\alpha(q_b) = ((P^W)^{(1-\alpha)/\alpha} + (P_W^L/(1-P_L^W/P^W))^{(1-\alpha)/\alpha})^\alpha.$$

By symmetry, $v_\alpha(q_c)$ can be found, and the result follows.   ∎

B   Appendix II: Proof of Theorem 2

The maxmin problem is equivalent to the following LP problem

$$\text{maximize } \nu,$$
$$q^W P^W + q P_L^W \geq \nu, q P_W^L + q^L P^L \geq \nu, \ q^W + q + q^L = 1, q^W, q, q^L \geq 0. \tag{27}$$

Let for while the component q of the strategy q be fixed and optimal, while Then, component q^W and q^L might vary. Since $q^W + q^L = 1 - q$ the optimal q^W can be found as a solution of the following problem:

$$\text{maximize } \nu,$$
$$q^W P^W + q P_L^W \geq \nu, q P_W^L + (1-q)P^L - q^W P^L \geq \nu, q^W \in [0, 1-q]. \tag{28}$$

First we look for the optimal channel sharing strategy, i.e., when $q > 0$. Figure 5 illustrates that the solution of LP problem (28) can be found as an intersection of the corresponding lines.

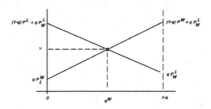

Fig. 5. Solution of the LP problem

Thus, the channel sharing solution holds if and only if $(1-q)P^L + qP_W^L > qP_L^W$ and $(1-q)P^W + qP_L^W > qP_W^L$. These inequalities are equivalent to

$$(1-q)P^L > q(P_L^W - P_W^L) \text{ and } (1-q)P^W > q(P_W^L - P_L^W). \tag{29}$$

Since either $P_L^W \geq P_W^L$ or $P_W^L > P_L^W$, then one of the conditions (29) always hold. Without loss of generality we assume that

$$P_L^W \geq P_W^L. \tag{30}$$

Then, conditions (29) are equivalent to

$$q \leq 1/(1 + (P_L^W - P_W^L)/P^L). \tag{31}$$

Let us switch on to finding the optimal ν and q^W. By Fig. 5 and (28),

$$q^W P^W + q P_L^W = q P_W^L + (1 - q)P^L - q^W P^L = \nu. \tag{32}$$

Thus,

$$q^W = (P^L - (P^L + P_L^W - P_W^L)q)/(P^L + P^W). \tag{33}$$

Thus, by (30), q^W is decreasing in q, and

$$\nu = (P^W P^L + (P^L P_L^W + P^W P_W^L - P^L P^W)q)/(P^W + P^L). \tag{34}$$

So, (33) and (34) give channel sharing solution of (27) for a fixed q, and (31) is the condition such solution holds. Note that, by (34), ν is increasing in q if $P^L P_L^W + P^W P_W^L > P^L P^W$, and ν is decreasing in q otherwise. Thus, if (30) holds, then the channel sharing solution exists ($q > 0$) if and only if $P^L P_L^W + P^W P_W^L > P^L P^W$, and then $q = 1/(1 + (P_L^W - P_W^L)/P^L)$. Substituting this q into (33) implies $q^W = 0$. Thus, $q^L = 1 - q$. The case of the channel on/off optimal strategy can be considered similarly, and the result follows. ∎

References

1. Barbagallo, P., McElgunn, T.: 3Gpp Set to Study LTE over Unlicensed Technology, Sowing New Spectrum Battles. Bloomberg BNA, 10, September 2014. http://www.bna.com/3gpp-set-study-n17179894699/
2. Cavalcante, A.M., et al.: Performance evaluation of LTE and Wi-Fi coexistence in unlicensed bands. In: IEEE 77th Vehicular Technology Conference (VTC Spring), pp. 1–6 (2013)
3. Abinader, F.M., et al.: Enabling the coexistence of LTE and Wi-Fi in unlicensed bands. IEEE Comm. Mag. **52**, 54–61 (2014)
4. Liu, F., Bala, E., Erkip, E., Yang, R.: A framework for femtocells to access both licensed and unlicensed bands. In: International Symposium Modeling and Optimization in Mobile, Ad Hoc and Wireless Networks (WiOpt), pp. 407–411 (2011)

5. Hajmohammad, S., Elbiaze, H., Ajib, W.: Fine-tuning the femtocell performance in unlicensed bands: case of WiFi Vo-existence. In: International Wireless Communications and Mobile Computing Conference (IWCMC), pp. 250–255 (2014)

6. Hajmohammad, S., Elbiaze, H.: Unlicensed spectrum splitting between femtocell and WiFi. In: IEEE International Conference Communication (ICC), pp. 1883–1888 (2013)

7. Jian, Y., Shih, C.-F., Krishnaswamy, B., Sivakumar, R.: Coexistence of Wi-Fi and LAA-LTE: experimental evaluation, analysis and insights. In: IEEE International Conference Communication (ICC), pp. 10387–10393 (2015)

8. Cano, C., Leith, D.: Coexistence of WiFi and LTE in unlicensed bands: a proportional fair allocation scheme. In: IEEE International Conference Communication (ICC), pp. 10350–10355. IEEE (2015)

9. Sagari, S., Seskar, I., Raychaudhuri, D.: Modeling the coexistence of LTE and WiFi heterogeneous networks in dense deployment scenarios. In: Proceedings of IEEE International Conference Communication (ICC), pp. 10363–10368 (2015)

10. Raychaudhuri, D., Baid, A.: NASCOR: network assisted spectrum coordination service for coexistence between heterogeneous radio systems. IEICE Trans. Comm. **E97–B**, 251–260 (2014)

11. Sagari, S., Baysting, S., Sahay, D., Seskar, I., Trappe, W., Raychaudhuri, D.: Coordinated dynamic spectrum management of LTE-U and Wi-Fi networks. In: IEEE International Conference on Dynamic Spectrum Access Network (DySPAN), pp. 209–220 (2015)

12. Huaizhou, S., Prasad, R.V., Onur, E., Niemegeers, I.: Fairness in wireless networks: issues, measures and challenges. IEEE Commun. Surv. Tutorials **16**, 5–24 (2014)

13. Mo, J., Walrand, J.: Fair end-to-end window-based congestion control. IEEE/ACM Trans. Networking **8**, 556–567 (2000)

14. Kim, S.: Multi-objective power control algorithm for femtocell networks. Wireless Pers. Commun. **75**, 2281–2288 (2014)

15. Altman, E., Avrachenkov, K., Garnaev, A.: Generalized α-fair resource allocation in wireless networks. In: 47th IEEE Conference on Decision and Control (CDC), pp. 2414–2419 (2008)

16. Altman, E., Avrachenkov, K., Garnaev, A.: Fair resource allocation in wireless networks in the presence of a jammer. Perform. Eval. **67**, 338–349 (2010)

17. Garnaev, A., Sagari, S., Trappe, W.: Fair allocation of throughput under harsh operational conditions. In: Jonsson, M., Vinel, A., Bellalta, B., Tirkkonen, O. (eds.) MACOM 2015. LNCS, vol. 9305, pp. 108–119. Springer, Heidelberg (2015)

18. Yerramalli, S., Jain, R., Mitra, U.: Coalitional games for transmitter cooperation in MIMO multiple access channels. IEEE Trans. Sign. Proc. **62**, 757–771 (2014)

19. Altman, E., Avrachenkov, K., Garnaev, A.: Jamming in wireless networks: the case of several jammers. In: International Conference on Game Theory for Networks (GameNets 2009), pp. 585–592 (2009)

20. Garnaev, A., Hayel, Y., Altman, E.: A Bayesian jamming game in an OFDM wireless network. In: 10th International Symposium on Modeling and Optimization in Mobile, Ad Hoc and Wireless Networks (WIOPT), pp. 41–48 (2012)

21. Altman, E., Avrachenkov, K., Garnaev, A.: Transmission power control game with SINR as objective function. In: Altman, E., Chaintreau, A. (eds.) NET-COOP 2008. LNCS, vol. 5425, pp. 112–120. Springer, Heidelberg (2009)

22. Garnaev, A., Trappe, W., Kung, C.-T.: Dependence of optimal monitoring strategy on the application to be protected. In: IEEE Global Communications Conference (GLOBECOM), pp. 1054–1059 (2012)

23. Garnaev, A., Trappe, W.: One-time spectrum coexistence in dynamic spectrum access when the secondary user may be malicious. IEEE Trans. Inf. Forensics Secur. **10**, 1064–1075 (2015)
24. Garnaev, A., Trappe, W.: A bandwidth monitoring strategy under uncertainty of the adversary's activity. IEEE Trans. Inf. Forensics Secur. **11**, 837–849 (2016)
25. Marina, N., Arslan, G., Kavcic, A.: A power allocation game in a four node relaynetwork: an upper bound on the worst-case equilibrium efficiency. In: Conference onTelecommunications (ICT), pp. 1–6 (2008)
26. Garnaev, A., Baykal-Gursoy, M., Poor, H.V.: Incorporating attack-type uncertainty into network protection. IEEE Trans. Inf. Forensics Secur. **9**, 1278–1287 (2014)

Is Bayesian Multi-armed Bandit Algorithm Superior?: Proof-of-Concept for Opportunistic Spectrum Access in Decentralized Networks

Sumit J. Darak[1(✉)], Amor Nafkha[2], Christophe Moy[2], and Jacques Palicot[2]

[1] IIIT-Delhi, Okhla Phase III, New Delhi 110020, India
sumit@iiitd.ac.in
[2] SCEE, CentraleSupélec/IETR, CS 47601, 35576 Cesson-Sévigné Cedex, France

Abstract. Poor utilization of an electromagnetic spectrum has led to surge of interest in paradigms such as cognitive radio, unlicensed LTE etc. Such paradigms allow opportunistic spectrum access in the vacant frequency bands of the licensed spectrum. Though various spectrum detectors to check the status of frequency bands (i.e., vacant or occupied) have been studied, the selection of the frequency band from wideband spectrum is a challenging problem especially in the decentralized network. In this paper, a testbed for analyzing the performance of decision making policies (DMPs) for identifying optimum frequency band in the decentralized network is presented. Furthermore, experimental results using real radio signals show that the proposed DMP using Bayesian multi-armed bandit algorithm leads to 7–12 % improvement in an average spectrum utilization over existing DMPs. Added advantages of 6–20 % lower switching cost and 30–46 % fewer collisions make the proposed DMP energy-efficient.

Keywords: Cognitive radio · Multi-armed bandit algorithm · Opportunistic spectrum access · Usrp testbed

1   Introduction

Further boosting the average utilization of an electromagnetic spectrum has led to a surge of interest in paradigms such as cognitive radio (CR), device-to-device (D2D) communications, unlicensed LTE (LTE-U) etc. from the academia as well as industrial partners [1–4]. Such paradigms allow opportunistic spectrum access (OSA) in the vacant licensed bands with the constraint of zero interference to the active licensed users. For example, an underlay inband unlicensed D2D communications allow direct communication between secondary (or unlicensed) users (SUs) over identified vacant licensed subband(s) without the need of base stations or access points for establishing the communication link. Also, CR and LTE-U allow SUs to access vacant TV white and 5 GHz spectrum, respectively. Though most of the existing works focus on the centralized networks, decentralized networks would be an efficient choice over the centralized approach due to

© ICST Institute for Computer Sciences, Social Informatics and Telecommunications Engineering 2016
D. Noguet et al. (Eds.): CROWNCOM 2016, LNICST 172, pp. 104–115, 2016.
DOI: 10.1007/978-3-319-40352-6_9

advantages such as ease of implementation, robustness to link or node failures, no communication overhead and lower delay [1, 2]. Furthermore, decentralized app-roach is preferred choice for public safety networks and proximity-aware social networking services.

Decision making policies (DMPs) are desired for OSA in the decentralized networks in order to: (1) Enable SUs to identify optimum vacant subbands, (2) Minimize collisions among SUs, and (3) Keep the subband switching cost (SSC) as low as possible. Here, SSC stands for the total penalty incurred in terms of delay, power, hardware reconfiguration and protocol overhead when SU switches from one frequency subband to another. From the energy efficiency perspective, SSC and number of collisions should be as low as possible. Design of such DMP for decentralized CRNs is a challenging task and one of the objective the work presented in this paper.

In this paper, USRP based testbed for analyzing the performance of DMPs for OSA in decentralized networks is presented. To the best of our knowledge, the proposed testbed is the first proof-of-concept which compares the performance of various DMPs using real radio signals. Furthermore, experimental results using real radio signals show that the proposed DMP using Bayesian multi-armed bandit algorithm leads to 7–12 % improvement in an average spectrum utilization over existing DMPs. Added advantages of 6–20 % lower switching cost and 30–46 % fewer collisions make the proposed DMP energy-efficient.

The paper is organized as follows. The detailed literature review of DMPs for decentralized CRNs is given in Sect. 2. The proposed DMP is presented in Sect. 3 followed by the proposed testbed description in Sect. 4. The experimental results are discussed in Sect. 5. Section 6 concludes the paper.

2   Literature Review: Decision Making Policies

Various DMPs have been proposed for decentralized CRNs [3–9]. These DMPs consist of MAB algorithms such as frequentist approach based upper confidence bound (UCB) algorithm and its extensions, $\varepsilon-$greedy, optimization based Kullback-Leibler UCB (KL–UCB) to estimate subband statistics (e.g., probability of being vacant, transition probability from vacant state to occupied state and vice-versa etc.) [5, 10]. Such algorithms are based on exploration-exploitation trade-off where they explore all subbands multiple number of times before settling down to optimum subband with respect to desired statistics. These algorithms are asymptotically optimal with logarithmic regret that is the best one can expect when there is no prior information about subband statistics. Such optimality guarantees that the estimated statistics of all subbands, and not just optimum subband, are closed to their actual statistics [5, 10].

Another challenging task of the DMP is to orthogonalize SUs to different frequency bands and minimize collisions between them. This is a challenging problem considering SUs do not share any information with each other. In ρ^{rand} DMP with M SUs [4], each SU randomly and independently chooses the rank, $R(k) \in \{1, 2, ..M\}$ in the beginning. In subsequent time slots, underling MAB algorithm calculates the quality index for each subband. Then, the SU with the rank $R(k)$ chooses the subband with the $R(k)^{th}$ best quality index. Another DMP in [6] follows time division fare share approach where the rank of each SU is rotated in circular fashion between 1 to M to allow an equal access to the optimum subbands among all SUs. In [4,6], a new rank is randomly and independently chosen for SU experiencing collision. Though both DMPs are asymptotically optimal with logarithmic regret, SSC of [6] increases linearly with t. The SSC can be minimized using the DMP, ρ^{rand}, in [4], where the rank is changed only when corresponding SU collides with other SUs. The performance of the ρ^{rand} DMP is further improved in [7]. In [7], the range for rank i.e. $1 \leq R(k) < N$, $\forall k$ is made wider to minimize the number of collisions. In [8], variable filtering architecture and its integration with tunable subband access DMP, ρ^{t_rand}, is proposed that takes into account tunable bandwidth requirements of SUs. Still, the average SSC of [7,8] is quite high. Recently, Bayesian MAB algorithms such as Bayes-UCB and Thompson Sampling have become more popular and are proved to be efficient than the other MAB algorithms [9,10]. However, experimental analysis of these MAB algorithms using real radio signals and non-ideal detectors has not been done yet.

3   Proposed Decision Making Policy

The proposed DMP consists of subband statistic estimation using Bayes-UCB algorithm and subband access scheme for orthogonalization of SUs. The proposed DMP is discussed in detail next.

3.1   System Model

Consider the slotted CRN consisting of multiple primary users and M SUs. The wideband spectrum of bandwidth, B, is divided into N uniform subbands of bandwidth, B_{cmin}. Hence, $B_{cmin} = (B/N)$. The status of each subband (i.e., vacant or occupied) is independent of the status of other subbands. For a given subband, vacancy statistic depends on the underlining traffic model which can be either independent and identically distributed (i.i.d.) model or Markovian model. In case of i.i.d. model, the status of $i^{th}, i \in \{1, 2, ...N\}$ subband depends on its $P_{vac}(i)$ distribution and is independent of its status in previous time slots. In case of Markovian model, i^{th} subband switches its state from being vacant to occupied and vice versa according to a discrete Markov process with the probabilities of $p_{vo}(i)$ and $p_{ov}(i)$, respectively. Then, using Markov chain analysis, steady-state probabilities of subband being vacant, denoted as $P_{vac}(i), i \in \{1, 2, ...N\}$, are given by [11]

$$P_{vac}(i) = \frac{p_{ov}^i}{p_{ov}^i + p_{vo}^i}, \ \forall i \tag{1}$$

Basic assumptions made in this paper for SUs in decentralized CRNs are:

1. Infrastructure less decentralized CRN where all SUs employ the same DMP but do not exchange information with other SUs.

2. SU can sense only one subband in each time slot.
3. $P_{vac}(i)$, $p_{vo}(i)$ or $p_{ov}(i)$, $i \in \{1, 2, ...N\}$ are unknown to SUs.
4. All SUs must sense the chosen frequency band at the start of each time slot irrespective of sensing outcomes in the previous time slots.

At the beginning of each time slot, DMP of SU chooses the subband for sensing. Let $N_k(t) \in \{1, 2, .., N\}$ be the subband chosen by k^{th} SU in time slot t. The analog-front end and digital front-end of SU filter the chosen subband, down-convert it to baseband and passed it to the spectrum detector. If the subband is vacant, it is assumed that the SU transmits over that subband. When multiple SUs transmit on the same subband i.e., $N_k(t) = N_j(t)$ for any $k \neq j$, collision occurs leading to failed transmission. Otherwise, it is assumed that SU transmits successfully. Let $\Delta_k(t)$ be instantaneous reward of kth SU in time slot t and is given by,

$$\Delta_k(t) = \begin{cases} 1 & \text{No collission} \\ 0 & \text{Collision} \end{cases} \tag{2}$$

Let $r_k(t)$ be the total number of successful transmissions by k^{th} SU and is given by,

$$r_k(t) = r_k(t-1) + \Delta_k(t) \tag{3}$$

Let $S^*(t)$ and $S(t)$ denote the total number of successful transmissions by genie-aided DMP (i.e. the DMP where $P_{vac}(i)$, $\forall i$ are known a priori and central unit allocates distinct subband to each SUs) and decentralized DMP, respectively. Then, total loss in terms of transmission opportunities, $U(t)$, up to time t is given by Eq. 4 and should be as small as possible.

$$U(t) = S^*(t) - S(t) = \sum_{k=1}^{M} \sum_{v=1}^{t} \mathbb{E}[r_k^*(v) - r_k(v)] \tag{4}$$

In addition, SSC and number of collisions, $C(t)$, given by Eqs. 5 and 6, respectively, should be as minimum as possible.

$$SSC(t) = \sum_{k=1}^{M} \sum_{v=2}^{t} \mathbb{E}[\mathbf{1}_{N_k(v) \neq N_k(v-1)}] \tag{5}$$

$$C(t) = \sum_{k=1}^{M} \sum_{v=1}^{t} \Delta_k(v) \tag{6}$$

3.2   Subband Statistic Estimation

The Bayes-UCB algorithm employed in the proposed DMP of kth SU is given in Algorithm 1. Here, H is horizon size, $T_k(i, t)$ indicates the number of times the subband i is chosen by k^{th} SU up to time t, $S_k(i, t)$ indicates the number of times out of $T_k(i, t)$, the subband i is observed as vacant by k^{th} SU and $R(k)$ is the rank of kth SU. The rank calculation is discussed later in Sect. 3.3. Initially, all subbands are sensed once as shown in Steps 1–6 of Algorithm 1. Then, at each time slot, $t > N$, Bayes-UCB algorithm calculates quality index, $q(i,t), \forall i$ for each subband as shown in Step 12 of Algorithm 1. This is done by calculating quantile of order i for a given beta distribution. Based on the values of $q(i,t), \forall i$, SU chooses the subband according to its rank $R(k)$ as shown in Step 14. After getting the status of the chosen subband from spectrum detector, parameters $S_k(N_k(t),t)$ and $T_k(N_k(t),t)$ are updated as shown in Steps 15–18. Note that, $S_k(i,t) \le T_k(i,t), \forall i, \forall k$.

Algorithm 1: Subband Selection and Statistic Estimation Using Bayes-UCB Algorithm for k^{th} Secondary User

---

Parameters: $N, R(k), i \in \{1,2,..,N\}, H$
Input: $S_k(i,t-1), T_k(i,t-1), \forall i$
Output: $N_k(t), S_k(i,t), T_k(i,t), \forall i$

1. **if** $(any(T_k(:,t-1) == 0))$
2. $N_k(t) = i$ s.t. $T_k(i,t-1) = 0$
3.   **if** $N_k(t)$ is vacant
4. $S_k(N_k(t),t) = S_k(N_k(t),t) + 1$
5.   **end**
6. $T_k(N_k(t),t) = T_k(N_k(t),t) + 1$
7. **else**
8.   **for** $t = N + 1$ **to H do**
9. $S_k(:,t) = S_k(:,t-1)$
10. $T_k(:,t) = T_k(:,t-1)$
11.   **for** $i = 1$ **to N do**
12.   **Compute**
$$q(i,t) = Q\left\{1-\tfrac{1}{t}; Beta[S_k(i,t)+1, T_k(i,t)-S_k(i,t)+1]\right\}$$
13.   **end**
14.   Select the band $N_k(t) = i$
                      s.t. $q(i,t)$ is $R(k)^{th}$ max. value of $q(:,t)$
15.   **if** $N_k(t)$ is vacant
16. $S_k(N_k(t),t) = S_k(N_k(t),t) + 1$
17.   **end**
18. $T_k(N_k(t),t) = T_k(N_k(t),t) + 1$
19.   **end**
20. **end**

Bayes-UCB algorithm is asymptotically optimal which means that

$$\limsup_{t\to\infty} \frac{\mathbb{E}[T_k(i,t)]}{\ln t} \ge \frac{1}{KL(P_{vac}(i), P_{vac}(i^*))}, \quad \forall i, \forall k \tag{7}$$

where KL stands for the Kullback-Leibler divergence. Equation 7 indicates that Bayes-UCB algorithm achieves optimal balance between exploration of all subbands and exploitation of optimum subband. Also, it guarantees that learned subbands statistics are closed to their actual values. Another advantage of Bayes-UCB algorithm is that SSC is low through empirical observations. This means that Bayes-UCB algorithm chooses the same band consecutively more number of times than other MAB algorithms like UCB. This feature might be advantageous for accurate estimation of transition probabilities, i.e., P_{vo} and P_{ov}. However, usefulness of transition probabilities for DMP is out of scope of this paper. Next, subband access scheme is presented.

3.3   Subband Access in Multi-user Decentralized Network

For any DMP, the upper bound on $U(t)$ in Eq. 4, is given by [4],

$$U(t) \leq P_{vac}(1^*) \left\{ \sum_{k=1}^{M} \sum_{i \in Z\_worst} \mathbb{E}[T_k(i,t)] + \mathbb{E}[C(t)] \right\} \tag{8}$$

where $P_{vac}(1^*)$ is the highest vacancy statistics among N subbands, Z_worst is set of all subbands excluding first M subbands when arranged according to decreasing values of P_{vac} and $C(t)$ are the number of collisions when SU chooses any of the first M subbands. For lower $U(t)$, subband ordering at each SU should be accurate and rank selection must be orthogonal. Hence, for OSA in a multi-user decentralized network, accurately estimation of subband statistics alone is not sufficient. In addition, DMP needs to avoid collisions, $C(t)$, among active SUs.

In the proposed DMP, modified randomization based subband access scheme, proposed in [8], has been used. The proposed subband access scheme in [8] is based on ρ_{rand} [4]. The novelty of the proposed scheme is that the upper limit of rank is tunable based on subband statistics compared fixed value of rank (equal to M) in ρ_{rand} [4]. This leads to fewer collisions and hence, higher number of transmission opportunities and lower SSC compared to ρ_{rand} [4].

4   Proposed USRP Testbed

The proposed USRP testbed is shown in Fig. 1 and is an significant extension of the testbed in [13]. It consists of two units: (1) Left hand side unit is primary user traffic generator, and (2) Right hand side unit acts as DMP of secondary user(s). Both the units are discussed in detail next.

4.1   Primary User Traffic Generator

The chosen design environment for the primary user traffic generation is GNU Radio Companion (GRC) and the hardware platform is made of a USRP from Ettus Research. The main reason for choosing GRC is the precise control on each parameter of the transmission chain compared to other environments. The proposed primary user traffic generator is shown in Fig. 2. In the beginning, number of frequency bands, traffic model (i.i.d. or Markovian) and corresponding subband statistics are taken from the user using the block named Traffic Model in

Fig. 1. Proposed USRP based testbed for analyzing the performance of DMPs using real radio signals and non-ideal spectrum detectors.

Fig. 2. The transmission bandwidth, which is restricted by bandwidth of analog front-end of USRP, is divided into N subbands. In each time slot, masking vector of size N is generated by Traffic Model block based on given subband statistics. This masking vector can have 1 or 0 values where 1 and 0 indicate that corresponding band is occupied and vacant, respectively. Next step is mapping data to be transmitted on sub-carriers of occupied bands. The data modulation used is a differential QPSK modulation with Gray encoding. This is followed by sub-carrier mapping using OFDM and transmission via USRP. In the proposed tested, number of sub-carriers, center frequency and transmission bandwidth are 256, 433.5 MHz and 1 MHz, respectively. For demonstration purpose, each time slot duration is one second so that it can be followed by human eye. However, it can be reduced to the order of milliseconds and will have no direct effect on the performance of DMP.

4.2   Secondary User with Decision Making Policy

The chosen design environment for the SU terminal is Matlab/Simulink and USRP from Ettus Research. USRP is tuned to receive signal of bandwidth 1 MHz centered at 433.5 MHz. The received signal is then down-sampled, digitized and passed to the DMP implemented using Simulink. The DMP uses MAB algorithm to select single subband in each time slot. The chosen subband is sensed using energy detector. Note that energy detector is not ideal and sensing error may occur [12]. If subband is sensed as vacant, it is assumed that SU transmits over the chosen subband. If multiple SUs choose the same subband, it is assumed that both users suffer collision and fail to transmit any data. In case of multiple SUs, each user is independently implemented in Simulink with their respective DMP. In existing works, sensing is assumed as perfect which is not true in real radio conditions. Thus, proposed testbed with non-ideal spectrum detectors will enable to study performance of DMPs in the presence of sensing errors. Note

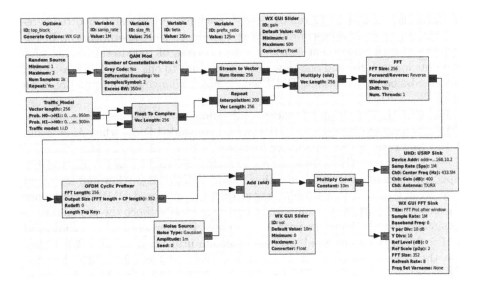

Fig. 2. Proposed primary user traffic generator.

that performance comparisons of various detectors and their effect on DMPs is not discussed here due to brevity of the paper.

4.3   Synchronization

The synchronization between transmitter and receiver is an important aspect of slotted decentralized network infrastructure considered in this paper. For demonstration purposes, synchronization has been achieved by switching first band from occupied to vacant states or vice-versa in each time slot. This enables SUs to detect the transitions between OFDM symbols as well as to synchronize the energy detection phase on an entire OFDM symbol of the primary traffic. In a real OSA scenario, SU should be able to synchronize with PU network via synchronization signals or pilot carriers. Note that the synchronization band in the proposed approach is not wasted because DMP does not restrict it as synchronization band but consider it as possible option for data transmission.

5   Experimental Results and Analysis

In this section, experimental comparison of various DMPs on the proposed testbed, discussed in Sect. 4, is presented. For $N = 8$ and $B = 1\,\mathrm{MHz}$, we have $B_{cmin} = 125\,\mathrm{KHz}$. In case of i.i.d. rewards, two different P_{vac} distributions, denoted as case 1^1 and case 2^2 are considered. Similarly, P_{vo} and P_{ov} distributions

---

[1] P_{vac} :- $[.50\ .10\ .20\ .30\ .40\ .60\ .80\ .90]$.
[2] P_{vac} :- $[.50\ .05\ .95\ .10\ .80\ .60\ .40\ .75]$.

for Markovian rewards are given by case 3^3. Each numerical result reported here-after is the average of values obtained over 15 independent experiments on USRP testbed and each experiment consider a time horizon of 1000 iterations i.e. 1000 time slots for each SU and one time slot corresponds to one second. It is assumed that all SUs employ the same DMP but do not exchange any information with others.

For Case 1, Fig. 3a and b show total number of successful transmissions, $S(t)$, in percentage for various DMPs w.r.t. genie-aided DMP when $M = 2$ and $M = 4$, respectively. For simplicity, policies UCB($\alpha = 2$)+ρ_{rand} [4], UCB($\alpha = 0.5$)+ρ_{rand} [4] and KLUCB+ρ_{rand} are hereafter referred to as E1, E2 and E3, respectively. Here, α is an exploration factor of UCB algorithm and it should be between 0.5 to 2. It can be observed that the proposed DMP offers higher transmission opportunities compared to existing DMPs. Since the probability of collision among SUs increases with M, $S(t)$ in % is lower when $M = 4$ compared to $M = 2$. On the other hand, average spectrum utilization for $M = 4$ is higher than the same when $M = 2$. For instance, average spectrum utilization due to licensed users was only 47 % for Case 1. Due to OSA with $M = 2$, average spectrum utilization can be increased to 58 %, 63 %, 66 % and 70 % using E1, E2, E3 and proposed DMPs, respectively. In case of $M = 4$, average spectrum utilization can be increased to 70 %, 76 %, 78 % and 81 % using E1, E2, E3 and proposed DMP, respectively.

Figures 4 and 5 show total number of successful transmissions, $S(t)$, in percentage for various DMPs w.r.t. genie-aided DMP in Case 2 and Case 3, respectively. Improvements in the average spectrum utilization, similar to Case 1, can also be seen for Case 2 and 3. To conclude, proposed DMP leads to higher avearge spectrum utilization and hence, superior compared to existing DMPs for various traffic distributions.

As discussed in Sect. 1, SSC should be as minimum as possible for making SU terminals energy efficient. In Fig. 6 (a), the SSC of different DMPs are compared

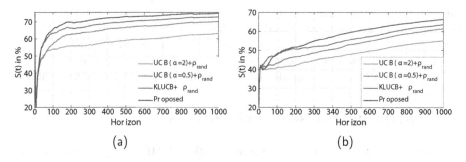

(a)   (b)

Fig. 3. Comparisons of average $S(t)$ in % of different DMPs with respect to the genie-aided DMP in Case 1 for (a) $M = 2$ and (b) $M = 4$.

---

[3] P_{vo} :- [.50 .05 .10 .20 .30 .40 .50 .60]
P_{ov} :- [.50 .80 .70 .60 .50 .40 .30 .20].

for subband distributions in Case 1, Case 2 and Case 3. It can be observed that SSC of the proposed DMP is low. Numerically, average SSC of the proposed DMP is 20 %, 6 % and 2 % lower than that of E1, E2 and E3, respectively. In additions to SSC, the number of collisions of DMP should be as minimum as possible. This is because, collision leads to waste of the energy required for transmission of corresponding data and it may be higher than the energy required for subband switching. In Fig. 6 (b), the number of collisions suffered by all SUs are compared for subband distributions in Case 1, Case 2 and Case 3. Numericaly, SUs employing proposed DMP suffers 46 %, 30 % and 13 % less number of collisions than SUs employing policies E1, E2 and E3, respectively. Thus, lower SSC and collisions of the proposed DMP makes it energy efficient and suitable for resource constrained battery operated SU terminals.

In terms of computational complexity, KLUCB is based on optimization approach and is the most computationally complex. The complexity of Bayes-UCB is slightly lower than that of UCB [10]. Based on experimental results, we argue that proposed DMP using Bayesian MAB algorithm for OSA in multi-user decentralized network is not only superior but also energy efficient.

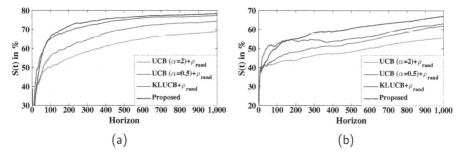

Fig. 4. Comparisons of average $S(t)$ in % of different DMPs with respect to the genie-aided DMP in Case 2 for (a) $M = 2$ and (b) $M = 4$.

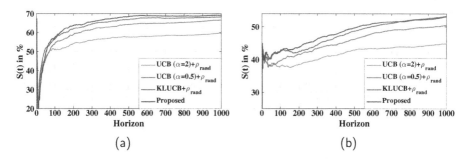

Fig. 5. Comparisons of average $S(t)$ in % of different DMPs with respect to the genie-aided DMP in Case 3 for (a) $M = 2$ and (b) $M = 4$.

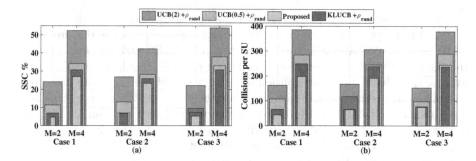

Fig. 6. (a) Comparison of SSC for various P_{vac} distributions, (b) Comparison of number of collisions for various P_{vac} distributions.

6 Conclusion and Future Works

A USRP based testbed for experimentally analyzing the performance of decision making policies (DMPs) for opportunistic spectrum access (OSA) in the decentralized cognitive radio networks has been proposed. To the best of our knowledge, the proposed testbed is the first proof-of-concept which compares the performance of various DMPs using real radio signals. Furthermore, experimental results showed that the proposed DMP designed using Bayesian multi-armed bandit (MAB) algorithm offers superior performance over existing policies in terms of average spectrum utilization, subband switching cost as well as number of collisions. Thus, we argue that Bayesian MAB algorithm based DMPs are superior for OSA in the decentralized networks. Future work involves study of the effect of various spectrum detectors on the performance of DMPs and realization of actual data transmission on the chosen frequency band.

Acknowledgments. The authors would like to thank Department of Science and Technology (DST), Government of India for INSPIRE fellowship in support of this work.

References

1. Asadi, A., Wang, Q., Mancuso, V.: A survey on device-to-device communication in cellular networks. IEEE Commun. Surv. Tutorials **16**(4), 1801–1819 (2014)
2. Phunchongharn, P., Hossain, E., Kim, D.I.: Resource allocation for device-to-device communications underlaying LTE-advanced networks. IEEE Wirel. Commun. **20**(4), 91–100 (2013)
3. Xu, Y., Anpalagan, A., Wu, Q., Shen, L., Gao, Z., Wang, J.: Decision-theoretic distributed channel selection for opportunistic spectrum access: strategies, challenges and solutions. IEEE Commun. Surv. Tutorials **15**(4), 1689–1713 (2013)
4. Anandkumar, A., Michael, N., Tang, A., Swami, A.: Distributed algorithms for learning and cognitive medium access with logarithmic regret. IEEE J. Sel. Areas Commun. **29**(4), 731–745 (2011)

5. Auer, P., Cesa-Bianchi, N., Fischer, P.: Finite-time analysis of the multiarmed bandit problem. Mach. Learn. **47**(2), 235–256 (2002)
6. Liu, K., Zhao, Q.: Distributed learning in multi-armed bandit with multiple players. IEEE Trans. Signal Process. **58**(11), 5667–5681 (2010)
7. Zandi, M., Dong, M.: Learning-stage based decentralized adaptive access policy for dynamic spectrum access. In: IEEE International Conference on Acoustics, Speech and Signal Processing (ICASSP), pp. 5323–5327, Vancouver, Canada, May 2013
8. Darak, S.J., Zhang, H., Palicot, J., Moy, C.: Efficient decentralized dynamic spectrum learning and access scheme for multi-standard multi-user cognitive radio networks. In: 11th International Symposium on Wireless Communication Systems (IEEE ISWCS 2014), pp. 271–275, Barcelona, Spain, August 2014
9. Darak, S.J., Zhang, H., Palicot, J., Moy, C.: An efficient policy for D2D communications and energy harvesting in cognitive radios: go bayesian! In: 23rd European Signal Processing Conference (EUSIPCO), pp. 1236–1240, Nice, France, August 2015
10. Kaufmann, E., Cappé, O., Garivier, A.: On bayesian upper confidence bounds for bandit problems. In: 15th International Conference on Artificial Intelligence and Statistics, pp. 592–600, La Palma, Canary Islands, April 2012
11. Hoang, D.T., Niyato, D., Wang, P., Kim, D.I.: Opportunistic channel access and RF energy harvesting in cognitive radio networks. IEEE Trans. Sel. Areas Commun. **32**(11), 1–14 (2014)
12. Bahamou, S., Nafkha, A.: Noise uncertainty analysis of energy detector: bounded and unbounded approximation relationship. In: 21st European Signal Processing Conference (EUSIPCO), pp. 1–4, Marrakech, Morocco, September 2013
13. Robert, C., Moy, C., Wang, C.X.: Reinforcement learning approaches and evaluation criteria for opportunistic spectrum access. In: IEEE International Conference on Communications (ICC), pp. 1508–1513, Sydney, Australia, June 2014

Minimum Separation Distance Calculations for Incumbent Protection in LSA

Markku Jokinen[1(✉)], Marko Mäkeläinen[1], Tuomo Hänninen[1],
Marja Matinmikko[2], and Miia Mustonen[2]

[1] Centre for Wireless Communications, University of Oulu, Oulu, Finland
{markku.jokinen,marko.makelainen,tuomo.hanninen}@ee.oulu.fi
[2] VTT Technical Research Centre of Finland, Oulu, Finland
{marja.matinmikko,miia.mustonen}@vtt.fi

Abstract. In this paper, we consider minimum separation distance calculations from the perspective of a real-life Licensed Shared Access (LSA) system in the 2.3 GHz band in Europe. In the LSA system, an LTE network shares spectrum resources with incumbent users, such as programme making and special events (PMSE) users, which need to be protected from harmful interference. Plenty of potential resources are available, in case the incumbent activity is occasional or localized. The sharing scenario requires realistic separation distances to be calculated to protect the incumbents. The minimum separation distances were calculated using methods presented in the ECC report on compatibility studies on 2.3 GHz band, but by using the parameters from the real-life LSA test network. With this work, we bridge the gap between theoretical research for incumbent protection and practical LSA deployment. In the process of defining new separation distances, discrepancies were found in the original example calculations.

Keywords: Minimum separation distance · Exclusion zone · Protection zone · Licensed Shared Access (LSA)

1 Introduction

As mobile traffic keeps increasing, new ways of finding more resources need to be established. One fundamental resource in mobile communication is spectrum, but since many different systems are allocated to dedicated frequency bands this resource is becoming scarce. On the other hand this allocated spectrum can be under utilized, hence spectrum sharing can be one solution to the lack of resources.

One of the emerging concepts of spectrum sharing is Licensed Shared Access (LSA) [1] that introduces additional licensed users on a shared basis. When applied to the mobile broadband, a mobile network operator (MNO) can share spectrum with different kinds of incumbent users in a licensed manner with quality of service (QoS) guarantees for all involved. In Europe, regulation framework [1] is ready for the 2.3–2.4 GHz for national deployment and standardization

© ICST Institute for Computer Sciences, Social Informatics and Telecommunications Engineering 2016
D. Noguet et al. (Eds.): CROWNCOM 2016, LNICST 172, pp. 116–128, 2016.
DOI: 10.1007/978-3-319-40352-6_10

is ongoing to be used by both licensed and incumbent users. Incumbents in this band vary depending on national deployments and include e.g. aeronautical telemetry and programme making and special events (PMSE) applications. Research efforts are on-going to protect the incumbents, see e.g. [2]. Trialing this kind of new system is important for verifying usability and operation in practice. In Finland the LSA concept has been extensively trialed [3–5] in Core+ and Core++ projects [6] using Finnish LSA trial environment.

In the LSA concept protecting incumbent is the most important issue while ensuring good operational conditions for MNOs. There are several methods for incumbent protection. First method is to preserve a certain partition of the frequency band to the incumbent use and allow the operation of MNOs in other parts of the frequency band. Another method is to allow the shared use of the whole band by both systems. This method requires geographical separation between the systems [7]. In this paper, we concentrate on sharing scenario between PMSE and MNO users. The separation can be defined by calculating a minimum separation distance [8] between the PMSE user and the closest MNO network element or by defining a protection zone as an aggregate effect of the MNO network elements. To expand the generic work of [8], we apply the minimum separation distance calculations to the incumbent protection in LSA. We decided to use minimum separation distance calculations, because the amount of required network information and the calculational complexity is lower than with the protection zone method. Minimum separation distance calculations are a foundation of the protection zone method, thus giving us a good starting point for more advanced implementations.

2   LSA Concept and Trial Environment

The LSA concept allows the introduction of additional licensed users on bands currently used by incumbents on a shared basis. Only two additional blocks, LSA Controller and LSA Repository [9], are needed on top of the commercial LTE network for the LSA concept. Figure 1 illustrates LSA management system as a part of LSA architecture. Each MNO willing to operate on the LSA frequencies should have its own LSA Controller for managing its networks according to incumbent activity. The LSA Repository contains information about the operation of the incumbent as well as of the LSA Controllers and LSA Licenses for specific areas. The incumbent reports its location, frequency, operation time and type of incumbent to the LSA Repository. The LSA Controller receives this information from the LSA Repository and controls the MNO network accordingly. For making the decision of the spectrum usage, the LSA Controller uses information of the MNO network layout. The first implementation of the LSA concept is the Finnish LSA trial environment [3,4] that consists of commercial LTE base stations (BS), both macro and small cells, operating on 2.3 GHz TDD band. BSs are connected to the conventional LTE core network and Operations, Administration and Management (OAM) system. The core network provides internet connectivity to user equipments (UEs), the OAM side consists of NetAct OSS providing a single system for managing LTE network.

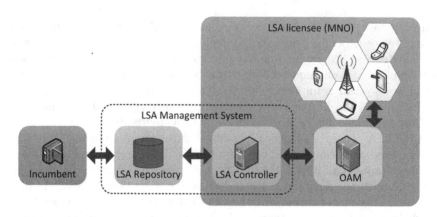

Fig. 1. LSA system arcitecture.

To protect the incumbent user from harmful interference, the LSA Controller needs to adjust the LTE network to achieve a given interference power, which depends on factors such as path loss and transmit power. This can be accomplished by setting the minimum separation distance between the MNO and the PMSE users to guarantee interference free operation at the same time and on the same frequency resource. Minimum separation distance can be expanded to an exclusion zone, if a circle with a radius of the separation distance is drawn around the incumbent receiver. These minimum separation distances calculations have been introduced in ECC report 172 [8]. However, the calculations were revised to reflect more realistic heterogeneous network environment used also in the Finnish LSA trial environment. The revised calculations lead to more realistic exclusion zones and makes the sharing scenario more efficient, not wasting spatial resources, but still guaranteeing interference free operation.

3   Minimum Separation Distance Calculation

For two systems to operate in same geographical area and same frequency band, an interference threshold needs to be set. An interfering transmitter generates a signal and the victim receiver needs to have sufficient protection against it to continue its operations. This protection can be achieved if there is a sufficient separation in the spatial domain between the interfering transmitter and the victim receiver. This means that the path loss between two systems is high enough. The other possibility is to have separation in frequency; transmit and receive filters can suppress the signal to a low enough level. More generally, a combination of both mechanisms is present in the system.

The ECC report 172 [8] introduces sharing scenarios for mobile broadband and incumbents in the 2.3 GHz band. In many European countries, an incumbent is a PMSE user for which the report introduces three different use cases: 1. Cordless camera link, which consists of a hand-held camera transmitter and a small portable receiver, 2. Mobile video link, where a transmitter is on top of a motorcycle and a receiver is carried by a helicopter, 3. Portable video link, which consists of a two-man camera team transmitter and a truck receiver. Scenarios 1. and 3. are located in an urban environment, scenario 2. takes place in a rural environment. The selection of the scenario has an effect on the used propagation model, therefore having significant effect on achieved separation distance. In our study, the coexistence scenarios studied involve LTE TDD (BS or UE) transmitters and video link receiver on the other end. Since the PMSE service has primary status on using the spectrum, the LTE system should not create interference against it. The systems are assumed to be deployed either in the same channel (co-channel case), in channels directly adjacent to each other (adjacent channel case), or with a guard band (alternate channel case). Figure 2 illustrates different channel cases [10], where B is the channel bandwidth and f_c is the center frequency.

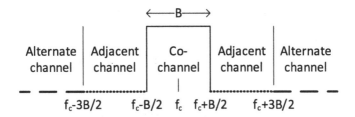

Fig. 2. Measurement mask normalized to channel bandwidth.

For calculating the minimum separation distance [8], first the median minimum coupling loss (MCL) is calculated, which defines the minimum required pathloss between an interferer and a victim receiver. It is calculated with,

$$MCL_{50} = P_t + G_t - G_{td} - G_{fe} + G_r - G_{rd} - IC - G_b, \quad (1)$$

where P_t [dBm] is transmitted power, G_t [dBi] is transmit antenna gain, G_{td} [dB] is transmit antenna directivity loss, G_{fe} [dB] is transmit antenna feeder loss, G_r [dBi] is receive antenna gain, G_{rd} [dB] is receive antenna directivity loss, IC [dBm] is interference criterion and G_b [dB] is bandwidth mitigation factor. Antenna directivity loss is caused if transmitter and receiver antennas are not pointed directly at each other. This can be considered in both vertical and horizontal directions. Feeder loss is mainly caused by cable attenuation in the base station tower. IC is the maximum allowable received interference level, where the interference limit is assumed to be 6 dB under thermal noise of the receiver. Value of 6 dB is commonly used in coexistence studies involving video links and MNO terminals. Thermal noise N [dBm] can be calculated with,

$$N = -174 + 10\log(B_r) + F, \quad (2)$$

where F [dB] is receiver noise figure and B_r is receiver bandwidth in Hz, all the later formulas consider bandwidths in MHz.

If two systems are not operating in the same channel e,g, they are in adjacent or alternate channels, not all transmitted power is effecting the victim. In the co-channel case all transmitted energy is received, hence $P_t = P_{MAX}$ [dBm], where P_{MAX} is maximum output power. In the adjacent channel case

$$P_t = \max(P_{MAX} - ACLRr; 10\log(B_t \cdot 10^{ACLRa/10})), \qquad (3)$$

where the adjacent channel leakage ratio $ACLRr$ [dB] is a relative limit, compared to P_{MAX}, $ACLRa$ [dBm/MHz] is an absolute limit for LTE transmitter on adjacent channel. The less stringent limit [11] is used since we consider the worst-case scenario of interference. In the alternate channel case

$$P_t = 10\log(B_t \cdot 10^{I_{sp}/10}), \qquad (4)$$

where I_{sp} [dBm/MHz] is a maximum absolute interference emission density in guard band and B_t is transmission bandwidth. Numerical values of $ACLR$ and I_{sp} can be found in Tables 6, 7, 20 and 21 in [8]. If the transmitter bandwidth is higher than the receiver bandwidth, only part of the transmitted energy is received. That is modeled with G_b a bandwidth mitigation factor. For the co-channel case

$$G_b = \max(0; 10\log(B_t/B_r)). \qquad (5)$$

For the adjacent channel case the specific mitigation factor is subtracted from the previous equation. These values are derived from the transmitter emission masks and are presented in Table 23 in [8]. For the alternate channel case $G_b = 0$.

In the presence of fading, MCL_{50} limits the received interference under desired threshold only 50 % of the time. Fading statistics is taken into account to limit interference power under the threshold 95 % of time. Correction term for MCL_{50} is

$$MCL_{95} = MCL_{50} + \sigma \cdot \sqrt{2} \cdot \text{erf}^{-1}(2 \cdot 0.95 - 1), \qquad (6)$$

where erf^{-1} is the inverse error function, which results in the constant multiplier for σ that is the distance dependent standard deviation given in [12].

According to the MCL_{95}, the minimum separation distance can be calculated by using different propagation models depending on the calculation scenario. Different versions of Modified Hata propagation model and the Free space propagation model [12] was used in the calculations.

Effect of the directional antennas was modeled using ITU-R F.1336-2 [13] approach, because depending on antenna heights, distance and tilt angle, different antenna directivity loss is achieved. Effect of geometry between interferer and victim systems is illustrated in Fig. 3. The PMSE receiver antenna is assumed to be parallel with the surface of the earth. In this case, the angle α is used to define the receiver antenna directivity loss G_{rd}, meaning that the main beam of the receiver antenna is not pointing towards the transmitter. In the LTE BS antennas are usually tilted downwards, meaning that the tilt angle defines the

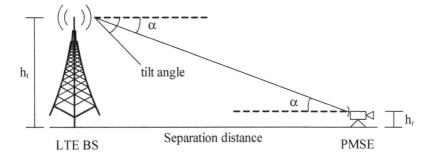

Fig. 3. Geometry between interfering transmitter and victim receiver.

main beam direction of the antenna. In this case, the effective angle for calculating G_{td} is $\alpha-$tilt angle. It should be noted, that the PMSE receiver in scenario 2 is located in the helicopter and is above the LTE BS antennas. In the horizontal domain, we always assume the worst case scenario where the maximum antenna gain is considered on both the interferer and the victim side.

Since there are distance dependent values that effect the MCL, like antenna directivity loss and fading standard deviation, calculation of the MCL is performed iteratively. In the iteration process we first give an initial guess for the distance value, and calculate the MCL accordingly. Then the MCL can be used to calculate the minimum separation distance and that distance can be used for fine tuning distance dependent values and recalculating the MLC. This iteration process is continued until saturation is achieved, meaning that two consecutive iterations give the same result with a three decimal accuracy. If the iterations are not saturating, meaning that the calculations are oscillating between two values, then average of consecutive iterations is used for the next iteration.

4   Verification of the Calculations

The minimum separation distances, obtained according to the previously presented methods, were calculated with different input parameters using Matlab model that was first verified by repeating calculations from [8].

Table 1 presents calculated values for the Cordless camera link use case. Documented values from ECC report 172 Table 25 [8] are reproduced in Table 2. Parameters used in the calculations are presented in Table 3. It should be noted that these parameters (as well as the separation distances in Tables 1 and 2) represent the typical small cell BS values, not those of a macro BS. In the calculations, Urban sub-case of Modified Hata propagation model was used. By comparing our results to the results given in the report, we can see that they match with high accuracy. Therefore it can be concluded that the created calculations are inline with those presented in [8]. This was true for all three use cases.

In a couple of scenarios there are differences in the results. They are marked with * in Table 2. An analysis of the results revealed that the discrepancies occur

Table 1. Cordless camera link use case, calculated results

interference scenario	victim bandwidth B_r		Interfering system and bandwidth B_t					
			LTE TDD BS			LTE TDD UE		
			20Mhz	10MHz	5Mhz	20Mhz	10MHz	5Mhz
co-channel	20 MHz	MCL (dB)	162.35	162.35	162.35	156.79	156.79	156.79
		d (km)	3.236	3.236	3.236	0.609	0.609	0.609
	10 MHz	MCL (dB)	162.35	165.32	165.32	156.79	159.80	159.80
		d (km)	3.236	3.930	3.930	0.609	0.741	0.741
	5 MHz	MCL (dB)	162.35	165.32	168.30	156.79	159.80	162.81
		d (km)	3.236	3.930	4.775	0.609	0.741	0.902
adjacent	20 MHz	MCL (dB)	129.37	127.77	127.77	139.91	139.91	139.91
		d (km)	0.375	0.337	0.337	0.202	0.202	0.202
	10 MHz	MCL (dB)	130.67	130.06	130.06	141.19	141.97	141.97
		d (km)	0.408	0.392	0.392	0.220	0.231	0.231
	5 MHz	MCL (dB)	131.51	130.98	131.85	142.16	142.63	143.91
		d (km)	0.431	0.416	0.441	0.234	0.241	0.262
alternate	20 MHz	MCL (dB)	130.68	128.62	125.54	129.96	125.40	119.03
		d (km)	0.408	0.357	0.292	0.105	0.095	0.086

in the adjacent channel scenarios. There are discrepancies in both the LTE BS and the LTE UE scenarios, recreating these results with our model means different changes in both scenarios. In the LTE BS scenarios, to replicate documented results, values of B_t and B_r need to be switched in Eqs. 2, 3 and 5. Also, from Eq. 5 the max-operator is left out, which limits the values to the positive side. Additionally, the specific mitigation factor needs to be added to G_b instead of subtracted from it, as it should be. In the LTE UE scenario to produce values marked with *, the specific mitigation factor is used as G_b value instead of subtracting it from G_b. As a conclusion, it is clear that there are some discrepancies in the calculations of the original report. Fortunately the differences are within an order of 1.5 dB and 40 m, thus the effect is not significant. Hence, by back tracking discrepancies in the original report, we can conclude that our code can be used to calculate separation distances more suitable for our trial network.

5   Results for LSA Concept and Trial Environment

Next the minimum separation distance calculations are applied to the LSA concept in the 2.3 GHz band for sharing between LTE and incumbent PMSE by using parameters from the Finnish LSA trial environment. Suburban below rooftop propagation model was used for the Cordless camera link and for the Portable video link use cases. This channel model represents our trial environment more accurately than Urban channel model used in reference calculations. Free space propagation model was used for Mobile video link use case. Separation distances were calculated also for a small cell scenario, where BS and UE are located inside of a building and are influencing the incumbent user outside. In the indoor to outdoor case, propagation models need to be modified. It is specified in [12] that an external wall creates additional (L_{we}) 10 dB attenuation to the signal and increases deviation caused by fading by (σ_{add}) 5 dBs.

Table 2. Cordless camera link use case, results from [8]

interference scenario	victim bandwidth B_r		Interfering system and bandwidth B_t					
			LTE TDD BS			LTE TDD UE		
			20Mhz	10MHz	5Mhz	20Mhz	10MHz	5Mhz
co-channel	20 MHz	MCL (dB)	162.3	162.3	162.3	156.8	156.8	156.8
		d (km)	3.236	3.236	3.236	0.609	0.609	0.609
	10 MHz	MCL (dB)	162.3	165.3	165.3	156.8	159.8	159.8
		d (km)	3.236	3.953	3.953	0.609	0.744	0.744
	5 MHz	MCL (dB)	162.3	165.3	168.3	156.8	159.8	162.8
		d (km)	3.236	3.953	4.780	0.609	0.744	0.900
adjacent	20 MHz	MCL (dB)	129.4	129.4*	129.4*	139.9	139.9	139.9
		d (km)	0.373	0.373*	0.373*	0.203	0.203	0.203
	10 MHz	MCL (dB)	128.7*	130.1	130.1	140.7*	142.0	142.0
		d (km)	0.359*	0.392	0.392	0.213*	0.231	0.231
	5 MHz	MCL (dB)	129.8*	131.0	131.9	141.8*	143.2*	143.9
		d (km)	0.385*	0.417	0.442	0.229*	0.250*	0.263
alternate	20 MHz	MCL (dB)	130.6			130.0		
		d (km)	0.408			0.106		

Table 3. Cordless camera link calculation parameters

Parameter	symbol	unit	LTE TDD BS	LTE TDD UE
Maximum transmit power	P_{max}	dBm	24	23
Tx antenna height	h_t	m	15	1.5
Rx antenna height	h_r	m	1.5	1.5
Transmit bandwidth	B_t	MHz	20, 10, 5	
Receive bandwidth	B_r	MHz	20, 10, 5	
Spurious emission	I_{sp}	dBm/MHz	−30	−30
Relative ACLR	ACLRr	dB	45	30
Absolute ACLR	ACLRa	dBm/MHz	−32	−30
Tx antenna gain (max)	dBi	G_t	17	0
Rx antenna gain (max)	dBi	G_r	16	16
Feeder loss	G_{fe}	dB	3	0
Rx noise figure	F	dB	4	4
Tx antenna tilt	tilt	degree	3	0
3dB vertical beamwidth	θ_3	degree	3	-
Center frequency	f_c	MHz	2310	2310

The results of the separation distance calculations are presented in Table 4 through Table 6. The results are calculated with parameters relevant to the trial environment, presented in Table 7. In the result tables, both macro cell and small cell scenarios are presented for both the BS and the UE scenarios. All these scenarios have been considered with three different incumbent use cases.

Table 4. Cordless camera link use case results for trial environment

			Interfering system and bandwidth B_t			
			macro cell		small cell	
interfenrece scenario	victim bandwidth B_r		LTE BS	LTE UE	LTE BS	LTE UE
			20 Mhz	20 MHz	20 Mhz	20 MHz
co-channel	8 MHz	MCL (dB)	188.26	151.79	151.52	144.17
		d (km)	42.88	0.971	1.275	0.591
adjacent	8 MHz	MCL (dB)	144.09	133.68	118.69	125.9
		d (km)	3.712	0.297	0.090	0.095
alternate	8 MHz	MCL (dB)	133.24	129.91	125.63	114.54
		d (km)	1.796	0.232	0.235	0.079

Table 5. Mobile video link use case results for trial environment

			Interfering system and bandwidth B_t			
			macro cell		small cell	
interfenrece scenario	victim bandwidth B_r		LTE BS	LTE UE	LTE BS	LTE UE
			20 Mhz	20 MHz	20 Mhz	20 MHz
co-channel	8 MHz	MCL (dB)	180.19	143.79	143.52	135.92
		d (km)	10182.2	154.2	149.5	62.32
adjacent	8 MHz	MCL (dB)	135.23	115.67	101.42	107.68
		d (km)	57.50	6.053	1.173	2.413
alternate	8 MHz	MCL (dB)	123.51	109.69	106.15	101.3
		d (km)	14.93	3.042	2.024	1.158

Table 6. Portable video link use case results for trial environment

			Interfering system and bandwidth B_t			
			macro cell		small cell	
interfenrece scenario	victim bandwidth B_r		LTE BS	LTE UE	LTE BS	LTE UE
			20 Mhz	20 MHz	20 Mhz	20 MHz
co-channel	8 MHz	MCL (dB)	202.22	165.79	165.52	157.92
		d (km)	85.09	4.805	5.749	2.872
adjacent	8 MHz	MCL (dB)	157.54	137.68	127.81	131.95
		d (km)	18.50	0.765	0.488	0.526
alternate	8 MHz	MCL (dB)	146.49	133.00	129.78	128.91
		d (km)	8.829	0.563	0.556	0.431

Parameters used in the calculations are following the format from [8,12,13], but are specific for the trial environment.

Parameters presented in Table 7 have multiple values for Rx antenna height and gain. These present the values for different use cases, the first one is for the Cordless camera link, the second is for the Mobile video link and the third is for the Portable video link use case. Differences in the parameters are due to the use case definitions. In the Cordless camera link case the receiver antenna is directional disk or Yagi standing on the ground. In the Mobile video link case,

Table 7. Calculation parameters for macro cell and small scenarios

Parameter	symbol	unit	macro BS	macro UE	small BS	small UE
Maximum Tx power	P_{max}	dBm	43	21	24	21
Tx antenna height	h_t	m	40	1.5	2.5	1.5
Rx antenna height	h_r	m	1.5, 150, 5			
Transmit bandwidth	B_t	MHz	20	20	20	20
Receive bandwidth	B_r	MHz	8	8	8	8
Spurious emission	I_{sp}	dBm/MHz	-30	-30	-30	-30
Relative ACLR	ACLRr	dB	45	30	45	30
Absolute ACLR	ACLRa	dBm/MHz	-15	-	-32	-
Tx antenna gain (max)	G_t	dBi	17	0	5	0
Rx antenna gain (max)	G_r	dBi	13, 5, 27			
Feeder loss	G_{fe}	dB	0.4	0	0.4	0
Rx noise figure	F	dB	4	4	4	4
Tx antenna tilt	tilt	degree	3	0	0	0
3dB vertical beamwidth	θ_3	degree	7	-	-	-
Center frequency	f_c	MHz	2380	2380	2380	2380
Wall attenuation	L_{we}	dB	-	-	10	10
Additional deviation	σ_{add}	dB	-	-	5	5

an omnidirectional receiver antenna is mounted on a helicopter levitating 150 m above ground. In the Portable video link case, a parabolic disk receiver antenna is mounted on top of a truck. In addition to these changes, different propagation models were used in the calculations. This time, the Tx and the Rx bandwidths were fixed according to the values used in the trial environment. It can be noted, that the macro cell BS has the highest transmit power and antenna gain, leading to the highest separation distances seen in the following paragraphs.

Results of the Cordless camera link use case are presented in Table 4. The co-channel scenario naturally results in the highest separation distances. When a macro cell BS is the interfering transmitter, almost 43 Km of separation distance is required. Separation distance is reduced to around 1.3 Km with a small cell BS interferer. The UE scenarios result in under 1 Km separation distance. In small cell scenarios both the BS and the UE are located inside, hence the external wall effect is taken into account. In the adjacent channel scenarios, the separation distance is under 4 Km and in the alternate channel scenario under 2 Km.

In the Mobile video link use case, separation distances are significantly higher than in the Cordless camera link use case. The propagation model in Mobile video link use case is Free space, which has lower attenuation than Hata models. Results of the Mobile video link use case are presented in Table 5. The MCL values are in the same scale as in the Cordless camera link use case, meaning that most of the increase in the separation distance is caused by a different propagation model. The results show up to 10000 Km of separation distance in the co-channel scenario which is not realistic and is due to the assumptions taken in the model (e.g. lack of considering the curvature of the earth). In the co-channel scenario, the UE and the small cell scenarios lead to separation distances between 150 to 60 Km. In the adjacent channel scenario, up to 58 Km of separation distance is achieved and the highest separation distance in the alternate channel case is almost 15 Km in the macro BS scenario.

In the Portable video link use case, the use of a highly directive receiver antenna and the assumption that the receiver and transmitter are pointing to each other increase the MCL values compared to the previous use cases. This means that also the separation distances are higher compared to the Cordless camera link use case. Results for the Portable video link use case are presented in Table 6. In the case of macro cell BS, the co-channel scenario separation distance is 85 Km, in the adjacent channel scenario 19 Km and in the alternate channel scenario 9 Km. All the other scenarios have significantly shorter separation distances varying from 6 Km to 400m.

6   Conclusion and Future Work

In this paper we have studied the problem of incumbent protection in LSA and calculated the minimum separation distances from LTE to incumbent PMSE in the LSA trial environment with realistic parameters. Calculations were done according to the principles presented in the ECC report 172 [8], the ERC report 68 [12] and the Recommendation ITU-R F.1336-2 [13]. Using these principles, numerical values for different PMSE system co-existence use cases were produced. By applying more realistic separation distances to the system, spatial resources can be used without causing interference to the incumbent user.

The LSA Controller can use parameters achieved from the current network deployment and from the PMSE system to adaptively calculate separation distances. This will make the LSA Controller more flexible to cope with the changes in environment of different network deployments and multiple types of PMSE. This is a useful feature when expanding the LSA deployment to a larger scale.

One future direction is to consider the incumbent protection with other methods, like extending this work to consider multiple interference sources and to calculate the protection zone for accumulated interference. Taking into account a mobile network layout consisting of multiple spatially separated BSs and UEs which are transmitting simultaneously on the same frequency band, the aggregate field strength may need to be considered instead of the minimum separation distance. However, adding more complexity to the interference calculations will lead to increased amount of information needed as well as more time needed for interference calculations. This will make it more challenging for example to follow changing location of an incumbent with mobility.

Acknowledgments. This work has been done in the CORE++ research project within the 5th Gear programs of Tekes - the Finnish Funding Agency for Innovation. The authors would like to acknowledge the CORE++ project consortium: VTT Technical Research Centre of Finland, University of Oulu, Centria University of Applied Sciences, Turku University of Applied Sciences, Nokia, PehuTec, Bittium, Anite, Fairspectrum, Finnish Defence Forces, Finnish Communications Regulatory Authority, and Tekes.

References

1. ECC Report 205, Licensed Shared Access (LSA), February 2014. http://www.erodocdb.dk/Docs/doc98/official/pdf/ECCREP205.PDF
2. Pérez, E., Friederichs, K-J., Lobinger, A., Redana S., Viering, I., Naranjo, J.D., Optimization of authorised/licensed shared access resources. In: 9th International Conference on Cognitive Radio Oriented Wireless Networks (CROWNCOM) (2014)
3. Palola, M., Matinmikko, M., Prokkola, J., Mustonen, M., Heikkilä, M., Kippola, T., Yrjölä, S., Hartikainen, V., Tudose, L., Kivinen, A., Paavola, J., Heiska, K.: Live field trial of licensed shared access (LSA) concept using LTE network in 2.3 GHz band. In: International Symposium on Dynamic Spectrum Access Networks (DYSPAN) (2014)
4. Palola, M., Rautio, T., Matinmikko, M., Prokkola, J., Mustonen, M., Heikkilä, M., Kippola, T., Yrjölä, S., Hartikainen, V., Tudose, L., Kivinen, A., Paavola, J., Okkonen, J., Mäkeläinen, M., Hänninen, T., Kokkinen, H.: Licensed shared access (LSA) trial demonstration using real LTE network. In: 9th International Conference on Cognitive Radio Oriented Wireless Networks (CROWNCOM) (2014)

5. Matinmikko, M., Palola, M., Mustonen, M., Heikkilä, M., Kippola, T., Yrjölä, S., Hartikainen, V., Tudose, L., Kivinen, A., Kokkinen, H., Mäkeläinen, M.: Field trial of licensed shared access (LSA) with enhanced LTE resource optimization and incumbent protection. In: International Symposium on Dynamic Spectrum Access Networks (DYSPAN) (2015)
6. The CORE++ project web page. http://core.willab.fi/
7. CEPT Report 58, Technical sharing solutions for the shared use of the 2300–2400 MHz band for WBB and PMSE, July 2015. http://www.erodocdb.dk/Docs/doc98/official/pdf/CEPTREP058.PDF
8. ECC Report 172, Broadband Wireless Systems Usage in 2300–2400 MHz, March 2012. http://www.erodocdb.dk/Docs/doc98/official/pdf/ECCREP172.PDF
9. ETSI TR 103 113 V1.1.1, Electromagnetic compatibility and Radiospectrum Matters (ERM); System Reference document (SRdoc); Mobile broadband services in the 2300 MHz–2400 MHz frequency band under Licensed Shared Access regime, July 2013. http://portal.etsi.org/webapp/WorkProgram/Report_WorkItem.asp?WKI_ID=39874
10. ETSI EN 302 064-1 V1.1.2, Electromagnetic compatibility and Radiospectrum Matters (ERM); Wireless Video Links (WVL) operating in the 1,3 GHz to 50 GHz frequency band; Part 1: Technical characteristics and methods of measurement, July 2004. http://portal.etsi.org/webapp/WorkProgram/Report_WorkItem.asp?WKI_ID=21026
11. ETSI TS 136 104 V9.13.0, LTE; Evolved Universal Terrestrial Radio Access (E-UTRA); Base Station (BS) radio transmission and reception (3GPP TS 36.104 version 9.13.0 Release 9), November 2012. http://portal.etsi.org/webapp/workprogram/Report_WorkItem.asp?WKI_ID=39947
12. ERC Report 68, Monte-Carlo simulation methodology for the use in sharing and compatibility studies between different radio services or systems. http://www.erodocdb.dk/docs/doc98/official/pdf/Rep068.pdf
13. Recommendation ITU-R F.1336-2, Reference radiation patterns of omnidirectional, sectoral and other antennas in point-to-multipoint systems for use in sharing studies in the frequency range from 1GHz to about 70GHz. http://www.itu.int/dms_pubrec/itu-r/rec/f/R-REC-F.1336-2-200701-S!!PDF-E.pdf

Mobile Content Offloading in Database-Assisted White Space Networks

Suzan Bayhan[1(✉)], Gopika Premsankar[2], Mario Di Francesco[2],
and Jussi Kangasharju[1]

[1] Department of Computer Science, University of Helsinki, Helsinki, Finland
bayhan@hiit.fi, jakangas@cs.helsinki.fi
[2] Department of Computer Science, Aalto University, Espoo, Finland
{gopika.premsankar,mario.di.francesco}@aalto.fi

Abstract. Mobile data offloading leverages more affordable or even free network capacity to reduce the traffic experienced by cellular operators through their limited over-the-air resources. One way to harvest free capacity is to employ the white space, namely, frequencies that are assigned to licensed users but are not actively utilized, as long as no harmful interference is generated. In this article, we characterize the benefits of harnessing node contacts for mobile content offloading through dynamic spectrum access assisted by a white space database (WSDB). We take a content-centric approach and model the selection of distributors among the subscribers of each content served through a base station. We formulate an optimization problem to maximize the offloading gain based on realistic settings. We show that such a problem is NP-hard and devise efficient heuristics for practical mobile data offloading. Our results show that the offloading gain allowed by white space is significant even when WSDB data are inaccurate.

Keywords: White spaces · Dynamic spectrum access · Mobile opportunistic offloading · Content delivery · White space database · WSDB

1 Introduction

Mobile *data offloading* is a method to move traffic from the cellular network through other means, such as local area networks or device-to-device communications. It has emerged as a promising solution to decrease the load on mobile networks [1]. As WiFi is densely deployed, on-the-spot offloading to local wireless networks when the user is under coverage provides a significant decrease in the mobile operator traffic [2]. Mobile communications could also be postponed until users reach an area covered by a WiFi access point through the so-called delayed WiFi offloading. This is an option for delay-tolerant traffic, as long as the time spent without WiFi connectivity is short [2]. However, WiFi offloading may be restricted by the capacity of the backhaul [3] which is often subject to data caps for private WiFi networks. Motivated by these concerns, offloading to

© ICST Institute for Computer Sciences, Social Informatics and Telecommunications Engineering 2016
D. Noguet et al. (Eds.): CROWNCOM 2016, LNICST 172, pp. 129–141, 2016.
DOI: 10.1007/978-3-319-40352-6_11

mobile opportunistic networks has been proposed [4,5]. It leverages the capacity of short-distance communications without relying on any infrastructure and entails almost no monetary cost. As this mode is driven by contacts between mobile nodes, it may fail to provide guaranteed delays, making it a better fit for delay-tolerant traffic. However, unlicensed bands such as the Industrial Scientific and Medical (ISM) are already congested, thus possibly incurring in a low transmission capacity for opportunistic offloading.

A different approach to address the spectrum capacity crunch is to employ the *white space*, namely, the spectrum that is licensed to *primary users* (PU) while being spatiotemporally unused. Offloading mobile data to unused PU channels is called *white space offloading* [6,7] and some existing solutions have explicitly targeted proximity-based communications in such a context. Among them, Cui et al. [7] presented a model to leverage WiFi and white spaces instead of cellular communications, with focus on power efficiency and channel assignment under delay constraints. Ding et al. [8] proposed using TV bands for device-to-device communications by creating location-specific white space databases with the help of "big spectrum data" collected by the mobile crowd.

In this article, we characterize the benefits of mobile data offloading through dynamic spectrum access assisted by a white space database (WSDB). We take a content-centric approach and model the selection of distributors among the subscribers of content served through a base station. We then formulate an optimization problem to maximize the offloading gain based on realistic settings. We show that such a problem is NP-hard and devise efficient heuristics for practical mobile data offloading. Our results show that the offloading gain allowed by white space is significant even when WSDB data are inaccurate.

The key contributions of this article are the following.

- We consider database-assisted white space access in realistic settings. While WSDBs are expected to provide accurate and up-to-date information on the incumbents, some flaws (e.g., bogus entries and incorrect device locations) have been discovered due to several reasons (e.g., unsynchronized WSDBs and manual entry of device information) [9]. We explicitly include the factors affecting the availability and the reliability of the WSDB in our model.
- We provide a general and flexible framework for content-driven mobile data offloading. Our model supports two different options: mobile opportunistic offloading through ISM bands and white space offloading via unoccupied PU channels retrieved from the WSDB.
- We propose several heuristics with different levels of complexity to improve the offloading gain. Some of them focus on the number of distributors for each content based on its size and popularity, whereas more sophisticated ones aim at identifying nodes with high offloading potential. Our experiments using realistic user mobility demonstrate that white space enables offloading 67 % more capacity compared to a purely opportunistic approach when the information in the WSDB is accurate. Even in the presence of inaccuracies, the offloading gain is still higher than 47 %.

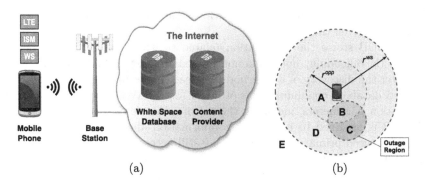

Fig. 1. (a) Reference architecture and (b) offloading regions: opportunistic offloading takes place in zones A and B, white space offloading in zone D, while no offloading is possible in zones C and E.

To the best of our knowledge, our work is the first one focusing on white space offloading for mobile content delivery. In fact, existing solutions in the literature addressed either wireless capacity of white space networks [8] or distributed dynamic access schemes [10]. In contrast, we propose a content delivery framework for white space offloading in database-assisted networks that jointly utilizes opportunistic contacts for offloading. Our solution also distinguishes itself from the state of the art on mobile opportunistic offloading. For instance, Li et al. [5] presented optimal offloading in mobile opportunistic networks through careful selection of distributors. Even though our approach is somewhat similar, we focus on the offloading gain rather than on the distribution delay and buffer constraints.

2   System Model

Our reference architecture is the cellular network illustrated in Fig. 1a. A base station (BS) is connected to both a content provider and a white space database (WSDB). A set $\mathcal{N}$ of mobile users (through their respective devices) is also part of the network and requests content as well as spectrum availability through the BS. We assume that the overhead[1] associated with such requests (and the related responses) is negligible. Each mobile device is equipped with three radio transceivers: one for cellular connectivity, one for communications in the ISM bands, and one for white space access. We denote the ranges of the white space and the ISM radios by r^{ws} and r^{opp}, respectively (Fig. 1b). These ranges may differ as they depend on the actual frequencies employed. In the following, we consider TV bands for white space offloading; as a consequence, $r^{ws} > r^{opp}$. Moreover, we assume that the connection to the BS can be used independently

---

[1] We analyze the WSDB querying delay and its impact on offloading capacity in Sect. 4, based on our measurements from the Google Spectrum Database.

from the others, while only one of the white space and the ISM interfaces can be active at a given time.

The WSDB stores the information related to white spaces. In contrast to opportunistic offloading, nodes cannot immediately start offloading to white spaces after the discovery of a peer node. Instead, nodes first consult the WSDB by using a database communication protocol (e.g., PAWS [11]) for the permitted operation parameters, including the list of available channels. Even though the WSDB is assumed to have perfect information about white space utilization, unregistered primary users (PU) may still access certain frequencies, thus resulting in interference. To this end, we model the probability of unsuccessful communications due to PU collisions on a given frequency as p^{un}. We assume that the availability of white space is such that all requests can be accommodated without competition between users. We also assume that mobile devices may not be able to reach the BS from certain regions in the nominal coverage range of the BS, e.g., due to shadowing. We call these regions *outage areas* and denote the probability that a mobile device lies in such a region with p^{sh}.

We describe the different content in the network through the set $\mathcal{C} = \{c_1, \cdots, c_k, \cdots, c_K\}$. Each specific content c_k is characterized by its size l_k and its delivery deadline T_k. The set of mobile nodes subscribing to c_k (i.e., the *subscribers*) is denoted as $\mathcal{S}_k$. Without loss of generality, we assume that c_1 is the most popular and c_K is the least popular content. Each mobile user subscribes to one content only. The BS serves the content requests by the users in its coverage area through the *distributors* $\mathcal{D}_k$, each responsible for content c_k. The BS transfers the allocated content to distributors through its F frequencies. The distributors, in turn, deliver the cached content to rest of the nodes (i.e., $\mathcal{S}_k \setminus \mathcal{D}_k$) as long as it is valid. Mobile nodes request the content directly from the BS as soon as the related deadline expires. The BS selects the distributors and announces the association between them and the cached content to the network. As a consequence, subscribers know from which node to request their content.

We model the inter-contact time between pair of nodes through an exponential distribution with parameter $\lambda_{i,j}^{opp}$ for the opportunistic radio and $\lambda_{i,j}^{ws}$ for the white space interface. We assume that contacts are long enough to completely transfer a content item.

3   White Space Offloading

A subscriber node n_i fetches content c_k in one of the three modes detailed next.
– *Mobile opportunistic* (or *ISM*) *offloading:* Let n_j be a distributor for c_k: $n_j \in \mathcal{D}_k$, and $d(n_i, n_j)$ denote the Euclidean distance between n_i and n_j. Node n_i receives the content from n_j if it is in the range of n_j's ISM radio interface, i.e., the $d(n_i, n_j) \leqslant r^{opp}$. Given that the inter-contact time between n_i and n_j is exponentially distributed with parameter $\lambda_{i,j}^{opp}$, opportunistic offloading is possible if n_i has a contact with n_j during the lifetime of c_k. More formally, we state opportunistic offloading probability as:

$$p_{i,j,k}^{opp} = 1 - e^{-\lambda_{i,j}^{opp} T_k}. \tag{1}$$

– *White space offloading:* Node n_i receives the content from n_j through the white space only if it cannot get the content by opportunistic offloading. This case is possible only if $r^{opp} < d(n_i, n_j) \leqslant r^{ws}$. In other words, opportunistic offloading is preferred over white space offloading due to the entailed cost and possibly poorer performance, e.g., inaccurate WSDB data. In this case, the distributor first consults the BS to get an available white space channel for offloading. Therefore, this mode is possible only when n_j has an uplink channel to the BS. In Fig. 1b, n_j sends the content to n_i using white space offloading only in region D. Recall that a node may be under outage with probability p^{sh} due to shadowing or other channel impairments. Moreover, even if the BS assigns a channel for its use, the transmission may fail as the assigned channel may be occupied by an unregistered PU. By considering all these cases, we calculate the probability of white space offloading $p_{i,j,k}^{ws}$ as follows:

$$p_{i,j,k}^{ws} = (e^{-\lambda_{i,j}^{opp}T_k} - e^{-\lambda_{i,j}^{ws}T_k})(1 - p^{sh})(1 - p^{un}), \tag{2}$$

where the first term represents the probability that the distributor node is in the white space offloading range but not close enough for opportunistic offloading. Under our assumption that $r^{ws} > r^{opp}$, the contact rates are such that $\lambda_{i,j}^{ws} > \lambda_{i,j}^{opp}, \forall i, j$. Based on that, we calculate $p_{i,j,k}$ which is the probability that n_i gets c_k from n_j in one of the two offloading modes before T_k as:

$$p_{i,j,k} = p_{i,j,k}^{opp} + p_{i,j,k}^{ws}. \tag{3}$$

– *No offloading:* Node n_i receives the content directly from the BS in two cases: (i) it is selected as a distributor (i.e., $n_i \in \mathcal{D}_k$) and gets the content just after the related request, or (ii) it could not receive the content from any of the distributors during time T_k (precisely, the BS serves the content just after T_k). We then express the probability of getting the content from the BS as:

$$p_{i,k} = 1 - \prod_{j \in \mathcal{D}_k} (1 - p_{i,j,k}). \tag{4}$$

We define *offloading gain* for a content item c_k as the traffic saved by offloading which would otherwise be delivered by the BS through the cellular network. Let $\mathbf{Y} = [y_{i,k}]$ denote the subscriber matrix where $y_{i,k} = 1$ indicates that n_i requests c_k. The BS decides which nodes to select as distributors for each item c_k based on $\mathbf{Y}$, $\Lambda = [\lambda_{i,j}]$, p^{sh}, and p^{un}. Let $\mathbf{X} = [x_{j,k}]$ represent the decision variables where $x_{j,k} = 1$ stands for n_j being selected as distributor for c_k. We define the set of subscribers as $\mathcal{S}_k = \{n_i \,|\, n_i \in \mathcal{N}\}$ and the set of distributors as $\mathcal{D}_k = \{n_j \,|\, n_j \in \mathcal{S}_k\}$, where $S_k = |\mathcal{S}_k|$ and $D_k = |\mathcal{D}_k|$. Note that $D_k \subseteq S_k, \forall k$. We can now formulate the offloading gain (i.e., the traffic saved by either opportunistic or white space offloading) maximization problem as:

$$\max_{\mathbf{X}} \sum_{k=1}^{K} l_k \left(S_k - D_k - \sum_{i \in \mathcal{S}_k \setminus \mathcal{D}_k} (1 - p_{i,k}) \right) \tag{5}$$

subject to the following constraints:

$$S_k = \sum_{i=1}^{N} y_{i,k} \qquad \forall k \in \mathcal{C} \tag{6}$$

$$D_k = \sum_{i=1}^{N} x_{i,k} \qquad \forall k \in \mathcal{C} \tag{7}$$

$$p_{i,k} = 1 - \prod_{j \in \mathcal{D}_k} (1 - p_{i,j,k}^{opp} - p_{i,j,k}^{ws}) \quad \forall i \in \mathcal{S}_k, k \in \mathcal{C} \tag{8}$$

$$p_{i,j,k}^{opp} = 1 - e^{-x_{j,k}\lambda_{i,j}^{opp}T_k} \quad \forall i, j \in \mathcal{N}, k \in \mathcal{C} \tag{9}$$

$$p_{i,j,k}^{ws} = (e^{-x_{j,k}\lambda_{i,j}^{opp}T_k} - e^{-x_{j,k}\lambda_{i,j}^{ws}T_k})(1 - p^{sh})(1 - p^{un}) \quad \forall i, j \in \mathcal{N}, k \in \mathcal{C} \tag{10}$$

$$p_{i,k} = 0 \qquad \forall i \in \mathcal{N} \setminus \mathcal{S}_k, k \in \mathcal{C} \tag{11}$$

$$x_{i,k} \leqslant y_{i,k} \qquad \forall i \in \mathcal{N}, k \in \mathcal{C} \tag{12}$$

$$\sum_{k \in \mathcal{C}} D_k \leqslant F \tag{13}$$

$$x_{i,k} \in \{0, 1\} \qquad \forall i \in \mathcal{N}, k \in \mathcal{C}. \tag{14}$$

The number of subscribers and distributors are described by Eqs. (6) and (7), respectively. Distributors can offload content only when in contact with the related subscribers. As contacts are stochastic, each user receives the content with a certain probability from the selected distributors. In detail, the probability that n_i receives content c_k is expressed by Eq. (8), while the probability for each offloading mode is described by Eqs. (9) and (10). The constraint in Eq. (11) ensures that only offloading to the subscribers of a given content is taken into account. The constraint in Eq. (12) guarantees that distributors are selected only from the set of subscribers of that content. As the BS has only F frequencies, Eq. (13) ensures that the number of selected distributors is smaller than or equal to F. Finally, Eq. (14) signifies that the decision variables are binary.

3.1 Heuristics

The optimization problem introduced earlier is a variant of the 0–1 knapsack problem: the total number of frequencies F corresponds to the knapsack capacity while nodes are items to be packed. The utility of each node depends on its capacity to deliver content to the other unselected nodes. Precisely, our optimization formulation is a more general version of the target set selection problem shown to be NP-hard in [4]. As a consequence, we introduce several heuristics – with varying computational complexity – to leverage node and (or) content diversity and obtain a high offloading gain.

Random selection (RAND). The BS randomly selects F nodes. Let $p(n_i)$ denote the probability that n_i is selected as distributor. In this case, $p(n_i)$ is set

Algorithm 1. IBOS

1: $\mathbb{D} = \emptyset$ and set $p_{j,i,k} = 0$ for all i, j, k
2: **for** $n_i \in \mathcal{N}$ **do**
3:     Get the content id k where $n_i \in \mathcal{S}_k$
4:     **for** $n_j \in \mathcal{S}_k$ **do**
5:         Calculate $p_{j,i,k}$ as in Eq. (3)
6: **for** $f = 1$ to F **do**
7:     $U(n_i) = l_k \sum_j p_{j,i,k}$ for all $n_i, n_j \in \mathcal{N} \setminus \mathbb{D}$
8:     Select $n_o = \arg\max U(n_i)$ where $n_i \in \mathcal{N} \setminus \mathbb{D}$
9:     **if** $U(n_o) > 0$ **then**
10:         $\mathbb{D} = n_o \cup \mathbb{D}$ and assign f to n_o for content delivery
11:         Set $p_{o,i,k} = 0$ for all $n_i \in \mathcal{N} \setminus \mathbb{D}$
12:     **else**
13:         return $\mathbb{D}$
14: return $\mathbb{D}$

---

to $\min(1, F/N)$. This heuristic has a complexity of $O(F)$ and does not employ content diversity or node diversity. We use RAND for comparison purposes only.

Content diversity (CD). This two-step approach explicitly considers both content size and popularity, different from RAND. In the first step, the BS determines the number of distributors for each content, i.e., $D_k \propto S_k l_k$. After D_k is decided, D_k nodes are randomly selected from $\mathcal{S}_k$ in the second step. This approach ignores the differences among nodes and it does not consider content lifetimes. We calculate $p(n_i)$ for $n_i \in \mathcal{S}_k$ as $p(n_i) = \min(1, D_k/S_k)$, where

$$D_k = F \frac{S_k l_k}{\sum_{m \in \mathcal{C}} S_m l_m}.$$

The complexity of this approach is $O(K)$.

Content and node diversity (CND). CND differs from CD in the second step to better harness the diversity among nodes. For a given c_k, nodes in $\mathcal{S}_k$ are evaluated according to their capacity to offload that content item to the other subscribers. The probability of selection is proportional to the utility of n_i, defined as $U(n_i) = \sum_{j \in \mathcal{S}_k} p_{j,i,k}$. Specifically, the probability $p(n_i)$ is:

$$p(n_i) = \min\left(1, F \frac{S_k l_k}{\sum_{m \in \mathcal{C}} S_m l_m} \frac{U(n_i)}{\sum_{j \in \mathcal{S}_k} U(n_j)}\right).$$

The complexity of this approach is $O(N^2)$.

Iterative Best Offloader Selection (IBOS). This heuristic does not select all F distributors at once, but rather applies the iterative approach detailed in Algorithm 1. Let $\mathbb{D} = \{\mathcal{D}_k\}$ be the set of distributors at the current iteration. First, IBOS initializes $\mathbb{D} = \emptyset$. At each iteration, it calculates the utility of each node as $U(n_i) = l_k \sum_{j \in \mathcal{S}_k} p_{j,i,k}$. After sorting the nodes according to $U(n_i)$, IBOS adds the node with the highest utility, i.e., $n_o = \arg\max U(n_i)$ to the distributors set: $\mathbb{D} = n_o \cup \mathbb{D}$. Next, it sets $p_{o,i,k} = 0$ for all n_i that are candidates to be selected as distributors in the next iteration. This iteration is necessary to better identify the contribution of candidate nodes to offload data to the remaining unselected nodes. IBOS re-calculates $U(n_i)$ according to the updated $p_{j,i,k}$ and follows the same iterations until F number of nodes are selected as distributors or the maximum utility equals to zero. In fact, selecting new nodes as distributors is not expected to increase the offloaded traffic even if there are still some unassigned frequencies. The resulting complexity is $O(FN^2)$.

Improved IBOS (IBOS+). Different from IBOS, this heuristic stores a vector $P(\mathbb{D}) = [p_{i,k}]$ to keep track of each node's probability of receiving the content from the current set of distributors $\mathbb{D}$. Then, after a new node is selected as distributor, it updates $P(\mathbb{D})$ according to Eq. (4) and adds nodes with $p_{i,k}$ higher than some predefined probability (*safety threshold*) to the set of *safe nodes* $\mathbb{A}$. Nodes in $\mathbb{A}$ are then excluded in the calculation of the utility, i.e., in line 7 of Algorithm 1: $n_j \in \mathcal{N} \setminus (\mathbb{D} \cup \mathbb{A})$. Hence, IBOS+ selects nodes that can reach those without a high probability of getting the content from the already selected distributors. The complexity of IBOS+ is the same as that of IBOS, namely, $O(FN^2)$.

4   Performance Evaluation

We developed a custom simulator in Python to carry out our experiments. We used as input mobility traces generated through the ONE simulator [12], which are based on pedestrian paths extracted from real roads in the city of Helsinki. Pedestrians walk with a speed of $[0.5, 1.5]$ m/s and wait at a reached location for $[1, 4]$ minutes before moving towards their next destination. We recorded the contacts among $N = 200$ pedestrians using two transmission ranges $r^{opp} = 20$ m and $r^{ws} = 100$ m, according to the relatively higher range of white spaces [7]. Node contacts (e.g., their start and end times) lasted for three hours. We derived the average pairwise contact rate $\lambda_{i,j}$ from the trace and assumed that the related information is available to the BS for CND, IBOS, and IBOS+.

We drew content popularity from a Weibull distribution with parameters $k = 0.513$ and $\lambda = 6010$ according to [13]. We considered content sizes and lifetimes uniformly distributed between $[2, 5]$ MB and $[1, 3]$ hours, respectively.

4.1   Offloading Capacity

In our first set of experiments, we study the impact of white space availability on the offloading capacity irrespective of the offloading algorithm employed. To this end, we define *effective offloading capacity* as the product between the channel bandwidth and the time during which a node can transmit to its peer. As Fig. 2a shows, this period is equal to the remaining time after peer discovery and before the contact ends for opportunistic offloading. In case of white spaces, offloading starts only after the WSDB returns the data and it is successful only

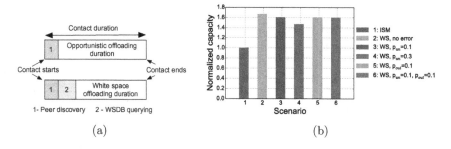

Fig. 2. (a) Offloading duration and (b) effective offloading capacity.

if the transmission does not collide with an unregistered (active) PU. As peer discovery is common to both modes, we ignore the related time loss in the results.

To obtain a realistic estimate for WSDB query overhead, we developed a mobile application that connects to the Google Spectrum Database to query the spectrum availability at some predetermined (urban as well as rural) locations in the US. We recorded the round trip time of the WSDB requests/responses and subtracted the related values from the contact duration to derive the effective offloading capacity in the white spaces. According to our observations from 10,000 queries, 90 % of delays are below 4.2 s. This is consistent with the delays below 3 s reported in [14], especially considering that our queries were run from a mobile device located in Helsinki. Regarding contact durations, we used the contact trace generated by the ONE simulator. At the beginning of each contact, we simulated the chance of outage as well as that of PU collision.

As Fig. 2b shows, exploiting white spaces in addition to ISM channels significantly improves the effective offloading capacity: 67 % when there is no outage and the WSDB information is reliable; 47–60% under different levels of unreliability, i.e., the outages and PU collision probabilities associated with the scenarios 3–6 reported in the figure.

4.2   Content Offloading

We now investigate the impact of the number of content items K on the offloaded traffic for different values of the frequency bands F available at the BS (Fig. 3). For clarity, we only describe[2] the results obtained for IBOS+ when $p^{sh} = p^{un} = 0.1$. Figure 3a shows that when K increases, the offloaded traffic decreases for all schemes. For a low number of contents, opportunistic offloading is very efficient. For instance, if there is a single item, ten nodes ($F = 10$) deliver 86 % of the traffic through opportunistic offloading. For white spaces, the offloaded traffic reaches 92 %. For $K = 100$, ISM-only mode can offload only 6 % of the traffic, as opposed to the 11 % value obtained when white spaces are employed for offloading. This behavior is due to the increasing content diversity. In other words, the probability that two random nodes subscribe to the same content is

---

[2] A detailed comparison of the different heuristics is provided in the next subsection.

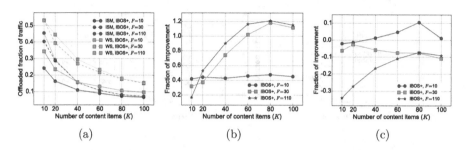

Fig. 3. (a) Offloaded traffic fraction, improvement in (b) offloaded traffic and (c) mean content delivery delay of white spaces compared to ISM.

lower when K increases. Consequently, the chance of offloading decreases too. Next, we observe better performance under white space offloading. Figures 3b and c illustrate the related fraction of improvement in performance compared to ISM-only offloading. Considering all three sub-plots, we can see that for low K, white spaces approximately allow a 20–40 % improvement in offloaded traffic as opportunistic offloading finds sufficiently many contacts for the distribution of say $K = 10$ contents. However, Fig. 3c shows that white spaces speed up the delivery very significantly in this operating region. With increasing content diversity, the benefit of white spaces becomes more apparent. For instance, the relative improvement is above 100 % for $K = 100$ and $F = 30$. For $F = 10$, the improvement is around 45 % due to the low number of distributors.

In summary, white spaces provide the most significant gains when: there are many diverse content items and sufficient frequencies; content items are less diverse but the number of frequencies available at the BS is limited, i.e., only a small fraction of the nodes can be selected as distributors. The improvement lies in the offloaded traffic in the first case, whereas it consists of reduced delivery delay in the latter case.

4.3   Comparison of Heuristics

We finally evaluate and compare our heuristics as a function of the fraction of distributors (Fig. 4). We define the fraction of distributors as the ratio of number of distributors that the BS can select to the total number of subscribers. Note that the BS may select less nodes than the allowed fraction for IBOS and IBOS+, as these algorithms stop when the remaining nodes are not expected to further increase the offloaded traffic. The number of selected nodes is indeed equal to the number of available frequencies for the rest of the algorithms.

First, we note that with increasing F, all schemes can initially offload more traffic by employing more distributors. However, after a certain number of frequencies is reached (e.g., $F = 50$ corresponding to 0.25 fraction of distributors in Fig. 4a), CD, CND, and RAND redundantly select nodes as distributors. Since IBOS and IBOS+ stop allocating distributors when the maximum utility of a

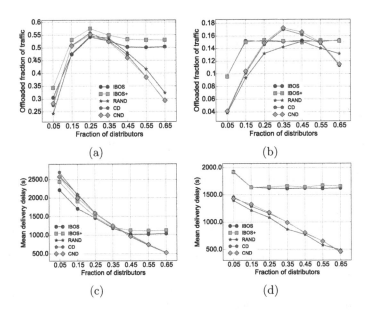

Fig. 4. Offloaded traffic for (a) 10 and (b) 100 content items. Delivery delay for (c) 10 and (d) 100 content items.

selected node is zero, these schemes can still maintain a high offloading performance. Second, IBOS+ outperforms others for $K = 10$ while IBOS is the second best scheme only when F is either low or high. In between, other naive schemes have higher offloading capacity. We expect a typical setting to have low to moderate F; hence, the best scheme is IBOS+ in these conditions. However, CD is a sensible choice when contact statistics are inaccurate or not even available.

In contrast to Fig. 4a, b shows that there is almost no difference in the performance of IBOS and IBOS+. This result can be explained through the impact of the safety threshold in IBOS+ (set to 0.9 in our experiments). As content diversity is higher in this scenario (i.e., $K = 100$), the probability that a random contact results in content offloading is much lower too. Hence, IBOS+ cannot add nodes to the safety set, which effectively reduces IBOS+ to IBOS. Moreover, our model assumes exponentially distributed inter-contact times and calculates the expected offloading probabilities based on such a model. However, the contact traces do not necessarily exhibit this property.

Figures 4c and d illustrate the mean content delivery delay for the nodes that receive the content before the deadline, i.e., the distributors and the nodes that are served by these distributors. IBOS and IBOS+ not only improve offloading capacity but also help faster content delivery for $K = 10$ and low F. IBOS outperforms IBOS+, which is followed by naive schemes until $F = 90$. After this point, CD, CND and even RAND perform better as they assign all frequencies to the distributors without considering the offloading capacity. For the same reason, Fig. 4d shows that CD, CND and RAND obtain lower delays while the delivery delay of IBOS and IBOS+ remains almost the same with increasing F. This could be seen as a trade-off between offloaded traffic fraction and delivery delay. However, we conclude that IBOS and IBOS+ are the best choices when timely delivery is needed to guarantee user satisfaction.

5   Conclusion

In this article, we leverage white spaces in addition to ISM bands for mobile content offloading. We specifically model database-assisted white space networks in which WSDB data may not be accurate. We then devise a general framework for content offloading through a content-centric approach. Specifically, we formulate an optimization problem to maximize the offloading gain through the selection of distributors. We show that solving such a problem is computationally hard, then propose several practical heuristics and analyze their offloading performance. Our results demonstrate that the availability of white space significantly increases the offloaded traffic, especially when there are many content items. When there are a few items, mobile opportunistic offloading provides a high gain, comparable to that for white spaces. In this case, white space offloading enables faster content delivery. As a future work, we seek to find an approximate solution to our optimization problem. We also plan to better analyze the impact of content and mobility characteristics on the offloading performance.

Acknowledgments. We thank Aleksandr Zavodovski for developing the mobile application for querying the WSDB. This work has been supported by the Academy of Finland under grant numbers 284806, 284807 and 278207.

References

1. Aijaz, A., Aghvami, H., Amani, M.: A survey on mobile data offloading: technical and business perspectives. IEEE Wirel. Commun. **20**(2), 104–112 (2013)
2. Lee, K., Lee, J., Yi, Y., Rhee, I., Chong, S.: Mobile data offloading: how much can WiFi deliver?. In: ACM CoNEXT (2010)
3. Golrezaei, N., Molisch, A.F., Dimakis, A.G., Caire, G.: Femtocaching and D2D collaboration: a new architecture for wireless video distribution. IEEE Commun. Mag. **51**(4), 142–149 (2013)
4. Han, B., et al.: Mobile data offloading through opportunistic communications and social participation. IEEE Trans. Mob. Comput. **11**(5), 821–834 (2012)
5. Li, Y., et al.: Multiple mobile data offloading through DTNs. IEEE Trans. Mob. Comput. **13**(7), 1579–1596 (2014)
6. Mancuso, A., Probasco, S., Patil, B.: Protocol to access white-space (PAWS) databases: Use cases and requirements. IETF, RFC 6953, May 2013
7. Cui, P., Tonnemacher, M., Rajan, D., Camp, J.: WhiteCell: energy-efficient use of unlicensed frequency bands for cellular offloading. In: IEEE DySPAN (2015)
8. Ding, G., Wang, J., Wu, Q., Yao, Y.D., Song, F., Tsiftsis, T.A.: Cellular-base-station assisted D2D communications in TV white space. IEEE J. Sel. Areas Commun. **34**(1), 107–121 (2016)

9. National Association of Broadcasters, Emergency motion for suspension of operations and petition for rulemaking, March 2015
10. Khoshkholgh, M.G.: Connectivity of cognitive D2D communications underlying cellular networks. IEEE JSAC **33**(1), 81–98 (2015)
11. Chen, V., Das, S., Zhu, L., Malyar, J., McCann, P.: Protocol to access whitespace (PAWS) databases, RFC 7545, May 2015
12. Keränen, A., Ott, J., Kärkkäinen, T.: The ONE simulator for DTN protocol evaluation. In: Simutools (2009)
13. Cheng, X., Dale, C., Liu, J.: Statistics and social network of YouTube videos. In: International Workshop on Quality of Service (IWQoS) (2008)
14. Majid, A.Y.: Better mobility support for radio spectrum white space-enabled devices, in MS Thesis, Delft University of Technology (2015)

Neighbours-Aware Proportional Fair Scheduler for Future Wireless Networks

Charles Jumaa Katila[✉], Melchiorre Danilo Abrignani, and Roberto Verdone

University of Bologna, Radio Networks Laboratory,
Viale Risorgimento 2, Bologna, Italy
{charlesjumaa.katila,danilo.abrignani,roberto.verdone}@unibo.it

Abstract. In this paper, we present an uplink scenario where primary and secondary users coexist on the same set of radio resources. The primary users rely solely on a centralised scheduler within the base station for the assignment of resources, and the secondary users rely on an unslotted Carrier Sense Multiple Access (CSMA) protocol for channel access. We propose a novel centralised scheduling algorithm, Neighbours-Aware Proportional Fair (N-PF), which considers the uplink channel state conditions and the number of secondary users neighbouring each primary user in the aggregate scheduling metric. Through simulations we demonstrate that N-PF outperforms the chosen benchmark algorithm, Proportional Fair (PF), in terms of packet delivery rate while maintaining fairness.

Keywords: Proportional Fair · Neighbours-Aware · Primary users · Secondary users · Unslotted CSMA · Packet delivery rate · Fairness

1 Introduction

One of the practical challenges in the design of future wireless networks will be the presence of interference. Because the spectrum resource will remain limited, numerous primary users (hereafter denoted as scheduled nodes) and secondary users (hereafter denoted as uncoordinated nodes) will have to coexist on the same set of radio resources, resulting in enormous interference on communication links and consequently network performance degradation. Advanced medium access schemes can play a significant role towards achieving efficient utilization of radio resources and hence, current research activities on Medium Access Control (MAC) protocols are of paramount importance.

In this paper we present an uplink scenario where scheduled nodes and uncoordinated nodes coexist on the same pool of radio resources within a cell. Both groups of nodes transmit to a common base station (BS) but unlike scheduled nodes, the uncoordinated nodes do not have a global reference time and therefore, they are totally asynchronous with the base station and with each other. To access the channel, the uncoordinated nodes rely on an unslotted Carrier Sense Multiple Access with Collisions Avoidance (CSMA/CA) protocol, while

© ICST Institute for Computer Sciences, Social Informatics and Telecommunications Engineering 2016
D. Noguet et al. (Eds.): CROWNCOM 2016, LNICST 172, pp. 142–153, 2016.
DOI: 10.1007/978-3-319-40352-6_12

the scheduled nodes rely on a centralised scheduling algorithm located within the base station for radio resources assignment. This scenario could be applicable in many different network instances of future generation wireless networks e.g., 5G and beyond. We propose and evaluate through simulation a novel centralised scheduling algorithm, which outperforms the baseline algorithm, i.e., proportional fair (PF), in terms of packet delivery rate, while maintaining fairness.

In wireless networks, MAC protocols are classified into two main groups: contention-based and contention-free MAC protocols. The contention-based MAC protocols are distributed in nature and suffer from packet collisions. Nodes whose packets collide, perform a random backoff before attempting to access the channel again for retransmission of the lost frames. Such protocols include ALOHA [1], slotted ALOHA [2] and CSMA/CA family of protocols [3]. On the other hand, the contention free MAC protocols are mainly coordinated in nature involving a centralised master entity which develops and allocates orthogonal or non-orthogonal radio resources according to some policies defined by the scheduling algorithms. Schedules assigned to users can either be in time, frequency, space, code or combination of more than one resource dimension. The conventional scheduling algorithms include: Round-Robin (RR), Earliest Deadline First (EDF) [4], Maximum Throughput (MT), and Proportional Fair (PF) [5]. Each scheduling algorithm aims at maximizing/minimizing some network performance metrics such as fairness measure, sum throughput, power consumption, latency, etc., subject to some constraints.

In this paper, we contribute to the performance of centralised uplink scheduling algorithms by proposing a novel algorithm, called Neighbour-Aware proportional Fair (N-PF), which takes into account both channel state conditions and the number of uncoordinated nodes neighbouring each of the scheduled nodes in the aggregate scheduling metric. To maximize packet delivery rate of the scheduled nodes, N-PF, prioritises users with large subsets of uncoordinated neighbours and good channel conditions. In fact, in the presence of capture effect, good uplink channel conditions for the scheduled nodes results in high packet capture probability (p_c), since p_c depends on the Signal to Interference Ratio (SIR). Similarly, a large subset of uncoordinated neighbours belonging to a given scheduled node results in high transmission success probability, because all the uncoordinated nodes in the subset can sense the scheduled transmissions in progress and refrain from accessing the channel.

The main contributions of this paper can be summarised as: (a) we study a new problem where scheduled nodes coexist on the same pool of radio resources with uncoordinated nodes; (b) we propose and evaluate through simulation a novel scheduling algorithm for the scenario, N-PF, which takes into account the relative channel quality metric, and the relative neighbourhood metric accounting for the presence of uncoordinated nodes in the cell; (c) we evaluate through simulation the impact of CSMA parameters (e.g., Clear Channel Assessment (CCA) threshold, NB_{max}, and backoff exponent (BE)) on the benchmark and the proposed algorithms.

The rest of the paper is organised as follows: Sect. 2 discusses related literature, Sect. 3 describes the system model, Sect. 4 describes the benchmark and the proposed scheduling algorithms, Sect. 5 describes simulator setup and the numerical results and finally Sect. 6 provides conclusions.

2   Related Literature

In the past, most MAC protocols for wireless networks have been designed to work in an environment where all users on the same set of radio resources rely exclusively on a centralised scheduling algorithms for resource assignments or contention based MAC protocols for channel access. However, a few studies in literature have been carried out on hybrid MAC protocols which combine features of ordinary TDMA and contention MAC based schemes. The Probabilistic TDMA (PTDMA) in [6], is an hybrid MAC protocol for a single-hop wireless LAN. PTDMA adapts the behaviour of the MAC between TDMA and CSMA according to the level of contention in the network. In [7], DrxMAC, an hybrid MAC protocol for low power and resource constrained devices is discussed. Drx-MAC is a slotted TDMA protocol with in-slot carrier sensing, and more than one device can be assigned the same slot. In [8], the authors proposed an hybrid MAC protocol for heterogeneous Machine to Machine (M2M) networks which combine features of contention based and TDMA schemes. [9] Proposes a spectrum-aware cluster-based energy-efficient routing scheme with an hybrid MAC which combine CSMA and TDMA schemes, but the two schemes operate on non conflicting set of radio resources. In IEEE 802.15.4 standard [10], the MAC protocol for beacon-enabled mode uses slotted CSMA/CA as the default channel access scheme, but the coordinator optionally assigns granted time slots to some nodes based on need.

In our work we consider an hybrid scenario, where the scheduled nodes rely on a centralised scheduling scheme for resources assignments, while the uncoordinated nodes rely on an unslotted CSMA/CA protocol for the channel access. Both groups of users coexist on the same set of radio resources. To the best of our knowledge, N-PF is the first dynamic centralised scheduling algorithm to account for the uncoordinated neighbours of the scheduled nodes as a part of aggregate scheduling metric in an hybrid scenario.

3   The System Model

We consider an uplink scenario in a single square cell of side 1 km, consisting of K scheduled nodes $\{j = 1, ..., K\}$, M uncoordinated nodes $\{i = 1, ..., M\}$, a single base station (BS) placed at the center of the cell, and a single frequency TDMA channel. Scheduled nodes are synchronized with the BS and they rely on a scheduler located within the BS for radio resources assignment, while uncoordinated nodes are asynchronous with the BS and they rely on a CSMA/CA MAC protocol for channel access. All nodes are randomly and uniformly distributed within the cell as shown in Fig. 1. Moreover, the scheduled and uncoordinated nodes transmit towards the BS on the same set of radio resources. Radio resources are in form of TDMA slots (hereafter referred as slots). A single frame is divided

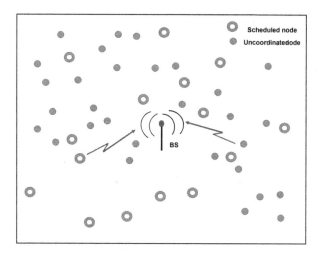

Fig. 1. The system model.

into T slots $\{s = 0, ..., T - 1\}$ of unit length, each of which is subdivided into t equal-sized sub-slots $\{ss = 0, ..., t - 1\}$.

All devices generate packets of equal length (L sub-slots) according to a Poisson arrival process with arrival rate λ. For the scheduled nodes, when a new packet arrives it is buffered until the base station grants the user a slot to transmit the packet, while in the case of uncoordinated nodes, when a new packet arrives it is buffered until the device successfully contends for a transmission opportunity and finishes transmission of the packet or unsuccessfully contents for a transmission opportunity and drops the packet. Each of the scheduled nodes can only be assigned at maximum a single unique slot in a given frame to transmit a single packet. When assigned to a given slot index s, the node starts transmission at the beginning of that slot. On the other hand uncoordinated nodes can start transmission at any instant along the time-line when the channel access attempt is successful. Depending on the number of packets in the buffer, uncoordinated nodes can attempt transmission of more than one packet, but for each packet the normal CSMA procedure has to be performed.

Let i be a network user connected to the base station, i is affected by path-loss according to the model given as

$$P_{Li}(d)(dB) = k_0(dB) + k_1 \log_{10} d(i, BS)(dB) - \gamma_i(dB) \qquad (1)$$

where k_0 and k_1 are constants depending on the propagation environment and the channel frequency, $d(i, BS)$ is the distance between user i and the base station. In linear scale, γ_i is an exponentially distributed component accounting for Rayleigh fading effect on the link. A packet is considered to have been correctly received if for the entire packet transmission time both the Signal to Noise Ratio (SNR) and the SIR are above the respective system thresholds as given by

$$SNR > \xi \text{ and } SIR > \alpha \qquad (2)$$

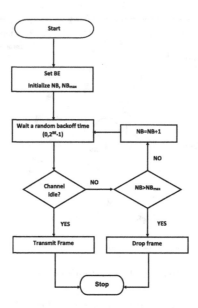

Fig. 2. The CSMA/CA protocol.

For simplicity, uncoordinated node i is considered to be a neighbour of the scheduled node j only if i can hear transmissions of j. Let $\mathcal{U}_{j_n} = \{1, 2, ..., n\}$ denote the subset of all uncoordinated nodes neighbouring j. The properties of $\mathcal{U}_{j_n}$ i.e. cardinality of the subset and its elements change according to the time coherence of the channel because of Rayleigh fading effect on links. Therefore $\mathcal{U}_{j_n}$ has a minimum and a maximum cardinality of 0 and M respectively.

The CSMA protocol implemented in our model is represented by Fig. 2. In the protocol BE is set to a fixed value. Clear Channel Assessment (CCA) is performed using Energy Detection (ED) technique. As shown in the figure the channel access attempt fails when the channel is sensed to be busy in all backoff stages up to the maximum stage (NB_{max}).

4   Benchmark and Proposed Algorithms

4.1   Proportional Fair Scheduling Algorithm

Wireless networks are characterized by time varying channel conditions, which are independent for different users. The proportional fair algorithm is designed to take advantage of multiuser diversity, while maintaining comparable long term throughput for all users. Let $R_j(s)$ denote the instantaneous data rate that user j can achieve at time instant s, and $T_j(s)$ be the average throughput for user j up to time slot s. The proportional fair scheduler selects the user, denoted as j^* with the best relative channel quality according to the metric $R_j(s)/T_j(s)$ for transmission. The average throughput $T_j(s)$ for all the users is updated as

$$T_j(s+1) = \begin{cases} (1-\beta)T_j(s) + \beta R_j(s), \ j = j^* \\ (1-\beta)T_j(s), \qquad\qquad j \neq j^* \end{cases} \tag{3}$$

where $0 \leq \beta \leq 1$ and $1/\beta$ is the time constant of the exponential moving average. By changing β the scheduler can trade off between the throughput of the system and temporal fairness among the users. In this paper, R_j is computed according to the normalised Shannon capacity formula as $\log_2(1 + SNR)$.

4.2   The Proposed Scheduling Algorithm

Extending the PF algorithm to account for the number of uncoordinated nodes neighbouring each of the scheduled nodes in a scenario where uncoordinated nodes coexists with scheduled nodes on the same set of radio resources, can lead to significant improvement in the performance of the algorithm. At time instant s, our proposed algorithm, N-PF, selects the user, denoted as j^*, with the best aggregate scheduling metric given as

$$\frac{R_j(s)}{T_j(s)} * \left(\frac{1}{\Omega_j}\right)^{\rho} \tag{4}$$

where $\rho \geq 0$ is an optimization constant used by the scheduler to emphasize or de-emphasize relative neighbourhood metric Ω_j during scheduling. For $\rho = 0$, the algorithm turns to be the PF algorithm. For higher values of ρ, the metric Ω_j becomes predominant. For a given scheduled node j, the metric Ω_j is given by

$$\Omega_j = \begin{cases} 1 - \left(\frac{n_j(s)}{M}\right), \ M > 0 \ \& \ n_j(s) \neq M \\ b, \qquad\qquad M > 0 \ \& \ n_j(s) = M \\ 1, \qquad\qquad M = 0 \end{cases} \tag{5}$$

where $n_j(s)$ is the number of uncoordinated neighbours of scheduled node j at time instant s, M is the total number of uncoordinated nodes deployed within the cell and b is an arbitrarily small positive constant.

5   Simulator Setup and Numerical Results

5.1   Simulator Setup

A C++ simulator is used to evaluate the performance of N-PF algorithm. The simulator implements the system model as described in Sect. 3. A single TDMA frame is divided into 10 slots, and each slot is subdivided into 200 sub-slots. The scheduling algorithm runs at the beginning of each new frame. We assume that the BS and all users have omnidirectional antennas. Default parameters considered in our simulations are summarised in Table 1.

A single simulation consists of 1000 frames. Results are averaged over 10 different scenarios, characterised by different nodes' positions the area.

Table 1. Default simulation parameters

Parameter	Value	Parameter	Value
Transmit power	$30dBm$	SNR threshold (ξ)	5 dB
β	0.1	SIR threshold (α)	3 dB
k_o	$41.7dB$	BS height	20 m
k_1	3.0	NB_{max}	10
λ	$1\ packet/frame$	CCA threshold	$-85dBm$
M	40 nodes	CCA duration	8 sub-slots
K	30 nodes	Contention Window (CW)	31 sub-slots
Packet length	50 sub-slots		

5.2   Performance Metrics

1. **Jain Index (JI)** [11], given as

$$Jain\ Index = \left(\sum_{j=1}^{K} x_j\right)^2 / \left(K\sum_{j=1}^{K} x_j^2\right) \tag{6}$$

where x_j is the average number of radio resource units allocated to user j within an interval of 1000 frames.

2. **Packet Delivery Rate (PDR)** is given by

$$PDR = \frac{n^o\ of\ successful\ packets}{n^o\ of\ transmitted\ packets} * 100 \tag{7}$$

3. **Blocking Rate (BR)**: if we let U_A be the number of unsuccessful channel access attempts and T_A be the total number of channel access attempts, BR is then given by

$$BR = \frac{U_A}{T_A} * 100 \tag{8}$$

5.3   Results

Where not indicated, default parameters in Table 1 should be assumed. Figure 3 shows packet delivery rate (PDR) metric for the scheduled nodes versus ρ with different number of uncoordinated nodes deployed. From the figure, PDR increases with increasing ρ and decreases with increasing number of uncoordinated nodes (M). The former trend is due to the fact that, for higher values of ρ the scheduler selects the scheduled nodes with best Ω_j metric which results in minimizing collision loss probability. The latter trend is attributed to the fact that, packet collision losses increase with increasing M.

Figure 4 shows how the JI varies with the neighbour metric coefficient (ρ). According to the figure, the JI slightly increases with increasing ρ for values

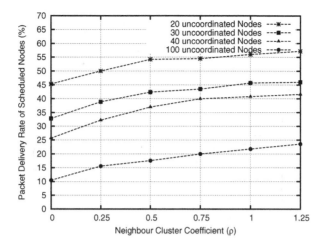

Fig. 3. Packet delivery rate of the scheduled nodes with $K = 30$ and different values of uncoordinated nodes (M).

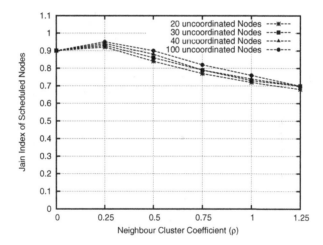

Fig. 4. Jain index of the scheduled nodes with $K=30$ and different values of M.

of ρ between 0 and 0.25. This effect can be attributed to the additional randomness introduced in the scheduling algorithm by Ω_j metric. Above $\rho = 0.5$, the JI decreases with increasing ρ because the Ω_j component becomes predominant. The decreasing performance in fairness is compensated by an improved performance in PDR as shown in Fig. 3.

Figure 5 shows the blocking rate (BR) metric versus ρ for the uncoordinated nodes. BR increases with increasing ρ and M. As shown in the figure, BR slightly increases with increasing ρ because the scheduler selects users with the best Ω_j metric and as a consequence more uncoordinated nodes are blocked

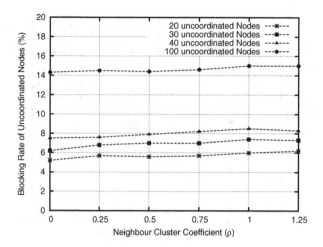

Fig. 5. Packet blocking rate of the uncoordinated nodes with $K = 30$ and different values of M.

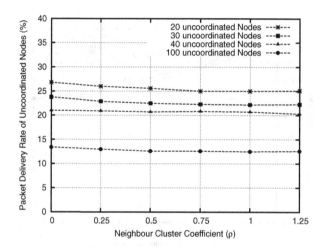

Fig. 6. Packet delivery rate of the uncoordinated nodes with $K = 30$ and different values of M.

from accessing the channel when scheduled nodes are transmitting. On the other hand, BR increases with increasing M because of increasing contention which result in many uncoordinated nodes reaching NB_{max} and consequently dropping packets. Likewise, according to Fig. 6 the PDR of the uncoordinated nodes slightly decrease with increasing ρ because of increasing packet collision loss probability.

Figure 7 shows the impact of packet length on packet delivery rate. The PDR of scheduled nodes increases with decreasing packet length. This is because the probability of packet collisions decreases with decreasing packet transmission time.

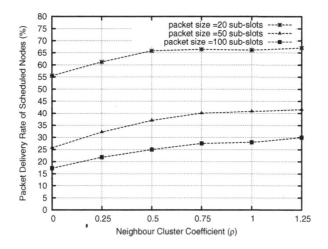

Fig. 7. Packet delivery rate of the scheduled nodes with $K = 30$ and $M = 40$ and different packet sizes.

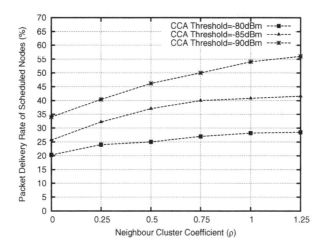

Fig. 8. Packet delivery rate of the scheduled nodes with $K = 30$, $M = 40$ and different values of CCA thresholds.

Figures 8, 9 and 10 show the impact of CSMA parameters on the scheduler i.e. BE, CCA threshold and NB_{max}. Packet delivery rate of the scheduled nodes increases with decreasing CCA threshold and increasing BE because these parameters have an impact of reducing packet collisions. On the other hand, the delivery rate slightly increases with decreasing NB_{max} because decreasing NB_{max} results in an increased blocking rate of uncoordinated nodes and hence decreased collisions.

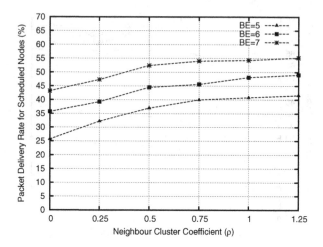

Fig. 9. Packet delivery rate of the scheduled nodes with $K = 30$ and $M = 40$ and different values of **BE**.

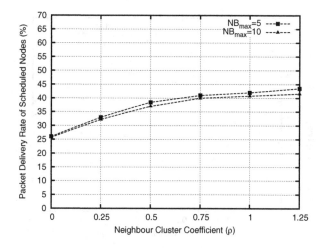

Fig. 10. Packet delivery rate of the scheduled nodes with $K = 30$ and $M = 40$ and different values of NB_{max}.

6   Conclusion

This paper presents a novel centralised scheduling algorithm for a scenario where scheduled nodes and uncoordinated nodes coexist on the same pool of radio resources. The proposed algorithm takes into account relative channel quality metric and relative neighbourhood metric in order to maximize packet delivery rate while maintaining fairness. Performance evaluations through simulations have demonstrated that, with respect to the benchmark algorithm, the proposed algorithm: (i) improves performance of scheduled nodes in terms of packet delivery rate rate; (ii) for small values of ρ, it improves the performance of scheduled nodes in terms of Jain index of fairness; (iii) maintains comparable network performance in terms of blocking rate.

References

1. Abramson, N.: The ALOHA system - another alternative for computer communications. In: Proceedings of the Fall Joint Computer Conference (1970)
2. Roberts, L.G.: ALOHA packet system with and without slots and capture. Comput. Commun. Rev. **5**(2), 28–42 (1975)
3. Wang, Y., Garcia-Luna-Aceves, J.J.: Performance of collision avoidance protocols in single-channel ad hoc networks. In: Proceedings of IEEE Tenth International Conference on Network Protocols (ICNP) 2002, pp. 68–77 (2002)
4. Pang, Q., Bigloo, A., Leung, V.C.M., Scholefield, C.: Service scheduling for general packet radio service classes. In: Proceedings of WCNC, New Orleans, LA, vol. 3, pp. 1229–1233 (1999)
5. Wengerter, C., Ohlhorst, J., von Elbwart, A.G.E.: Fairness and throughput analysis for generalized proportional fair frequency scheduling in OFDMA. In: 2005 IEEE 61st VTC, vol. 3, pp. 1903–1907 (2005)
6. Ephremides, A., Mowafi, O.A.: Analysis of a hybrid access scheme for buffered users-probabilistic time division. IEEE Trans. Softw. Eng. **SE–8**(1), 52–61 (1982)
7. Bergamini, L., Corbellini, G., Mangold, S.: Resource-constrained medium access control protocol for wearable devices. In: 2014 IEEE 10th International Conference on Wireless and Mobile Computing, Networking and Communications (WiMob), pp. 634–641, 8–10 October 2014
8. Liu, Y., Yuen, C., Cao, X., Hassan, N.U., Chen, J.: Design of a scalable hybrid MAC protocol for heterogeneous M2M networks. IEEE Internet Things J. **1**(1), 99–111 (2014)
9. Shah, G.A., Akan, O.B.: Spectrum-aware cluster-based routing for cognitive radio sensor networks. In: 2013 IEEE International Conference on Communications (ICC), pp. 2885–2889, 9–13 June 2013
10. IEEEStd. 802.15.4: Wireless Medium Access Control(MAC) and Physical Layer (PHY) Specifications for Low Data Rate Wireless. IEEE Std 802.15.4-2006 (2006)
11. Jain, R., Chiu, D., Hawe, W.: A Quantitative Measure Of Fairness And Discrimination For Resource Allocation In Shared Computer Systems. DEC Research Report TR-301, September 1984

Performance Analysis of Dynamic Spectrum Allocation in Multi-Radio Heterogeneous Networks

Yongjae Kim[1], Yonghoon Choi[2], and Youngnam Han[1(✉)]

[1] Department of Electrical Engineering,
Korea Advanced Institute of Science and Technology (KAIST), Daejeon, Korea
{yongjaekim,ynhan}@kaist.ac.kr
[2] Department of Electrical Engineering, Chonnam National University,
Gwangju, Korea
yh.choi@jnu.ac.kr

Abstract. In heterogeneous networks, multi-radio access technologies (RATs) can coexist for a variety of traffic demands and it is called multi-RAT network. Also, cognitive radio enable to use white space of frequency band, and thus spectrum resources can be dynamically allocated. This paper analyzes an effect of multi-radio access (MRA) users, who simultaneously exploit multi-RATs, on network performance where dynamic spectrum allocation (DSA) is performed. Multi-dimensional Erlang loss (MDEL) model, which is based on queueing, is suitable to describe behaviors of single radio access users in multi-RAT networks under the performing DSA. Based on the MDEL model, extended MDEL model is proposed to investigate the effect of MRA users. As MRA users increase, blocking probability, utilization, and expected processing time of a user in the multi-RAT networks deteriorate, since the MRA users require multiple spectrum resources at a time. Numerical results verify the performance degradation resulted from the MRA users under the DSA and FSA scenarios.

Keywords: Dynamic spectrum allocation · Heterogeneous networks · Multi-radio access technology networks · Multi-radio access user

1 Introduction

In heterogeneous networks (HetNets), users who exploit multi-radio access technologies (RATs) become widely common and many wireless applications for multi-RAT have been developed. Hence, spectrum scarcity is one of important issues in HetNets and many researches have been studied the scarcity problem. A large portion of allocated spectrum is used sporadically and the utilization of allocated spectrum has a geographical variation with a high variance in time [1]. Zhao *et al.* [2] show that the depletion of the spectrum is the result of current fixed spectrum allocation (FSA) policy rather than physical scarcity of frequency

© ICST Institute for Computer Sciences, Social Informatics and Telecommunications Engineering 2016
D. Noguet et al. (Eds.): CROWNCOM 2016, LNICST 172, pp. 154–165, 2016.
DOI: 10.1007/978-3-319-40352-6_13

bands. For compensation of underutilized spectrum, J. Mitola *et al.* [3] propose a concept of cognitive radio (CR). The CR enables user equipment to dynamically access the spectrum. This is known as dynamic spectrum access.

Dynamic spectrum access is categorized as opportunistic spectrum access model, spectrum sharing model and dynamic exclusive use model [2,4,5]. Liang *et al.* [4] provide an overview on CR networking. They review physical, medium access control and network layer involved in a CR design. Zhang *et al.* [5] focus on spectrum sharing problem in the view point of convex optimization. Moreover, Kliks *et al.* [6] consider flexible pluralistic licensing concept for 5G wireless networks from spectrum sharing point of view. Akyildiz *et al.* [7] classify CR technologies according to different functionalities such as sensing, decision, sharing, and mobility. Dynamic spectrum allocation (DSA) which belongs to the dynamic exclusive use model denotes that spectrum is exclusively distributed by a central entity for spectrum utilization.

DSA has been considered in many researches for spectrum utilization or revenue. Auction or trading methods of spare frequency bands are introduced in [8–10]. Subramanian *et al.* [8] apply a centralized entity called spectrum broker in multi-RAT network based on dynamic auctions. Each base station (BS) bids for channels depending on their demands. Their objective is to maximize the overall revenue subject to interference in the networks. They exploit greedy algorithm for bidding. However, the greedy algorithm is not an optimal solution. Thus binary integer programming is introduced for spectrum allocation [9] to compensate for drawback of the greedy algorithm. They employ an interference graph based on interference constraints. This scheme obtains an optimal set of binary decisions on whether to allocate or not the channels to BSs. The optimal result is found in accordance multiple objectives of maximization of total revenue and spectrum efficiency. Le *et al.* [10] propose a scheme in which the adjacent cells lease the spectrum to each other to maximize the revenue of HetNets. This scheme can maximize profit of operators and solve inter-system interference issue. Game theory, which is an efficient method for resource optimization algorithms, is applied to design a spectrum trading algorithm. Zhang *et al.* [11] investigate joint subchannel and power allocation in cognitive small cell networks. They formulate resource allocation as a cooperative Nash bargaining game, and near optimal solutions are derived by relaxing variables and using Lambert-W function. Nash bargaining resource allocation algorithm is developed and show to converge to a Pareto-optimal equilibrium. In [12], DSA framework for multi-RAT network is proposed. The available frequency band is divided into sharable spectrum blocks. These blocks are heuristically distributed to each RAT according to the amount of traffic load. Choi *et al.* [13] explore the benefit of multiple transmissions by multi-RATs over a single transmission by a single RAT. The optimal solution is founded with respect to band selection and power allocation using a distributed joint allocation algorithm which is proposed for parallel multi-radio access (MRA) scheme to maximize system capacity. In [13], they consider that each RAT, which is based on orthogonal frequency division multiple access (OFDMA), can allocate the scalable spectrum bandwidth size to users. However, the scalable spectrum bandwidth results in high computation complexity to OFDMA systems.

This paper analyzes the effect of MRA and single radio access (SRA) users in multi-RAT networks under the DSA and FSA scenarios. Under the DSA scenario, total spectrum resources can be distributed to each RAT according to traffic loads, whereas the amount of assigned spectrum to each RAT is fixed in the FSA policy. Therefore, the spectrum resources can be more efficiently used in case of DSA. Each RAT is based on OFDMA allocates subchannels to MRA or SRA users. To analyze an effect of MRA users, mathematical models which describe behaviors of MRA users in multi-RAT networks under DSA and FSA are proposed as queueing models. Using the proposed models, blocking probability (BP), utilization, and processing time of a user are evaluated according to the proportion of MRA users.

2   Effect of Multi-Radio Access

This paper considers a region covered by a set of different N BSs, that is, N RATs are in the region [12], and the multi-RATs belong to same or different network operators. This is called multi-RAT network.

Every RAT is assumed to adopt OFDMA and frequency band of each RAT is divided into multiple subchannels. Suppose that the multi-RAT network has C_T subchannels. Based on the FSA policy, the subchannels are evenly distributed among multi-RATs, i.e., C_T/N. On the other hand, subchannels which are in a multi-RAT network can be shared among multi-RATs where DSA is performed. In this case, it is possible that one of RATs uses all C_T subchannels when the RAT has huge traffic loads. The spectrum scarcity problem can be mitigated by performing DSA. The subchannels are assigned to each RAT according to traffic load when DSA is performed.

When an SRA user access to a RAT, the RAT assigns a subchannel to the user, and therefore an SRA user occupies a subchannel at a time. Similarly, when an MRA user access to multi-RAT, each RAT assigns a subchannel. Hence, multiple subchannels are assigned to the MRA user during access to multi-RATs. When a RAT has no remaining subchannel, a user is dropped, and this phenomenon is known as *blocking*.

2.1   Single Radio Access Users in Multi-Radio Access Technology Network

In this subsection, let us consider a case that only SRA users are in a multi-RAT network. An SRA user occupies a subchannel of the RAT which the SRA user accesses. As mentioned above, based on the FSA policy, each RAT has C_T/N subchannels. On the contrary, under DSA, RAT_k user can use RAT_ls subchannel, because subchannels can be shared.

Multi-dimensional Erlang loss (MDEL) model [14] is composed of multiple Erlang loss models ($M/M/c/c$). The multiple Erlang loss models can share their subchannels, therefore total subchannels are shared in the MDEL model. Hence, the multi-RAT network, where DSA is performed, can be modeled as the MDEL model.

The number of subchannels that users occupy is denoted by c_k, where k is index of RAT, $k \in (1, N)$. Then, $\sum_{k=1}^{N} c_k \leq C_T$. When state of MDEL model is defined as the number of users who occupy subchannels in the network, state space is expressed as

$$S = \left\{ \mathbf{c} = (c_1, \cdots, c_N) \,\middle|\, \forall c_k \geq 0, \; \sum_{k=1}^{N} c_k \leq C_T \right\}, \tag{1}$$

where $\mathbf{c} = (c_1, c_2, \cdots, c_N)$ denotes the state and $k \in (1, N)$. The state space is an N-dimensional Euclidean space. Transition rates are as follows:

$$r(\mathbf{c}, \mathbf{c}') = \begin{cases} \lambda_k, & \text{if } \mathbf{c}' = \mathbf{c} + e_k \\ c_k \mu_k, & \text{if } \mathbf{c}' = \mathbf{c} - e_k \end{cases}, \tag{2}$$

where $\mathbf{c}, \mathbf{c}' \in S$, $k \in (1, N)$, and e_k denotes standard basis of N-dimensional Euclidean space. Arrival and departure process assumed Poisson process, and thus λ_k and μ_k are rate values of exponential distribution. In steady-state, balance equation can be derived by using (1) and (2).

$$p_{(c_1, \cdots, c_N)} \cdot \sum_{k=1}^{N} \left[I \left(\sum_{i=1}^{N} c_i < C_T \right) \lambda_k + c_k \mu_k \right]$$
$$= \sum_{k=1}^{N} \left[\lambda_k p_{(c_1, \cdots, c_k+1, \cdots, c_N)} + (c_k + 1) \mu_k p_{(c_1, \cdots, c_k+1, \cdots, c_N)} \right], \tag{3}$$

where $p_{(c_1, c_2, \cdots, c_N)}$ represents steady-state probability of the state $(c_1, c_2, \cdots, c_N)$ and $I(A)$ denotes the indicator function of event A. Also, if $\mathbf{c} = (c_1, c_2, \cdots, c_N) \notin S$, then $p_{(c_1, c_2, \cdots, c_N)} = 0$. Using (3) and normalization condition $\sum_{\mathbf{c} \in S} p_{(c_1, c_2, \cdots, c_N)} = 1$, the steady-state probabilities are found as follows:

$$p_{(0, \cdots, 0)} = \left[\sum_{c_1=0}^{C_T} \sum_{c_2=0}^{C_T-c_1} \cdots \sum_{c_N=0}^{C_T-\sum_{i=1}^{N-1} c_i} \prod_{k=1}^{N} \frac{\lambda_k{}^{c_k}}{c_k! \mu_k{}^{c_k}} \right]^{-1},$$

$$p_{(c_1, \cdots, c_N)} = \begin{cases} p_{(0, \cdots, 0)} \cdot \prod_{k=1}^{N} \frac{\lambda_k{}^{c_k}}{c_k! \mu_k{}^{c_k}}, & (c_1, \cdots, c_N) \in S \\ 0, & (c_1, \cdots, c_N) \notin S \end{cases}. \tag{4}$$

2.2   Multi-Radio Access Users in Multi-Radio Access Technology Network

Multi-Radio Access Technology Network with Dynamic Spectrum Allocation: The state space of extended MEDL model is same with the MDEL model.

To describe the behaviors of MRA users in multi-RAT network, the MDEL model is extended. Two state transitions are added to the extended MDEL model. Let $1 - \gamma_k$ be the proportion of MRA users, and γ_k be the proportion of SRA users. Then the transition rates are given as follows:

$$r(\mathbf{c}, \mathbf{c}') = \begin{cases} \gamma_k \lambda_k, & \text{if } \mathbf{c}' = \mathbf{c} + e_k \\ \gamma_k c_k \mu_k, & \text{if } \mathbf{c}' = \mathbf{c} - e_k \\ \sum_{k=1}^{N} (1 - \gamma_k)\lambda_k, & \text{if } \mathbf{c}' = \mathbf{c} + \sum_{k=1}^{N} e_k \\ \sum_{k=1}^{N} (1 - \gamma_k)c_k \mu_k, & \text{if } \mathbf{c}' = \mathbf{c} - \sum_{k=1}^{N} e_k \end{cases} \tag{5}$$

where $\mathbf{c}, \mathbf{c}' \in S$ and $k \in (1, N)$. The third line of (5) denotes that subchannels are assigned to an MRA user. An MRA user simultaneously uses N RATs, and thus N subchannels are assigned. The fourth line of (5) denotes that an MRA user releases assigned subchannel. Figure 1 shows state transitions of multi-RAT network under DSA when MRA users are considered. Using (1) and (5), balance equation is derived in steady-state as (6). Normalization condition and (6), the steady-state probabilities can be computed using numerical approach [15].

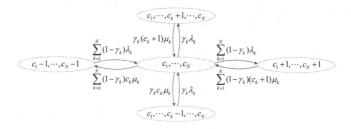

Fig. 1. State transition diagram for behaviors of MRA user in multi-RAT system

$$p_{(c_1, \cdots, c_N)} \left[\sum_{k=1}^{N} I\left((c_1, \cdots, c_k + 1, \cdots c_N) \in S\right) \gamma_k \lambda_k + \sum_{k=1}^{N} I\left((c_1, \cdots, c_k - 1, \cdots c_N) \in S\right) \gamma_k c_k \mu_k \right.$$

$$\left. + \sum_{k=1}^{N} I\left((c_1 + 1, \cdots c_N + 1) \in S\right)(1 - \gamma_k)\lambda_k + \sum_{k=1}^{N} I\left((c_1 - 1, \cdots c_N - 1) \in S\right)(1 - \gamma_k)c_k \mu_k \right]$$

$$= I\left((c_1 + 1, \cdots c_N + 1) \in S\right) p_{(c_1 + 1, \cdots, c_N + 1)} \left(\sum_{k=1}^{N} (1 - \gamma_k)(c_k + 1)\mu_k \right)$$

$$+ I\left((c_1 - 1, \cdots c_N - 1) \in S\right) p_{(c_1 - 1, \cdots, c_N - 1)} \left(\sum_{k=1}^{N} (1 - \gamma_k)\lambda_k \right)$$

$$+ \sum_{k=1}^{N} \left[I\left((c_1, \cdots, c_k - 1, \cdots c_N) \in S\right) p_{(c_1, \cdots, c_k - 1, \cdots, c_N)} \gamma_k \lambda_k \right.$$

$$\left. + I\left((c_1, \cdots, c_k + 1, \cdots c_N) \in S\right) p_{(c_1, \cdots, c_k + 1, \cdots, c_N)} \gamma_k(c_k + 1)\mu_k \right] \tag{6}$$

Multi-Radio Access Technology Network with Fixed Spectrum Allocation: Difference between DSA and FSA in the multi-RAT network is whether there are sharable subchannels. Under the FSA policy, every RAT exclusively uses subchannels and there are no sharable subchannels. If there is a gap in the traffic loads among multi-RATs, the FSA is inefficient resources management method compared with DSA.

In case of FSA, the state space is expressed as follows:

$$S = \left\{ \mathbf{c} = (c_1, \cdots, c_N) \middle| 0 \leq c_k \leq \frac{C_T}{N}, \ k \in (1, N) \right\}. \tag{7}$$

Figure 2 shows the difference of state spaces and state transitions between DSA and FSA. In Fig. 2, solid arrows denote the transitions associated with SRA users, whereas dashed arrows denote the transitions associated with MRA users. The state transitions and transition rates are same with DSA as described in (5) and Fig. 1, respectively.

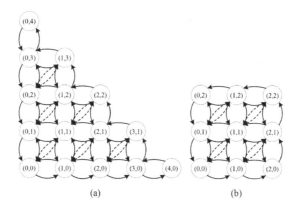

(a) (b)

Fig. 2. State transition diagram of extended MDEL model under (a) DSA and (b) FSA, when $C_T = 4$ and $N = 2$

3 Performance Evaluations

3.1 Blocking Probability

Let the probability of the number of total users in the network be π_j, and then it is computed as follows:

$$\pi_j(\boldsymbol{\gamma}, \boldsymbol{\rho}) = \sum_{\mathbf{c} \in S_j} P_{(c_1, \cdots, c_N)}(\boldsymbol{\gamma}, \boldsymbol{\rho}), \tag{8}$$

where $S_j = \left\{ \mathbf{c} = (c_1, \cdots, c_N) \in S \middle| \forall c_k \geq 0, \sum_{k=1}^{N} c_k = j \right\}$, $k \in (1, N)$, $\boldsymbol{\gamma} = (\gamma_1, \cdots, \gamma_N)$, and $\boldsymbol{\rho} = (\rho_1, \cdots, \rho_N) = \left(\frac{\lambda_1}{\mu_1}, \cdots, \frac{\lambda_N}{\mu_N} \right)$. In addition, BP which is defined as a probability that an arrival user is blocked because a RAT has no remaining subchannels is derived by using PASTA theorem, i.e.,

$$BP(\boldsymbol{\gamma}, \boldsymbol{\rho}) = \sum_{\mathbf{c} \in S_B} P_{(c_1, \cdots, c_N)}(\boldsymbol{\gamma}, \boldsymbol{\rho}) = \pi_{C_T}(\boldsymbol{\gamma}, \boldsymbol{\rho}), \tag{9}$$

where $S_B = \left\{ (c_1, \cdots, c_N) \in S \middle| \sum_{k=1}^{N} c_k = C_T \right\}$.

3.2　Utilization

The expected value of the number of total users in the network is given as $L(\boldsymbol{\gamma}, \boldsymbol{\rho}) = \sum_{j=1}^{C_T} j \cdot \pi_j(\boldsymbol{\gamma}, \boldsymbol{\rho})$, and thus utilization is defined as

$$U(\boldsymbol{\gamma}, \boldsymbol{\rho}) = \frac{L(\boldsymbol{\gamma}, \boldsymbol{\rho})}{C_T}. \tag{10}$$

3.3　Processing Time

Assume that a blocked user tries to access a RAT again until the user succeed in accessing the RAT. Using above BPs, processing time $T(\boldsymbol{\gamma}, \boldsymbol{\rho})$ for a user is derived as follows:

$$T(\boldsymbol{\gamma}, \boldsymbol{\rho}) = \Pr[\mathbf{c}^1 \notin S_B] \cdot T_s(\mu)$$
$$+ \sum_{n=2}^{\infty} \Pr[\mathbf{c}^2, \cdots, \mathbf{c}^{n-1} \in S_B, \mathbf{c}^n \notin S_B] \cdot (T_s(\mu) + \sum_{m=1}^{n-1} T_i(\lambda)), \tag{11}$$

where $\mathbf{c}^\tau$ represents the state after τ-th state transition, and random variables T_s and T_i denote service time and inter-arrival time, respectively. $\Pr[\mathbf{c}^1 \notin S_B]$ and $\Pr[\mathbf{c}^2, \cdots, \mathbf{c}^{n-1} \in S_B, \mathbf{c}^n \notin S_B]$ denote the probability that a user is not blocked at the first arrival and the probability that a user in not blocked n-th arrival after $(n-1)$-th blocked. In (11), the probability that user is sequentially blocked is too small, and therefore (11) can be rewritten as follows:

$$T(\boldsymbol{\gamma}, \boldsymbol{\rho}) \cong \Pr[\mathbf{c}^1 \notin S_B] \cdot T_s(\mu) + \Pr[\mathbf{c}^1 \in S_B, \mathbf{c}^2 \notin S_B] \cdot (T_s(\mu) + T_i(\lambda)). \tag{12}$$

The first term of right-hand side in (12) can be rewritten $(1 - BP(\boldsymbol{\gamma}, \boldsymbol{\rho})) \cdot T_s(\mu)$, and then it is also expressed according to which RAT user:

$$(1 - BP(\boldsymbol{\gamma}, \boldsymbol{\rho})) \cdot T_s(\mu) = \sum_{k=1}^{N} Pr[\text{departure user} \in \text{RAT}_k] \cdot (1 - BP(\boldsymbol{\gamma}, \boldsymbol{\rho})) \cdot T_s(\mu_k)$$
$$= \sum_{k=1}^{N} \frac{\mu_k}{\sum_{i=1}^{N} \mu_i} \cdot (1 - BP(\boldsymbol{\gamma}, \boldsymbol{\rho})) \cdot T_s(\mu_k). \tag{13}$$

In addition, $\Pr[\mathbf{c}^1 \in S_B, \mathbf{c}^2 \notin S_B]$ of (12) is expressed as follows:

$$
\begin{aligned}
\Pr[\mathbf{c}^1 \in S_B,\ \mathbf{c}^2 \notin S_B] &= \Pr[\mathbf{c}^2 \notin S_B \,|\, \mathbf{c}^1 \in S_B] \cdot \Pr[\mathbf{c}^1 \in S_B] \\
&= \Pr[R_s(\mu)\ \text{of at least one of}\ C_T < T_i(\lambda)] \cdot BP(\boldsymbol{\gamma}, \boldsymbol{\rho}) \\
&\overset{(a)}{=} (1 - \Pr[R_s(\mu) > T_i(\lambda)])^{C_T} \cdot BP(\boldsymbol{\gamma}, \boldsymbol{\rho}) \\
&\overset{(b)}{=} (1 - \Pr[T_s(\mu) > T_i(\lambda)])^{C_T} \cdot BP(\boldsymbol{\gamma}, \boldsymbol{\rho}),
\end{aligned}
\tag{14}
$$

where $R_s(\mu)$ represents remaining time of a user. The equality (a) results from the independent-identically distributed (i.i.d.) condition and (b) results from the memoryless property of arrival/departure process. $\Pr[T_s(\mu) > T_i(\lambda)]$ in (14) can be expressed according to which RAT user:

$$
\Pr[T_s(\mu) > T_i(\lambda)] = \sum_{k=1}^{N} \sum_{l=1}^{N} \Pr[T_s(\mu_l) > T_i(\lambda_k)] \cdot Pr[\text{arrival user} \in \text{RAT}_k]
$$

$$
\cdot \Pr[\text{departure user} \in \text{RAT}_l]
$$

$$
= \sum_{k=1}^{N} \sum_{l=1}^{N} \frac{\lambda_k}{\lambda_k + \mu_l} \cdot \frac{\lambda_k}{\sum_{i=1}^{N} \lambda_i} \cdot \frac{\mu_l}{\sum_{i=1}^{N} \mu_i}.
\tag{15}
$$

After some manipulations, distribution of processing time of a user can be presented as (16) where the number of RATs N is 2. From (16), moment generating function of T is calculated as follows:

$$
\begin{aligned}
f_T(t) = \sum_{i=1}^{2} \sum_{j=1}^{2} &\left[\beta_{01}\beta_{02} \left(\frac{\alpha_{02}\alpha_{ij}}{\alpha_{ij}\mu_2 - \alpha_{02}\lambda_i} \right)^2 + \sum_{k=0}^{2} \sum_{l=1}^{2} \frac{\alpha_{ij}\alpha_{kl}\omega_{kl}}{\alpha_{ij}\mu_l - \alpha_{kl}\lambda_i} \right] \eta_{ij} e^{-\frac{\lambda_i}{\alpha_{ij}}t} \\
+ \sum_{i=0}^{2} \sum_{j=1}^{2} &\left[\alpha_{ij}\omega_{ij} \left(\sum_{k=1}^{2} \sum_{l=1}^{2} \frac{\alpha_{kl}\eta_{kl}}{\alpha_{ij}\lambda_k - \alpha_{kl}\mu_j} \right) \right. \\
+ I(i = 0,\ j = 2)&\beta_{01}\beta_{02} \left. \left(\sum_{k=1}^{2} \sum_{l=1}^{2} \frac{\alpha_{ij}\alpha_{kl}\eta_{kl}}{\alpha_{ij}\lambda_k - \alpha_{kl}\mu_j} \cdot t - \left(\frac{\alpha_{ij}\alpha_{kl}}{\alpha_{ij}\lambda_k - \alpha_{kl}\mu_j} \right)^2 \eta_{kl} \right) \right] e^{-\frac{\mu_j}{\alpha_{ij}}t},
\end{aligned}
\tag{16}
$$

where $\alpha_{ij} = \begin{cases} \dfrac{\mu}{\sum_{k=1}^{2} \mu_k}(1 - BP(\boldsymbol{\gamma}, \boldsymbol{\rho})), & i = 0 \\[2ex] \left[1 - \left(\dfrac{\lambda_i}{\lambda_i + \mu_j} \cdot \dfrac{\lambda_i}{\sum_{k=1}^{2} \lambda_k} \cdot \dfrac{\mu_j}{\sum_{k=1}^{2} \mu_k} \right)^{C_t} \right] \cdot BP(\boldsymbol{\gamma}, \boldsymbol{\rho}), & i \neq 0 \end{cases}$,

$\beta_{ij} = \dfrac{\alpha_{ij}\mu_j}{\prod_{k=0, k\neq i}^{2}(\alpha_{ij} - \alpha_{kj})}$,

$\omega_{ij} = \begin{cases} \left(\dfrac{\sum_{k=0}^{2} \alpha_{k,j+1}\beta_{k,j+1}}{\mu_j} - \sum_{k=1}^{2} \dfrac{\alpha_{k,j+1}^2\beta_{k,j+1}}{\alpha_{k,j+1}\mu_j - \alpha_{ij}\mu_{j+1}} \right) \dfrac{\beta_{ij}\mu_j}{\mu_{j+1}}, & i = 0, j = 1 \\[3ex] \sum_{k=1}^{2} \dfrac{\alpha_{k,j-1}\beta_{k,j-1}}{\alpha_{ij}\mu_{j-1} - \alpha_{k,j-1}\mu_j}, & i = 0, j = 2 \\[3ex] \sum_{k=1}^{2} \dfrac{\alpha_{k,j+1}\beta_{k,j+1}}{\alpha_{ij}\mu_{j+1} - \alpha_{k,j+1}\mu_j} \alpha_{ij}\beta_{ij}, & i \neq 0, j = 1 \\[3ex] \sum_{k=1}^{2} \dfrac{\alpha_{k,j-1}\beta_{k,j-1}}{\alpha_{ij}\mu_{j-1} - \alpha_{k,j-1}\mu_j} \alpha_{ij}\beta_{ij}, & i \neq 0, j = 2 \end{cases}$,

$$
\eta_{ij} = \begin{cases} \dfrac{(-1)^{j+1}\alpha_{ij}\lambda_i\lambda_{i+1}}{\prod_{m=1}^{2}\left(\sum_{n=1}^{2}(-1)^{n+1}\alpha_{mn}\right)} \sum_{k=1}^{2} (-1)^{k+1}\dfrac{\alpha_{i+1,k}}{\alpha_{ij}\lambda_{i+1}-\alpha_{i+1,k}\lambda_i}, & i = 1 \\[4mm] \dfrac{(-1)^{j+1}\alpha_{ij}\lambda_{i-1}\lambda_i}{\prod_{m=1}^{2}\left(\sum_{n=1}^{2}(-1)^{n+1}\alpha_{mn}\right)} \sum_{k=1}^{2} (-1)^{k+1}\dfrac{\alpha_{i-1,k}}{\alpha_{ij}\lambda_{i-1}-\alpha_{i-1,k}\lambda_i}, & i = 2 \end{cases},
$$

$$i \in \{0,1,2\},\ j \in \{1,2\}.$$

$$
\Phi_T(s) = \prod_{i\in\{1,3,5\}} \left(\frac{\mu_1}{\mu_1 - \alpha_i s}\right) \cdot \prod_{j\in\{2,4,6\}} \left(\frac{\mu_2}{\mu_2 - \alpha_j s}\right) \cdot
$$
$$
\prod_{k\in\{3,4\}} \left(\frac{\lambda_1}{\lambda_1 - \alpha_k s}\right) \cdot \prod_{l\in\{5,6\}} \left(\frac{\lambda_2}{\lambda_2 - \alpha_l s}\right). \tag{17}
$$

Expected processing time is obtained as follows:

$$
E[T(\boldsymbol{\gamma},\boldsymbol{\rho})] = \frac{\sum_{i\in\{1,3,5\}}\alpha_i}{\mu_1} + \frac{\sum_{j\in\{2,4,6\}}\alpha_j}{\mu_2} + \frac{\sum_{k\in\{3,4\}}\alpha_k}{\lambda_1} + \frac{\sum_{l\in\{5,6\}}\alpha_l}{\lambda_2}. \tag{18}
$$

4   Numerical Results

The objective of numerical analysis is to examine how MRA users affect the performance metrics given in terms of BP, utilization, and expected processing time. For numerical results, the parameter configuration is set as follows: $\lambda_1 = 3$ [users/s], $\lambda_2 = 0.6, 0.8, 1.0$ [users/s], $\mu_1 = 1.5$ [users/s], $\mu_2 = 1$ [users/s], $C_t = 6$, $N = 2$ and $\gamma_1 = \gamma_2 = \gamma$.

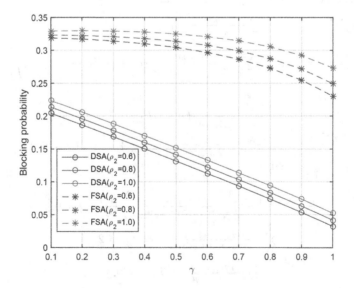

Fig. 3. Blocking probability of DSA and FSA under different γ

Figure 3 presents BPs according to the change in γ under DSA or FSA scenarios. When traffic loads of RAT_2 are low, the BPs are always low in both DSA and FSA cases. If traffic loads are high, the BPs also grow up regardless of which scenario. In addition, BPs always decrease as γ becomes large, because MRA users occupy more subchannels than SRA users at a time. At low γ region, there is no difference of BPs according to γ under FSA, whereas at high γ region, the difference of BPs become large. It describes that DSA is more sensitive with respect to the effects of MRA users than FSA.

Figure 4 shows the utilization of DSA and FSA in accordance with the change in γ. In FSA scenario, the utilization becomes low when MRA users are densely deployed. As γ increases, the utilization also increases. This is because an MRA user need multiple subchannels, whereas an SRA user only needs a subchannel. On the contrary, there is no considerable change according to γ in the utilization of DSA scenario compared to FSA, because DSA takes on a role of load balancing.

Figure 5 illustrates the expected time of processing a user according to different γ under DSA and FSA scenarios. In (18), α_i, $i \in \{3, 4, 5, 6\}$ can be approximated by the terms of $BP(\boldsymbol{\gamma}, \boldsymbol{\rho})$, therefore expected processing time is given by

$$E[T(\boldsymbol{\gamma}, \boldsymbol{\rho})] \cong C_1 \cdot BP(\boldsymbol{\gamma}, \boldsymbol{\rho}) + C_2, \tag{19}$$

where $C_1 = 2 \left[\frac{1}{\mu_1 + \mu_2} \left(1 + \frac{\mu_1^2 + \mu_2^2}{\mu_1 \mu_2} \right) + \frac{\lambda_1 + \lambda_2}{\lambda_1 \lambda_2} \right]$, and $C_2 = \frac{2}{\mu_1 + \mu_2}$. From (19), it is noticed that $E[T(\boldsymbol{\gamma}, \boldsymbol{\rho})] \propto BP(\boldsymbol{\gamma}, \boldsymbol{\rho})$.

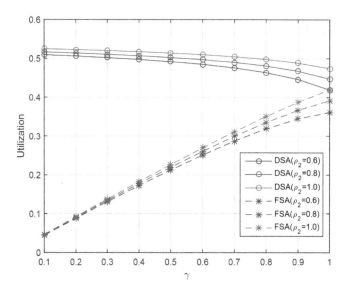

Fig. 4. Utilization of DSA and FSA under different γ

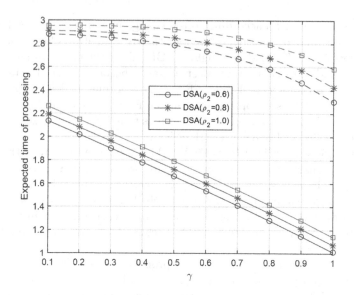

Fig. 5. Expected time of procssing a user under different γ

5   Conclusion

In this paper, an effect of MRA users on performance metrics such as the BP, the utilization, and the expected processing time was analyzed in multi-RAT networks under DSA and FSA scenarios. For this, an analytic model which is extended from MDEL model was provided to describe the behaviors of MRA users. Numerical results verified that DSA gives better performance than FSA and the BP and the expected processing time are improved as the number of SRA users increases.

References

1. Akyildiz, I.F., Lee, W.Y., Vuran, M.C., Mohanty, S.: NeXt generation/dynamic spectrum access/cognitive radio wireless networks: a survey. Comput. Netw. **50**(13), 2127–2159 (2006)
2. Zhao, Q., Sadler, B.M.: A survey of dynamic spectrum access: signal processing, networking, and regulatory policy. IEEE Signal Proc. Mag. **24**(3), 79–89 (2007)
3. Mitola, J., Maguire, G.Q.: Cognitive radio: making software radios more personal. IEEE Pers. Commun. **6**(4), 13–18 (1999)
4. Liang, Y.C., Chen, K.C., Li, G.Y., Mahonen, P.: Cognitive radio networking and communications: an overview. IEEE Trans. Veh. Technol. **60**(7), 3386–3407 (2011)
5. Zhang, R., Liang, Y.C., Cui, S.: Dynamic resource allocation in cognitive radio networks. IEEE Signal Proc. Mag. **27**(3), 102–114 (2010)
6. Kliks, A., Holland, O., Basaure, A., Matinmikko, M.: Spectrum and license flexibility for 5G networks. IEEE Commun. Mag. **53**(7), 42–29 (2015)

7. Akyildiz, I.F., Lee, W.Y., Vuran, M.C., Mohanty, S.: A survey on spectrum management in cognitive radio networks. IEEE Commun. Mag. **46**(4), 40–48 (2008)
8. Subramanian, A.P., Al-Ayyoub, M., Gupta, H., Das, S.R., Buddhikot, M.M.: Near-optimal dynamic spectrum allocation in cellular networks. In: 3rd IEEE Symposium on New Frontiers in Dynamic Spectrum Access networks (DySPAN), pp. 1–11, October 2008
9. Lee, S., Lee, H.: Dynamic spectrum allocation based on binary integer programming under interference graph. In: 23rd International Symposium on Personal Indoor and Mobile Radio Communications (PIMRC), pp. 226–231, September 2012
10. Le, V., Feng, Z., Bourse, D., Zhang, P.: A cell based dynamic spectrum management scheme with interference mitigation for cognitive networks. Wirel. Pers. Commun. **49**(2), 275–293 (2008)
11. Zhang, H., Jiang, C., Beaulieu, N.C., Chu, X., Wang, X., Quek, T.Q.S.: Resource allocation for cognitive small cell networks: a cooperative bargaining game theoretic approach. IEEE Trans. Wireless Commun. **14**(6), 3481–3493 (2015)
12. Alsohaily, A., Sousa, E.S.: Dynamic spectrum management in multi-radio access technology (RAT) cellular systems. IEEE Wirel. Commun. Lett. **3**(3), 249–252 (2014)
13. Choi, Y., Kim, H., Han, S., Han, Y.: Joint resource allocation for parallel multi-radio access in heterogeneous wireless networks. IEEE Trans. Wirel. Commun. **9**(11), 3324–3329 (2010)
14. Melikov, A., Ponomarenko, L.: Multidimensional Queueing Models in Telecommunication Networks. Springer, Heidelberg (2014)
15. Chapra, S.C., Raymond, P.C.: Numerical Methods for Engineers. McGraw-Hill, New York (2012)

Secondary User QoE Enhancement Through Learning Based Predictive Spectrum Access in Cognitive Radio Networks

Anirudh Agarwal$^{(\boxtimes)}$, Shivangi Dubey, Ranjan Gangopadhyay,
and Soumitra Debnath

Department of ECE, The LNM Institute of Information Technology, Jaipur, India
alanirudh@gmail.com, shivangidbl@gmail.com,
ranjan_iitkgp@yahoo.com, soumitra_deb@yahoo.com

Abstract. Quality of experience (QoE) of a secondary spectrum user is mainly governed by its spectrum utilization, the energy consumption in spectrum sensing and the impact of channel switching in a cognitive radio network. It can be enhanced by prediction of spectrum availability of different channels in the form of their idle times through historical information of primary users' activity. Based on a reliable prediction scheme, the secondary user chooses the channel with the longest idle time for transmission of its data. In contrast to the existing method of statistical prediction, the use and applicability of supervised learning based prediction in various traffic scenarios have been studied in this paper. Prediction accuracy is investigated for three machine learning techniques, artificial neural network based Multilayer Perceptron (MLP), Support Vector Machines (SVM) with Linear Kernel and SVM with Gaussian Kernel, among which, the best one is chosen for prediction based opportunistic spectrum access. The results highlight the analysis of the learning techniques with respect to the traffic intensity. Moreover, a significant improvement in spectrum utilization of the secondary user with reduction in sensing energy and channel switching has been found in case of predictive dynamic channel allocation as compared to random channel selection.

Keywords: Machine learning · Dynamic spectrum access · Prediction · Spectrum utilization · Channel switching

1 Introduction

With the ever-increasing need for spectral resources, it becomes necessary for a secondary user (SU) to smartly and efficiently access the resources of idle channel primary radio systems without creating harmful interference to the licensed users, which is possible through Cognitive Radio (CR) technology. However, random spectrum sensing by an SU can result in a bad channel selection, as the channel might be heavily used by the existing primary user (PU) during sensing. Moreover, it might result into multiple unnecessary channel switch which would create delay in the SU data transmission [1]. Therefore, conservation of spectrum sensing energy and reduction in channel switching become really important for efficient dynamic spectrum access

© ICST Institute for Computer Sciences, Social Informatics and Telecommunications Engineering 2016
D. Noguet et al. (Eds.): CROWNCOM 2016, LNICST 172, pp. 166–178, 2016.
DOI: 10.1007/978-3-319-40352-6_14

(DSA) and better QoE of SU, thereby improving the spectrum utilization of a CR user (terms, CR user and SU, are used interchangeably in this paper). This in turn suggests the need of spectrum prediction before sensing and channel allocation, where SU predicts the primary activity and sense only if the primary predicted state is idle. In this way, SU would perform prediction based sensing, starting with the channel having longest idle time and if the predicted state is busy, would switch to the next longer idle time channel for improved opportunistic spectrum access.

Machine learning (ML) proves to be a powerful tool for a CR system opening up versatile DSA applications viz. spectrum sensing, spectrum occupancy modeling, spectrum prediction, traffic pattern prediction, spectrum scheduling etc. [2]. The primary advantage of ML over other statistical models is that it does not require a-priori knowledge of the distributions under consideration. In the context of CR, ML techniques are generally used for signal classification, feature extraction, spectrum prediction [3–5] etc. For CR, mainly artificial neural networks (ANN) and Support Vector Machines (SVM) have been investigated in case of supervised ML [6]. But application specific work in reference to DSA and channel allocation based on learning has not been explored in sufficient detail in the existing literature. Moreover, most of the mentioned papers are restricted to one possible traffic model only.

Tumuluru et al. [7] has done the spectrum prediction based on MLP in Poisson traffic and has shown some improvement in SU spectrum utilization and reduction in sensing energy but the essence of channel switching has not been considered. In this paper, we have analyzed the performance of three supervised ML techniques e.g. artificial neural network based Multilayer Perceptron, Support Vector Machines with Linear Kernel (LSVM) and SVM with Gaussian Kernel (GSVM), for the prediction of primary activity as governed by several well known network traffic models namely, Poisson, Interrupted Poisson (IP) and Self-similar (SS) traffic. These traffic models reasonably capture the traffic characteristics that exist in most of the types of the wireless networks. The performance analysis of the prediction techniques is done in accordance with the statistical variation of the primary user data traffic. ML technique with highest prediction accuracy in estimating the average length of OFF period of primary in a single channel, is chosen for predicting the primary activity in multiple channel scenario. SU spectrum utilization, spectrum sensing energy and channel switching have been taken into account for claiming better QoE of CR user.

The paper is organized as follows: Sect. 2 discusses about the system model and methodology. A brief description of traffic models and various ML techniques are provided in Sect. 3. The performance analysis and the results are discussed in Sect. 4. Finally, Sect. 5 concludes the paper.

2 System Model and Methodology

In our model, for simplicity, we assume initially, one SU, targeting a single channel of PU whose channel state information (CSI) is used by the system for learning based prediction of future primary activity with three ML techniques in different network traffics. This model is subsequently used in the scenario of multiple PU channels where the CR would utilize the information of the average length of the OFF period $(\overline{L_{OFF}})$ of

PU activity, predicted by an accurate ML technique. This process is repeated for the available channels and the channel having longest $\overline{L_{OFF}}$ is chosen, thereby resulting in an improved prediction based CR sensing-transmission strategy for DSA. It is also assumed that the sensing information is highly accurate. The channel allocation methodology has been explained by the flow graph in Fig. 1.

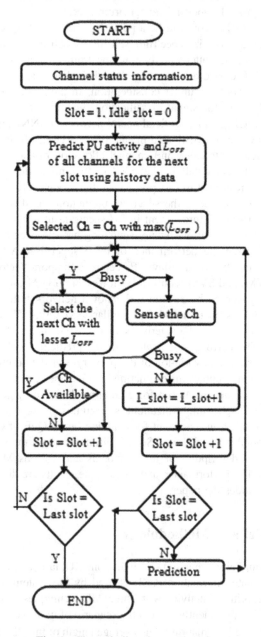

Fig. 1. Channel allocation methodology

3 Data Traffic Models and ML Prediction Techniques

3.1 Data Traffic Models

In this work, we have considered three different traffic models, i.e. Poisson traffic, Interrupted Poisson (IP) traffic and Self-similar (SS) traffic, for characterizing the statistics of a PU channel. Poisson traffic is one of the most widely used traffic model used to model the traditional voice data. The IP traffic is a good representation of data found in computers, e-mails, etc., i.e., there is heavy traffic for some time and then no traffic for some time. Self-similarity is a well known feature in the Internet traffic. SS traffic is characterized by long range dependence of traffic, burstiness and high correlation over varying time scales.

3.2 ML Prediction Techniques

In this sub-section, a brief description of three ML prediction techniques used in the present study is provided.

3.2.1 Multilayer Perceptron Neural Network Based Prediction

An MLP is a feed-forward network of simple neurons called *perceptrons*. It consists of three or more layers (an input and an output layer with one or more hidden layers) of nodes in a directed graph. Each node excluding the nodes at the input layer is a computing unit i.e. perceptron. The perceptron computes single output from multiple real-valued inputs by forming a linear combination of their input weights and then putting the output through some nonlinear activation function. Mathematically, this can be written as:

$$y = \varphi\left(w^T x + b\right) \tag{1}$$

where y is the output vector, φ is the activation function, w is the weight vector, x is the input vector and b is the bias.

The activation function is often chosen to be the logistic sigmoid or tangent hyperbolic. The number of hidden layers and the number of neurons in each layer vary according to the application [7].

MLP networks are typically used in supervised learning problems, that can be further solved by the *back-propagation algorithm (BPA)* [8]. The signal flow graph for MLP is presented in Fig. 2, where the circles in blue, excluding the input layer, denote neurons.

For the implementation of the MLP algorithm, we have used MATLAB Neural Network Toolbox with 4 inputs in the input layer, two hidden layers consisting of 15 and 20 neurons respectively and one neuron in the output layer. We have used Tangent Sigmoid and Purelin as the activation functions respectively for the hidden layer and the output layer neurons. Both learning rate and momentum constant for the gradient descent method in BPA are taken as 0.2 and 0.9 respectively. These values are chosen after rigorous cross-verification for optimality.

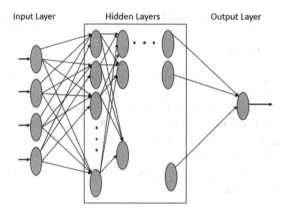

Fig. 2. Signal flow graph for MLP

3.2.2   Linear and Gaussian Support Vector Machines Based Prediction

SVM is a maximum margin discriminative classifier which means that, it learns a decision boundary that maximizes the distance between samples of the two classes, given a kernel. The distance between a sample and the learned decision boundary can be used to make the SVM a "soft" classifier. In the present implementation, we have used linear kernel based SVM and Gaussian kernel based SVM. The training feature and response vectors can be represented as $Z = (T_i, x_i)$ where $T_i \in \{-1, 1\}$. The two classes are separated by a hyperplane denoted as $H: w^T x + b = \varepsilon$, where w is the weight vector, b is the bias and $\varepsilon = \sum_{i=1}^{m} \varepsilon_i$ is a slack variable vector whose 1-norm is the penalty term. The hyperplanes which separate the two classes are given by:

$$T_i = \left\{ \begin{array}{ll} 1 \ when \ w^T x_i + b > 1 - \varepsilon \\ -1 \ when \ w^T x_i + b < -1 + \varepsilon \end{array} \right\} \tag{2}$$

With a soft margin, the optimization problem for the SVM can be defined as follows:

$$\min_{(w,b,\varepsilon) \in R^{n+1+m}} \left(\frac{1}{2} \|w\|^2 + C \sum_{i=1}^{m} \varepsilon_i^2 \right) \tag{3}$$

$$s.t. T_i (w^T x_i + b) > 1 - \varepsilon,$$
$$for \ i = 1, 2, \ldots, m.$$

where $C > 0$ is a regularization parameter that balances the weight of the penalty term $\sum_{i=1}^{m} \varepsilon_i$ and the margin maximization term $\frac{1}{2} \|w\|^2$ [9].

For training and testing purposes, we have utilized the widely used software tool i.e. LIBSVM [10], integrated and compiled in MATLAB, where the algorithm is iterated until the minimum tolerance value (taken as 0.0001 in this work) is achieved.

4   Performance Evaluation, Results and Discussion

For evaluating the performance of the prediction based technique for the reliable prediction of $\overline{L_{OFF}}$ estimate for different traffic scenarios, we have used two performance measures i.e. probability of error in predicting the busy state, $Prob_{err}(busy\ state)$, and the mean square error in predicting the average length of the OFF duration of primary activity, $MSE\ of\ \overline{L_{OFF}}$. The significance of $Prob_{err}(busy\ state)$ lies in the interference caused by the CR user to PU. More $Prob_{err}(busy\ state)$ would lead to more tendency of CR user to violate the interference constraint. However, the overall utilization efficiency of SU with better channel allocation strategy, is governed by $MSE\ of\ \overline{L_{OFF}}$. Moreover, the prediction accuracy and $MSE\ of\ \overline{L_{OFF}}$ for all the ML techniques are calculated and depicted in Table 1. The traffic intensity for an ON-OFF random data traffic is defined as:

$$\rho = \frac{T_{ON}}{T_{ON} + T_{OFF}} \tag{4}$$

where T_{ON} and T_{OFF} are respectively the average time for which the primary user is busy and idle.

Table 1. Comparison of ML Techniques for different training lengths and traffic intensities in 3 traffic scenarios.

Percentage of training data	Traffic intensity	ML technique	Poisson		IP		SS	
			MSE	PA	MSE	PA	MSE	PA
30 %	0.5	MLP	2.032e-1	91.558	2.034e-1	87.040	1.195e-1	85.104
		LSVM	8.828e-7	92.850	2.233e-6	87.470	5.817e-7	85.398
		GSVM	2.012e-1	91.890	2.055e-1	87.290	6.597e-7	85.368
	0.8	MLP	5.360e-4	92.645	5.223e-3	89.950	2.140	86.746
		LSVM	7.382e-8	92.931	1e-7	90.003	4.082e-7	87.068
		GSVM	1.088e-7	92.919	7.330e-7	89.988	8.850e-7	87.054

The mean-square error in predicting the average length of the OFF duration of primary is calculated as:

$$MSE\ of\ L_{OFF} = \frac{1}{N}\sum_{n=1}^{N}\left(\overline{L_{OFF-pre}}(n) - \overline{L_{OFF-org}}(n)\right)^2 \tag{5}$$

where N is the total number of simulation intervals, $\overline{L_{OFF-pre}}(n)$ is the average length of OFF duration of PU activity in predicted data at the nth simulation interval and similarly, $\overline{L_{OFF-org}}(n)$ is the average length of OFF duration of PU activity in original data at the nth simulation interval. The other performance metric, the prediction accuracy (PA), is defined as:

$$PA = (1 - P_e) \times 100 \tag{6}$$

where P_e is the overall probability of error, i.e. when the busy state is predicted as idle and vice versa, in the prediction by an ML technique. For prediction analysis, 30 % of the primary user data is used for training while the rest of the data is utilized for testing the trained model. The total number of traffic slots for primary data is taken as 50000 in this work. Moreover, as the characteristics of the particular data traffic might change with time, we have evaluated the performance of all the considered parameters after averaging over sufficient number (50 in this work) of simulation intervals.

In a multi-channel scenario, two cases have been analyzed for the performance, i.e. random sensing, where CR user randomly senses the channels and check the status at every slot, and prediction based sensing, where CR user predicts the channel status of all the channels with the help of slot history information. In the second case, the channels are prioritized in the decreasing order of their $\overline{L_{OFF}}$, then the CR user starts priority based sensing of the channel among the channels with predicted idle status. Moreover, we have considered 10 licensed channels of different traffic characteristics, as described in Table 2, and a single SU in the slotted-time mode. It is assumed that at every slot, SU has the sensing information of all the channels.

Table 2. Different Channels for Primary User System

Channel number	Mean inter arrival time	Traffic intensity
1	22	0.760
2	20	0.500
3	18	0.700
4	16	0.625
5	12	0.600
6	10	0.700
7	10	0.600
8	20	0.500
9	25	0.525
10	22	0.500

The QoE of SU is characterized using three performance measures:

1. *Spectrum Utilization Improvement (α_{imp})*

   The spectrum utilization (α) is defined as the fraction of slots in the system over a finite duration of time, for which number of slots are detected as idle by the CR user. So, spectrum utilization improvement (in %) due to prediction is given by

$$\alpha_{imp} = \frac{\alpha_{PS} - \alpha_{RS}}{\alpha_{RS}} \times 100 \tag{7}$$

where α_{RS} and α_{PS} are respectively the spectrum utilizations by CR in the case of random sensing and prediction based sensing.

2 *Reduction in Channel Switching (β_{red})*

The percentage reduction in channel switching due to spectrum prediction can be expressed as

$$\beta_{red} = \frac{\beta_{RS} - \beta_{PS}}{\beta_{RS}} \times 100 \tag{8}$$

where β_{RS} and β_{PS} are number of times CR has to switch the channel in case of random sensing and prediction based sensing respectively.

3 *Reduction in Sensing Energy (ξ_{red})*

Sensing energy is reduced in case of prediction based sensing, because in random sensing, CR user has to sense all the slots while through prediction, it senses only when the state of the channel is predicted to be idle. Here, we have assumed that one unit of sensing energy is required to sense one slot. The percentage reduction in sensing energy is given by

$$\xi_{red} = \frac{\xi_{RS} - \xi_{PS}}{\xi_{RS}} \times 100 \tag{9}$$

where ξ_{RS} and ξ_{PS} denote the product of unit sensing energy and corresponding number of idle slots sensed by CR in case of random sensing and prediction based sensing.

Figures 3 and 4 depict the probability of error in predicting the busy state of the primary for different traffic types utilizing the three learning schemes. It can be observed that for all types of data traffics, the probability of error in predicting the busy state decreases as we increase the traffic intensity. As ρ increases, the number of times

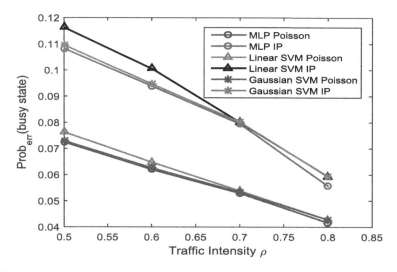

Fig. 3. Prob$_{err}$(busy state) vs. ρ for different data traffic using different ML techniques.

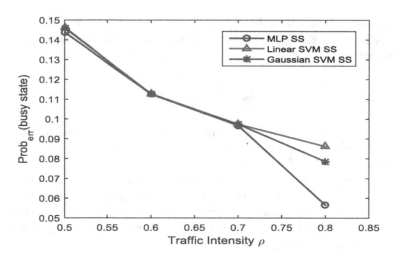

Fig. 4. Prob$_{err}$(busy state) vs. ρ for SS traffic using different ML techniques.

the channel being idle tends to decrease. It may be noted that MLP prediction technique performs slightly better than the other two in this case.

Figure 5 shows the variation of *MSE of* $\overline{L_{OFF}}$, for Poisson and IP traffic against the traffic intensity, ρ. The decreasing nature is attributed from the fact that, there are less number of transitions from busy state to idle and vice-versa with increase in ρ. This leads to more dependency of future states on the present and previous states, thereby suggesting a decrease in prediction error and an improvement in the prediction

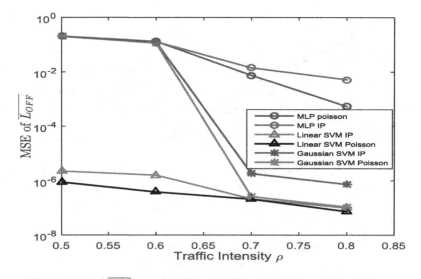

Fig. 5. MSE of $\overline{L_{OFF}}$ vs. ρ for different traffic using different ML techniques.

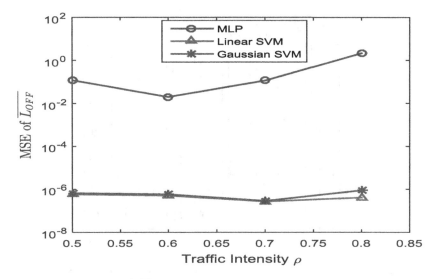

Fig. 6. MSE of $\overline{L_{OFF}}$ vs. ρ for SS traffic using different ML techniques.

accuracy. However, in this case, LSVM based prediction turns out to be the best in terms of $\overline{L_{OFF}}$.

With the same reasoning, similar pattern was expected for bursty SS traffic in Fig. 6. But it is observed that the prediction is not so accurate for high traffic intensity. This may be due to the heavy burstiness and strong OFF period correlation in the data traffic. Nevertheless, LSVM is found to perform uniformly and reliably for the SS traffic.

Table 1 provides a comparison of the performance of ML techniques under various simulated conditions in terms of the MSE of $\overline{L_{OFF}}$ and prediction accuracy of the algorithm under consideration. It is clearly observed that for any data traffic, LSVM outperforms both the other techniques.

Figures 7 and 8 show the comparison of random sensing and prediction based sensing on the basis of improvement in spectrum utilization and reduction in channel switching respectively. It can be seen in Fig. 7 that the number of times a channel used by a CR user for a duration of time, is far more in the case of prediction based sensing than that in random sensing. This is due to the fact that through prediction, a CR can find more idle slots in a channel thereby utilizing it optimally. Moreover, the percentage of channel switching would also be decreased as shown in Fig. 8. In prediction based sensing, CR remains in the best predicted channel, i.e. the channel with longest $\overline{L_{OFF}}$, for more time and switches to the next prioritized channel only when the state of the present channel is predicted to be busy, unlike in random sensing where CR senses any channel randomly and is supposed to switch, if it is found to be busy in sensed slot.

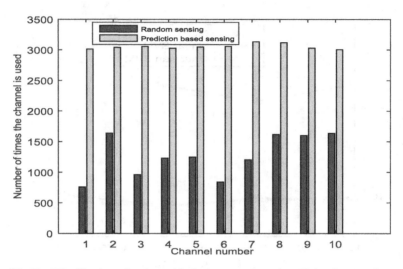

Fig. 7. SU utilization of a channel in both cases of sensing (Color figure online)

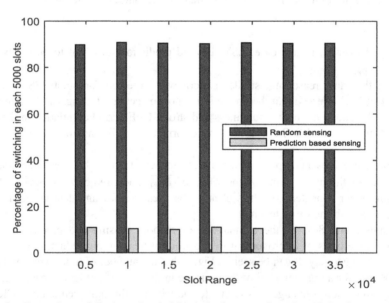

Fig. 8. Comparison of % channel switching in both cases of sensing (Color figure online)

In Table 3, an improvement in spectrum utilization due to prediction ranging from around 88 % to 146 % considering different number of channels, has been found. Channel switching is reduced by more than 83 %. Finally, Table 4 shows the reduction in sensing energy for a channel with $\rho = 0.5$ and for different mean inter-arrival times. Energy is saved in the case of prediction because there is no sensing operation when the predicted status of the slot is busy.

Table 3. Comparison of α_{imp} and β_{red} due to prediction for different number of channels

Number of channels	α_{imp} (%)	β_{red} (%)
3	88.5394	87.4291
5	124.0066	85.1706
7	146.3669	83.9832
9	145.1946	86.7339
10	139.7345	88.1905

Table 4. ξ_{red} in a channel due to prediction for $\rho = 0.5$ and different mean inter-arrival time

Mean inter-arrival time	ξ_{red} (%)
10	57.1265
12	55.9721
14	55.0063
16	54.2976
18	54.1090
20	53.5461
22	53.2832

5 Conclusion and Future Work

In this work, the importance of machine learning spectrum prediction is highlighted in the context of CR for efficient DSA. LSVM is found to consistently predict the primary $\overline{L_{OFF}}$ in data traffics with high accuracy. This technique is further used in multi-channel CR scenario and prediction based sensing is done over the channels which are prioritized on the basis of decreasing $\overline{L_{OFF}}$. A significant improvement in spectrum utilization, decrease in channel switching frequency and reduction in spectrum sensing energy have been found thereby providing improved quality of experience to CR user for opportunistic spectrum access.

However, this preliminary work needs further extension for multiple SU's without assuming perfect detection. Moreover, some advanced prediction algorithms based on deep belief need to be exploited for further QoE enhancement of the SU in multiple channel radio access in a CR system.

Acknowledgement. This work has been carried out as a part of the project "Mobile Broadband Service Support over Cognitive Radio Networks", sponsored by Information Technology Research Academy (ITRA), Department of Electronics and Information Technology (DeitY), Ministry of Communications & IT, Govt. of India.

References

1. Hoyhtya, M., Pollin, S.: Improving the performance of cognitive radios through classification, learning, and predictive channel selection. Adv. Electron. Telecommun. **2**, 28–38 (2011)
2. Azmat, F., Chen, Y., Member, S., Stocks, N., Mar, N.I.: Analysis of Spectrum Occupancy Using Machine Learning Algorithms, 0–22
3. Xing, X., Jing, T., Cheng, W., Huo, Y., Cheng, X.: Spectrum prediction in cognitive radio networks. IEEE Wirel. Commun. **20**, 90–96 (2013)
4. Lee, J., Park, H.-K.: Channel prediction-based channel allocation scheme for multichannel cognitive radio networks. J. Commun. Networks. **16**, 209–216 (2014)
5. He, A., Member, S., Bae, K.K., Newman, T.R., Gaeddert, J., Member, S., Kim, K., Menon, R., Morales-tirado, L., Neel, J.J., Zhao, Y., Reed, J.H., Tranter, W.H., Fellow, L.: Survey of artificial intelligence for cognitive radios. IEEE Trans. Vech. Tech. **59**, 1578–1592 (2010)
6. Bkassiny, M., Member, S., Li, Y., Member, S.: A Survey on Machine-Learning Techniques in Cognitive Radios, pp. 1–53
7. Tumuluru, V.K., Wang, P.W.P., Niyato, D.: A Neural Network Based Spectrum Prediction Scheme for Cognitive Radio (2010)
8. Rumelhart, D.E., Hinton, G.E., Williams, R.J.: Learning representations by back-propagating errors. Cog. mod. **5**, 3 (1988)
9. Burges, C.J.C.: A tutorial on support vector machines for pattern recognition. Data Mining Knowl. Discov. **2**(2), 121–167 (1998)
10. http://www.csie.ntu.edu.tw/∼cjlin/libsvm

Sensing Based Semi-deterministic Inter-Cell Interference Map in Heterogeneous Networks

Fatima Zohra Kaddour, Dimitri Kténas$^{(\boxtimes)}$, and Benoît Denis

Cea-Leti, Minatec Campus, 17 Rue des Martyrs, 38054 Grenoble, France
dimitri.ktenas@cea.fr

Abstract. This paper describes a low computational complexity semi-deterministic Inter-Cell Interference (ICI) map construction procedure. The built Interference Map (IM) gives the ICI level at each pixel of a two-dimensional area, based on an initialization map and ICI levels measured by collaborative User Equipments (UEs). In a first step, the initialization map is obtained with an analytical location-dependent ICI prediction model based on the Poisson Point Process (PPP) framework, where a priori deterministic information about the indoor/outdoor UE status can be injected. The analytical interference map is then updated following a self-learning approach, after spatially interpolating the gap sensed by the UEs with respect to analytical predictions in their visited positions. Two conventional spatial interpolation techniques are thus considered under regular and irregular sensing grids: Inverse Distance Weighting (IDW) and kriging, where exponential and Von Kàrmàn variograms are used. In order to show the benefits of the IM initialization, the performance is compared to that of traditional approaches (i.e., direct spatial interpolation of the ICI measured values), while varying the density of sensing positions.

Keywords: Inter-Cell Interference (ICI) · Interference Map (IM) · Self-learning · Spatial interpolation · Kriging · Long Term Evolution

1 Introduction

The emerging wireless communication standard Long Term Evolution Advanced (LTE-A) and beyond introduces significant technological enhancements to meet the ambitious requirements set by the Third Generation Partnership Project (3GPP) in terms of high data rate and high spectral efficiency.

To satisfy this continuous growth, the network is brought closer to the users by making the cells smaller [1]. However, a cell densification generates a high Inter-Cell Interference (ICI) level that should be mitigated in order to preserve high performance and satisfy the User Equipment (UE) Quality of Service (QoS). ICI estimation has been identified as a priority research topic in the recent literature. Some studies assume that the ICI follows a Gaussian distribution (by invoking the central limit theorem) and derive the ICI statistical parameters accordingly, as given in [2]. However, higher layer algorithms (e.g., handover, resource

© ICST Institute for Computer Sciences, Social Informatics and Telecommunications Engineering 2016
D. Noguet et al. (Eds.): CROWNCOM 2016, LNICST 172, pp. 179–191, 2016.
DOI: 10.1007/978-3-319-40352-6_15

scheduling, coordinated base stations' transmission activity...) usually require more accurate channel estimations/predictions. Accordingly, current researches focus on the construction of refined Interference Maps (IMs) that aim at representing the ICI level as a function of the UE location in 2D spaces. Such maps are even more welcome and beneficial in Heterogeneous Networks (HetNets), which consist of different layers (i.e., macro and micro cells).

IMs can be designed using: (i) fingerprinting techniques (deterministic approach), or (ii) (location-dependent) ICI prediction models (e.g., stochastic approach). The high resolution fingerprinting approach consists in collecting real ICI levels with a drive test to measure the Received Signal Strength (RSS) at each location of the considered area [3,4]. This method needs a pre-training step, which may be too greedy in terms of time, or alternatively, blind crowd-sourcing (i.e., performing calibration "on the fly" as UEs physically sense and report their ICI conditions). The latter method requires that a minimum amount of field measurements are collected to converge properly and ensure uniform coverage. Thus it may introduce significant latency in the system (e.g., before being able to take any reliable IM-based decision). To overcome this drawback, low resolution fingerprinting (where the ICI levels can also be collected from collaborating UEs that send their locations and RSS measurements to the eNB) is exploited by estimating the ICI level of the non-visited locations using well-known spatial interpolation techniques [5,6]. The performance thus depends mainly on the accuracy of the used spatial interpolation technique. However, it may also suffer from similar latency issues under too sparse sensing. Moreover, both high and low resolution fingerprinting approaches can just account for an instantaneous picture of the ICI conditions (most likely, corresponding to distant sensing epochs) but not for varying base stations activity over time. On the contrary, ICI prediction models derive an analytical expression for the average ICI level as a function of UEs locations based on a priori stochastic assumptions (typically, regarding the spatial distribution of fixed interfering base stations) [2,7]. They represent a flexible alternative to cope with HetNets complexity. This may come at the expense of a tolerated mapping accuracy degradation (e.g., local prediction biases), somehow resulting from spatial interference averaging or even from deliberate approximations made for the sake of tractability (e.g., about distance-dependent integration bounds) in a few particular proposals [8].

In this study, we introduce a new mixed semi-deterministic method to dynamically build the IM. First of all, interference values are predicted by a stochastic location-dependent ICI analytical model proposed in our previous work [8] in a 2D area. Then prediction errors, which account for the deviation between the previous IM predictions and the deterministic UE measurements at the visited positions (i.e., at the discret sensing positions), are spatially interpolated to get a global prediction error surface. The latter can be further used to correct and update the final IM in a self-learning way (See Fig. 1). In comparison with classical approaches, where the UE measurements are directly interpolated, the proposed solution is expected to converge more rapidly[1] towards the actual ICI conditions under spatially sparse sensing conditions. In the two compared IM approaches, conventional techniques are considered for the respective spatial interpolation steps, namely the Inverse Distance Weighted (IDW) technique and the Kriging technique with both exponential and Von Kàrmàn variogram models [9,10].

Numerical results are provided for illustration, based on deterministic Ray-Tracing radio simulations in a representative urban scenario and for different rates of regularly and irregularly drawn active UEs.

Overall, the novelty of this study lies in (i) the practical description of the IM initialization out of the predictive model in [8], while introducing local partitioning of the 2D prediction space around physical sensing positions, (ii) the coupling of this IM initialization with additional deterministic Indoor/Outdoor information and finally, (iii) the dynamic IM update depending on sensing positions through the spatial interpolation of prediction errors.

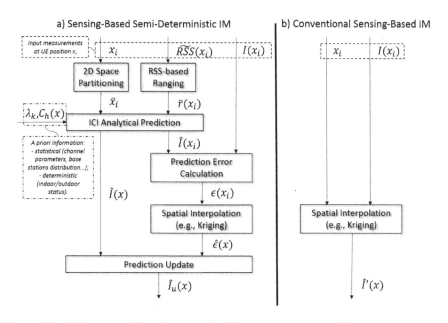

Fig. 1. Block diagrams of both proposed (a) and conventional (b) sensing-based IM approaches.

The paper is organized as follows. In Sect. 2 we define the system model. The semi-deterministic interference map construction steps are detailed in Sect. 3.

---

[1] The so-called "convergence speed" is just intended here in terms of the overall amount of collected UE measurements, indifferently of their acquisition conditions (i.e., synchronous measurements at distributed static UEs, asynchronous measurements under UE(s) mobility).

Section 4 summarizes the simulation parameters and the numerical results. Finally, Sect. 5 concludes the paper.

2   System Model

Our goal is to build a semi-deterministic interference map, where the IM is first initialized according to an analytical location-dependent ICI estimation model. Then a self-learning procedure allows an update of the IM according to the prediction errors obtained regarding the ICI levels sensed by the UEs in the visited positions. The analytical location-dependent ICI estimation model is based on stochastic approach, where in K-tier HetNets, the network's nodes are modeled according to an independent Poisson Point Process (PPPs). Let K be the total number of tiers in the network. Thus, the nodes positions of each tier k are modeled with an independent homogeneous PPP $\varphi_k = \{y_1, y_2, \dots\}$, $k = 1, \dots, K$ of intensity λ_k.

The Orthogonal Frequency Division Multiple Access (OFDMA) technique is considered for resource allocation. Thereby, the intra-cell interference is canceled and the total interference received by a user u on the downlink is caused only by the nodes transmitting on the same frequency band. However the macro and small cell base stations can be allowed to transmit in the same frequency band (i.e., co-channel deployment), or in separate frequency band. When the macro and small cell BSs operate in different frequency bands, the ICI is called a co-tier interference, which means that the interference is generated between network elements on/belonging to the same layer. The macro cell users and the small cell users are interfered only by the macro cell base stations and the small cell base stations, respectively. Unlike the separate frequency band allocation, in a co-channel deployment, the users are interfered by all the network nodes without distinction between tiers. In this case, the interference is called a cross-tier interference.

In [8], both co-tier and cross-tier location-dependent ICI estimation model are derived. However, in this paper, we focus only on the co-tier interference. Without loss of generality, the same methodology can be used to apply the semi-deterministic approach in a cross-tier interference map construction.

The analytical model considers various sources of channel variations. A standard path loss function that depends on the distance $r_j = \|y_j\|$ between the user[2] and the interfering base station j is used. First (of all) the path loss function is classically expressed by $Pl(r_j) = l\ r_j^{-\gamma}$, where l and $\gamma > 2$ are respectively the reference (constant) path loss and the path loss exponent. Their values depend on the considered scenario (i.e., macro or small cell scenario), and can be instantiated by using the corresponding model specified in [11,12]. In addition, Rayleigh fast fading effects of the form $h \sim \exp(1)$ are taken into account. In [13], it has been shown that shadowing has a considerable impact on the ICI level. Thus, in our study a log-normal shadowing $\chi = 10^{\frac{X}{10}}$ such that $X \sim \mathcal{N}(\mu, \sigma)$ is

---

[2] For straightforward mathematical analysis, the user is assumed to be located at the origin.

considered while deriving the ICI estimation model. μ and σ are, respectively, the shadowing mean and standard deviation in dB. We depict as μ_k and σ_k the specific shadowing parameters associated with the kth tier. Based on these radio environment parameters, a location-dependent analytical model of the downlink ICI level is given by:

$$\hat{I}(x) = \left(\frac{\pi \lambda_k^{(e)} \Phi(r, R_{ob}, 4)}{2\sqrt{\eta}\operatorname{erfc}^{-1}(0.5)} \right)^2 \tag{1}$$

where $\Phi(r, R_{ob}, 4) = \arctan(R_{ob}) - \arctan(r)$, r is the true distance between the UE and the first interfering eNB, $R_{ob} \simeq 3r$ is the radius of the observing area (i.e., the distance beyond which the interfering signal is considered negligible) and $\eta = \frac{1}{P_k l}$, where p_k is the node's transmission power of tier k. For more details regarding this model, please refer to [8].

Using the latter equation as a practical prediction tool in a real system (while assuming that eNB positions are unknown and/or non-disclosed to the UE), for any particular occupied position x_i of the 2D area of interest, $r(x_i)$ could be practically estimated as $\widetilde{r}(x_i)$ relying on the RSS perceived from the closest interfering eNB, $\widetilde{RSS}(x_i)$. Then the set of neighboring positions $\bar{x}_i$ around each occupied position x_i defines a region where one can assume in first approximation that $\widetilde{r}(x_i)$ is an acceptable common distance to the closest interfering eNB, $\forall x \in \bar{x}_i$. For instance, those regions, which do not necessarily coincide with eNB coverage cells, can be defined in a Voronoi sense around the sensing points (ensuring at least that any point in the region is closer to the sensing point leading to the common r approximation). Accordingly, the entire 2D area can be partitioned into small sub-areas of constant nominal ICI levels depending on the physically visited positions, as illustrated within the simplified 1D scenario of Fig. 2. Even under relatively low sensing points density, anyway much lower than the density simulated hereafter while evaluating the proposed solution, the assumption of a "common closest interfering eNB per region" has been reasonably validated by additional simulations (not shown here due to the limited number of pages). Extra deterministic a priori information, $C_h(x) = \{Indoor, Outdoor\}$, regarding the environment status at any position x can be used to account for lower perceived ICI levels in indoor zones, by adding an extra power penalty on top of the initial analytical prediction, while assuming Non Line of Sight (NLoS) outdoor channel parameters by default. Overall, the combination of analytical predictions, space partitioning and deterministic information leads to a piece-wise IM initialization $\hat{I}(x)$, $\forall x$ in the 2D area of interest.

3 Inter-Cell Interference Map Construction and Updates

So as to get more reliable information, the previous IM initialization is updated following a self-learning procedure. For this sake, one can perform either a statistical shape analysis such as the Procrustes analysis, which defines the required transformations to be applied to the initial map based on the observed shape deviations (i.e., scaling, rotation...) [15], or more simple interpolation techniques. As defined in the 3GPP standard, the active UEs report their measurements to the eNB. Accordingly, the observed "gap" between the measured ICI level and the estimated level can be evaluated at all the visited sensing points x_i, as illustrated in Fig. 2. Since the active UEs can be sparsely (i.e., under low deployment density) and/or non-uniformly distributed in the geographic area of interest, the IM is thus herein updated by spatially interpolating the perceived prediction errors as follows:

$$\epsilon(x_i) = \hat{I}(x_i) - I(x_i) \tag{2}$$

where, $\hat{I}(x_i)$ and $I(x_i)$ are respectively the estimated and the observed ICI levels at the sensing position x_i.

Finally, the corrected ICI value $\hat{I}_u(x)$ is defined by:

$$\hat{I}_u(x) = \hat{I}(x) - \hat{\epsilon}(x) \tag{3}$$

where, $\hat{I}(x)$ and $\hat{\epsilon}(x)$ are respectively the theoretical ICI level obtained by the stochastic location-dependent ICI model over partitioned regions and the interpolated gap at any position (i.e., including non-visited points).

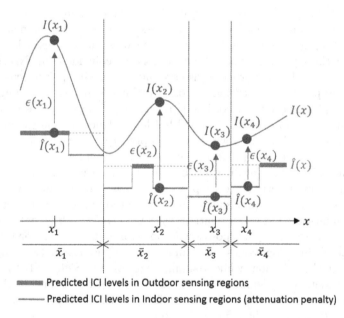

Predicted ICI levels in Outdoor sensing regions
Predicted ICI levels in Indoor sensing regions (attenuation penalty)

Fig. 2. Analytical ICI prediction errors based on visited UE sensing positions.

So as to estimate the gap $\hat{\epsilon}(x)$ at any location x, we consider classical spatial interpolation techniques that are widely used in the context of radio environment cartography [5,16,17]. More specifically, we focus on the well-known IDW and Kriging methods. These two approaches rely on a set of neighborhood observed gap values, through weighted linear combinations.

For IDW, the ICI gap values at unobserved positions are expressed as:

$$\hat{\epsilon}(x) = \sum_{i=1}^{N} \lambda(x_i)\epsilon(x_i) \qquad (4)$$

where

$$\lambda(x_i) = \frac{d(x, x_i)^p}{\sum_{i=1}^{N} d(x, x_i)^{-p}} \qquad (5)$$

is the Shepard's weighting function, given according to the distance between the location of the unobserved gap value x and the location of the observed gap values x_i, N is the number of considered observed gap values, and p is a real positive power (usually p is set to 4 or 6).

Unlike the IDW technique, the Kriging weights are based on the spatial correlation between the considered observations, which can be reflected by the covariance matrix C such that,

$$C(\tau)\lambda = C(0) \qquad (6)$$

where $C(\tau)$ is the covariance matrix between the sensing positions x_i, $\tau = d(x_i, x_j)$ with $1 \le i \le N$ and $1 \le j \le N$. and $C(0)$ is the covariance vector between the interpolation position x and the sensing positions x_i.

An alternative way to describe the spatial relationship of the observed values is the variogram (or semi-variogram). Contrarily to the covariance matrix, the variogram can be calculated even if the mean of the observation values is unknown. Thus, the variogram is more convenient to describe the spatial relationship inside a data set and is described as,

$$\gamma(\tau) = \frac{1}{2}Var[\epsilon(x_i) - \epsilon(x_i + \tau)] \qquad (7)$$

Under the hypothesis of first and second order stationarity, the semi-variogram is given as:

$$\gamma(\tau) = C(0) - C(\tau) \qquad (8)$$

The experimental semi-variogram is calculated from the given data and therefore a discrete and often irregularly sampled function. The experimental variogram can then be approximated through a continuous parameter model. The most frequently used in the radio environment cartography are:

- the exponential model defined as:

$$\gamma(\tau) = n + c(1 + \exp(\frac{-3\tau}{r_g})) \qquad (9)$$

With, n is the nugget, c is the sill and r_g is the range of the variogram (i.e. the correlation length).
- the Von Kàrmàn model defined as:

$$\gamma(\tau) = \frac{c}{2^{\nu-1}\Gamma(\nu)}(\frac{\tau}{r_g})^{\nu} K_{\nu}(\frac{\tau}{r_g}) \qquad (10)$$

where Γ is the gamma function, K_{ν} is the modified Bessel function of the second kind of order $0 \le \nu \le 1$ [18].

4   Simulation Parameters and Performance Analysis

In the following simulations, the active UEs' ICI measurements (i.e., at visited positions x_i), as well as the true generalized ICI level $I(x)\forall x$, are taken from a realistic interference map obtained by ray-tracing techniques [14] over a mixed Indoor/Outdoor urban area of 1 km². Both the IDW and the Kriging (using exponential and Von Kàrmàn variogram models) interpolation techniques are first compared. We consider that the sensing task (i.e., ICI measurements) is performed by active UEs. Thus, the number of sensing positions is proportional to the population density and the rate of active UEs in the area. Regular and irregular sensing positions are considered. We assume that the studied area is in the city of Paris (France) with a population density of 21564 pop. per km² [19], which are equipped with mobile phones that can collaborate to the sensing procedure during their active mode, with an active UE rate of 45 %.

As a first step, comparing the so-called "true" IM over the whole area (i.e., obtained by ray-tracing deterministic simulations) with that obtained through low resolution fingerprinting (i.e., using a spatial interpolation over the deterministic ICI values at the sensing positions) based on the IDW and the Kriging methods, we obtain the Cumulative Distribution Functions (CDFs) of ICI level estimation error illustrated in Fig. 3. From the latter, we notice that the ICI error obtained with regular sensing positions is close to that of irregular sensing positions, whatever the interpolation technique. At the 50-percentile, IDW generates an ICI error of 6 dB, whereas kriging generates an error of about 2 dB independently of the used variogram model. The better performance of the kriging method is due to the weighting strategy, which relies on data correlation. It is worth noting that using the exponential variogram model generates lower ICI prediction errors. However the performance gap obtained with the two variogram models is very small (in the order of 0.01 dB). Based on this numerical results, in the following we focus only on the kriging method with the exponential variogram.

Next, in order to show the benefits from IM initialization, we compare the performance of the new proposed algorithm (i.e., IM update procedure) with the conventional low resolution fingerprinting technique that directly interpolates the sensed ICI values (see Fig. 1). We just assume perfect knowledge of the distance to the closest eNB at any point of the map for simplicity, considering the relatively high spatial density of sensing UEs (see the discussion on space partitioning validity in Sect. 2). The comparison is based on the CDF of ICI error with respect to the "true" IM (i.e., obtained by ray-tracing deterministic simulations) (see Fig. 4). As said before, the type of sensing grid has a relatively low impact on the performance under the considered practical UE density. The ICI obtained with the regular and the irregular sensing positioning is indeed very small. However, the performance of the updated IM is higher. At 90-percentile, the ICI error obtained with the updated IM (i.e., using the theoretical IM at the initialization step) is about 5 dB, whereas the latter is about 9 dB in case of low resolution fingerprinting. The a priori information provided by both the analytical IM and the indoor/outdoor map used in the initialization step allows

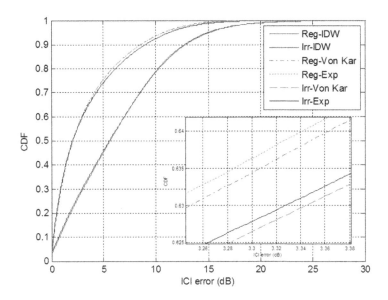

Fig. 3. CDF of ICI error in regular sensing positions

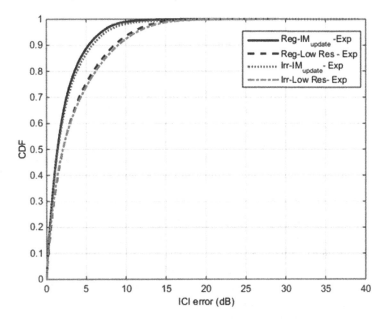

Fig. 4. CDF of ICI error: low resolution fingerprinting and updated IM

to smooth the observed gap values in comparison with the deterministic ICI values, while capturing even better local spatial correlation effects.

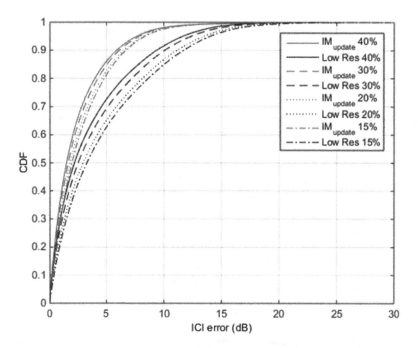

Fig. 5. CDF of ICI error: updated IM and Low resolution IM vs active UE rate (Colour figure online)

An active UE rate of 45 % may be high in practice and thus, not sufficiently realistic. Thus, the performance of the proposed updated IM is also studied with regards to different values in terms of active UEs rate. It is expected that under lower spatial density, the performances of both the updated IM and the low resolution fingerprinting techniques will decrease. We assume an irregular sensing positioning, where the updated IM and the low resolution fingerprinting IM are both generated with the kriging method based on the exponential variogram. Figure 5 shows the CDF of the corresponding ICI errors. The red and black curves represent the CDFs of ICI error obtained with the updated IM (i.e., the proposed algorithm) and the low resolution IM, respectively. The active UE rates are shown with different line types. We can notice that when the active UE rate decreases, the ICI error increases independently of the used technique for the IM construction. However, the updated IM generates a low ICI error compared to the low resolution fingerprinting technique. In addition, the gap introduced when the active UE rate decreases is higher in case of low resolution fingerprinting compared to the updated IM. At 90-percentile, when the active rate decreases from 40 % to 15 %, the ICI error of the updated IM increases from 5.92 dB to 6.63 dB, whereas the ICI error of the low resolution fingerprinting increases from 9.17 dB to 11.65 dB. In fact, even at low active UE rates (e.g., 15 %), the proposed algorithm outperforms the low resolution technique. This can be explained by the presence of an initialization step and by the update

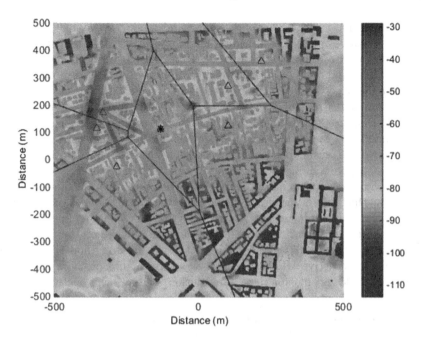

Fig. 6. Semi-deterministic IM based on exponential variogram: irregular sensing positions with active UE rate of 15 %

procedure according to the gap between the ICI level observed at the sensing positions and the theoretical IM at the same positions. It is also worth noting that this improvement is achieved with relatively low computational complexity. The additional computation cost inherent to our algorithm is related to the gap computation at the N sensing positions with a computational complexity of $\mathcal{O}(N)$. However, the complexity of (Eq. 4) can be decreased by considering only the closest neighboring sensing positions (i.e., choose the sensing positions within a radius $R \simeq 50$ m for the closest neighboring sensing positions). Figure 6 shows the resulting IM obtained with the proposed algorithm using the kriging interpolation with the exponential variogram, where an active UEs rate of 15 % is considered. The ICI levels are represented with a color code. Since the eNBs are located outdoor, the IM shows a low generated ICI level in indoor, a higher ICI level in outdoor, especially in the streets where UEs are exposed to more than one eNB in Line of Sight (LoS).

5   Conclusion

To build an IM, a stochastic-based location-dependent ICI estimation model that considers the shadowing, the fast fading and the path loss, was first developed. However, the analytical IM may be practically challenging to get reliable a priori information in real systems, due to the possibly low representativeness of the statistical model parameters in practical operating environments. Combining the latter analytical IM with space partitioning and indoor/outdoor deterministic information enables piece-wise IM initialization when starting from scratch. To build a reliable IM, we propose an update of the analytical IM in a self-learning procedure. The update is performed by a spatial interpolation technique based on the ICI gap observed by collaborative UE at the sensing positions considered either regular and irregular. Two spatial interpolation techniques are studied: (i) Inverse Distance Weighting (IDW), and (ii) Kriging where exponential and Von Kàrmàn variogram model are investigated. The performance analysis shows that, unlike to the sensing positioning type, where the regular and irregular positioning sensing give a similar ICI estimation error, the choice of the spatial interpolation technique impacts more the ICI estimation performance. The numerical results show that the IDW is less reliable than the kriging spatial interpolation technique, since in the latter case the weights are based on the inputs correlations. The best ICI estimation is given with the kriging based on the exponential variogram. In addition, the performance of the proposed method is studied as a function of the active UE rate, where the performance of the updated IM degrades more slowly that for the low resolution fingerprinting method.

Acknowledgement. The research leading to this paper has been supported by the Celtic-Plus project SHARING (project number C2012/1-8).

References

1. Bhushan, N., et al.: Network densification: the dominant theme for wireless evolution into 5G. IEEE Commun. Mag. **52**(2), 82–89 (2014)
2. Pinto, P., et al.: A stochastic geometry approach to coexistence in heterogeneous wireless networks. IEEE JSAC **27**(7), 1268–1282 (2009)
3. Brunel, L., et al.: Inter-cell interference coordination and synchronization based on location information. In: Proceedings of WPNC 2010, April 2010
4. Rodrigues, M.L., et al.: Fingerprinting-based radio localization in indoor environments using multiple wireless technologies. In: Proceedings of IEEE PIMRC 2011, September 2011
5. Darakanath, R.C., et al.: Modeling of interference map for licensed shared access in LTE-advanced networks supporting carrier aggregation. In: Proceedings of IFIP Wireless Days 2013 (2013)
6. Kim, S.-J., et al.: Cooperative spectrum sensing for cognitive radios using Kriged Kalman filtering. IEEE J. Sel. Top. Sig. Process. **5**(1), 23–36 (2011)
7. Babaei, A., et al.: Interference statistics of a poisson field of interferers with random puncturing. In: Proceedings of IEEE MILCOM 2011, November 2011
8. Kaddour, F.Z., et al.: Downlink interference analytical predictions under shadowing within irregular multi-cell deployments. In: Proceedings of IEEE ICC 2015, June 2015
9. Shepard, D.: A two-dimensional interpolation function for irregularly-spaced data. In: Proceedings of ACM National Conference, pp. 517–524 (1968)
10. Krige, D.G.: A statistical approach to some mine valuations and allied problems at the Witwatersrand, Master's Thesis of the University of Witwatersrand (1951)

11. Evolved Universal Terrestrial Radio Access (E-UTRA); Further advancements for E-UTRA physical layer aspects (Release 9), 3Gpp. TR 36.814, v. 9.0.0 (2010)
12. Small Cell Enhancements for E-UTRA and E-UTRAN- Physical Layer Aspects (Release12), 3Gpp. TR 36.872, v. 12.1.0 (2013)
13. Kaddour, F.Z., et al.: A new method for inter-cell interference estimation in uplink SC-FDMA networks. In: Proceedings of IEEE VTC-Spring 2012, May 2012
14. Brau, M., et al.: Assessment of 3D network coverage performance from dense small-cell LTE. In: Proceedings of IEE ICC 2012, pp. 6820–6824, June 2012
15. Kendall, D.G.: Survey of the statistical theory of shape. Stat. Sci. **4**(2), 87–99 (1989)
16. Alaya-Feki, A., et al.: Informed spectrum usage in cognitive radio networks: interference cartography. In: Proceedings IEEE PIMRC 2008, September 2008
17. Seung-Jun, K., et al.: Cooperative spectrum sensing for cognitive radios using Kriged Kalman filtering. IEEE J. Sel. Top. Sig. Process. **5**(1), 24–36 (2011)
18. Sidler, R.: Kriging and Conditional Geostatistical Simulation Based on Scale-Invariant Covariance Models, Diploma Thesis, Institute of geophysics Department of Earth Science, October 2003
19. INSEE Recensement de 2012, Série historique des résultats du recensement de la population (2012)

Simultaneous Uplink and Downlink Transmission Scheme for Flexible Duplexing

Adrian Kliks[(✉)] and Paweł Kryszkiewicz

Faculty of Electronics and Telecommunications,
Poznan University of Technology, Poznan, Poland
{adrian.kliks,pawel.kryszkiewicz}@put.poznan.pl

Abstract. The idea of adaptive usage of uplink frequency resources for other purposes, such as downlink data transmission, has attracted researchers for many years. This paper discusses the concept of applying a simultaneous uplink and downlink transmission scheme for flexible duplexing, where the middle part of the so-called uplink component carrier is used for downlink data delivery. The realization of such an idea is based on the adaptive change of the transmit/reception mask at the base stations and/or mobile devices. Beside a theoretical analysis, this paper provides the calculation results of the interference rise in the system.

Keywords: Flexible duplexing · Non-contiguous transmission · Interference · LTE/LTE-A · 5G systems

1 Introduction

The problem of asymmetric traffic, typical for modern wireless networks, can be solved in various ways depending on the applied duplexing schemes between uplink (UL) and downlink (DL) data delivery. In time division duplexing (TDD), a fragment of the frequency spectrum is utilized in both directions of data transmission (i.e., from base station (BS) to mobile terminal (MT), and from MT to BS), and the split between UL and DL is done in the time domain. As the wireless standards typically provide a very detailed time hierarchy (i.e., transmitted bits are organized into frames, bursts, chunks, etc.), it can be foreseen that asymmetric traffic can be easily managed by the application of adaptive allocation of more time slots to that direction which needs to serve higher traffic. Such an approach has been discussed in, e.g., [1–4]

Contrarily, in the classical frequency division duplexing mode (FDD), the data between BS and MT can be delivered continuously in the time domain in both directions if needed, but the split between UL and DL is realized in the spectrum domain, i.e., dedicated fragments of the frequency spectrum are assigned to each transmission direction. In such a case, the problem of asymmetric traffic can be solved by allowing data transmission in a selected direction in both bands. In other words, UL band can be utilized for DL transmission and vice-versa. Some interesting discussion can be found in, e.g., [5,6]. It is also

© ICST Institute for Computer Sciences, Social Informatics and Telecommunications Engineering 2016
D. Noguet et al. (Eds.): CROWNCOM 2016, LNICST 172, pp. 192–203, 2016.
DOI: 10.1007/978-3-319-40352-6_16

worth noticing here that the frequency bands dedicated for UL and DL transmission are separated by a dedicated frequency gap guaranteeing enough isolation between the transmit and receive signals. This isolation is required, as the problem of efficient canceling of strong interference at the reception interface from the transmit one is not mature enough today, although much progress has been achieved in the area of wireless full-duplex transmission [7,8]. These observations have to be considered while realizing the concept of adaptive utilization of both frequency bands for data delivery only in one direction.

In our work we concentrate on FDD solutions with particular attention given to LTE/LTE-A systems. Our motivation behind such a selection is to provide new technological solutions, while keeping the backward compatibility with current standards. In other words, one may notice that many of the existing wireless standards (including practical deployments of LTE) are based on FDD solutions. Thus, it is highly expected that the existing infrastructure be utilized as effectively as possible, and flexible usage of frequency bands in such a scenario can be one of the interesting solutions. Moreover, we assume that the amount of traffic in the DL direction is much higher than in the opposite direction, as this represents a typical situation in crowded areas where the role of the dominating service is played by mobile video streaming [9]. Clearly, an opposite situation is also possible (one can consider mass events where many persons decide to upload the photos or videos), but it is not as popular as the previous one. Finally, in such a context, the selection of LTE-based systems is natural, as this is the technology that permanently supersedes 3G systems in many places in the world. However, the application of flexible duplexing in FDD-LTE is not straight-forward, due to the continuous transmission of control signals in PUCCH in the UL band (PUCCH stands for physical uplink control channel). It means that (potentially) every time there is a useful signal present in the uplink band and it is not possible to allocate the whole UL band for downlink transmission without interfering to the base station.

One of the possible solutions in such a case is to apply the TDD mode in the uplink band [1]. In other words, the whole DL band is utilized for delivering data from BS to MT only (so the classic FDD transmission scheme is kept in the DL band), but the UL band is split in time into equal time slots which can be adaptively assigned to UL or DL depending on the current traffic in both directions. In this context, it is worth to mention further developments in this topic, known as the TDD Enhanced Interference Management and Traffic Adaptation (eIMTA) [10,11], also standardized in Release 12 of LTE [12]. Here, the split between the uplink and downlink data is considered to be flexible and modified adaptively to the observed traffic.

In the approach discussed in this paper, we propose to use the uplink bands in a highly flexible way, so that the split between uplink and downlink traffic depends mainly on the current user demands and assumed priorities. We consider simultaneous data transmission in both uplink and downlink directions, implementing advanced adaptive transmission/reception filtering for out-of-band attenuation. In general, this technique allows us to utilize the middle part of the uplink component carrier for DL transmission. Clearly, such a transmission scheme results in an interference rise observed inside the serving and all surrounding cells. Thus, in order to evaluate the proposed scheme, we have analyzed its impact on the interference boost observed by other users (mobile terminals or base stations). Based on that analysis, we propose to apply simple localization techniques for further enhancements in the proposed scheme, as well as we discuss the backward compatibility of that study. The key idea of the paper is to check the possibility of utilizing a fragment of all uplink resources for downlink transmission, even if the rest of these uplink resources are occupied. This approach is highly flexible, and can be used as a solution for advanced spectrum utilization in 5G networks. Moreover, the proposed technique can also guarantee backward compatibility with 4G systems. Thus, in order to check the correctness of the proposed approach (i.e., simultaneous uplink and downlink transmission) and its backward compatibility, we intentionally modeled a 4G-like scenario where uplink control data are transmitted at the edges of the uplink band. Nevertheless, this scenario can easily be generalized.

The paper is structured as follows. In the next section, we briefly remind selected techniques proposed for flexible duplexing, and present a few potential application scenarios including the key limiting factors of the LTE/LTE-A systems. We also discuss the main idea of simultaneous data transmission and the method of pre-calculating the transmit/reception masks. The analysis is followed by a presentation of the obtained numerical results.

2   Flexible Duplexing for 4G and 5G

2.1   TDD Mode Applied in Uplink Band

The concept of flexible utilization of unused frequency resources in FDD-based wireless communication systems is one of the immediate solutions that appeared during the discussion on the effective management of asymmetric traffic in current and future networks. As introduced in the previous section, various schemes have been considered so far (please see the discussion in, e.g. [1]), but the most focus has been put on TDD-based solutions. In such a case, the UL band is utilized following the time division duplex mode, where for certain time slots, the UL channel is used for downlink data transmission. Clearly, such a scheme results in an interference increase observed by other system users (for example, by those located in the surrounding cells). In order to minimize the impact of this phenomenon, quite often, (almost) ideal synchronization between cells is considered. This issue is illustrated in Fig. 1, where a frame consisting of 10 subframes is shown. In the upper part, the classic case is shown, where the whole UL band is used only for UL transmission. In such a case, one can ideally assume the lack of interference between the neighboring cells (in the figure, we used the names Cell A and Cell B). In the middle part of the figure, some arbitrarily selected subframes are used for DL data transmission, causing interference to the adjacent cells. The problem is more severe if there are some synchronization problems between the cells, as shown in the bottom part of the figure.

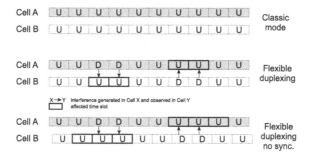

Fig. 1. Examples of flexible duplexing schemes in TDD mode

2.2   Considered Use Cases

There are various practical use cases considered in the context of flexible duplexing that can be found in the literature. In order to illustrate the idea we show two of them below. First, as discussed in the previous subsection, one can apply flexible duplexing at the macro-cell level. Thus, it is the macro base station that decides to assign some of the slots for downlink transmission, and such a transmission scheme is unique within the whole cell. The interference is then observed in the neighboring cells - please see Fig. 2. Another case includes the presence of small cells, i.e., there is more flexibility in assigning the subframes for UL or DL transmission, as this decision can be made either at the macro, or small-cell level. In the latter case, interference is induced into both serving and neighboring sites, as shown in Fig. 3. Without a loss of generality, we concentrate on the two macro-cell scenario, having in mind that the scheme proposed in this paper can be immediately applied to other cases with few or even no modifications.

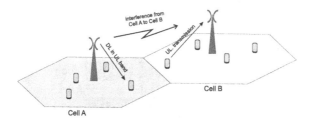

Fig. 2. Two macro cell scenario

2.3   Proposed Simultaneous Flexible Duplexing Scheme

Let us now consider an FDD-LTE-based system operating in both macro cells (hereafter denoted as Cell A and Cell B) with the frequency reuse factor close to unity; the application of the soft frequency reuse can be considered as well, but it does not influence the idea investigated here. In such a case, the uplink

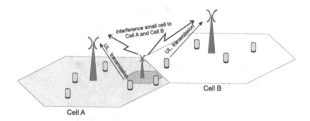

Fig. 3. Two macro cell and one small cell scenario

component carrier is used, broadly speaking, for user data delivery (realized in PUSCH, standing for physical uplink shared channel) and for uplink control information transmission (typically performed via PUCCH). It is important to notice that the PUCCH data are transmitted using small frequency segments located on the borders of the component carrier. In the considered scenario, we focus on the case when the UL channel (in Cell B) is used only for conveying control information, and there are no user data to be delivered to the base station. At the same time, Cell A is using the middle part of its own component carrier for downlink transmission, causing some interference rise. The idea is illustrated in Fig. 4, where the first seven subframes are managed in the way described above. However, we foresee that the proposed scheme can be extended to a more flexible situation where the fragment of the UL band will also be used for user data transmission (as shown in the last four subframes); this is a topic for further investigation. Please notice that we have also intentionally illustrated the presence of inference observed within the serving cell (Cell A), as the resource blocks used for PUCCH delivery in Cell A will be affected by the DL transmission in the middle of the band. The key concept proposed by us is to apply advanced, adaptive spectrum shaping algorithms originally considered to be used in non-

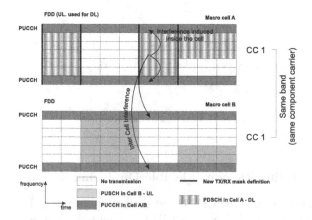

Fig. 4. Proposed scheme for simultaneous flexible duplexing

contiguous multicarrier transmission schemes [13]. These solutions can guarantee a significant reduction of unwanted out-of-band emission even in a very narrow frequency band at a reasonable complexity. Moreover, these algorithms can be applied at the beginning of a frame, allowing for a precalculation of the required spectrum masks (filter shapes). The moments when the new spectrum masks have to be changed within the cell are indicated by solid bold vertical lines in Fig. 4.

2.4   Synchronization Problems

As stated, in the considered scenario, the DL data signal in Cell A will be transmitted adjacently (in frequency) to the UL control signal in Cell B. As such, there will be some interference between UL (SC-FDMA, single-carrier frequency division multiple access) and DL (OFDMA), depending on the guaranteed synchronization level between both systems:

− full frequency and time synchronization - there is no inter-carrier interference,
− only frequency or only time synchronization - there is inter-carrier interference.

It is well known that both systems utilize the same subcarrier spacing and cyclic prefix (CP) length. Both systems could have the same reference oscillator frequency, synchronized using, e.g., GPS. On the other hand, according to LTE modulation specification, subcarriers in UL (after DFT processing) are moved by 0.5 subcarrier spacing in comparison to the DL signal in order to obtain minimum peak-to-average power Ratio (PAPR). As such, DL and UL subcarriers are not orthogonal to each other, i.e., inter-carrier interference occurs. One solution would be to shift 5G transmission by 0.5 subcarrier spacing to achieve subcarriers orthogonality and rely on the carrier frequency offset estimation at the receiver.

Another issue is related to the time synchronization of frames between Cells A and B. Typically, all mobile terminals attached to Cell B have different *timing advance* values (due to different propagation times) in order to synchronize their UL signal at the Cell B base station. However, in a practical case, DL transmission within Cell A is not synchronized or only partially synchronized with Cell B UL, and interference occurs. On the other hand the protection of Cell A UEs (DL) from interference caused by Cell B UEs (UL) cannot be achieved by means of synchronization (many sources of interference with different propagation delays). Other solutions have to be considered, such as the above-mentioned filtering.

2.5   Sources of Interference in Non-orthogonal OFDM/SC-OFDM Systems

The interference power in the considered scenario at the SC-OFDM receiver (Cell B base station/eNodeB) comes from two sources [14]:

- *Out-of-band (OOB) radiation of the DL/UL transmitter* - this is mostly the effect of sinc-like subcarrier spectrum. As it has been shown in [14], each subcarrier in the time domain can be represented by a complex sinusoid windowed using a rectangular window. However, there are a number of spectrum shaping methods designed in order to reduce OOB radiation power, e.g., [15,16]. The simplest one is to use so-called *guard subcarriers* [14] by modulating subcarriers closest to the currently utilized spectrum bandwidth with zeros (i.e., subcarriers close to the PUCCH band in Cell B are zeroed). Another source of OOB radiation is the high PAPR of an OFDM transmission carried using a nonlinear front-end. However, it can be effectively reduced using PAPR minimization or predistortion techniques [17].
- *Limited selectivity of OFDM/SC-OFDM receiver* - in a case when there is no user data transmission in the uplink (only the control channel is present), the middle part of the UL component carrier is empty, and only the resource blocks at the border convey useful data. However, the DFT operation is carried at the receiver on N consecutive incoming samples cut out from a stream of incoming samples. As such, time-domain windowing with N-length rectangular window is used. The reception filter characteristic of a single subcarrier has a sinc-like shape. If there is a DL signal (transmitted as stated above in the middle of the UL band) observed by the Cell B receiver, high power sidelobes of a sinc function will cause interference on the used subcarrier (conveying control information). In order to overcome this problem, time-domain windowing or filtering should be applied at the receiver.

3   Simulation Results

In order to evaluate the proposed scheme, we would like to measure the impact of the interference induced to other users due to the application of the proposed flexible duplexing scheme (i.e., when the advanced spectrum shaping proposed in [15,16] is implemented). We consider the presence of various receivers in the system, ones that are equipped with the proposed spectrum shaping algorithms, and others that can be treated as classical LTE devices (base stations or mobile terminals). The idea here is to guarantee backward compatibility with the existing devices.

3.1   Power Spectral Density Analysis

First, SC-OFDM (UL) and OFDM (DL) occupying a maximum of 20 MHz bandwidth are considered with IFFT/FFT of size $N = 2048$. Both systems transmit their signals in dedicated time slots (0.5ms duration each) composed of 7 OFDM/SC-OFDM symbols. While the first symbol in a slot utilizes 160 samples with CP, the rest are 144 samples long, each. It is assumed that there is no synchronization, neither in the time, nor in the frequency domain (as it is the most challenging scenario, as presented in the previous section). According to [19], the maximal number of Resource Blocks (RBs) in the considered scheme is

100. In the case of no data transmission in UL, only PUCCH is transmitted on both ends of the available band, i.e., RBs indexed 50 and -50. In the proposed scheme, unused resource blocks in the middle can be utilized by Cell A for its DL transmission. In Fig. 5, normalized PSDs of UL and DL signals are shown using solid lines. Cell A transmission utilizes RBs with indexes $\{-48, ..., -1, 1, ..., 48\}$, i.e., a contiguous band around the DC subcarrier (0th RB is not used). Signals are normalized to have equal received power per utilized RB. It can be observed that the OOB radiation of DL transmission in-band of UL transmission equals about -25 dB (solid, blue line). However, as presented previously, OOB radiation can be decreased by the application of advanced spectrum shaping methods (like Optimized Cancellation Carriers Selection, OCCS, discussed in [15]). In the presented plot, 20 OCCS subcarriers have been used, i.e., about 1.8 % out of all used subcarriers are devoted to OOB reduction. It is visible that a significant OOB radiation reduction can be obtained (solid, black line).

On the other hand, the values of the effective signal and interference power observed at the SC-OFDM receiver (i.e. after passing the FFT block) are shown with dashed lines. It can be noticed that the application of the proposed flexible duplexing scheme with no dedicated filtering, i.e., where standard receivers (std. RX) at the Cell B and standard transmitters (std. TX) at Cell A are used, results in the highest effective interference power, i.e., about -22 dB. The modification of only one side of the system (either the transmitter using the OCCS method, or the receiver by applying 256-samples-long Hanning windowing before FFT operation) does not significantly decrease the interference power. However, if both methods are combined (dashed, black line) the interference power is decreased significantly giving about -45 dB.

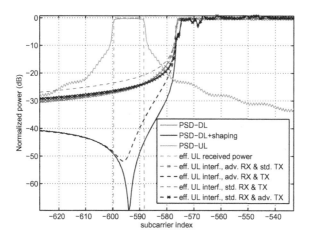

Fig. 5. Normalized PSDs of UL and DL transmitted signals, effective interference/useful signal power at Cell B base station (receiver). Only lower frequency edge is shown

3.2   Adjacent Channel Interference Ratio Analysis

In this section we analyze the adjacent channel interference ratio (ACIR) as a good metric used for the assessment of the ratio of wanted power to the interference power from the other bands. Mathematically, ACIR is the function of the adjacent channel leakage ratio (ACLR, used to characterize the transmitter) and the adjacent channel selectivity (ACS, used to characterize the receiver), i.e., $\text{ACIR} = \frac{1}{\text{ACLR}^{-1} + \text{ACS}^{-1}}$. As it has been mentioned, another possibility to decrease interference power at the Cell B receiver is to use Guard Subcarriers (GSs) [14]. Turning off DL subcarriers closest to the utilized UL band increases both ACLR and ACS. Let us note that such an approach is compliant with the existing LTE TX/RX technology, although it decreases the achievable rate (as some subcarriers are not utilized). The efficiency of this approach is considered here. In Fig. 6, ACIR values of the standard and advanced TX/RX technologies are shown as the functions of frequency separation between the UL and DL signals. ACIR is defined as the ratio of Cell-A-originated transmitted power observed at the antenna of Cell B base station to the power of effective interference at Cell B RX, i.e., power observed at the utilized subcarriers after the FFT block (in resource blocks used for PUCCH transmission). It is shown that when the PDSCH and PUCCH overlaps (frequency separation equal 0), ACIR equals about 20 dB. The situation changes rapidly when guard subcarriers are used. Even in the case of a standard transmitter and receiver, the introduction of a single empty resource block between the UL and DL bands (frequency separation equal to 2) increases ACIR to 42 dB. In the case of advanced TX and RX utilization (with the spectrum shaping algorithms discussed previously), the frequency

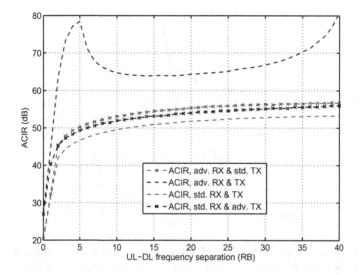

Fig. 6. Adjacent Channel Interference Ratio vs UL-DL frequency separation (0 RBs means that PUCCH and PDSCH overlap) with standard and advanced TX/RX.

separation of 2 RBs results in ACIR equal to 63.47 dB. It means that by proper signal processing, e.g., spectrum shaping at the transmitter [15,16] and windowing at the receiver [14] together with the application of guard subcarriers, a significant ACIR increase can be achieved. It is worth explaining why the curve representing ACIR for an advanced transmitter and receiver rises steeply for low frequency separation, and then it falls down. Such behavior is observed due to the specificity of the OCCS method which reduces the OOB most significantly in the adjacent subcarriers.

3.3 Influence on Transmission Opportunity in Neighboring Cell

In the most rigorous approach, Cell A can transmit according to the proposed scheme only if Cell B's transmission is not deteriorated (in practice, some deterioration should be acceptable). According to Table 8.2.1.1-6 in [20], the minimum SINR that should allow for transmission using QPSK modulation and $\frac{1}{3}$ coding rate is -0.4 dB. However, PUCCH reception should even be possible for lower SINR values, namely, -3.8 dB. Assuming the proposed scheme is used when only PUCCH is transmitted in Cell B's UL (although in general, other scenarios can be considered as well), it is visible that interference plus noise power can be increased by 3.4 dB without decreasing the effective Cell B radius. In order to evaluate this issue, the effective cell radius has been calculated using the COST 231 model for carrier frequency $f = 2$ GHz, base station antenna height $H_A = 30$ m, mobile terminal antenna height $h_{UE} = 1.5$ m, and mobile terminal and BS gains $G_{UE} = 0$ dBi and $G_{BS} = 18$ dBi, respectively (according to [21]). Assuming the mobile terminal transmit power is $P_{UE} = 23$ dBm and thermal noise power in 300 K increased by a Noise Figure (NF) of 5 dB [21], Cell B radius equals $R_A = 0.83\,km$.

For the same system parameters, the interference plus noise power can be increased by 3.4 dB while transmitting PUCCH (instead of PUSCH), as discussed above. It can be calculated that the effective interference power from Cell A TX to Cell B RX should be equal or lower than -93 dBm. In Fig. 7, the minimum required ACIR value is plotted as a function of distance between the cell centers for different propagation parameters between these BSs. In the worst-case scenario, both BSs have the Line-of-Sight propagation condition (received power decreases with $n = 2$ power of the distance), and a maximum antenna gain of 18 dBi at both BSs is assumed. Even in such a case, the required distance between Cell A BS and Cell B BS equals only about 327 m for ACIR equal to 78.4 dB (advanced processing at both TX and RX), which is less than the calculated cell radius.

In the case of some reflections (received power decreases with $n = 2.5$ power of the distance), the required distance equals, e.g., 576 m for ACIR=50 dB that can be achieved even without any improvements at the TX/RX side. If the received power decreases with the third distance power ($n = 3$) and because of tilt, the antennas do not have maximal gains in their directions, i.e., $G_{AB} = G_{BA} = 10$ dBi, the required ACIR decreases even further.

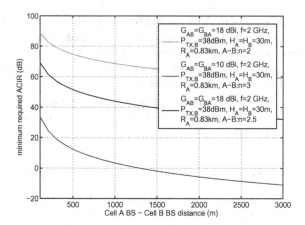

Fig. 7. Minimum required ACIR vs distance between Cell A and Cell B BS for different system parameters.

4   Conclusions

In this work, we have evaluated the possibility of simultaneous UL and DL data transmission in the flexible duplexing mode by the application of advanced spectrum shaping algorithms. Based on the achieved results, one can state that such a transmission scheme is possible, and the proposed technical solutions guarantee that the assumed maximum level of allowable interference is not exceeded. In some cases (by the application of guard subcarriers) even backward compatibility can be achieved. It means that the unused frequency resources in the UL channel can be utilized simultaneously for DL transmission, leading to better spectrum utilization. As the results are promising, one should consider investigating a situation where not only PUCCH is present in the UL band, but user data as well.

Acknowledgments. The work has been funded by the EU H2020 project COHERENT (contract no. 671639).

References

1. Wan, L., Zhou, M., Wen, R.: Evolving LTE with Flexible Duplex. In: IEEE Globecom Workshops (GC Wkshps), pp. 49–54, 9–13 December 2013. doi:10.1109/GLOCOMW.2013.6824960
2. Liu, H., Jiao, Y., Gao, Y., Sang, L., Yang, D.: Performance evaluation of flexible duplex implement based on radio frame selection in LTE heterogeneous network. In: 22nd International Conference on in Telecommunications (ICT), pp. 308–312, 27-29 April 2015
3. 3Gpp. R1–120196, Performance evaluation of dynamic TDD reconfiguration, Samsung, February 2012

4. 3Gpp. R1–120059, Evaluation of TDD traffic adaptive DL-UL reconfiguration in isolated cell scenario, Huawei, HiSilicon, January 2012
5. Soh, Y.S., Quek, T.Q.S., Kountouris, M., Caire, G.: Cognitive hybrid division duplex for two-tier femtocell networks. IEEE Trans. Wirel. Commun. 12(10), 4852–4865 (2013)
6. Pirinen, P.: Challenges and possibilities for flexible duplexing in 5G networks. In: CAMAD 2015, Guildford, UK, 7–9 September 2015
7. Choi, J., Jain, M., Srinivasan, K., Levis, P., Katti, S.: Achieving single channel, full duplex wireless communication. In: Proceedings of the ACM MobiCom Conference, pp. 1–12 (2010)
8. Sabharwal, A., Schniter, P., Guo, D., Bliss, D.W., Rangarajan, S., Wichman, R.: In-band full-duplex wireless: challenges and opportunities. IEEE JSAC 32(9), 1637–1652 (2014)
9. CISCO White paper, Cisco Visual Networking Index: Global Mobile Data Traffic Forecast Update, 2014–2019, 3 February 2015. http://www.cisco.com/c/en/us/solutions/collateral/service-provider/visual-networking-index-vni/white_paper_c11-520862.pdf
10. Pauli, V., Li, Y., Seidel, E.: Dynamic TDD for LTE-A and 5G, white paper, Nomor Research GmbH, Munich, Germany, September 2015
11. 3GPP, TR 36.828 - Evolved Universal Terrestrial Radio Access (E-UTRA) Further Enhancements to LTE Time Division Duplex (TDD) for Downlink-Uplink (DL-UL) Interference Management and Traffic Adaptation (Release 11) (2012)
12. 3GPP, TS 36.300 - Evolved Universal Terrestrial Radio Access (E-UTRA) and Evolved Universal Terrestrial Radio Access Network (E-UTRAN) Overall Description (Release 12) (2015)
13. Bogucka, H., Kryszkiewicz, P., Kliks, A.: Dynamic spectrum aggregation for future 5G communications. IEEE Commun. Mag. 53(5), 35–43 (2015). doi:10.1109/MCOM.2015.7105639
14. Weiss, T., Hillenbrand, J., Krohn, A., Jondral, F.K.: Mutual interference in OFDM-based spectrum pooling systems. IEEE Veh. Technol. Conf. (VTC) 4, 1873–1877 (2004)
15. Kryszkiewicz, P., Bogucka, H.: Out-of-band power reduction in NC-OFDM with optimized cancellation carriers selection. IEEE Commun. Lett. 17(10), 1901–1904 (2013)
16. Kryszkiewicz, P., Bogucka, H.: Flexible quasi-systematic precoding for the out-of-band energy reduction in NC-OFDM. In: IEEE Wireless Communications and Networking Conference (WCNC), pp. 209–214, April 2012
17. Kryszkiewicz, P., Kliks, A., Bogucka, H.: Obtaining Low Out-Of-Band Emission Level of an NC-OFDM Waveform in an SDR Platform, Inter-national Symposium on Wireless Communication Systems, Brussels, Belgium, 25–28 August 2015
18. Bogucka, H.: Directions and recent advances in PAPR reduction methods. In: IEEE International Symposium on Signal Processing and Information Technology, pp. 821–827, August 2006
19. Dahlman, E., Parkvall, S., Skold, J.: 4G: LTE/LTE-Advanced for Mobile Broadband: LTE/LTE-Advanced for Mobile Broadband. Elsevier Science, Burlington (2011)
20. 3Gpp. TS 36.104 V12.5.0 (2014–09), Technical Specification Group Radio Access Network; Evolved Universal Terrestrial Radio Access (E-UTRA); Base Station (BS) radio transmission and reception (Release 12), September 2014
21. CEPT Report 40: Report from CEPT to European Commission in response to Task 2 of the Mandate to CEPT on the 900/1800 MHz bands

Networking Protocols for Cognitive Radio

FTA-MAC: Fast Traffic Adaptive Energy Efficient MAC Protocol for Wireless Sensor Networks

Van-Thiep Nguyen, Matthieu Gautier$^{(\boxtimes)}$, and Olivier Berder

IRISA, University of Rennes 1, Lannion, France
{van-thiep.nguyen,matthieu.gautier,olivier.berder}@irisa.fr

Abstract. This paper presents the FTA-MAC (Fast Traffic Adaptive MAC) protocol, a novel energy-efficient MAC protocol based on asynchronous duty cycling for Wireless Sensor Networks (WSNs). In FTA-MAC protocol, the communication between the sensor nodes is initiated by sending a wake-up beacon from the receiving node. The latter fast adapts its *wake-up interval* according to the traffic rate in order to reduce the *idle listening* of the transmitting nodes, which results in reducing the energy consumption. FTA-MAC protocol is implemented and evaluated in OMNeT++/Mixim network simulator. Simulation results show that FTA-MAC outperforms state-of-the-art protocols under fixed and variable traffic rates and also with multiple concurrent nodes. Thanks to this better wake-up schedule, the lifetime of a sensor node is increased 1.5 to 2 times.

Keywords: Energy-efficient MAC protocol · Traffic aware · Wireless sensor networks · Wake-up interval

1 Introduction

Wireless sensor networks (WSNs) received tremendous attention in recent years because of the evolution of sensor devices as well as wireless communication technologies. The sensor devices connected to the Internet of Things (IoT) [1] can be used to enable remote health monitoring, emergency notification or other observation systems. A problem of such sensor devices is the limited lifetime because of the limitation of the battery capacity. To tackle this problem, one solution is to use an energy efficient protocol to control the behavior of each device in the network. Another challenge in these systems is to deal with variable traffic that can appear in applications such as remote health monitoring. In this type of systems, the sensor nodes are used to monitor the patient physiological parameters such as pulse rate, body temperature, etc., which are periodically transmitted to the center node (e.g. a smart phone). However when the observed patient changes its activity, the sensor nodes can also modify their data rate. Therefore, a new communication management mechanisms are required allowing these devices to adapt with respect to the variable traffic.

© ICST Institute for Computer Sciences, Social Informatics and Telecommunications Engineering 2016
D. Noguet et al. (Eds.): CROWNCOM 2016, LNICST 172, pp. 207–219, 2016.
DOI: 10.1007/978-3-319-40352-6_17

The Medium Access Control (MAC) layer allows a sensor node to efficiently share the wireless medium with others in the network. In this layer, the main causes of energy consumption are the idle listening, the overheads, the overhearing and the collisions [2]. Thus, in order to reduce the energy, these factors need to be optimized but there are trade-offs between them. For example, reducing idle listening and collisions requires extra synchronization and overheads, whereas reducing the synchronization and overheads causes the waste of energy in collisions. In the context of energy-efficient MAC protocols, an important mechanism for reducing the energy consumption is duty cycling. In this technique, the radio is turned on periodically, switching between awake and sleeping states.

The recent duty cycling MAC protocols can be grouped into two types: synchronous and asynchronous. While synchronous duty cycling MAC protocols (such as SMAC [3], TMAC [4]) reduce the energy consumption by synchronizing the sleep and wake-up times of sensor nodes, asynchronous duty cycling protocols do not require any synchronization period and can be categorized into two groups: sender initiated (e.g. BMAC [5], XMAC [6], WiseMAC [7]) and receiver initiated (PW-MAC [8], RICER [9], RI-MAC [10], TAD-MAC [11]). In asynchronous protocols, the receiver/sender sends a Wake-up Beacon (WB) to schedule a rendez-vous with other nodes for receiving/sending data frames.

While many of these protocols have been proposed to reduce energy consumption, few of them address the case of WSNs with variable traffics. Indeed, the traffic in WSNs can be variable depending on the requirements of the applications. The senders can change their traffic rate many times and without any preview.

In this paper, a new asynchronous duty cycling energy efficient MAC protocol, called Fast Traffic Adaptive MAC (FTA-MAC) protocol, is presented. Typically, in a duty cycling protocol, the optimal protocol has the schedule such that the sender and receiver wake up at same time, as it is the case for synchronous protocols but at the price of less flexibility and extra overheads, leading to energy waste. To approach this optimal schedule, FTA-MAC protocol adapts the wake-up interval of the receiver toward the transmission rate of the sender. To deal with variable traffic, FTA-MAC provides an adaptive mechanism at the receiver side to quickly reach a steady state value of wake-up interval according to the new traffic rate of sender.

The contributions of our work are:

- designing of a novel energy-efficient MAC (FTA-MAC) protocol for WSNs with variable traffics. FTA-MAC minimizes the energy consumption of the sensor nodes by reducing their idle listening time.
- implementing FTA-MAC in OMNeT++/Mixim network simulator for a star network topology. The performance of FTA-MAC are compared with TAD-MAC and RICER under fixed and variable traffic rates.

The rest of this paper is organized as follows. Section 2 describes the related works on receiver-initiated and adaptive MAC protocols. The main contribution of the paper is presented in Sects. 3 and 4, which describe the design of

FTA-MAC protocol and its evaluation through network simulations, respectively. Finally, the paper ends with the conclusion in Sect. 5.

2   Related Works

One of the early MAC protocols based on duty cycling technique with initiation by the receiver is RICER proposed in [9]. A receiving node wakes up with a constant period I_{WU} and initiates the communication by transmitting a short WB (with a duration T_b) and by monitoring the channel for a fixed duration T_l. If there is no response, this node goes back to sleep. A source node with data to transmit wakes up, stays awake and waits for the WB from the destination. Once receiving a WB, it sends its data frame. The communication ends with an acknowledgement (ACK) sends from the destination after correctly receiving the data frame. Two other versions of RICER were proposed in [12] to avoid data collisions.

The receiver-initiated mechanism has been also applied to asynchronous duty cycling in RI-MAC (Receiver-Initiated MAC) [10]. In RI-MAC, the WB is used to initiate the communication. In some cases, the receiver is already active and sends WB before the sender wakes up so the latter misses it. An mechanism called *beacon-on-request* aims to avoid this bad situation. The sender sends a special WB which contains the receiver address to specify who is supposed to receive the following $DATA$ frame. The receiver answers a beacon to enable the sender to send $DATA$ immediately. RI-MAC works with dynamic traffic load by increasing or decreasing the number of nodes but does not work for the variable traffic.

Proposed in [8], PW-MAC (Predictive-Wakeup MAC) aims at optimizing the idle listening. PW-MAC schedules the rendez-vous for a receiving node and a transmitting node by using a pseudo-random number generator in each node. The receiver includes the parameters of its generator in the WB. Based on these values, the sender can predict the next wake-up of the receiver so that the sender can wake up shortly before the receiver. Although this mechanism reduces the idle listening and energy, it is not suitable for networks with variable traffic. If the sender needs to change its traffic rate, it must buffer the data frame and waits until the receiver wakes up. The traffic latency will increase and frames will be dropped if the buffer is full.

One protocol that includes a mechanism for variable traffic is TAD-MAC (Traffic-Aware Dynamic MAC) protocol, presented in [11]. In TAD-MAC, the receiver periodically sends a WB and considers the variable traffic by adapting its wake-up interval to the traffic rate of each individual sender. In order to be reactive to the traffic change, the receiver wakes up twice more frequently than the sender. This leads to energy waste due to unnecessary wake-ups of the receiver. Unlike in RI-MAC, the WB of TAD-MAC contains a specific node address to identify the destination address. This mechanism allows the reduction of wireless collisions.

FTA-MAC and TAD-MAC protocols are both based on the Traffic Status Register (TSR) technique introduced in the next section. However, FTA-MAC proposes an algorithm which allows the receivers to estimate the data rate of senders and to wake up just in time to reduce the *idle listening* of senders.

3 Fast Traffic Adaptive MAC Protocol

3.1 Traffic Status Register (TSR) Based Protocol

In a network under variable traffic, the traffic estimation is an important part. With their limited computation capacity, the sensor nodes require a light-weight mechanism to estimate the traffic. The *TSR* technique is first introduced in [11] to store the status information of the traffic. When a receiver wakes up to send a *WB*, if it receives a data response, the *TSR* corresponding to the transmitted node is filled with a BIT 1, in contrast, the *TSR* is filled with a BIT 0. The new status (either BIT 1 or BIT 0) is inserted into the *TSR* at the first index by shifting left one bit. In a star network, the receiver contains a list of *TSR* to track the traffic status of all the potential senders, as described in Fig. 1. This is a blind way to keep the track of traffic load without any information about the data rate of the transmitter. This blind estimation is not a quantitative evaluation of the data rate but rather a relative measure of either the traffic is increasing (which is represented by multiple consecutive ones in the *TSR*) or decreasing (which is represented by multiple zeros in the *TSR*). In this context, the sequence [10101010...] seems the best trade-off between the optimal sequence [11111111...] (*i.e.* each wake-up of the node is followed with a successful data reception), and the too frequent wake-up [10001000...] due to several reasons:

- Sequence of [11111111...] is an ideal sequence which means that each time the receiver wakes up, it receives data. Because of the blind estimation, it always has probability of missing data as soon as the traffic rate increases (at the transmitter side). The wake-up interval of the receiver will remain at lower traffic rate.

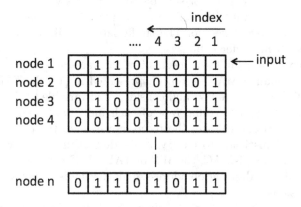

Fig. 1. Traffic Register Status (*TSR*) structure.

– Sequence of [10001000...] is another extreme as the receiver wakes up too often and wastes energy by sending unnecessary WB.

Thus, [10101010...] sequence is selected as a steady state in TAD-MAC protocol. The sender periodically wakes up to send data. This period is called interval of transmission (I_{Tx}) and the receiver adapts it wake-up interval (I_{WU}) according to the I_{Tx} of sender. With the sequence [10101010...], the receiver wakes up twice more frequently than the sender. It means that I_{WU} is equal to $0.5I_{Tx}$. The sender wakes up and waits for WB from the receiver in a limited interval of time (t_{WBmax}) and the t_{WBmax} is always less than I_{Tx}.
Because of the blind estimation, a drawback of the TSR technique is that the receiver can converge to a wrong value even if the sequence is still the best trade-off sequence [10101010...] with the condition that t_{WBmax} is less than $0.5I_{Tx}$. This problem is described more clearly in Fig. 2 where the receiver converges to a wrong value: I_{WU} is equal to $1.5I_{Tx}$.

Moreover, the speed of convergence depends on the length of the TSR and the pair values of I_{WU_0} (the initial value of I_{WU}) and I_{Tx}. In some worst case, the receiver takes a long time to reach the steady state and loses much energy. The blind estimation without any additional information from the sender nodes is a good idea but it still has many limitations.

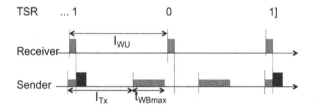

Fig. 2. Wrong convergence of TAD-MAC protocol (I_{WU} converges to $\frac{3}{2} I_{Tx}$ instead of $\frac{1}{2} I_{Tx}$).

3.2   FTA-MAC Protocol Design

FTA-MAC is a novel protocol based on the duty cycling technique. The communications between the sensor nodes are initiated by the receiving node and the principle is described in Fig. 3. The figure is divided into two phases: before and after the convergence. On the one hand, the receiver periodically sends a WB frame to notify its wake-up to the other nodes. On the other hand, the sender periodically wakes up with a period I_{Tx} which depends on its data rate. Before sending the data, it waits for the WB from the receiver. This period is called *idle listening* time (t_{idle}) which is the most energy consumption activity in receiver-initiated MAC protocols. Once receiving the WB, the sender sends the data after sensing the medium channel during an interval of time called t_{CCA} (Clear Channel Assessment). The communication session ends with an acknowledgement (ACK) message from the receiver to the sender after successfully receiving the data frame.

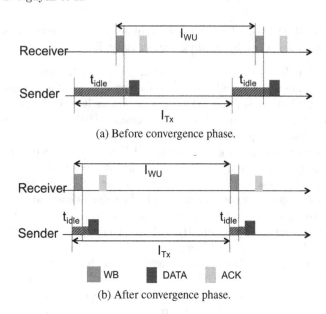

(a) Before convergence phase.

(b) After convergence phase.

Fig. 3. Behaviour of FTA-MAC protocol.

To reduce t_{idle}, the receiver attempts to wake up according to the I_{Tx} of each sender. The wake-up interval (I_{WU}) of the receiver is adapted itself to the data transmission rate of each transmitting node by collecting the statistics of its data traffic. To this aim, a TSR is used to store the recent history of data transmission, as for TAD-MAC protocol.

The second phase after convergence indicates that the receiver has adapted its I_{WU} to the I_{Tx} of the senders in such a way that the idle listening time is reduced. This phase is described in Fig. 3b, where the t_{idle} of the source node is reduced. The adaptive mechanism of I_{WU} is explained in Fig. 4 and I_{WU} is calculated by the following adaptive function:

$$
I_{WU}(i+1) = \begin{cases} I_{WU}(i) + n_0(i).t_{ref} & \text{if } TSR(i) = 0 \\ \frac{\sum_{j=k}^{i} I_{WU}(j) + t_{idle_k} - t_{idle_i}}{N_{WW}+1} & \text{if } TSR(i) = 1 \end{cases} , \tag{1}
$$

where $n_0(i)$ is the number of bits 0 in the TSR, t_{ref} is the system clock factor, i and k indexes stand for the two last moments when the receiving node received data from the sender, t_{idle_k} and t_{idle_i} are the sender idle listening time for the two last times it received a WB and N_{WW} is the number of wake-ups without receiving WB. The three values t_{idle_k}, t_{idle_i} and N_{WW} are calculated at the sender side and sent to the receiver in the data frame, whose structure is described in the next section.

The variable N_{WW} stores the statistic of the received WB in order to deal with the TSR wrong convergence problem described in Fig. 2. Each time the sender wakes up without receiving the WB, the variable N_{WW} is increased. This information is sent to the receiver at the next data frame and then is reset to 0.

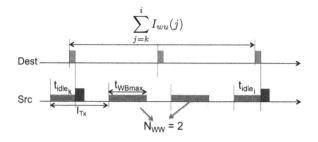

Fig. 4. Adaptive function to calculate I_{WU}.

In FTA-MAC, the algorithm is designed to converge toward a steady state value of I_{WU}, where the *TSR* contains a sequence $[11111111\ldots]$. This sequence is the best sequence of the *TSR*, *i.e.* each time the receiver wakes up, it receives the data from the sender. This sequence also specifies that the I_{WU} has converged. As discussed in Sect. 3.1, the sequence $[111111111\ldots]$ has the latent problem that the receiver can not detect the increase of traffic rate. But in FTA-MAC protocol, the adaptive function is provided the essential information (t_{idle}, N_{WW}) of the sender so the receiver can detect as soon as the traffic rate changes (either increases or decreases) and the receiver does not need to wake up more frequently than the sender.

3.3 Frame Structure

The frame structure proposed for FTA-MAC is described in Fig. 5. The first byte is used for the frame information in which the 3 first bits specify the type of frame. In FTA-MAC, there are 3 types of frame: *WB* (000), *DATA* (001) and *ACK* (010). The next 5 bits are reserved for further utilization. Based on the type of frame, the 4 or 8 next bytes are used for address information. The *WB* frame is broadcast so the destination address does not need to be defined. In this case, the frame size of *WB* is 7 bytes and 11 bytes for *ACK* because these frames have not data.

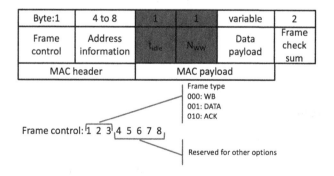

Fig. 5. New frame structure of FTA-MAC.

As discussed in Sect. 3.2 above, the sender will include the 2 variables t_{idle} and N_{WW} in the data frame. The 2 first bytes in the payload part are reserved for these variables so the size of the *DATA* frame is greater than 13 bytes. For the simulations of the next section, the size of *DATA* frame is set to 16 bytes.

4   Performance Evaluation

In this section, the evaluation of performance of FTA-MAC protocol is presented. This protocol is implemented in the OMNeT++ network simulator using the framework MiXiM for WSNs. For comparison purpose, TAD-MAC and RICER protocols have also been implemented. This section addresses first point-to-point communications, then the same simulation framework is used to validate the protocol for multi-sender communications.

4.1   Simulation Topology

OMNeT++ is an object-oriented modular event network simulator and it provides the infrastructure for writing simulations with the component architecture. MiXiM is a modeling framework for wireless network (WSNs, body area network, ad-hoc network, vehicular network, etc.).

Each sensor node is simulated with a single omni-directional antenna and using the simple path loss model of radio propagation. Table 1 summarizes the key parameters which are used to simulate the radio. Most of these parameters are from the data sheet of CC2420 radio chip [13], which is used in popular WSN testbed systems such as MICAz [14], TelosB [15] and PowWow [16]. These parameters are used to calculate the energy consumption in the Physical layer of MiXiM.

Table 1. Configuration of radio parameters.

Bitrate	250 kbit/s
RX current	18.8 mA
TX current	17.4 mA
Sleep current	0.03 mA
P_{Tx}	1mW

4.2   Convergence Speed

In a network with variable traffics, the convergence speed is an important factor to evaluate the performance of a MAC protocol. The convergence speed corresponds to the number of wake-ups of the receiver needed to converge toward the good *TSR* register, *i.e.* [1111...] for FTA-MAC and [1010...] for TAD-MAC.

Table 2. Configuration of MAC protocol parameters.

TSR length	4
t_{WBmax}	500 ms
t_{CCA}	.5 ms
I_{WU_0}	0.1s - 2s

When the receiver reaches the steady state, the t_{idle} of the sender is minimized such that the faster the receiver converges toward the steady state, the less energy the sender consumes. In order to evaluate the convergence speed of FTA-MAC and TAD-MAC, the first simulation scenario is configured with a fixed traffic rate of 2 frames per second for a point-to-point communication. Table 2 summarizes the MAC protocol parameters which are used in the simulations. These parameters are the values used for TAD-MAC protocol in [11]. The variable I_{WU_0} is tuned from 0.1 s to 2 s to study the impact of the initial wake-up interval on the convergence speed of these protocols.

The simulation results in Fig. 6 show that FTA-MAC protocol reaches the steady state value faster than TAD-MAC and the convergence speed is nearly independent of the variable I_{WU_0}. The convergence speed is constant when the value of I_{WU_0} is greater or equal than the traffic rate (I_{Tx}=0.5 s for a data rate of 2 frames/s). On the contrary, the convergence speed of TAD-MAC depends on the value of I_{WU_0}. Even with the best value of I_{WU_0} (0.3 s) for TAD-MAC, the latter converges slower than FTA-MAC.

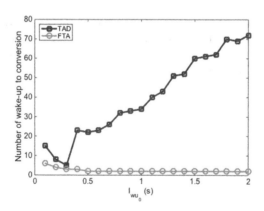

Fig. 6. I_{WU} convergence speed as a function of I_{WU_0}.

4.3 Energy Evaluation for Variable Traffics

One of the advantages of FTA-MAC compared with TAD-MAC is the fast convergence speed toward a steady state value of I_{WU}. This advantage can be tra-

Table 3. Simulation parameters for variable traffic.

Simulation time	2000 s
Traffic rate	1 frame/s - 10 frames/s
I_{WU_0} (FTA &TAD)	0.3 s
I_{WU} (RICER)	0.05s / 0.1s
Number of changes	from 0 to 30
Number of random simulations	100

duced in a significant energy gain for sending and receiving a frame in a network with variable traffics. The traffic rate changes by increasing or decreasing the I_{Tx} of the traffic generator of the sender. In this simulation, a change corresponds to a new random value for I_{Tx} uniformly distributed between 100 ms and 1000 ms. The number of changes is varied from 0 to 30 for a simulation time of 2000 s. All of these parameters are specified in Table 3. The results for variable traffic are shown in Figs. 7 and 8, where each average value is calculated from the results of 100 random simulations. FTA protocol is compared with TAD-MAC and also with the commonly used RICER protocol. RICER has no adaptation mechanism and is tuned by its wake-up interval I_{WU}.

Figure 7 shows the average energy consumption for sending and receiving a frame which is calculated at both sender and receiver sides for a successful transmission. The energy consumption of TAD-MAC is 1.6 to 2 times more than the one of FTA-MAC and this gain increases with the number of changes of I_{Tx}. The FTA-MAC protocol saves more energy than TAD-MAC because of the fast convergence speed. For each change of the traffic rate, the receiver needs to adapt its I_{WU} according to the new steady state value of I_{Tx}. The fast convergence speed reduces not only the energy consumption of the receiver but also of the sender by reducing the energy wasted during the *idle listening* period.

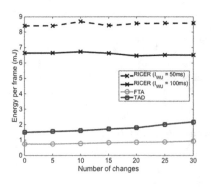

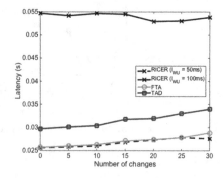

Fig. 7. Energy consumption per frame under variable traffic.

Fig. 8. Average latency of frame under variable traffic.

The adaptation of the receiver toward the data rate of the sender allows FTA-MAC to outperform also RICER protocol not only under the variable traffic but also with fixed traffic. Indeed, Fig. 7 shows that FTA-MAC uses nearly 7 times less energy consumption than RICER ($I_{WU} = 100$ ms) and 9 times less than RICER ($I_{WU} = 50$ ms).

The other factor used to evaluate the performance of the MAC protocols is the frame latency. The results are shown in Fig. 8. Although the RICER sensor node wakes up more frequently, FTA-MAC protocol has a lower latency. To gain 0.002 s less latency than FTA-MAC at 30 changes of traffic rate, RICER sensor node with a $I_{WU} = 50$ ms needs to wake up 10 times more frequently and thus consumes much more energy (about 9 times more).

4.4   Evaluation for Multi Sender Networks

When the communication is extended to multiple concurrent senders, data collisions can occur and decrease the performance of a protocol. Thus this kind of networks is also addressed to evaluate the performance of FTA-MAC. A star topology with a number of senders increasing from 1 to 9 is used and a traffic rate of 2 frames per second is set for all senders. When increasing the number of nodes, wake-up collisions will appear more frequently. Simulation results are given in Figs. 9 and 10, showing the average latency and the energy consumption of FTA-MAC, TAD-MAC and RICER3 (a version of RICER with a mechanism to avoid data collision). The latency and the energy consumption of FTA-MAC slightly increase while these factors for TAD-MAC and RICER3 sharply increase with the number of senders. In RICER3, the number of slots are set according to the number of senders to avoid data collision. Thus, when many senders concurrently send data to the receiver, some nodes will take a long time to wait their WB (*i.e.* the t_{idle} is increased). For TAD-MAC protocol, the data collision increases the time to converge toward the steady state. This issue increases the latency and the energy consumption in TAD-MAC.

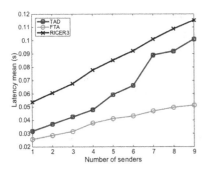 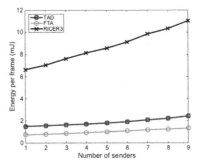

Fig. 9. Average latency of frame with multiple concurrent senders.

Fig. 10. Energy consumption per frame with multiple concurrent senders.

5 Conclusion

In this research, a novel adaptive energy efficient MAC protocol is introduced. The FTA-MAC protocol presents an algorithm that allows the receiver to adapt its I_{WU} according to the transmission rates of the sender. With the fast convergence speed, the receiver is well scheduled to wake up closely to the sender so reducing its idle listening. Thus the energy consumption in a sensor node is decreased 2 times compared to TAD-MAC protocol and 9 times compared to RICER protocol. This improvement is even more interesting for communications with variable traffics and also for a star network topology with concurrent multi-senders. The further work will extend FTA protocol to multi-hop networks, where the impact of idle listening on the energy consumption increases with the number of hops.

References

1. Definition of the Internet of Things (IoT). http://iot.ieee.org/definition.html. Accessed on 29 Nov 2015
2. Bachir, A., Dohler, M., Watteyne, T., Leung, K.K.: MAC essentials for wireless sensor networks. IEEE Commun. Surv. Tutorials **12**, 222–248 (2010)
3. Ye, W., Heidemann, J., Estrin, D.: An energy-efficient MAC protocol for wireless sensor networks. In: Proceedings of the 21st IEEE International Conference on Computer Communications (INFOCOM), pp. 1567–1576, June 2002
4. van Dam, T., Langendoen, K.: An Adaptive energy-efficient MAC protocol for wireless sensor networks. In: Proceedings of the 1st ACM International Conference on Embedded Networked Sensor Systems (SenSys), pp. 171–180, November 2003
5. Polastre, J., Hill, J., Culler, D.: Versatile low power media access for wireless sensor networks. In: Proceedings of the 2nd ACM International Conference on Embedded Networked Sensor Systems (SenSys), pp. 95–107, November 2004
6. Buettner, M., Yee, G.V., Anderson, E., Han, R.: X-MAC: a short preamble MAC protocol for duty-cycled wireless sensor networks. In: Proceedings of the 4th ACM International Conference on Embedded Networked Sensor Systems (SenSys), pp. 307–320, October 2006
7. El-Hoiydi, A., Decotignie, J.-D.: WiseMAC: an ultra low power MAC protocol for the downlink of infrastructure wireless sensor networks. In: Proceedings of the 9th International Symposium on Computers and Communications (ISCC), pp. 244–251, June 2004
8. Tang, L., Sun, Y., Gurewitz, O., Johnson, D.B.: PW-MAC: an energy-efficient predictive-wakeup MAC protocol for wireless sensor networks. In: Proceedings of the 30th IEEE International Conference on Computer Communications (INFOCOM), pp. 1305–1313, April 2011
9. Lin, E.-Y. A., Rabaey, J.M.: Power-efficient rendezvous schemes for dense wireless sensor networks. In: Proceedings of the 11th IEEE International Conference on Communications (ICC), pp. 3769–3776, June 2004
10. Sun, Y., Gurewitz, O., Johnson, D.B.: RI-MAC: a receiver-initiated asynchronous duty cycle MAC protocol for dynamic traffic loads in wireless sensor networks. In: Proceedings of the 6th ACM International Conference on Embedded Networked Sensor Systems (SenSys), pp. 1–14, November 2008

11. Alam, M.M., Berder, O., Menard, D., Sentieys, O.: TAD-MAC: traffic-aware dynamic MAC protocol for wireless body area sensor networks. IEEE J. Emer. Sel. Topics Circuits Syst. **2**(1), 109–119 (2012)

12. E.-Y. Lin, J. Rabaey, S. Wiethoelter, Wolisz, A.: Receiver initiated rendezvous schemes for sensor networks. In: Proceedings of the 48th IEEE Global Telecommunications Conference (GLOBECOM), vol. 5, pp. 3122–3128, December 2005

13. CC2420 Single-Chip 2.4GHz. http://www.ti.com/product/CC2420/technical documents. Accessed on 29 Nov 2015

14. MEMSIC, Inc - WSN Nodes — MPR2400CB. http://www.memsic.com/wireless-sensor-networks/MPR2400CB. Accessed on 29 Nov 2015

15. MEMSIC, Inc - WSN Nodes — TPR2420. http://www.memsic.com/wireless-sensor-networks/TPR2420. Accessed on 29 Nov 2015

16. Power Optimized Hardware, Software FrameWork for Wireless Motes. http://powwow.gforge.inria.fr/. Accessed on 29 Nov 2015

Threshold Based Censoring of Cognitive Radios in Rician Fading Channel with Perfect Channel Estimation

M. Ranjeeth and S. Anuradha$^{(\boxtimes)}$

National Institute of Technology, Warangal 506004, Telangana, India
ranjithamamidi2001@gmail.com, anuradha@nitw.ac.in

Abstract. In this paper, we have discussed about the performance analysis and comparison between hard decision (majority rule) logic and soft decision (maximal ratio combining) logic on cooperative spectrum sensing network using censoring scheme. We are assuming that both sensing channel and reporting channel are affected by Rician fading. Due to presence of fading in the channels, the information received at the fusion center is erroneous. Threshold based censoring scheme is used to eliminate the heavily faded cognitive radios in reporting channel. Majority logic and maximal ratio combining (MRC) schemes are applied individually at fusion center to make final decision about primary user. Finally, the performance is evaluated in terms of missed detection probability (Q_m) &total error probability $(Q_m + Q_f)$ using majority logic and MRC rule at fusion center. Simulations are performed with perfect channel estimation by varying the network parameters like probability of false alarm (P_f), S-channel SNR, R-channel SNR, Rician fading parameter (K) and number of CR users (N). Comparison table between majority logic and MRC rule is provided to know which fusion rule performs better under Rician fading.

Keywords: Co-operative spectrum sensing · Censoring · Energy detection · Majority logic · Maximal ratio combining · Rician fading

1 Introduction

The emerging new technologies and applications of wireless communication increase the uses of radio spectrum. To make good usage of spectrum, unused spectrum band also should be used in proper manner. Various reports on spectrum utilization have shown that the spectrum is inefficiently utilized. This means that there is ample opportunity to find many vacant spaces in the radio spectrum. In [1], to tackle the problem of spectrum underutilization they proposed a flexible spectrum models rather than fixed spectrum models. The main aim of the spectrum sensing (SS) is to detect the spectrum holes when the information about primary signal is unknown. Energy detection (ED) is relevant detection scheme that is used when the primary signal is unknown, because of most simple and non-coherent detector [2]. In spectrum sensing, detection performance is limited due to single cognitive radio (CR) user is present in the network, moreover due to fading and shadowing present in the environment. To overcome these drawbacks

© ICST Institute for Computer Sciences, Social Informatics and Telecommunications Engineering 2016
D. Noguet et al. (Eds.): CROWNCOM 2016, LNICST 172, pp. 220–231, 2016.
DOI: 10.1007/978-3-319-40352-6_18

co-operative spectrum sensing (CSS) [3] concept is introduced, where primary user (PU) is detected by multiple numbers of CRs, hence, detection probability of PU increases. These multiple number of CRs senses the PU individually and stores the information with them using sensing channel (S-channel). Fusion center (FC) collects the sensing information stored by individual CRs through reporting channel (R-channel). FC makes final decision about PU using hard decision (majority logic) rule and soft decision (MRC rule) rule individually [4].

However, most of the literature work exist on CSS is done by assuming noiseless R-channels [5]. But, in practical situations R-channels may not be noiseless (ideal) channels. There may be a chance to occur fading effect in R-channel due to this information passed through the R-channel gets affected and information received at FC also erroneous. Due to fading effect, the R-channel links gets heavily faded; this type of radio links has to be eliminated to improve the detection probability and to decrease system complexity. So, Threshold based censoring scheme is used to eliminate heavily faded R-channel radio links. The idea of censoring scheme is initiated and implemented on CSS in cognitive radio network in 2007 [6]. In [7] sensor network with impact of channel estimation error is described. Threshold-based censoring scheme is applied on CR users, whose R-channel fading coefficients exceed a pre-determined threshold. In [8], censoring scheme is used to overcome the drawback of overhead traffic at FC. By

Table 1. Comparison between majority logic and MRC fusion rules with perfect channel estimation using various network parameters at threshold value C_{th} = 1.0.

Parameters		Perfect channel		
		Q_m	Q_f	Total Error
1.	R-SNR = −9 dB (Majority)	0.0512	0.0395	0.0907
	R-SNR = − 9 dB (MRC)	0.0288	0.0233	0.0521
	R-SNR = −7 dB (Majority)	0.0264	0.0217	0.0481
	R-SNR = −7 dB (MRC)	0.0178	0.0127	0.0305
2.	S-SNR = 20 dB (Majority)	0.0264	0.0217	0.0481
	S-SNR = 20 dB (MRC)	0.0178	0.0127	0.0305
	S-SNR = 15 dB (Majority)	0.0753	0.0677	0.1430
	S-SNR = 15 dB (MRC)	0.0651	0.0615	0.1266
3.	P_f = 0.05 (Majority)	0.0264	0.0217	0.0481
	P_f = 0.05 (MRC)	0.0178	0.0127	0.0305
	P_f = 0.0005 (Majority)	0.0472	0.0397	0.0869
	P_f = 0.0005 (MRC)	0.0395	0.0386	0.0781
4.	N = 30 (Majority)	0.0512	0.0395	0.0907
	N = 30 (MRC)	0.0288	0.0233	0.0521
	N = 15 (Majority)	0.1340	0.0839	0.2179
	N = 15 (MRC)	0.0739	0.0743	0.1482
5.	K = 6 (Majority)	0.0393	0.0307	0.0700
	K = 6 (MRC)	0.0241	0.0127	0.0368
	K = 3 (Majority)	0.0512	0.0395	0.0907
	K = 3 (MRC)	0.0288	0.0233	0.0521

using this paper our contribution to the existing literature is that both S-channel and R-channel are affected by Rician fading and Threshold based censoring scheme is applied in R-channel to eliminate the heavily faded radio links. Performance comparison between majority logic and MRC rule is provided by varying several network parameters using perfect channel estimation in Table 1. Finally, we have investigated the performance using Q_m and $Q_m + Q_f$ with perfect channel estimation by varying the network parameters: S-channel SNR, R-channel SNR, false alarm probability (P_f), Rician fading parameter (K) and number of available CR users (N).

The rest of the paper is arranged as follows. The energy detection (ED) system model, CSS network, majority rule and MRC logic are discussed in Sect. 2. The discussion about simulation results are presented in Sect. 3, conclusions are provided in Sect. 4.

2 System Model

The below Fig. 1 represent the energy detection (ED) block diagram. The signal from the transmitter is first received i.e. x(t), it passed through the band pass filter (BPF) to limit the noise variance of the signal, then filtered signal is given to non–linear device called square law device. Due to filter action the filtered signal is band-limited and signal has a flat spectral density. The output of the square law device is given to the integrator block; it measures the energy of the received signal for a particular duration of time interval (T). Integrator output (Y), is compared with a pre-defined fixed threshold (λ) to make final decision about the PU [3]. The output of the threshold block produces two hypothesis H_0 & H_1 represent the absence and presence of PU. The test statistics can be given as:

$$T(y) = \frac{1}{N}\sum_{n=1}^{N} \{y(n)\}^2 \tag{1}$$

The received signal y(n) can be shown as below:

$$y(n) = \begin{cases} w(n) & H_0 \\ h * s(n) + w(n) & H_1 \end{cases} \tag{2}$$

Where s(n) is transmitted signal, w(n) is the normal distributed Additive White Gaussian Noise (AWGN) and 'h' is fading coefficient of Rician channel.

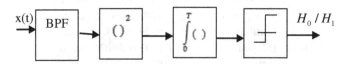

Fig. 1. Energy Detection block diagram

The expressions for probability of false alarm P_f and detection probability P_d over AWGN channels are given respectively [3],

$$P_d = pr(y > \lambda \,|\, H_1) = Q_m\left(\sqrt{2\gamma}, \sqrt{\lambda}\right) \tag{3}$$

$$P_d = pr(y > \lambda \,|\, H_0) = \frac{\Gamma(u, \frac{\lambda}{2})}{\Gamma(u)} \tag{4}$$

where $\Gamma(a, b)$ is the incomplete gamma function [9] and $Q_m(a, b)$ is the generalized Marcum Q-function [10].

The ED compares the output of integrator Y_k with pre-defined detection threshold (λ) then decides the presence of PU. The sensing information about the PU is sent to the FC through R-channel using BPSK signal modulation under Rician faded environment. The signal from k-th CR user received at the FC is:

$$y_k = m_k h_k + n_k \tag{5}$$

where m_k represent the bit energy of the each CR, in case of BPSK signal that bit energy values indicated as ($\sqrt{E_b}$ and $-\sqrt{E_b}$), corresponding H_1 & H_0, respectively. We are considering that R-channel is Rician faded and that fading coefficient h_k is modeled as a zero-mean complex Gaussian random variable with variance, σ^2, i.e., $h_k \sim CN(0, \sigma^2)$ and noise present in the channel is considered as zero-mean complex Gaussian noise with variance, σ_n^2, $n_k \sim CN(0, \sigma_n^2)$. Let us consider that n_k & h_k are mutually independent. Fading coefficients h_k present in the R-channel are estimated by minimum mean square estimation (MMSE) strategy, then as follows [11]:

$$\hat{h}_k = E[h_k \,/\, y_k] = y_k \frac{\sqrt{E_b}}{E_b + \sigma_n^2} \tag{6}$$

$$\hat{h}_k = E[h_k \,/\, y_k] = h_k \frac{E_b}{E_b + \sigma_n^2} + n_k \frac{\sqrt{E_b}}{E_b + \sigma_n^2} \tag{7}$$

In the above equations, $\hat{h}_k$ represents the estimated value of fading coefficient (h_k) after using MMSE estimation and E_b represents the bit energy of R-channel radio link.

Finally, the channel estimation error $\tilde{h}_k$ can be calculated by taking the difference between the actual fading coefficient and the estimated channel coefficient i.e. $\tilde{h}_k = h_k - \hat{h}_k$. In case of perfect channel, both actual fading coefficient & estimated fading coefficient are equal, hence estimation error is zero. So, estimation error does not show any effect on missed detection probability. Censoring scheme is applied on the estimated channel coefficients. CRs which are greater than predefined threshold, those estimated channel coefficients are selected to transmit through the R-channel to the FC. All these estimated channel coefficients are gathered at FC, it uses majority logic, MRC rule to know the activity of primary user.

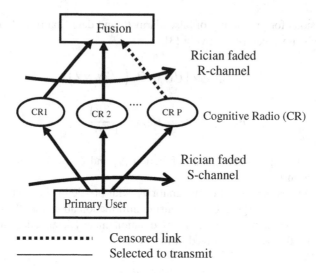

Fig. 2. Cooperative spectrum sensing network with censoring

2.1   Rician Fading

In Fig. 2, we are considering both S-channel and R-channel are affected by Rician fading. Due to Rician fading effect all the radio links present in the network are follows Rician distribution. Then fading coefficients can be generated as [3];

$$|h_k| = |X_1 + jX_2| = \sqrt{X_1^2 + X_2^2} \tag{8}$$

$$X_1 \sim (v, sigma) \qquad X_2 \sim (0, sigma)$$

$$v = \sqrt{\frac{K}{1+K}} \qquad sigma = \sqrt{\frac{1}{2(1+K)}} \tag{9}$$

Where 'K' is the Rician fading parameter.

Rician distribution PDF given [3],

$$f_y(x) = \frac{K+1}{\overline{\gamma}} \exp\left(-K - \frac{(K+1)\gamma}{\overline{\gamma}}\right) * I_0\left(2\sqrt{\frac{K(1+K)}{\overline{\gamma}}}\right) \tag{10}$$

Where $\overline{\gamma}$ is the average SNR. The average detection probability for Rician channel $(P_{d,Ric})$, can be obtained as:

$$\overline{P}_{dRic}\big|_{u=1} = Q\left(\sqrt{\frac{2K\overline{\gamma}}{k+1+\overline{\gamma}}}, \sqrt{\frac{\lambda(K+1)}{k+1+\overline{\gamma}}}\right) \tag{11}$$

2.2 Majority Logic Fusion Rule

In hard decision fusion rule, FC receives the one bit information (either '0' or '1') from all CRs. With the aid of censoring scheme in the R-channel, out of 'P' number of CR users only 'k' numbers of CRs are received at FC. These 'k' numbers of CRs send their information to the FC through the R-channel. The receive decision denoted by

$$u_k = \begin{cases} 1 & \text{if the received decision is } H_1 \\ 0 & \text{if the received decision is } H_0 \end{cases} \tag{12}$$

where $k \in \{1, 2, \ldots \ldots P\}$. Finally, all the information is collected at FC and majority logic rule can be applied at the fusion center by using below expressions [7];

$$u_0 = \Gamma(u_1 \ldots \ldots u_p) = \begin{cases} H_1 & \text{if } \sum_{k=1}^{p} u_k \geq \frac{P}{2} + 1 \\ H_0 & \text{if } \sum_{k=1}^{p} u_k < \frac{P}{2} + 1 \end{cases} \tag{13}$$

2.3 Maximal Ratio Combining

In case of soft decision fusion rule all the CRs send their sensing information in the form of energy values to the FC without performing any local decision. MRC logic can be obtained by simplifying LRT fusion rule [9]:

$$\wedge = \prod_{k=1}^{P} \frac{f(y_{k,d} \mid H_1)}{f(y_{k,d} \mid H_0)} \tag{14}$$

$$= \prod_{k=1}^{p} \frac{P_{d_k} + (1 - P_{d_k}) e^{\frac{-4\sqrt{E_b}}{\sigma_w^2} \text{Re}(y_{k,d} h_k^*)}}{P_{d_k} + (1 - P_{f_k}) e^{\frac{-4\sqrt{E_b}}{\sigma_w^2} \text{Re}(y_{k,d} h_k^*)}} \tag{15}$$

$$\sigma_w^2 = E_b \sigma_h^2 + \sigma_n^2 = \frac{E_b \sigma_n^2}{E_b + \sigma_n^2} + \sigma_n^2 \tag{16}$$

If the channel is error free i.e. perfect channel then, $\sigma_h^2 = 0$ then above Eq. (16) reduces to $\sigma_w^2 = \sigma_n^2$

taking logarithm on both sides to (15) and after simplification, it reduces to

$$\wedge_1 = \log(\wedge) = \frac{2\sqrt{E_b}}{\sigma_w^2} \sum_{k=1}^{K} (P_{d_k} - P_{f_k}) \text{Re}(y_{k,d} h_k^*) \tag{17}$$

If the selected CRs have identical local performances, $\wedge_1$ can be simplified further as follows:

$$\wedge_{MRC} = \sum_{k=1}^{K} \text{Re}(y_{k,d} h_k^*) \tag{18}$$

Where h_k^* is the complex conjugate of $\hat{h}_k$. At FC decision can be taken in favor of H_0 or H_1 by comparing $\wedge_{MRC}$ with the threshold zero.

3 Threshold Based Censoring

In case of threshold based censoring scheme, a particular CR can be transmitted if the amplitude of estimated R-channel coefficient is greater than censoring threshold (C_{th}). The cumulative distribution function (CDF) of Rician distributed random variable in terms of C_{th} can be derived as

$$F_{Ric}(C_{th}) = 1 - Q_1\left(\frac{v}{\sigma}, \frac{C_{th}}{\sigma}\right); \ C_{th} \geq 0 \tag{19}$$

where Q_1 is the Marcum-Q function, $v = \sqrt{\frac{K}{1+K}}, \sigma = \sqrt{\frac{\Omega}{2(1+K)}}$ and 'K' is the Rician fading parameter. Now the probability of selecting a CR in Rician faded R-channel can be derived as

$$p_{ric} = \text{Pr}(|\hat{h}_k| > C_{th}) \tag{20}$$

$$1 - F_{ric}(C_{th}) = Q_1\left(\frac{v}{\sigma}, \frac{C_{th}}{\sigma}\right) \tag{21}$$

Now by using binomial distribution, the probability of selecting 'k' CRs from 'P' CRs can be calculated as follows:

$$P(k) = \binom{P}{k} p^k (1-p)^{(P-k)} \tag{22}$$

In above Eq. (22), $P(k)$ represents the probability of selecting k CRs, P is the total number of CR users and k is the CRs which are gathered at FC.

4 Results and Discussion

Figure 3 is drawn between the missed detection probability Q_m Vs censoring threshold value C_{th} with perfect channel estimation using threshold based censoring scheme. Following network parameters are considered to simulate this graph: number of CR users N = 30, probability of false alarm P_f = 0.05, S-channel SNR = 20 dB, Rician

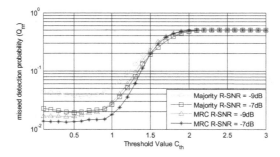

Fig. 3. Missed detection probability versus censoring threshold for various values of R-channel SNR with perfect channel estimation using majority and MRC logic rules at fusion center.

fading parameter (K) = 3 and R-channel SNR = -9 dB & -7 dB in the presence of Rician fading. As C_{th} value increases, Q_m obtain an optimum value, after that it increases with C_{th}, later it attains constant value after reaching a certain value of C_{th}, this is due to changing the value of probability mass function with the threshold value for each CR. We have considered two various values of R-channel SNR (-9 dB & -7 dB), as the R-channel SNR increases, Q_m value decreases in both MRC and majority logic rules because of fading effect present in the R-channel decreases, so that more number of CR users get selected and all these selected CRs passes their sensing information to the FC. After certain value of threshold, Q_m became constant in perfect channel because FC select best R-channel links which are having lowest probability of getting rejected. For a particular value of C_{th} = 0.8, R-channel SNR = -7 dB and MRC logic rule is used at FC instead of majority logic, the Q_m value reduced by 31.42 %.

From above lines we can conclude that MRC achieves lower value of Q_m than majority logic. If R-channel SNR increases from -9 dB to -7 dB, Q_m value decreases by 46.7 % with majority logic and it decreases by 25.1 % with MRC rule at C_{th} = 0.8.

In Fig. 4, comparison between MRC and majority logic with perfect channel estimation is provided for various values of S-channel SNR. Network parameters: N = 30, P_f = 0.05, R-channel SNR = -7 dB, K = 3 and S-channel SNR = 15 dB & 20 dB are used to simulate this graph under Rician fading. As the S-channel SNR increases, Q_m value decreases in both majority and MRC logic fusion rule because of fading effect and noise value in S-channel is decreases. As the fading effect decreases in the S-channel, CRs sense the PU activity more effectively hence, Q_m value decreases. For a particular value of C_{th} = 0.8 and S-channel SNR = 20 dB, Q_m values with majority logic and MRC fusion rules are 0.0215 and 0.0144 respectively. MRC logic achieves least value of Q_m compared to the majority logic. When S-channel SNR value is varying from 15 dB to 20 dB, Q_m value decreases by 68.2 % with majority logic and it decreases by 76.1 % with MRC rule at C_{th} = 0.8. For a particular value of S-SNR = 20 dB and C_{th} = 0.8, the Q_m value reduced by 31.4 % with MRC rule compared to the majority logic rule.

Figure 5 shows that graph between Q_m versus C_{th} value for various values of probability of false alarm P_f with perfect channel estimation under Rician fading. N = 30, R-channel SNR = -7 dB, P_f = 0.05 & 0.0005, (K) = 3 and S-channel

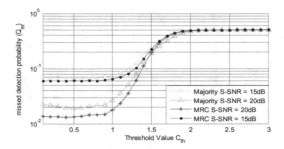

Fig. 4. Missed detection probability versus censoring threshold for various values of S-channel SNR with perfect channel estimation using majority and MRC logic rules at fusion center.

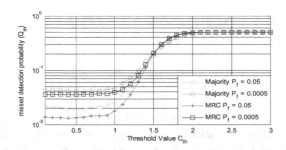

Fig. 5. Missed detection probability versus censoring threshold for various values of probability of false alarm with perfect channel estimation using majority and MRC logic rules at FC.

SNR = 20 dB are considered as network parameters to get Fig. 5. In this case, Q_m achieves constant value of 0.5056 for both majority logic and MRC fusion rule after it reaches $C_{th} = 2.0$. As P_f value increases from 0.0005 to 0.05, Q_m value decreases in both majority and MRC logic rule. If P_f value increases (from Eq. (4)), the detection threshold value (λ) is decreases for each CR, it improves the detection probability value. As P_f increases, there is a possibility of spectrum utilization becomes low and number of missing opportunities decreases this leads to increases in detection probability. As P_f value varying from 0.0005 to 0.05, Q_m value decreases by 61 % with MRC fusion rule and it decreases by 49.6 % with majority rule at $C_{th} = 0.8$. For a particular value of $P_f = 0.05$ and $C_{th} = 0.8$, the missed detection probability value is reduced by 31.4 % with MRC rule compared to the majority logic rule.

In Fig. 6, the performance comparison between majority logic and MRC fusion strategies are evaluated with perfect channel estimation by varying number of CR users (N). The performance is evaluated for $P_f = 0.05$, different values of number of CRs N = 15 & 30, K = 3, R-channel SNR = −7 dB and S-channel SNR = 20 dB. As the number of CR users increases, cooperation among the users increases, hence, (Q_m) value decreases in both MRC and majority logic rule. As the number of CR users are increases, the sensing information received from each CR at FC about PU increases, hence, detection performance is increases. For a particular value of $C_{th} = 0.8$ and N = 15, missed detection probability is more with majority logic (0.1102) compared to

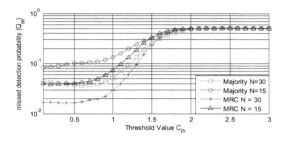

Fig. 6. Missed detection probability versus censoring threshold for various number of CRs with perfect channel estimation using majority and MRC logic rules at fusion center.

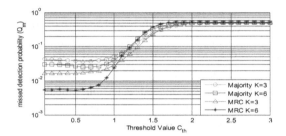

Fig. 7. Missed detection probability versus censoring threshold for various values of Rician fading parameter with perfect channel estimation using majority and MRC logic rules at FC.

MRC logic rule (0.0496), (Q_m) value reduced by 54.9 % with MRC rule compared to the majority logic rule.

Next, we have considered the effect of Rician fading parameter (K) on missed detection probability (Q_m) in Fig. 7. The following factors are used to simulate this graph: N = 30, P_f = 0.05, (K) = 3 & 6, R-channel SNR = −9 dB and S-channel SNR = 20 dB. As the fading parameter (K) increases, Q_m value decreases in both MRC logic rule and majority logic fusion rule because of fading effect and noise value decreases in S-channel and R-channel. As the fading parameter increases, the dominant multipath component power (means line of sight (LOS) wave power) increases, hence, information received at FC will be more perfect; this will increase the detection probability. For a particular value of C_{th} = 0.8 and fading parameter increases from K = 3 to K = 6, Q_m value decreases by 61.5 % with MRC logic rule and it decreases by 32.4 % with majority logic rule. Here also MRC logic has less value of Q_m compared to majority logic. At certain instant, if K = 6 and C_{th} = 0.8, (Q_m) value reduced by 71.8 % with MRC rule compared to the majority logic.

In Fig. 8, Rician fading parameter (K) and number of CRs (N) are considered as variable parameters to evaluate the performance of total error probability $(Q_m + Q_f)$. As 'K' value is increases from 3 to 6 & 'N' value is decreases from 30 to 15, $(Q_m + Q_f)$ decreases by 61.2 % with majority logic and it decreases by 62.2 % with MRC rule with perfect channel estimation at (C_{th}) = 1.0. Though 'K' value increases

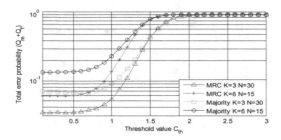

Fig. 8. Total error probability versus censoring threshold for various values of fading parameter (K) and number of CR users (N) with perfect channel estimation using majority and MRC logic rules at fusion center.

and 'N' value decreases, $(Q_m + Q_f)$ value increases because of cooperation among the users are decreases. MRC achieves lower value of total error compared to majority logic. For a particular case, for K = 3, N = 30 and MRC rule is used at FC instead of majority rule, $(Q_m + Q_f)$ value reduced by 42.5 % at C_{th} = 1.0.

The above Table 1 shows that missed detection probability (Q_m),false alarm probability (Q_f) and total error probabilities values and comparison between majority and MRC logic by varying different network parameters like: R-channel SNR, S-channel SNR, probability of false alarm, CR users (N) and Rician fading value (K) with perfect channel estimation. All the above tabulated values are considered at C_{th} = 1.0 under Rician fading channel. The performance of MRC logic is better in terms of missed detection and total error probabilities than majority logic fusion rule. MRC logic achieves lower error values compared to majority logic with perfect channel estimation.

5 Conclusion

In this paper, censoring of cognitive radios (CRs) under Rician fading using energy detection has been investigated in CSS. MRC rule and majority logic fusion rules are applied at FC to decide about primary user activity with perfect channel estimation. The performance is evaluated using missed detection probability (Q_m) & total error probability $(Q_m + Q_f)$. We also observed the effect on (Q_m) & $(Q_m + Q_f)$ by varying the network parameters like: S-channel SNR, R-channel SNR, probability of false alarm (P_f) and Rician fading parameter (K) with perfect channel estimation. Comparison between majority logic and MRC rule is provided and also observed that majority logic is having higher missed detection probability than MRC rule. Finally, we can conclude that the detection performance is increases by using censoring of CRs. This work is useful to reduce the complexity of CSS network, useful to eliminate the unwanted CRs which having lowest threshold value and traffic overhead problem can be avoided.

References

1. Federal Communication Commission: Spectrum Policy Task Force. Report ET Docket no. 02-135 (2002)
2. Digham, F.F., Alouini, M.-S., Simon, M.K.: On the energy detection of unknown signals over fading channels. In: Proceedings of IEEE International Conference on Communications (ICC 2003), pp. 3575–3579, May 2003
3. Ghasemi, A., Sousa, E.S.: Collaborative spectrum sensing for opportunistic access in fading environments. In: 2005 Proceedings of 1st IEEE Symposium on New Frontiers in Dynamic Spectrum Access Networks, pp. 131–136. Baltimore, USA (2005)
4. Teguig, D., Scheers, B., Le Nir, V.: Data fusion schemes for cooperative spectrum sensing in cognitive radio networks. In: Military Communication and Information System Conference (MCC), Poland, pp. 1–7, October 2012
5. Duan, J., Li, Y.: Performance analysis of co-operative spectrum sensing in different fading channels. In: Proceedings of IEEE International Conference on Computer Engineering and Technology, ICCET 2010, pp. v3-64–v3-68, June 2010
6. Sun, C., Zhang, W., Letaief, K.B.: Cooperative spectrum sensing for cognitive radios under bandwidth constraints. In: IEEE Wireless Communication and Networking Conference, WCNC 2007, pp. 1–5 (2007)
7. Ahmadi, H.R., Vosoughi, A.: Impact of channel estimation error on decentralized detection in bandwidth constrained wireless sensor networks. In: Proceedings of IEEE MILCOM 2008, San Diego, USA, pp. 1–7 (2008)
8. Lunden, J., Koivunen, V., Huttunen, A., Poor, H.V.: Censoring for collaborative spectrum sensing in cognitive radios. In: Proceedings of the 41st Asilomar Conference on Signals, Systems, and Computers, Pacific Grove, CA, USA, pp. 772–776 (2007)
9. Gradshteyn, I.S., Ryzhik, I.M.: Table of Integrals, Series, and Products, 5th edn. Academic Press, Cambridge (1994)
10. Nuttall, A.H.: Some integrals involving the QM-function. IEEE Trans. Inf. Theory $21(1)$, 95–96 (1975)
11. Ahmadi, H.R., Vosoughi, A.: Impact of channel estimation error on decentralized detection in bandwidth constrained wireless sensor networks. In: Proceedings of IEEE MILCOM 2008, pp: 1–7, November 2008
12. Ferrari, G., Pagliari, R.: Decentralized detection in sensor networks with noisy communication links. In: Devoli, F., Pelezzo, S., Zappetore, S. (eds.) Distibuted Cooperative Laboratories: Networking, Instrumentation, and Measurements, pp. 233–249. Springer, New York (2006)

Wireless Network Virtualization: Opportunities for Spectrum Sharing in the 3.5 GHz Band

Marcela M. Gomez[✉] and Martin B.H. Weiss[✉]

School of Information Sciences, University of Pittsburgh,
Pittsburgh, PA 15260, USA
{mmg62,mbw}@pitt.edu

Abstract. In this paper, we evaluate the opportunities that Wireless Network Virtualization (WNV) can bring for spectrum sharing by focusing on the regulatory framework that has been deployed by the Federal Communications Commission (FCC) for the 3.5 GHz band. We pair this innovative regulatory approach with another novel arrangement, Wireless Network Virtualization, and thus assess the resulting opportunities from the perspectives of regulation, technology and economics. To this end, we have established a comprehensive foundation for further exploration and development of virtualized networks that would provide significant opportunities for enabling and enhancing current sharing arrangements.

Keywords: Wireless Network Virtualization · Spectrum sharing

1 Introduction

The complexity of managing electromagnetic spectrum is not purely technical. There are crucial economic and regulatory implications that determine whether an alternative for making more efficient use of this resource would be beneficial or detrimental. Therefore, we perform an analysis that goes beyond the existing technical barriers and extends along three axes: regulation, technology and economics.

In this work, we focus on the 3.5 GHz band and its regulation as well as the innovative technology of Wireless Network Virtualization (WNV) to explore the opportunities and challenges in introducing sharing opportunities. Our study focuses on one particular approach of WNV that is built on resource pooling. Thus, we will study the characteristics of resource pools, the interaction between user types (Incumbents, Priority Access and General Authorized Access users) and how economic considerations drive the definition of networks and the resulting types of competition. We expect that this comprehensive analysis will permit us to solidify the basis for further deployment of an appropriate virtualization environment for spectrum sharing.

This paper is organized as follows: the regulatory framework for the 3.5 GHz band is presented in Sect. 2; Sect. 3 includes a description of WNV and the particular approach that will be considered in this work; Sect. 4 includes a technical

© ICST Institute for Computer Sciences, Social Informatics and Telecommunications Engineering 2016
D. Noguet et al. (Eds.): CROWNCOM 2016, LNICST 172, pp. 232–245, 2016.
DOI: 10.1007/978-3-319-40352-6_19

analysis, which presents the two models that could be adapted to the opportunities offered by regulation in the 3.5 GHz band; Sect. 5 analyzes three important aspects associated with Economics, which target at framing our model within this context, and finally, Sects. 7 and 8 present our conclusions and future work, respectively.

2   3.5 GHz Band: Current Status

To date, the 3.5 GHz band in the U.S. has been allocated to federal services (e.g.,DoD radar systems), Fixed Satellite Service (FSS) and, for a finite period, to grandfathered terrestrial wireless operations in the 3650–3700 MHz band [1]. The Federal Communications Commission (FCC) and the National Telecommunications and Information Agency (NTIA) have made a significant effort toward opening this band for shared operations between federal and commercial users. The FCC has referred to this band as an "innovation band," given that the main objective is to enable new spectrum access models that allow the use of modern technologies, thus enabling a move away from legacy spectrum management categories: Federal vs. Non-Federal; Licensed vs. Unlicensed and Carrier vs. Private [1]. The basis of this new spectrum sharing scheme is a three-tiered model for spectrum access, with each tier holding a different level of priority: Incumbent Access, Priority Access and General Authorized Access (GAA). Some important characteristics of these tiers include [2]:

- Incumbent users comprise federal services and some legacy satellite and wireless operations. These users have superior spectrum rights over Priority Access and GAA users at all times and in all areas.
- The Priority Access tier consists of seven channels of 10 MHz each, which can be assigned to Priority Access Licensees. These licensees will have more predictable spectrum access than GAA users. Nevertheless, Priority Access Licenses (PALs)[1] will be granted as long as the demand is greater than the supply in the area of interest. If that is not the case, the entire band will be allocated for GAA use.
- General Authorized Access (GAA) will be granted by rule. In this way, GAA users could potentially access the entire 150 MHz band in areas where PALs have not been issued (or are not in use) and up to 80 MHz where PALs are in use. It is important to note; however, that GAA users will not be protected from interference from other Citizens Broadband Radio Service (CBRS) users.

Through the aforementioned characteristics, it is expected that this three-tiered approach will enable the adaption of spectrum use to market and user demands. Figure 1 illustrates the tentative bandplan, proposed by the FCC, for the 3.5 GHz band.

---

[1] PALs are defined as an authorization to use a 10 MHz channel in a single census tract for three years. These licenses will be assigned in up to 70 MHz of the 3550–3650 MHz portion of the band [2].

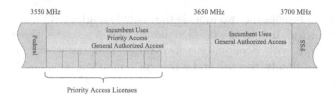

Fig. 1. Tentative bandplan under the 3.5 GHz sharing framework.

Sharing in the 3.5 GHz band will be enabled by a Spectrum Access System (SAS). According to [2], "[t]he SAS serves as an advanced, highly automated frequency coordinator across the band. It protects higher tier users from those beneath and optimizes frequency use to allow maximum capacity and coexistence for both GAA and Priority Access users". In other words, the SAS is an entity that will be in charge of authorizing spectrum access to CBRS users in any frequency and location. Additionally, the SAS is in charge of providing Priority Access Licensees and GAA users with alternative spectrum when they have been displaced by users with higher priorities [3]. In general terms, the SAS should fulfill the *automated frequency assignment* task that will enhance the band management flexibility pursued with this sharing scheme. With the flexible access model developed for this band, the FCC aims at creating a versatile band which will permit to adapt to market as well as technological opportunities [2]. Figure 2 summarizes some important details regarding this three-tiered sharing framework.

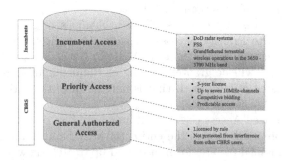

Fig. 2. Three-tier sharing framework

3   Wireless Network Virtualization: The Technology of Choice

From the regulatory approach presented in the previous section, we infer that flexibility for innovation is a key policy objective. Nevertheless, for innovation to be successful we should not only contemplate regulatory flexibility; in fact, we also require that technology allows for adding such flexibility to the network. Along these lines, we find that there is a significant link between Wireless Network Virtualization (WNV) and adding technical flexibility to networks and systems.

Through virtualization, different components of the network are partitioned, combined, sliced and abstracted to create virtual instances of the network. Further, each type of partition, combination or abstraction will yield distinct types of virtual networks giving us the impression that we are working with a *new* network, different from the original [4]. For benefit to be extracted, the virtualization process should be transparent to the users of a virtual network, thus making them oblivious to the underlying virtualization process. As a result, multiple virtual networks operate on one single network, each serving specific purposes and utilizing distinct technologies. Furthermore, co-existing virtual networks may be different from each other [5,6], or as stated in [8], Mobile Network Virtualization "promises multiple personality network elements in terms of virtual ownership by multiple operators. That means multiple networks running virtually (i.e., logically) and concurrently within one physical network equipment or hardware". Notably, this would call for an important degree of isolation embedded in the virtualized systems, which will permit a sound co-existence of virtual entities.

With the adequate application of virtualization technologies, we would be able to devise improved alternatives for the use, sharing and assignment of existing resources [7]. This could provide a degree of flexibility that would aid in maximizing the spectrum access and management options on the operator side. Several alternatives for the application and deployment of WNV have been explored. However, given the characteristics of the new sharing framework for the 3.5 GHz band, we consider virtualization from the perspective of *resource pooling*. This approach requires multiple entities/providers to share their resources in a pool and then make them accessible to alternative users/providers. To elaborate on the resource pooling concept, the authors in [9,10] have compared it to the Cloud (in a computer science context), given that, in principle, it gives us the illusion of an infinite amount of resources, which are available on demand without the need to incur in high upfront commitments and actually permitting users to pay for them on a short-term basis or as needed. Focusing on the idea of *access on demand*, we could expect that the users who have access to the pool will be allowed to choose the resources that are most suitable for a particular service, but which may belong to different incumbents or access tiers.

Centering our attention on spectrum, the objective of pooling this resource is to "enhance spectral efficiency by overlaying a new mobile radio system on an existing one without requiring any changes to the actual licensed system" [11]. Thus, the deployment of spectrum pools would imply a different resource allocation system, where the existing and new hardware can be operated transparently, or in other words, as if there were no other system concurrently present in the same frequency range [11]. In this manner, we can merge the key concepts behind WNV and the creation of resource pools and present them as important alternatives for providing enhanced spectrum access and sharing opportunities [10–13].

4 Technical Design

In this section, we aim at providing a technical overview of the creation of spectrum pools. We will present a *local* and a *global* architecture construct, which will permit us to illustrate some of the benefits that can be derived from virtualization.

Local Approach. In the local approach, we point out potential benefits of the construction and operation of a resource pool within the 3.5 GHz band *only*. From the regulatory approach presented in [2] and as shown in Fig. 1, the assets available for conforming the resource pool are the following:

- 3550–3650 MHz band: 0–70 MHz for PALs and 30–100 MHz for GAA
- 3650–3700 MHz band: 50 MHz for GAA

For the design of this approach we have explored the actual responsibilities of the SAS. Note that, at the basis, the SAS is in charge of the automated allocation of resources (i.e., spectrum access management). Nevertheless, in a virtualized environment, we consider the option of the SAS outsourcing part of its spectrum pool management duties to an external entity known as the *Virtual Network Builder (VNB)*. The VNB is an intermediate entity in charge of aggregating spectrum (and perhaps additional network resources) and offering it to its own customers (i.e., Service Providers). For aggregating spectrum, the VNB should negotiate access with the SAS, and at the same time, it should be aware of the expected demand of the SPs with whom it works. In this context, the SAS would treat the VNBs as large spectrum users or operators. As such, VNBs would auction for PALs from the SAS and compete with other Priority Access and GAA users under the same rules. In a broad sense, this is consistent with the notion of *polycentric governance* described in [24]. This structure is portrayed in Fig. 3.

Given that the VNB should account for the resources to serve the aggregate requirements of its customers, the demand from the VNB should be significantly larger than that of individual entities. When posting bids for PALs, the VNB operations could lead to two important consequences: (1) the VNB can compete with other large stakeholders (e.g., Verizon, AT&T) in terms of the amount that the latter are able to pay for obtaining a license; (2) it is likely that the 'demand greater than supply' constraint for PAL assignment will be met given the aggregate demand that the VNB carries. In this light, this *local* approach provides opportunities for enhancing the sharing arrangements.

As shown in Fig. 3, there is a certain hierarchy among the different entities that belong to this type of network. Indeed, we could associate specific tasks and behaviors to each layer: The SAS would be considered as the regional spectrum access coordinator. It is in charge of the automated process of assigning licenses to the entities in the layer below and, in turn, it is accountable to the regulator (i.e., the FCC) and incumbents in the layer above. The next layer consists of the VNBs or large Network Operators who will negotiate spectrum access directly with the SAS. These will be entities that require larger spectrum assignments

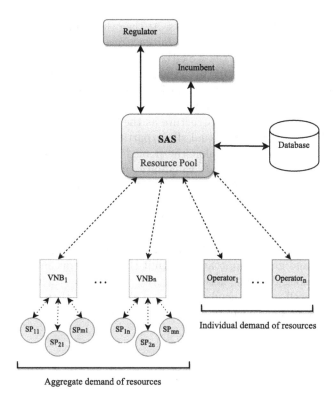

Fig. 3. Virtual network builders as part of the sharing scheme in the 3.5 GHz band

than smaller SPs. The final layer of the hierarchy will be composed of individual SPs who will require spectrum from VNBs or from large Network Operators (as in the case of Mobile Virtual Network Operators (MVNOs)).

We could expect this localized approach to evolve into a virtualized one, especially if we consider pooling resources that belong to multiple providers and we make them available to additional SPs. This type of arrangement can be explained through our global approach, where the virtualization options can be further explored.

Global Approach. A global approach represents a more complex arrangement that targets at adding flexibility to the network and incrementing the opportunities for new entrants. In this scheme, we envision the resources of the 3.5 GHz band as *one* of the multiple inputs to the resource pool. Hence, we would have various frequency bands, licensed and unlicensed, available in the pool, which would represent more possibilities for the VNB to aggregate resources and thus satisfy the service requirements of a larger range of users.

The changes in the architecture under the global approach are shown in Fig. 4. In this case, multiple resource providers (RPs) make their resources available to the pool, which is managed by the VNBs. At the other end, we have various SPs requesting resources from the pool via interactions with the VNBs. Note that the VNBs have also access to the 3.5 GHz band via interactions with the SAS.

The virtualization process in this scenario would be complete when we envision the pool as a set of spectrum and infrastructure resources which can be seamlessly accessed by the RPs and SPs. For this purpose, through WNV, RPs could be utilizing the same infrastructure as the one they are making available in the pool, just on different virtual slices/partitions. If virtualization is properly deployed, we could fully exploit the pooled resources given that we would have the illusion of higher *virtual* availability while preserving the fixed *physical* resources. The VNB would be in charge of aggregating resources upon SPs' demand, which will in turn depend on the specific service that each SP intends to provide. Note that at the basis we would still have physical resources, which are partitioned in different forms. In this way, we would expect the SPs to be compatible and capable of using the virtualized resources offered by the VNB.

These local and global approaches permit the incorporation of WNV, redistribute tasks among different network entities and rely on their interactions to enhance the overall sharing environment. In the section that follows, we look into relevant economic aspects that could help us further evaluate the feasibility of the virtualized approaches.

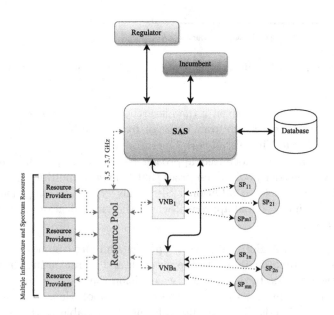

Fig. 4. Generalized approach for sharing and virtualization.

5   Economic Evaluation

5.1   The Innovative Architecture from an Economics Perspective

Innovation has driven significant changes, not only in the technological field, but also on the markets developed to sustain and spread that innovation. In order to place our virtualization ideas within the appropriate context, we would like to point out some significant similarities between our study and the work developed by Hagel and Seeley-Brown in [14].

From the various proposals presented in [14], we find an important similarity between our virtualized approaches and the concept of reverse markets. In such markets, customers can seek the greatest possible value from a broad set of providers which are available at an appropriate time and place. Reverse markets have further led to the design of *process networks*, which are in charge of mobilizing "highly specialized companies across more than one level of an extended business process" [14]. Process networks adopt a *pull* model "where resources are flexibly provided in response to a specific market demand" [14]. When the network needs cannot be easily determined in advance, operators and providers could create platforms permitting them to mobilize their resources readily. This model further suggests a different means to deal with uncertainty given that it can "help people come together and innovate by drawing on a growing array of specialized and distributed resources" [15]. In this light, the ultimate benefit from process networks and pull systems, in terms of uncertainty, would be the possibility of not seeing it as a threat, but as an opportunity to innovate [15].

In this context, we could also associate the characteristics of the VNB with that of a *process orchestrator*, which is an entity in charge of organizing and managing process networks. Some of its duties include determining the eligibility of an entity to participate in the process network; defining the role of each participant in particular process implementation and ensuring that each participant performs as expected and is rewarded accordingly [14]. The orchestrators should focus on expanding the range of participants and creating strong relationships among them. In this way, more specialized skills are accessible, and at the same time, the collaborating parties can build their capabilities faster [15].

To summarize, the local and global models we present in this work adapt to the pull system studied in [14], given that it explores the possibility of generating supply from the aggregation of (specialized) resources belonging to different entities. Additionally, it aims at managing local resource assignment by means of a general orchestrator, which in our models corresponds to the Virtual Network Builder. Since we are dealing with a framework in which different entities (SPs) are providing a service with the aggregation of resources belonging to other operators (RPs), we envision a service-based type of competition. In this way, it is important to shed some light on the nuances, opportunities and challenges of switching from a traditional facility-based competition to service-based competition.

5.2   Facility-Based vs. Service-Based Competition

When we analyze facility (or infrastructure)-based competition and contrast it with service-based competition, we are not facing a "black or white" type of situation. Instead, we can find a wide range of possibilities and arrangements between these two poles. This has important implications in terms of the complexity of the strategies adopted by incumbents and entrants and the regulatory schemes that are optimal.

At the core of these competition decisions, we have a set of trade-offs that incumbents and entrants should take into account. Indeed, each user will decide to enter in either arrangement depending on the level of profitability that it represents. For instance, incumbents should evaluate the benefit from investing in their own infrastructure and share it with new entrants versus the possible threat of competing with new market entrants who possess their own market infrastructure. New entrants, on the other hand, should determine how limited their competitiveness will be in the market if they are subject to the lease arrangements provided by the incumbents, and at the same time, they should contrast those limitations with the investment required for deploying their own infrastructure (i.e., opportunity cost of technology adoption) [16,17].

Referring to a traditional view of networks, we find that it widely favors facility-based competition and sees service-based competition as the stepping stone for the rise of the first. Nevertheless, if we adopt the process networks perspective presented in Subsect. 5.1, we could envision models and systems that successfully operate under service-based competition. Furthermore, when adapting our virtualization considerations, a wider array of resource usage models can be considered, which not only represents additional service opportunities for the new entrants, but also decreases the threat that these users can pose to the incumbents, e.g., threat caused by new entrants providing the same service as the incumbent. Moreover, the aggregation and assignment activities of the VNB could make the negotiation process easier for entrants and incumbents, thus reducing the associated costs. In this way, we would obtain positive conditions for a successful switch toward service-based competition.

5.3   Value Chains vs. Value Networks

According to [12], "[t]he value chain includes all the activities that exist as a direct result of usage of the cellular network. The purpose of creating the chain is to understand where the costs are incurred and the revenue is generated". Generally, a value chain is associated with a particular network operator or incumbent, and it will help to determine the activities that will be more profitable. Due to the significant changes in spectrum sharing arrangements, technology use and service availability, we can expect that the traditional value chain will shift to new perspectives in which, not only an incumbent's view on how to derive value from its resources and make profits is considered; instead, we might be interested in a new approach which encompasses the interactions of multiple users for generating valued services.

We have already evidenced examples that portray significant changes in the structure of value chains, such as the appearance of MVNOs, the evolution of Wi-Fi which has turned its hotspots into important complements of regular mobile networks, and also the creation of over-the-top services. From these examples, one can notice that different parts of the value chain that generate revenue, can be actually controlled by entities different from those that have deployed and control the parts associated with the highest costs [12]. In this way, as value chains continue to evolve, it is possible to observe how various value chains become intertwined for the creation of more complex networks where different entities are simultaneously involved in more than one value chain. We can refer to these as *value networks*.

A value network presents multiple entry and exit points, which increase the complexity of operations for all the members involved [20]. Additionally, it is expected that this network will be formed by "different actors drawn form a range of industries that collectively provide goods and services to the end users" [20]. For this purpose, these industries should show a higher level of specialization in particular activities, instead of managing the overall production of services. Furthermore, the companies involved are expected to dynamically evolve and perhaps specialize and gain expertise in additional areas. Hence, for the final service provision, relationships among multiple, specialized companies should be established [20].

This new notion of specialization and interaction among entities, calls for the modification of the boundaries of a company, which is evidently accompanied by a corresponding trade-off: value of specialization versus the transaction costs associated with external suppliers [20]. In this light, for setting their boundaries,

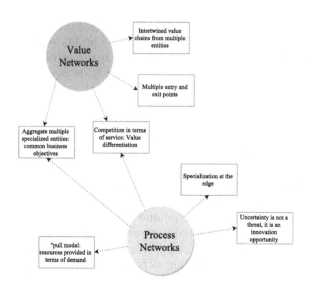

Fig. 5. Similarities between process and value networks.

firms should consider a balance between facing low transaction costs from internal production of services, thus lower agency costs and the economies of scale derived from obtaining resources from external entities [20].

Ultimately, the interaction of multiple users proposed by the value network approach permits us to study a firm's relationship with other network members and thus understand where value lies in the network and how it is created by multiple parties; how the activities of a firm will affect the network and how other members are likely to respond [21].

From the concepts presented in this section, we can find the relationship between value networks and process networks, which are illustrated in Fig. 5. Both mechanisms envision the aggregation of specialized entities to provide valued services, targeting at the deployment of service-driven networks and the accompanying type of competition.

6   Putting Things Together

Analyzing the network presented in Fig. 4, as a whole, we can point out important details that map to the concepts presented throughout this paper.

The entities in this network may have different degrees of specialization in multiple areas. In turn, these entities share their resources with others, thus promoting the development and provision of additional, perhaps more specialized services. This creates intertwined value chains as there is greater value extracted from a set of resources initially owned and used by a reduced group of incumbents or RPs. Additionally, this translates in a wider array of services provided throughout the network, which defines it as a service-based competition environment.

From the perspective of the RPs, there are increased opportunities for analyzing whether participation in the pool results in a profitable arrangement. This presents them with options to continue to participate, increase their participation or exit the network. The SPs at the other end of the network will generate a dynamic demand, dependent on the type of service that lies at the core of their business model. This represents less restrictions in terms of resource access and thus definition of the service to provide.

In a traditional system-based competition model, each SP would need to negotiate with every RP from which it requires resources. This is not a practical solution in terms of transaction costs, and possible restrictions in the establishment of leasing agreements with RPs. In the network we study, both RPs and SPs will negotiate resource access with a single entity: the VNB. In fact, the VNB will aggregate the required type and amount of resources based on the demand of the SPs, which is expected to be service-specific and dynamic. At the same time, the VNB should be in charge of providing the appropriate compensation to the RPs and/or negotiating with the SAS depending on the type of resources accessed.

Note that the flexible management of the resources belonging to the pool responds to the utilization of an enabling technology such as wireless network virtualization. In this way, the co-existence of multiple RPs and SPs would be ensured. It is evident that there is a greater degree of flexibility stemming from this network when compared with traditional system-based or facility-based competition arrangements. In the case of the latter, we can expect higher transaction costs associated with negotiations, given that specific leasing agreements should be developed among particular RPs and SPs, on a one-to-one basis. In the virtualized case, the negotiation is done through the VNB, which reduces the resulting overhead and allows for the seamless negotiation with multiple entities at a time. However, when designing the negotiation mechanisms between the VNBs and the SPs, we should take into account a framework that reduces agency costs, thus deterring strategic behaviors which could affect the overall welfare in the system.

7 Conclusions

We propose the incorporation of WNV to the sharing framework defined by the FCC for the 3.5 GHz band. The analysis we present does not reflect regulatory and technical considerations only, it also explores additional economic factors, which play a key role for the deployment of successful sharing models.

The studied fields pose important challenges and opportunities for the sharing model we devise. In this way, we have been able to find some benefits that could stem from embedding virtualization as the technical enabler for sharing approaches. Indeed, WNV would permit to add technical flexibility to the network, which is required to accomplish the regulatory flexibility that the current regulation seeks. Additionally, we have pointed out how the addition of a new entity, the Virtual Network Builder, could allow for the distribution of the functionality that has been assigned entirely to the SAS. In the model we propose, it is likely that smaller entrants will have higher opportunities to access spectrum. This results from having a VNB in charge of aggregating the demand from multiple users and posting bids for spectrum access. In this way, the VNBs could be better competitors in the market than smaller entities alone, and their possibilities to win resources in an auction may be significantly enhanced.

We found several similarities between the characteristics and objectives of process networks and those of value networks. When adapting these concepts to our model, we expect virtualization to allow for a seamless aggregation of resources from multiple entities thus permitting to exploit the specialization of network entities at their edge. This would provide an avenue for achieving common or service-differentiated business objectives, which could lead to appealing service-based competition opportunities taking place in current telecommunications market scenarios. Overall, our analyzed framework suggests that in an environment where multiple users with varied levels and areas of specialization come together to innovate, we could actually derive opportunities instead of threats from the uncertainty of sharing.

8   Future Work

In our efforts to extend our work, we consider it important to delve into details regarding how rights are adapted to these novel sharing schemes and, how social concepts and constructs influence the deployment of accurate models. Following the study presented in [22], we expect bundles of rights to be redefined in virtualized scenarios, which will in turn have a significant impact on the model design, outcomes and evaluation.

From a social perspective, our analysis of process and value networks has shed light on the interaction of multiple entities in order to achieve common and service-differentiated business objectives. In turn, these entities will be sharing assets, which could be mapped to the *common-pool resource* definition[2]. Keeping this in mind, and as explored by Ostrom in [23], we could expect *collective-action problems* to arise under our virtualization scenarios. As pointed out by Ostrom, a possible solution is the adoption of polycentric governance approaches, which implies the development of systems of governmental and non-governmental organizations working at multiple scales. The authors in [24] have already explored the inclusion of CPR concepts and polycentric governance to the design of the SAS and how this would help define facilitating conditions for the development of successful systems. In this way, we consider that analyzing CPR and Polycentric governance notions would provide us with a richer view on how to design our virtualization system.

Acknowledgments. This work has been supported in part by the U.S. National Science Foundation under Grant 1443796.

References

1. Leibovitz, J.: Breaking down barriers to innovation in the 3.5 GHz band (2015). https://www.fcc.gov/blog/breaking-down-barriers-innovation-35-ghz-band
2. FCC: Report and Order and Second Further Notice of Proposed Rulemaking. In the matter of the Commission's rules with regard to commercial operations in the 3550–3650 MHz band (2015). https://apps.fcc.gov/edocs_public/attachmatch/FCC-15-47A1.pdf
3. Marshall, P.: Spectrum Access System: Managing three tiers of users in the 3550–3700 MHz band. http://wireless.fcc.gov/workshops/sas_01-14-2014/panel-1/Marshall-Google.pdf
4. Wang, A., Iyer, M., Dutta, R., Rouskas, G.N., Baldine, I.: Network virtualization: technologies, perspectives, and frontiers. J. Lightwave Tech. **31**(4), 523–537 (2013)
5. Wen, H., Tiwary, P.K., Le-Ngoc, T.: Wireless Virtualization. SpringerBriefs in Computer Science. Springer, New York (2013)

---

[2] According to [23], "[c]ommon-pool resources are systems that generate finite quantities of resource units so that one person's use does subtract from the quantity of resource units available to others. Most common-pool resources are sufficiently large that multiple actors can simultaneously use the resource system and efforts to exclude potential beneficiaries are costly".

6. Zaki, Y., Zhao, L., Goerg, C., Timm-Giel, A.: LTE mobile network virtualization. Mob. Netw. Appl. **16**(4), 424–432 (2011)
7. Gomez, M.M., Cui, L., Weiss, M.B.H.: Trading wireless capacity through spectrum virtualization using LTE-A. TPRC Conf. Paper (2014)
8. Panchal, J.S., Yates, R., Buddhikot, M.M.: Mobile network resource sharing options: performance comparisons. IEEE Trans. Wireless Comm. **12**(9), 4470–4482 (2013)
9. Forde, T.K., Macaluso, I., Doyle, L.: Exclusive sharing and virtualization of the cellular network. In: New Frontiers in Dynamic Spectrum Access Networks (DyS-PAN) (2011)
10. Forde, T., Doyle, L.: Cellular clouds. Telecom. Policy **37**(2–3), 194–207 (2013)
11. Weiss, T.A., Jondral, F.K.: Spectrum pooling: an innovative strategy for the enhancement of spectrum efficiency. IEEE Comm. Mag. **42**(3), S8–14 (2004)
12. Doyle, L., Kibilda, J., Forde, T.K., DaSilva, L.: Spectrum without bounds, networks without borders. Proc. IEEE. **102**(3), 351–365 (2014)
13. Hua, S., Liu, P., Panwar, S.S.: The urge to merge: when cellular service providers pool capacity. In: IEEE International Conference on Communications (2012)
14. Hagel, J., Brown, J.S.: The Only Sustainable Edge: Why Business Strategy Depends on Productive Friction and Dynamic Specialization. Harvard Business Press, Boston (2005)
15. Brown, J.S., Hagel, J.: The next frontier of innovation. McKinsey Q. **3**, 82–91 (2005)
16. Baranes, E., Bourreau, M.: An economist's guide to local loop unbundling. Comm. Strat. **57**, 13 (2005)
17. Bourreau, M., Doğan, P.: Service-based vs. facility-based competition in local access networks. Info. Economics and Policy **16**(2), 287–306 (2004)
18. Cave, M.: Snakes and ladders: unbundling in a next generation world. Telecom. Policy. **34**(1), 80–85 (2010)
19. Garrone, P., Zaccagnino, M.: Seeking the links between competition and telecommunications investments. Telecom. Policy **39**(5), 388–405 (2015)
20. Li, F., Whalley, J.: Deconstruction of the telecommunications industry: from value chains to value networks. Telecom. Policy **26**(9), 451–472 (2002)
21. Peppard, J., Rylander, A.: From value chain to value network: insights for mobile operators. Eur. Manage. J. **24**(2), 128–141 (2006)
22. Cui, L., Gomez, M.M., Weiss, M.B.H.: Dimensions of cooperative spectrum sharing: rights and enforcement. In: New Frontiers in Dynamic Spectrum Access Networks (DySPAN) (2014)
23. Ostrom, E.: Polycentric systems as one approach for solving collective-action problems. Indiana University, Bloomington: School of Public & Environmental Affairs Research Paper. no. 2008-11 (2008)
24. Weiss, M.B.H., Lehr, W., Acker, A., Gomez, M.M.: Socio-technical considerations for Spectrum Access System (SAS) design. In: IEEE International Symposium on Dynamic Spectrum Access Networks (DYSPAN) (2015)

Distributed Topology Control with SINR Based Interference for Multihop Wireless Networks

Maryam Riaz[(✉)], Seiamak Vahid, and Klaus Moessner

University of Surrey, Guildford GU2 7XH, UK
m.riaz@surrey.ac.uk

Abstract. In this paper, a distributed approach to topology control (TC) is proposed where the network topology is established considering interference and routing constraints. This optimization problem however involves link scheduling and power assignment under SINR constraint, which is an NP hard problem. Opting for heuristics rather than exact approach, the proposed algorithms in the literature, either cannot guarantee the quality of the solution, or approximate the interference (protocol interference model) rather than using realistic SINR models. There is also lack of distributed exact/approximation approaches which can reduce complexity and provide practical solutions. Here, we propose a distributed approximation algorithm using column generation (CG) with knapsack transformation on the SINR constraint. Particle Swarm Optimization (PSO) is integrated with CG, to provide robust initial feasible patterns. The results show that, DCG-PSO with knapsack transformation increase the solvable instances three fold in terms of number of nodes, in comparison to the state-of-art approaches. The links are scheduled with less power, shorter scheduling lengths and reduces the computation time at lower penalty cost.

Keywords: Topology control · Scheduling · SINR · Approximation · Distributed

1 Introduction

Wireless Multihop networks have wide range of application in todays world [1,2] such as the military field communication and hot-spots for daily use. These networks can be deployed independently or can also co-exist with fixed infrastructure. Thus forming an integral part of the structure for future networks, which is considered a large dynamic mesh network. While the application of such networks is increasing with the advent of new applications, the three fundamental aspects: energy efficiency, connectivity and receiver centric interference mitigation are becoming more and more important. In order to address these fundamental challenges many solutions have been proposed in the literature, such as, power aware MAC layers [3] and location based routing for connectivity [4]. Despite the considerable amount of work in these directions, the proposed approaches not only increase the complexity of the layers but are also not able to accommodate

© ICST Institute for Computer Sciences, Social Informatics and Telecommunications Engineering 2016
D. Noguet et al. (Eds.): CROWNCOM 2016, LNICST 172, pp. 246–258, 2016.
DOI: 10.1007/978-3-319-40352-6_20

effectively all three fundamental challenges. Here we have considered topology control (TC) [1,2]. TC addresses these challenges on the link layer level and reduces the complexity of other layers such as MAC and routing layer. Generally, TC mainly comprises of power control [5] and/or scheduling [6,8] at the link layer. Here, we have considered scheduling with power control to implement the optimization framework. The proposed approach is a distributed approximation algorithm, providing fault tolerance, energy efficiency and mitigating interference. The interference is based on SINR rather than protocol model thus giving a realistic measurement of interference [9,11,12] and fault tolerance is attained by subjecting to the explicit k-vertex connectivity constraint [10,15], that connects each node to k other nodes. Here, the complexity of finding the solution is due to the SINR constraint which makes the problem NP complete [7,13]. While the maximum number of nodes cannot exceed six in order to find the optimal solution with SINR constraint [14], the approximation approaches in literature are also confined to a maximum of 18 to 30 links [19,20]. In most cases, either the SINR approximations such as node degree [13] or protocol model are used or the heuristics are opted instead of exact solutions. Although heuristic approaches offer less computation time and larger solvable instances than exact solutions, unlike the optimal (exact) approaches, heuristics cannot guarantee the quality of the solution: that is to remain within upper and lower bounds. On the other hand, the approximation approaches offer better overall solutions, by adhering to the lower or upper bound by a constant or dependent factor. s Here, mixed integer linear programming (MILP) is used with distributed CG method [16]. All the transmitters run the distributed algorithm simultaneously while utilizing the local information only. Depending on the network a centralized entity can provide the global information however the algorithm can run without it too. The proposed method provides a distributed topology with links offering minimum scheduling length with power control for Spatial Time Division Multiple Access (STDMA) multihop networks. In summary, our main contributions are as follows.

- We present a novel MILP formulation to minimize the scheduling length and total power of the network under routing, power and k connectivity constraints.
- A novel distributed approximation algorithm based on CG, DCG-PSO is proposed, where we transform the SINR constraint to knapsack problem and solve the pricing problem in two stages, which reduces the complexity of the problem. The feasible link set is increased and interference set is decreased while the upper and lower bounds are made efficient in each iteration.
- As CG performance is influenced by the initialization, instead to random initialization which may effect the stabilization of CG, we integrate distributed Particle Swarm Optimization (DPSO) to provide better initial feasible solutions.

The proposed DCG-PSO results in a 5 times increase in the number of nodes than the state of the art [18,19], less computation time and better solutions.

The resulting topology consists of links with less scheduling length, less power consumption and more spatial reuse. The DCG-PSO is the first approximation algorithm to the best of our knowledge that can solve up to 80 nodes, while supporting simultaneously k-connectivity, minimum transmit power and receiver centric interference mitigation.

The rest of the paper is organized as follows. At first, the network model is explained in Sect. 2. In Sect. 3, the formulation of optimization framework and algorithm is illustrated. The evaluation and results are given in Sect. 4, followed by conclusions in the Sect. 5.

2   Network Model

We consider a multihop STDMA wireless network consisting of N nodes with $(i,j)\varepsilon L$ directed links. The free space model is used for the channel gain G_{ij} calculation. The $G_{ij} = \varphi.d_{ij}^{-\alpha}$ where, d_{ij} is the distance between two nodes i and j, α is the path loss exponent and φ is the uniform random perturbation. P^i is the power of a node i while the value of the power can be continuous or integer value. The following formulation can be applied to CSMA given the respective changes are made in formulation.

2.1   Feasible Access Patterns

The feasible pattern is a subset of links transmitting simultaneously subjected to the given constraints. Here, for the links L in the network, a set of links $S \subseteq L$ that are simultaneously active such that no links share a node is called matching. If the links in a matching S are concurrently activated such that the minimum SINR requirement is met, then such a matching is called feasible matching or pattern. Here, the minimum SINR requirement, i.e given the SINR is provided by (1), the links within a matching satisfy the specified SINR threshold. In (1), γ is the SINR threshold and n_o is the noise The SINR threshold can be different for each link, here for simplicity purposes we have assumed same threshold for all links.

$$\frac{P_i G_{ij}}{n_o + \sum_{j,i\neq m}(P_m G_{jm})} \geq \gamma \tag{1}$$

The power of a link needs to be high enough to meet the minimum k connectivity and data rate requirement such that QoS does not get effected. Scheduling can be defined in a number of ways. Here, schedule Q is the index collection such that $Q = (S^q, \tau^q, q\epsilon Q_I)$, the scheduling length is measured in terms of number of slots as well as total duration in seconds. The length of schedule is $\tau^q \geq 0$, and represents the duration in the matching S, while $Q_I \subseteq Z^+$ is a large and finite set. The traffic demand vector is f_{ij} for the link (i,j) and is given in mbps. Each transmission frame length is divided into slots $q\epsilon Q_I$ and the within each slot the matching is active for the duration τ^q. The given demand is completely transferred in the frame length.

The flow conservation balance is the multihop routing constraint here. Flow conservation is applied per session $c \epsilon C$, where one session c comprises of a source, a destination and the demand. In the demand flow constraint, the f_{ij}^c is the demand flow for link i, j and session c. The $f_i^c = k$ when i node is the origin of session c, $f_i^c = -k$ if i is destination of c and zero otherwise. It means all the relaying nodes have $f_i^c = 0$.

3   Optimization Framework

3.1   Centralized Algorithm: Column Generation (CG)

The centralized algorithm is based on CG, it takes global information about the network into account and executes centralized knapsack algorithm for the formulation. The CG comprises of a master and a pricing problem, and it reduces the complexity of the problem by focusing on the variables that can have potential in improving the existing solution. CG is an iterative approach where it attempts to search for reduced cost variables in each iteration. In order to do so, the master problem (MP) solves its constraints and passes the dual variable to the pricing problem (PP) which comprises of constraints with exponential complexity. If a variable with reduced cost is found it is added to the optimal solution. The following is the MP formulation.

$$minimize \sum_{1 \leq q \leq |Q|} \tau^q \tag{2}$$

subjected to:

$$\sum_{j \epsilon N} f_{ji}^c - \sum_{i \epsilon N} f_{ij}^c = f_i^c \tag{3}$$

$$\sum_{i \epsilon N} f_{ij}^c \leq 1 \tag{4}$$

$$f_{ij}^c \epsilon (0, 1) \tag{5}$$

$$\sum_{1 \leq q \leq |Q|} u_{ij} \tau^q \geq f_{ij} \tag{6}$$

$$\tau^q > 0 \tag{7}$$

Here, the master problem of CG consists of the flow conservation constraint with the objective function of minimizing the scheduling length. The flow conservation and disjointed node is ensured by (3) and (4) respectively. The capacity constraint is (6) where u_{ij} is the Shannons capacity. Among all possible feasible solutions of an optimization problem, only a subset of such solutions/variables, known as basic variables participate in getting the optimal solution while the rest of non-basic variables can be discarded. The master problem (MP) is thus transformed into a restricted master problem (RMP) which considers a subset of initial feasible scheduling patterns. The set of all possible feasible matching is Q while its subset is $Y \subseteq Q$. The RMP consisting of (8)(9) and (3-5)(7) is provided below.

$$minimize \sum_{q \epsilon Y} \tau^q \qquad (8)$$

subjected to:

Constraints (3),(4),(5),(7)

$$\sum_{q \epsilon Y} u_{ij} \tau^q \geq f_{ij} \qquad (9)$$

3.2   Distributed Algorithm

The CG as described above is a centralized approach thus providing solution based on global information. Such techniques can result in high computational complexity and significant overhead. In order to provide a solution which can be implemented in small to large multi hop networks here we provide the distributed approach for CG. Although every node is capable to decide on the local information only, however in the presence of a centralized entity such as access points in mesh networks or base stations in case of multi hop cellular networks the global information is used for better efficiency. The distributed RMP (same formulation as centralized RMP), takes into account the local information and runs at each transmission node simultaneously. While the pricing problem is executed as per individual nodes situation. The pricing problem (PP) from (10)–(16) consists of mainly the constraints with exponential complexity, which is the SINR constraint here. The objective of the PP is to maximize the reduced cost and upon finding a reduced cost, the new pattern is added to RMP to contribute in finding the optimal solution.

$$maximize_{(i,j) \epsilon E} u_{ij} v_{ij} x_{ij} \qquad (10)$$

subjected to:

$$P_i G_{ij} x_{ij} + M(x_{ij} - 1) - \gamma \sum_{j, i \neq m} (P_m G_{jm} x_{jm}) \geq \gamma n_o \qquad (11)$$

$$\sum_{i \epsilon N} x_{ij} + \sum_{j \neq i, j \epsilon N} x_{ji} \leq 1 \qquad (12)$$

$$x_{ij} \epsilon (0, 1) \qquad (13)$$

$$0 \leq P_i \leq P_{max} i \epsilon N \qquad (14)$$

$$\sum_{i \epsilon N} x_i \leq k - 1 \qquad (15)$$

$$x_i \epsilon (0, 1) \qquad (16)$$

The objective of PP is (10), where v_{ij} is the dual variable of the constraint (9) in RMP that is provided to PP. The inequality in (11) is the SINR constraint where M is a large integer which linearises the constraint. This method is known as Big M method. The constraint (12) ensures that each node either transmits or receives at a time, here x_{ij} is a decision variable representing an edge such that $x_{ij} = 1$ if link (i,j) is active, otherwise $x_{ij} = 0$. The k connectivity constraint (15) is ensuring minimum of $k-1$ links to provide fault tolerance.

In case of power control, P_i also becomes the decision variable with an additional constraint on the values of power (14). These constraints on the value of power depends on the nature of value i.e. continuous or integer. The M is a large integer which linearises the SINR constraint, this method is known as the 'big M approach. However in an attempt to calculate an individual value for each link, here the value of M is taken as:

$$M_{ij} = \gamma \left(\eta_o + \sum_{m \neq i,j} P_m G_m j \right) \tag{17}$$

The distributed algorithm is executed at all transmitters simultaneously.

3.3   SINR Transformation

Although the big M approach makes unnecessary constraints redundant, its values are not optimized, as it introduces numerical discrepancy in linear formulation. Here, we first transform the explicit SINR constraint to knapsack problem. This allows minimization of the time required to reach the convergence point and avoids the numerical complexity related to M. The generalized form of a knapsack problem with objective (18) and constraint (19) is as under:

$$maximize \sum_{i \in N} a_i x_i \tag{18}$$

subjected to:

$$\sum_{i \in N} c_i x_i \leq B \tag{19}$$

Such that x_i can have binary or finite integer range. The a_i is profitable cost while c_i is the weight and B is simply a constraint constant. Given that link (i,j) is active, upon applying the knapsack transformation, the SINR constraint can be represented as following.

$$P_i G_{ij} - \gamma \sum_{j,i \neq m} (P_m G_{jm} x_{jm}) \geq \gamma \eta_o \tag{20}$$

$$\sum_{j,i \neq m} (P_m G_{jm} x_{jm}) \leq P_i G_{ij}/\gamma - \eta_o \tag{21}$$

Here, $i,j \in N$. If, we take $P_m G_{jm} = c_{jm}$ and $P_i G_{ij}/\gamma - \eta_o = B$ then by taking $u_{ij} = a_{ij}$, we can transform PP to knapsack. Here, the transformation to knapsack simplifies the problem as it eliminates the explicit SINR constraints (11) removes the numerical discrepancies induced by big M method and represents the problem at hand as set cover. The set cover removes the unnecessary constraints and converges towards the optimal more effectively.

3.4 Algorithm Description

The algorithm starts by running RMP problem and call the algorithm for pricing problem in each iteration. Here in RMP algorithm, at first the transmitting node broadcasts its power and data demand. After establishing the neighbourhood information, PSO provides initial feasible patterns Y and power values P_i. The RMP problem is solved by each transmitting node simultaneously and the dual variable v_{ij} associated with each link is then provided to PP. The algorithm for PP returns the new column variable and power values. This loop run till the cost $u_{ij}v_{ij} \leq 1$. The pricing problem is solved by two main sub-functions such that end results impacts the bounds and optimal feasible results. For the PP algorithm the dual of primal formulation is considered. As, the dual of the knapsack problem is the covering problem, so by transforming the primal pricing problem with knapsack constraint as stated above is converted to covering problem. Here, the vertex set is formed based on the nodes in violation of SINR. The number of nodes in the interference set is decreased by adjusting the power and allowing the subsets of links that can be tolerated for simultaneously transmission. As a result the cardinality of interference set and feasible set decreases and increases respectively. The optimal k connectivity is calculated. The vertex cover number serves as the high priority to SINR and allow to choose optimal k per node where $k \geq 2$. The final scheduled links form the network topology where the link is removed as its demand is met and new links are added. Traditionally, the topology control is triggered when a link vanishes or added, here the link demand is the main criteria.

3.5 Initial Feasible Solutions: DPSO

CG is usually initialized by a set of feasible solution/pattern which is taken as a single link in the network. As the performance of CG is significantly influenced by the initial settings, here distributed particle swarm optimization (DPSO) is chosen to provide better initial feasible pattern and initial power levels. It not only provides various patterns but also reduces the number of iterations to provide stable solution. The DPSO is executed by each transmitter and in take only the local information. The formulation of DPSO involves defining the particle of the population, velocity and position update technique. Here, the particles represent nodes in the network, while forming a matrix that represents the power allocated to each node forms. If we represent the particle as $x_{i,j}$, where $(i,j)\epsilon L$, then $v_{i,j}^t$ and $d_{i,j}^t$ is the particle's velocity and distance at iteration t. The local and global best are calculated, represented as $x_{i,j}^{lo}$ and $x_{i,j}^*$ respectively. The velocity and position update equations are provided here. In the formulation below, $\varsigma_{soc} \epsilon R^+$ and $\varsigma_{cog} \epsilon R^+$ are acceleration coefficients for social and cognition effect while $w_{in} \epsilon [0, 1]$ is the inertia coefficient that controls the velocity.

$$v_{i,j}^{t+1} = w_{in} v_{i,j}^t + \varsigma_{soc}(x_{i,j}^{lo} - d_{i,j}^t) + \varsigma_{cog}(x_{i,j}^* - d_{i,j}^t) \tag{22}$$

$$x_{i,j}^{t+1} = v_{i,j}^{t+1} + d_{i,j}^t \tag{23}$$

4   Evaluation and Results

In this section, we evaluate the efficiency of the proposed optimization framework which provided additional insights on topology control. Here, we have used MATLAB, together with CPLEX 12.5v as an optimal LP solver. The nodes are uniformly distributed, forming an initial random topology and the weight of the link G_{ij} is calculated where $\alpha = 3$ and $\vartheta = [0.8, 1.2]$. The values for k is $[2, |N| - 1])$ here, although the values can be used within the $2 \leq k \leq 20$ range. The SINR threshold is set to 10, noise is set at $10^{(-6)}$ and the maximum power is 0.1 Watts. Here, none of the links share a common node as stated in Sect. 2. The STDMA based network with varying number of nodes from 5 till 80 nodes are taken into consideration. In total of 5 instances of each network size are considered and 250 monte carlo simulation are run for the results.

4.1   Approximation Solution

In this section, we discuss the performance of proposed DCG-PSO knapsack algorithm in terms of average transmission length in terms of number of slots as well as total length in seconds. This discussion is followed by the analysis of the approach with and without power control, computational complexity and then the system level analysis of the algorithm in comparison to similar state of the art techniques. First we determine the minimum transmission length needed to fulfil a given traffic demand over the links. The transmission length is calculated in terms og number of slots, however, the transmission length in seconds can be calculated. At first the proposed distributed algorithm DCG-PSO is compared with centralized version CG-PSO in Fig. 1. As the Fig. 1 shows that the distributed approach provides transmission length with almost a constant gap from the centralized approach. It is observed that the gap can increase slightly at higher number of links, due to the fact that accumulative interference in presence of multicast scenario can increases thus requiring more number of slots to fulfil the traffic demand.

The comparison of DCG-PSO in terms of average transmission length with state of the art is shown in Fig. 2. In this comparison the probing based algorithm [17], ϵ bounded approach [18] and CG distributed based approach in [19] is compared to DCG-PSO. The transmission length provided by DCG-PSO is very low. While the transmission length provided by probing is the largest, the rest of the approaches computed almost the results as DCG-PSO. However all the approaches except DCG-PSO cannot compute for higher number of links. It is

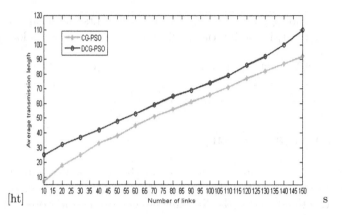

Fig. 1. Average transmission length by centralized and distributed algorithm

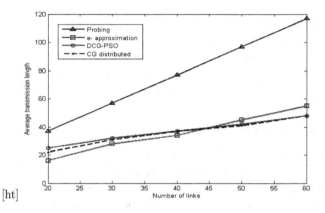

Fig. 2. Comparison of proposed distributed algorithm in terms of average transmission length with state of the art

due to the exponentially increasing computational complexity in terms of run time and number of iterations to converge. In order to compare the convergence, the number of iterations required by each algorithm is illustrated in Fig. 3. The number of iterations is highest in case of [19] while other approaches cannot find convergence at relatively higher number of links as illustrated in Table 2.

The spatial reuse is one in case of TDMA which means only one link is activate in one time slot, in case of STDMA spatial reuse can be greater than or equal to one. The analysis and comparison of DCG-PSO with centralized approach shows that relatively less power is needed while improving the spatial reuse as shown in Fig. 4. The comparison shows that spatial reuse is much closer to centralized approach when distributed algorithm (DCG-PSO) with power control is executed. The increasing spatial reuse is because the total sum of power of the network decreases significantly in case of power control and less power is needed as the number of nodes increases, resulting in increase in simultaneous

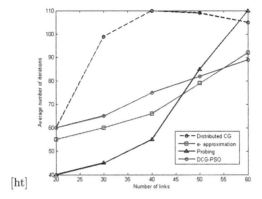

Fig. 3. Comparison of number of iterations of proposed distributed algorithm DCG-PSO and state of the art

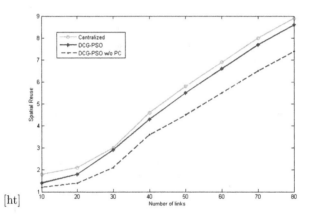

Fig. 4. Spatial Reuse of proposed centralized and distributed approach with and without power control

activation of the links. However, the power control formulation has higher run time and is significant if the density of node is relatively higher, as illustrated in Table 1. This is because, the power control formulation involves extra decision variables and spares networks usually need higher power for connectivity while having larger margin to avoid interference. In Table 1, the average computational time is illustrated for solving a network of 25 nodes and total of 250 instances are simulated. In the Table 2 below the total power of the network for DCG-PSO in comparison to state of the art is provided.

As our CG procedure is initialized by particle swarm instead of greedy and/or single link configuration, the number of iterations to attain the objective value has been reduced and better objective values are obtained. This is due to the fact that CG is sensitive to initial values which can then affect algorithm stability upon each iteration. The average percentage cost penalty, which is defined as

Table 1. Column generation objective values and run time

Technique	Average computational time	Solved instances
DCG-PSO	21.1 s	245
DCG-PSO w/o PC	11 s	248
ϵ- approximation	200 s	180
Distributed CG	587 s	178
Probing	11.12 s	239

Table 2. Total power of the network by DCG-PSO, ϵ approximation and distributed CG approaches

No. of Nodes	Avg. No. of Links	DCG-PSO w/o PC	DCG-PSO w PC	e approximation	CG distributed
10	18	45.0	32.50	31.4	59.40
30	87	82.51	60.90	62.41	No solution found
50	280	502.95	410.61	No solution found	No solution found
80	650	731.52	698.01	No solution found	No solution found

the difference between the optimal O_{opt} and the findings of the algorithm O_{algo}, that is: $(O_{algo} - O_{opt})/O_{opt}$, reduces as the bounding interval and reduced in range. The penalty cost of DCG-PSO is 20 percent at most in worse case. This reduction is due to PSO based initialization and the knapsack transformation, which allows for solving the pricing problem through solving inequalities. The cost penalty also shows that: with increasing search space, the greedy based approach tends to have higher cost penalty than PSO based. The lower bound of RMP is calculated by finding the dual variables di and vijof constraint (3) and (9) If D is a traffic demand vector for the links in Y and optimal value is z then $LB = di.D/1 - z$. The upper bound is also calculated at each iteration, thus contributing into fast convergence.

5   Conclusion

We have considered the problem of network topology based on minimum scheduling length with power control for STDMA multihop network, subject to SINR based interference. We opted for distributed approximation algorithm based on CG. The SINR constraint is transformed such that it reduces complexity. Thus DCG-PSO can solve larger instances, that is at least three fold increase in solvable instances in terms of number of nodes as compared to literature. DCG-PSO provide network topology with shorter scheduling length and minimum power consumption, it has a profound effect on spatial reuse and has minimum penalty cost. Thus, to the best of our knowledge DCG-PSO is the first approximation algorithm to provide the solution for 80 nodes while considering SINR,

k-connectivity and power consumption. The evaluation of proposed algorithm with realistic propagation models, CSMA, effect on throughput and routing is part of our future work.

References

1. Haghpanahi, M., Kalantari, M., Shayman, M.: Topology control in large-scale wireless sensor networks: between information source and sink. Ad Hoc Netw. **11**(3), 975–990 (2013)
2. Ao, X., Yu, F.R., Jiang, S., Guan, Q., Leung, V.C.M.: Distributed cooperative topology control for WANETs with opportunistic interference cancelation. IEEE Trans. Veh. Technol. **63**(2), 789–801 (2014)
3. Zuo, J., Dong, C., Nguyen, H.V., Ng, S.X., Yang, L.-L., Hanzo, L.: Cross-layer aided energy-efficient opportunistic routing in ad hoc networks. IEEE Trans. Commun. **62**(2), 522–535 (2014)
4. Parissidis, G., Karaliopoulos, M., Spyropoulos, T., Plattner, B.: Interference-aware routing in wireless multihop networks. IEEE Trans. Mob. Comput. **10**(5), 716–733 (2011)
5. Santi, P.: Topology control in wireless ad hoc and sensor networks. ACM Comput. Survey **37**(2), 164–194 (2005)
6. Sardellitti, S., Barbarossa, S., Swami, A.: Optimal topology control and power allocation for minimum energy consumption in consensus networks. IEEE Trans. Signal Process. **60**(1), 383–399 (2012)
7. Fathi, M., Taheri, H., Mehrjoo, M.: Cross-layer joint rate control and scheduling for OFDMA wireless mesh networks. IEEE Trans. Veh. Technol. **59**(8), 3933–3941 (2010)
8. Yong, D., Chen, W., Li, X.: An adaptive partitioning scheme for sleep scheduling and topology control in wireless sensor networks. IEEE Trans. Parallel Distrib. Syst. **20**(9), 1352–1365 (2009)
9. Zengmao, C., et al.: Aggregate interference modelling in cognitive radio networks with power and contention control. IEEE Trans. Commun. **60**(2), 456–468 (2012)
10. Burt, C., Chan, Y., Sonenberg, N.: Exact models for the k-connected minimum energy problem. In: Zheng, J., Mao, S., Midkiff, S.F., Zhu, H. (eds.) ADHOCNETS 2009. LNICST, vol. 28, pp. 392–406. Springer, Heidelberg (2010)
11. Cardieri, P.: Modelling interference in wireless ad hoc networks. IEEE Commun. Surveys Tutorials **12**(4), 551–572 (2010)
12. Fussen, M., Wattenhofer, R., Zollinger, A.: Interference arises at the receiver. In: 2005 International Conference on Wireless Networks, Communications and Mobile Computing, vol. 1, pp. 427–432, 13-16 June 2005
13. Burkhart, M., et al.: Does topology control reduce interference? In: Proceedings of the 5th ACM International Symposium on Mobile Ad Hoc Networking and Computing 2004, pp. 9–19. ACM, Roppongi Hills (2004)
14. Yan, G., Hou, J.C., Hoang, N.: Topology control for maintaining network connectivity and maximizing network capacity under the physical model. In: The 27th Conference on Computer Communications, INFOCOM 2008. IEEE (2008)
15. Moraes, R.E.N., Ribeiro, C.C., Duhamel, C.: Optimal solutions for fault-tolerant topology control in wireless ad hoc networks. IEEE Trans. Wireless Commun. **8**(12), 5970–5981 (2009)

16. Bjrklund, P., Vrbrand, P., Yuan, D.: A column generation method for spatial TDMA scheduling in ad hoc networks. Ad Hoc Netw. **2**(4), 405–418 (2004)
17. Chen, C.-C., Lee, D.-S.: A joint design of distributed QOS scheduling and power control for wireless networks. In: Proceedings of the 25th IEEE International Conference on Computer Communications, INFOCOM 2006, pp. 1–12, April 2006
18. Li, M., Salinas, S., Li, P., Huang, X., Fang, Y., Glisic, S.: Optimal scheduling for multi-radio multi-channel multi-hop cognitive cellular networks. IEEE Trans. Mob. Comput. **14**(1), 139–154 (2015)
19. Phunchongharn, P., Hossain, E.: Distributed robust scheduling and power control for cognitive spatial-reuse TDMA networks. IEEE J. Sel. Areas Commun. **30**(10), 1934–2012 (1946)
20. Kompella, S., Wieselthier, J.E., Ephremides, A., Sherali, H.D., Nguyen, G.D.: On optimal SINR-based scheduling in multihop wireless networks. IEEE/ACM Trans. Netw. **18**(6), 1713–1724 (2010)

PHY and Sensing

A Comparison of Physical Layers for Low Power Wide Area Networks

Yoann Roth[1]($\boxtimes$), Jean-Baptiste Doré[1], Laurent Ros[2], and Vincent Berg[1]

[1] CEA, LETI, MINATEC Campus, 38054 Grenoble, France
{yoann.roth,jean-baptiste.dore,vincent.berg}@cea.fr
[2] Univ. Grenoble Alpes, GIPSA-Lab, 38000 Grenoble, France
laurent.ros@gipsa-lab.grenoble-inp.fr

Abstract. In Low Power Wide Area networks, terminals are expected to be low cost, low power and able to achieve successful communication at long range. Communication should be low rate compared to cellular mobile networks. Most of the current technologies dedicated to Machine-to-Machine communication rely on the use of a spreading factor to achieve low levels of sensitivity. To contain power consumption at the terminal side, the cost of complexity should be paid by the receiving side. We propose to use turbo processing schemes as potential physical layers. We compare these schemes to a standard and an industrial solution, and show that a significant gain in sensitivity can be achieved, and very energy efficient scheme can be designed by mixing turbo processing and orthogonal modulation concepts.

Keywords: Low Power Wide Area · LPWA · Internet of Things · PHY-layer · FSK · Turbo Code · Low rate · Low SNR · Machine-to-Machine

1 Introduction

The Internet-of-Things (IoT) is a vast concept, where every object is expected to be connected. Amongst the foreseeable applications for these Machine-to-Machine (M2M) communications, many require the design of a new Low Power Wide Area (LPWA) network [1]. While recent advances (3G, 4G) sought for higher data rate and spectral efficiency, this kind of network is expected to be low rate, low power and long range. Packet sizes should be from a few bytes to hundreds maximum. Terminal power consumption and sensitivity working levels are critical issues, relying on the choice of an efficient and robust physical layer. We consider a star network topology, where each node is connected to a base-station. We focus on the uplink communication.

To achieve low sensitivity levels, a widely used technique is the low complexity repetition code. Indeed, repeating the information by a Spreading Factor (SF) λ at the transmitter can offer a gain of $10\log_{10}(\lambda)$ dB in sensitivity at the receiver side. This technique offers a good compromise between sensitivity and complexity and many systems rely on it, as for instance in Direct Sequence

© ICST Institute for Computer Sciences, Social Informatics and Telecommunications Engineering 2016
D. Noguet et al. (Eds.): CROWNCOM 2016, LNICST 172, pp. 261–272, 2016.
DOI: 10.1007/978-3-319-40352-6_21

Spread Sprectrum (DSSS). Another natural choice for low data rate and low power applications is the use of M-ary orthogonal modulation. Increasing the size of the alphabet gives a gain in energy efficiency, while reducing the spectral efficiency [2]. This property can be extended to Orthogonal Sequence Spread Spectrum (OSSS), and using high orders of alphabet can bring low levels of sensitivity.

Some solutions have been proposed to answer the need of a new LPWA network. A recent proposal [3,4] is based on the combination of OSSS with a low complexity block code. Also, efforts have been made to design the standard 802.15.4k using the DSSS technique combined with a convolutional code [5], which can be considered for LPWA networks. These two schemes are low complexity and straightforward solutions. However, since low power consumption is expected mainly at the terminal side, the cost of complexity can be paid by the receiver side, *i.e.* the base station. More elaborate decoding schemes can then be used, such as turbo processing [6], and lead to a better energy efficiency of the system. We consider the widely known Turbo Code (TC) standardized by the Universal Mobile Telecommunications System (UMTS) [7,8] combined with a common modulation and a SF as a potential physical layer for LPWA. Our recent work [9] pointed out that if the repetition scheme allows a decrease of the required Signal-to-Noise Ratio (SNR) of a specific modulation, it does not increase the energy efficiency of the system (which can be expressed in terms of energy per bit to noise spectral density ratio, denoted E_b/N_0). This is a consequence of the information rate reduction implied by the use of the SF. Redundancy should be used more efficiently by the receiver, by means of turbo processing. The scheme proposed in [9], dedicated to LPWA networks, combines orthogonal modulation with turbo processing, and achieves very promising level of performance.

In this paper, we propose a comparison of four different schemes: the 802.15.4k standard [5], OSSS with a Hamming code, the TC UMTS from [7,8] with a common modulation (Binary Phase Shift Keying (BPSK) or M-ary orthogonal modulation), and the Turbo-FSK scheme proposed in [9]. All of them perform at a low spectral efficiency, aim a high energy efficiency, and would be suitable for uplink communication in the context of LPWA networks. We focus on the physical layer, and assume perfect synchronization and Additive White Gaussian Noise (AWGN). All architectures are described, and simulations are performed. To the best of our knowledge, comparison of these potential LPWA networks have never been done. Also, the use of the UMTS Turbo Code and a modulation have never been considered before as a potential uplink scheme for LPWA.

The paper is organized as follows. The different schemes and their architectures are introduced Sect. 2. Performance comparison are presented Sect. 3. In Sect. 4, results are discussed and Sect. 5 concludes the paper.

2 System Model

This section is dedicated to the description of the four conceivable physical layers previously mentioned. For each one, a short introduction is followed by the presentation of the block diagram of the transmitter. As receivers is rarely standardized, a receiver scheme has been proposed for each physical layer.

2.1 IEEE 802.15.4k

The IEEE 802.15.4k is a standard for local and metropolitan area networks, and is part of the Low-Rate Wireless Personal Area Networks (LR-WPANs) [5]. It aims at low energy critical infrastructure monitoring networks. This standard supports three physical layer (PHY) modes: DSSS with BPSK or Offset-Quadrature Phase Shift Keying (O-QPSK), or Frequency Shift Keying (FSK). DSSS with BPSK modulation is adapted to more constrained situations, and will be presented here. Standard specifications allow the use of a spreading factor value λ from 16 to 32768.

Block diagrams for transmitter and receiver are given in Fig. 1. The transmitter is composed of a Forward Error Correction (FEC) block, defined to be the convolutional code of rate 1/2, with generators polynomial [171 133] (in octal) and constraint length $k = 7$. After encoding, interleaving is done to ensure diversity at the reception side. Differential encoding is applied, and encoded data is "repeated" by the use of a binary direct sequence of size λ. BPSK modulation is then used to transmit the signal through the channel. The normalized spectral efficiency of the physical layer, expressed in bits/s/Hz, is

$$\eta_1 = \frac{1}{2\lambda}. \tag{1}$$

The receiver executes the reverse operations of the transmitting side. After despreading the signal (executing the mean weighted by the elements of the binary direct sequence), we use a soft differential BPSK decoder. Knowing how data was interleaved at the transmitter side, the deinterleaving operation allows the soft input Viterbi decoder to decode the data and to finally retrieve the information bits.

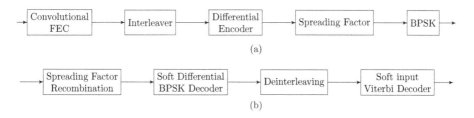

Fig. 1. Standard IEEE 802.15.4k: (a) transmitter architecture, (b) receiver architecture.

2.2   Orthogonal Modulation with Block Code

It is a known fact that orthogonal modulations reach channel capacity for large size of alphabet [2]. An original solution relies on the use of orthogonal modulations with a simple channel code [4,10]. Spreading operation is replaced by increasing the alphabet size, thus lowering the spectral efficiency but also increasing the energy efficiency. As we assumed ideal synchronization and AWGN channel, performance does not depend on the orthogonal alphabet choice (*e.g* FSK, OSSS, Pulse Position Modulation, ...).

The transmitter and receiver architectures are given in Fig. 2. After applying a simple Hamming $(7, 4, 3)$ block code, coded bits are interleaved and then mapped to an orthogonal alphabet of size λ. The spectral efficiency is defined as

$$\eta_2 = \frac{\log_2(\lambda)}{\lambda} \frac{4}{7}. \tag{2}$$

At the receiving side, soft demodulation is performed. The output is the log likelihood ratio (LLR) of the bits. It is defined as

$$L(b_n) = \log \frac{p(b_n = 1)}{p(b_n = 0)}. \tag{3}$$

The principle of soft decoding is to compute the *A Posteriori* Probabilities (APP) of the codewords, *i.e.* for each codeword the probability of having a given codeword knowing the observation, and then compute the LLR of the information bits, knowing the APP for all the possible codewords. The elements of the orthogonal alphabet are denoted $\boldsymbol{c}^i$ with $i \in [0, \lambda - 1]$ and $\boldsymbol{c}^i = [c_0^i, \ldots c_{\lambda-1}^i]$. For each $\boldsymbol{c}^i$, the associated information word is denoted $\boldsymbol{b}^i = [b_0^i, \ldots b_{\log_2(\lambda)-1}^i]$. We consider one noisy received codeword $\boldsymbol{y}$, and $\boldsymbol{d}$ its associated decoded information word. The APP can be expressed as

$$p(\boldsymbol{c}^i | \boldsymbol{y}) = \frac{p(\boldsymbol{y} | \boldsymbol{c}^i) \, p(\boldsymbol{c}^i)}{p(\boldsymbol{y})}, \tag{4}$$

after applying Bayes' law. $p(\boldsymbol{c}^i)$ is the *a priori* probability of having the codeword, which here is equal to $1/\lambda$. As $\boldsymbol{y}$ and $\boldsymbol{c}^i$ are vector with λ elements, the APP becomes:

$$p(\boldsymbol{c}^i | \boldsymbol{y}) = \frac{1}{\lambda p(\boldsymbol{y})} \prod_{m=0}^{\lambda-1} p(y_m | c_m^i). \tag{5}$$

Considering the complex AWGN case,

$$p(y_m | c_m^i) = \frac{1}{2\pi\sigma^2} \exp\left\{ -\frac{1}{2\sigma^2} \|y_m - c_m^i\|^2 \right\}, \tag{6}$$

where σ is the noise variance and $\|.\|$ is the Euclidean norm. y_m and c_m^i are complex numbers, hence (5) can be expressed

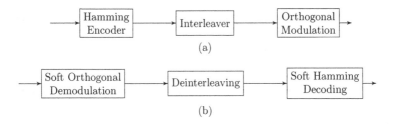

(a)

(b)

Fig. 2. Orthogonal modulation and Hamming code: (a) transmitter architecture, (b) receiver architecture.

$$p(\boldsymbol{c}^i|\boldsymbol{y}) = \frac{C}{\lambda p(\boldsymbol{y})} \exp \left\{ \frac{1}{\sigma^2} \sum_{m=0}^{\lambda-1} \mathrm{Re}\left(y_m.\overline{c_m^i}\right) \right\}, \tag{7}$$

where C is a constant that will be canceled out in further computations, and $\overline{c_m^i}$ the complex conjugate of c_m^i. LLR of the information bits can be computed using the expression

$$L(d_n|\boldsymbol{y}) = \log \sum_{\substack{i=0, \\ b_n^i=1}}^{\lambda-1} p(\boldsymbol{c}^i|\boldsymbol{y}) - \log \sum_{\substack{i=0, \\ b_n^i=0}}^{\lambda-1} p(\boldsymbol{c}^i|\boldsymbol{y}), \tag{8}$$

where the index $b_n^i = 1$ (resp. $b_n^i = 0$) for the first sum (resp. the second sum) signifies that the sum is done over the codewords $\boldsymbol{c}^i$ that encode an information word for which the value of bit b_n^i is 1 (resp. 0). The factor before the exponential in (7) is suppressed at this step.

Once LLR have been computed, deinterleaving is done. Soft Hamming decoding can then be executed to get the decoded information bits.

2.3   Standard Modulation with Turbo Coding

3G and 4G standards both rely on a turbo code for forward error correction [7]. The TC used is a parallel concatenation of two identical recursive systematic codes of rate 1/2, making the code rate equal to 1/3 (or slightly less when the termination bits needed to close the trellis are considered). The constituent code has a constraint length of $k = 4$, and its feedback and generator polynomials are respectively [1315] (in octal). This channel code allows any system to achieve an energy efficiency gain relatively high, thanks to iterative soft decoding at the receiver side. Complexity at the transmitter side is rather low. We consider this code combined with a modulation (BPSK or M-ary FSK) as a potential physical layer for LPWA network, or at least as an interesting element of comparison.

The architecture of the transmitter is depicted Fig. 3. The Parallel-to-Serial Conversion has three inputs: the information bits, the parity bits of the first Recursive Systematic Coder (RSC), and the parity bits of the second RSC which input has been interleaved. After serial conversion, a SF of value λ is applied, and then encoded data is modulated. The spectral efficiency of this physical layer is

$$\eta_3 = \frac{\eta_{\mathrm{mod}}}{3\lambda},\tag{9}$$

where η_{mod} is the normalized spectral efficiency of the modulation used, defined as

$$\eta_{\mathrm{mod}} = \begin{cases} 1 & \text{if BPSK is used,} \\ \frac{\log_2(M)}{M} & \text{if } M\text{-ary orthogonal modulation is used.} \end{cases}$$

The receiver suggested is described Fig. 4. After the noisy channel, the SF Recombination module averages λ consecutive symbols. Signal is then demodulated, *i.e.* LLR are computed. Serial-to-Parallel Conversion will give the systematic bits (*i.e.* uncoded information bits), and parity bits of each encoder. Each decoder is then fed by the systematic bits (which are interleaved for the second decoder) and parity bits corresponding to each RSC. *Maximum A Posteriori* (MAP) rule is used to decode the trellis, using the Bahl, Cocke, Jelinek and Raviv (BCJR) algorithm [11]. The max-log approximation can be used, implying a small loss in performance for an interesting reduction of complexity. For each decoder, the output of the other decoder is used as *a priori* input after interleaving. Several iterations of this process are performed, and a decision can be made at the end of each iteration.

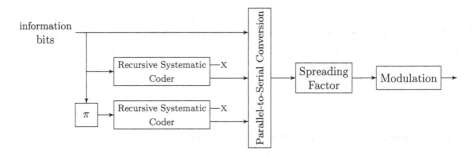

Fig. 3. Modulation and Turbo Code UMTS: transmitter side.

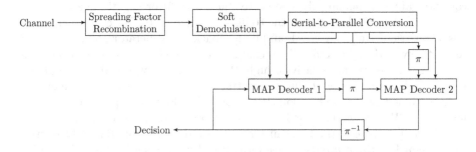

Fig. 4. Modulation and Turbo Code UMTS: receiver side.

2.4 Turbo-FSK

A recent scheme has been proposed, using orthogonal modulations and turbo decoding at the receiver side [9]. This scheme allows for a better use of redundancy than in pure repetition scheme, and achieves an interesting energy efficiency gain. The transmitter low complexity makes it suitable for uplink communication in the LPWA context.

Transmitter architecture is given in Fig. 5. The structure is composed of λ stages, each one encoding an interleaved version of the input bits. Information bits are gathered into P groups of r bits, and Convolutional-FSK encoding is applied, as described Fig. 6. For every group of r bits, parity is computed and accumulated in the memory. The $r + 1$ bits are then mapped to a codeword of the FSK alphabet, which size is $M = 2^{r+1}$. Thanks to the accumulator, every consecutive symbol is linked to the previous one. The output of each encoder is then a set of P FSK codewords: this scheme mixes coding and modulation in the same process. A Parallel-to-Serial Conversion is then done to send the FSK codewords through the channel. The spectral efficiency is defined by

$$\eta_4 = \frac{\log_2(M) - 1}{\lambda M}, \tag{10}$$

The receiving side, depicted in Fig. 7, consists of Serial-to-Parallel Conversion, to reconstruct the λ stages emitted at the first place. A Soft FSK Detector is used to determine the probabilities of each possible codewords. This step can be done using the Fast Fourier Transform (FFT) algorithm. These probabilities

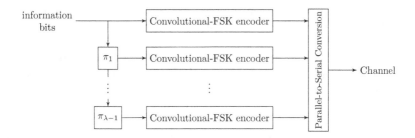

Fig. 5. The Turbo-FSK transmitter.

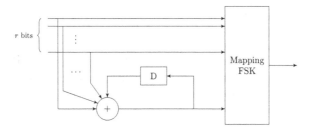

Fig. 6. The Convolutional-FSK encoder.

are then fed to the APP decoder, which will use them as channel observation, while output of the other decoders will be used as *a priori* information. BCJR algorithm is used to decode the trellis. Orthogonality of the transmitted codewords is one of the key feature of this scheme, which offers a great performance gain compared to simple repetition schemes. Detailed explanations about the transmitter and the receiver's computations can be found in [9].

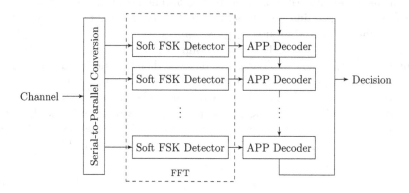

Fig. 7. The Turbo-FSK receiver.

3   Simulation Results

To perform a comparison between the schemes presented in Sect. 2, we perform simulations for the AWGN channel, with coherent reception. Random interleavers are used, except for the TC UMTS where the interleaver specified by the 3G standard is considered. Packet size is set to 1024 bits (or 128 bytes), which is rather small but appropriate for the LPWA context. For iterative schemes, 10 decoding iterations are performed, since performance improvement is rather small for larger numbers of iterations. MAP algorithm is used.

For the sake of fairness, the different schemes should be compared with an equal normalized spectral efficiency η. This physical feature can be adjusted thanks to the definition of the parameter λ, which is related to either a repetition factor or an alphabet size. The normalized spectral efficiency of a configuration of the Turbo-FSK (32-FSK with $\lambda = 4$ stages, with $\eta_4 \simeq 1/32 = 3.125 \cdot 10^{-2}$) is taken as reference. The adjusted spectral efficiencies, the parameters λ, the modulations, the FEC chosen and the binary code rate are summarized in Table 1. For the third scheme, two different modulations are chosen: BPSK (physical layer 3) and 32-FSK (physical layer 3bis), hence the total of 5 physical layers. To obtain the SNR, we consider the straightforward conversion

$$\text{SNR} = \eta \frac{E_b}{N_0}. \tag{11}$$

Table 1. Parameters used for comparison. Termination bits for convolutional codes are considered when computing the normalized spectral efficiency values, implying a lower value than expected. For Turbo-FSK, no binary code rate can be defined, since coding and modulation are mixed.

PHY-layer	1	2	3	3bis	4
Modulation	DBPSK	128-Orthog	BPSK	32-FSK	32-FSK
FEC	CC [171 133]	Hamming	TC [13 15]	TC [13 15]	Turbo-FSK
Binary code-rate	1/2	4/7	1/3	1/3	-
λ	16	128	11	2	4
$\eta \ (\cdot 10^{-2})$	3.125	3.120	3.019	2.594	3.113

Figure 8 shows the Bit Error Rate (BER) performance versus SNR for the 5 physical layers. All schemes have interesting performance at a very low SNR, and will work at low sensitivity levels, which was the goal to achieve. Scheme 1 shows the typical performance of a convolutional code. Scheme 2 uses the Hamming code, which is less powerful than the convolutional code of scheme 1, but offers better performance when combined with a relatively large size of alphabet (128). Performance clearly shows the gain of turbo processing. Scheme 3, corresponding to the Turbo Code UMTS with BPSK modulation and $\lambda = 11$, achieves a BER of 10^{-5} for a SNR of -14.2 dB, hence outperforming scheme 2 by 4.5 dB. Scheme 3bis, for which 32-FSK modulation is used, suffers from a performance loss compared to 3. Scheme 4 outperforms all the other schemes, showing the beneficial impact of mixing convolutional encoding and FSK modulation as proposed in [9]. It should be noted that both schemes 3bis and 4 use the same size of orthogonal alphabet, with yet a gain greater than 1.3 dB at a BER of 10^{-4} for the Turbo-FSK scheme.

Packet Error Rate (PER) performance versus the SNR is depicted Fig. 9. Schemes involving turbo processing are again shown to be far more efficient than the other schemes, with a gain superior to 3.5 dB for a PER of 10^{-3}.

The ultimate Shannon's limit [12] is the maximal transmission rate with arbitrarily small bit-error probability, for a given SNR and a given bandwidth. Comparing schemes to this limit thus shows how efficiently the channel resource is used. The limit, expressed in terms of normalized spectral efficiency versus E_b/N_0, is represented in Fig. 10, along with the performance of the 5 physical layers that are here presented. M-ary orthogonal modulation is also added for reference. Since we choose a specific value of normalized spectral efficiency, all performance corresponding to the schemes are approximately on the same horizontal line. Scheme 4 is the closest scheme to the E_b/N_0 Shannon's limit as it is only 2.3 dB away, showing it to be the scheme with the best energy efficiency. We also give the performance of scheme 3 but without any repetition (thus $\lambda = 1$). As repetition does not influence performance in E_b/N_0, both schemes have the same energy efficiency (same vertical line). Comparing scheme 4 to the uncoded M-ary orthogonal modulation, performance shows that turbo processing is a

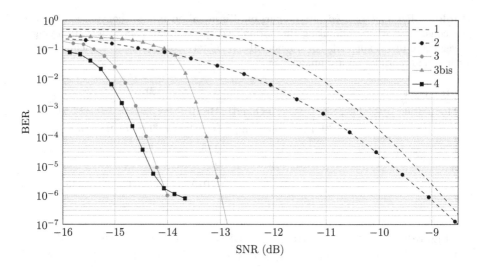

Fig. 8. BER performance comparison in SNR. For schemes using turbo processing (plain curves), 10 iterations have been performed and the block size is set to 1024 bits. Spectral efficiency is roughly equal to $3.113 \cdot 10^{-2}$ bits/s/Hz.

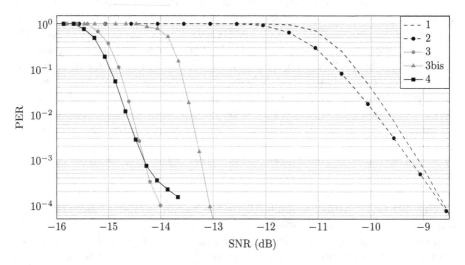

Fig. 9. PER performance comparison in SNR. Packet size is 1024 bits. For schemes using turbo processing (plain curves), 10 iterations have been performed and the block size is set to 1024 bits. Spectral efficiency is roughly equal to $3.113 \cdot 10^{-2}$ bits/s/Hz.

way to get closer to Shannon's limit with a reasonable size of alphabet, hence a reasonably low spectral efficiency.

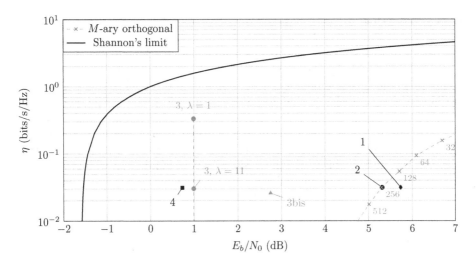

Fig. 10. Performance comparison considering spectral efficiency versus E_b/N_0. The given E_b/N_0 values are the ones for which the BER is 10^{-5}. For schemes using turbo processing, 10 iterations have been performed and the block size is set to 1024 bits.

4 Discussion

As presented in Sect. 3, schemes involving turbo processing have clearly the potential to reach low sensitivities. All gains presented in the previous section can be associated to the gain in sensitivity: using scheme 4 instead of scheme 1 with the same bandwidth will lead to the same data rate but sensitivity will be 5 dB lower. Same performance can thus be achieved with a transmitted power divided by 3.1 (in linear scale) or distance between transmitter and receiver increased by a factor 1.7 (considering a free space path loss). In an other way, for a same level of sensitivity, schemes based on repetition will have to use a higher SF, thus a longer signal for an equivalent bandwidth. This can imply harder constraints for synchronization, in terms of phase noise or channel coherence time.

The performance improvements are done at the expense of an increased complexity. However, the cost of complexity of the turbo processing-based schemes is paid by the receiver side, *i.e.* the base station, which can be considered to have unlimited resource. This is not often suggested for the LPWA context, but more advanced processing can be considered, leading to better performance.

As mentioned in Sect. 1, we assumed perfect synchronization. Since every communication system needs to be synchronized to work properly, this aspect must be taken into account. This is one of the major issues concerning low SNR communication, that still needs to be solved.

5   Conclusion

The design of a new LPWA network is a major issue addressed in the context of the IoT. The uplink scheme is critical, as terminals must have low power consumption and high reliability. We compared several schemes as potential candidates, and have shown that if repeating the information is a good way to lower the sensitivity, more complex systems can allow for higher energy efficiency gain. Simulation results show that turbo processing is an efficient way to achieve a gain in sensitivity. Amongst the studied turbo schemes, Turbo-FSK is the most efficient, and demonstrates the positive impact of mixing turbo processing, M-ary orthogonal modulation and low spectral efficiency. However, synchronization for this ranges of SNR and complexity remain open questions.

References

1. Rebbeck, T., Mackenzie, M., Afonso, N.: Low-powered wireless solutions have the potential to increase the M2M market by over 3 billion connections. Analysys Mason, September 2014
2. Proakis, J.: Digital Communications. Communications and Signal Processing, 3rd edn. McGraw-Hill, New York (1995)
3. LoRa Alliance. https://www.lora-alliance.org/. Accessed 30 April 2016
4. Seller O., Sornin, N.: Low power long range transmitter, US Patent 20140219329 A1 August 2014
5. 802.15.4k: Low-rate wireless personal area networks (LR-WPANs) amendment 5: physical layer specifications for low energy, critical infrastructure monitoring networks. In: IEEE Standard for Local and Metropolitan Area Networks, pp. 1–149, August 2013
6. Berrou, C., Glavieux, A., Thitimajshima, P.: Near Shannon limit error-correcting coding, decoding: Turbo-codes 1. In: IEEE International Conference on Communications (ICC), Geneva, vol. 2, pp. 1064–1070, May 1993
7. LTE Evolved Universal Terrestrial Radio Access (E-UTRA): Multiplexing and Channel Coding. 3GPP TS 36.212, V12.6.0, Release 12, pp. 12–15, October 2015
8. Valenti, M.C., Sun, J.: The UMTS turbo code and an efficient decoder implementation suitable for software defined radios. Int. J. Wireless Inf. Networks 8, 203–216 (2001)
9. Roth, Y., Dore, J.-B., Ros, L., Berg, V.: Turbo-FSK: a new uplink scheme for low power wide area networks. In: 2015 IEEE 16th International Workshop on Signal Processing Advances in Wireless Communications (SPAWC), pp. 81–85, June 2015
10. SX1272 from Semtech, datasheet. http://www.semtech.com/wireless-rf/rf-transceivers/sx1272/. Accessed 30 April 2016
11. Bahl, L., Cocke, J., Jelinek, F., Raviv, J.: Optimal decoding of linear codes for minimizing symbol error rate (corresp.). IEEE Trans. Inf. Theory 20(2), 284–287 (1974)
12. Shannon, C.: A mathematical theory of communication. Bell Syst. Tech. J. 27(3), 379–423 (1948)

A Novel Sequential Phase Difference Detection Method for Spectrum Sensing

Shaojie Liu[(✉)], Zhiyong Feng, Yifan Zhang, Sai Huang, and Dazhi Bao

Key Laboratory of Universal Wireless Communications, Ministry of Education,
Wireless Technology Innovation Institute (WTI), Beijing University of Posts
and Telecommunications, Beijing 100876, People's Republic of China
liushaojie@bupt.edu.cn

Abstract. As traditional spectrum sensing approaches in cognitive radio network unable to deal with the contradiction between accuracy and complexity, a novel sequential spectrum detection based on phase difference (SPDD) is proposed in this paper to achieve good performance with less complexity. The variance of phase difference of signal is utilized as the statistics to detect the signal under a realistic Rayleigh fading channel. Moreover, a variable sample size of proposed algorithm is conducted to minimize the complexity while maintained an acceptable performance. Simulation shows that our SPDD method yields about 2 dB gain over the conventional sequential energy detection. In addition, when the cutoff sample number is set to 1000, a substantial efficiency improvement is obtained compared to the fixed sample detection scheme.

Keywords: Cognitive radio · Spectrum sensing · Phase difference · Sequential detection · Gaussian noise

1 Introduction

With the increasing scarcity of spectrum resources, measurement shows the average utilization rate of current spectrum below 3 GHz is merely 5.2 %, which unveils that the spectrum resources are heavily underutilized [1]. Cognitive Radio (CR) is proposed to sense radio environment and utilize vacant spectrum to improve the spectrum utilization [2]. The most important function in CR is to determine whether the primary user (PU) is present or not, which is called spectrum sensing. If CR determines the PU is absent, then the secondary user (SU) can access the licensed bands. The process of detection of spectrum holes is the key enabler for efficient spectrum utilization in CR.

Various spectrum sensing approaches have been proposed in previous literatures, such as energy detection (ED) [3], cyclostationarity based detection [4], matched filter detection [5], multitaper spectrum estimation [6] and sequential energy detection [7]. Matched filtering detection has the optimal performance while detailed information of PU signal such as pulse shaping is required. In practice, it is generally impractical to get priori feature information of PU.

© ICST Institute for Computer Sciences, Social Informatics and Telecommunications Engineering 2016
D. Noguet et al. (Eds.): CROWNCOM 2016, LNICST 172, pp. 273–283, 2016.
DOI: 10.1007/978-3-319-40352-6_22

Cyclostationarity based detection does not require detailed information of PU signal and has robust performance under the low signal to noise ratio (SNR). However, the high computational complexity restricts its widespread usage on energy-constrained devices. For the simplicity of energy detection (ED), it was popularized in the context of IEEE 802.22 cognitive radio networks [8]. However, a longer length of samples of ED is necessary to maintain an acceptable performance at a low SNR. There implements a sequential approach to energy detection to deliver a significant improvement compared with the fixed sample size detection, which is called sequential energy detection (SED) [7]. Since the thresholds setting of SED cannot determined without noise power, we must estimate the noise in advance. Thus it will bring complexity increase and performance deterioration because of the noise uncertainty. Moreover, it also suffers from huge performance degradation at low SNR. In [9] we have proposed a phase difference sensing method, which improves the detection performance compared to energy detection. Simultaneously, it needs a large fixed number samples for detection and thus the complexity increases inevitably.

To solve the contradiction between complexity and accuracy, this paper formulates a novel sequential phase difference detection method (SPDD). In [10,11] R.F. Pawula and F. Adachi have derived the phase difference distribution of the noise-perturbed signal. We have known that there is a obvious difference in the phase difference distribution between noise-perturbed signal and Gaussian noise through plenty of researches [9]. The sequential test sensitivity to the primary signal phase difference addressed in this paper can obviously yield great performance improvement and have a promising future. In addition, compared to the conventional sequential energy detection, the SPDD sensing scheme is immune to the noise uncertainty because the noise power is not required to set the thresholds.

The rest of the paper is organized as follows. In Sect. 2, we formulate the model of spectrum sensing and phase difference distributions of the noise-perturbed signal and Gaussian noise. The algorithm of SPDD is described and the corresponding performance is analyzed in Sect. 3. We provide the simulation results in Sect. 4 and conclude in Sect. 5.

2   System Model and Phase Difference

2.1   System Model

The spectrum sensing problem can be formulated as per Eq. (1) for $n = 1, 2...$

$$
\begin{aligned}
H_1 : \quad & r(n) = hs(n) + w(n) \\
H_0 : \quad & r(n) = w(n).
\end{aligned}
\tag{1}
$$

where, $r(n)$ is the received signal, h is the instantaneous channel gain, $s(n)$ is the transmitted signal of PU, and $w(n)$ is the Additive White Gaussian Noise (AWGN). H_1 represents that the PU signal is present while H_0 indicates that there is only noise. It is assumed that $s(n)$ is independent identically distributed and $s(n)$ and $w(n)$ are mutually independent.

Instead of the amplitude squares of the received signal as test statistics in conventional sequential energy detection, we focus on the phase difference between two adjacent samples of received signal. We propose a Sequential Probability Ratio Test (SPRT) formulation of the phase difference for detection. Unlike the most sensing method which always have a fixed number of samples to be received before calculating the test statistic, here the samples will be received sequentially, and the likelihood ratio $T(\widetilde{\Delta\theta_n})$ can be calculated as

$$T(\widetilde{\Delta\theta_n}) = \frac{f(\widetilde{\Delta\theta_n}|H_1)}{f(\widetilde{\Delta\theta_n}|H_0)}, \tag{2}$$

where $\Delta\theta_n$ means phase difference between two adjacent samples, $\widetilde{\Delta\theta_n} = [\Delta\theta_1 \Delta\theta_2 ... \Delta\theta_n]$. As $\Delta\theta_i$ is independent identically distributed under H_1 and H_0 hypotheses, we can deduce

$$T(\widetilde{\Delta\theta_n}) = \prod_{i=1}^{n} \frac{f(\Delta\theta_i|H_1)}{f(\Delta\theta_i|H_0)}. \tag{3}$$

The explanation for the independence of the $\Delta\theta_i$ will be given later.

2.2   Phase Difference

In [9] the phase θ_n of the received signal $r(n)$ can be calculated as

$$\theta'_n = \begin{cases} \arctan \frac{\text{Im}(r(n))}{\text{Re}(r(n))}, & \text{Re}(r(n)) \geq 0 \\ \arctan \frac{\text{Im}(r(n))}{\text{Re}(r(n))} + \pi, & \text{Re}(r(n)) < 0, \end{cases} \tag{4}$$

$$\theta_n = \theta'_n \bmod 2\pi. \tag{5}$$

Where $\text{Re}(r(n))$ and $\text{Im}(r(n))$ mean the real and imaginary part of $r(n)$ respectively, and $(\bullet) \bmod 2\pi$ can make the phase θ fall between 0 and 2π. Then, the phase difference $\Delta\theta$ mentioned in Eq. (2) can be obtained as

$$\Delta\theta_n = (\theta_{n+1} - \theta_n) \bmod 2\pi. \tag{6}$$

2.3   Phase Difference Distribution

We know that the phase θ_n of Gaussian noise follows a uniform distribution, which means $\theta_n \sim U(0, 2\pi)$. According to our assumption, the instantaneous phases of Gaussian noise are all mutually independent and also identically distributed. Then $\Delta\theta'_n = \theta_{n+1} - \theta_n$ follows a triangular distribution from -2π to 2π, which can be shown that

$$f_{\Delta\theta'_n}(\Delta\theta'_n) = \begin{cases} \frac{1}{2\pi} + \frac{\Delta\theta'_n}{4\pi^2}, & -2\pi \leq \Delta\theta'_n < 0 \\ \frac{1}{2\pi} - \frac{\Delta\theta'_n}{4\pi^2}, & 0 \leq \Delta\theta'_n \leq 2\pi. \end{cases} \tag{7}$$

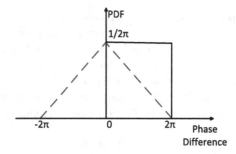

Fig. 1. Distribution of phase difference for Gaussian noise

Since $\Delta\theta_n = (\Delta\theta'_n) mod 2\pi$, we can express the distribution of phase difference of Gaussian noise $f^n_{\Delta\theta_n}(\Delta\theta_n)$ as

$$f^n_{\Delta\theta_n}(\Delta\theta_n) = f_{\Delta\theta'_n}(\Delta\theta_n) + f_{\Delta\theta'_n}(\Delta\theta_n - 2\pi) = \frac{1}{2\pi}. \qquad (8)$$

As shown in Fig. 1, it indicates the phase difference $\Delta\theta \sim U(0, 2\pi)$. It seems that $\Delta\theta_{n+1}$ and $\Delta\theta_n$ may be correlated because both of them refer to θ_{n+1}, but the correlation between them can be eliminated after $(\bullet) mod$. So we assume $\Delta\theta_i$ is independent identically distributed, and this assumption matches very well with our simulation results. The phase difference cumulative distribution of the signal perturbed by Gaussian noise has been given as formula (17) in [11], the formula can be written as

$$F_{\Delta\theta_n}(\Delta\theta_n) = \frac{1}{4\pi} \int_{-\pi/2}^{\pi/2} e^{-E} \left[\frac{W \sin\alpha}{E} + Q \right] dt, \qquad (9)$$

where,

$$E = U - V\sin t - W\cos\beta\cos t$$

$$U = \frac{1}{2}(SNR_{n+1} - SNR_n)$$

$$V = \frac{1}{2}(SNR_{n+1} + SNR_n)$$

$$W = \sqrt{SNR_{n+1}SNR_n} = \sqrt{U^2 - V^2}$$

$$Q = \frac{\rho\sin\Delta\theta_n - \lambda\cos\Delta\theta_n}{1 - (\rho\cos\Delta\theta_n + \lambda\sin\Delta\theta_n)\cos t}$$

$$\alpha = (\Delta\phi_n - \Delta\theta_n) mod 2\pi$$

$\Delta\phi_n$ represents the phase difference between nth and $(n + 1)$th sampling point, $\rho + \lambda i$ means complex correlation of Rayleigh fading signal and noise. It should be noted that SNR_n is the instantaneous SNR of the nth received sampling signal, and SNR_n will not change for phase modulation through amplitude and QAM modulation. The received sample signal is considered to be constant, thus we can assume $SNR_{n+1} = SNR_n = \gamma$.

In addition, the phase difference can be assumed to be independent with each other just as noise. For received continuous wave, if the Gaussian noise is without fading, $Q = 0$. Here we discuss the case of Rayleigh fading, thus $\rho + j\lambda = \sqrt{\rho^2 + \lambda^2}e^{j\Delta\phi_n} = \frac{\gamma e^{j\Delta\phi_n}}{\gamma+1}$. The performance of SPDD algorithm shows significant improvement in comparison to the conventional algorithm at extremely low SNR, then we derive $e^{-E} \approx 1$. Therefore, the form of $F_{\Delta\theta_n}(\Delta\theta_n)$ can be rewritten as

$$
\begin{aligned}
&F_{\Delta\theta_n}(\Delta\theta_n) \\
&= \frac{1}{4\pi} \int_{-\pi/2}^{\pi/2} \left[\frac{\sin\alpha}{1-\cos\alpha\cos t} + \frac{\gamma\sin\alpha}{\gamma(1-\cos\alpha\cos t)+1}\right]dt \\
&= \frac{\sin\alpha}{\pi|\sin\alpha|}\arctan\left|\cot\frac{\alpha}{2}\right| \\
&\quad + \frac{\sin\alpha}{\pi\sqrt{\left(1+\frac{1}{\gamma}\right)^2-\cos^2\alpha}}\arctan\sqrt{\frac{(\gamma+1)+\gamma\cos\alpha}{(\gamma+1)-\gamma\cos\alpha}}.
\end{aligned}
\tag{10}
$$

After simplifying, we have

$$
F_\alpha(\alpha) = \frac{1}{2} + \frac{\alpha}{2\pi} + \frac{\sin\alpha G(\alpha)}{2\pi H(\alpha)},
\tag{11}
$$

in which,

$$
G(\alpha) = \frac{\pi}{2} + \arcsin\frac{\gamma\cos\alpha}{\gamma+1}
$$

$$
H(\alpha) = \sqrt{\left(1+\frac{1}{\gamma}\right)^2 - \cos^2\alpha}
$$

Since $\alpha = (\Delta\phi_n - \Delta\theta_n)$ mod 2π, we can use the formula Eq. (11) to get the derivation. Finally, the distribution of phase difference of the received signal perturbed by Gaussian noise can be obtained as

$$
\begin{aligned}
f_{\Delta\theta_n}^s(\Delta\theta_n) &= \frac{1}{2\pi} + \frac{\cos\alpha G(\alpha)}{2\pi H(\alpha)} - \frac{\cos\alpha \sin^2\alpha G(\alpha)}{2H^3(\alpha)} \\
&\quad - \frac{\gamma\sin^2\alpha}{2\pi(\gamma+1)H(\alpha)\sqrt{1-\frac{\gamma^2\cos^2\alpha}{(\gamma+1)^2}}}.
\end{aligned}
\tag{12}
$$

Here $\Delta\phi_n = \frac{\pi}{2}$, which means that sampling rate is set to four times of the residual carrier frequency. It is obvious that the phase difference distribution of the received signal perturbed by Gaussian noise is quite different with that of noise, which can be utilized to detect PU signal sequentially. As shown in Fig. 2, the curves represent the phase difference distribution when SNR is 0 dB, -5 dB, -10 dB. With the decrease of SNR, the distribution of phase difference of pure signal converges to linear distribution $\frac{1}{2\pi}$, which is the distribution of phase difference for Gaussian noise. This indicates the proposed SPDD is not only reasonable in theory but also feasible in practice.

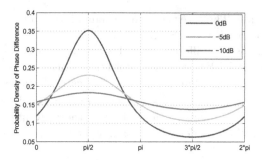

Fig. 2. Phase difference distribution (Color figure online)

3   SPDD Algorithm and Analysis

3.1   SPDD Algorithm

Through the description above, the distribution of phase difference between signal perturbed by noise and Gaussian noise is completely different. Therefore, we propose a novel SPDD algorithm to improve detection efficiency. It is assumed that the received signal is perturbed by noise and signal Rayleigh fading is slow, then using formula (3), (8), (12) we rewrite the likelihood ratio $T(\widetilde{\Delta\theta_n})$ as

$$T(\widetilde{\Delta\theta_n}) = \prod_{i=1}^{n} \frac{f^s_{\Delta\theta_i}(\Delta\theta_i)}{f^n_{\Delta\theta_i}(\Delta\theta_i)} = (2\pi)^n \prod_{i=1}^{n} f^s_{\Delta\theta_i}(\Delta\theta_i). \tag{13}$$

where, $f^s_{\Delta\theta_i}(\Delta\theta_i)$ denotes the distribution of phase difference when the PU signal is present while $f^n_{\Delta\theta_i}(\Delta\theta_i)$ indicates that of only noise. The proposed SPDD calculates the likelihood ratio $T(\widetilde{\Delta\theta_n})$ sequentially, and thus the statistic test $T(\widetilde{\Delta\theta_n})$ is compared with two thresholds, the upper threshold B and the lower threshold A. If the likelihood radio is above B then PU is present, if it is below A then the PU is absent, else accept new samples. The decision rule can be given as

$$D = \begin{cases} H_1, & T(\widetilde{\Delta\theta_n}) \geq B \\ Accept\ New\ Sample & A < T(\widetilde{\Delta\theta_n}) < B \\ H_0, & T(\widetilde{\Delta\theta_n}) \leq A. \end{cases} \tag{14}$$

where D means the sensing decision, and the SPDD algorithm is described as the following **Algorithm 1**.

We can iteratively update the $T(\widetilde{\Delta\theta_n})$ as

$$\begin{aligned} \ln T(\widetilde{\Delta\theta_{n+1}}) &= (n+1)\ln(2\pi) + \sum_{i=1}^{n+1} \ln f^s_{\Delta\theta_i}(\Delta\theta_i) \\ &= n\ln(2\pi) + \sum_{i=1}^{n} \ln f^s_{\Delta\theta_i}(\Delta\theta_i) + \ln(2\pi) + \ln f^s_{\Delta\theta_{n+1}}(\Delta\theta_{n+1}) \\ &= \ln T(\widetilde{\Delta\theta_n}) + \ln(2\pi) + \ln f^s_{\Delta\theta_{n+1}}(\Delta\theta_{n+1}). \end{aligned} \tag{15}$$

Algorithm 1. Sequential Phase Difference Detection Algorithm

Require: The decision threshold A and B
Ensure: $D \in H_0, H_1$
1: Calculate the distribution of phase difference of the received sample signal using the formula(12);
2: Calculate $T(\widetilde{\Delta\theta_n})$ to get the decision D by the formula (13), (14);
3: **if** $T(\widetilde{\Delta\theta_n}) \geq B$ **then**
4:      $H_1 \leftarrow D$,declaring PU is present;
5: **else**
6:     **if** $T(\widetilde{\Delta\theta_n}) \leq A$ **then**
7:         $H_0 \leftarrow D$,declaring PU is absent;break
8:     **else**
9:         Accept new sample, go to step1 and update $T(\widetilde{\Delta\theta_n})$;
10:     **end if**
11: **end if**

The formula (15) provides a natural simplification for SPDD implementation and avoids approximations caused by the threshold setting and others. The computational complexity to update $T(\widetilde{\Delta\theta_n})$ can be reduced substantially by using this iterative method as well.

3.2   Threshold Setting

These thresholds A and B are calculated by using the Wald approximations. [12] has proved that the SPRT has the minimum average expected sample size amongst the class of all sequential and fixed size likelihood ratio for a given fixed P_d and P_f, which provides a theoretical basis for our algorithm.

$$A = \frac{1 - P_d}{1 - P_f} \quad \& \quad B = \frac{P_d}{P_f}. \tag{16}$$

The P_d and P_f used to set the thresholds mean the detection probability and false alarm probability are called Design Values, which are set to fixed values before detection. The thresholds and the likelihood radio $T(\widetilde{\Delta\theta_{n+1}})$ are determined without the demand of noise power because the distribution of phase difference of noise is uniformly distributed despite of noise power. Therefore, the proposed SPDD is completely immune to the noise uncertainty problem, which means considerable superiority over the sequential energy detection.

3.3   Performance Analysis

The stopping sample size of the sequential phase difference detection is a random variable, and the relative performance gained by SPDD can be characterized by comparing the detection duration with the conventional fixed sample size detection. In the case of the hypotheses H_1, we can derive from [12] the following lower bounds for the expected values of the termination time as (18) where $\beta = 1 - P_d$,

$$P[\ln T(\widetilde{\Delta\theta_n} \,|H_1) \leq \ln A] = \beta,$$
$$P[\ln T(\widetilde{\Delta\theta_n} \,|H_1) \geq \ln B] = 1 - \beta. \tag{17}$$

Let $\xi = \frac{f_{\Delta\theta}^s(\Delta\theta)}{f_{\Delta\theta}^n(\Delta\theta)}$, with the assumption of negligible overshoot of the test statistic, then the expectation of the average sample number can be approximatively given as

$$E\,[N|H_1] = \frac{1}{E[\ln \xi|H_1]} \left(\frac{A(B-1)}{B-A} \ln A + \frac{B(1-A)}{B-A} \ln B \right). \tag{18}$$

Similarly, in the case of the hypotheses H_0,

$$E[N|H_0] = \frac{1}{E_0[\ln \xi|H_0]} \left(\frac{B-1}{A-1} \ln A + \frac{1-A}{B-A} \ln B \right). \tag{19}$$

Using (8) and (12),

$$E[\ln \xi \,|H_1] = 2\pi E[\ln f_{\Delta\theta}^s(\Delta\theta) \,|H_1],$$
$$E[\ln \xi \,|H_0] = 2\pi E[\ln f_{\Delta\theta}^s(\Delta\theta) \,|H_0]. \tag{20}$$

Unfortunately, as the complicated formulation of $f_{\Delta\theta}^s(\Delta\theta)$, an exhaustive mathematical formulation of the average sample size of SPDD is intractable. But we can get that the distribution of the stopping sample size depends on the P_d, P_f and the SNR of our sensing system. In practical detection, it may occur the situation that the sample size is too excessive to degrade the performance at extremely low probability. In order to avoid this situation, we set a cut-off number M, which is much larger than the average sample number. The result of SPDD will be decided as H_0 when the system sample number comes to M and the decision has not been made. Otherwise, SPDD only needs to store N samples. Thus the computational complexity is $O(N)$, which is a great advantage compared to other more sophisticated schemes such as cyclostationary detection.

4   Simulation Analysis

In this section, Monte Carlo Simulation is conducted to analyze the performance (2000 experiments for each point) of the proposed scheme. The influence of SNR, different kinds of modulated signals and Design Values on the performance of SPDD are analyzed carefully and rigorously here.

Figure 3 shows the relationship of detection probability P_d of SPDD with SNR for several basic modulation signals, when the Design Values $P_d = 0.9, P_f = 0.01$ and the cut-off sample number is set to 1000. It can be observed that the SNR of Sine Wave, BPSK, 16QAM and 2FSK are $-9.3\,\mathrm{dB}$, $-9.2\,\mathrm{dB}, -9.2\,\mathrm{dB}$ and $-8.5\,\mathrm{dB}$ correspondingly when the actual detection probability is 0.9. The P_d curves for kinds of modulated signals are similar, which demonstrates that SPDD

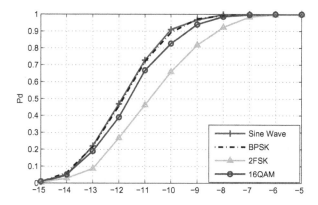

Fig. 3. Detection probability P_d for Sine Wave, BPSK, 16QAM and 2FSK (Color figure online)

is robust with respect to modulation mode. Generally, the sample symbols in one modulation are independently distributed and occur with an equal probability. Therefore, the opposite influence on phase difference of the sequences $r(i)r(j)$ and $r(j)r(i)$ can be offset by each other.

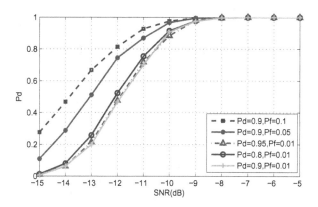

Fig. 4. Detection probability P_d for different design values (Color figure online)

Figure 4 illustrates the receiver operating characteristic (ROC) curves of sine wave for different design values, where shows the effect of SNR on the actual P_d obtained via Monte Carlo simulation for increasing design values P_d and P_f. It can be seen that the proposed SPDD matches its design specifications upto a SNR of -10 dB, below which the actual detection probability P_d precipitously declines. Also, the actual P_d is influenced by thresholds setting, which are determined by design values P_d and P_f.

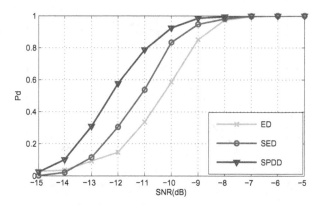

Fig. 5. Detection probability P_d for SPDD, SED and ED (Color figure online)

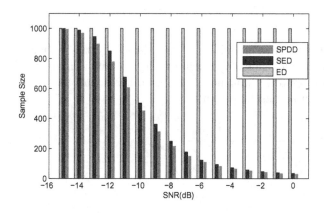

Fig. 6. Sample size for SPDD, SED and ED (Color figure online)

Figure 5 compares the ROC curves of SPDD, conventional sequential energy detection(SED) and energy detection (ED), where the design values of SPDD, SED $P_d = 0.9, P_f = 0.01$ and the signal used is continuous carrier. Moreover, the corresponding stopping sample number are compared in Fig. 6, where the cut-off sample number of SPDD, SED is set to 1000. It shows substantial performance improvement in the SPDD by analyzing the distribution of phase difference. The actual P_d equals 1 till $SNR = -7\,\text{dB}$ and the P_d is still > 0.9 upto $-9\,\text{dB}$. Meanwhile, SPDD can achieve 2 dB gain compared with conventional SED and ED, which demonstrates that SPDD has a significant performance improvement. The ratio of the average sample size of SPDD test to the sample size of fixed sample ED test reaches 0.5 while the ratio between SPDD and SED reaches about 0.95 when the actual P_d are both 0.9. It is obvious that the proposed SPDD is more efficient and flexible.

5 Conclusions

In this paper, the novel SPDD spectrum sensing scheme based on phase difference has been shown to deliver substantial efficiency gain over the conventional detection methods based on fixed sample size and amplitude statistics, which are susceptive to the noise uncertainty and inefficient. Through careful analysis, the distribution of phase difference between two adjacent received signal samples of signal perturbed by noise and Gaussian noise can be utilized to sense signal. Moreover, the Iterative Probabilistic Update method proposed above has been developed to robustly evaluate the likelihood ratio and thus bring significant performance improvement. Meanwhile, the detection thresholds setting of SPDD are free from noise power, therefore the proposed scheme is immune to noise uncertainty.

Acknowledgment. This work was supported by the National Natural Science Foundation of China (61227801), the National Key Technology R&D Program of China (2014ZX03001027-003).

References

1. McHenry, M.A.: NSF Spectrum Occupancy Measurements Project Summary. Shared Spectrum Company Report (2005)
2. Haykin, S.: Cognitive radio: brain-empowered wireless communications. IEEE J. Sel. Areas Commun. **23**(2), 201–220 (2005)
3. Urkowitz, H.: Energy detection of unknown deterministic signals. Proc. IEEE **55**(4), 523–531 (1967)
4. Gardner, W.A., Spooner, C.M.: Signal interception: performance advantages of cyclic-feature detectors. IEEE Trans. Commun. **40**(1), 149–159 (1992)
5. Price, R., Abramsom, N.: Detection theory. IEEE Trans. Inf. Theory **7**(3), 135–139 (1961)
6. Thomson, D.J.: Spectrum estimation and harmonic analysis. Proc. IEEE **70**(9), 1055–1096 (1982)
7. Kundargi, N., Tewfik, A.: A performance study of novel sequential energy detection methods for spectrum sensing. In: IEEE International Conference on Acoustics Speech and Signal Processing, pp. 3090–3093 (2010)
8. Cordeiro, C., Challapali, K., Birru, D.: IEEE 802.22: an introductionto the first wireless standard based on cognitive radios. J. Commun. **1**(1), 38–47 (2006)
9. Yang, J., Yan, X., Gao, M.F., Lian, H., Feng, Z.Y., Zhang, Y.F.: A novel spectrum sensing scheme based on phase difference. In: IEEE WCNC, pp. 264-268 (2014). doi:10.1109/WCNC.2014.6951978
10. Pawula, R.F., Rice, S.O., Roberts, J.H.: Distribution of the phase angle between two vectors perturbed by Gaussian noise. IEEE Trans. Commun. **30**(8), 1828–1841 (1982)
11. Pawula, R.F.: Distribution of the phase angle between two vectors perturbed by Gaussian noise II. IEEE Trans. Veh. Technol. **50**(2), 576–583 (2001)
12. Wald, A.: Sequential Analysis. Dover Publications, New York (2004)

A Simple Formulation for the Distribution of the Scaled Largest Eigenvalue and Application to Spectrum Sensing

Hussein Kobeissi[1,2]([✉]), Youssef Nasser[3], Amor Nafkha[1], Oussama Bazzi[2], and Yves Louët[1]

[1] SCEE/IETR, CentraleSupélec of Rennes, Rennes, France
{hussein.kobeissi,Amor.Nafkha,Yves.Louet}@centralesupelec.fr,
hussein.kobeissi.87@gmail.com
[2] Department of Physics and Electronics, Faculty of Science 1,
Lebanese University, Hadath, Beirut, Lebanon
obazzi@ul.edu.lb
[3] ECE Department, AUB, Bliss Street, Beirut, Lebanon
youssef.nasser@aub.edu.lb

Abstract. Scaled Largest Eigenvalue (SLE) detector stands out as the best single-primary-user detector in uncertain noisy environments. In this paper, we consider a multi-antenna cognitive radio system in which we aim at detecting the presence/absence of a Primary User (PU) using the SLE detector. We study the distribution of the SLE as a large number of samples are used in detection without constraint on the number of antennas. By the exploitation of the distributions of the largest eigenvalue and the trace of the receiver sample covariance matrix, we show that the SLE could be modeled as a normal random variable. Moreover, we derive the distribution of the SLE and deduce a simple yet accurate form of the probability of false alarm. Hence, this derivation yields a very simple form of the detection threshold. The analytical derivations are validated through extensive Monte Carlo simulations.

Keywords: Scaled largest eigenvalue detector · Spectrum sensing · Wishart matrix

1 Introduction

In Cognitive Radio (CR) networks, Spectrum Sensing (SS) is the task of obtaining awareness about the spectrum usage. Mainly it concerns two scenarios of detection: (i) detecting the absence of the Primary User (PU) in a licensed spectrum in order to use it and (ii) detecting the presence of the PU to avoid interference. Hence, SS plays a major role in the performance of the CR as well as the performance of the PU networks that coexist. In this context, an extreme importance for a CR network is to have an optimal SS technique with high probability of accuracy in uncertain environments. The Scaled Largest Eigenvalue detector (SLE) is an efficient technique that is proved to be the optimal

© ICST Institute for Computer Sciences, Social Informatics and Telecommunications Engineering 2016
D. Noguet et al. (Eds.): CROWNCOM 2016, LNICST 172, pp. 284–293, 2016.
DOI: 10.1007/978-3-319-40352-6_23

detector under Generalized Likelihood Ratio (GLR) criterion and noise uncertainty environments [1,2].

SLE is among the detectors that use the eigenvalues of the receiver sample covariance matrix. Such detectors are known as the eigenvalue based detectors and includes, in addition to SLE, other detectors like the Largest Eigenvalue detector (LE) and the Standard Condition Number detector (SCN) [3–7]. In a scenario with perfect knowledge of the noise power, the LE detector is the optimal detector [5]. However, in practical systems the noise power may not be perfectly known. In this case, the SLE and SCN detectors outperform the LE detector due to their blind nature. Moreover, the SLE is proved to be the optimal detector under GLR criterion [1,2] and outperforms the SCN detector.

Even with its importance, existing results on the statistics of the SLE, defined as the ratio of the largest eigenvalue to the normalized trace of the sample covariance matrix, are relatively limited. These results are based on tools from random matrix theory [2,8,9] and Mellin transform [9–11]. SLE was proved, asymptotically, to follow the LE distribution (i.e. Tracy-Widom (TW) distribution) [2]. However, a non-negligible error still exists and new form is derived based on TW distribution and its second derivative [8]. Using Mellin transform, The distribution of the SLE was derived by the exploitation of the distribution of LE and the distribution of the trace [9–11]. The complexity in the form of SLE distribution in these results motivated us to find a simpler form.

In this paper, we are interested in finding a simple form for the Cumulative Density Function (CDF) and Probability Density Function (PDF) of the SLE. We consider the following hypotheses: (i) $\mathcal{H}_0$: there is no primary user and the received signal is only noise; and (ii) $\mathcal{H}_1$: the primary user exists. Our work is concentrated under the $\mathcal{H}_0$ hypothesis which is common to all CR systems, i.e. there are no constraints on the PU signal, number of PUs and the channel conditions. Probability of False-alarm (P_{fa}), defined as the probability of detecting the presence of PU when it does not exist, is also considered. We prove that the SLE can be modeled as a normal random variable and a simple form of the detection threshold is derived. In the following, we summarize the contributions of this paper:

- Derivation of the distribution of the trace of a complex sample covariance matrix.
- Derivation of the distribution of the SLE detector.
- Derivation of a simple form for the correlation coefficient between the largest eigenvalue and the trace.
- Derivation of a simple form for the P_{fa} and the threshold for detection.

The rest of this paper is organized as follows. Section 2 studies the system model. In Sect. 3, we recall the distribution of the LE and we derive the distribution of the trace of sample covariance matrix. The distribution of the SLE is considered in Sect. 4. We derive its distribution and formulate the correlation coefficient between the LE and the trace. The false alarm probability and the threshold are also addressed. Theoretical findings are validated by simulations in Sect. 5 while the conclusion is drawn in Sect. 6.

2   System Model

Consider a multi-antenna cognitive radio system and denote by K the number of received antennas. Let N be the number of samples collected from each antenna, then the received sample from antenna $k = 1 \cdots K$ at instant $n = 1 \cdots N$ under the two hypotheses is given by

$$\mathcal{H}_0 : \; y_k(n) = \eta_k(n), \tag{1}$$
$$\mathcal{H}_1 : \; y_k(n) = s(n) + \eta_k(n), \tag{2}$$

with $\eta_k(n)$ is a complex circular white Gaussian noise with zero mean and unknown variance σ_η^2 and $s(n)$ is the received signal sample including the channel effect.

After collecting N samples from each antenna, the received signal matrix, $\boldsymbol{Y}$, is given by:

$$\boldsymbol{Y} = \begin{pmatrix} y_1(1) & y_1(2) & \cdots & y_1(N) \\ y_2(1) & y_2(2) & \cdots & y_2(N) \\ \vdots & \vdots & \ddots & \vdots \\ y_K(1) & y_K(2) & \cdots & y_K(N) \end{pmatrix}, \tag{3}$$

Without loss of generality, we suppose that $K \le N$ then the sample covariance matrix is given by $\boldsymbol{W} = \boldsymbol{Y}\boldsymbol{Y}^\dagger$ where $\dagger$ is the Hermitian notation. Denote the eigenvalues of $\boldsymbol{W}$ by $\lambda_1 \ge \lambda_2 \ge \cdots \ge \lambda_K > 0$.

Under $\mathcal{H}_0$, the received samples are complex circular white Gaussian noise with zero mean and unknown variance σ_η^2. Consequently, the sample covariance matrix is a central uncorrelated complex Wishart matrix denoted as $\boldsymbol{W} \sim \mathcal{CW}_K(N, \sigma_\eta^2 \mathbf{I}_K)$ where K is the size of the matrix, N is the number of Degrees of Freedom (DoF), and $\sigma_\eta^2 \mathbf{I}_K$ is the correlation matrix. $\mathbf{I}$ and '$\sim$' denote the identity matrix and 'distributed as' respectively.

3   Distributions of the Largest Eigenvalue and of the Trace

This section considers the distributions of the LE and of the trace under $\mathcal{H}_0$ hypothesis. We prove that the LE and the trace follow Gaussian distributions for which the mean and variance are formulated. Since the SLE does not depend on the noise power, we suppose, in this section, that $\sigma_\eta^2 = 1$. Based on results of this section, we derive the distribution of the SLE in the next section.

3.1   Distribution of the LE

Let λ_1 be the maximum eigenvalue of $\boldsymbol{W}$ under $\mathcal{H}_0$ and denote the centered and scaled version of λ_1 of the central uncorrelated Wishart matrix $\mathbf{W} \sim \mathcal{CW}_K(N, \mathbf{I}_K)$ by:

$$\lambda_1' = \frac{\lambda_1 - a(K, N)}{b(K, N)} \tag{4}$$

with $a(K, N)$ and $b(K, N)$, the centering and scaling coefficients respectively, are defined by:

$$a(K, N) = (\sqrt{K} + \sqrt{N})^2 \tag{5}$$

$$b(K, N) = (\sqrt{K} + \sqrt{N})(K^{-1/2} + N^{-1/2})^{\frac{1}{3}} \tag{6}$$

then, as $(K, N) \to \infty$ with $K/N \to c \in (0,1)$, λ_1' follows a Tracy-Widom distribution of order 2 (TW2) [12]. However, it was shown that, for a fixed K and as $N \to \infty$, λ_1 follows a normal distribution [13]. The mean and the variance of λ_1 could be approximated using TW2 and they are, respectively, given by:

$$\mu_{\lambda_1} = b(K, N)\mu_{TW2} + a(K, N), \tag{7}$$

$$\sigma^2_{\lambda_1} = b^2(K, N)\sigma^2_{TW2}, \tag{8}$$

where $\mu_{TW2} = -1.7710868074$ and $\sigma^2_{TW2} = 0.8131947928$ are, respectively, the mean and variance of TW distribution of order 2. This approximation is very efficient and it achieves high accuracy for K as small as 2 [13].

3.2 Distribution of the Trace

For a fixed K, as $N \to \infty$ the LE converges to a Gaussian distribution. On the other hand, let $T = \sum \lambda_i$ be the trace then the following theorem holds:

Theorem 1. *Let T be the trace of $\mathbf{W} \sim \mathcal{CW}_K(N, \mathbf{I}_K)$. Then, as $N \to \infty$, T follows a Gaussian distribution as follows:*

$$P(\frac{T - NK}{\sqrt{NK}} \leq x) = \frac{1}{\sqrt{2\pi}} \int_{-\infty}^{x} e^{-\frac{u^2}{2}} du, \tag{9}$$

Proof. Let us write:

$$T = tr(\mathbf{Y}\mathbf{Y}^\dagger) = \sum_{i=1}^{K} \left[\sum_{j=1}^{N} |y_{i,j}|^2 \right] \tag{10}$$

with y_{ij} are independent circularly symmetric complex standard normal random variables $(y_{i,j} \sim \mathcal{CN}(0,1))$. Accordingly, the square of the norm, $|y_{i,j}|^2$, is exponentially distributed with unit mean and unit variance. Hence, by Central Limit Theorem (CLT), as $N \to \infty$ the term in the square bracket of (10) follows Gaussian distribution and T is the sum of Gaussians.

To the best of the authors' knowledge, the result in Theorem 1 is new. Let $T_n = \frac{1}{K}T$ be the normalized trace, then T_n, following Theorem 1, is Normally distributed with mean and variance given respectively by:

$$\mu_{T_n} = N, \tag{11}$$

$$\sigma^2_{T_n} = N/K, \tag{12}$$

4 SLE Detector

Let W be the sample covariance matrix at the CR receiver, then the SLE of W is defined by:

$$X = \frac{\lambda_1}{\frac{1}{K}\sum_{i=1}^{K}\lambda_i} = \frac{\lambda_1}{T_n} \tag{13}$$

Denoting by α the decision threshold, then the P_{fa} is given by:

$$P_{fa} = P(X \geq \alpha/\mathcal{H}_0) = 1 - F_X(\alpha), \tag{14}$$

where $F_X(.)$ is the CDF of X under $\mathcal{H}_0$ hypothesis. If the expression of the P_{fa} is known, then a threshold could be set according to a required error constraint. Hence, it is important to have a simple and accurate form for the distribution of X.

4.1 SLE Distribution

Under $\mathcal{H}_0$, both the LE and the normalized trace follow the Gaussian distribution as $N \to \infty$ which is realistic in practical spectrum sensing scenarios. Herein, we show that the SLE is normally distributed when both the LE and the normalized trace follows the normal distribution as stated by the following theorem:

Theorem 2. *Let X be the SLE of* $\mathbf{W} \sim \mathcal{CW}_K(N, \sigma_\eta^2 \mathbf{I}_K)$. *Then, for a fixed K and as* $N \to \infty$, *X follows a normal distribution with CDF and PDF, respectively, given by:*

$$F_X(x) = \Phi\left(\frac{x\mu_{T_n} - \mu_{\lambda_1}}{\sqrt{\sigma^2_{\lambda_1} - 2xc + x^2\sigma^2_{T_n}}}\right) \tag{15}$$

$$f_X(x) = \frac{\mu_{T_n}\sigma^2_{\lambda_1} - c\mu_{\lambda_1} + (\mu_{\lambda_1}\sigma^2_{T_n} - c\mu_{T_n})x}{(\sigma^2_{\lambda_1} - 2xc + x^2\sigma^2_{T_n})^{\frac{3}{2}}}\phi\left(\frac{x\mu_{T_n} - \mu_{\lambda_1}}{\sqrt{\sigma^2_{\lambda_1} - 2xc + x^2\sigma^2_{T_n}}}\right) \tag{16}$$

with

$$\Phi(v) = \int_{-\infty}^{v}\phi(u)du \quad \textit{and} \quad \phi(u) = \frac{1}{\sqrt{2\pi}}e^{-\frac{u^2}{2}} \tag{17}$$

where μ_{λ_1}, μ_{T_n} *and* $\sigma^2_{\lambda_1}$, $\sigma^2_{T_n}$ *are, respectively, the mean and the variance of* λ_1 *and* T_n *given by (7), (11) and (8), (12) respectively. The parameter* $c = \sigma_{\lambda_1}\sigma_{T_n}\rho$ *where* ρ *is the correlation coefficient between* λ_1 *and* T_n.

Proof. Let λ_1 and T_n be two normally distributed random variables with μ_{λ_1}, μ_{T_n}, $\sigma^2_{\lambda_1}$ and $\sigma^2_{T_n}$ their means and variances and let ρ be their correlation coefficient. Denote by $g(\lambda, t)$ the joint density of λ_1 and T_n then the PDF of X is $f_X(x) = \int_{-\infty}^{+\infty}|t|g(xt,t)dt$ and the result is found in [14], however, since W is positive definite then $Pr(T_n > 0) = 1$ and the CDF of X could be written as:

$$F_X(x) = Pr(\lambda/t < x) = Pr(\lambda_1 - xt < 0) \tag{18}$$

and thus, CDF is given by (15) and the PDF is its derivative in (16) [15].

4.2   Correlation Coefficient ρ

Theorem 2 gives the form of the distribution of the SLE as a function of the mean and the variance of λ_1 and T_n as well as the correlation coefficient ρ usually not negligible especially for small K. In this section, we will give a simple analytical form to calculate the correlation coefficient, ρ, between the largest eigenvalue and the trace of Wishart matrix based on the mean of the SLE. In the following, we calculate the mean of SLE in two different ways such that a simple form for ρ could be derived.

Mean of SLE Using Independent Property: Using (13) and the property that the SLE and the trace are independent as proved in [16], then the mean of λ_1 could be written as:

$$E[\lambda_1] = E\big[X \times T_n\big] = E[X] \cdot E[T_n] \tag{19}$$

where $E[.]$ stands for expectation operator.

Recall that the mean of λ_1 and the mean of T_n are given respectively by (7) and (11), then based on (19), the mean of the SLE is given by:

$$\mu_X = \frac{\mu_{\lambda_1}}{\mu_{T_n}} = \frac{b(K,N) \cdot \mu_{TW2} + a(K,N)}{N} \tag{20}$$

Mean of SLE Using Its Distribution: Using SLE distribution, it is difficult to find numerically the mean of the SLE, however, it turns out that a simple and accurate approximation could be found.

An approximation of the mean of the ratio $(u + Z_1)/(v + Z_2)$ could be found where u and v are positive constants and Z_1 and Z_2 are two independent standard normal random variables. It is based on approximating formula for $E[1/(v + Z_2)]$ when $v + Z_2$ is normal variate conditioned by $Z_2 > -4$ and $v + Z_2$ is not expected to approach zero as follows [15]:

$$E\left[\frac{1}{v + Z_2}\right] = \frac{1}{1.01v - 0.2713} \tag{21}$$

By using the transformation of the general ratio of two normal random variable λ_1/T_n into the ratio $(u + Z_1)/(v + Z_2)$, which has the same distribution, we have:

$$\frac{\lambda_1}{T_n} \sim \frac{1}{q}\left(\frac{u + Z_1}{v + Z_2}\right) + s \tag{22}$$

with $s = \rho\frac{\sigma_{\lambda_1}}{\sigma_{T_n}}$, $v = \frac{\mu_{T_n}}{\sigma_{T_n}}$ and

$$u = \frac{\mu_{\lambda_1} - \rho\frac{\mu_{T_n} \cdot \sigma_{\lambda_1}}{\sigma_{T_n}}}{(\pm\sigma_{\lambda_1}\sqrt{1 - \rho^2})} \tag{23}$$

$$q = \frac{\sigma_{T_n}}{(\pm\sigma_{\lambda_1}\sqrt{1 - \rho^2})} \tag{24}$$

where one chooses the $\pm$ sign so that u and v have the same sign (i.e. positive). Consequently, the left-side and the right-side of (22) have the same mean. Therefore the mean of the SLE could be approximated as follows:

$$\mu_X = \frac{\mu_{\lambda_1} - \delta\,\mu_{T_n}}{\theta} + \delta \tag{25}$$

with $\delta = \rho\frac{\sigma_{\lambda_1}}{\sigma_{T_n}}$ and $\theta = 1.01\mu_{T_n} - 0.2713\sigma_{T_n}$.

Correlation Coefficient ρ: Using (25), then ρ, after some algebraic manipulation, is given by:

$$\rho = \frac{\sigma_{T_n}}{\sigma_{\lambda_1}} \cdot \frac{\theta\,\mu_X - \mu_{\lambda_1}}{\theta - \mu_{T_n}} \tag{26}$$

where μ_{λ_1}, μ_{T_n} and μ_X are respectively the means of the LE, the normalized trace and the SLE given by (7), (11) and (20) respectively. σ_{λ_1} and σ_{T_n} are respectively the standard deviations of the LE and the normalized trace and are the square root of (8) and (12) respectively.

4.3   Performance Probabilities and Threshold

Using (14) and (15), then P_{fa} is given by:

$$P_{fa}(\alpha) = Q\left(\frac{\alpha\mu_{T_n} - \mu_{\lambda_1}}{\sqrt{\sigma^2_{\lambda_1} - 2\alpha c + \alpha^2\sigma^2_{T_n}}}\right) \tag{27}$$

where $Q(.)$ is the Q-function. Based on (27), we can derive a simple and accurate form for the threshold as a function of the means and variances of the LE and T_n and the correlation coefficient between them as well as the false alarm probability. That is, for a target false alarm probability, $\hat{P}_{fa}$, the equation of the threshold of the SLE detector will be:

$$\alpha = \frac{\mu_{12} - \beta^2\rho\sigma_{12} + \beta\sqrt{m_v - 2\rho\mu_{12}\sigma_{12} + \beta^2\sigma^2_{12}(\rho^2 - 1)}}{\mu^2_{T_n} - \beta^2\sigma^2_{T_n}} \tag{28}$$

where $\mu_{12} = \mu_{\lambda_1}\mu_{T_n}$, $\sigma_{12} = \sigma_{\lambda_1}\sigma_{T_n}$, $m_v = \mu^2_{T_n}\sigma^2_{\lambda_1} + \mu^2_{\lambda_1}\sigma^2_{T_n}$ and $\beta = Q^{-1}(\hat{P}_{fa})$ with $Q^{-1}(.)$ is the inverse Q-function.

5   Numerical Validation

In this section, we discuss the analytical results through Monte-Carlo simulations. We validate the theoretical analysis presented in Sects. 3 and 4. The simulation results are obtained by generating 10^5 random realizations of Y. The inputs of Y are complex circular white Gaussian noise with zero mean and unknown variance σ^2_η.

Table 1 shows the accuracy of the analytical approximation of the correlation coefficient (ρ) of the SLE in (26). The results are shown for $K = \{2, 4, 50\}$

Table 1. The empirical and approximated value of the correlation coefficient ρ for different values of K and N.

$K \times N$	2×500	4×500	2×1000	4×1000	50×1000
ρ-Emperical	0.849	0.6974	0.839	0.6915	0.3353
ρ-Analytical	0.8548	0.6957	0.8623	0.6967	0.3356

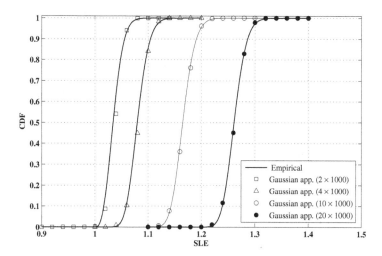

Fig. 1. Empirical CDF of the SLE and its corresponding Gaussian approximation for different values of K with $N = 1000$.

antennas and $N = \{500, 1000\}$ samples per antenna. Table 1 shows that the accuracy of this approximation is higher as the number of antenna increases, however, we can also notice that we have very high accuracy even when $K = 2$ antennas. Also, as expected, it is easy to notice that the correlation between the largest eigenvalue and the trace decreases as the number of antenna increases, however, this correlation could not be ignored even if the number of antenna is large.

Figure 1 shows the empirical CDF of the SLE and its corresponding Gaussian approximation given by Theorem 2. The results are shown for $K = \{2, 4, 10, 20\}$ antennas and $N = 1000$ samples per antenna. Results show a perfect match between the empirical results and our Gaussian formulation. The slight difference in the case $K = 2$ is due to the use of an approximation for the mean and variance of the largest eigenvalue as mentioned in Sect. 3.1. If the exact mean and variance of the LE are used, better results would be expected. However, the results in this paper combine between accuracy and simplicity.

Figures 2 shows the accuracy of the proposed false alarm form proposed in (27). We have considered multi-antenna CR with different number of antennas that aim to sense the spectrum for a time corresponding to $N = 500$ samples.

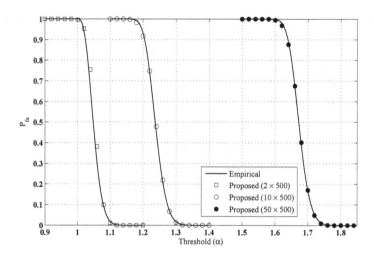

Fig. 2. Empirical probability of false alarm for the SLE detector and its corresponding proposed form in (27) for different values of K with $N = 500$ samples.

The considered number of antennas is as small as $K = 2$ and as large as $K = 50$. Simulation results show a high accuracy in our proposed form which increases as K increases. In addition to the accuracy, the form given in (27) is a simple Q-function equation.

6 Conclusion

In this paper, we have considered the SLE detector due to its optimal performance in uncertain environments. We proved that the SLE could be modeled as a normal random variable and we derive its CDF and PDF. The false alarm probability and the threshold were also considered as we derive new simple and accurate forms. These forms are simple function of the means and variances of the LE and the trace as well as the correlation function between them. Simple forms for the mean, the variance and the correlation coefficient are provided. As a result, this paper provides a simple form for the false alarm probability and the threshold for the SLE detector under relatively large number of samples. However, this constraint is always satisfied in spectrum sensing. Simulation results have shown that the proposed simple forms achieve high accuracy. In addition, results have shown that the correlation between the largest eigenvalue and the trace decreases as the number of antenna increases but it could not be ignored even for large number of antennas. However, the approximation of the correlation coefficient, derived in this paper, shows high accuracy.

Acknowledgment. This work was funded by a program of cooperation between the Lebanese University and the Azem & Saada social foundation (LU-AZM) and by Cen-traleSupélec (France).

References

1. Wang, P., Fang, J., Han, N., Li, H.: Multiantenna-assisted spectrum sensing for cognitive radio. IEEE Trans. Veh. Technol. **59**(4), 1791–1800 (2010)
2. Bianchi, P., Debbah, M., Maida, M., Najim, J.: Performance of statistical tests for single-source detection using random matrix theory. IEEE Trans. Inform. Theory **57**(4), 2400–2419 (2011)
3. Cardoso, L., Debbah, M., Bianchi, P., Najim, J.: Cooperative spectrum sensing using random matrix theory. In: Proceedings of the IEEE International Symposium on Wireless Pervasive Computing (ISWPC), Greece, pp. 334–338, May 2008
4. Zeng, Y., Liang, Y.C.: Eigenvalue-based spectrum sensing algorithms for cognitive radio. IEEE Trans. Commun. **57**(6), 1784–1793 (2009)
5. Nadler, B., Penna, F., Garello, R.: Performance of eigenvalue-based signal detectors with known and unknown noise level. In: 2011 IEEE International Conference on Communications (ICC), pp. 1–5, June 2011
6. Zhang, W., Abreu, G., Inamori, M., Sanada, Y.: Spectrum sensing algorithms via finite random matrices. IEEE Trans. Commun. **60**(1), 164–175 (2012)
7. Kobeissi, H., Nasser, Y., Bazzi, O., Louet, Y., Nafkha, A.: On the performance evaluation of eigenvalue-based spectrum sensing detector for mimo systems. In: XXXIth URSI General Assembly and Scientific Symposium (URSI GASS), pp. 1–4, August 2014
8. Nadler, B.: On the distribution of the ratio of the largest eigenvalue to the trace of a wishart matrix. J. Multivar. Anal. **102**(2), 363–371 (2011)
9. Wei, L., Tirkkonen, O.: Analysis of scaled largest eigenvalue based detection for spectrum sensing. In: 2011 IEEE International Conference on Communications (ICC), pp. 1–5, June 2011
10. Wei, L., Tirkkonen, O., Dharmawansa, K.D.P., McKay, M.R.: On the exact distribution of the scaled largest eigenvalue. CoRR abs/1202.0754 (2012)
11. Wei, L.: Non-asymptotic analysis of scaled largest eigenvalue based spectrum sensing. In: 2012 4th International Congress on Ultra Modern Telecommunications and Control Systems and Workshops (ICUMT), pp. 955–958, October 2012
12. Johansson, K.: Shape fluctuations and random matrices. Comm. Math. Phys. **209**(2), 437–476 (2000)
13. Tirkkonen, O., Wei, L.: Exact and asymptotic analysis of largest eigenvalue based spectrum sensing. In: Foundation of Cognitive Radio Systems, InTech, pp. 3–22 (2012)
14. Hinkley, D.V.: On the ratio of two correlated normal random variables. Biometrika **56**(3), 635–639 (1969)
15. Marsaglia, G.: Ratios of normal variables. J. Stat. Softw. **16**(4), 1–10 (2006)
16. Besson, O., Scharf, L.: Cfar matched direction detector. IEEE Trans. Signal Process. **54**(7), 2840–2844 (2006)

Doppler Compensation and Beamforming for High Mobility OFDM Transmissions in Multipath

Kalyana Gopala$^{(\boxtimes)}$ and Dirk Slock

EURECOM, Sophia-Antipolis, France
{gopala,slock}@eurecom.fr

Abstract. The paper focuses on the use of receive beamforming (BF) for high speed train (HST) scenario under independent Doppler for the different multipath components in a Rician Fading environment. To combat ICI, we not only null out the ICI in the frequency domain, but do a pre-processing in the time domain via frequency correction (demodulation) to maximise the signal part at the output of the FFT. To obtain a suitable demodulation frequency, location aware and location agnostic approaches are considered. Cyclic prefix (CP) based estimation method is also considered as part of location information agnostic approach. In the location-aware approach, a technique that uses both the LoS and dominant scatterer information is also proposed. The paper then provides the optimal weights for the maximisation of the SINR criterion from a theoretical and practical perspective. In the case of linear approximation of the channel variation, the ICI is shown to be a rank 1 interferer and hence can be nulled out with just 2 receiver antennas. Finally, all the methods are compared via simulations. We conclude that in an LTE OFDM system simple, low complex, location agnostic BF schemes are very effective against ICI even with just two receive antennas.

Keywords: ICI · Doppler · Beamforming · SINR · OFDM

1 Introduction

The current LTE design has been done to support speeds of up to 500 kmph but this is yet to be seen in practical demonstrations [1]. As is well known, the high Doppler in these environments violates the orthogonality requirement for the OFDM, resulting in ICI. While the lower data rate transmissions are not impacted by Doppler, the higher data rates are severely impacted. An analysis of the SINR due to ICI is clearly useful to bring out this dependency [2,3]. In [4], one can clearly see the presence of multipath components with independent Doppler values. Interestingly, these multipath components appear with the same delay (based on the sampling rates) and have amplitudes comparable to that of the LoS path. In [5], the Digital Video Broadcast (DVB) scenario is considered with multiple LoS paths from adjacent base stations. Here, individual LoS paths

© ICST Institute for Computer Sciences, Social Informatics and Telecommunications Engineering 2016
D. Noguet et al. (Eds.): CROWNCOM 2016, LNICST 172, pp. 294–306, 2016.
DOI: 10.1007/978-3-319-40352-6_24

are extracted by beamforming on the spatial signatures, and then individually corrected for the Doppler. The results are shown for strong LoS path. In practice, the multipath components would also have significant power and would arrive with their independent Doppler values. In another recent publication [6], ICI is analyzed using a two path model, but does not attempt to improve the SINR due to ICI. In [7], Taylor series approximation of the time varying channels is exploited for Doppler compensation with multiple receive antennas. In this paper, we consider a Rician fading channel with multipath components, each of which have individual Doppler frequencies. The optimal beamforming weights that maximise the SINR in the frequency domain is derived assuming full knowledge of the individual multipaths. It is then shown how the theoretical analysis can be effectively used to obtain the optimal weights in a practical scenario.

Using the simple time-varying nature of the channel it is shown that to maximise the SINR, it is important to maximise the signal part at the output of the FFT at the receiver. Hence, to combat ICI, we not only null out the ICI in the frequency domain, but do a pre-processing in the time domain via frequency correction (demodulation) to maximise the signal part at the output of the FFT. Several criterion are considered to obtain the suitable choice for this frequency. One of them is to choose a demodulation frequency that maximises the signal power in the frequency domain. We also derive an approximate expression to derive this frequency. A well known technique to estimate the carrier frequency offset in OFDM is to cross-correlate the cyclic prefix(CP) of a symbol with it's repeated segment. Intuitively, this technique tries to minimize the time variance of the channel, and hence we analyze this correlation technique as a candidate to estimate the demodulation frequency. These methods are contrasted against the location aided approach of using the LoS frequency. We also suggest a multiple demodulation approach based on the Doppler frequencies corresponding to the strongest paths, and derive the SINR expression in this case.

The rest of the paper is organized as follows. We first present the system model in Sect. 2. Section 3 motivates the need for an optimal demodulation frequency and details the various approaches to obtain an appropriate demodulation frequency. The theoretical SINR analysis and a practical beamformer for the same is given in Sect. 4.1. Section 4.2 derives the SINR for the case of multiple demodulation frequencies. In Sect. 4.3 we derive the beamformer weights when the channel variation due to ICI is approximately linear as in [7–9]. This is followed by simulation results for multipath scenario in Sect. 5. Finally, conclusions are given in Sect. 6. In the following discussions, an underscore on a variable indicates that it is a vector. Capital letters are used for frequency domain representation.

2   System Model

Consider a single input multiple output (SIMO) system with N_r receive antennas on a train moving with velocity v. A typical LTE OFDM framework is chosen with N subcarriers and sampling rate f_s. We consider a Rician multipath channel with all the multipath components appearing in the same time sample as that of the LoS path. In other words, there is negligible delay spread in the system. Let L be the total number of paths and θ_i be the direction of arrival (DoA) of each path (LoS and multipaths). The Doppler frequency of the individual multipaths is then given by

$$f_i = \frac{v \cos(\phi_i)}{c} f_c \qquad (1)$$

where c is the velocity of light, f_c is the carrier frequency and ϕ_i is related to θ_i and direction of travel of the train. For every sample index n in the time domain,

$$\underline{y}_n = \underline{h}_n x_n + \underline{\nu}_n \qquad (2)$$

The time varying channel h_n can be expressed as

$$\underline{h}_n = \sum_{i=0}^{L-1} A_i \underline{a}(\theta_i) e^{j2\pi \epsilon_i \frac{n}{N}} \qquad (3)$$

where A_i, $\underline{a}(\theta_i)$ are the amplitudes and the spatial signature of the individual channel paths. ϵ_i refers to the Doppler frequency divided by the subcarrier spacing of the OFDM symbol.

$$\epsilon_i = \frac{f_i}{\frac{fs}{N}} \qquad (4)$$

Assuming perfect symbol synchronization, the FFT of the received symbol is taken for each of the antennas.

FFT output for the lth subcarrier, in the frequency domain is

$$\underline{Y}_l = \sum_{i=0}^{L-1} A_i \underline{a}(\theta_i) \frac{1}{\sqrt{N}} \sum_{n=0}^{N-1} x_n e^{-j2\pi(l-\epsilon_i)\frac{n}{N}} + \frac{1}{\sqrt{N}} \sum_{n=0}^{N-1} \nu_n e^{-j2\pi l \frac{n}{N}} \qquad (5)$$

Let X_l be the data on subcarrier l. i.e.,

$$X_l = \frac{1}{\sqrt{N}} \sum_{n=0}^{N-1} x_n e^{-j2\pi l \frac{n}{N}} \qquad (6)$$

Let $\underline{G}$ be the weights used on the lth subcarrier to extract the data on that subcarrier.

$$\hat{X}_l = \underline{G}^H \underline{Y}_l; \qquad (7)$$

This can be expanded as,

$$\hat{X}_l = X_l \underline{G}^H \sum_{i=0}^{L-1} A_i \underline{a}(\theta_i) Q(\epsilon_i) + \underline{G}^H \sum_{i=0}^{L-1} A_i \underline{a}(\theta_i) \sum_{m \neq l} X_m Q(m-l+\epsilon_i) + \underline{G}^H \hat{\underline{\nu}}_n \qquad (8)$$

The first term in Eq. (8) corresponds to the signal component. For obvious reasons, we now define

$$H = \sum_{i=0}^{L-1} A_i \underline{a}(\theta_i) Q(\epsilon_i) \qquad (9)$$

The rest of the terms in Eq. (8) are the ICI components and the noise. The loss of orthogonality is captured by the function Q which is defined as

$$Q(m - l + \epsilon_i) = \frac{1}{N} \sum_{k=0}^{N-1} e^{j2\pi(m-l+\epsilon_i)\frac{k}{N}} \tag{10}$$

3   Choice of Demod Frequency

From Eq. (9), it is obvious that the signal power is dependent on the individual Doppler frequencies. We now provide an alternative argument for the same. The time varying channel model in the time domain can be expressed as

$$\underline{h}(t) = \underline{h}_0 + \underline{h}_1(t) \tag{11}$$

where h_0 is the (time-)average value of $h(t)$ over the OFDM time span, and $h_1(t) = h(t) - h_0$ has average value zero. h_0 (a constant signal) and $h_1(t)$ (with zero average) are orthogonal signals over the OFDM symbol duration. Upon sampling, we obtain

$$\underline{h}_n = \underline{h}_0 + \underline{h}_{1,n} \tag{12}$$

Orthogonality ensures that

$$\sum_{n=0}^{N-1} |\underline{h}_n|^2 = N|\underline{h}_0|^2 + \sum_{n=0}^{N-1} |\underline{h}_{1,n}|^2 \tag{13}$$

FFT output for the lth subcarrier, in the frequency domain is

$$\underline{Y}_l = \underline{h}_0 \frac{1}{\sqrt{N}} \sum_{n=0}^{N-1} x_n e^{-j2\pi l \frac{n}{N}} + \frac{1}{\sqrt{N}} \sum_{n=0}^{N-1} \underline{h}_{1,n} x_n e^{-j2\pi l \frac{n}{N}} + \frac{1}{\sqrt{N}} \sum_{n=0}^{N-1} \underline{\nu}_n e^{-j2\pi l \frac{n}{N}} \tag{14}$$

Using Eq. (6),

$$\underline{Y}_l = X_l \underline{h}_0 + \sum_{k=0}^{N-1} X_k \Psi_{k,l} + \hat{\underline{\nu}}_n \tag{15}$$

where

$$\Psi_{k,l} = \frac{1}{N} \sum_{n=0}^{N-1} \underline{h}_{1,n} e^{j2\pi(k-l)\frac{n}{N}} \tag{16}$$

Thus, as the mean of $\underline{h}_{1,n}$ is zero, $\Psi_{k,k}$ is zero for all k. Hence, the desired signal part is a function of $\underline{h}_0$ and $\underline{h}_{1,n}$ contributes only to ICI. Thus $\underline{h}_0$ is a measure of the signal power and $\underline{h}_{1,n}$ is a measure of the ICI power. Now, applying a demod frequency on the OFDM symbol is equivalent to multiplying the time domain channel in the following manner.

$$\tilde{\underline{h}}_n = \underline{h}_n e^{-j2\pi \epsilon \frac{n}{N}} \tag{17}$$

As this only impacts the phase of the channel, $\sum_{n=0}^{N-1} |\underline{h}_n|^2 = \sum_{n=0}^{N-1} |\tilde{\underline{h}}_n|^2$. But multiplication with the demod frequency impacts the signal power and ICI power. In addition, from Eq. 13, when the signal power increases, the ICI power decreases and vice versa. Thus, the demod frequency can be used effectively to improve the signal power to ICI ratio.

Let us assume that for every OFDM symbol, the receiver demodulates the time domain signal by multiplying sample n with a factor $e^{-j2\pi n \frac{f_d}{f_s}}$. The effective channel can now be represented as

$$\underline{h}_n = \sum_{i=0}^{L-1} A_i \underline{a}(\theta_i) e^{j2\pi(\epsilon_i - \epsilon)\frac{n}{N}} = \sum_{i=0}^{L-1} A_i \underline{a}(\theta_i) e^{j2\pi \acute{\epsilon}_i \frac{n}{N}} \tag{18}$$

where $\epsilon = \frac{f_d}{\frac{f_s}{N}}$ and $\acute{\epsilon}_i = \epsilon_i - \epsilon$.

Equation (9) also gets modified such that the terms ϵ_i get replaced by $\acute{\epsilon}$. The determination of the Demod frequency can be done from a location-aware stand point or from a location agnostic standpoint. We list different options here and later compare them in the Sect. 5.

3.1 Signal Power Maximisation

At low SNR, the noise terms would dominate over the ICI. Hence, the beam-forming weights would reduce to a maximal ratio combiner (MRC) and in this scenario, that the best ϵ would be the one that maximises the signal power. Let this be denoted by ϵ_{mrc}. The results of the following analysis will be used in simulations section to benchmark other techniques.

$$\max_\epsilon ||H^2|| = \max_\epsilon \sum_{p=0}^{Nr-1} \sum_{i=0}^{L-1} \sum_{j=0}^{L-1} A_i A_j^* a_p(\theta_i) a_p^*(\theta_j) Q(\acute{\epsilon}_i) Q^*(\acute{\epsilon}_j) \tag{19}$$

Note that

$$
\begin{aligned}
Q(\acute{\epsilon}_i) Q^*(\acute{\epsilon}_j) &= e^{-j\pi\acute{\epsilon}_i(1-\frac{1}{N}))} \frac{sin(\pi\acute{\epsilon}_i)}{N sin(\pi\frac{\acute{\epsilon}_i}{N})} e^{j\pi\acute{\epsilon}_j(1-\frac{1}{N})} \frac{sin(\pi\acute{\epsilon}_j)}{N sin(\pi\frac{\acute{\epsilon}_j}{N})} \\
&= e^{-j\pi(\epsilon_i - \epsilon_j)(1-\frac{1}{N})} \frac{sin(\pi\acute{\epsilon}_i)}{N sin(\pi\frac{\acute{\epsilon}_i}{N})} \frac{sin(\pi\acute{\epsilon}_j)}{N sin(\pi\frac{\acute{\epsilon}_j}{N})}
\end{aligned}
\tag{20}
$$

Thus, only the sinusoid terms involve ϵ. As the typical values of ϵ_i are small, using the first order Taylor series approximation,

$$\frac{sin(\pi\acute{\epsilon}_i)}{N sin(\pi\frac{\acute{\epsilon}_i}{N})} = \frac{\pi\acute{\epsilon}_i - (\pi\acute{\epsilon}_i)^3/6}{N * \pi\frac{\acute{\epsilon}_i}{N}} = 1 - (\pi\acute{\epsilon}_i)^2/6 \tag{21}$$

Thus, (20) becomes,

$$
\begin{aligned}
Q(\acute{\epsilon}_i) Q^*(\acute{\epsilon}_j) &\approx e^{(-j\pi(\epsilon_i - \epsilon_j)(1-\frac{1}{N})} (1 - (\pi\acute{\epsilon}_i)^2/6)(1 - (\pi\acute{\epsilon}_j)^2/6) \\
&\approx e^{(-j\pi(\epsilon_i - \epsilon_j)(1-\frac{1}{N})} (1 - (\pi\acute{\epsilon}_i)^2/6 - (\pi\acute{\epsilon}_j)^2/6)
\end{aligned}
\tag{22}
$$

Taking the derivative of (22) with respect to ϵ gives

$$\frac{\partial(Q(\acute{\epsilon}_i)Q^*(\acute{\epsilon}_j))}{\partial\epsilon} \approx e^{(-j\pi(\epsilon_i-\epsilon_j)(1-\frac{1}{N})}\left(-(\pi(\epsilon_i-\epsilon))/3 - (\pi(\epsilon_j-\epsilon))/3\right)$$
$$= \frac{\pi}{3}e^{(-j\pi(\epsilon_i-\epsilon_j)(1-\frac{1}{N})}(2\epsilon - \epsilon_i - \epsilon_j) \tag{23}$$

Using this in (19) and equating to zero,

$$\epsilon_{mrc,approx} = \frac{\sum_{p=0}^{Nr-1}\sum_{i=0}^{L-1}\sum_{j=0}^{L-1}A_iA_j^*a_p(\theta_i)a_p^*(\theta_j)e^{-j\pi(\epsilon_i-\epsilon_j)(1-\frac{1}{N})}(\epsilon_i+\epsilon_j)}{2\sum_{p=0}^{Nr-1}\sum_{i=0}^{L-1}\sum_{j=0}^{L-1}A_iA_j^*a_p(\theta_i)a_p^*(\theta_j)e^{-j\pi(\epsilon_i-\epsilon_j)(1-\frac{1}{N})}} \tag{24}$$

3.2 Cyclic Prefix Based Estimation

A well known technique to estimate the carrier frequency offset in OFDM is to cross-correlate the cyclic prefix (CP) of a symbol with it's repeated segment in the time domain. Intuitively, this technique tries to minimize the time variance of the channel, and hence we analyze this correlation technique as a candidate to estimate the demodulation frequency. The time domain correlation in the absence of noise is expressed as

$$\sum_{n=0}^{N_{cp}-1}\sum_{p=0}^{Nr-1}\sum_{i=0}^{L-1}A_ia_p(\theta_i)e^{j2\pi\epsilon_i\frac{n+N}{N}}\sum_{j=0}^{L-1}A_j^*a_p^*(\theta_j)e^{-j2\pi\epsilon_j\frac{n}{N}} \tag{25}$$

where N_{cp} is the length of the CP.

When the time variance is zero, Eq. (25) becomes real. Hence, the demodulation frequency ϵ is chosen such that the cross correlation becomes a real number.

3.3 LoS Doppler as Demodulation Frequency

Assuming sufficient number of receive antennas, we can exploit the knowledge of the Doppler frequency of the LoS component based on the location information. This method becomes optimal at high SNR, when the ICI becomes the dominant component compared to noise. In other words, the signal to interference ratio (SIR) is of interest here. It can be seen that, given sufficient number of antennas to remove the non-LoS multipath components, the best approach is to fully compensate for the Doppler of the LoS component. The optimal beamforming weights would essentially zero force the rest of the multi path components.

3.4 Doppler of LoS Path and Dominant Scatterer as Demodulation Frequencies

In scenarios where there is a significant LoS and a dominant scatterer with known Doppler, we propose a receiver with multiple Demod frequencies. Thus, the input at the receive antennas is first demodulated with the LoS frequency, then with the frequency of the dominant scatterer. The two sets of demodulated data are used to perform beamforming. The SINR analysis for this is given in Sect. 4.2.

4   Receive Beamformer Design

In this section, we analyze the SINR for different scenarios and also derive the corresponding optimal beamforming weights.

4.1   Beamformer for Single Demod

The ICI on the lth subcarrier from the mth subcarrier would be $H_{l,m}^{ICI}X_m$, where

$$H_{l,m}^{ICI} = \sum_{i=0}^{L-1} A_i \underline{a}(\theta_i)Q(m - l + \acute{\epsilon}_i) \tag{26}$$

The overall ICI power from all the subcarriers after weighting with G would be

$$\underline{G}^H\{\sum_{m \neq l} H_{l,m}^{ICI}(H_{l,m}^{ICI})^H\}\underline{G}\sigma_x^2 \tag{27}$$

where $\sigma_x^2 = E[|X_m|^2]$. The SINR for the lth carrier can now be written as

$$SINR_l = \frac{\underline{G}^H HH^H \underline{G}\sigma_x^2}{\underline{G}^H\{\sum_{m \neq l} H_{l,m}^{ICI}(H_{l,m}^{ICI})^H\sigma_x^2 + \sigma_n^2 I_{Nr}\}\underline{G}} \tag{28}$$

where σ_n^2 is the noise variance and I_{Nr} is the identity matrix of size Nr

This the well known Rayleigh quotient problem and the optimal weights G for this scenario are given by

$$\underline{G}_{opt,l} = R_l^{-1}H \tag{29}$$

where R_l is defined as

$$R_l = \{\sum_{m \neq l} H_{l,m}^{ICI}(H_{l,m}^{ICI})^H\sigma_x^2 + \sigma_n^2 I_{Nr}\} \tag{30}$$

The ICI power in the above equations is dependent on the subcarrier location due to the presence of guard bands in the LTE symbol structure. For sake of simplicity, we can assume that all subcarriers of OFDM symbol are occupied. In that case, it can be easily seen that the impact of ICI is uniform across all subcarriers and we can drop the subscript l for R and the optimal $\underline{G}$. The optimal SINR is then given by

$$SINR_{opt} = H^H R^{-1}H \tag{31}$$

From Theory to Practice. The optimal weights were derived assuming full knowledge of the individual channel paths. However, we can take advantage of the form of the optimal weights to derive it in a practical receiver which would not have complete knowledge of all the individual channel taps.

Let us now look at the term $\underline{G}_{opt}$ more carefully. The factor $\{\sum_{i=0}^{L-1} A_i \underline{a}(\theta_i)Q(\acute{\epsilon}_i)\}$ is nothing but the channel value that will be observed on the subcarrier minus the impact of ICI and noise. Let $\widehat{H}$ be the estimated channel in the frequency domain with P pilot subcarriers. Then

$$\widehat{H} = H + \widetilde{H} \qquad (32)$$

where $\widetilde{H} \sim \mathcal{CN}(0, \frac{1}{P}R)$. Hence, with sufficient number of pilots, $\widehat{H}$ becomes a good approximation for H. Now, if we can estimate R in (29), we would be able to compute the $\underline{G}_{opt}$. We now look at the relationship between R and $R_{yy} = E(\underline{y}_l \underline{y}_l^H)$.

$$R_{yy} = HH^H \sigma_x^2 + R \qquad (33)$$

Now, by the Matrix Inversion Lemma [10], $R_{yy}^{-1}H \propto R^{-1}H$. Hence, R_{yy} can readily be obtained by averaging across the subcarriers as follows

$$R_{yy} = \frac{1}{N} \sum_{l=1}^{N} \underline{y}_l \underline{y}_l^H \qquad (34)$$

Thus a practical receiver can use the following as optimal weights

$$\underline{G}_{opt,estimated} = R_{yy}^{-1}\widehat{H}. \qquad (35)$$

4.2 SINR Analysis with Multiple Demodulation

There would be scenarios when the strength of one of the multi path components could become comparable to that of the LoS path. Again, with the knowledge of location, one can have information of both the LoS component as well as the dominant scatterer at a given location. This motivates using two different demodulation frequencies corresponding to the LoS component and the strongest scatterer. Thus, the input at the receive antennas is first demodulated with the LoS frequency, then with the frequency of the dominant scatterer. We now analyse this algorithm on the same lines as in Sect. 4.

Let ϵ_a and ϵ_b be the demodulation frequencies being used for the LoS and dominant scatterer respectively. Let $\acute{\epsilon}_{i,a} = \epsilon_i - \epsilon_a$ and $\acute{\epsilon}_{i,b} = \epsilon_i - \epsilon_b$. For every subcarrier, we obtain $2Nr$ number of equations in the frequency domain as follows

$$\begin{bmatrix} \underline{Y}_{l,a} \\ \underline{Y}_{l,b} \end{bmatrix} = \begin{bmatrix} \sum_{i=0}^{L-1} A_i \underline{a}(\theta_i) \frac{1}{\sqrt{N}} \sum_{n=0}^{N-1} x_n e^{-j2\pi(l-\acute{\epsilon}_{i,a})\frac{n}{N}} + \frac{1}{\sqrt{N}} \sum_{n=0}^{N-1} \underline{\nu}_n e^{-j2\pi(l-\acute{\epsilon}_{i,a})\frac{n}{N}} \\ \sum_{i=0}^{L-1} A_i \underline{a}(\theta_i) \frac{1}{\sqrt{N}} \sum_{n=0}^{N-1} x_n e^{-j2\pi(l-\acute{\epsilon}_{i,b})\frac{n}{N}} + \frac{1}{\sqrt{N}} \sum_{n=0}^{N-1} \underline{\nu}_n e^{-j2\pi(l-\acute{\epsilon}_{i,b})\frac{n}{N}} \end{bmatrix} \qquad (36)$$

Following the same steps as in Sect. 4, we see that

$$\begin{aligned} H &= \begin{bmatrix} \sum_{i=0}^{L-1} A_i \underline{a}(\theta_i) Q(\acute{\epsilon}_{i,a}) \\ \sum_{i=0}^{L-1} A_i \underline{a}(\theta_i) Q(\acute{\epsilon}_{i,b}) \end{bmatrix} \\ H_{l,m}^{ICI} &= \begin{bmatrix} \sum_{i=0}^{L-1} A_i \underline{a}(\theta_i) Q(m-l+\acute{\epsilon}_{i,a}) \\ \sum_{i=0}^{L-1} A_i \underline{a}(\theta_i) Q(m-l+\acute{\epsilon}_{i,b}) \end{bmatrix} \end{aligned} \qquad (37)$$

The AWGN noise terms across the two sets of demodulation output in the frequency domain are correlated. The correlation C_{2Nr} is given by

$$C_{2Nr} = \sigma_n^2 \begin{bmatrix} I_{Nr} & Q(\epsilon_b - \epsilon_a)I_{Nr} \\ Q(\epsilon_a - \epsilon_b)I_{Nr} & I_{Nr} \end{bmatrix} \tag{38}$$

The SINR for the lth carrier can now be written as

$$SINR_l = \frac{\underline{G}^H H H^H \underline{G} \sigma_x^2}{\underline{G}^H \{ \sum_{m \neq l} H_{l,m}^{ICI} (H_{l,m}^{ICI})^H \sigma_x^2 + C_{2Nr} \} \underline{G}} \tag{39}$$

The optimal weights G for this scenario is given by

$$\underline{G}_{opt} = R^{-1} H \tag{40}$$

where $R = \{ \sum_{m \neq l} H_{l,m}^{ICI} (H_{l,m}^{ICI})^H \sigma_x^2 + C_{2Nr} \}$

4.3   Beamformer with Linear Approximation for Channel Variation

Equation (16) can be simplified if the channel variations are assumed to be linear. Several prior publications [7–9] have also done similar approximation.

$$\underline{h}_n = \underline{h}_0 + \left(n - \frac{N-1}{2} \right) \underline{h}_1 \tag{41}$$

Equation (15) can now be modified as

$$\underline{Y}_l = X_l \underline{h}_0 + \underline{h}_1 \sum_{k=0}^{N-1} X_k \Xi_{k,l} + \hat{\underline{\nu}}_n \tag{42}$$

where

$$\Xi_{k,l} = \frac{1}{N} \sum_{n=0}^{N-1} \left(n - \frac{N-1}{2} \right) e^{j2\pi(k-l)\frac{n}{N}} \tag{43}$$

As $\Xi_{k,l}$ is a constant, we see immediately that the ICI is approximately rank 1. Hence $N_r = 2$ receivers should be sufficient to cancel the ICI. Applying this to Eq. (29) (as in [7]),

$$G_{opt,linear} = \left(\underline{h}_1 \underline{h}_1^H \sum_{k=0}^{N-1} |\Xi_{k,l}|^2 + \sigma_n^2 I_{N_r} \right)^{-1} \underline{h}_0 \tag{44}$$

Using the matrix inversion lemma, this can be simplified to

$$G_{opt,linear} = \left(1 - \frac{\underline{h}_1 \underline{h}_1^H \frac{\sum_{k=0}^{N-1} |\Xi_{k,l}|^2}{\sigma_n^2}}{1 + \underline{h}_1^H \underline{h}_1 \frac{\sum_{k=0}^{N-1} |\Xi_{k,l}|^2}{\sigma_n^2}} \right)^{-1} \underline{h}_0 \tag{45}$$

5   Simulation Results

Here we consider an OFDM system with 1024 subcarriers and subcarrier spacing of 15 KHz as in LTE. In addition to the LoS, 3 additional multipaths are considered. The Doppler offsets for all the paths in Hertz are [1080 −1080 758 220]. The relative tap strengths in dB are [0 0 −11 −0.7]. Such a channel with two equal strength paths but differing Doppler is motivated by the observations in [4]. Channel estimation is done using pilots with a spacing of 12 subcarriers. As a flat frequency channel is considered, the channel estimates are averaged across the pilots.

Figure 1 plots the theoretical SINR for $N_r = 2$ with CP based Demod frequency offset estimation (Eq. (29)) and compares this against a Maximum Ratio Combiner receiver that does not factor in the ICI. Also shown is a single antenna receiver (No BF). The huge performance gap clearly brings out the need for optimal Beam Forming with ICI in the SIMO receiver. The MRC receiver and the single antenna receiver do not apply the Demod frequency correction.

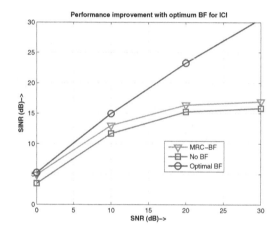

Fig. 1. Performance gains with optimum BF for $N_r = 2$

Figure 2 compares the different approaches to estimate the demodulation frequency over multiple channel realizations at an SNR of 30 dB. The data points corresponding to "Matlab Search for SINR" are obtained by determining the demodulation frequency that along with the corresponding optimal weights maximises the SINR. This was obtained by performing an exhaustive search in Matlab over the range of possible Doppler values. The data points "Matlab search for SIG PWR" were obtained by determining the demodulation frequencies that maximise the signal component in the frequency domain. This was again performed through exhaustive search in Matlab. The same was also obtained using Eq. (24) and is given by the data points "Approximate freq est". Finally, the demodulation frequencies obtained from CP correlation are also given. It can be

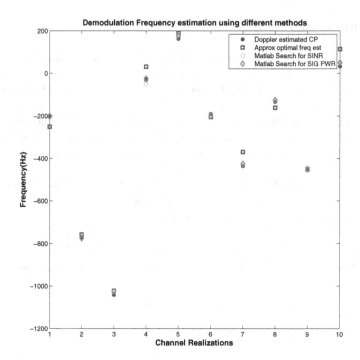

Fig. 2. Comparison of methods to estimate demodulation frequency

seen that the CP based correlation method and the signal power maximisation method closely match the SINR maximisation criterion. The approximation for the signal power maximisation is limited by the first-order Taylor series approximation, but is still a useful approach. This establishes the CP based estimation as a useful practical approach to improve the SINR under ICI.

In Fig. 3, for the same scenario as in Fig. 1, different receivers are compared. The "Optimal BF" curve implements Eq. (29) with CP based demod frequency estimation. The "Estimated BF" curve implements Eq. (35) and the demod frequency is estimated using the CP. The "Approx BF" corresponds to the method in Eq. (45) with CP based demod. The $\underline{h}_1$ is computed as in [7]. The SINR with the above mentioned three methods almost overlap. The "Approx BF, no Demod correction" refers to implementation of Eq. (45) without the application of the CP based Demod frequency correction to highlight the need for the Demod frequency correction. Also given are the curves with LoS Doppler used for Demod frequency correction. Clearly, in this scenario, the performance is suboptimal with this choice as there are scattered paths with significant signal strengths. The Multiple Demod, though computationally more expensive, performs slightly better than the optimal receiver with only a single Demod.

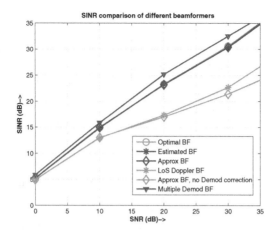

Fig. 3. Comparison of SINR with different receivers for $N_r = 2$

6 Conclusion

The paper focuses on ICI compensation for high Doppler scenarios. The need
for an optimal Demod frequency correction is clearly motivated. This is followed by a theoretical analysis of the ICI for a multipath scenario and determines the optimal weights to maximise the SINR. We then proceed to show
that these theoretical weights can indeed be obtained in a practical scenario and
show the equivalence via simulations. While there have been publications on ICI
for OFDM systems with multiple receivers [7–9], we focus on both the Demod
frequency as well as the beamforming weights. We propose and analyse different approaches for obtaining the optimal demod frequency that include both
location-aware and location information agnostic approaches. As a further extension, the performance is analyzed when two demodulation frequencies are used
instead of one. It is shown through simulations that CP based Demod frequency
estimation is a very practical way of obtaining the near-optimal Demodulation
frequency. With the help of the simulations, we conclude that for a typical LTE
OFDM scenario, good performance can be achieved with low complexity beamforming techniques even at 450 kmph in the presence of significant scatterers
with independent Doppler frequencies with just $N_r = 2$ receivers.

Acknowledgements. EURECOM's research is partially supported by its industrial
members: ORANGE, BMW, SFR, ST Microelectronics, Symantec, SAP, Monaco Telecom, iABG, by the EU FP7 NoE NEWCOM# and H2020 projects ADEL, HIGHTS.

References

1. Merz, R., Wenger, D., Scanferla, D., Mauron, S.: Performance of LTE in a high-velocity environment: a measurement study. In: 4th Workshop on All Things Cellular: Operations, Applications and Challenges (co-located with ACM SIGCOMM), August 22, Chicago, IL, USA, pp. 47–52. ACM, New York (2014). http://doi.acm.org/10.1145/2627585.2627589
2. Faulkner, M., Wilhelmsson, L., Svensson, J.: Low-complex ICI cancellation for improving doppler performance in OFDM systems. In: IEEE Vehicular Technology Conference, September 2006
3. Peng, Y., Wenbo, W., Kim, Y.I.: Performance analysis of OFDM sytem over time-selective fading channels. In: IEEE Wireless Communications and Networking Conference, April 2009
4. Kaltenberger, F., Byiringiro, A., Arvanitakis, G., Ghaddab, R., Nussbaum, D., Knopp, R., Berbineau, M., Cocheril, Y., Philippe, H., Simon, E.P.: Broadband wireless channel measurements for high speed trains. In: ICC 2015, IEEE International Conference on Communications, 8–12 June 2015, London, United Kingdom. ROYAUME-UNI, London, June 2015. http://www.eurecom.fr/publication/4464
5. Yang, L., Ren, G., Zhai, W., Qiu, Z.: Beamforming based receiver scheme for DVB-T2 system in high speed train environment. IEEE Trans. Broadcast. **59**, 146–154 (2013)
6. Zhang, C., Fan, P., Xiong, K.: Downlink resource allocation for the high-speed train and local users in OFDMA systems. In: IEEE Wireless Communications and Networking Conference (WCNC), March 2015
7. Serbetli, S.: Doppler compensation for mobile ofdm systems with multiplereceive antennas. In: 2012 IEEE 19th Symposium on Communications and Vehicular Technology in theBenelux (SCVT), pp. 1–6. IEEE (2012)
8. Mostofi, Y., Cox, D.: ICI mitigation for pilot-aided OFDM mobile systems. IEEE Trans. Wirel. Commun. **4**(2), 765–774 (2005)
9. Schellmann, M., Thiele, L., Jungnickel, V.: Low-complexity doppler compensation in mobile SIMO-OFDM systems. In: 2008 42nd Asilomar Conference on Signals, Systems and Computers, pp. 1015–1019, October 2008
10. Henderson, H.V., Searle, S.R.: On deriving the inverse of a sum of matrices. SIAM Rev. **23**, 53–60 (1981)

Frequency Agile Time Synchronization Procedure for FBMC Waveforms

Jean-Baptiste Doré$^{(\boxtimes)}$ and Vincent Berg

CEA-Leti Minatec, 17 rue des Martyrs, 38054 Grenoble Cedex 9, France
jean-baptiste.dore@cea.fr

Abstract. Access to multiple non-contiguous spectrum bands for non-synchronous users is foreseen as essential by some transmission scenarios of the 5th generation of cellular networks (5G). Filterbank Multicarrier modulation (FBMC) has been envisaged because of its ability to give an efficient answer to these requirements. This paper introduces and evaluates the performance of a frequency agile time synchronization solution adapted to FBMC asynchronous transmission on fragmented spectrum. The detection of the start of burst is solely made in the frequency domain assuming that the fast Fourier transform of the receiver is not synchronized. The proposed algorithms have a particular interest for real time implementation when Machine Type Communication applications are considered. This latter scenario is sometimes considered as one of the most challenging for 5G.

Keywords: FBMC · OFDM · Synchronization

1 Introduction

The current generation of cellular physical layers such as used by Long Term Evolution (LTE) and LTE-Advanced have been optimized to deliver high bandwidth pipes to wireless users but require the orthogonality between users within a single cell. More precisely, in case of Orthogonal Frequency Division Multiplexing (OFDM), users should be synchronized in time to avoid inter-user interference. As a consequence LTE networks rely on Timing Advance (TA) value adjustment procedure to manage the synchronization between users.

The expected explosion of Machine-Type-Communication (MTC) is posing new challenges. Fast dormancy necessary for this type of service to save battery power has resulted in significant control signaling growth especially for the TA procedure. Furthermore, as the availability of large amounts of contiguous spectrum is getting more and more difficult to guarantee, the aggregation of non-contiguous frequency bands has to be considered to increase the efficiency of the network.

Therefore relaxed synchronization and access to fragmented spectrum are key parameters for future generations of wireless networks [1]. This requirement of spectrum agility has encouraged the study of alternative multicarrier

© ICST Institute for Computer Sciences, Social Informatics and Telecommunications Engineering 2016
D. Noguet et al. (Eds.): CROWNCOM 2016, LNICST 172, pp. 307–318, 2016.
DOI: 10.1007/978-3-319-40352-6_25

waveforms such as Filterbank Multicarrier (FBMC) to provide better adjacent channel leakage performance without compromising spectral efficiency [2]. Frequency Division Multiple Access (FDMA) schemes have already been considered for FBMC modulations [3], and access to fragmented spectrum has been envisaged [4]. When licensed spectrum access is considered, fragmented spectrum is often a consequence of dynamic access to the frequency block. The phenomenon is expected to be further exacerbated when noncontiguous asynchronous dynamic access is considered particularly when messages get shorter as is sometimes the case for MTC applications.

When flexible asynchronous transmission on fragmented spectrum is considered, time synchronization of the multicarrier symbol is not straightforward. Time domain processing is usually considered for multicarrier receivers [5,6]. A signal a priori known to the receiver is initially sent to synchronize the packet. A preamble built on a known sequence (e.g. Zadoff-Chu LTE) or alternatively a training symbol sequence that is repeated n times [6] can be used. When multiple users access the same resource each preamble should be orthogonal to each other. For FDMA, frequency orthogonality usually applies. However, if the synchronization is performed in the time domain, it is difficult to guarantee orthogonality between the users with a reasonable level of complexity, particularly in the considered fragmented spectrum scenario. It requires a per-user filtering process during the synchronization procedure making the time synchronization process ill-adapted to flexible asynchronous fragmented access.

A frequency agile time synchronization solution adapted to the context is investigated in this paper. We propose a frame format associated with a frequency-domain time-synchronization processing particularly adapted to flexible multiuser scenario on fragmented spectrum. Section 2 describes the system model, the frame format and the block diagram of the proposed receiver. In Sect. 3 we derive a time synchronization algorithm and performance is evaluated and discussed. Section 4 concludes the paper.

2   System Model

2.1   FBMC and Proposed Receiver Architecture

A multicarrier system can be described by a synthesis/analysis filter bank, i.e. a transmultiplexer structure. The synthesis filter bank is composed of a set of parallel transmit filters. The analysis filter bank consists of the parallel matched receive filters. The most widely used multicarrier technique is OFDM, based on the use of the inverse and forward Fast Fourier Transform (FFT) for the analysis and the synthesis filter banks. The prototype filter of OFDM is a rectangular window whose size is equal to the duration of the FFT. At the receiver, perfect signal recovery is possible under ideal channel conditions thanks to the orthogonality of the prototype filters. Nevertheless under more realistic multipath channels, a data rate loss has to be paid with the mandatory introduction of a Cyclic Prefix (CP), longer than the impulse response duration of the channel. FBMC waveforms utilize a more advanced prototype filter designed to give a better frequency localization of the subcarriers. The prototype filter used in this paper is based on the frequency sampling technique [7]. This technique gives the advantage of using a closed-form representation that includes only a few adjustable design parameters.

The most significant parameter is the duration of the impulse response of the prototype filter also called overlapping factor, K. The impulse response of the prototype filter is given by [7]:

$$h(t) = G_P(0) + 2 \sum_{k=1}^{K-1} (-1)^k G_P(k) \; cos \left(\frac{2\pi k}{KN} (t+1) \right) \qquad (1)$$

where $G_P(0..3) = \left[1, \; 0.97195983, \; \frac{1}{\sqrt{2}}, \; 1 - G_P(1)^2 \right]$ for an overlapping factor of $K = 4$ and N is the number of carriers. The larger the overlapping factor K, the more localized the signal will be in frequency. Adjacent carriers significantly overlap with this kind of filtering. In order to keep adjacent carriers orthogonal, real and pure imaginary values alternate on successive carrier frequencies and on successive transmitted symbols for a given carrier at the transmitter side. In order to maximize spectral efficiency of the Offset QAM (OQAM) modulation the symbol period T is halved. The well-adjusted frequency localization of the prototype filter guarantees that only adjacent carriers interfere with each other. This allows for a more flexible operation than OFDM for FDMA access, i.e.: non synchronous flexible frequency division multiple access.

In [4], the authors describe a high performance receiver architecture denoted FS-FBMC (Frequency Spreading FBMC). One advantage of this architecture is that frequency domain time synchronization may be performed independently of the position of the FFT [4]. This is realized by combining timing synchronization with channel equalization. The proposed receiver architecture is depicted in Fig. 1. A free running FFT of size KN is processed every blocks of $N/2$ samples

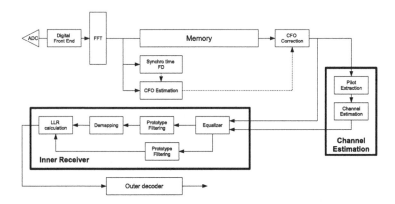

Fig. 1. Block diagram of the proposed receiver based on FS-FBMC principle. The time domain synchronization is performed in the frequency domain(box labelled "Synchro time FD").

generating KN points, i.e. if $\mathbf{r_m}$ is the mth received vector, a KN-point FFT is computed for samples $k = (n + m \times N/2)$ with $n = 0, 1, \ldots, NK - 1$. These successive KN points are stored in a memory unit.

Assuming that the detection of a start of burst is then achieved on the frequency domain (i.e. at the output of the FFT) using a priori information from the preamble. The channel coefficients may be estimated on the pilot subcarriers of the preamble. Once the channel coefficients are estimated on all the active subcarriers, a one-tap per subcarrier equalizer is applied before filtering by the FBMC prototype filter. Demapping and Log-Likelihood Ratio (LLR) computation complete the inner receiver architecture. A soft-input Forward Error Correction (FEC) decoder finally recovers the original message.

The asynchronous frequency domain processing of the receiver combined with the high stop-band attenuation of the FBMC prototype filter provides a flexible receiver architecture that allows for multiuser asynchronous reception. FFT and Memory Unit are common modules, while the remaining of the receiver should be duplicated as many times as the number of parallel asynchronous users the system may tolerate. The originality of this paper lies in the fact that the time domain synchronization procedure is performed in the frequency domain and well adapted to this flexible scenario.

2.2   Burst Format

In order to keep a flexible frequency and time block allocation, a preamble based burst approach is considered. Synchronization and channel estimation is performed using a training sequence known to the receiver. Its structure has been defined and is illustrated in Fig. 2. It is composed of a preamble of duration P-FBMC symbols (P is set to 4 in Fig. 2). The preamble has been designed to accurately detect the start of the burst and gives an estimate of the channel

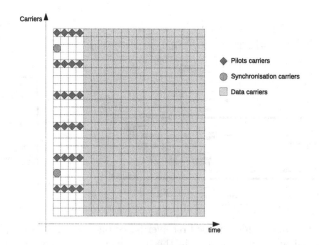

Fig. 2. Considered FBMC burst structure. (Color figure online)

frequency response while preserving the localization properties of the FBMC signal. It is mainly composed of pilot carriers spaced every D active carriers for the whole duration of the preamble (D is set to 4 in Fig. 2). The pilot carriers are designed so that the signal transmitted on each pilot carrier is constant for the duration of the preamble. Synchronization carriers are added on the first multicarrier symbol. These are designed to accurately estimate the start of burst as we will see after. By implementing at the receiver all the baseband signal processing functions in the frequency domain, the proposed scheme may be extended without loss of generalization to the aforementioned multiuser asynchronous environment.

In the following section, we will propose and evaluate signal processing algorithm for the detection of burst.

3   Time Synchronization

Time synchronization is the process by which the start of the burst is identified. In the proposed architecture, all the receiver signal processing operations are realized in the frequency domain. Consequently, the synchronization procedure should inform the other components of the receiver which block of samples, at the output of the FFT, corresponds to the beginning of the burst. Based on the structure of the preamble, a two-steps algorithm is realized. A coarse synchronization is performed followed by a fine synchronization procedure designed to accurately detect the start of the burst.

3.1   Coarse Time Synchronization

In this section, the metric for coarse frame detection in the frequency domain is introduced. A measure of the energy on the filtered carriers at specific carrier location is proposed. The decision metric is expressed as follows:

$$D_F(m) = \sum_{k \in \Omega_s} |y(k,m)|^2 \tag{2}$$

where Ω_s is the subset of carriers used for coarse synchronization and $y(k,m)$ is the received sample after prototype filtering at carrier index k and symbol m. The proposed algorithm is based on a classical hypothesis testing. We introduce the following two hypotheses:

- H_0 : only noise is received
- H_1 : signal plus noise is received

The decision whether a burst has been detected or not is realized on the rule:

$$D_F(m) > T \tag{3}$$

where T defines a threshold value. When this rule is true, the index m gives the reference time for the beginning of the burst. Two measures are usually studied for the characterization of such an algorithm: the probability of false alarm P_{fa} and the probability of detection P_d. The probability of false alarm informs about the probability of a burst detection when only noise is received. The probability of detection gives the probability that a burst is detected when hypothesis H_1 occurs. The probability of false alarm is evaluated by analyzing the probability density function of the decision metric when H_0 occurs.

The analytic derivation of the false alarm probability as a function of the noise variance σ_n^2 and the cardinal of Ω_S, M, is well known and is given by [8]:

$$P_{fa}(T, M, \sigma_n^2) = \frac{\Gamma\left(M, \frac{T}{\sigma_n^2}\right)}{\Gamma(M)} \tag{4}$$

where T is the threshold and $\Gamma(x, y)$ is the upper incomplete gamma function:

$$\Gamma(x, y) = \int_y^{+\infty} t^{x-1} e^{-t} dt \tag{5}$$

This expression assumes independent noise following a normal distribution. A closed form expression for the probability of detection can be derived as [8]:

$$P_d(T, M, \sigma_s^2, \sigma_n^2) = Q_M\left(\sqrt{\frac{\sigma_s^2}{\sigma_n^2 M}}, \sqrt{\frac{T}{\sigma_n^2}}\right) \tag{6}$$

where σ_s^2 is the power of the useful signal. $Q(x, y)$ is the generalized Marcum-Q function.

We have evaluated the receiver operating characteristics (ROC) for the proposed detector when $M = 32$ and for various Carrier to Noise (C/N) ratio ($= \sigma_s^2 / \sigma_n^2$). The size of the FFT is set to 256, and the synchronization preamble is built with 32 actives carriers. Simulations have been done assuming an AWGN (Additive white Gaussian Noise) channel. The C/N varies with a step of 1 dB starting from -5 dB. Due to the proposed receiver architecture and the synchronization process that is realized in the frequency domain, the FFT window is not synchronized with the emitted signal. Simulations have been done for a FFT position mismatch of $N/4$ samples. This is the worst case scenario as the signal energy leakage between consecutive FFT block is at its peak [4]. The coarse detector performance is given in Fig. 3. The probability of false alarm and the probability of misdetection are less than 10^{-2} for a C/N of 1 dB.

The main issue in classical detection problem resides in the determination of the decision threshold. The bottom graph in Fig. 3 illustrates the variation of the false alarm probability (resp. the misdetection probability) for various decision threshold values and for various C/N. For any given C/N a specific threshold that leads to the targeted false alarm probability can be derived (similarly for the misdetection probability).

In practice, it is not straightforward to compute the decision threshold as no information about hypothesis H_0 or H_1 is available. In order to propose a simpler practical realization, we modified the decision metric. We suggest computing

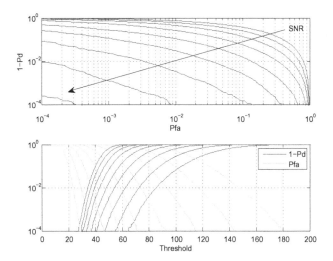

Fig. 3. Top: ROC curves for the proposed energy detector when $M = 32$ and for various SNR (resp C/N). The C/N varies with a step of 1 dB and starts at -5 dB. Bottom: variation of the false alarm probability and the misdetection probability in function of the threshold. (Color figure online)

on the same FFT block two physical observations in parallel. Energy level is computed at some carrier locations while an estimation of the noise is performed on other (null) carrier locations. Based on these two observations, we propose to build a metric by computing a likelihood ratio. We divide the power estimation at the carrier location by the power estimation at the null carrier positions. When H_0 occurs, the metric should tend to 1, and when H_1, the metric should be much greater than 1. The proposed metric can be defined as:

$$D_R(m) = \frac{|\Omega_n|}{|\Omega_s|} \frac{\sum_{k \in \Omega_s} |y(k,m)|^2}{\sum_{k \in \Omega_n} |y(k,m)|^2} \qquad (7)$$

where Ω_n is the set of null carriers and $|\Omega_n|$ (resp. $|\Omega_s|$) is the number of element of the set Ω_n (resp. Ω_s) with $\Omega_s = M$. The analytic computation of a threshold is not straightforward to determine in this case because the metric is a division of chi-squared distribution. Consequently, simulations have been done to estimate the performances of the proposed receiver.

Here again, results have been simulated for the worst case scenario, i.e. a FFT misalignment of $N/4$ samples. Active carriers of the preamble are equally spaced every 4 carriers. The estimation of the noise is realized at the center of two successive synchronization carriers. We depict in Fig. 4 (a) the ROC curves when $M = 32$ and for various C/N.

Compared to the classical energy detection algorithm, the proposed scheme exhibits worse performance. This can be explained by the nature of the statistical test. The results show that for a C/N of 4 dB, the probability of false alarm and

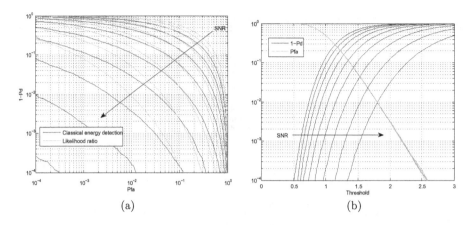

(a)   (b)

Fig. 4. (a) ROC curves for the proposed normalized energy detector using when $M = 32$ and for various C/N. The C/N varies with a step of 1 dB ans starts at -4 dB. (b) Variation of the false alarm probability and the misdetection probability for the proposed normalized energy detector using when $M = 32$ and for various C/N. The C/N varies with a step of 1 dB starting at -4 dB. (Color figure online)

the probability of misdetection is of around 10^{-2}. The advantage of this metric is depicted in Fig. 4 (b). The variation of the probability of false alarm and the probability of misdetection for various C/N as a function of a threshold is given. Here again a FFT with a misalignment of $N/4$ samples (worst case) is assumed.

We remark that thanks to the normalization, the decision threshold hardly depends on the SNR level (the variation is very small). For any C/N, a threshold equal to 2, leads to a false alarm probability smaller than 3.10^{-3}. The corresponding misdetection probability is lower than 10^{-2} for a C/N greater than 3 dB. Even if this metric suffers from a performance loss compared to the classical energy detection, the simplicity of the threshold determination makes this algorithm particularly interesting for practical realization of the coarse detection.

The evaluation of the performance has been done assuming an AWGN channel. In practice, fading channel affects the carrier. The knowledge of the fading positions are not available at the synchronization step: It is possible that at a given carrier position, only noise is present. Consequently when the number of active carriers is small performance of the proposed detectors is not sufficient.

The first improvement for the proposed algorithm consists of allocating more synchronization carriers. We illustrate in Fig. 5 the probability of misdetection when the false alarm probability is equal to 10^{-2} for various C/N values and various number of synchronization carriers. When we increase M, the number of synchronization carrier involved into the decision metric, performance is better. It should be mentioned that, in such a case, the threshold depends on M.

As dynamic spectrum usage is envisaged and possible multifrequency fragment, M can vary. Therefore it is interesting to establish a relationship between M and the synchronization threshold. We depict in Fig. 6 the variations of the

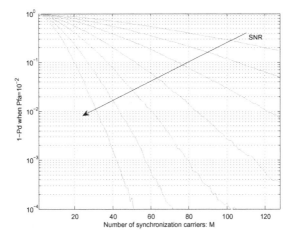

Fig. 5. Probability of misdetection when the false alarm probability is equal to 10^{-2} for various C/N values and various number of synchronization carriers. The C/N varies with a step of 1 dB starting at -4 dB.

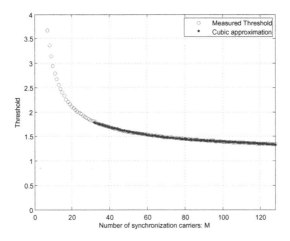

Fig. 6. Threshold values for a false alarm probability of 10^{-2} as a function of the number of synchronization carriers. (Color figure online)

threshold when the false alarm probability is set to 10^{-2} for different values of M. The threshold is a decreasing function of the number of synchronization carriers. It is useful for a flexible receiver to dynamically determine the threshold value according to the considered bandwidth for transmission. Under the assumption that the number of carriers dedicated to synchronization is in the range $[32\,128]$, we approximated the threshold value by a cubic function as follows:

$$T = -0.00000068M^3 + 0.000218M^2 - 0.024M + 2.358 \tag{8}$$

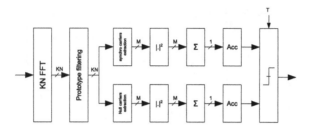

Fig. 7. Block diagram of the proposed coarse synchronization module.

This approximation allows for a simple and flexible computation of the threshold depending on the spectrum configuration.

The previous results have demonstrated that when the number of synchronization carriers is relatively small, performance is not really acceptable, especially if a margin on the C/N should be added to allow for performance over fading channels. If only a small fragment of spectrum is available for the transmission it is not possible to combine an important number of synchronization carriers (we have to keep in mind that null carriers without interference are required for noise estimation). The algorithm can be enhanced by exploiting the structure of the preamble and more precisely the repetition of the pilot carrier over time. The accumulation of different observations leads to better performance; this is effectively equivalent to increasing parameter M by time diversity.

The block diagram of the proposed synchronization module is depicted in Fig. 7. After the filtering in the frequency domain by the prototype filter, the power of each FFT block at synchronization (resp. null) carrier position is estimated and accumulated over successive blocks (Acc module). The two metrics are then compared using the decision threshold T to identify a coarse estimation of the start of the burst.

During the coarse detection, all the active carriers could be used for the energy estimation. More precisely, if we refer to the burst structure depicted in Fig. 2, the combination of the pilot and synchronization carriers may be used for the initial synchronization process. The coarse synchronization procedure is followed by a fine synchronization step designed to precisely identify the start of the burst.

3.2   Fine Time Synchronization

Fine time synchronization gives a refined position of the start of the burst. This step is essential for the signal processing operations that follow in the receiver chain such as channel estimation or equalization.

The proposed algorithm for fine time synchronization is based on the same principle as the one for the coarse time synchronization. The principle is to detect a power signature located at the beginning of the burst. While coarse

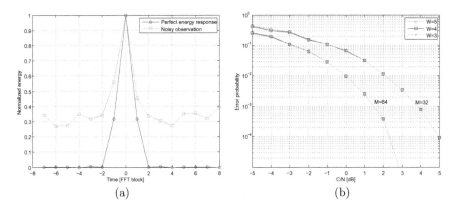

Fig. 8. (a) Example of an energy signature for fine time synchronization. (b) Error probability of the proposed detector for various C/N. (Color figure online)

synchronization relies on the pilot carriers, only the synchronization carriers are used for fine synchronization (Fig. 2). Coarse synchronization is indeed optimized for misdetection and false alarm, while fine synchronization evaluates the most probable for the start of burst event instant. An example of the energy signature of a fine preamble sequence build on equi-distributed positions of the synchronization carriers is given in Fig. 8 (a).

Once coarse synchronization is acquired, we propose to lock the synchronization on the position corresponding to a maximum of energy. Therefore, we propose to evaluate the signal energy at the synchronization carrier location (see Fig. 2) using the following rule:

$$z(m) = \sum_{k \in \Omega_{Sync}} |y(k,m)|^2 \tag{9}$$

$$\hat{m} = \max_{m \in W} z(m) \tag{10}$$

where W is the set of position to test and Ω_{Sync} is the set of synchronization carriers indexes. It should be mentioned that the fine time synchronization can be directly applied. However, the use of a robust coarse synchronization based on a higher number of observations (the number of pilot carrier is higher than the number of synchronization carrier) gives better performance.

The performance of the proposed detector has been evaluated by simulation over an AWGN channel. We depict in Fig. 8 (b) the error probability for various C/N. An error is assumed when the position at the output of the detector is not correct. These results show that the size of the window (W) has a limited impact on the performances especially at high C/N region. When the number of carriers used for the energy detection increases, performance also increases. Maximizing the number of synchronization carriers while maintaining orthogonality between the pattern is crucial to guarantee good performance.

4 Conclusion

In this work, a frequency agile time domain synchronization algorithm adapted to dynamic fragmented spectrum scenarios in a multiuser context has been addressed. The detection of the start of multicarrier frame algorithm is based on detecting the power signature of a preamble sequence. The procedure is divided into two steps, a coarse followed by a fine acquisition step. Performance and practical realization have been discussed and evaluated. An efficient architecture with an associated threshold determination method adapted to the dynamic scenario has been proposed. The method not only performs well for low C/N but is also very adapted to dynamic access per user as its decision threshold can be efficiently predicted according to the amount of occupied spectrum.

Acknowledgment. This work has been carried out in the frame of the PROFIL project, which is funded by the French National Research Agency (ANR) under the contract number ANR-13-INFR-0007-02.

References

1. Wunder, G., et al.: 5GNOW: Challenging the LTE design paradigms of orthogonality and synchronicity. In: 2013 IEEE 77th Vehicular Technology Conference (VTC Spring) (2013)
2. Noguet, D., Gautier, M., Berg, V.: Advances in opportunistic radio technologies for TVWS. EURASIP J. Wirel. Commun. Netw. **2011**(1), 170 (2011). http://jwcn.eurasipjournals.com/content/2011/1/170
3. Medjahdi, Y., Terre, M., Le Ruyet, D., Roviras, D.: On spectral efficiency of asynchronous OFDM/FBMC based cellular networks. In: IEEE 22nd International Symposium on Personal Indoor and Mobile Radio Communications (PIMRC), pp. 1381–1385, September 2011
4. Doré, J., Berg, V., Cassiau, N., Kténas, D.: FBMC receiver for multi-user asynchronous transmission on fragmented spectrum. EURASIP J. **2014**, 1–41 (2014). Special Issue on Advances in Flexible Multicarrier Waveform for Future Wireless Communications
5. Speth, M., Fechtel, S., Fock, G., Meyr, H.: Optimum receiver design for wireless broad-band systems using OFDM. I. IEEE Trans. Commun. **47**(11), 1668–1677 (1999)
6. Schmidl, T., Cox, D.: Robust frequency and timing synchronization for OFDM. IEEE Trans. Commun. **45**(12), 1613–1621 (1997)
7. Bellanger, M., et al.: FBMC physical layer: a primer (2010). http://www.ict-phydyas.org
8. Salehi, M., Proakis, J.: Digital Communications. McGraw-Hill Education, New York (2007)

IEEE 1900.7-2015 PHY Evaluation on TVWS Scenarios

Dominique Noguet$^{(\boxtimes)}$ and Jean-Baptiste Doré

CEA Leti, Minatec Campus, 17 rue des Martyrs, 38054 Grenoble, France
dominique.noguet@cea.fr
http://www-leti.cea.fr

Abstract. White space communications have attracted a particular interest after some national regulators have authorized TVWS unlicensed secondary usage. The IEEE 1900.7 working group has defined a specific air interface tailored to these applications. The standard, is based on an Filter Bank Multi Carrier (FBMC) physical layer and a CSMA-CA MAC sublayer. In this paper, the Modulation and Coding Schemes (MCS) of the IEEE 1900.7-2015 are considered to assess simulated expectations on throughput and coverage under various typical TVWS deployment scenarios: Fixed Long Range Access, campus WLAN (in Fixed Urban/Suburban environments), Indoor Femtocells (a TVWS low power transceiver for WLAN-like applications). Expected rate and coverage maps are provided for these scenarios.

Keywords: IEEE 1900.7-2015 · TV white space (TVWS) · Dynamic spectrum access · Physical layer · FBMC

1 Introduction

In the recent years, there has been a worldwide concern related to spectrum shortage. As an example, in June 2010, the White House issued a Presidential Memorandum stating that the National Telecommunications and Information Administration (NTIA) shall collaborate with FCC to make available a total of 500 MHz of Federal and non-federal spectrum over the next 10 years, suitable for both mobile and fixed wireless broadband use. One of the means to make new spectrum available is through sharing, and the Digital Switch Over (DSO) in TV bands, which has resulted in making the so-called TV White Space (TVWS) UHF spectrum available, was the first actual example where such a mechanism has been allowed. TVWS availability depends on TV broadcast frequency usage profile, thus changing across time and space. TVWS usage relies on unlicensed secondary Dynamic Spectrum Access (DSA) under the principle on a non-harmful interference with incumbent users [1].

In the USA, the FCC proposed rules for the Unlicensed Operation in the TV Broadcast Bands [2], with the final set of rules in 2009 [3] and an additional memorandum in 2010 [4], along with a notice in 2011 [5]. In Japan, MIC has

© ICST Institute for Computer Sciences, Social Informatics and Telecommunications Engineering 2016
D. Noguet et al. (Eds.): CROWNCOM 2016, LNICST 172, pp. 319–329, 2016.
DOI: 10.1007/978-3-319-40352-6_26

published rules for secondary operation in TV white space [6]. In Europe, the Ofcom UK was the first regulator to establish rules for TVWS usage [7,8]. Subsequently, Ofcom organized a nation-wide trial where several technologies were tested [9]. Following up the approval of the European Parliament and Council the first Radio Spectrum Policy Programme (RSPP) in March 2012 [10], the European Commission released a Communication [11] in which shared use of TV White Spaces in the 470–790 MHz band is identified as a major opportunity. Then, regulatory actions have taken place in this area. In a first report, European Regulation (CEPT) has defined technical considerations for TVWS operation in Europe [12]. Then, ECC Report 159 established technical and operational requirements for the possible operation of cognitive radio systems in the "white spaces of the [470–790] MHz band" [13]. CEPT thoroughly addressed the way forward in European TV White Spaces, assessing both geolocation database and spectrum sensing as enabling technologies and setting out technical requirements for the use of TVWS. Finally, ETSI established some requirements for TVWS equipment in answer to the R&TTE Directive and delivered the Harmonized EN 301 598 [14].

Alongside these regulatory harmonization initiatives, a set of standardization bodies developed technology standards to facilitate the implementation of TVWS radios. First, the CogNea consortium developed the ECMA 392 TVWS standard for WLAN-like application [15]. Then, IEEE 802.22 [16], and more recently IEEE 802.11af [17] issued standards for WRAN and WLAN applications respectively. Finally, IEEE 802.15.4m has been developed to address WPAN operation in the TVWS [18].

All broadband TVWS standards (ECMA 392, IEEE 802.22, IEEE 802.11af) are based on Cyclic Prefix - Orthogonal Frequency Division Multiplexing (CP-OFDM) physical layer (PHY), often inherited from previous standard developments such as IEEE 802.16e or IEEE 802.11a. These standards have been adapted to make them suitable for TVWS operation. Although this approach was the best to guarantee fast market readiness, it had some difficulties to face the TVWS specific requirements in terms of interference control [1]. For this reason, IEEE DySPAN Standards Committee (formerly known as SCC41) created an ad hoc group on White Space (WS) Radio in March 2010 "to consider interest in, feasibility of, and necessity of developing standard defining radio interface (medium access control and physical layers) for white space radio system". Subsequently, the 1900.7 working group on "Radio Interface for White Space Dynamic Spectrum Access Radio Systems Supporting Fixed and Mobile Operation" [19] was created.

The IEEE 1900.7-2015 standard [20] is the result of a clean slate technology analysis where the working group tried to identify the most suitable technology to TVWS requirements. The chosen technology is based on a Filter Bank Multi Carrier (FBMC) PHY and a contention based CSMA-CA MAC, in order to cover the scenarios described in [21]. The technology benefits of the IEEE 1900.7-2015 FBMC PHY vs classical OFDM approached have been addressed in a recent paper [21].

In this paper, our focus is to put the performance of the IEEE 1900.7-2015 PHY into the perspective of typical TVWS scenarios as identified by the European QoSMOS project in [22], namely Rural Broadband Access and indoor TVWS Femtocells. The paper investigates the standard modulation and coding schemes (MCS) and express throughput and coverage for these scenarios.

2   Scenarios and Path Loss Models

2.1   TVWS Rural Broadband Access

In this scenario, a fixed base station serves customer premises equipment (CPE) to deliver content access (cf. Fig. 1). This is the first scenario associated to TVWS as this was the first use case considered in standardization pioneering work, such as the one of IEEE 802.22 wireless regional area network (WRAN). The good propagation properties of sub-GHz UHF frequency and the authorized EIRP for fixed TVWS stations by the FCC (4 W) provide TVWS with good assets for this scenario as illustrated in Fig. 1.

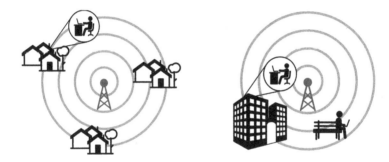

Fig. 1. Rural broadband access (left) and campus (right) scenarios

From a technical viewpoint this scenario also encompasses the campus WLAN use case. For long range communication in UHF bands, the ITU-R P.1546 [24] propagation model used by broadcasters is appropriate. However, in order to use closed form expressions in the path loss model, [23] suggests using the Okamura-Hata model with antenna hights correction that fits the empirical model of ITU-R P.1546 quite well. This will simplify the performance evaluation provided hereafter. Hence, under this scenario we consider the light urban scenario for which the path loss (PL) model is given by:

$$PL_{Urban}(d) = 9.55 + 26.16 \log f - 13.82 \log h_B - C_H + (44.9 - 6.55 \log h_B) \log d \tag{1}$$

Where f is the central frequency (in MHz), h_B is the antenna height of the fixed base station, d the distance between the base station and the CPE, C_H is a correction factor depending on the receiver antenna height h_M defined by:

$$C_H = 0.8 + (1.1 \log f - 0.7)h_M - 1.56 \log f \tag{2}$$

Where h_M is the height (in meters) of the CPE antenna. In the case of suburban deployments, PL shall be corrected as:

$$PL_{Suburban}(d) = PL_{Urban}(d) - 2(\log(f/28))^2 - 5.4 \qquad (3)$$

2.2   TVWS Indoor Femtocell

This scenario considers an indoor access point such as in classical WLAN scenarios for homes of larger buildings. UHF good propagation conditions makes it possible to expect indoor to outdoor coverage as illustrated in Fig. 2.

Fig. 2. Indoor scenario with indoor to outdoor coverage

In this case ITU-R recommendation P.1238 [23] is more suitable than the Okamura-Hata model.

$$PL_{Indoor}(d) = 20 \log f + N \log d + L_f(n) - 28 \qquad (4)$$

Where N is the power loss coefficient. In the LOS case, $N = 20$. In the case of inner-wall penetration, N shall be increased and we will use $N = 35$ as recommended in [22]. Propagation through floors is modelled using the floor penetration loss factor $L_f(n)$, which is recommended to be 9, 19 and 24 dB for one, two and three floors, respectively, at 900 MHz. Although it is expected that TVWS frequencies experience lower penetration losses, these values will be used as a conservative estimate.

3   IEEE 1900.7-2015 PHY Layer

3.1   Overview

The IEEE 1900.7-2015 PHY uses an FBMC modulation, which has been preferred to CP-OFDM, due to superior behavior with regards to adjacent channel leakage ratio (ACLR) and flexibility performance [21,24]. FBMC was introduced in the 60 s by [25,26]. It is a multi-carrier approach, but unlike CP-OFDM where the *sinc* kernel filter ensures subcarriers orthogonality with respect to the FFT operator, FBMC uses a prototype filter that filters each sub carrier. A proper

Table 1. IEEE 1900.7-2015 PHY modes

Mode	N	Intercarrier spacing	2 MHz channel	8 MHz channel
4 K	4096	3.75 KHz	504 (1.86 MHz)	2016 (7.56 MHz)
1 K	1024	15.00 KHz	124 (1.86 MHz)	504 (7.56 MHz)
0.5 K	512	30.00 KHz	64 (1.92 MHz)	252 (7.56 MHz)
0.25 K	256	60.00 KHz	32 (1.92 MHz)	124 (7.44 MHz)

design of the prototype filter enables to trade time and frequency localization, and thus to control ACLR [27]. Because the prototype filter's response spreads each subcarrier over several adjacent subcarriers, another dimension is used to maintain orthogonality. In the framework of IEEE 1900.7-2015, the offset QAM (OQAM) approach is used. OQAM consists in a complex to real conversion where real and imaginary parts of each complex symbol are multiplexed in consecutive time samples into Pulse Amplitude Modulation (PAM) symbols. In order for this pre-processing not to impact data rate, the PAM symbols are up-sampled by a factor of 2. Then, the output real numbers are multiplied by an offset QAM sequence to form a new complex symbol: $h_{k,l} = (-1)^{k.l}(j)^{k+l}e_{k,l}$. These symbols are then filtered through a polyphase network. The IEEE 1900.7-2015 standard specifies two different sizes for the prototype filter: K = 3 or K = 4 governing the level of protection of adjacent channels. Also, for the sake of flexibility, several modes are proposed (Table 1). They consider different number of carriers, and two channelization modes (2 or 8 MHz). Thus, IEEE 1900.7-2015 can be used in any country authorizing TVWS operation and can cover medium to broadband channels with flexible bandwidth (2, 4, 6, 8, 10... MHz). Of course the use of non-contiguous fragments is possible also.

The block diagram of the IEEE 1900.7-2015 transmitter is shown in Fig. 3. The transmitter architecture is composed of two main elements: forward error correction block and data mapping and modulation block. Forward error correc-

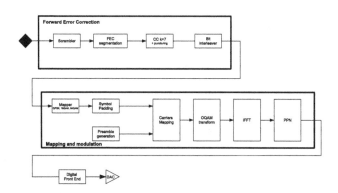

Fig. 3. IEEE 1900.7-2015 PHY transmitter block diagram

Table 2. MCS parameters

MCS	Modulation	Coding rate	Peak throughput (2 MHz) [Mbps]	Peak throughput (8 MHz) [Mbps]
0	BPSK	1/2	0.93	3.78
1	QPSK	1/2	1.86	7.56
2	QPSK	3/4	2.79	11.34
3	16QAM	1/2	3.72	15.12
4	16QAM	3/4	5.58	22.68
5	64QAM	2/3	7.44	30.24
6	64QAM	3/4	8.37	34.02
7	64QAM	5/6	9.30	37.80

tion (FEC) is implemented using a standard convolutional encoder. The code may be punctured to support variable encoding rates. The convolutional code is segmented by blocks of fixed size. The trellis is closed at the beginning and the end of each FEC block. The output of the encoder is forwarded to a bit interleaver of size multiple of the output length of the encoder.

The second block maps and modulates the encoded bits to the multicarrier modulation. The coded data are mapped to QPSK, 16QAM or 64QAM modulation symbols, depending on the desired data rate. Symbols are then padded to make the transmitted burst length an integer multiple of multicarrier symbols. The generated block of QAM symbols is mapped to active carriers and modulated to an offset-QAM before being transformed into a time domain signal using FBMC waveform. A modulation and coding scheme (MCS) index is defined to describe the combination of the modulation and coding schemes that are used when transmitting data (Table 2. Summarizes the possible MCS and the associated peak throughput (Mode 1K), where sampling rate is set to 15.36 Msps and subcarrier spacing is set to 15 KHz as in LTE.

3.2   Receiver Architecture

Most of the published FBMC receiver architectures are based on PolyPhase Network (PPN) receivers [28]. In this case, the filter bank process is applied in the time domain before the frequency domain transform (FFT) using a polyphase filter. It reduces the size of the FFT but makes the receiver less tolerant to large channel delay spread or synchronization mismatch of the FFT. Therefore, this strategy is not well adapted to scenarios with large delay spread. In [29] the authors describe a high performance receiver architecture denoted FS-FBMC (Frequency Spreading FBMC). One advantage of this architecture comes from the fact that time synchronization may be performed in the frequency domain independently of the position of the FFT [30]. This is realized by combining time synchronization with channel equalization. Moreover, good performance for channel exhibiting large delay spread is achieved [31]. As suggested in [31], it gives flexibility in the choice of the intercarrier spacing to support high Doppler spread in combination with robustness against fading channel. Finally, it was shown that FS-FBMC implementation could result in limited extra complexity compared to its OFDM counterpart [28] (typically 30 % additional computational logic). Such a receiver architecture will be considered in the following for performance evaluation.

4   Performance Evaluation

4.1   Parameters and Approach

The evaluation of the cell coverage using IEEE 1900.7-2015 PHY layer is realized assuming the mode 1K, 2 MHz and 8 MHz bandwith, and the parameters of the draft standard (sampling rate $= 15.36$ Msps, subcarrier spacing $= 15$ KHz). We have first evaluated by simulations the minimal required signal to noise ratio (SNR) to ensure a 10^{-3} bit error rate (BER) for all the MCS. The channel estimation and synchronization are assumed perfect. The performance are given for an additive white gaussian noise (AWGN) channel and a standardized channel: the Extended Typical Urban (ETU) channel. This channel model is a 9 tap channel with a delay spread up to 5us. Taps and associated relative power are given in Table 3.

Table 3. Extended Typical Urban (ETU) channel

Tap index [sample]	0	1	2	3	4	8	25	35	77
Relative power [dB]	−1	−1	−1	0	0	0	−3	−5	−7

Table 4 gives the SNR value for a BER of 10^{-3} over AWGN and ETU channels. In the ETU channel case, the resulting BER at the output of the receiver has been evaluated and averaged for 2000 channel realizations assuming a perfect synchronization, perfect channel estimation and a "FS based" receiver structure.

Table 4. Required SNR on AWGN and ETU channels for each MCS (@BER $= 10^{-3}$)

MCS	1	2	3	4	5	6	7
$SNR_{AWGN}[dB]$	4.2	7.6	10.0	14.2	17.6	20.0	21.2
$SNR_{ETU}[dB]$		7.33	12.76	12.82	18.60	20.92	23.68	36.00

4.2   Simulation Results

Performance evaluations have been carried out for the scenarios introduced in Sect. 2. In the case of the rural broadband access, transmit power has been set to 36 dBm (4 W). The path loss has been computed to maintain the minimal SNR required in the case of AWGN and ETU channel models. In the following, channel estimation and synchronization are assumed perfect. One can expect a typical implementation loss of 1dB in actual implementation. Then the maximal range has been computed according to Eq. 1 for urban environment and to Eq. 3 for suburban environment. Figure 4 shows these range estimates for a 2 MHz and a 8 MHz profile. Results are provided for MCS 1 to MCS 7.

In the case of the indoor scenario, the TX power was limited to 20 dBm (100 mW). Minimal SNR in the AWGN case is considered, bearing in mind that strong attenuation factors are already introduced in the distance calculation based on Eq. 4. Figure 5 shows these range estimates for a 2 MHz and a 8 MHz profile, considering 3 different situations: same floor, through 1 floor, through 2 floors. Results are provided for MCS 1 to MCS 7.

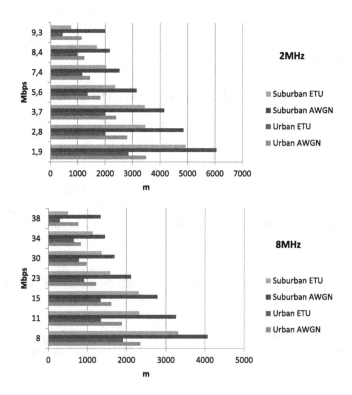

Fig. 4. Max. data rate [Mbps] vs. max. range [m] in rural broadband scenario (Color figure online)

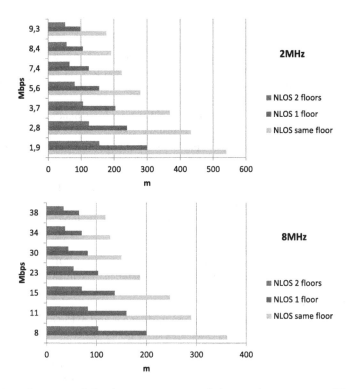

Fig. 5. Max. data rate [Mbps] vs. max. range [m] in indoor scenario (Color figure online)

5   Conclusion

In this paper, communication range are estimated for the 7 MCS of the IEEE 1900.7-2015 standard. These evaluations are provided for two typical TVWS scenarios: rural broadband access with 4 W Tx power and indoor femto-cells with 100 mW Tx power.

Simulations show that in the rural broadband access case, a range of 2 km at 8 Mbps to 23 Mbps depending on channel conditions, with an 8 MHz bandwidth (equivalent to one single European TV channel) can be reached. This makes IEEE 1900.7-2015 technology very suitable for this scenario.

In the indoor case, where TX power is limited to 100 mW (similar to the Tx power of WiFi), range can reach 100 m at 8 Mbps when crossing 2 floors and 38 Mbps when on the same floor (considering 8 MHz bandwidth), making IEEE 1900.7-2015 suitable for indoor and indoor to outdoor connectivity, even in large buidling. This makes IEEE 1900.7-2015 suitable for low CAPEX large building or campus coverage (e.g. shopping malls, airport, university campus, libraries).

References

1. Noguet, D., Gauthier, M., Berg, V.: Advances in opportunistic radio technologies for TVWS. EURASIP J. Wirel. Commun. Netw. **2011**, 1–12 (2011)
2. FCC proposed rule: unlicensed operation in the TV broadcast bands. Technical report, US Federal Register, vol. 69, no. 117, June 2004
3. FCC final rule: unlicensed operation in the TV broadcast bands. Technical report, US Federal Register, vol. 74, no. 30, February 2009
4. FCC: In the matter of unlicensed operation in the TV broadcast bands: additional spectrum for unlicensed devices below 900 MHz and in the 3 GHz band. Technical report, Second memorandum opinion and order, Septembre 2010
5. FCC notice: Unlicensed operation in the TV broadcast bands. Technical report, US Federal Register, vol. 76, no. 26, February 2011
6. Harada, H.: Status report on usage of TV white space in Japan. Technical Report 802.11-12/677r0, IEEE 802.11-12/677r0, May 2012
7. OFCOM: Digital dividend: cognitive access, statement on licence exempting cognitive devices using interleaved spectrum. Technical report, OFCOM, July 2009
8. OFCOM: TV white spaces - approach to coexistence. Technical report, OFCOM, Septembre 2013
9. OFCOM: TV White Spaces Pilot, December 2014
10. Decision no 243/2012/eu of the european parliament and of the council of 14 March 2012 establishing a multiannual radio spectrum policy programme. Text with EEA relevance (2012)
11. Communication from the Commission to the European Parliament, the Council, the European Economic and Social Committee and the Committee of the Promoting the shared use of radio spectrum resources in the internal market (COM/2012/0478) (2012)
12. Report: Technical considerations regarding harmonisation options for the Digital Dividend. Technical report 24, CEPT (2008)
13. ECC Report: Technical and operational requirements for the possible operation of cognitive radio systems in the "white spaces" of the frequency band 470–790 MHz. Technical Report 159, CEPT, January 2011
14. ETSI: White Space Devices (WSD); Wireless Access Systems operating in the 470 MHz to 790 MHz frequency band; Harmonized EN covering the essential requirements of article 3.2 of the R&TTE Directive. Technical report EN 301 598, ETSI, April 2014
15. ECMA-392 standard: MAC and PHY for operation in TV white space. Technical report, ECMA (2009)
16. IEEE standard 802.22: Cognitive wireless ran medium access control (MAC) and physical layer (PHY) specifications: policies and procedures for operation in the TV bands. Technical report, IEEE, July 2011
17. IEEE standard 802.11af: Wireless LAN medium access control (MAC) and physical layer (PHY) specifications: television white spaces (TVWS) operation. Technical report, IEEE, December 2013
18. IEEE standard 802.15.4m: IEEE Standard for Local and metropolitan area networks - Part 15.4: Low-Rate Wireless Personal Area Networks (LR-WPANs) - Amendment 6: TV White Space Between 54 MHz and 862 MHz Physical Layer. Technical report, IEEE, March 2014
19. IEEE P1900.7 PAR: Radio interface for white space dynamic spectrum access radio systems supporting fixed and mobile operation. Technical report, IEEE, June 2011

20. IEEE: 1900.7-2015 - IEEE Standard for Radio Interface for White Space Dynamic Spectrum Access Radio Systems Supporting Fixed and Mobile Operation (2015). https://standards.ieee.org/findstds/standard/1900.7-2015.html
21. Noguet, D., Doré, J.-B., Miscopein, B.: Preliminary performance evaluation of the FBMC based future IEEE 1900.7 Standard. In: Wireless VITAE, Hyderabad, India, December 2015
22. Lehne, P.H., Grondalen, O., MacKenzie, R., Noguet, D., Berg, V.: Mapping cognitive radio system scenarios into the TVWS context. J. Signal Process. Syst. **73**, 227–242 (2013)
23. ITU-R: Propagation data and prediction methods for the planning of indoor radio-communication systems and radio local area networks in the frequency range 900 MHz to 100 GHz. Recommendation ITU-R P.1238-7. Technical report, ITU, Geneva (2012)
24. Filin, S., Noguet, D., Doré, J.-B., Mawlawi, B., Holland, O., Zeeshan Shakir, M., Harada, H., Kojima, F.: IEEE 1900.7 standard for white space dynamic spectrum access radio systems. In: IEEE Conference on Standards for Communications and Networking, Tokyo, Japan, October 2015
25. Chang, R.W.: Synthesis of band-limited orthogonal signals for multichannel data transmission. Bell Syst. Tech. J. **45**, 1175–1796 (1966)
26. Saltzberg, B.R.: Performance of an efficient parallel data transmission system. IEEE Trans. Commun. Technol. **15**, 805–813 (1967)
27. Farhang-Boroujeny, B.: OFDM versus filter bank multicarrier. IEEE Signal Process. Mag. **28**, 92–112 (2011)
28. Berg, V., Doré, J.-B., Noguet, D.: A flexible radio transceiver for TVWS based on FBMC. Elsevier J. Microprocess. Microsyst. (MICPRO) **38**, 743–753 (2014)
29. Bellanger, M.: FS-FBMC: a flexible robust scheme for efficient multi-carrier broadband wireless access. In: Globecom Workshops, pp. 192–196, December 2012
30. Berg, V., Doré, J.-B., Noguet, D.: A flexible FS-FBMC receiver for dynamic access in the TVWS. In: International Conference on Cognitive Radio Oriented Wireless Networks (CROWNCOM), Oulu, Finland, June 2014
31. Doré, J.-B., Berg, V., Cassiau, N., Ktenas, D.: FBMC receiver for multi-user asynchronous transmission on fragmented spectrum. EURASIP J. Wirel. Commun. Netw. **2014**, 1–20 (2014)

LRS-G^2 Based Non-parametric Spectrum Sensing for Cognitive Radio

D.K. Patel[1(✉)] and Y.N. Trivedi[2]

[1] Institute of Engineering and Technology, Ahmedabad University, Gujarat, India
dkpcad@gmail.com
[2] Institute of Technology, Nirma University, Gujarat, India
yogesh.trivedi@nirmauni.ac.in

Abstract. In this paper, a novel non-parametric spectrum sensing scheme in cognitive radio (CR) is proposed based on robust Goodness of Fit (GoF) test. The proposed scheme uses likelihood ratio statistics (LRS-G^2), from which goodness of fit test is derived. The test is applied assuming different types of primary user (PU) signals such as static or constant, single frequency sine wave and Gaussian signals, whereas different types of channels such as additive white Gaussian noise (AWGN), block fading and time-varying channels. Considering a real time scenario, uncertainty in noise variance is also assumed. The performance of the proposed scheme is shown using receiver operating characteristics (ROC) and it is compared with energy detection (ED) and prevailing GoF based sensing techniques such as Anderson-Darling (AD) sensing, Order Statistic based sensing and Kolmogrov-Smirnov (KS) sensing. It is shown that the proposed scheme outperforms all these prevailing schemes.

Keywords: Spectrum sensing · Goodness of fit test · Likelihood ratio statistic · Noise uncertainty · Time-varying channel (AR1)

1 Introduction

The opportunistic spectrum access plays an important role to improve spectral efficiency in wireless communications. It becomes achievable using cognitive radio (CR) [1]. One of the most important task in CR is spectrum sensing in which the presence of licensed user, also known as a primary user (PU), is to be detected. If PU is absent in the spectrum, then unlicensed user, also known as a secondary user (SU), can use the spectrum. However, SU has to vacate the spectrum as soon as PU becomes active. Therefore, spectrum sensing technique should take less time with higher detection accuracy. However, the spectrum sensing function is suffered by various factors such as multi-path fading, receiver's uncertainty, interference, etc. Hence, design of a spectrum sensing algorithm is a challenging problem [2,3].

There are two categories for spectrum sensing. First is parametric sensing in which CR uses some known information of PU to sense its presence. In the second

© ICST Institute for Computer Sciences, Social Informatics and Telecommunications Engineering 2016
D. Noguet et al. (Eds.): CROWNCOM 2016, LNICST 172, pp. 330–341, 2016.
DOI: 10.1007/978-3-319-40352-6_27

category of non-parametric sensing, CR does not have any information about PU. A latter approach is realistic in the current scenario of wireless communications as different PU use distinct bandwidth, modulation and coding schemes. In this category, energy detection (ED) based sensing [4] is the simplest one for spectrum sensing due to its low complexity. To improve the performance of ED sensing, antenna diversity [5] or modified ED [6] is used. However, the assumption of having perfect information about distribution of noise at the CR becomes very crucial at the low signal-to-noise ratio (SNR) of the PU signal or time varying nature of wireless channel. In such circumstances, the performance of the ED degrades drastically and results in SNR wall [7]. Therefore, it is of interest to develop a non-parametric sensing algorithm, which provides better performance at low SNR with less number of observations and false alarm probabilities.

Recently, some goodness of fit (GoF) based sensing schemes have been proposed in the category of non-parametric sensing. In this kind of sensing, we determine cumulative distribution function (CDF) of the received observations. This empirical CDF is compared with known CDF of noise, or we test the null hypothesis (H_0), where H_0 denotes absence of PU. Deviation of empirical CDF from the known CDF of noise (F_0) decides presence of PU or hypothesis H_1 [8]. Based on this, [9] has proposed Anderson Darling (AD) sensing, which outperforms ED based sensing at low SNR assuming an additive white Gaussian noise (AWGN) channel. In [10], ordered statistics (OS) based sensing has been proposed. This method outperforms both AD and ED based sensing at low SNR. Furthermore, based on Kolmogorov-Smirnov (KS) GoF test, [11] has proposed KS sensing. The KS sensing outperforms ED based sensing in AWGN channel. In addition to this, based on sequential KS test, [12], has proposed sequential KS sensing scheme in dispersive MIMO channel. In this paper, we propose new GoF based sensing scheme called as likelihood ratio statistics (LRS-G^2), which outperforms OS, AD, KS and ED sensing in AWGN channel with Gaussian noise assumption under H_0.

The above-mentioned papers on GoF have assumed PU as a constant signal in AWGN channel. However, [13] has investigated performance of AD sensing with different PU signals such as independent and identically distributed (*i.i.d*) Gaussian and single frequency sine signals. Under both these PU signals, ED sensing outperforms the AD sensing. We will show that the proposed LRS-G^2 scheme outperforms ED sensing in this condition also.

The assumption of known variance of noise is very crucial in ED sensing. The change of noise variance deteriorates the performance of ED sensing method. In [14], blind AD sensing scheme has been proposed in block fading channel with constant PU signal. This scheme does not require any information about variance of noise. This blind AD outperforms ED sensing significantly. Our proposed scheme without having knowledge of variance of noise, we call it as Blind LRS-G^2, outperforms Blind AD and ED based sensing methods.

In [14], GoF based sensing has been used assuming a quasi-static channel. However, in a practical scenario, the channel is time-varying. Hence, it is of interest to evaluate the performance of GoF test in a time-varying channel. We have shown performance of the proposed scheme assuming a time-varying channel which is modeled using autoregressive (AR) process. The proposed scheme shows significant improvement in the performance compared to AD and ED sensing.

In [15], authors have proposed likelihood ratio test based sensing scheme under AWGN channel environment. They proposed sensing scheme which out-performs AD and ED based sensing for the system model proposed in [9]. In this paper, we propose a GoF sensing based on a robust normality test [16]. The authors of the paper have used different weighting functions in quadratic equation of the test statistic, called as Zhang's statistic (Z_c), and proposed powerful omnibus GoF test for the Gaussian distribution under H_0. It gives the highest statistical power in comparison with the other GoF test such as AD, KS and Cramer-von-Mises(CvM) tests.

The sampling distribution of the Zhang statistic (Z_c), is mathematically intractable, so it is unattainable to derive the close form expression of the false alarm probability (P_f) and probability of detection (P_d). Hence, we use extensive Monte Carlo Simulations to evaluate the sensing performance of the proposed scheme. We have shown that the proposed LRS-G^2 outperforms all the available GoF based sensing methods and ED sensing method in various scenarios such as different structures of PU, different channel conditions and unknown variance of noise.

The rest of the paper is organized as follows. Section 2 presents the system model and the problem of spectrum sensing as GoF testing using LRS is formulated in Sect. 3. In Sect. 4, the LRS-G^2 sensing algorithm is proposed under known and unknown assumption of noise uncertainty. The simulation results are presented in Sect. 5. Finally, the paper is concluded in Sect. 6.

2   System Model

Let $\mathbf{y} = [y_1, y_2, ..., y_n]^T$ be a vector of n observations of PU, received at CR, where $n \geq 1$. We assume that all the received observations are real as considered in [9,10,14], and each y_i is represented as,

$$y_i = \sqrt{\rho} h_i s_i + w_i, \quad i = 1, 2, 3, \cdots \cdots n, \tag{1}$$

where ρ is the received SNR, h_i represents the channel coefficient. In (1), $w_i \sim \mathcal{N}(0, \sigma^2)$, where $1 \leq i \leq n$, denotes gaussian noise samples and s_i denotes symbol of PU, which can be assumed as constant one or $i.i.d.$ Gaussian as $s_i \sim \mathcal{N}(0, 1)$ or single frequency sine signal as defined in [13]. The CDF of w_i is denoted by $F_0(w)$. The PU signal as a single carrier frequency (f_c) in the discrete version of sine signal can be represented as,

$$s_i = \sqrt{2} sin \left(\frac{2\pi}{k} i + \theta \right), \tag{2}$$

where θ is an initial phase and $k = \frac{f_s}{f_c}$ is the ratio of the sampling frequency (f_s) to the carrier frequency (f_c). The value of k is assumed to be six. Without loss of generality, we assume that all n observations are in ascending order. It means $y_1 \leq y_2 \leq \cdots \leq y_n$.

We assume three different models for channel coefficient h_i.

- AWGN channel: In this case, h_i is assumed to be one and noise distribution is Gaussian with mean zero and variance σ^2.
- Block fading channel: In this case, $h_i \sim \mathcal{N}(0,1)$, however it remains constant during a block of n symbols.
- Time-varying channel: In this case, $h_i \sim \mathcal{N}(0,1)$, however it varies with time in a block of n symbols. This channel is generated using first ordered autoregressive ($AR1$) process,

$$h_i = ah_{i-1} + \sqrt{1-a^2}v_i, \quad 0 \leq a \leq 1 \tag{3}$$

where v_i denotes $i.i.d$ as Gaussian with mean zero and variance one. In (3), a indicates correlation coefficient between consecutive symbols i.e. $a = E[h_{i-1}^* h_i]$, where $E[\cdot]$ represents expectation operator. Here, $a = 1$ and $a = 0$ denote a constant (block fading) channel and an independent channel respectively. The value of a is determined using Jake's autocorrelation function [17] as $a = J_0(2\pi f_d T_s)$, where f_d and T_s denote doppler frequency in Hz and symbol time in seconds respectively.

3   Goodness of Fit Based Sensing Using Likelihood Ratio Statistics (LRS-G^2)

In GoF based sensing, we test the received observations whether they are drawn from null hypothesis (H_0) or not. We assume that the CDF of Gaussian noise under H_0 is known and denoted by $F_0(t)$. In literature, null hypothesis testing algorithms are classified in two ways, Pearson's Chi-squared test and empirical distribution function (EDF) test. The AD, KS and CvM tests are under the category of EDF tests. In [18], authors have proposed a new hypothesis test based on power divergence statistics for null-hypothesis testing as,

$$2nI^\lambda = \frac{2n}{\lambda(\lambda+1)} \left\{ F_n(t) \left[\frac{F_n(t)}{F_0(t)} \right]^\lambda \right.$$
$$\left. + [1 - F_n(t)] \left[\frac{1-F_n(t)}{1-F_0(t)} \right]^\lambda - 1 \right\} \tag{4}$$

where, λ represents a parameter for selection of goodness of fit test, n and $F_n(t)$ denote number of received observations and empirical CDF respectively.

By selecting $\lambda = 1$, (4) represents Pearson's Chi-squared test statistics ($\mathbb{X}^2$) as,

$$\mathbb{X}^2 = \frac{n[F_n(t) - F_0(t)]^2}{F_0(t)[1 - F_0(t)]} \tag{5}$$

and $\lambda = 0$, (4) represents Likelihood Ratio Statistics (LRS-G^2) as,

$$G^2 = 2n \left\{ F_n(t) log \frac{F_n(t)}{F_0(t)} + [1 - F_n(t)] log \frac{1 - F_n(t)}{1 - F_0(t)} \right\} \tag{6}$$

In [16], authors have proposed a parametrization approach to construct a generalized omnibus GoF tests for a specified distribution (F_0) under hypothesis H_0 as normal distribution using different weight functions. They have proposed general test statistics called as Z statistics using,

$$Z = \int_{-\infty}^{\infty} z_t \, w(t) \, dt, \tag{7}$$

where z_t indicates a type of goodness of fit test statistics and $w(t)$ denotes weighting function. The power of any goodness of fit test depends on these two parameters z_t and $w(t)$.

Let $z_t = X^2$ as shown in (5). Then, (7) can be expressed as

$$Z = \int_{-\infty}^{\infty} \frac{n[F_n(t) - F_0(t)]^2}{F_0(t)[1 - F_0(t)]} \, w(t) \, dt \tag{8}$$

Substituting the distinct weighting functions $w(t) = F_0(t)$, $w(t) = n^{-1}F_0(t)[1 - F_0(t)]$ and $w(t) = F_0(t)[1 - F_0(t)]$ in (8), the Z statistics represent AD, KS and CvM statistics respectively as discussed in [8]. Using these AD, KS and CvM statistics, different spectrum sensing schemes have been proposed in [9,11,12,19].

The authors of [16] have proposed powerful omnibus tests. To derive such test, they used LRS-G^2 by substituting (6) into (7) in place of z_t,

$$Z = \int_{-\infty}^{\infty} G^2 \, w(t) \, dt$$
$$= \int_{-\infty}^{\infty} 2n \left\{ F_n(t) log \frac{F_n(t)}{F_0(t)} + [1 - F_n(t)] log \frac{1 - F_n(t)}{1 - F_0(t)} \right\} w(t) dt \tag{9}$$

By using different weight functions ($w(t)$) in (9) as mentioned below, Z produces Z_k, Z_a and Z_c statistics called as Zhang's omnibus statistics.

For $w(t) = 1$, Z approaches Z_k statistic, which is expressed as

$$Z_k = \max_{1 \leq i \leq n} \left(\left(i - \frac{1}{2} \right) log \left\{ \frac{i - \frac{1}{2}}{nF_0(y_{(i)})} \right\} + \left(n - i + \frac{1}{2} \right) log \left\{ \frac{n - i + \frac{1}{2}}{n\{1 - F_0(y_{(i)})\}} \right\} \right) \tag{10}$$

For $w(t) = F_n(t)^{-1}\{1 - F_n(t)\}^{-1}$, Z approaches Z_a statistic, which is expressed as

$$Z_a = -\sum_{i=1}^{n} \left[\frac{log\{F_0(y_{(i)})\}}{n - i + \frac{1}{2}} + \frac{log\{1 - F_0(y_{(i)})\}}{i - \frac{1}{2}} \right] \tag{11}$$

For $w(t) = F_0(t)^{-1}\{1 - F_0(t)\}^{-1}$, Z approaches Z_c statistic, which is expressed as

$$Z_c = \sum_{i=1}^{n} \left[log\left\{ \frac{F_0(y_{(i)})^{-1} - 1}{(n - \frac{1}{2})/(i - \frac{3}{4}) - 1)} \right\} \right]^2 \tag{12}$$

We choose above mentioned statistics and use it for hypothesis testing considering different conditions for channels and PU. The effect of the different Zhang statistics [16] on the detection performance of SU are discussed in Sect. 5.

4   LRS-G^2 Spectrum Sensing Algorithm

The problem of spectrum sensing as a null-hypothesis testing problem is defined as [9],

$$H_0 : F_Y(y) = F_0(y)$$
$$H_1 : F_Y(y) \neq F_0(y) \tag{13}$$

For LRS-G^2 sensing, we use statistics defined in (12) to measure distance between $F_Y(y)$ and $F_0(y)$. Let $F_n(y)$ be the empirical cumulative distribution function (ECDF) of the received observations which can be expressed as,

$$F_n(y) = \frac{|\{i - \frac{1}{2} : y_i \leq y, 1 \leq i \leq n\}|}{n} \tag{14}$$

where $|.|$ indicates cardinality.

4.1   LRS-G^2 Sensing Without Noise Uncertainty

We assume that the noise power is known a priori. The noise under H_0 is $w_i \sim \mathcal{N}(0, \sigma^2)$. Here, we assume that $\sigma^2 = 1$.

First, for the detection of PU at the CR, the value of threshold (ξ) is selected so that the false alarm probability (P_f) is at a desired level (α) as,

$$\alpha = \mathbb{P}\{ Z_c > \xi | H_0 \} \tag{15}$$

To find ξ, it is worth mentioning that the distribution of Z_c under H_0 is independent of the $F_0(y)$. Hence, after applying the probability integration transform (PIT) for available observations,

$$Z_c = \int_0^1 2n\left\{ F_Z(z)log\frac{F_Z(z)}{z} + [1 - F_Z(z)] \right.$$

$$\left. \times log\frac{1 - F_Z(z)}{1 - z} \right\} z^{-1} \{1 - z\}^{-1} dz, \tag{16}$$

where $z = F_0(y)$ and $F_Z(z_i)$ denotes ECDF of the transformed observations z_i, where $z_i = F_0(y_i)$ for $1 \leq i \leq n$. All statistics of observations are independent and uniformly distributed over $[0, 1]$. As shown in [9] for AD sensing, the distribution of A^2 is independent of the $F_0(y)$. The same is also true for the distribution of Z_c. As given in [16], the value of ξ is determined for a specific value of P_f. For example, when $P_f = 10^{-3}$ and $n = 50$, then the value of ξ is 31.707.

Second, sort all the received observations in ascending order. Then, we get

$$y_{(1)} \leq y_{(2)} \leq \cdots \leq y_{(n)}. \tag{17}$$

Third, calculate the test statistics (Z_c) using (12) as,

$$Z_c = \sum_{i=1}^{n} \left[log \left\{ \frac{u_i^{-1} - 1}{(n - \frac{1}{2})/(i - \frac{3}{4}) - 1)} \right\} \right]^2 \tag{18}$$

where $u_i = F_0(y_{(i)})$.

At last, compare the value of (18) with ξ. If $Z_c > \xi$, then reject the null hypothesis H_0 in favor of the presence of PU signal. Otherwise, declare that the PU is absent. Compute performance metric as Probability of Detection (P_d) with a given value of P_f. Furthermore, the detection probability (P_d) is computed theoretically as,

$$\begin{aligned} P_d &= \mathbb{P}\{ Z_c > \xi | H_1 \} \\ &= 1 - F_{Z_c, H_1}(\xi) \end{aligned} \tag{19}$$

4.2   LRS-G^2 Sensing with Noise Uncertainty

In this case, LRS-G^2 sensing method is used considering an uncertainty in the variance of noise, we call it Blind LRS-G^2 sensing.

Recently, [14] has proposed the Blind AD sensing method, where noise uncertainty was considered. Authors of the papers have considered the spectrum sensing problem as Student's t-distribution testing problem. We have used the same approach by replacing AD test with the proposed Zhang test in LRS-G^2 sensing. The summary of the algorithm is as follows:

Step:1 Select an integer m, where $m > 1$ and it is a factor of n. Divide all the samples $Y = \{y_i\}_{i=1}^{n}$ into $g = \frac{n}{m}$ groups, where m number of received observations are there in one group [14].

Step:2 For the jth group $(j = 1, 2, 3 \cdots \cdots g)$, calculate T_j,

$$T_j = \frac{\overline{Y_j}}{S_j/\sqrt{m}}, j = 1, 2 \cdots, g \tag{20}$$

where $\overline{Y_j}$ is mean and S_j^2 is variance of the received observations in the jth group,

$$\overline{Y_j} = \sum_{k=0}^{m-1} \frac{Y_{mj-k}}{m} \ and \ S_j^2 = \sum_{k=0}^{m-1} \frac{(Y_{mj-k} - \overline{Y_j})^2}{m} \tag{21}$$

Step:3 Find the threshold ξ for a given probability of false alarm P_f using (15).
Step:4 Sort T_j in ascending order. Hence, we get

$$T_{(1)} \leq T_{(2)} \leq \cdot\cdot \leq T_{(g)}$$

Step:5 Calculate the required test statistic Z_c for each group as shown in (18)
by replacing $y_{(i)}$ by $T_{(j)}$.
Step:6 If $Z_c < \xi$, then reject null hypothesis H_0 i.e. If $T_j \sim N(0, \sigma^2)$, then T_j
is Student's t-distributed variable with $m - 1$ degrees. It shows the absence of
PU. Compute P_d for the fixed value of P_f. Repeat the above-mentioned steps
for other values of P_f.

5   Simulation Results

In this section, we have presented receiver operating characteristics (ROC) i.e.
plot of P_d versus P_f for different values of SNR for the proposed LRS-G^2 sensing
method using simulations. We have also presented P_d versus SNR for lower values
of P_f. We have considered three types of channels such as AWGN, block fading
and time-varying channels using auto regressive process ($AR1$). model. We have
also considered three types of PU such as constant, single frequency sine wave
and $i.i.d$ Gaussian with mean zero.

In AWGN channel environment, Z_c, Z_k and Z_a provide similar detection
performance. So, we choose the Z_c statistic for taking decision at secondary
user (SU). However, in fading channel, Z_k statistic provides better performance
Therefore, we choose Z_k statistic for block fading and time varying channel.
Furthermore, we have considered the noise uncertainty and shown its effect on
detection performance by varying SNR.

Finally, we have compared all our results with prevailing GoF sensing such
as AD, KS, OS and ED schemes.

Figure 1 shows the ROC for the proposed LRS-G^2 method in comparison
with prevailing GOF sensing schemes at SNR $= -4$ dB, $n = 30$ and constant
PU signal. It can be seen that the proposed technique outperforms all under
AWGN channel. To observe the performance of the proposed scheme at lower
value of P_f such as 0.01, we have shown P_d versus SNR with $n = 30$ under AWGN
channel in Fig. 2. At SNR $= -8dB$, the detection probabilities of 0.7293, 0.5505,
0.4026, 0.3206 and 0.0195 are achieved for LRS-G^2, KS, OS, AD and ED sensing
respectively.

Considering the PU signal as a discrete sinusoidal signal or independent and
identically distributed Gaussian signal [13], Fig. 3 shows ROC for the proposed
scheme along with AD and ED sensing at an SNR of $-5dB$ and $n = 30$. It can be
seen that the proposed scheme outperforms both the AD and ED sensing in both
the PU signals. Furthermore, it can be seen that the ED sensing outperforms GoF
based AD sensing, however proposed GoF based LRS-G^2 scheme outperforms
ED sensing. It proves that the LRS-G^2 scheme is robust against the nature of
PU signal.

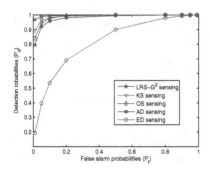

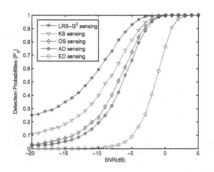

Fig. 1. ROC for different sensing schemes in AWGN channel for constant PU signal at SNR = −4dB and $n = 30$.

Fig. 2. P_d versus SNR for different sensing schemes in AWGN channel for constant PU signal at $P_f = 0.01$ and $n = 30$.

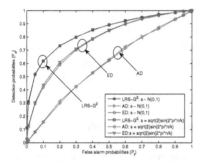

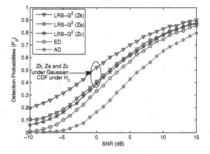

Fig. 3. ROC with different types of PU signals at SNR = −5dB and $n = 30$ in AWGN channel.

Fig. 4. P_d versus SNR in block fading channels with PU Signal as single frequency sine signal at $P_f = 10^{-3}$.

So far, we have shown performance of the proposed scheme in AWGN channel with different PU signals. Further, in Fig. 4, the detection performance of LRS-G^2 is shown under block fading channel with PU signal as single frequency sine signal with $n = 30$. We have also presented performance for LRS-G^2 sensing taking all Zhang test statistics as derived in [16]. The ED outperforms AD sensing. Interestingly, we can observe that the LRS-G^2 with Z_k, Z_a and Z_c outperform ED and AD sensing under fading environment.

Now, we consider blind LRS-G^2 with uncertainty in noise, i.e. the noise variance (σ^2) is unknown. We assume that the channel (h) is block fading and PU signal is constant [14]. In Fig. 5, we have shown P_d versus SNR for $P_f = 0.05$ with $m = 4$ and $n = 32$. It can be seen that uncertainty in noise degrades the performance as expected. We have also presented performance of AD sensing and blind AD sensing (for $m = 4$ and $m = 2$) along with performance of ED sensing with known variance of noise. It can be seen that the blind LRS-G^2 outperforms AD and ED sensing with known variance also. In Fig. 6, we have shown ROC for the proposed scheme assuming PU signal as a single frequency sine signal and channel is time-varying modeled by $AR1$ process. The ROC for

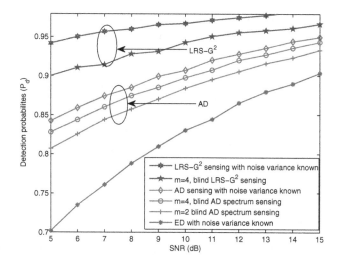

Fig. 5. P_d versus SNR in block fading channels with noise uncertainty for constant PU signal at $P_f = 0.05$ and $n = 32$.

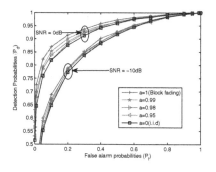

Fig. 6. ROC for LRS-G^2 sensing with different correlation coefficient (a) at $n = 30$ in time-varying channel.

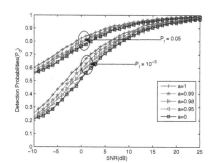

Fig. 7. P_d versus SNR for LRS-G^2 sensing with different values of correlation coefficient (a) at $n = 30$.

LRS-G^2 sensing is presented for different values of correlation coefficient (a) such as 1, 0.99, 0.98, 0.95, 0 at $n = 30$ and SNR of 0dB and -10dB. It can be seen that performance improves as the value of a increases towards unity. In Fig. 7, we have shown P_d versus SNR for $P_f = 0.05, 0.001$ for the same values of n and a. From the results, shown in Figs. 6 and 7, we can say that LRS-G^2 sensing improves P_d when the channel is block faded $(a = 1)$. However, as the value of a decreases, the performance degrades as the channel becomes time-varying.

6 Conclusion

In this paper, a novel non-parametric spectrum sensing scheme based on likelihood ratio statistics using goodness of fit test has been proposed. The detection performance is presented using ROC assuming various types of primary user signals as well as different channel conditions. Furthermore, the adverse effect of noise uncertainty is also shown on the performance. The ROC for ED and prevailing GoF based sensing schemes such as AD, OS and KS are compared with the proposed one. The ED based sensing usually outperforms traditional GoF based sensing schemes when PU signal is not static. However, the proposed GoF based scheme outperforms ED as well as all these GoF based sensing. In case of time-varying channel, the performance of the proposed scheme degrades as the channel changes from slow time varying to fast time varying.

References

1. Haykin, S.: Cognitive radio: brain-empowered wireless communications. IEEE J. Sel. Areas Commun. **23**(2), 201–220 (2005)
2. Yücek, T., Arslan, H.: A survey of spectrum sensing algorithms for cognitive radio applications. IEEE Commun. Surv. Tutorials **11**(1), 116–130 (2009)
3. Axell, E., Leus, G., Larsson, E.G., Poor, H.V.: Spectrum sensing for cognitive radio: State-of-the-art and recent advances. IEEE Sig. Process. Mag. **29**(3), 101–116 (2012)
4. Urkowitz, H.: Energy detection of unknown deterministic signals. Proc. IEEE **55**(4), 523–531 (1967)
5. Digham, F.F., Alouini, M.S., Simon, M.K.: On the energy detection of unknown signals over fading channels. IEEE Trans. Commun. **55**(1), 21–24 (2007)
6. Chen, Y.: Improved energy detector for random signals in gaussian noise. IEEE Trans. Wirel. Commun. **9**(2), 558–563 (2010)
7. Tandra, R., Sahai, A.: Snr walls for signal detection. IEEE J. Sel. Top. Sig. Process. **2**(1), 4–17 (2008)
8. D'Agostino, R.B.: Goodness-of-Fit-Techniques, vol. 68. CRC Press, Boca Raton (1986)
9. Wang, H., Yang, E.H., Zhao, Z., Zhang, W.: Spectrum sensing in cognitive radio using goodness of fit testing. IEEE Trans. Wirel. Commun. **8**(11), 5427–5430 (2009)
10. Rostami, S., Arshad, K., Moessner, K.: Order-statistic based spectrum sensing for cognitive radio. IEEE Commun. Lett. **16**(5), 592–595 (2012)
11. Arshad, K., Moessner, K.: Robust spectrum sensing based on statistical tests. IET Commun. **7**(9), 808–817 (2013)
12. Zhang, G., Wang, X., Liang, Y.C., Liu, J.: Fast and robust spectrum sensing via kolmogorov-smirnov test. IEEE Trans. Commun. **58**(12), 3410–3416 (2010)
13. Nguyen-Thanh, N., Kieu-Xuan, T., Koo, I.: Comments and corrections comments on "spectrum sensing in cognitive radio using goodness-of-fit testing". IEEE Trans. Wirel. Commun. **11**(10), 3409–3411 (2012)
14. Shen, L., Wang, H., Zhang, W., Zhao, Z.: Blind spectrum sensing for cognitive radio channels with noise uncertainty. IEEE Trans. Wirel. Commun. **10**(6), 1721–1724 (2011)
15. Teguig, D., Scheers, B., Le Nir, V., Horlin, F.: Spectrum sensing method based on the likelihood ratio goodness of fit test under noise uncertainty. Int. J. Eng. Res. Technol. (IJERT) **3**(9), 488–494 (2014)
16. Zhang, J., Wu, Y.: Likelihood-ratio tests for normality. Comput. Stat. Data Anal. **49**(3), 709–721 (2005)

17. Jakes, W.C., Cox, D.C.: Microwave Mobile Communications. Wiley-IEEE Press, Piscataway (1994)
18. Cressie, N., Read, T.R.: Multinomial goodness-of-fit tests. J. R. Stat. Soc. Ser. B (Methodol.) **46**, 440–464 (1984)
19. Kieu-Xuan, T., Koo, I.: Cramer-von mises test based spectrum sensing for cognitive radio systems. In: Wireless Telecommunications Symposium (WTS), pp. 1–4. IEEE (2011)

On Convergence of a Distributed Cooperative Spectrum Sensing Procedure in Cognitive Radio Networks

Natalia Y. Ermolova[✉] and Olav Tirkkonen

Aalto University, 00076 Aalto, Finland
{natalia.ermolova,olav.tirkkonen}@aalto.fi

Abstract. In this work, we analyze a distributed cooperative spectrum sensing scheme where N secondary users (SUs) of a cognitive wireless network try to agree about the primary user presence (absence) by iterative interchanging individual opinions (states) over an unreliable wireless propagation medium. It is assumed that the SUs update their personal states based on the "K-out-of-N" rule, and the interchange session fails, if the consensus has not been reached within a fixed number of iterations. The problem of forming a joint opinion becomes challenging because a SU makes its personal decision based on local observations distorted by a wireless propagation medium. This fact may cause a disorder. In this paper, we formulate sufficient conditions of reaching the agreement on the basis of local observations.

Keywords: Cognitive radio networks · Distributed spectrum sensing · Social wireless networks · Wireless propagation

1 Introduction

Spectrum sensing (SS) is a crucial function of cognitive radio since it provides secondary users (SUs) information about spectrum availability and preserves primary users (PUs) from interference coming from unlicensed spectrum users. In order to improve the SS quality in the wireless medium characterized by fading, interference, and path-loss effects, cooperative SS (CSS) schemes employing SU spatial diversity have been proposed [1,2]. A large amount of research has been devoted to analyzing and designing CSS algorithms, and the most works on the topic considered centralized schemes where a fusion center makes a joint decision on the basis of local decisions or/and measurements [3,4]. In [5], a distributed CSS algorithm was analyzed where the SUs attempted to reach a joint decision on the PU presence via interchange of their individual measurements, which were received undistorted at each node.

In contrast to the absolute majority of previous works on CSS, in this paper, we consider a distributed CSS scheme where the SUs try to reach the agreement on the PU presence by interchanging their personal binary opinions (yes/no) via an unreliable propagation medium. Such scenarios are typical in wireless

© ICST Institute for Computer Sciences, Social Informatics and Telecommunications Engineering 2016
D. Noguet et al. (Eds.): CROWNCOM 2016, LNICST 172, pp. 342–350, 2016.
DOI: 10.1007/978-3-319-40352-6_28

networks where the nodes have also social ties, and a dedicated control channel is used for opinion interchange. Trying to reach the agreement, the SUs update their personal opinions (states) based on the "K-out-of-N" rule. But each SU changes its opinion based on only the local observations of the network state, which are different for different users since the wireless medium distorts the transmitted binary signals in a random manner. Therefore the above distributed procedure may result in a disorder (divergence).

In this paper, we obtain sufficient conditions assuring the stochastic convergence of the presented distributed algorithm.

2 System Model

2.1 Model of Opinion Interchange

We consider a secondary network comprising N nodes operating in a finite area. The SUs cooperate to form a joint decision on the PU presence. We denote the network state vector

$$\mathbf{x}(t) = \{x_1(t), x_2(t), \ldots, x_N(t)\} = \{\underbrace{+1, \ldots +1}_{\mathbf{s}^+(t)}, \underbrace{-1, \ldots -1}_{\mathbf{s}^-(t)}\} \tag{1}$$

where the random variate (RV) $x_i(t)$ corresponds to the opinion of the i th SU (yes/no) on the PU presence at the t th iteration, $0 \leq t \leq T$, where T is a fixed integer. The initial state $\mathbf{x}(0)$ is formed based on individual spectrum sensing, after which the SUs start to update their opinions following the "K-out-of-N" rule as

$$x_i(t+1) = \text{Sign} \left[x_i(t) + \sum_{j \neq i} x_{i,j}(t) + N - 2K \right] \tag{2}$$

where $x_{i,j}(t)$ is the state of node j observed at node i. Taking into account that the binary opinion $x_j(t)$ can be interpreted either correctly or incorrectly, $x_{i,j}(t)$ can be represented as

$$x_{i,j}(t) = w_{i,j}(t)x_j(t) \tag{3}$$

where $w_{i,j}(t)$ is a two-point RV taking on the value $+1$ with the probability of correct bit detection $P_{\text{cd}_{i,j}}$ and taking on the value -1 with the probability $(1 - P_{\text{cd}_{i,j}})$. Obviously, $w_{i,j}$ follows a Bernoulli distribution [6] with the success probability equal to $P_{\text{cd}_{i,j}}$, which can be defined as the probability of correct bit detection of binary phase shift keying as [7]

$$P_{\text{cd}_{i,j}} = 1 - Q\left(\sqrt{2\gamma_{i,j}}\right) \tag{4}$$

where $Q(.)$ is the Gaussian Q function, and $\gamma_{i,j}$ is the signal-to-noise ratio (SNR) characterizing the transmission path between the nodes j and i.

In this work, we assume that $x_i(t+1) = x_i(t)$ if the sum in the brackets in (2) is zero. The transform $\mathbf{x}(0) \rightarrow \mathbf{x}(1) \rightarrow \ldots \rightarrow \mathbf{x}(T)$ represents the stochastic dynamics of the considered system. We obtain below sufficient conditions assuring the convergence of $\mathbf{x}(0)$ to a consensus in a probabilistic sense.

2.2 Model of Wireless Propagation

In this work, we model wireless propagation by taking into account fading and path-loss (PL) effects. We apply a bounded PL model that can be represented as [8]

$$
\gamma_{pl} = \begin{cases} 1, R < R_0 \\ \left(\frac{R}{R_0}\right)^{-\kappa}, R \geq R_0 \end{cases}
\tag{5}
$$

where R is the transmitter-receiver distance, κ is the path-loss exponent, and R_0 is a path-loss constant.

A gamma distribution models fading effects as

$$
f_{\gamma_f}(x) = \frac{x^{m-1}}{\Gamma(m)\theta^m}\exp\left(-\frac{x}{\theta}\right)
\tag{6}
$$

where m and θ are the respective shape and scale parameters, and $\Gamma(.)$ is the gamma function. In fading channels, m is inversely proportional to the amount of fading. This model represents the channel power gains in Nakagami-m small-scale fading, as well as it is used as a substitute for composite Nakagami-m-log-normal shadowing fading [9].

3 Convergence to Consensus

In view of (1)–(2), the agreement in the considered network means that the network state is either $\mathbf{x}^+ = \{\underbrace{1, 1, \ldots, 1}_{N}\}$ or $\mathbf{x}^- = \{\underbrace{-1, -1, \ldots, -1}_{N}\}$. From the point of view of convergence analysis, $\mathbf{x}^+$ and $\mathbf{x}^-$ are equivalent, and we will concentrate on the convergence to $\mathbf{x}^+$, and the convergence will be understood in the sense of the ϵ-convergence given by Definition 1.

Definition 1. A state $\mathbf{x}(0) = \{\underbrace{1, 1, \ldots, 1}_{n}\underbrace{-1, -1, \ldots, -1}_{m=N-n}\}$ converges to $\mathbf{x}^+$ if the probability $Pr\left\{\cap_{i=1}^{N}\{x_i(t) = 1\}\right\} \geq 1 - \epsilon$, where ϵ is a predetermined number.

From the state update Eq. (2), it is seen that there are two reasons that may affect the convergence to the consensus: the initial state $\mathbf{x}(0)$ and statistics of $w_{i,j}$. We give below more details about them.

3.1 Distribution of Network Initial State

The network initial state $\mathbf{x}(0)$ is defined by results of individual SS. For example, in the case of energy detection, the probabilities of correct detection P_{d_i} and false alarm P_f at node i can be defined as [10]

$$
P_{d_i} = Q_u\left(\sqrt{2\gamma_i}, \sqrt{2\lambda}\right),
\tag{7}
$$

and

$$
P_f = \frac{\Gamma(u, \lambda/2)}{\Gamma(u)}
\tag{8}
$$

where u is the product of the observation time and signal bandwidth, $\Gamma(a,x) = \int_x^\infty t^{a-1}\exp(-t)\,dt$ is the upper incomplete gamma function, $Q_u\left(\sqrt{2\gamma_i}, \sqrt{2\lambda}\right)$ is the generalized Marcum Q function [11], γ_i denotes the signal-to-noise ratio (SNR) at the node i, and λ is the detector threshold.

In the scenario considered in this work, we assume that u and λ are the same for all SUs, and thus the false probability is the same for all SUs, while the received SNR γ_i is obviously defined by the channel gain and distance between the PU and node i. Then the probability of obtaining less than M indications (I) of PU presence $P_I(M) = Pr\{I \leq M\}$ is the probability of less than M successes in N independent and non-identical (i.n.d.) trials where the success probability of i th trial is P_{d_i}. This probability is defined by the cumulative distribution function (CDF) $F_{\mathcal{B}_P}(N, \mathbf{p})$ of the Poisson binomial distribution $\mathcal{B}_P$ [12], where $\mathbf{p} = \{P_{d_1}, \ldots, P_{d_N}\}$. In each interchange epoch, the CDF $F_{\mathcal{B}_P}(N, \mathbf{p})$ is random, and it is defined by a concrete realization of γ_i, $i = 1, \ldots, N$. $P_I(M)$ can be averaged over the channel and node location statistics.

In Fig. 1, we show simulation results for the complementary CDFs $Pr\{I > N/2\}$ and $Pr\{I > 2N/3\}$ for the SUs uniformly distributed over a circle of the radius $R_{\max}$ and PU located at the origin. The network and propagation parameters are: $m = 1.7$ and $m = 3.5$, $\kappa = 2.6$, $R_0 = 0.1R_{\max}$. We assume that the probability of false detection $Q_f = 0.1$, and the product of the observation time and signal bandwidth $u = 2$.

Actually, the estimates in Fig. 1 characterize the average (over the channel statistics and operating area) CSS performance for either centralized or distributed scenarios where a decision is made via "K-out-of N" rule on the basis of perfect (undistorted) SU decisions that, however, are made by taking into

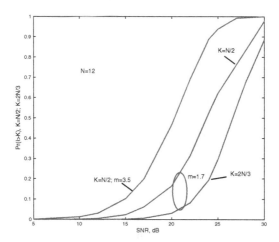

Fig. 1. Complementary CDF, $Pr\{I > K\}$ ($N = 12$, $K = N/2$ and $K = 2N/3$), versus the PU SNR. SUs are uniformly distributed over a circle, and the PU is located at the origin.

account imperfections imposed by the wireless propagation medium on individual SS. Both scenarios correspond to cases of a very high SNR of the control channel: $SNR_c \to \infty$. This condition assures that for $\forall\, i, j$, $P_{cd_{i,j}} \approx 1$, and even under a distributed scenario, all nodes make decisions on the basis of the same information. We focus below on distributed scenarios where the wireless propagation medium distorts the individual opinions.

3.2 Statistical Properties of Node Weights

The node weights $w_{i,j}$ are two-point RVs, and thus their statistical properties are defined by Bernoulli distribution [6]. We assume that $w_{i,j}$ may differ in different interchange epochs, but they are constants at a fixed interchange epoch. It is seen that generally $w_{i,j}$ differ for different i and j due to different (random) propagation conditions between different node pairs caused by random fading and node locations. Thus, any sum of $w_{i,j}$ follows the Poisson binomial distribution. Moreover, we note that

$$D_i = \left(\sum_{j \in J} w_{i,j} - \sum_{k \in K} w_{i,k} \right) \overset{d}{=}$$

$$\Sigma_i = \left(\sum_{j \in J} w_{i,j} + \sum_{k \in K} w'_{i,k} \right) \tag{9}$$

where $\overset{d}{=}$ means equal in distribution, and $w'_{i,j}$ is a two-point RV: $w'_{i,j} = +1$ with the probability $(1 - P_{cd_{i,j}})$, and $w'_{i,j} = -1$ with the probability $P_{cd_{i,j}}$. Thus D_i also follows the Poisson binomial distribution with the average success probability

$$\bar{p}_i = \frac{1}{|J| + |K|} \left(\sum_{j \in J} P_{cd_{i,j}} + \sum_{k \in K} (1 - P_{cd_{i,j}}) \right) \tag{10}$$

3.3 Sufficient Conditions of Convergence

We introduce node subsets $S^+(0) = \{k : x_k(0) = 1\}$ ($|S^+(0)| = n$) and $S^-(0) = \{k : x_k(0) = -1\}$ ($|S^-(0)| = m = N - n$). For the sake of simplicity, we omit below the iteration index t and assume that each SU makes its decision following the opinion of the majority, that is $K = N/2$. Also for the sake of simplicity and without loss of generality, we assume that $(N-1)$ is even (otherwise we had just to use the corresponding integer parts). We suppose that the control channel is designed in such a way that $P_{cd_{i,j}} > 0.5$ for $\forall i, j$ since otherwise the probability of incorrect opinion reception is larger than that of correct reception.

It is seen from (2) that starting from $t = 1$ the components of $\mathbf{x}(t)$ become dependent RVs. Thus, a question is, which values of n, m, and the success probabilities $P_{cd_{i,j}}$ can guarantee the ϵ-convergence? Sufficient conditions of ϵ-convergence can be formulated via Proposition 1.

Proposition 1. *The network is ϵ-convergent if*

$$n > m,$$ (11)

and for each node $x_i \in \mathbf{S}^+$,

$$\bar{p}^{(+)} \geq \left\lceil \frac{(N-1)\left(1 - I_{\epsilon/N}^{-1}\left[(N-1)/2 + 2, (N-1)/2 - 1\right]\right)}{n-1} \right.$$

$$\left. - \frac{(N-n)(1 - \bar{p}_i^{(-)})}{n-1} \right\rceil,$$ (12)

while for each node $x_i \in \mathbf{S}^-$,

$$\bar{p}_i^{(+)} \geq \max\left\{ \left\lceil 0.5 + \frac{2}{N-1} \right\rceil ; \right.$$

$$\left\lceil \frac{(N-1)\left(1 - I_{\epsilon/N}^{-1}\left[(N-1)/2 - 1, (N-1)/2 + 2\right]\right)}{n} \right.$$

$$\left. \left. - \frac{(N-n-1)(1 - p_i^{(-)})}{n} \right\rceil \right\}$$ (13)

where $\bar{p}_i^{(+)}$ is the average success probability for the neighborhood of node $i \in \mathbf{S}^+$, $\bar{p}_i^{(-)}$ is the average success probability for the neighborhood of node $i \in \mathbf{S}^-$, and $I_r^{-1}()$ is the inverse regularized beta function [11].

Proof. Let E_i be the event of $x_i = 1$. Then $Pr\left\{\cap_{i=1}^N E_i\right\} = 1 - Pr\left\{\cup_{i=1}^N \bar{E}_i\right\}$, where $Pr\left\{\cup_{i=1}^N \bar{E}_i\right\}$ is the probability that at least one of E_i is not true. By Boole's inequality,

$$Pr\left\{\cap_{i=1}^N E_i\right\} \geq 1 - \sum_{i=1}^N Pr\left\{\bar{E}_i\right\}.$$ (14)

Thus, conditions assuring $Pr\left\{\bar{E}_i\right\} \leq \epsilon/N$ for $\forall i$ guarantee the ϵ-convergence. If $x_i \in \mathbf{S}^+$, then the probability that it will change the opinion is

$$P^+ = Pr\left\{ \underbrace{\left(\sum_{j \in \mathbf{S}_i^+, j \neq i} w_{i,j} - \sum_{k \in \mathbf{S}_i^-} w_{i,k} \right)}_{\Sigma_i, i \in \mathbf{S}^+} < -1 \right\},$$ (15)

and the probability that a node $x_i \in \mathbf{S}^-$ will not change the opinion is

$$P^- = Pr \left\{ \underbrace{\left(\sum_{j \in \mathbf{S}_i^+} w_{i,j} - \sum_{k \in \mathbf{S}_i^-, k \neq i} w_{i,k} \right)}_{\Sigma_i, i \in \mathbf{S}^-} \leq +1 \right\}. \tag{16}$$

The RV Σ_i in (15)–(16) follow the Poisson binomial distribution. Bounds on the CDF $U \sim \mathcal{B}_P$ can be obtained due to Hoeffding as [12]

$$Pr\{U \leq M\} \leq \sum_{k=0}^{M} \binom{N-1}{k} \bar{p}^k (1 - \bar{p})^{N-1-k} \tag{17}$$

iff $\bar{p} \geq (M + 1)/(N - 1)$, where $\bar{p}$ is the average success probability. On the right-hand side of (17) we observe the CDF $F(N - 1, \bar{p})$ of ordinary binomial distribution with the parameters $(N - 1)$ and $\bar{p}$ defined as [6]

$$F_{\mathrm{B}}(N - 1, \bar{p}) = I_{1-\bar{p}}(N - 1 - M, M + 1) \tag{18}$$

where $I_r(a, b)$ is the regularized beta function [11].

Then using (15) and (17)–(18) as well as taking into account that $P_{\mathrm{cd}_{i,j}} > 0.5$ and $\bar{p} = (n - 1)\bar{p}^{(+)} + (N - n)(1 - \bar{p}^{(-)})$, we conclude that $P^+ \leq F_{\Sigma_i}((N - 1)/2 - 2) \leq \epsilon/N$ if (11)–(12) hold.

Similarly, one can show that $P^- \leq \epsilon/N$ if (11), (13) hold. In this case, $M = (N - 1)/2 + 1$ in (17), and $\bar{p} = n\bar{p}^{(+)} + (N - n - 1)(1 - \bar{p}^{(-)})$.

It is possible to formulate stricter sufficient conditions of ϵ-convergence for both $x_i \in \mathbf{S}^+$ and $x_i \in \mathbf{S}^-$.

Corollary 1. *The network is ϵ-convergent if for $\forall$ x_i, $i = 1, \ldots, N$, (11) holds and*

$$\bar{p}_i^{(+)} > \max \left\{ \left[\frac{(N - n)\bar{p}_i^{(-)} + n - \frac{N+1}{2} + 2}{n - 1} \right]; \right.$$
$$\left[\frac{(N - 1)\left(1 - I_{\epsilon/N}^{-1}[(N - 1)/2 - 1, (N - 1)/2 + 2]\right)}{n - 1} \right.$$
$$\left. \left. - \frac{(N - n)(1 - \bar{p}_i^{(-)})}{n - 1} \right] \right\} \tag{19}$$

Proof. Aiming at formulating joint convergence conditions for all nodes, we note that Σ_i for $i \in \mathbf{S}^+$ defined by (15) is the sum of $(n - 1)$ i.n.d. Bernoulli RVs, each with the success probability larger than 0.5 and m i.n.d. Bernoulli RVs, each with the success probability less than 0.5. In Σ_i for $i \in \mathbf{S}^-$ specified by (16), the number of the i.n.d. Bernoulli RVs with the success probability larger than 0.5 is n, and the number of the i.n.d. Bernoulli RVs with the success probability less than 0.5 is $(m - 1)$. This is due to (9)–(10) and $P_{\mathrm{cd}_{i,j}} > 0.5$. Then it is easy to show that under other equal conditions (that is under equal $P_{\mathrm{cd}_{i,j}}$), conditions assuring $Pr\{\Sigma_i, i \in \mathbf{S}^+ \le +1\} \le \epsilon/N$ guarantee also that $P^+ < \epsilon/N$ and $P^- \le \epsilon/N$.

The validity of (11)–(13) and (19) is defined by many factors such as the cardinality N of the node set, shape and size of the operating area, node distribution in the area, control channel reliability (that is the SNR and coding used). In Fig. 2, we show graphs presenting the probability $Pr(P_\epsilon)$ that (19) holds at the first iteration. The operating area and parameters of wireless propagation medium are described in Subsect. 3.1. The results presented in Fig. 2 show that the wireless propagation medium affects significantly the validity of sufficient conditions. At the same time, we emphasize that (11)–(13) represent sufficient conditions of ϵ-convergence, and (19) (valid for $\forall\, x_i$) represents rather strict conditions.

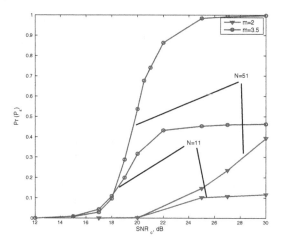

Fig. 2. Graphs of probability of validity of (19) under considered scenario versus the $\mathrm{SNR_c}$ of the control channel. The PU SNR $= 30\,\mathrm{dB}$.

4   Conclusion

In this work, we analyzed a distributed cooperative spectrum sensing algorithm where the SUs tried to reach an agreement about the PU presence/absence by interchanging their personal opinions via an untrustworthy propagation medium. Such scenarios are typical in scenarios where the SUs have also social ties implemented via a dedicated control channel. Under conditions that the SUs can make their decision only on the basis of local observations that can be misinterpreted, we obtained sufficient conditions of convergence to the consensus. Our numerical results showed that propagation conditions affect significantly the validity of the derived sufficient conditions.

Acknowledgment. This work was supported by the Finnish Funding Agency for Technology and Innovations under the project "Exploiting Social Structure for Cooperative Mobile Networking"

References

1. Lataief, K.B., Zhang, W.: Cooperative communications for cognitive radio networks. Proc. IEEE **97**, 878–893 (2009)
2. Akyildiz, I.F., Lo, B.F., Balakrishnan, R.: Cooperative spectrum sensing in cognitive radio networks: A survey. Phy. Commun. **4**, 40–62 (2011)
3. Zhang, W., Mallik, R.K., Ben Letaief, K.: Optimization of cooperative spectrum sensing with energy detection in cognitive radio networks. IEEE Trans. Wirel. Commun. **8**, 5761–5766 (2009)
4. Nallagonda, S., Roy, S.D., Kundu, S.: Performance evaluation of cooperative spectrum sensing scheme with censoring of cognitive radios in Rayleigh fading channel. Wirel. Person. Commun. **70**, 1409–1424 (2013)
5. Li, Z., Yu, R., Huang, M.: A distributed consensus-based cooperative spectrum-sensing scheme in cognitive radio. IEEE Trans. Veh. Techn. **59**, 383–393 (2010)
6. Papoulis, A.: Probability, Random Variables and Stochastic Processes. McGraw-Hill, New York (1994)
7. Goldsmith, A.: Wireless Communications. Cambridge University Press, New York (2005)
8. Rappaport, T.: Wireless Communications: Principles and Practice, 2nd edn. Prentice Hall PTR, Upper Saddle River (2001)
9. Ermolova, N.Y.: Analysis of OFDM error rates over nonlinear fading radio channels. IEEE Trans. Wirel. Commun. **9**, 1855–1860 (2010)
10. Digham, F.F., Alouini, M.-S., Simon, M.K.: On the energy detection of unknown signals over fading channels. IEEE Trans. Commun. **55**, 21–24 (2007)
11. The wolfram functions site. http://functions.wolfram.com/
12. Hoeffding, W.: On the distribution of the number of successes in independent trials. Annals Math. Statist. **27**, 713–721 (1956)

Simple and Accurate Closed-Form Approximation of the Standard Condition Number Distribution with Application in Spectrum Sensing

Hussein Kobeissi[1,2(✉)], Amor Nafkha[1], Youssef Nasser[3], Oussama Bazzi[2], and Yves Louët[1]

[1] SCEE/IETR, CentraleSupélec of Rennes, Rennes, France
{hussein.kobeissi,Amor.Nafkha,Yves.Louet}@centralesupelec.fr
[2] Department of Physics and Electronics, Faculty of Science 1, Lebanese University, Beirut, Lebanon
obazzi@ul.edu.lb, hussein.kobeissi.87@gmail.com
[3] ECE Department, AUB, Bliss Street, Beirut, Lebanon
youssef.nasser@aub.edu.lb

Abstract. Standard condition number (SCN) detector is a promising detector that can work effectively in uncertain environments. In this paper, we consider a Cognitive Radio (CR) with large number of antennas (eg. Massive MIMO) and we provide an accurate and simple closed form approximation for the SCN distribution using the generalized extreme value (GEV) distribution. The approximation framework is based on the moment-matching method and the expressions of the moments are approximated using bi-variate Taylor expansion and results from random matrix theory. In addition, the performance probabilities and decision threshold are also considered as they have a direct relation to the distribution. Simulation results show that the derived approximation is tightly matched to the condition number distribution.

Keywords: Standard condition number · Spectrum sensing · Wishart matrix · Massive MIMO

1 Introduction

Cognitive Radio (CR) is being the technology that provides solution for the scarcity and inefficiency in using the spectrum resource. For the CR to operate effectively and to provide the required improvement in spectrum efficiency, it must be able to effectively detect the presence/absence of the Primary User (PU) to avoid interference if it exists and freely use the spectrum in the absence of the PU. Thus, Spectrum Sensing (SS), being responsible for the presence/absence detection process, is the key element in any CR guarantee.

Several SS techniques were proposed in the last decade, however, Eigenvalue Based Detector (EBD) has been shown to overcome noise uncertainty challenges

© ICST Institute for Computer Sciences, Social Informatics and Telecommunications Engineering 2016
D. Noguet et al. (Eds.): CROWNCOM 2016, LNICST 172, pp. 351–362, 2016.
DOI: 10.1007/978-3-319-40352-6_29

and performs adequately even in low SNR conditions. It presents an efficient way for multi-antenna SS in CR [1,2] as it does not need any prior knowledge about the noise power or signal to noise ratio. EBD is based on the eigenvalues of the received signal covariance matrix and it utilises results from random matrix theory. It detects the presence/absence of the PU by exploiting receiver diversity and includes the Largest Eigenvalue detector, the Scaled Largest Eigenvalue detector, and the Standard Condition Number (SCN) detector [1–6].

The SCN is defined as the ratio of maximum to minimum eigenvalues. The SCN detector compares the SCN of the sample covariance matrix with a certain threshold. This threshold was set according to Marchenco-Pastur law (MP) in [1], however, it is not related to any error constraints. In [2], the authors have provided an approximate relation between the threshold and the False-Alarm Probability (P_{fa}) by exploiting the Tracy-Widom distribution (TW) for the maximum eigenvalue while maintaining the MP law for the minimum eigenvalue. This work was further improved in [3,4] by using the Curtiss formula for the distribution of the ratio of random variables. In these two cases, the threshold could not be computed online and Lookup Tables (LUT) should be used instead. The exact distribution of the SCN was, also, derived in [5] for 2 antennas and in [6] for 3 antennas, however, it is very complicated to extend this work for CR with more number of antennas.

In this paper, we are interested in finding a simple approximation for the SCN detector that allows the system to dynamically compute its threshold online. For this purpose, we propose to asymptotically approximate the SCN distribution with the Generalized Extreme Value (GEV) distribution by matching the first three central moments. This approximation yields a simple and accurate closed form expression for the SCN detector. Accordingly, the threshold could be simply computed. The main contributions of this paper are summarized as follows:

- derivation of the asymptotic central moments of the extreme eigenvalues.
- derivation of an asymptotic approximated form of the central moments of the SCN from that of the extreme eigenvalues.
- proposition of a new simple and asymptotic closed form approximation of the SCN detector using the central moments.

The rest of this paper is organized as follows. Section 2 provides the system model, the SCN detector and hypotheses analysis. In Sect. 3, we give the asymptotic mean, variance and skewness of the extreme eigenvalues under $\mathcal{H}_0$ and $\mathcal{H}_1$ hypotheses. The asymptotic mean, variance and skewness of the SCN are derived in Sect. 4. Then, we propose a new asymptotic approximation for the SCN detector. Theoretical findings are validated by simulations in Sect. 5 while the conclusion is drawn in Sect. 6.

2   Standard Condition Number Detector

2.1   System Model

Consider a CR equipped by K receiving antennas. After collecting N samples from each antenna, the received signal matrix, $\boldsymbol{Y}$, is given by:

$$Y = \begin{pmatrix} y_1(1) & y_1(2) & \cdots & y_1(N) \\ \vdots & \vdots & \ddots & \vdots \\ y_K(1) & y_K(2) & \cdots & y_K(N) \end{pmatrix}, \tag{1}$$

where $y_k(n)$ is the baseband sample at antenna $k = 1 \cdots K$ and instant $n = 1 \cdots N$.

Two hypotheses exist: (i) $\mathcal{H}_0$: there is no PU and the received sample is only noise; and (ii) $\mathcal{H}_1$: the PU exists (single PU case is considered in this paper). The received vector, at instant n, under both hypotheses is given by:

$$\mathcal{H}_0 : \ y_k(n) = \eta_k(n), \tag{2}$$
$$\mathcal{H}_1 : \ y_k(n) = h_k(n)s(n) + \eta_k(n), \tag{3}$$

with $\eta_k(n)$ is a complex circular white Gaussian noise with zero mean and unknown variance σ_η^2, $h_k(n)$ is a the channel coefficient between the PU and antenna k at instant n, and $s(n)$ stands for the primary signal sample modeled as a zero mean Gaussian random variable with variance σ_s^2. Without loss of generality, we suppose that $K \leq N$ and the channel is considered constant during the sensing time for simplicity.

2.2 SCN Detector

Let $W = YY^\dagger$, with $\dagger$ denotes the Hermitian notation, be the sample covariance matrix and denote by $\lambda_1 \geq \lambda_2 \geq \cdots \geq \lambda_K > 0$ its ordered eigenvalues. Then the SCN of W, defined as the ratio of the maximum to minimum eigenvalues, is given by:

$$X = \frac{\lambda_1}{\lambda_K}. \tag{4}$$

Denoting by α the decision threshold, then the probability of false alarm (P_{fa}), defined as the probability of detecting the presence of PU while it does not exist, and the detection probability (P_d), defined as the probability of correctly detecting the presence of PU, are, respectively, given by:

$$P_{fa} = P(X \geq \alpha / \mathcal{H}_0) = 1 - F_0(\alpha), \tag{5}$$
$$P_d \ = P(X \geq \alpha / \mathcal{H}_1) = 1 - F_1(\alpha), \tag{6}$$

where $F_0(.)$ and $F_1(.)$ are the Cumulative Distribution Functions (CDF) of X under $\mathcal{H}_0$ and $\mathcal{H}_1$ hypotheses respectively. If the expressions of the P_{fa} and/or P_d are known, then a threshold could be set according to a required error constraint. For a given threshold, $\hat{\alpha}$, the SCN detector algorithm could be summarized as follows:

1- compute λ_1 and λ_K of $W = YY^\dagger$.
2- evaluate the SCN as $X = \lambda_1/\lambda_K$.
3- decide according to $X \underset{\mathcal{H}_0}{\overset{\mathcal{H}_1}{\gtrless}} \hat{\alpha}$.

2.3   Hypotheses Analysis

$\mathcal{H}_0$ **hypothesis:** By considering $\mathcal{H}_0$ hypothesis, the received samples are complex circular white Gaussian noise with zero mean and unknown variance σ_η^2. Consequently, the sample covariance matrix is a central uncorrelated complex Wishart matrix denoted as $W \sim \mathcal{CW}_K(N, \sigma_\eta^2 I_K)$ where K is the size of the matrix, N is the number of Degrees of Freedom (DoF), and $\sigma_\eta^2 I_K$ is the correlation matrix and I denotes the identity matrix. The symbol '$\sim$' stands for distributed as.

$\mathcal{H}_1$ **hypothesis:** By considering $\mathcal{H}_1$ hypothesis, the single PU sample is Gaussian and the channel is constant during sensing time. Consequently, the sample covariance matrix is a non-central uncorrelated complex Wishart matrix denoted as $W \sim \mathcal{CW}_K(N, \sigma_\eta^2 I_K, \Omega_K)$ where Ω_K is a rank-1 non-centrality matrix[1].

Let $\widehat{\Sigma}_K$ be the correlation matrix defined as:

$$\widehat{\Sigma}_K = \sigma_\eta^2 I_K + {}^{\Omega_K}/_N, \tag{7}$$

and denote by $\sigma = [\sigma_1, \sigma_2, \cdots, \sigma_K]^T$ its vector of eigenvalues. Then W, under $\mathcal{H}_1$, could be modeled as a central semi-correlated complex Wishart matrix denoted as $W \sim \mathcal{CW}_K(N, \widehat{\Sigma}_K)$ [7]. Since Ω_K is a rank-1 matrix, then $\widehat{\Sigma}_K$ belongs to the class of spiked population model with all but one eigenvalue of $\widehat{\Sigma}_K$ are still equal to σ_η^2 while σ_1 is given by:

$$\sigma_1 = \sigma_\eta^2 + {}^{\omega_1}/_N, \tag{8}$$

with ω_1 is the only non-zero eigenvalue of Ω_K. Denote the channel power by σ_h^2 and the signal to noise ratio by $\rho = \frac{\sigma_s^2 \sigma_h^2}{\sigma_\eta^2}$, then it can be easily shown that:

$$\omega_1 = tr(\Omega_K) = NK\rho. \tag{9}$$

3   Assymptotic Moments of λ_1 and λ_K

This section considers the statistical analysis of the extreme eigenvalues (λ_1 and λ_K) of the sample covariance matrix (W) by considering both hypotheses. Since SCN is not affected by the noise power, let $\sigma_\eta^2 = 1$ and define the Asymptotic Condition (AC) and the Critical Condition (CC) as follows:

$$\mathbf{AC}: \qquad (K, N) \to \infty \text{ with } K/N \to c \in (0, 1), \tag{10}$$

$$\mathbf{CC}: \qquad \rho > \rho_c = \frac{1}{\sqrt{KN}}. \tag{11}$$

---

[1] The non-centrality matrix is defined as $\Omega_K = \Sigma_K^{-1} MM^\dagger$ where Σ_K and M are respectively the covariance matrix and the mean of Y defined as $\Sigma_K = E[(Y - M)(Y - M)^\dagger]$ and $M = E[Y]$.

3.1 $\mathcal{H}_0$ Hypothesis

Let $\lambda_1^{\mathcal{H}_0}$ and $\lambda_K^{\mathcal{H}_0}$ be the maximum and minimum eigenvalue of $\boldsymbol{W}$ under $\mathcal{H}_0$ respectively, then:

Distribution of $\boldsymbol{\lambda_1^{\mathcal{H}_0}}$: Denote the centered and scaled version of $\lambda_1^{\mathcal{H}_0}$ of the central uncorrelated Wishart matrix $\boldsymbol{W} \sim \mathcal{CW}_K(N, \boldsymbol{I}_K)$ by:

$$\lambda_1' = \frac{\lambda_1^{\mathcal{H}_0} - a_1(K, N)}{b_1(K, N)} \tag{12}$$

with $a_1(K, N)$ and $b_1(K, N)$, the centering and scaling coefficients respectively, are defined by:

$$a_1(K, N) = (\sqrt{K} + \sqrt{N})^2 \tag{13}$$
$$b_1(K, N) = (\sqrt{K} + \sqrt{N})(K^{-1/2} + N^{-1/2})^{\frac{1}{3}} \tag{14}$$

then, as AC is satisfied, λ_1' follows a TW distribution of order 2 (TW2) [8].

Distribution of $\boldsymbol{\lambda_K^{\mathcal{H}_0}}$: Denote the centered and scaled version of $\lambda_K^{\mathcal{H}_0}$ of the central uncorrelated Wishart matrix $\boldsymbol{W} \sim \mathcal{CW}_K(N, \boldsymbol{I}_K)$ by:

$$\lambda_K' = \frac{\lambda_K^{\mathcal{H}_0} - a_2(K, N)}{b_2(K, N)} \tag{15}$$

with $a_2(K, N)$ and $b_2(K, N)$, the centering and scaling coefficients respectively, are defined by:

$$a_2(K, N) = (\sqrt{K} - \sqrt{N})^2 \tag{16}$$
$$b_2(K, N) = (\sqrt{K} - \sqrt{N})(K^{-1/2} - N^{-1/2})^{\frac{1}{3}} \tag{17}$$

then, as AC is satisfied, λ_K' follows a TW2 [9].

Central Moments of $\boldsymbol{\lambda_1^{\mathcal{H}_0}}$ and $\boldsymbol{\lambda_K^{\mathcal{H}_0}}$: The mean, variance and skewness of λ_1' and λ_K' are that of the TW2. They are given by $\mu_{TW2} = -1.7710868074$, $\sigma_{TW2}^2 = 0.8131947928$ and $\mathcal{S}_{TW2} = 0.2240842036$ respectively [10]. Accordingly, using (12), the mean, variance and skewness of $\lambda_1^{\mathcal{H}_0}$ are, respectively, given by:

$$\mu_{\lambda_1^{\mathcal{H}_0}} = b_1(K, N)\mu_{TW2} + a_1(K, N), \tag{18}$$
$$\sigma^2_{\lambda_1^{\mathcal{H}_0}} = b_1^2(K, N)\sigma_{TW2}^2, \tag{19}$$
$$\mathcal{S}_{\lambda_1^{\mathcal{H}_0}} = \mathcal{S}_{TW2}, \tag{20}$$

and using (15), the mean, variance and skewness of $\lambda_K^{\mathcal{H}_0}$ are, respectively, given by:

$$\mu_{\lambda_K^{\mathcal{H}_0}} = b_2(K, N)\mu_{TW2} + a_2(K, N), \tag{21}$$
$$\sigma^2_{\lambda_K^{\mathcal{H}_0}} = b_2^2(K, N)\sigma_{TW2}^2, \tag{22}$$
$$\mathcal{S}_{\lambda_K^{\mathcal{H}_0}} = -\mathcal{S}_{TW2}. \tag{23}$$

3.2 $\mathcal{H}_1$ Hypothesis

Let $\lambda_1^{\mathcal{H}_1}$ and $\lambda_K^{\mathcal{H}_1}$ be the maximum and minimum eigenvalue of W under $\mathcal{H}_1$ respectively, then:

Distribution of $\lambda_1^{\mathcal{H}_1}$: Denote the centered and scaled version of $\lambda_1^{\mathcal{H}_1}$ of the central semi-correlated Wishart matrix $\mathbf{W} \sim \mathcal{CW}_K(N, \widehat{\boldsymbol{\Sigma}}_K)$ by:

$$\lambda_1'' = \frac{\lambda_1^{\mathcal{H}_1} - a_3(K, N, \sigma)}{\sqrt{b_3(K, N, \sigma)}} \tag{24}$$

with $a_3(K, N)$ and $b_3(K, N)$, the centering and scaling coefficients respectively, are defined by:

$$a_3(K, N, \sigma) = \sigma_1(N + \frac{K}{\sigma_1 - 1}) \tag{25}$$

$$b_3(K, N, \sigma) = \sigma_1^2(N - \frac{K}{(\sigma_1 - 1)^2}) \tag{26}$$

then, as AC and CC are satisfied, λ_1'' follows a standard normal distribution ($\lambda_1'' \sim \mathcal{N}(0, 1)$) [11]. However, if CC is not satisfied, then $\lambda_1^{\mathcal{H}_1}$ follows the TW2 distribution of $\lambda_1^{\mathcal{H}_0}$ as AC is satisfied [11]. Accordingly, the PU signal has no effect on the eigenvalues and could not be detected.

Distribution of $\lambda_K^{\mathcal{H}_1}$: As mentioned in [12], when $\widehat{\boldsymbol{\Sigma}}_K$ has only one non-unit eigenvalue such that CC is satisfied, then only one eigenvalue of $\mathbf{W}$ will be pulled up. In other words, and as could be deduced from [13, Proof of Lemma 2], the rest $K - 1$ eigenvalues of $\mathbf{W}$ ($\lambda_2^{\mathcal{H}_1}, \cdots, \lambda_K^{\mathcal{H}_1}$) has the same distribution of the eigenvalues of $\mathbf{W} \sim \mathcal{CW}_{K-1}(N, \mathbf{I}_{K-1})$ under $\mathcal{H}_0$ hypothesis.

Denote the centered and scaled version of $\lambda_K^{\mathcal{H}_1}$ of the central semi-correlated Wishart matrix $\mathbf{W} \sim \mathcal{CW}_K(N, \widehat{\boldsymbol{\Sigma}}_K)$ by:

$$\lambda_K'' = \frac{\lambda_K^{\mathcal{H}_1} - a_2(K - 1, N)}{b_2(K - 1, N)} \tag{27}$$

with $a_2(K, N)$ and $b_2(K, N)$ are, respectively, given by (16) and (17). Then, as the AC and CC are satisfied, λ_K'' follows a TW2.

Central Moments of $\lambda_1^{\mathcal{H}_1}$ and $\lambda_K^{\mathcal{H}_1}$: The mean, variance and skewness of $\lambda_1^{\mathcal{H}_1}$ are, due to (24), given respectively by:

$$\mu_{\lambda_1^{\mathcal{H}_1}} = a_3(K, N, \sigma), \tag{28}$$

$$\sigma^2_{\lambda_1^{\mathcal{H}_1}} = b_3(K, N, \sigma), \tag{29}$$

$$S_{\lambda_1^{\mathcal{H}_1}} = 0, \tag{30}$$

and using (27), the mean, variance and skewness of $\lambda_K^{\mathcal{H}_1}$ are respectively given by:

$$\mu_{\lambda_K^{\mathcal{H}_1}} = b_2(K-1,N)\mu_{TW2} + a_2(K-1,N), \tag{31}$$

$$\sigma^2_{\lambda_K^{\mathcal{H}_1}} = b_2^2(K-1,N)\sigma^2_{TW2}, \tag{32}$$

$$S_{\lambda_K^{\mathcal{H}_1}} = -S_{TW2}. \tag{33}$$

As a result, this section provides a simple form for the central moments of the extreme eigenvalues. These moments are used, in the next section, to derive an approximation for the mean, the variance and the skewness of the SCN under both hypotheses.

4   Approximating the SCN Distribution

This section approximates the asymptotic distribution of the SCN by the GEV distribution using moment matching. First, we consider both detection hypotheses and we derive an approximation of the mean, the variance and the skewness of the SCN to be used in the next subsection for the approximation.

4.1   Asymptotic Central Moments of the SCN

The bi-variate first order Taylor expansion of the function $X = g(\lambda_1, \lambda_K) = \lambda_1/\lambda_K$ about any point $\theta = (\theta_{\lambda_1}, \theta_{\lambda_K})$ is written as:

$$X = g(\theta) + g'_{\lambda_1}(\theta)(\lambda_1 - \theta_{\lambda_1}) + g'_{\lambda_K}(\theta)(\lambda_K - \theta_{\lambda_K}) + O(n^{-1}), \tag{34}$$

with g'_{λ_i} is the partial derivative of g over λ_i.

Let $\theta = (\mu_{\lambda_1}, \mu_{\lambda_K})$, then it could be easily proved that:

$$E[X] = g(\theta), \tag{35}$$

$$E\left[(X - g(\theta))^2\right] = g'_{\lambda_1}(\theta)^2 E\left[(\lambda_1 - \theta_{\lambda_1})^2\right] + g'_{\lambda_K}(\theta)^2 E\left[(\lambda_K - \theta_{\lambda_K})^2\right] \\ + 2g'_{\lambda_1}(\theta)g'_{\lambda_K}(\theta)E\left[(\lambda_1 - \theta_{\lambda_1})(\lambda_K - \theta_{\lambda_K})\right], \tag{36}$$

$$E\left[(X - g(\theta))^3\right] = g'_{\lambda_1}(\theta)^3 E\left[(\lambda_1 - \theta_{\lambda_1})^3\right] + g'_{\lambda_K}(\theta)^3 E\left[(\lambda_K - \theta_{\lambda_K})^3\right] \\ + 3g'_{\lambda_1}(\theta)^2 g'_{\lambda_K}(\theta)E\left[(\lambda_1 - \theta_{\lambda_1})^2(\lambda_K - \theta_{\lambda_K})\right] \\ + 3g'_{\lambda_1}(\theta)g'_{\lambda_K}(\theta)^2 E\left[(\lambda_1 - \theta_{\lambda_1})(\lambda_K - \theta_{\lambda_K})^2\right], \tag{37}$$

where $E[.]$ stands for the expectation. Accordingly, we give the following theorems that formulate an approximation for the central moments of the SCN.

Theorem 1. *Let X be the SCN of $\mathbf{W} \sim \mathcal{CW}_K(N, \sigma_\eta^2 \mathbf{I}_K)$. The mean, the variance and the skewness of X, as AC is satisfied, can be tightly approximated using the mean, the variance and the skewness of the $\lambda_1^{\mathcal{H}_0}$ and $\lambda_K^{\mathcal{H}_0}$ as follows:*

$$\mu_X = \frac{\mu_{\lambda_1^{\mathcal{H}_0}}}{\mu_{\lambda_K^{\mathcal{H}_0}}} \tag{38}$$

$$\sigma_X^2 = \frac{\sigma_{\lambda_1^{\mathcal{H}_0}}^2}{\mu_{\lambda_K^{\mathcal{H}_0}}^2} + \frac{\mu_{\lambda_1^{\mathcal{H}_0}}^2 \sigma_{\lambda_K^{\mathcal{H}_0}}^2}{\mu_{\lambda_K^{\mathcal{H}_0}}^4} \tag{39}$$

$$\mathcal{S}_X = \frac{1}{\sqrt{\sigma_X^3}} \cdot \left[\frac{\sqrt{\sigma_{\lambda_1^{\mathcal{H}_0}}^3} \mathcal{S}_{\lambda_1^{\mathcal{H}_0}}}{\mu_{\lambda_K^{\mathcal{H}_0}}^3} - \frac{\sqrt{\sigma_{\lambda_K^{\mathcal{H}_0}}^3} \mu_{\lambda_1^{\mathcal{H}_0}}^3 \mathcal{S}_{\lambda_K^{\mathcal{H}_0}}}{\mu_{\lambda_K^{\mathcal{H}_0}}^6} \right] \tag{40}$$

Proof. The result follows (35), (36) and (37) while considering $\lambda_1^{\mathcal{H}_0}$ and $\lambda_K^{\mathcal{H}_0}$ asymptotically independent [14]. The mean, the variance and the skewness of $\lambda_1^{\mathcal{H}_0}$ and $\lambda_K^{\mathcal{H}_0}$ are given in Sect. 3.1.

Theorem 2. *Let X be the SCN of $\mathbf{W} \sim \mathcal{CW}_K(N, \widehat{\boldsymbol{\Sigma}}_K)$ where $\widehat{\boldsymbol{\Sigma}}_K$ has only one non-unit eigenvalue. The mean, the variance and the skewness of X, as the AC and CC are satisfied, can be tightly approximated using the mean, the variance and the skewness of the $\lambda_1^{\mathcal{H}_1}$ and $\lambda_K^{\mathcal{H}_1}$ as follows:*

$$\mu_X = \frac{\mu_{\lambda_1^{\mathcal{H}_1}}}{\mu_{\lambda_K^{\mathcal{H}_1}}} \tag{41}$$

$$\sigma_X^2 = \frac{\sigma_{\lambda_1^{\mathcal{H}_1}}^2}{\mu_{\lambda_K^{\mathcal{H}_1}}^2} + \frac{\mu_{\lambda_1^{\mathcal{H}_1}}^2 \sigma_{\lambda_K^{\mathcal{H}_1}}^2}{\mu_{\lambda_K^{\mathcal{H}_1}}^4} \tag{42}$$

$$\mathcal{S}_X = -\frac{\sqrt{\sigma_{\lambda_K^{\mathcal{H}_1}}^3} \mu_{\lambda_1^{\mathcal{H}_1}}^3 \mathcal{S}_{\lambda_K^{\mathcal{H}_1}}}{\sqrt{\sigma_X^3} \cdot \mu_{\lambda_K^{\mathcal{H}_1}}^6} \tag{43}$$

Proof. The result follows (35), (36) and (37) while considering $\lambda_1^{\mathcal{H}_1}$ and $\lambda_K^{\mathcal{H}_1}$ asymptotically independent [15]. The mean, the variance and the skewness of $\lambda_1^{\mathcal{H}_1}$ and $\lambda_K^{\mathcal{H}_1}$ are given in Sect. 3.2

4.2   Approximating the SCN Using GEV

Generalized Extreme Value (GEV) is a flexible 3-parameter distribution used to model the extreme events of a sequence of i.i.d random variables. These parameters are the location (δ), the scale (β) and the shape (ξ). In the following two propositions, we approximate the distribution of the SCN under $\mathcal{H}_0$ and $\mathcal{H}_1$ hypotheses respectively, however, the proof is omitted due to the lack of space.

Proposition 1. *Let X be the SCN of $\mathbf{W} \sim \mathcal{CW}_K(N, \sigma_\eta^2 \mathbf{I}_K)$ with defined skewness $-0.63 \leq \mathcal{S}_X < 1.14$. If AC is satisfied, then the CDF and PDF of X can be asymptotically and tightly approximated respectively by:*

$$F(x; \delta, \beta, \xi) = e^{-(1+(\frac{x-\delta}{\beta})\xi)^{-1/\xi}} \tag{44}$$

$$f(x; \delta, \beta, \xi) = \frac{1}{\beta}(1 + (\frac{x-\delta}{\beta})\xi)^{\frac{-1}{\xi}-1} e^{-(1+(\frac{x-\delta}{\beta})\xi)^{-1/\xi}} \tag{45}$$

where ξ, β and δ are defined respectively by:

$$\xi = -0.06393\mathcal{S}_X^2 + 0.3173\mathcal{S}_X - 0.2771 \tag{46}$$

$$\beta = \sqrt{\frac{\sigma_X^2 \xi^2}{g_2 - g_1^2}} \tag{47}$$

$$\delta = \mu_X - \frac{(g_1 - 1)\beta}{\xi} \tag{48}$$

where μ_X, σ_X^2 and $\mathcal{S}_X$ are defined in Theorem 1.

Proposition 2. *Let X be the SCN of $\mathbf{W} \sim \mathcal{CW}_K(N, \widehat{\mathbf{\Sigma}}_K)$ with defined skewness $-0.63 \leq \mathcal{S}_X < 1.14$ and $\widehat{\mathbf{\Sigma}}_K$ has only one non-unit eigenvalue. If AC and CC are satisfied, then the CDF and PDF of X can be asymptotically and tightly approximated by (44) and (45) respectively. The parameters ξ, β and δ are defined respectively by (46), (47) and (48) with μ_X, σ_X^2 and $\mathcal{S}_X$ are defined in Theorem 2.*

Now, given (5) and (6), Theorems 1 and 2 and Propositions 1 and 2, the false-alarm probability, the detection probability and the threshold are straightforward. For example, for a target false alarm probability $(\hat{\gamma})$, the threshold is given by:

$$\alpha = \delta + \frac{\beta}{\xi}\left(-1 + \left[-\ln(1 - \hat{\gamma})\right]^{-\xi}\right) \tag{49}$$

with δ, β and ξ given in Proposition 1.

5   Numerical Validation

In this section, we discuss the analytical results through Monte-Carlo simulations. We validate the theoretical analysis presented in Sects. 3 and 4. The simulation results are obtained by generating 10^5 random realizations of $\mathbf{Y}$. For $\mathcal{H}_0$ case, the inputs of $\mathbf{Y}$ are complex circular white Gaussian noise with zero mean and unknown variance σ_η^2 while for $\mathcal{H}_1$ case the channel is considered flat and the PU transmits a BPSK signal.

Table 1 shows the accuracy of the analytical approximation of the mean, the variance and the skewness of the SCN provided by Theorems 1 and 2. It can be easily seen that these Theorems provide a good approximation for the

Table 1. The Empirical and Approximated mean, variance and skewness of the SCN under $\mathcal{H}_0$ and $\mathcal{H}_1$ hypotheses using Theorems 1 and 2 respectively.

	$K \times N$	Empirical			Proposed App.		
		mean	variance	skewness	mean	variance	skewness
$\mathcal{H}_0$	50×500	3.3946	0.0117	0.2639	3.3975	0.0114	0.1652
$\mathcal{H}_1$		12.3363	0.4006	0.1710	12.3139	0.3906	0.0291
$\mathcal{H}_0$	100×500	6.3076	0.0386	0.2992	6.3126	0.0367	0.1740
$\mathcal{H}_1$		34.6345	3.2246	0.1618	34.5387	3.1154	0.0306
$\mathcal{H}_0$	50×1000	2.3386	0.0026	0.2339	2.3396	0.0026	0.1619
$\mathcal{H}_1$		9.6702	0.1205	0.1177	9.6612	0.1184	0.024

statistics of the SCN, however, it could be noticed that the skewness is not perfectly approximated. In fact, the skewness is affected by the slow convergence of the skewness of λ_K that must converge to $-\mathcal{S}_{TW2}$ (i.e. -0.2241) as AC is satisfied. For example, when $K = 50$, the empirical skewness increases from $\mathcal{S}_{\lambda_K} = -0.1504$ to $\mathcal{S}_{\lambda_K} = -0.1819$ as the number of samples increases from $N = 500$ to $N = 1000$. Comparing these results with SCN results in Table 1, one can notice that the empirical and approximated SCN skewness become closer as λ_K skewness converges to that of TW2. Accordingly, Theorems 1 and 2 are good approximations for the mean, the variance and the skewness of the SCN under both hypotheses. It is worth noting that one could approximate the SCN moments using second order bi-variate Taylor series to get a slightly higher accuracy, however, this will cost higher complexity and it is not necessary as shown in Table 1 and Figs. 1 and 2.

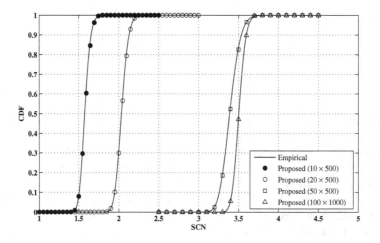

Fig. 1. Empirical CDF of the SCN and its corresponding proposed GEV approximation for different values of K and N under $\mathcal{H}_0$ hypothesis (i.e. false alarm probability).

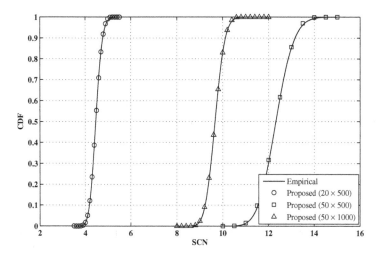

Fig. 2. Empirical CDF of the SCN and its corresponding proposed GEV approximation for different values of K and N under $\mathcal{H}_1$ hypothesis (i.e. detection probability).

Figure 1 shows the empirical CDF of the SCN and its corresponding GEV approximation given by Proposition 1. The results are shown for $K = \{10, 20, 50, 100\}$ antennas and $N = \{500, 1000\}$ samples per antenna. Results show a perfect match between the empirical results and our proposed approximation. Also, it could be noticed that the convergence of the skewness does not affect the approximation and thus the skewness in Theorem 1 holds for this approximation even though the convergence of the skewness of λ_K is slow. From SS perspective, the P_{fa} is in direct relation with this CDF and hence the P_{fa} is perfectly approximated.

Figure 2 shows the empirical CDF of the SCN and its corresponding GEV approximation given by Proposition 2. The results are shown for $K = \{20, 50\}$ antennas and $N = \{500, 1000\}$ samples per antenna and $SNR = -10dB$. Results show high accuracy in approximating the empirical CDF. Also, the difference in the skewness shown in Table 1 does not affect the approximation. Consequently, it could be concluded that the P_d is perfectly approximated.

Finally, it could be noticed that due to the large number of antennas considered in this paper, the proposed SS approximation could be directly applied to the Massive MIMO environment, a potential candidate in 5G.

6   Conclusion

In this paper, we have considered the SCN detector for large number of antennas and/or massive MIMO cognitive radios. We have derived the asymptotic mean, variance and skewness of the SCN using those of the extreme eigenvalues of the sample covariance matrix by means of bi-variate Taylor expansion. A new simple closed form approximation for the false-alarm and the detection probabilities was, also, proposed. This approximation is based on the extreme value theory distributions and uses results from random matrix theory. In addition to its simple form, simulation results show high accuracy of the proposed approximation for different number of antennas.

Acknowledgment. This work was funded by a program of cooperation between the Lebanese University and the Azem & Saada social foundation (LU-AZM) and by CentraleSupélec (France).

References

1. Cardoso, L., Debbah, M., Bianchi, P., Najim, J.: Cooperative spectrum sensing using random matrix theory. In: Proceedings of the IEEE International Symposium on Wireless Pervasive Computing (ISWPC), Greece, pp. 334–338, May 2008
2. Zeng, Y., Liang, Y.C.: Eigenvalue-based spectrum sensing algorithms for cognitive radio. IEEE Trans. Commun. **57**(6), 1784–1793 (2009)
3. Penna, F., Garello, R., Spirito, M.: Cooperative spectrum sensing based on the limiting eigenvalue ratio distribution in wishart matrices. IEEE Commun. Lett. **13**(7), 507–509 (2009)
4. Penna, F., Garello, R., Figlioli, D., Spirito, M.: Exact non-asymptotic threshold for eigenvalue-based spectrum sensing. In: Proceedings of the IEEE 4th International Conference CROWNCOM, Germany, pp. 1–5, June 2009
5. Zhang, W., Abreu, G., Inamori, M., Sanada, Y.: Spectrum sensing algorithms via finite random matrices. IEEE Trans. Commun. **60**(1), 164–175 (2012)
6. Kobeissi, H., Nasser, Y., Bazzi, O., Louet, Y., Nafkha, A.: On the performance evaluation of eigenvalue-based spectrum sensing detector for mimo systems. In: XXXIth URSI General Assembly and Scientific Symposium (URSI GASS), pp. 1–4, August 2014
7. Tan, W.Y., Gupta, R.P.: On approximating the non-central wishart distribution with wishart distribution. Commun. Stat. Theory Method **12**(22), 2589–2600 (1983)
8. Johansson, K.: Shape fluctuations and random matrices. Comm. Math. Phys. **209**(2), 437–476 (2000)
9. Feldheim, O.N., Sodin, S.: A universality result for the smallest eigenvalues of certain sample covariance matrices. Geom. Funct. Anal. **20**(1), 88–123 (2010)
10. Bornemann, F.: On the numerical evaluation of distributions in random matrix theory: A review with an invitation to experimental mathematics. Markov Proc. Relat. Fields **16**, 803–866 (2009)
11. Baik, J., Ben Arous, G., Pch, S.: Phase transition of the largest eigenvalue for non-null complex sample covariance matrices. Ann. Probab. **33**(5), 1643–1697 (2005)
12. Baik, J., Silverstein, J.W.: Eigenvalues of large sample covariance matrices of spiked population models. J. Multivar. Anal. **97**(6), 1382–1408 (2006)
13. Kritchman, S., Nadler, B.: Determining the number of components in a factor model from limited noisy data. Chemometr. Intell. Lab. Syst. **94**, 19–32 (2008)
14. Bornemann, F.: Asymptotic independence of the extreme eigenvalues of gaussian unitary ensemble. J. Math. Phy. **51**, 023514 (2009)
15. Hachem, W., Hardy, A., Najim, J.: A survey on the eigenvalues local behavior of large complex correlated wishart matrices. ARXIV to be published in the "Proceedings of the Journees MAS 2014" (September 2015)

Spectrum Sensing for Full-Duplex Cognitive Radio Systems

Abbass Nasser[1,3]($\boxtimes$), Ali Mansour[1], Koffi-Clement Yao[2], Hussein Charara[4], and Mohamad Chaitou[4]

[1] LABSTICC UMR CNRS 6285, ENSTA Bretagne, 2 Rue François Verny, 29200 Brest, France
abbass.nasser@ensta-bretagne.fr, mansour@ieee.org
[2] LABSTICC UMR CNRS 6285, Université de Bretagne Occidentale, 6 Av. Le Gorgeu, 29238 Brest, France
koffi-clement.yao@univ-brest.fr
[3] Computer Science Department, American University of Culture and Education, Beirut, Lebanon
[4] Computer Science Department, Lebanese University, Beirut, Lebanon

Abstract. Full-Duplex (FD) transceiver has been proposed to be used in Cognitive Radio (CR) in order to enhance the Secondary User (SU) Data-Rate. In FD CR systems, in order to diagnose the Primary User activity, SU can perform the Spectrum Sensing while operating. Making an accurate decision about the PU state is related to the minimization of the Residual Self Interference (RSI). RSI represents the error of the Self Interference Cancellation (SIC) and the receiver impairments mitigation such as the Non-Linear Distortion (NLD) of the receiver Low-Noise Amplifier (LNA). In this manuscript, we deal with the RSI problem by deriving, at the first stage, the relation between the ROC curves under FD and Half-Duplex (HD) (when SU stops the transmission while sensing the channel). Such relation shows the RSI suppression to be achieved in FD in order to establish an efficient Spectrum Sensing relatively to HD. In the second stage, we deal with the receiver impairments by proposing a new technique to mitigate the NLD of LNA. Our results show the efficiency of this method that can help the Spectrum Sensing to achieve a closed performance under FD to that under HD.

Keywords: Full-Duplex · Self-Interference Cancellation · Non-Linearity Distortion · Spectrum Sensing · Cognitive Radio

1 Introduction

Recently, the Full Duplex (FD) transmission has been introduced in the context of Cognitive Radio (CR) to enhance the Data-Rate of the Secondary (unlicensed) User (SU). In FD systems, SU can transmit and sense the channel at the same time. Classically, Half Duplex (HD) was used, therefore the SU should stop transmitting to sense the status of the Primary (licensed) User (PU). Recent

© ICST Institute for Computer Sciences, Social Informatics and Telecommunications Engineering 2016
D. Noguet et al. (Eds.): CROWNCOM 2016, LNICST 172, pp. 363–374, 2016.
DOI: 10.1007/978-3-319-40352-6_30

advancements in the Self-Interference Cancellation (SIC) make the application of FD in CR possible. Due to many imperfections, a perfect elimination of the self-interference cannot be reached in real world applications [1,2]. In CR, the SU makes a decision on the PU status using a Test Statistic (TS) [3]. This TS depends on the PU signal and the noise. Any residual interference from the SU signal can affect the TS norm and leads to a wrong decision about the presence of PU.

In wireless systems, the FD is considered as achieved if the Residual Self Interference (RSI) power becomes lower than the noise level. For that an important SIC gain is required (around 110 dB for a typical WiFi system [2]). This gain can be achieved using the passive suppression and the active cancellation. The passive suppression is related to many factors that reduce the Self Interference (SI) such as the transmission direction, the absorption of the metals and the distance between the transmitting antenna, T_x, and the receive antenna, R_x. The active cancellation reduce the Self-Interference (SI) by using a copy of the known transmitted signal. The estimation of channel coefficients becomes an essential factor in the active cancellation process. Any error in the channel estimation leads to decreasing the SIC gain.

Experimental results show that hardware imperfections such as the non-linearity of amplifiers and the oscillator noise are the main limiting performance factors [2,4–6]. Therefore, the SIC should also consider the receiver imperfections. The authors of [2] modify their method previously proposed in [7] to estimate the channel and the Non-Linearity Distortion (LND) of the receiver Low -Noise Amplifier (LNA). Their method requires two training symbol periods. During the first period, the channel coefficients are estimated in the presence of the NLD. The non-linearity of the amplifier is estimated in the second period using the already estimated channel coefficients. It is worth mentioned that the estimation of the NLD parameters in the second phase depends on the one of the channel coefficients done in the first phase. However the estimation of the channel coefficients in the first phase can be depending on unknown NLD parameters. To solve the previous dilemma, we propose hereinafter an estimation method of the NLD in such way that the estimation of the channel cannot be affected by the NLD.

[8–13] deal with the application of FD in CR. In [8–10,12], the RSI is modeled as a linear combination of the SU signal without considering hardware imperfections. In [10,13] the Energy Detection (ED) is studied in FD mode and the probability of detection, (P_d), and false alarm, P_{fa}, are found analytically. According to our best knowledge, there was no analytic relationship between the RSI, P_d and P_{fa} for both HD and FD mode.

This paper deals with the Spectrum Sensing in real world applications. At first we analytically address the impact of the RSI power on the detection process. For that objective, we derive a relation between the RSI power, the probabilities of detection and false alarm under HD and FD modes. Secondly, we analyze the NLD impact on the channel estimation and the Spectrum Sensing Performance. Hereinafter, a novel method is proposed to suppress the NLD of LNA without affecting the channel estimation process. Further, our proposed method outperforms significantly the method proposed in [2]. In addition, using our method, the receiver requires only one training symbol period to perform the estimation of the channel and the NLD estimation.

2 System Model

In the spectrum sensing context, we usually assume two hypothesis: H_0 (PU signal is absent) and H_1 (otherwise). In our works, we assume that PU signal and SU signals are wideband signals such as OFDM. Throughout this paper, uppercase letters represent frequency-domain signals and lower-case letters represent signals in time-domain. By focusing only on the additive receiver distortion which is dominated by the NLD of the LNA [2], the received signal can be modeled as follows:

$$Y_a(m) = HS(m) + W(m) + D(m) + \eta X(m) \tag{1}$$

H is the channel between the SU transmitter antenna T_x and the SU receive antenna R_x, $S(m)$ is the SU signal, $W(m)$ is an Additive White Gaussian Noise (AWGN), $D(m)$ represents the NLD of the LNA, $X(m)$ is image of the PU signal on R_x and $\eta \in \{0,1\}$ is the PU indicator ($\eta = 1$ under H_1 and $\eta = 0$ under H_0).

After the SIC and the circuit imperfections mitigation, the obtained signal, $\hat{Y}(m)$, can be presented as follows;

$$\hat{Y}(m) = \xi(m) + W(m) + \eta X(m) \tag{2}$$

Where $\xi(m)$ is the RSI and is defined as $\xi(m) = (H - \hat{H})S(m) + D(m) - \hat{D}(m)$. $\hat{H}$ and $\hat{D}(m)$ are the estimated channel and the NLD respectively.

Ideally $\hat{H} = H$ and $\hat{D}(m) = D(m)$, therefore Eq. (2) becomes: $\hat{Y}(m) = W(m) + \eta X(m)$, which corresponds to an HD mode. Any mistake in the cancellation process may lead to a wrong decision about the PU presence.

3 The RSI Effect on the Spectrum Sensing

In order to decide the existence of the PU, the Energy Detector (ED) compares the received signal energy, T, to a predefined threshold, λ.

$$T = \frac{1}{N} \sum_{m=1}^{N} |\hat{Y}(m)|^2 \underset{H_0}{\overset{H_1}{\gtrless}} \lambda \tag{3}$$

By assuming the *i.i.d* property of $\epsilon(n)$, $w(n)$ and $x(n)$, then $\xi(m)$, $W(m)$ and $X(m)$ become *i.i.d*. (See Appendix (A.1)). In this case, the distribution of T should asymptotically follow a normal distribution for a large number of samples, N, according to the central limit theorem. Consequently, the probabilities of False Alarm, P_{fa}^F, and the Detection, P_d^F, under the FD mode can be obtained as follows (See Appendix (A.2)):

$$P_{fa}^F = Q(\frac{\lambda - \mu_0}{\sqrt{V_0}}) = Q\left(\frac{\lambda - (\sigma_w^2 + \sigma_d^2)}{\frac{1}{\sqrt{N}}(\sigma_w^2 + \sigma_d^2)}\right) \tag{4}$$

$$p_d^F = Q(\frac{\lambda - \mu_1}{\sqrt{V_1}}) = Q\left(\frac{\lambda - (\sigma_w^2 + \sigma_d^2 + \sigma_x^2)}{\frac{1}{\sqrt{N}}(\sigma_w^2 + \sigma_d^2 + \sigma_x^2)}\right) \tag{5}$$

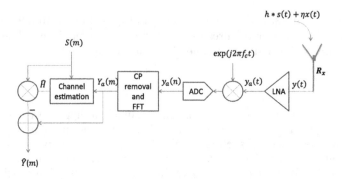

Fig. 1. Classical SIC circuit for OFDM receiver

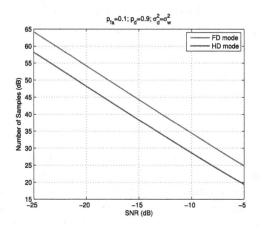

Fig. 2. The number of samples required to reach $P_d = 0.9$ and $P_{fa} = 0.1$ (Color figure online)

Where μ_i and V_i are the mean and the variance of T under H_i respectively, $i \in \{0; 1\}$, $\sigma_d^2 = E[|\xi(m)|^2]$ represents the RSI power, $\sigma_w^2 = E[|W(m)|^2]$ and $\sigma_x^2 = E[|X(m)|^2]$. The SNR, γ_x, is defined as: $\gamma_x = \frac{\sigma_x^2}{\sigma_w^2}$. If the SIC is perfectly achieved, i.e. $\sigma_d^2 = 0$, P_{fa}^F and P_d^F take their expressions under the HD mode.

Figure (2) shows the required number of samples to reach $P_d = 0.9$ and $P_{fa} = 0.1$ under the HD and FD modes for different values of SNR. In FD mode, we set $\sigma_d^2 = \sigma_w^2$ as the target values of σ_d^2 in digital communication. Figure (2) shows that the number of required samples slightly increases under the FD modes. For example if $\gamma_x = -5$ dB, then 85 samples are enough to reach the target $(P_d; P_{fa})$ under the HD mode while under FD mode, around 300 samples are needed.

Let us define the Probability of Detection Ratio (PDR), δ, for the same probability of false alarm under FD and HD modes, as follows:

$$\delta = \frac{P_d^F}{P_d^H} \quad with \quad P_{fa}^F = P_{fa}^H = \alpha \tag{6}$$

Where P_d^H and P_{fa}^H are the probabilities of detection and false alarm under HD respectively, $0 \leq \alpha \leq 1$ and $0 \leq \delta \leq 1$. As with an excellent SIC, the ROC can mostly reach in FD the same performance of HD. In order to show the effect of RSI on δ, let us define the RSI to noise ratio γ_d as follows:

$$\gamma_d = \frac{\sigma_d^2}{\sigma_w^2} \tag{7}$$

Using (4) and (6), the threshold, λ, can be expressed as follows:

$$\lambda = \left(\frac{1}{\sqrt{N}}Q^{-1}(\alpha) + 1\right)(\sigma_w^2 + \sigma_d^2) \tag{8}$$

By replacing (8) in (5), γ_d can be expressed as follows:

$$\gamma_d = \frac{(1+\gamma_x)Q^{-1}(\delta P_d^H) - Q^{-1}(\alpha) + \sqrt{N}\gamma_x}{Q^{-1}(\alpha) - Q^{-1}(\delta P_d^H)} \tag{9}$$

If $\delta = 1$, then we can prove that γ_d becomes zero, which means that the SIC is perfectly achieved. Figure (3) shows the curves of γ_d for various values of PDR, δ, with respect to the SNR, γ_x, for $P_d^H = 0.9$ and $\alpha = 0.1$. This figure shows that as δ increases γ_d decreases. To enhance the PDR, the selected SIC technique should mitigate at most the SI. In fact, for $\gamma_x = -5$ and a permitted loss of 1% (*i.e.* $\delta = 0.99$), γ_d is about $-15\,\mathrm{dB}$.

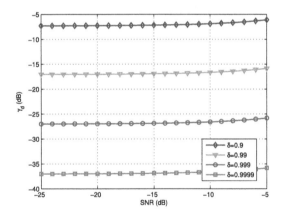

Fig. 3. Evolution of γ_d with respect to γ_x for various values of δ, $(P_{fa}^H \; ; \; P_d^H) = (0.1 \; ; \; 0.9)$

4 The Effect of the Amplifier Distortion on Spectrum Sensing

In real world applications, the full duplex transceiver seems hard to be achieved due to hardware imperfections: the non-linearity of the amplifiers, the quantization noise of the Analog to Digital Converter (ADC), the phase noise of the oscillator, *etc.* The NLD of LNA is an important performance limiting factor [2,4–7]. According to NI 5791 datasheet [14], the NLD power is of 45 dB below the power of the linear amplified component. A new efficient algorithm is proposed in this section, it shows more reliable performance than that proposed in [2] and make the channel estimation performed without the influence of NLD.

4.1 Estimation of the Non-linearity Distortion of LNA

The LNA output can be written as an odd degrees polynomial of the input signal [15]. The NLD stands for the degrees greater than one. By limiting to the third degree and neglecting the higher degrees power [16], the NLD component can be written as follows:

$$d(t) = \beta y^3(t) \tag{10}$$

Where β is the NLD coefficient. The estimation of β can be helpful to suppress the LNA output. In this case, the channel estimation is no longer affected by the NLD. The overall output signal of the LNA, $y_a(t)$, can be expressed as follows:

$$y_a(t) = \theta y(t) + \beta y^3(t) \tag{11}$$

θ and $y(t)$ are the power gain and the input signal of the LNA respectively. To estimate θ and β, by a and b respectively, one can minimize the following cost function:

$$J = E\left[\left(y_a(t) - (ay(t) + by^3(t))\right)^2\right] \tag{12}$$

By deriving J with respect to a and b we obtain:

$$\frac{\partial J}{\partial a} = 0 \Rightarrow aE[y^2(t)] + bE[y^4(t)] = E[y_a(t)y(t)] \tag{13}$$

$$\frac{\partial J}{\partial b} = 0 \Rightarrow aE[y^4(t)] + bE[y^6(t)] = E[y_a(t)y^3(t)] \tag{14}$$

Using Eq. (13) and (14), a linear system of equations can be obtained:

$$\begin{bmatrix} a \\ b \end{bmatrix} = A^{-1}B \tag{15}$$

Where:

$$A = \begin{bmatrix} E[y^2(t)] & E[y^4(t)] \\ E[y^4(t)] & E[y^6(t)] \end{bmatrix} ; \ B = \begin{bmatrix} E[y_a(t)y(t)] \\ E[y_a(t)y^3(t)] \end{bmatrix} \tag{16}$$

Once the non-linearity coefficient, β, is estimated, the non-linearity component can be subtracted from the output signal of the amplifier.

4.2 Numerical Results

Figure (4) shows the residual power of the NLD cancellation. The NLD power is fixed to 45 dB under the linear component [14]. This power is reduced to less than -300 dB after the application of our method. The method of [2] reduces the NLD

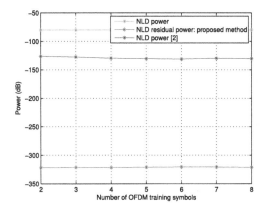

Fig. 4. The effect of the number of training symbols on the NLD residual power (Color figure online)

power by around 50 dB. β is estimated using various number of training symbols, N_e. In this simulation, OFDM modulations are used with 64 sub-carriers and a CP length equal to 16. The received power is fixed to -5 dBm and the noise power to -72 dBm [14]. As shown in Fig. (4), the residual power of NLD decreases with an increasing of N_e when the method of [2] is applied. However our method keeps a constant value of this power. Our technique outperforms significantly the method proposed in [2]. To show the impact of NLD on the channel estimation and the RSI power, Fig. (5) shows the power of $\hat{Y}(m)$ obtained in FD under H_0. The channel is estimated according to the method previously proposed by [17] as follows:

$$\hat{h} = IDFT\left\{ \frac{1}{N_e} \sum_{k=0}^{N_e} \frac{Y_a^k(m)}{Y^k(m)} \right\} \ and \ \hat{H} = DFT\{\hat{h}(1,..,n_{tap})\} \qquad (17)$$

Where $IDFT$ stands for the inverse discrete Fourier transform and n_{tap} is the channel order. The number of training symbols, N_e, is fixed to 4 symbols. To deal with a practical scenario, the number of sub-carrier is 64, the transmitted signal is of -10 dB (*i.e.* 20 dBm), and the noise floor is -102 dB (*i.e.* -72 dBm) [14]. The transceiver antenna is assumed to be omni-directional with 35 cm separation between T_x and R_x, so that a passive suppression of 25 dB is achieved [1]. According to the experimental results of [1], in a low reflection environment, 2 channel taps are enough to perform the SIC when the passive suppression is bellow 45 dB. Furthermore, the line of sight channel is modeled as Rician channel with K-factor about 20 dB. The non-line of sight component is modeled by a Rayleigh fading channel.

Figure (5) shows that our method leads to mitigate almost all the self interference, so that the power of $\hat{Y}(m)$ becomes very closed to the noise power. However, with the method of [2], the RSI power increases with the NLD power

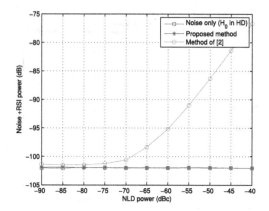

Fig. 5. The power of $\hat{Y}(m)$ under H_0 obtained after applying: (1) our proposed method, (2) the method of [2] is applied and (3) under HD mode

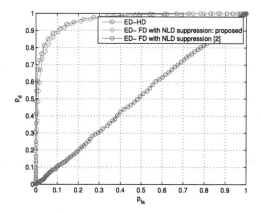

Fig. 6. The ROC curve after applying the proposed technique of NLD suppression (Color figure online)

because the NLD power is a limiting factor of the channel estimation which leads to a bad estimation of the channel.

To show the impact of the NLD on the Spectrum Sensing, Fig. (6) shows the ROC in various situations under $\gamma_x = -10$ dB. The simulations parameters in this figure are similar to those of Fig. (5), only the NLD power is 45 dB under the linear component according NI 5791 indications [14]. The method of [2] leads to a linear ROC, which means that no meaningful information about the PU status can be obtained. By referring to Fig. (5), the RSI power is of -82 dB for a NLD power of -45 dBc, which means that γ_d in this case is about 20 dB. This high RSI power leads to a harmful loss of performance (see Fig. (3)). From the other hand, our method makes the ROC in FD mode almost colinear with that of the ROC of HD mode, which means that all SI and receiver impairments is mitigated.

Figure (7) shows the PDR for a target $\alpha = 0.1$ and $P_d^H = 0.9$. The ratio δ increases with the SNR. At a low SNR of $-10\,\text{dB}$, δ becomes closed to 1, so that a negligible performance loss is happen. As the SNR decreases the detection process in FD mode becomes more sensitive to the RSI power.

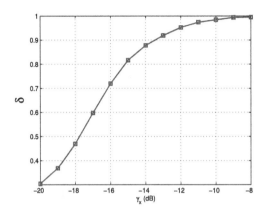

Fig. 7. The gain ratio: $\delta = \frac{P_d^F}{P_d^H}$ for different values of SNR

5   Conclusion

In this paper, we address the impact of the Residual Self Interference on the Spectrum Sensing for Full-Duplex Cognitive Radio. An analytic relation is derived relating the residual self interference with the probabilities of detection and false alarm under Full-Duplex and Half-Duplex modes. Furthermore, a new method is proposed to mitigate an important receiver impairment, which is the Non-Linear Distortion of the Low Noise Amplifier. This method shows its efficiency, leading the Spectrum Sensing performance in Full-Duplex mode to be closed to that under Half-Duplex mode.

A   Appendix

A.1   *i.i.d.* property in Frequency Domain

Let $r(n)$ be an *i.i.d.* time-domain signal. The DFT, $R(m)$, of the $r(n)$ is defined as follows:

$$R(m) = \sum_{n=1}^{L} r(n)e^{-j2\pi m \frac{n}{L}} \tag{18}$$

Where L is the number of samples of $r(n)$. According to the Central Limit Theorem (CLT), $R(m)$ follows asymptotically a Gaussian distribution for a large L. Based on [18], two normal variables are independent *iff* they are uncorrelated.

Let $C(m_1, m_2)$ the correlation of $R(m_1)$ and $R(m_2)$ $\forall$ $m_1 \neq m_2$.

$$C(m_1, m_2) = E[R(m_1)R^*(m_2)]$$

$$= E\left[\sum_{n_1,n_2=1}^{L} r(n_1)r^*(n_2)e^{-j2\pi\frac{n_1m_1-m_2n_2}{L}}\right]$$

$$= \sum_{n_1=n_2=1}^{L} E\left[|r(n_1)|^2\right]e^{-j2\pi\frac{(m_1-m_2)n_1}{L}}$$

$$+ \sum_{n_1\neq n_2=1}^{L} \underbrace{E\left[r(n_1)r^*(n_2)\right]}_{=0,\ since\ r(n)\ is\ i.i.d.} e^{-j2\pi\frac{(n_1m_1-n_2m_2)}{L}}$$

$$= E\left[|r(n_1)|^2\right] \underbrace{\sum_{n_1=1}^{L} e^{-j2\pi(m_1-m_2)\frac{n_1}{L}}}_{=0} = 0 \qquad (19)$$

$C(m_1, m_2) = 0$ $\forall$ $m_1 \neq m_2$, therefore $R(m_1)$ and $R(m_2)$ are uncorrelated and they become independent because of their Gaussianity distribution.

A.2   Probility of Detection and Probability of False Alaram

As by our assumption $\xi(m)$, $W(m)$ and $X(m)$, are asymptotically Gaussian i.i.d., then $\hat{Y}(m)$ is also Gaussian and i.i.d.. Therefore the TS, T, of Eq. (3) follows a normal distribution according to CLT for a large N. Under H_0 (i.e. $X(m)$ does not exist), the mean, μ_0, and the variance, V_0 of T can be obtained as follows:

$$\mu_0 = E[T] = E\left[\frac{1}{N}\sum_{m=1}^{N}|\xi(m)+W(m)|^2\right] = \sigma_w^2 + \sigma_d^2 \qquad (20)$$

$$V_0 = E[T^2] - E^2[T] = \frac{1}{N^2}E\left[\left(\sum_{m=1}^{N}|\hat{Y}(m)|^2\right)^2\right] - (\sigma_w^2+\sigma_d^2)^2$$

$$= \frac{1}{N^2}E\left[\sum_{m_1=m_2=1}^{N}|\hat{Y}(m_1)|^4\right]$$

$$+ \frac{1}{N^2}E\left[\sum_{m_1\neq m_2=1}^{N}|\hat{Y}(m_1)|^2\hat{Y}(m_2)|^2\right] - (\sigma_w^2+\sigma_d^2)^2$$

$$= \frac{1}{N^2}\sum_{m_1=m_2=1}^{N}E\left[|\hat{Y}(m_1)|^4\right] - \frac{1}{N}(\sigma_w^2+\sigma_d^2)^2 \qquad (21)$$

Since $\hat{Y}(m)$ is Gaussian, then its kurtosis $kurt(\hat{Y}(m))$ is zero.

$$kurt(\hat{Y}(m)) = E[|\hat{Y}(m)|^4] - E[\hat{Y}^2(m)] - 2E^2[|\hat{Y}(m)|^2] = 0 \qquad (22)$$

Assuming that the real and the imaginary parts of $\hat{Y}(m)$ are independent and of the same variance then $E[\hat{Y}^2(m)]$ becomes zero. Therefore: $E[|\hat{Y}(m)|^4] = 2E^2[|\hat{Y}(m)|]^2 = 2(\sigma_w^2 + \sigma_d^2)^2$. Back to Eq. (21), the variance, V_0 becomes:

$$V_0 = \frac{1}{N}(\sigma_w^2 + \sigma_d^2)^2 \tag{23}$$

By following the same procedure, μ_1 and V_1 can be obtained as follows under H_1 ($X(m)$ exists):

$$\mu_1 = \sigma_w^2 + \sigma_d^2 + \sigma_x^2 \tag{24}$$

$$V_1 = \frac{1}{N}(\sigma_w^2 + \sigma_d^2 + \sigma_x^2)^2 \tag{25}$$

References

1. Everett, E., Sahai, A., Sabharwal, A.: Passive self-interference suppression for full-duplex infrastructure nodes. IEEE Trans. Wirel. Commun. **13**, 680–694 (2014)
2. Ahmed, E., Eltawil, A.M.: Al-digital self-interference cancellation technique for full-duplex systems. IEEE Trans. Wirel. Commun. **7**, 291–294 (2015)
3. Yucek, T., Arslan, H.: A survey of spectrum sensing algorithms for cognitive radio applications. IEEE Commun. Surv. Tutorials **11**, 116–130 (2009). First Quarter
4. Eltawil, A.M., Ahmed, E., Sabharwal, A.: Rate gain region and design tradeoffs for full-duplex wireless communications. IEEE Trans. Wirel. Commun. **7**, 3556–3565 (2013)
5. Sahai, A., Patel, G., Dick, C., Sabharwal, A.: On the impact of phase noise on active cancelation in wireless full-duplex. IEEE Trans. Veh. Technol. **62**, 34494–34510 (2013)
6. Bliss, D.W., Hancock, T.M., Schniter, P.: Hardware phenomenological effects on cochannel full-duplex mimo relay performance. In: Proceedings of Asilomar Conference on Signals, Systems and Computers, pp. 34–39, April 2012
7. Ahmed, E., Eltawil, A.M., Sabharwal, A.: Self-interference cancellation with non-linear distortion suppression for full-duplex systems. In: Proceeding Asilomar Conference on Signals, Systems and Compututer, vol. II, pp. 1199–1203 (2013)
8. Afifi, W., Krunz, M.: Adaptive transmission-reception-sensing strategy for cognitive radios with full-duplex capabilities. In: International Symposium on Dynamic Spectrum Access Networks (DYSPAN), pp. 149–160 (2014)
9. Heo, J., Ju, H., Park, S., Kim, E., Hong, D.: Simultaneous sensing and transmission in cognitive radio. IEEE Trans. Wirel. Commun. **13**, 149–160 (2014)
10. Riihonen, T., Wichman, R.: Energy detection in full-duplex cognitive radios under residual self-interference. In: 9th International Conference on Cognitive Radio Oriented Wireless Networks (CROWNCOM), pp. 57–60, July 2014
11. Cheng, W., Zhang, X., Zhang, H.: Full-duplex spectrum-sensing and mac-protocol for multichannel nontime-slotted cognitive radio networks. IEEE J. Sel. Areas Commun. **33**(5), 820–831 (2015)
12. Afifi, W., Krunz, M.: Incorporating self-interference suppression for full-duplex operation in opportunistic spectrum access systems. IEEE Trans. Wirel. Commun. **14**, 2180–2191 (2015)

13. Cheng, W., Zhang, X., Zhang, H.: Full duplex spectrum sensing in non-time-slotted cognitive radio networks. In: The Military Communications Conference (Milcom), pp. 1029–1034. IEEE (2011)
14. NI 5791R: User Manual and Specifications. National Instruments, Austin (2013). http://www.ni.com/pdf/manuals/373845d.pdf
15. Schenk, T.: Rf Imperfections in High-rate Wireless Systems, Impact and Digital Compensation. Springer, New York (2008)
16. Razavi, B.: Design of Analog CMOS Integrated Circuits. McGraw-Hill, New York (2001)
17. Kang, Y., Kim, K., Park, H.: Efficient DFT-based channel estimation for ofdm systems on multipath channels. IET Commun. 1, 197–202 (2007)
18. Barkat, M.: Signal Detection and Estimation. Artech House, Norwood (2005)

Performance of an Energy Detector with Generalized Selection Combining for Spectrum Sensing

Deep Chandra Kandpal[✉], Vaibhav Kumar, Ranjan Gangopadhyay, and Soumitra Debnath

Department of ECE, The LNM Institute of Information Technology, Jaipur 302031, India
deepkandpal.lnmiit@gmail.com

Abstract. Diversity reception schemes are well-known to have the ability to mitigate the adverse effects of multipath wireless channels. This paper analyzes the performance of an energy detector with generalized selection combining (GSC) over a Rayleigh fading channel and compares the results with those of the conventional diversity combining schemes such as, maximal-ratio combining (MRC) and the selection combining (SC). Novel closed-form expressions have been derived for the average detection probability over the independently, identically distributed (i.i.d) diversity paths. Receiver operating characteristics (ROCs) and average detection probability versus SNR curves have been presented for different scenarios of interest.

Keywords: Cognitive radio · Spectrum sensing · Energy detection · Diversity combining · Generalized selection combining

1 Introduction

Cognitive radio has been well-recognized to offer smart solutions to meet the increasing bandwidth demands of emerging wireless services and communication devices by utilizing the licensed / license-free radio spectrum. Spectrum sensing is the key technology for the realization of opportunistic spectrum access (OSA), as it enables the secondary users (SUs) to reliably detect the *white spaces* and ensures the effective use of the vacant bands without causing any deleterious effect to the primary incumbent [10]. Among different spectrum sensing methods including energy detection, matched filtering, cyclostationary detector etc., energy detection is the most popular approach owing to the non-coherent structure as well as low computation and implementation cost. However, the performance of energy detector is highly susceptible to the variation of the detection threshold due to noise uncertainty and interference level [5]. The performance of energy detector based spectrum sensing system degrades further in the multipath fading and shadowing scenario.

© ICST Institute for Computer Sciences, Social Informatics and Telecommunications Engineering 2016
D. Noguet et al. (Eds.): CROWNCOM 2016, LNICST 172, pp. 375–384, 2016.
DOI: 10.1007/978-3-319-40352-6_31

Diversity combining schemes are known to have the distinct ability to mitigate the above harmful effects. A detailed analysis of the performance of energy detection based spectrum sensing for diversity reception has been presented in [3], for the composite shadow fading channel (K and K_G channels) where, the MRC based detector has been shown to outperform the SC based detector at the cost of increased system complexity. A moment generating function (MGF) based approach for the performance evaluation of energy detector with diversity reception in generalized fading channels (including η-μ, κ-μ, α-μ, K, G and K_G channels) has been presented in [1], where the authors have analyzed three different diversity combining schemes namely the MRC, the square law combining (SLC) and the square law selection (SLS) receivers, in which the MRC based receiver has been shown to provide the optimal detection.

One major deficiency of the MRC combining scheme is its sensitivity to channel estimation error which tends to be more vulnerable when the instantaneous SNR is low. In addition, the SC scheme makes the use of only one path out of L resolvable multipaths and hence fails to exploit the full diversity offered by the wireless multipath channel. In order to bridge the gap between the two extreme schemes (SC and MRC), the generalized selection combining (GSC) has been suggested [2], which is an adaptive combining scheme that *selects* L_c strongest resolvable paths (in terms of SNR) among the total L available paths and then coherently *combines* these L_c paths using the MRC scheme. The error probability analysis of GSC in different fading channels has received much research interest in the past years [6,7,9], and it has now been well-established as an alternative to both MRC and SC in terms of complexity and performance respectively. To the best of author's knowledge, the performance analysis of GSC in the context of spectrum sensing is still missing in the open literature.

In the present paper, we endeavor to analyze the performance of energy detection based spectrum sensing system using GSC in a Rayleigh fading channel. A closed-form expression for the average detection probability has been derived and the receiver operating characteristic (ROC) has been obtained by evaluating both the integral and the closed form expressions in order to verify the validity of the obtained results.

The rest of the paper is organized as follows: Sect. 2 briefly discusses the system model for energy detection with no diversity and with GSC and gives the tractable solution for the case of GSC. Numerical results have been presented in Sect. 3, followed by conclusions in Sect. 4. Appendix A, B and C are provided at the end of the paper in order to illustrate the derivation of the closed-form expression.

2   System Model

The received signal sample at a sensing node can be expressed as:

$$y[n] = \begin{cases} w[n], & H_0 \\ h[n]s[n] + w[n], & H_1 \end{cases} \tag{1}$$

where, $y[n]$, $h[n]$, $s[n]$ and $w[n]$ denote the nth sample of the signal received, channel fading coefficient, transmitted sample and the zero-mean additive white Gaussian noise (AWGN) with variance σ_w^2 respectively at the sensing node. H_0 and H_1 denote the null and the alternate hypotheses respectively, corresponding to the absence and the presence of the primary user (PU). At the sensing node the energy of the received signal is measured for a predefined bandwidth Ω over a period of time τ, provided $N = \Omega\tau \in \mathbb{Z}^+$, with $\mathbb{Z}^+$ being the set of positive integers. The received energy per sensing event is given as:

$$\Lambda = \sum_{n=0}^{N-1} \left[|y[n]|^2 \right] \tag{2}$$

The decision rule can be adopted as:

$$\begin{aligned} H_0 &: \Lambda < \lambda \\ H_1 &: \Lambda \geq \lambda \end{aligned} \tag{3}$$

Λ is also termed as the *test statistic* for the energy detector and follows a central chi-square distribution with $2N$ degrees of freedom under H_0 hypothesis, or a non-central chi-square distribution with $2N$ degrees of freedom under hypothesis H_1. λ in (3) is the predefined threshold. In order to analyze the performance of the sensing scheme, the probability of false alarm P_{fa} and the probability of detection P_d need to be evaluated. The parameters are defined as:

$$\begin{aligned} P_{fa} &= P\left[H_1 | H_0\right] \\ P_d &= P\left[H_1 | H_1\right] \end{aligned} \tag{4}$$

where, $P\left[\cdot|\cdot\right]$ denotes the conditional probability.

2.1   Energy Detection with No Diversity

For the case of energy detection without any diversity, P_{fa} and P_d are defined as [3]:

$$\begin{aligned} P_{fa} &= \frac{\Gamma\left(N, \lambda/2\right)}{\Gamma(N)} \\ P_d &= Q_N\left(\sqrt{2\gamma}, \sqrt{\lambda}\right) \end{aligned} \tag{5}$$

where, $\Gamma(\cdot)$ is the Gamma function, $\Gamma(s,x) = \int_x^\infty t^{s-1}\exp(-t)dt$ is the upper incomplete Gamma function, $Q_N(\cdot,\cdot)$ is the generalized Marcum-Q function and γ is the received SNR for the target signal. The expression for P_d in (5) represents the detection probability for the AWGN case. In the case of fading channel, where the fading coefficient varies, the average detection probability $\overline{P_d}$ is obtained by averaging $P_d(\gamma)$ over the statistics of the instantaneous channel SNR γ, i.e.,

$$\overline{P_d} = \int_0^\infty P_d(\gamma)f(\gamma)d\gamma \tag{6}$$

where, $f(\gamma)$ is the probability density function (PDF) of the channel SNR, with $\gamma = |h|^2 E_s / \sigma_w^2$ (SNR per received symbol) and E_s being the transmission energy per received symbol.

2.2   Energy Detection with GSC

In the case of energy detector with GSC scheme, the energy detector compares the received energy after combining the signals from L_c i.i.d. branches against a predefined threshold [3]. The nominal expressions for the instantaneous P_{fa} and P_d in this case remain the same at the output of GSC as for the AWGN channel as (5). The instantaneous SNR of the combiner output can be expressed as [2, (8)]:

$$\gamma_{\text{GSC}} = \sum_{i=1}^{L_c} \gamma_{i:L} \tag{7}$$

where, $\gamma_{i:L}$ is the instantaneous SNR of the ith received diversity path and $\gamma_{1:L} \geq \gamma_{2:L} \geq \ldots \geq \gamma_{L:L}$. The nominal expressions for the instantaneous false-alarm and detection probability in this case remain the same at the output of GSC as for the AWGN channel as (5) with γ replaced by γ_{GSC}. To get the average detection probability for the case of fading channel, $P_d(\gamma_{\text{GSC}})$ should be averaged over the statistics of the channel SNR, γ_{GSC}. Assuming that $\overline{\gamma}_{1:L} = \overline{\gamma}_{2:L} = \ldots = \overline{\gamma}_{L_c:L} = \overline{\gamma}$, where, $\overline{\gamma}_{i:L}$ is the average SNR of the ith received branch, the PDF of γ_{GSC} considering the i.i.d. Rayleigh fading diversity channels can be given as [2]:

$$f(\gamma_{\text{GSC}}) = \binom{L}{L_c} \left[\frac{\gamma_{\text{GSC}}^{L_c-1} e^{-\gamma_{\text{GSC}}/\overline{\gamma}}}{\overline{\gamma}^{L_c}(L_c-1)!} + \frac{1}{\overline{\gamma}} \sum_{l=1}^{L-L_c} (-1)^{L_c-l+1} \binom{L-L_c}{l} \left(\frac{L_c}{l} \right)^{L_c-1} \right.$$
$$\left. \cdot e^{-(\gamma_{\text{GSC}}/\overline{\gamma})} \left(e^{(-l\gamma_{\text{GSC}}/L_c\overline{\gamma})} - \sum_{m=0}^{L_c-2} \frac{1}{m!} \left(\frac{-l\gamma_{\text{GSC}}}{L_c\overline{\gamma}} \right)^m \right) \right] \tag{8}$$

The average detection probability for the energy detector based spectrum sensing with GSC in the fading case is obtained as:

$$\overline{P_d^{\text{GSC}}} = \int_0^\infty P_d(\gamma_{\text{GSC}}) f(\gamma_{\text{GSC}}) d\gamma_{\text{GSC}}$$
$$= A_1 + A_2 + A_3 \tag{9}$$

where,

$$A_1 = \int_0^\infty Q_N\left(\sqrt{2\gamma_{\text{GSC}}}, \sqrt{\lambda}\right) \binom{L}{L_c} \frac{\gamma_{\text{GSC}}^{L_c-1} e^{-\gamma_{\text{GSC}}/\overline{\gamma}}}{\overline{\gamma}^{L_c}(L_c-1)!} d\gamma_{\text{GSC}} \tag{10}$$

$$A_2 = \int_0^\infty Q_N\left(\sqrt{2\gamma_{\text{GSC}}}, \sqrt{\lambda}\right) \binom{L}{L_c} \frac{1}{\overline{\gamma}} \sum_{l=1}^{L-L_c} (-1)^{L_c-l+1} \binom{L-L_c}{l} \left(\frac{L_c}{l} \right)^{L_c-1}$$
$$\cdot e^{-(\gamma_{\text{GSC}}/\overline{\gamma})} e^{(-l\gamma_{\text{GSC}}/L_c\overline{\gamma})} d\gamma_{\text{GSC}} \tag{11}$$

$$A_3 = -\int_0^\infty Q_N\left(\sqrt{2\gamma_{\text{GSC}}}, \sqrt{\lambda}\right) \binom{L}{L_c} \frac{1}{\overline{\gamma}} \sum_{l=1}^{L-L_c} (-1)^{L_c-l+1} \binom{L-L_c}{l} \left(\frac{L_c}{l} \right)^{L_c-1}$$
$$\cdot e^{-(\gamma_{\text{GSC}}/\overline{\gamma})} \sum_{m=0}^{L_c-2} \frac{1}{m!} \left(\frac{-l\gamma_{\text{GSC}}}{L_c\overline{\gamma}} \right)^m d\gamma_{\text{GSC}} \tag{12}$$

Exact infinite series for (10) was proposed in [4, (9)], which does not qualify the conditions for a tractable solution. In the present paper, we derive the closed form solutions for A_1, A_2 and A_3. With the aid of Appendix A, the solution for (10) can be derived as:

$$A_1 = 2 \binom{L}{L_c} \frac{1}{\overline{\gamma}^{L_c}(L_c - 1)!} \left[G_{N-1} + \frac{\Gamma(L_c) \left(\frac{\lambda}{2}\right)^{N-1} \exp\left(-\frac{\lambda}{2}\right)}{2(N-1)! \left(1 + \frac{1}{\overline{\gamma}}\right)^{L_c}} \right.$$
$$\left. {}_1F_1\left(L_c; N; \frac{\lambda}{2}\frac{\overline{\gamma}}{1+\overline{\gamma}}\right) \right] \tag{13}$$

where, ${}_1F_1(\cdot)$ is the confluent hypergeometric function. The solution for (11) is obtained as [see Appendix B]:

$$A_2 = 2\binom{L}{L_c} \frac{1}{\overline{\gamma}} \sum_{l=1}^{L-L_c} (-1)^{L_c-l+1} \binom{L-L_c}{l} \left(\frac{L_c}{l}\right)^{L_c-1} \left[D_{N-1} + \right.$$
$$\left. \frac{(\lambda/2)^{N-1} \exp\left(-\frac{\lambda}{2}\right)}{2(N-1)! \left[\frac{1}{\overline{\gamma}}\left(1 + \frac{l}{L_c}\right) + 1\right]} {}_1F_1\left(1; N; \frac{\lambda}{2}\frac{\overline{\gamma}}{\frac{l}{L_c} + \overline{\gamma}}\right) \right] \tag{14}$$

In a similar fashion, the solution for (12) can be derived as [see Appendix C]:

$$A_3 = -2\binom{L}{L_c} \frac{1}{\overline{\gamma}} \sum_{l=1}^{L-L_c} (-1)^{L_c-l+1} \binom{L-L_c}{l} \left(\frac{L_c}{l}\right)^{L_c-1} \sum_{m=0}^{L_c-2} \frac{1}{m!} \left(\frac{-l}{L_c\overline{\gamma}}\right)^m$$
$$\cdot \left[J_{N-1} + \frac{\Gamma(m+1) \left(\frac{\lambda}{2}\right)^{N-1}}{2(N-1)!} \frac{e^{(-\lambda/2)}}{(1+\overline{\gamma}^{-1})^{m+1}} {}_1F_1\left(m+1; N; \frac{\lambda\overline{\gamma}}{2(1+\overline{\gamma})}\right) \right] \tag{15}$$

3 Numerical Results

The performance behavior of the energy detection based spectrum sensing system with the generalized selection combining is presented for different scenarios of interest by depicting the receiver operating characteristics (ROC) and $\overline{P_d^{GSC}}$ vs. $\overline{\gamma}$ curves.

Figure 1 shows the comparison of ROCs for $L = 4$, $L_c = 3$ and $N = 1$ with different values of $\overline{\gamma}$. For the verification of the derived closed-form expressions, the curves are drawn through integration as well as through the closed form expression. Furthermore, as expected, with the increase in average SNR per branch ($\overline{\gamma}$), the detection probability increases.

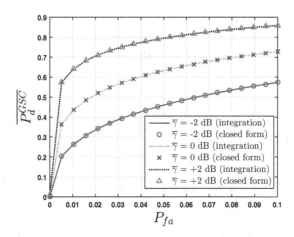

Fig. 1. ROCs for $N = 1$, $L = 4$, $L_c = 3$ and different values of $\overline{\gamma}$.

Figure 2 shows the effect of the chosen value of L_c on the overall detection performance. The value of L is taken as 6 and ROC plots have been shown for L_c varying from 1 (i.e., SC) to 6 (i.e., MRC). It is interesting to note that although $L_c = 6$ provides the best detection performance, the degradation in the performance with $L_c = 5$ as compared to the case $L_c = 6$ is almost negligible.

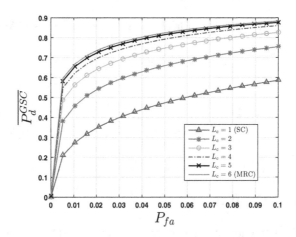

Fig. 2. ROCs for $N = 1$, $L = 6$, $\overline{\gamma} = 0$ dB and different values of L_c.

In Fig. 3, the variation of the detection probability versus the average SNR $\overline{\gamma}$ has been shown for three different values of target false alarm probability 0.01, 0.05 and 0.1.

Figure 4 shows the variation of the detection probability versus the average SNR for $L = 6$ and for different values of L_c varying from 1 to 6. It is important

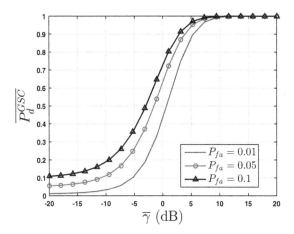

Fig. 3. Variation of detection probability versus $\overline{\gamma}$ with $N = 1$, $L = 4$ and $L_c = 3$.

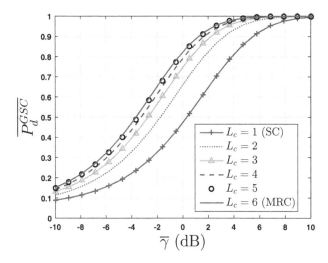

Fig. 4. Variation of detection probability versus $\overline{\gamma}$ with $N = 1$, $L = 6$ and $P_{fa} = 0.05$.

to note that for a lower value of the target P_{fa} (0.05 for the case), the detection performance for $L_c = 5$ and 6 are almost identical.

4   Conclusions

We study the performance of energy detector with generalized selection combining under the Rayleigh fading channel. Novel closed-form expressions are derived for the average detection probability. Numerical evaluation both through integration and the closed-form expression have been provided to validate the expected accuracy of the expression and to illustrate the behavior of the energy detector with GSC. The results confirm that the GSC receivers perform very well as compared to the MRC receivers for spectrum sensing, with a reasonable value of L_c and the associated reduction in system complexity.

Acknowledgments. The work has been carried out under the project, "Mobile Broadband Service Support over Cognitive Radio Networks," sponsored by Information Technology Research Academy (ITRA), Department of Electronics and Information Technology (DeitY), Govt. of India.

A   Appendix

Evaluation of A_1 in (10)

Using [8, (5)], A_1 can be written as:

$$A_1 = \binom{L}{L_c} \frac{1}{\overline{\gamma}^{L_c}(L_c-1)!} \int_0^\infty \left[1 - \exp\left(-\frac{2\gamma_{\mathrm{GSC}}+\lambda}{2}\right) \sum_{n=N}^\infty \left(\frac{\sqrt{\lambda}}{\sqrt{2\gamma_{\mathrm{GSC}}}}\right)^n \right.$$
$$\left. \cdot I_n\left(\sqrt{2\gamma_{\mathrm{GSC}}\lambda}\right) \right] \gamma_{\mathrm{GSC}}^{L_c-1} \exp\left(-\frac{\gamma_{\mathrm{GSC}}}{\overline{\gamma}}\right) d\gamma_{\mathrm{GSC}} \quad (16)$$

where, $I_n(\cdot)$ is the modified Bessel function of order n. Using transformation and change of variable, (16) can be written as:

$$A_1 = 2\binom{L}{L_c} \frac{1}{\overline{\gamma}^{L_c}(L_c-1)!} \int_0^\infty Q_N\left(\sqrt{2}\gamma_{\mathrm{GSC}}, \sqrt{\lambda}\right) \gamma_{\mathrm{GSC}}^{(2L_c-1)} \exp\left(-\frac{\gamma_{\mathrm{GSC}}^2}{\overline{\gamma}}\right) d\gamma_{\mathrm{GSC}}$$
$$= 2\binom{L}{L_c} \frac{1}{\overline{\gamma}^{L_c}(L_c-1)!} \cdot G_N \quad (17)$$

From [8, (29)], the above equation becomes equal to (13) where, G_1 can be defined as [8, (25)]:

$$G_1 = \frac{2^{L_c-1}(L_c-1)!}{(2/\overline{\gamma})^{L_c}} \left(\frac{\overline{\gamma}}{1+\overline{\gamma}}\right) \exp\left(-\frac{\lambda}{2(1+\overline{\gamma})}\right) \sum_{k=0}^{L_c-1} \epsilon_k \left(\frac{1}{1+\overline{\gamma}}\right)^k$$
$$\cdot L_k\left(-\frac{\lambda\overline{\gamma}}{2(1+\overline{\gamma})}\right) \quad (18)$$

where,

$$\epsilon_k \equiv \begin{cases} 1; & k < L_c - 1 \\ 1 + \dfrac{1}{\overline{\gamma}}; & k = L_c - 1 \end{cases} \quad (19)$$

and $L_k(\cdot)$ is the Laguerre polynomial of degree k.

B Appendix

Evaluation of A_2 in (11)

From (11), A_2 can be given as:

$$A_2 = \binom{L}{L_c} \frac{1}{\overline{\gamma}} \sum_{l=1}^{L-L_c} (-1)^{L_c-l+1} \binom{L-L_c}{l} \left(\frac{L_c}{l}\right)^{L_c-1} \int_0^\infty \left[1 - \exp\left(-\frac{2\gamma_{\text{GSC}}+\lambda}{2}\right)\right.$$

$$\cdot \sum_{n=N}^\infty \left(\frac{\sqrt{\lambda}}{\sqrt{2\overline{\gamma}}}\right)^n I_n\left(\sqrt{2\gamma_{\text{GSC}}\lambda}\right)\right] \exp\left[-\frac{\gamma_{\text{GSC}}}{\overline{\gamma}}\left(1+\frac{l}{L_c}\right)\right] d\gamma_{\text{GSC}}$$

$$= 2\binom{L}{L_c} \frac{1}{\overline{\gamma}} \sum_{l=1}^{L-L_c} (-1)^{L_c-l+1} \binom{L-L_c}{l} \left(\frac{L_c}{l}\right)^{L_c-1} \int_0^\infty Q_N\left(\sqrt{2\gamma_{\text{GSC}}},\sqrt{\lambda}\right)$$

$$\cdot \exp\left[-\frac{\gamma_{\text{GSC}}^2}{\overline{\gamma}}\left(1+\frac{l}{L_c}\right)\right] \gamma_{\text{GSC}} d\gamma_{\text{GSC}}$$

$$= 2\binom{L}{L_c} \frac{1}{\overline{\gamma}} \sum_{l=1}^{L-L_c} (-1)^{L_c-l+1} \binom{L-L_c}{l} \left(\frac{L_c}{l}\right)^{L_c-1} \cdot D_N \tag{20}$$

From [8, (29)], the above equation becomes equal to (14) where, D_1 can be defined as [8, (25)]:

$$D_1 = \frac{(\overline{\gamma}L_c)^2}{2(l+Lc)(1+L_c+\overline{\gamma}L_c)} \exp\left(-\frac{\lambda(l+L_c)}{2(1+L_c+\overline{\gamma}L_c)}\right) \cdot \left[\left(1+\frac{l+L_c}{\overline{\gamma}L_c}\right)\right] \tag{21}$$

In the above equation, it is important to note that the value of Laguerre polynomial for order 0 becomes 1.

C Appendix

Evaluation of A_3 in (12)

Following the same analogy as in Appendix A, (12) can be written as:

$$A_3 = -2\binom{L}{L_c} \frac{1}{\overline{\gamma}} \sum_{l=1}^{L-L_c} (-1)^{L_c-l+1} \binom{L-L_c}{l} \left(\frac{L_c}{l}\right)^{L_c-1} \sum_{m=0}^{L_c-2} \frac{1}{m!}\left(\frac{-l}{L_c\overline{\gamma}}\right)^m$$

$$\cdot \int_0^\infty Q_N\left(\sqrt{2\gamma_{\text{GSC}}},\sqrt{\lambda}\right) \exp\left(-\gamma_{\text{GSC}}^2/\overline{\gamma}\right) \gamma_{\text{GSC}}^{2m+1} d\gamma_{\text{GSC}}$$

$$= -2\binom{L}{L_c} \frac{1}{\overline{\gamma}} \sum_{l=1}^{L-L_c} (-1)^{L_c-l+1} \binom{L-L_c}{l} \left(\frac{L_c}{l}\right)^{L_c-1} \sum_{m=0}^{L_c-2} \frac{1}{m!}\left(\frac{-l}{L_c\overline{\gamma}}\right)^m \cdot J_N \tag{22}$$

From [8, (29)], the above equation becomes equal to (15) where, J_1 can be defined as [8, (25)]:

$$J_1 = \frac{2^m m!}{(2/\overline{\gamma})^{(m+1)}} \frac{\overline{\gamma}}{1 + \overline{\gamma}} \exp\left(-\frac{\lambda}{2(1 + \overline{\gamma})}\right) \sum_{k=0}^{m} \phi_k \left(\frac{1}{1 + \overline{\gamma}}\right)^k L_k \left(-\frac{\lambda \overline{\gamma}}{2(1 + \overline{\gamma})}\right)$$

(23)

where,

$$\phi_k \equiv \begin{cases} 1; & k < m \\ 1 + \dfrac{1}{\overline{\gamma}}; & k = m \end{cases}$$

(24)

References

1. Adebola, E., Annamalai, A.: Unified analysis of energy detectors with diversity reception in generalised fading channels. IET Commun. **8**(17), 3095–3104 (2014)
2. Alouini, M.S., Simon, M.K.: An mgf-based performance analysis of generalized selection combining over rayleigh fading channels. IEEE Trans. Commun. **48**(3), 401–415 (2000)
3. Atapattu, S., Tellambura, C., Jiang, H.: Performance of an energy detector over channels with both multipath fading and shadowing. IEEE Trans. Wirel. Commun. **9**(12), 3662–3670 (2010)
4. Cui, G., Kong, L., Yang, X., Ran, D.: Two useful integrals involving generalised marcum q-function. Electron. Lett. **48**(16), 1017–1018 (2012)
5. Hamdi, K., Zeng, X.N., Ghrayeb, A., Letaief, K.: Impact of noise power uncertainty on cooperative spectrum sensing in cognitive radio systems. In: 2010 IEEE Global Telecommunications Conference (GLOBECOM 2010), pp. 1–5, December 2010
6. Lu, C., Lans, W.: Approximate ber performance of generalized selection combining in nakagami-m fading. IEEE Commun. Lett. **5**(6), 254–256 (2001)
7. Ma, Y., Chai, C.C.: Unified error probability analysis for generalized selection combining in nakagami fading channels. IEEE J. Sel. Areas Commun. **18**(11), 2198–2210 (2000)
8. Nuttall, A.H.: Some Integrals Involving the Q_M-Functions. Naval Underwater Systems Center, New London (1974)
9. Theofilakos, P., Kanatas, A., Efthymoglou, G.: Performance of generalized selection combining receivers in k fading channels. IEEE Commun. Lett. **12**(11), 816–818 (2008)
10. Yucek, T., Arslan, H.: A survey of spectrum sensing algorithms for cognitive radio applications. Commun. Surv. Tutorials **11**(1), 116–130 (2009). IEEE

Modelling and Theory

Analysis of a Multicarrier Communication System Based on Overcomplete Gabor Frames

Alexandre Marquet[1,2](✉), Cyrille Siclet[1], Damien Roque[2], and Pierre Siohan[3]

[1] GIPSA-Lab, Univ. Grenoble Alpes, 38400 Grenoble, France
alexandre.marquet@gipsa-lab.grenoble-inp.fr
[2] Institut Supérieur de l'Aéronautique et de l'Espace (ISAE-SUPAERO),
Univ. Toulouse, 31055 Toulouse, France
[3] Orange–Labs, 35512 Cesson-Sevigné, France

Abstract. A multicarrier signal can be seen as a Gabor family whose coefficients are the symbols to be transmitted and whose generators are the time-frequency shifted pulse shapes to be used. In this article, we consider the case where the signaling density is increased such that inter-pulse interference is unavoidable.

Such an interference is minimized when the Gabor family used is a tight frame. We show that, in this case, interference can be approximated as an additive Gaussian noise. This allows us to compute theoretical and simulated bit-error-probability for a non-coded system using a quadrature phase-shift keying constellation. Such a characterization is then used in order to predict the convergence of a coded system using low-density parity check codes. We also study the robustness of such a system to errors on the received bits in an interference cancellation context.

Keywords: Multicarrier modulations · Faster-than-Nyquist signaling · Linear system · Optimal pulse-shapes · Gabor frames · Interference analysis · Interference cancellation · Low-density parity check codes

1 Introduction

In most of current communication systems, the linear part allows for perfect symbol reconstruction: the synthesis and analysis families used in the transmitter and the receiver form biorthogonal frames (also known as Riesz bases). In a single-carrier bandlimited scenario, this requires the Nyquist criterion to be respected [7]. In other words, the transmission rate must be lower than the bilateral bandwidth of the transmitted signal.

With an increasing need of spectral efficiency driven by overcrowded frequency bands, the main strategy relies on an increase of constellation size while keeping a constant transmission power, bandwidth and symbol rate (below the Nyquist limit). This choice induces a decrease of the distance between symbols, and the transmitted signal becomes more sensitive to noise, thus increasing bit-error-probability [5].

© ICST Institute for Computer Sciences, Social Informatics and Telecommunications Engineering 2016
D. Noguet et al. (Eds.): CROWNCOM 2016, LNICST 172, pp. 387–399, 2016.
DOI: 10.1007/978-3-319-40352-6_32

A more unusual way to improve spectral efficiency is to increase the symbol rate until the Nyquist criterion is overridden, leading to unavoidable inter-pulse interference (IPI). This idea has been proposed by J. Mazo under the denomination *"faster-than-Nyquist"* (FTN) [6]. He showed that an increase up to approximately 25 % of the Nyquist symbol rate keeps the minimal distance between symbols unchanged. As a consequence, considering the work of G.D. Forney on the optimal detection in presence of inter-symbol interference, one can preserve an acceptable bit-error-probability at the price of a greater computational complexity at the receiver side [5] (*e.g.*: maximum likelihood approaches...).

FTN transmission techniques can be extended to multicarrier modulations [9]. In this case, denoting F_0 the inter-carrier spacing and T_0 the multicarrier symbol duration, it can be shown that if $\rho = 1/(F_0T_0) > 1$ then the synthesis and analysis families, respectively used for transmission and reception, can no longer be biorthogonal but can still form overcomplete frames [3]. This leads to IPI both in time and/or frequency. Numerous studies focus on the realization of coded multicarrier FTN systems using, in particular, series or parallel concatenations [10] as well as turboequalization techniques [4]. Studies of these latter systems over additive white Gaussian noise (AWGN) channels show great performance, confirming their relevance, even if their intrinsic complexity makes their design and performance comparison particularly demanding in terms of simulation time.

In this article, we study a linear multicarrier system operating with overcomplete Gabor frames (*i.e.*: a generalization of an FTN system), as it plays a fundamental role in practical systems, including decision feedback and iterative structures (*e.g.*: turboequalizers). Our work includes guidelines for the design of such systems over an additive white Gaussian noise (AWGN) channel, only based on the linear part of the system. First of all, we focus on the determination of the expression of the bit-error-probability of our linear system provided that tight frames are used, as prescribed in [11] in order to maximize the signal to interference plus noise ratio (SINR). Secondly, we investigate the behavior of interference cancellation receivers in this context. Finally, we show how the bit-error-probability closed-form expression of the linear system can be used to guide the design of more complex structures (including iterative receivers such as turboequalizers).

This article is constructed as follows. Section 2 details the input-output relations of the system in presence of noise, based on the frame theory. This theoretical framework allows for the determination of the SINR and the theoretical bit-error-probability, based on the assumption of normality of the interference. Section 3 first aims for the study of the statistical properties of the interference in an empirical way, as to confirm the relevance of its Gaussian approximation. We then present bit-error-rate (BER) simulations aiming at the verification of our theoretical results, and an example of how our closed-form expression of the error probability can predict the performance of a coded system is presented through the simulation of a non-linear system using low-density parity check (LDPC) codes. The last simulation scenario analyzes the relevance of interference cancellation techniques in this communication context. Finally, conclusions and insights are presented in Sect. 4.

2   System Model

2.1   Input-Output Relationship in Presence of White Gausssian Noise

Let us denote $c = \{c_{m,n}\}_{(m,n)\in\Lambda} \in \ell_2(\Lambda)$ with $\Lambda \subset \mathbf{Z}^2$, a sequence of zero-mean, independent and identically distributed coefficients. Its variance is σ_c^2. The multicarrier signal is then written as:

$$s(t) = \sum_{(m,n)\in\Lambda} c_{m,n} g_{m,n}(t), \quad t \in \mathbf{R} \tag{1}$$

with $g = \{g_{m,n}\}_{(m,n)\in\Lambda}$ a Gabor family, with parameters $F_0, T_0 > 0$ and whose elements are given by the generator filter (also known as prototype) $g(t) \in \mathcal{L}_2(\mathbf{R})$ such as:

$$g_{m,n}(t) = g(t - nT_0)e^{j2\pi m F_0 t}. \tag{2}$$

As a result, the information carried by c is regularly spread in the time-frequency plane (Fig. 1) with a minimum distance F_0 in frequency and T_0 in time.

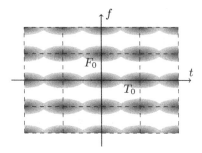

Fig. 1. Representation of a transmitted signal in the time-frequency plane. Here, the generator filter g and the parameters of the lattice allow for a separation in the frequency domain, but not in the time domain.

In a real case scenario, we usually have $\Lambda = \{0, \ldots, M-1\} \times \{0, \ldots, K-1\}$ where M, K are strictly positive integers representing respectively the number of subcarriers and the number of multicarrier symbols to be transmitted. Such a restriction to a finite signaling set induces the convergence of the sum in (1). Nevertheless it can still contain a large amount of terms, so it is important to make it stable. Denoting $\mathcal{H}_g = \overline{\text{Vect}}(g)$ the closure of the linear span of the family g^1, the stability of (1) is guaranteed when g is a Bessel sequence, which means that we can find an upper bound $B_g > 0$ such as:

$$\sum_{(m,n)\in\Lambda} |\langle g_{m,n}, x\rangle|^2 \leq B_g \|x\|^2, \quad \forall x \in \mathcal{H}_g \tag{3}$$

where $\langle\cdot,\cdot\rangle$ and $\|\cdot\|$ are the usual inner product and norm defined $\forall x, y \in \mathcal{L}_2(\mathbf{R})$ by

$$\langle x, y\rangle = \int_{-\infty}^{+\infty} x^*(t)y(t)\,\mathrm{d}t, \quad \|x\| = \sqrt{\langle x, x\rangle} \tag{4}$$

respectively, with x^* the complex conjugate of x. In order to retrieve the data c from the knowledge of $s(t)$, it is furthermore necessary (and sufficient) for g to be a linearly independent family. Hence g is a Riesz basis of $\mathcal{H}_g$, in other words a family for which we can find $0 < A_g \leq B_g$ such that:

$$A_g \|x\|^2 \leq \sum_{(m,n)\in\Lambda} |\langle g_{m,n}, x\rangle|^2 \leq B_g \|x\|^2, \; \forall x \in \mathcal{H}_g. \tag{5}$$

In this case, the density ρ of g is necessarily lower than or equal to one: $\rho = 1/(F_0 T_0) \leq 1$. On the contrary, in order to increase the spectral efficiency of the system (for a fixed number of bits per symbol), this article focuses on the case where $\rho > 1$. Thus, this increase in spectral efficiency is counterbalanced by an induced interference. In a linear receiver, this interference can be considered as an noise leading to an increased error probability. Indeed, when $\rho > 1$, g is necessarily a linearly dependent Gabor family, but it may be an overcomplete frame of $\mathcal{L}_2(\mathbf{R})$, i.e. a family for which (5) is valid not only for $x \in \mathcal{H}_g$, but for every $x \in \mathcal{L}_2(\mathbf{R})$. In this case, (1) is always stable and $\mathcal{H}_g = \mathcal{L}_2(\mathbf{R})$. However, g cannot be a basis of $\mathcal{L}_2(\mathbf{R})$.

A linear receiver is considered as a first stage of a more complete FTN system (necessarily non-linear in order to yield acceptable performance). In this context, the estimated symbols $\hat{c} = \{\hat{c}_{p,q}\}_{(p,q)\in\Lambda}$ are given by

$$\hat{c}_{p,q} = \langle \breve{g}_{p,q}, r\rangle, \quad \forall (p,q) \in \Lambda \tag{6}$$

where $\breve{g} = \{\breve{g}_{m,n}\}_{(m,n)\in\Lambda}$ is a reception family, $r(t) = s(t) + n(t)$ is the signal seen by the receiver and $n(t)$ is a zero-mean white complex circular noise independent from the symbols, and whose bilateral power spectral density is $\gamma_n(f) = 2N_0$ for $f \in \mathbf{R} : \mathrm{E}(n(t)) = 0$ and $\mathrm{E}(n^*(t)n(t')) = 2N_0\delta(t - t')$, with $\mathrm{E}(\cdot)$ the expectation operator.

---

[1] The closure of a normed vector space $\mathbf{E}$ contains all the elements of $\mathbf{E}$, together with its limit elements. For example, the closure of the set of the rational numbers is the set of the real numbers.

2.2 Interference and Noise Analysis

By rewriting (6), we can clearly identify the interference and noise terms:

$$\hat{c}_{p,q} = \underbrace{c_{p,q} \langle \check{g}_{p,q}, g_{p,q} \rangle}_{\tilde{c}_{p,q}:\text{useful signal}} + \underbrace{\sum_{(m,n)\in\Lambda\setminus\{(p,q)\}} c_{m,n} \langle \check{g}_{p,q}, g_{m,n} \rangle}_{i_{p,q}:\text{interference}} + \underbrace{\langle \check{g}_{p,q}, n \rangle}_{n_{p,q}:\text{noise}}. \tag{7}$$

We already showed in [11] that the SINR is maximized when $\check{g}$ and g are dual canonical ($A_g = 1/A_{\check{g}}$ and $B_g = 1/B_{\check{g}}$) tight ($A_g = B_g$ and $\check{g} = g/A_g$) frames. This leads to the following expressions:

$$E_s = \frac{1}{2}\sigma_c^2 \|g\|^2 = \frac{\sigma_c^2 A_g}{2\rho}, \tag{8}$$

$$\sigma_i^2 = \text{E}(|i_{p,q}|^2) = (\rho - 1)\sigma_c^2, \tag{9}$$

$$\sigma_n^2 = \text{E}(|n_{p,q}|^2) = 2\frac{\rho}{A_g}N_0 \tag{10}$$

with E_s the per-symbol energy, σ_i^2 the variance of the interference and σ_n^2 the variance of the filtered noise. The SINR is then written as

$$\text{SINR} = \frac{1}{\rho - 1 + \frac{N_0}{E_s}}. \tag{11}$$

We can see that the interference term $i_{p,q}$ is a random variable independent from the noise and corresponding to the sum of a large number of random variables $\tilde{c}_{m,n}$ which are zero-mean, independent, following the same type of law but with different variances $\sigma_{\tilde{c}_{m,n}}^2$:

$$\tilde{c}_{m,n} = c_{m,n} \langle \check{g}_{p,q}, g_{m,n} \rangle \text{ and } \sigma_{\tilde{c}_{m,n}}^2 = \sigma_c^2 |\langle \check{g}, g_{m-p,n-q} \rangle|^2. \tag{12}$$

All the conditions for applying the central limit theorem are thus not fulfilled but, as shown by our simulations in Subsect. 3.1, the Gaussian approximation is accurate for the sake of error-probability estimation. That is why in the following, we will assume the interference $i_{p,q}$ to be a normal zero-mean random variable independent from the noise. This is analogous to a case where the symbols would have been transmitted through an AWGN channel characterized by a signal-to-noise ratio given by (11). It is interesting to note that the noise term $n_{p,q}$ is zero-mean and Gaussian, but not necessarily white.

2.3 Theoretical Error Probability

We now restrict our analysis to the case where the symbols c are taken from a quadrature phase-shift keying (QPSK) constellation. In that case, given the fact that both the noise and the interference are considered Gaussian, the bit error probability for a transmission through an AWGN channel is given by

$$P_e = Q\left(\sqrt{\text{SINR}}\right) = Q\left(\sqrt{\frac{1}{(\rho - 1) + \frac{N_0}{2E_b}}}\right) \tag{13}$$

where $Q(\cdot)$ is the complementary cumulative distribution function of a standard normal distribution and $E_b = E_s/2$ the per-bit energy.

3   Simulations

3.1   Empirical Study of the Interference Term

In this part, we discuss the relevance of the Gaussian approximation of the interference. To this extent, we measure 3.6×10^6 realizations of the interference $i_{p,q}$ term by performing a transmission of $M = 64$ subcarriers over $K = 50000$ multicarrier symbols for different values of ρ, using a QPSK constellation and tight frames. The variance of the obtained samples is then normalized thus giving standardized versions of $i_{p,q}$ depending on ρ, whose empirical probability density functions and cumulative distribution functions (CDF) are comparable. The behavior described here has been observed to be similar with both the real and the imaginary part of $i_{p,q}$, and for various prototypes forming tight frames.

Considering a transmission over a noise-free perfect channel (SINR $= 1/(\rho-1)$), zero-mean, independent and identically distributed bits, and denoting $F_{i,\rho}(x)$ the complementary CDF (CCDF) of the interference given a density of ρ, we can express the bit-error-probability as

$$P_e(\rho) = 1 - F_{i,\rho}\left(\sqrt{\text{SINR}}\right) = 1 - F_{i,\rho}\left(\sqrt{\frac{1}{\rho-1}}\right). \tag{14}$$

In order to assess the Gaussian approximation, we compare the values of the functions $P_e(\rho)$ and $Q\left(\sqrt{\frac{1}{\rho-1}}\right)$ for various ρ on Fig. 2. Even though the interference cannot be characterized by a Gaussian distribution, we can see that the relative approximation error is negligible, except for ρ close to one, in this context of error probability estimation. Our simulations furthermore revealed that the Gaussian approximation then constitutes an upper bound for the bit-error-probability. This result ensures that the Gaussian approximation can be safely used for multicarrier FTN communication system design and engineering, provided that tight frames are used.

3.2   Linear System Performance

The simulations presented in this part consist in the transmission of $K = 5000$ multicarrier symbols over $M = 128$ subcarriers with a QPSK constellation. They were run for various prototypes. The prototypes maximizing the time-frequency localization (TFL) and minimizing the out-of-band energy (OBE) [8] form tight frames, as shown in [11] thanks to the Wexler–Raz theorem [3, Theorem 9.3.4]. It is as well the case for the square-root-raised-cosine (SRRC) with roll-off factor $\alpha = \rho - 1$ and the T_0-width rectangular (RECT$_{T_0}$) prototypes. When such a prototype is used both in transmission and reception, it is sufficient to set its norm to $1/\sqrt{\rho}$ in order to obtain dual canonical tight frames with $A_g = 1$.

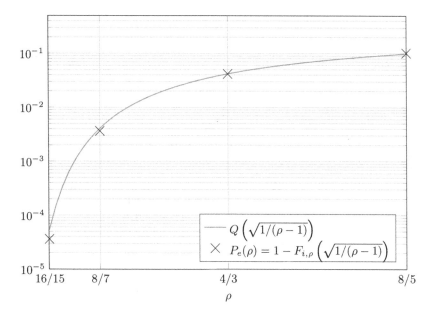

Fig. 2. Comparison of the real CCDF of the interference varying on ρ and its Gaussian approximation.

Although the rectangular prototype of width ρT_0 and T_0 are able to form dual frames, they are not canonical dual. In addition, the pair of frames produced is non-tight so that using it in transmission and reception does not lead to a pair of canonical dual tight frames.

Figure 3 exhibits the perfect prediction of the SINR by (11) when the prototypes used in transmission and reception form a dual canonical tight pair of frames. In addition, Fig. 4 confirms the accuracy of the expression of the bit-error-probability (13) and the relevance of the Gaussian approximation of the interference, although we can see its limits for strong E_b/N_0 ($\geq 14\,\mathrm{dB}$) and ρ close to 1 ($\rho = 16/15$).

In terms of performance, for this kind of non-coded multicarrier FTN system, Fig. 4 shows that the bit-error-rate (BER) rapidly rises with the density. We can also see on Fig. 5 that a lower-bound of the BER appears when the power of the noise becomes negligible compared to the one of the interference. In addition, and in accordance with the expression of the SINR, the performance gets worse if the frames used are not tight nor canonical dual. Theses results confirm the needs to develop non-linear detectors allowing for a more efficient IPI mitigation.

3.3   Use in a Coded System with Iterative Decoding

BER curves of systems using efficient coding schemes such as turbocodes or LDPC are characterized by a so-called "convergence threshold" [2] which is the E_b/N_0 value from which the coded system achieves better performance than the

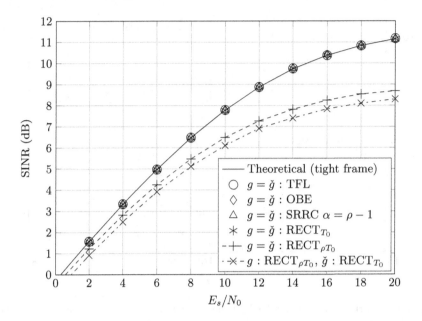

Fig. 3. SINR as a function of E_s/N_0, with $\rho = 16/15$.

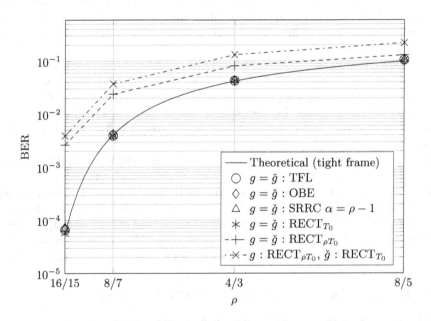

Fig. 4. BER as a function of ρ, with $E_b/N_0 = 20$ dB.

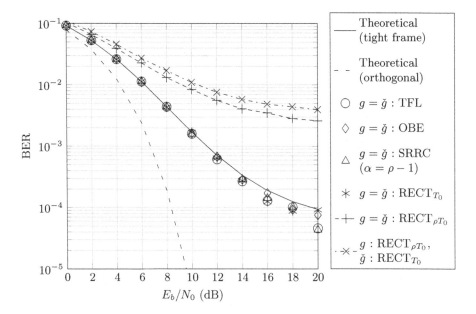

Fig. 5. BER as a function of E_b/N_0, with $\rho = 16/15$.

uncoded one. Given an AWGN channel, it is also possible to characterize the coded system with a curve presenting the BER at the output of the decoder (denoted as "output BER" - BER_{out}) varying with the BER at the input of the decoder (denoted as "input BER" - BER_{in}) as in Fig. 6. On this kind of curve, the convergence threshold is found at a given input BER. As a consequence, and thanks to the expression of the error probability (13), it is possible to determine the optimal density ρ allowing the coded system to converge given an arbitrary value of E_b/N_0.

As an example, Fig. 6 shows that a coded system using the LDPC code of rate 1/2 specified in the DVB-S2 specification [1] has its convergence threshold for an input BER of approximately 0.15. On Fig. 7, we can see that when used with a multicarrier FTN system using tight frames, the coded system converges as expected when the input BER goes below 0.15, at $E_b/N_0 = 2\,\text{dB}$.

3.4 Performance with Interference Cancellation

From the expression of the bit error probability (13) it is obvious that the FTN linear system shows worse performance compared to the orthogonal case. Besides, from the expression of the received signal (7), one can notice that the performance of the orthogonal system can be retrieved by removing the interference induced by the FTN system, allowing for an improvement of the spectral efficiency of the transmission while keeping the same BER.

Such an interference cancellation (IC) is usually performed by estimating the received symbols, then computing the interference term from these estimations

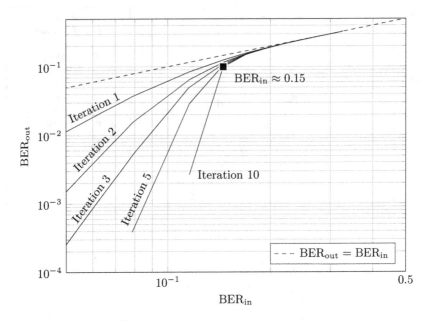

Fig. 6. Output BER as a function of the input BER for a rate=1/2 LDPC system. In this configuration, the convergence threshold is at an input BER of 0.15.

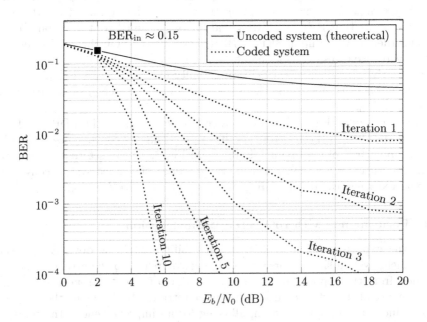

Fig. 7. BER as a function of E_b/N_0 using a rate 1/2 LDPC code, $\rho = 4/3$ and a TFL prototype for 10 iterations of the decoder.

and substract it to the received signal. Given that this estimation might not be perfect, it is interesting to assess the behavior of this system in presence of errors on the estimated symbols. To fulfill that purpose, we implemented the pseudo-genie receiver depicted by Fig. 8. The difference with a "true genie" receiver is that its knowledge of the transmitted symbols is corrupted by a binary symmetric channel inducing an error probability $P_{e,\text{genie}}$ on the bits used to compute the interference term.

Figure 9 presents the performance of this system, simulated by the transmission of $K = 5000$ multicarrier symbols over $M = 32$ subcarriers using a TFL prototype and a QPSK constellation. We can see that it is quite robust to the presence of errors on the bits used to compute and cancel the interference, which gives an insight on how non-linear receivers using interference cancellation (such as decision feedback or turboequalization) could efficiently prevent

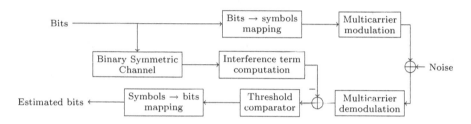

Fig. 8. Synoptic of the pseudo-genie IC receiver.

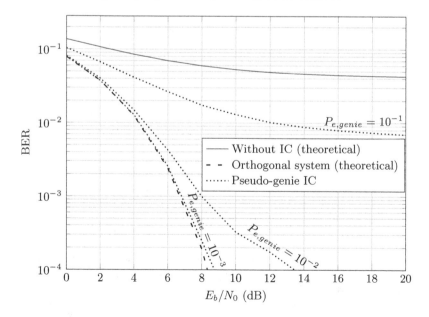

Fig. 9. BER as a function of E_b/N_0, with a pseudo-genie receiver, $\rho = 4/3$, a TFL prototype and $P_{e,\text{genie}} \in \{10^{-1}, 10^{-2}, 10^{-3}\}$.

inter-carrier interference. Although not presented here, we ran simulations with other prototypes yielding tight frames, and obtained similar results.

4 Conclusion

Through this article, we specified a linear multicarrier system based on the use of overcomplete Gabor frames, allowing an increase in signaling density in the time and/or the frequency domain and leading to a bidimensional FTN system. Consequently, an increase of the spectral efficiency beyond (bi)-orthogonal systems (for a given constellation size) yields interference between pulse-shapes. Such interference can be mitigated by the use of tight frames within the context of a linear system. Furthermore, we showed that interference cancellation based on noisy estimates of the transmitted symbols (pseudo-genie receiver) can lead to the same BER as orthogonal systems, but at a higher bitrate.

The results presented in this article allow the ability to relatively compare the performance of FTN multicarrier systems based on the parameters of their linear part (*e.g.*: time-frequency lattice density, transmission/reception filters...). Secondly, we showed how the knowledge of this performance can help the design of more complex receiver structures (*e.g.*: LDPC/turbodecoders, turboequalizers) by predicting their behavior.

Future work may consist in the analysis and efficient implementations of various multicarrier non-linear systems based on tight frames, and transmissions over more complex channels.

Acknowledgment. The authors would like to address a particular thank to Dr. Laurent Ros for his valuable advices and relevant remarks concerning this work.

References

1. ETSI: EN 302 307 Digital Video Broadcasting (DVB); Second generation framing structure, channel coding and modulation systems for Broadcasting, Interactive services, News Gathering and other broadband satellite applications (DVB-S2) (2009)
2. Berrou, C., Cavalec, K.A., Arzel, M., Glavieux, A., Jezequel, M., Langlais, C., Le Bidan, R., Saoudi, S., Battail, G., Boutillon, E., Saouter, Y., Maury, E., Laot, C., Kerouedan, S., Guilloud, F., Douillard, C.: Codes et turbocodes (sous la direction de Claude Berrou). Iris. Springer, Paris (2007)
3. Christensen, O.: Frames and bases: An introductory course. Birkhauser, Basel (2008)
4. Dasalukunte, D., Rusek, F., Owall, V.: Multicarrier faster-than-Nyquist transceivers: hardware architecture and performance analysis. IEEE Trans. Circ. Syst. I Regul. Pap. **58**(4), 827–838 (2011)
5. Forney, G.: Maximum-likelihood sequence estimation of digital sequences in the presence of intersymbol interference. IEEE Trans. Inf. Theory **18**(3), 363–378 (1972)
6. Mazo, J.E.: Faster than Nyquist signaling. Bell Syst. Tech. J. **54**, 1451–1462 (1975)

7. Nyquist, H.: Certain topics in telegraph transmission theory. Trans. Am. Inst. Electr. Eng. **47**(2), 617–644 (1928)
8. Pinchon, D., Siohan, P.: Closed-form expressions of optimal short PR FMT prototype filters. In: Proceedings of IEEE Global Telecommunications Conference GLOBECOM 2011 (2011)
9. Rusek, F., Anderson, J.: The two dimensional Mazo limit. In: Proceedings of International Symposium on Information Theory, ISIT 2005, pp. 970–974, September 2005
10. Rusek, F., Anderson, J.: Serial and parallel concatenations based on faster than Nyquist signaling. In: 2006 IEEE International Symposium on Information Theory, pp. 1993–1997, July 2006
11. Siclet, C., Roque, D., Shu, H., Siohan, P.: On the study of faster-than-Nyquist multicarrier signaling based on frame theory. In: 2014 11th International Symposium on Wireless Communications Systems (ISWCS), pp. 251–255, August 2014

Efficient Power Allocation Approach for Asynchronous Cognitive Radio Networks with FBMC/OFDM

Juwendo Denis[✉], Mylene Pischella, and Didier Le Ruyet

CEDRIC/LAETITIA Lab., Conservatoire National des Arts et Métiers, Paris, France
{juwendo.denis,mylene.pischella,didier.le_ruyet}@cnam.fr

Abstract. In this paper, we address the problem of power minimization under rate constraint for a multi-carrier-based underlay cognitive radio (CR) network. In fact, both primary users (PUs) and secondary users (SUs) employ either orthogonal frequency-division multiplexing (OFDM) or filter bank based multi-carrier (FBMC). The problem is formulated as a non-cooperative interference pricing-based game (NPBG) in order to circumvent the coupling primary interference constraint and also to propose distributed solutions. We provide a sufficient convergence condition to a Nash-equilibrium (NE) point for the modified water-filling algorithm. Moreover, we propose a distributed algorithm that always converges to a unique NE of the NPBG. Simulation analyses are then provided to demonstrate the efficiency of our proposed distributed algorithms. Furthermore, the simulations enhance the advantages of using FBMC as a modulation technique if compared to OFDM.

Keywords: FBMC · OFDM · Cognitive radio · Game theory · Resource allocation

1 Introduction

New paradigms that can enable efficient spectrum utilization emerge to anticipate shortages of radio spectrum in wireless networks that face increasing number of demand from users. Cognitive radio (CR) is proposed as an appealing technology capable of not only improving the spectrum utilization, it can also enhance efficiency of spectrum sharing in wireless networks. This can be done by dynamically allocate radio spectrum by permitting unlicensed users; the secondary users (SUs) to access the frequency band of the primary users (PUs).

There exist two paradigms for the operation of CR technology: opportunistic and concurrent spectrum access [1–3]. In an opportunistic spectrum access scenario, SUs are permitted to communicate only when the PUs are detected to be inactive. On the contrary, SUs are allowed to transmit simultaneously with the PUs in a concurrent spectrum access scenario, provided the quality of service (QoS) of the PUs is not degraded by the activity of the SUs. In this paper, we mainly focus on the second paradigm.

© ICST Institute for Computer Sciences, Social Informatics and Telecommunications Engineering 2016
D. Noguet et al. (Eds.): CROWNCOM 2016, LNICST 172, pp. 400–411, 2016.
DOI: 10.1007/978-3-319-40352-6_33

Multi-carrier modulation techniques such as the orthogonal frequency division multiplexing (OFDM) are eligible for the physical layer of CR networks [4]. For CR networks that experience asynchronous transmission due to lack of cooperation among the users, OFDM may sacrifice data rate transmission because of imperfect time and frequency synchronization. Its high spectral efficiency makes filter bank multi-carrier (FBMC) an alternative to conventional OFDM for transmission over CR networks.

For asynchronous multi-carrier-based CR networks, judicious resource allocation is required to mitigate the effect of inter-carrier interference. The problem of resource allocation for asynchronous underlay CR networks employing FBMC and OFDM was greatly studied over the past decade [5–7]. In [5], the authors addressed the downlink resource allocation for a CR network consisting of a single PU and a single SU. The same scenario as [5] was investigated in [6], yet by considering the uplink case. In [7], Shaat et al. proposed a modified water-filling solution to the problem of downlink rate maximization multi-cell CR network.

In this paper, we investigate the problem of sum power minimization subject to rate constraint for downlink asynchronous underlay CR networks with FBMC and OFDM. To the best of our knowledge, no other research group has addressed this issue. Motivated by [8,9], we formulate the problem as a non-cooperative interference pricing-based game (NPBG). In order to keep a distributed resource allocation, the couple primary interference constraint is embedded into the utility function by mean of pricing. The optimal power allocation strategy for each secondary BS is given by the modified water-filling. We provide a distributed sufficient convergence criterion for the water-filling algorithm. Furthermore, we theoretically demonstrate that our proposed NPBG converges to a Nash-equilibrium (NE) point whenever the convergence criterion is met. In addition to that, we proposed a new distributed algorithm (NDA) to solve the NPBG game. The NDA always converges to a unique NE point. The efficiency of the proposed methods is validated through extensive simulations results.

The rest of this paper is organized as follows: in Sect. 2, we describe the system model together with the problem formulation. In Sect. 3, we introduce a distributed convergence criterion for the water-filling. The new distributed approach that converges to a unique NE point is provided in Sect. 4. Numerical results highlighting some important features of our proposed schemes are given in Sect. 5. Finally, the work is concluded in Sect. 6.

2   System Model and Problem Formulation

Consider a multi-carrier based underlay spectrum sharing CR network that consists of $\mathcal{K}$ active primary users and $\mathcal{S}$ active secondary users. Each active PU and SU is formed by a single transmitter-receiver pair. The total spectrum is divided into L subcarriers. Each subcarrier has a bandwidth B. We consider a downlink transmission where all mobile terminals (MT) and BSs are equipped each with a single antenna. In this configuration, the PUs do not interfere with each other and have a fixed transmission power scheme regardless of the transmission strategy used by the SUs.

The systems that coexist in the network do not cooperate with each other. Moreover, there exists no synchronization neither between any two secondary BSs nor between a secondary BS and a primary BS. Lack of cooperation coupled with asynchronism will create inter-carrier interference which may have detrimental effect on the overall performance of the network. The study of inter-carrier interference was done in [10] where Medjahdi et al. quantified the number of subcarriers affected by interference generated from a given subcarrier. It was demonstrated in [10] that 17 and 3 neighbouring subcarriers suffered from inter-carrier interference in the case of OFDM and FBMC, respectively. The interference weight vector that was derived in [10] can be summarized as

$$
\begin{aligned}
V^{\mathrm{OFDM}} &= \left[\{705, 89.4, 22.3, 9.95, 5.6, 3.59, 2.5, 1.84, 1.12\} \times 10^{-3}\right] \\
V^{\mathrm{FBMC}} &= \left[8.23 \times 10^{-1}, 8.81 \times 10^{-2}\right]
\end{aligned}
\tag{1}
$$

The interference coefficients will be used throughout this work. In subsequent sections, the interference weight vector is denoted as $V = [V_0, \ldots, V_S]$ where $S = 1$ in the case of FBMC and $S = 8$ for OFDM.

Due to the distributed nature of CR network, all secondary MTs use single user detection i.e., interference caused by other SUs and the PUs are treated as noise. We assume that channel gains which include path loss and shadowing change sufficiently slowly to be considered unchanged during each scheduling interval. Perfect knowledge of channel state information (CSI) is available at each BS. The CSI between secondary BS and primary MT can be periodically measured by a band manager [11]. Also, the MTs can estimate the CSI and feed it back to their respective serving BS.

Denote P_s^l the power that the sth secondary BS allocates on the lth subcarrier. Let $\mathbf{P}_s \triangleq \left(P_s^1, \cdots, P_s^L\right)^{\top}$ be the power allocation vector of the sth secondary BS and $\mathbf{P}_{-s} \triangleq \{\mathbf{P}_j\}_{j \in \{1, \cdots, s-1, s+1, \cdots, S\}}$ the set of transmit power of all other secondary BSs. $\mathbf{P} = (\mathbf{P}_1, \cdots, \mathbf{P}_S)^{\top}$ denote all secondary BSs power vector. $G_{s,s}^l$ is the channel gain between secondary BS s and its served MT on subcarrier l. $G_{s,j}^l$ denote the channel gain between BS of SU s and MT of SU j on subcarrier l while $G_{s,k}^l$ is the channel gain between BS of SU s and receiver of the PU k within the lth subcarrier. The achievable data rate of the secondary MT s is given by

$$
\mathcal{R}_s(\mathbf{P}_s, \mathbf{P}_{-s}) = \sum_{l=1}^{L} B \log_2 \left(1 + \frac{P_s^l G_{s,s}^l}{\overline{N}_s^l + I_s^l}\right)
\tag{2}
$$

where

$$
I_s^l = \sum_{j \neq s}^{S} \sum_{l'=1}^{L} P_j^{l'} V_{|l-l'|} G_{j,s}^{l'}, \text{ and } \overline{N}_s^l = N_0 + \sum_{k=1}^{K} \sum_{l'=1}^{L} P_k^{l'} V_{|l-l'|} G_{k,s}^{l'}
$$

N_0 denotes the thermal noise on a subcarrier and $G_{k,s}^l$, the channel gain between the kth primary BS and the mobile terminal of SU s.

For underlay CR networks, secondary users can simultaneously with the PUs communicate on the same frequency band provided that the degradation induced on the QoS of the primary users is tolerable. This is captured by preventing the per subcarrier total interference caused by SUs activity to the kth PU from exceeding a predefined threshold.

$$
\sum_{s=1}^{S} \sum_{l'=1}^{L} P_s^{l'} G_{s,k}^{l'} V_{|l-l'|} \leq I_k^{l,\max}, \forall k, \forall l
\tag{3}
$$

The main bottleneck of (3) is that the interference constraint does include the power of all secondary users. Circumvent this coupling constraint is of great important since we ought to provide distributed solutions.

In this work, we formulate the transmission strategy of the SUs as a noncooperative game. Let $\mathcal{P}_s$ be the feasible set of the transmission strategy of secondary BS s.

$$\mathcal{P}_s(\mathbf{P}_{-s}) \triangleq \left\{ \mathbf{P}_s : \mathcal{R}_s(\mathbf{P}_s, \mathbf{P}_{-s}) \geq \widehat{\mathcal{R}}_s, \; P_s^l \geq 0, \forall l \right\}$$

where $\widehat{\mathcal{R}}_s$ is the rate constraint of the SU s. To deal with the coupling PUs interference constraint that is not incorporated in the feasible set, we introduce a pricing in the secondary utility function which is given by

$$\mathcal{U}(\mathbf{P}_s; \boldsymbol{\mu}) \triangleq \sum_{l=1}^{L} P_k^l + \sum_{k=1}^{\mathcal{K}} \sum_{l=1}^{L} \mu_k^l \sum_{l'=1}^{L} P_s^{l'} G_{s,k}^{l'} V_{|l-l'|} \tag{4}$$

by rearranging the terms on the left hand side, we have

$$\mathcal{U}(\mathbf{P}_s; \boldsymbol{\mu}) \triangleq \sum_{l=1}^{L} P_s^l + \sum_{l=1}^{L} P_s^l \left(\sum_{k=1}^{\mathcal{K}} G_{s,k}^l \left(\sum_{l' \in \mathcal{I}_{s,k}^l} \mu_k^{l'} V_{|l-l'|} \right) \right)$$

where $\mathcal{I}_{s,k}^l$ represents the set of subcarrier of kth primary BS that suffers from interferences generated by the lth subcarrier of the s-th secondary BS. $\boldsymbol{\mu} = (\boldsymbol{\mu}_1, \cdots, \boldsymbol{\mu}_{\mathcal{K}})^\top$ where $\boldsymbol{\mu}_k = (\mu_k^1, \cdots, \mu_k^L)^\top$ represents the vector prices set by the kth PU due to the activity of SUs. The prices are chosen such that the complementary conditions are met, i.e.,

$$\boldsymbol{\mu} \geq 0$$

$$\mu_k^l \left(\sum_{s=1}^{S} \sum_{l'=1}^{L} P_s^{l'} G_{s,k}^{l'} V_{|l-l'|} - I_k^{l,\max} \right) = 0, \; \forall k, l \tag{5}$$

Clearly, the prices set by the PUs aim to control the interference generated by the secondary users. It is straightforward to see that the vector of prices will be null whenever the interference generated by the SU is less than the interference threshold.

Denote $\mathcal{G} = \{ S, \{\mathcal{P}_s\}, \{\mathcal{U}_s\} \}$, the non-cooperative pricing-based game (NPBG). $S = \{1, 2, \cdots, \mathcal{S}\}$ represents the index set of the secondary BSs. $\mathcal{U}_s$ and $\mathcal{P}_s$ denote the utility function and the strategy space for secondary BS s, respectively.

In this paper, each SU selfishly minimizes its utility function while satisfying its rate constraint. More specifically, the game is formulated as

$$\text{NPBG}: \quad \min_{\mathbf{P}_s \in \mathcal{P}_s(\mathbf{P}_{-s})} \mathcal{U}(\mathbf{P}_s; \boldsymbol{\mu}), \quad \forall s \in S \tag{6}$$

For a fixed prices $\boldsymbol{\mu}^\star$, a strategy profile $\mathbf{P}^\star$ is said to be a pure-strategy Nash equilibrium (NE) if no single SU has the incentive to unilaterally change its own transmission power to achieve a lower utility function.

3   Convergence Criterion

For any fixed and non-negative pricing $\boldsymbol{\mu}$ and power allocation $\mathbf{P}_{-s}$, the optimal solution of problem (6) is given by the modified water-filling, i.e.,

$$P_s^l = \left[\frac{\nu_s \frac{B}{\ln 2}}{1 + \sum_{k=1}^{\mathcal{K}} G_{s,k}^l \left(\sum_{l' \in \mathcal{I}_{s,k}^l} \mu_k^{l'} V_{|l-l'|} \right)} - \frac{\overline{N}_s^l + I_s^l}{G_{s,s}^l} \right]^+ \tag{7}$$

where $[x]^+ \triangleq \max(x,0)$ and ν_s is the Lagrangian multiplier associated to the rate constraint. Let $\boldsymbol{\nu} = (\nu_1, \cdots, \nu_S)$, (7) can be compactly written as

$$\mathbf{P} = \Xi(\boldsymbol{\nu}, \boldsymbol{\mu}) - \mathbf{G}^{-1}\overline{\mathbf{N}} - \mathbf{G}^{-1}\overline{\mathbf{G}}\mathbf{P} \qquad (8)$$

where $\Xi(\boldsymbol{\nu}, \boldsymbol{\mu})$, $\mathbf{G}^{-1}$, $\overline{\mathbf{N}}$, $\overline{\mathbf{G}}$ are defined in (9). More specifically, $\mathbf{0}_s$ denotes a $L \times L$ zero entry matrix and $\overline{\mathbf{G}}$ is the interference matrix of the entire secondary system.

$$\mathbf{G} = \text{diag}\left(G_{1,1}^1, \cdots, G_{1,1}^L, \cdots, G_{S,S}^1, \cdots, G_{S,S}^L\right),$$

$$\overline{\mathbf{N}} = (\overline{N}_1^1, \cdots, \overline{N}_1^L, \cdots, \overline{N}_S^1, \cdots, \overline{N}_S^L)^\top, \quad \Xi(\boldsymbol{\nu}, \boldsymbol{\mu}) \triangleq (\xi(\nu_1, \boldsymbol{\mu}), \cdots, \xi(\nu_S, \boldsymbol{\mu}))^\top$$

$$\xi(\nu_s, \boldsymbol{\mu}) \triangleq \left(\frac{\nu_s \frac{B}{\ln 2}}{1 + \sum_{k=1}^K G_{s,k}^l \left(\sum_{l' \in \mathcal{I}_{s,k}^l} \mu_k^{l'} V_{|l-l'|}\right)}, \cdots, \right.$$

$$\left. \frac{\nu_s \frac{B}{\ln 2}}{1 + \sum_{k=1}^K G_{s,k}^L \left(\sum_{l' \in \mathcal{I}_{s,k}^L} \mu_k^{l'} V_{|L-l'|}\right)} \right)^\top$$

$$\overline{\mathbf{G}} \triangleq \begin{pmatrix} \mathbf{0}_1 & \overline{\mathbf{G}}_{12} & \cdots & \overline{\mathbf{G}}_{1S} \\ \overline{\mathbf{G}}_{21} & \mathbf{0}_2 & \cdots & \overline{\mathbf{G}}_{2S} \\ \vdots & \vdots & \ddots & \vdots \\ \overline{\mathbf{G}}_{S1} & \overline{\mathbf{G}}_{S2} & \cdots & \mathbf{0}_S \end{pmatrix},$$

$$\overline{\mathbf{G}}_{j,s} = \begin{pmatrix} G_{j,s}^1 V_0 & G_{j,s}^2 V_1 & \cdots & G_{j,s}^L V_{|L-1|} \\ G_{j,s}^1 V_1 & G_{j,s}^2 V_0 & \cdots & G_{j,s}^L V_{|L-2|} \\ \vdots & \vdots & \ddots & \vdots \\ G_{j,s}^1 V_{|L-1|} & G_{j,s}^2 V_{|L-2|} & \cdots & G_{j,s}^L V_0 \end{pmatrix}$$

$$\qquad (9)$$

At the nth iteration, for any fixed, $\boldsymbol{\mu}$, the modified water-filling function denoted as ψ can be expressed as

$$\mathbf{P}^{(n)} = \psi\left(\mathbf{P}^{(n-1)}, \boldsymbol{\mu}^{(n-1)}\right) = \Xi(\boldsymbol{\nu}, \boldsymbol{\mu}^{(n-1)}) - \mathbf{G}^{-1}\overline{\mathbf{G}} - \mathbf{G}^{-1}\overline{\mathbf{G}}\mathbf{P}^{(n-1)} \qquad (10)$$

Now, we proceed to define the price update.

Definition 1. *[9] Let* $\{n_k^{l,t(k,l)}\}_{t(k,l)=1}^\infty$ *of* $\{n\}_{n=1}^\infty$ *be a unique subsequence associated with* μ_k^l, *the asynchronous price update is defined as the price update approach in which each is updated only at time instances* $\{n_k^{l,t(k,l)}\}_{t(k,l)=1}^\infty$.

By using Definition 1, problem (6) can be solved by iteratively solving (10). The proposed approach is summarized as follow.

Algorithm 1. Iterative modified water-filling algorithm for solving (6)

1: **Input** A solution accuracy $\epsilon > 0$ and a feasible $\mathbf{P}^0$.
2: Initialize $\boldsymbol{\mu}^{(0)}, \boldsymbol{\nu}^{(0)}$, set $n = 0$ and let $t(k, l) = 1$;
3: **repeat**
4:    $n = n + 1$;
5:    Find $\mathbf{P}^n$ by using (10);
6:    For each secondary BS s, update $\nu_s^{(n)}$ by using bisection method.
7:    Asynchronous update of the interferences prices μ_k^l, $\forall k, l$

$$\mu_k^{l,(n)} = \begin{cases} \left[\mu_k^{l,(n-1)} + \delta \left(I_k^{l,(n)} - I_k^{l,\max}\right)\right]^+, & \text{if } n + 1 = n_k^{l,t(k,l)} \\ \mu_k^{l,(n-1)}, & \text{otherwise} \end{cases}$$

   If $n + 1 = n_k^{l,t(k,l)}$ then $t(k, l) = t(k, l) + 1$;
8: **until** $n > \mathcal{N}$ or $\left| \left(\mathcal{U}(\mathbf{P}_s^{(n)}; \boldsymbol{\mu}) - \mathcal{U}(\mathbf{P}_s^{(n-1)}; \boldsymbol{\mu})\right) / \mathcal{U}(\mathbf{P}_s^{(n-1)}; \boldsymbol{\mu}) \right| \leq \epsilon, \forall s$
9: **Output** $\mathbf{P}^n$.

---

where $\sum_{s=1}^{S} \sum_{l'}^{L} P_s^{l',n} G_{s,k}^{l'} V_{|l-l'|} = I_k^{l,n}$ and $\delta \in (0, 1)$ is a coefficient to control the convergence speed of the price update. The following lemma captures the convergence of the interference prices. See [9] for the proof.

Lemma 1. *The sequence of interferences prices* $\{\boldsymbol{\mu}^{(n)}\}_{n=1}^{\infty}$ *generated by Algorithm 1 converges, i.e.,*

$$\lim_{n \to \infty} \boldsymbol{\mu}^{(n)} = \boldsymbol{\mu}^{\star}$$

Now, we provide a sufficient criterion for convergence of the proposed Algorithm 1 to a unique NE point of the game $\mathcal{G}$. This is done in the following theorem.

Theorem 1. *The sequence* $\{\mathbf{P}^{(n)}\}_{n=1}^{\infty}$ *generated by the proposed Algorithm 1 converges to a unique NE regardless of the initial power allocation value if*

$$\sum_{j=1, j \neq s}^{S} \frac{\sum_{l'=1}^{L} V_{|l-l'|} G_{j,s}^{l'}}{G_{s,s}^l} \leq 1, \forall s, l \tag{11}$$

Proof: The proof of Theorem 1 follows the step of the proof of [9, Theorem 4]. Due to limited space, we leave the details for future publication. ∎

4   New Distributed Scheme

From Theorem 1, we notice that our proposed Algorithm 1 converges to a unique NE point only if the sufficient convergence condition is met. In this section, we propose a distributed algorithm that always convergences to a unique NE point of $\mathcal{G}$. This is done by providing a new distributed convergence criterion that can be embedded into problem (6).

The power allocation to solve (6) can be done by

$$P_s^l = \gamma_s^l \left(\frac{\overline{N}_s^l}{G_{s,s}^l} + \frac{\sum_{j \neq s}^{S} \sum_{l'=1}^{L} P_j^{l'} V_{|l-l'|} G_{j,s}^{l'}}{G_{s,s}^l} \right) \tag{12}$$

where γ_s^l is the signal-to-interference-plus-noise ratio (SINR) for secondary user s on subcarrier l. (12) can be compactly written as

$$\mathbf{P} = \mathbf{G}^{-1}\mathbf{\Gamma}\overline{\mathbf{G}}\mathbf{P} + \mathbf{G}^{-1}\mathbf{\Gamma}\overline{\mathbf{N}} \tag{13}$$

where $\mathbf{\Gamma} = \operatorname{diag}\left(\gamma_1^1, \cdots, \gamma_1^L, \cdots, \gamma_S^1, \cdots, \gamma_S^L\right)$. For a fixed SINR $\mathbf{\Gamma}$, at the nth iteration, the power allocation function ϕ is expressed as

$$\mathbf{P}^{(n)} = \phi\left(\mathbf{P}^{(n-1)}, \mathbf{\Gamma}\right) = \mathbf{G}^{-1}\mathbf{\Gamma}\overline{\mathbf{G}}\mathbf{P}^{(n-1)} + \mathbf{G}^{-1}\mathbf{\Gamma}\overline{\mathbf{N}} \tag{14}$$

Theorem 2. *The power allocation scheme (14) converges to a unique fixed point for any arbitrary starting point if*

$$\frac{\gamma_s^l\left(\sum_{j\neq s}^S \sum_{l'=1}^L V_{|l-l'|}G_{j,s}^{l'}\right)}{G_{s,s}^l} < 1, \quad \forall s, l \tag{15}$$

Proof: Given an arbitrary initial power $\mathbf{P}^{(0)}$, we have

$$\left\|\mathbf{P}^{(n+1)} - \mathbf{P}^{(n)}\right\| = \left\|\mathbf{G}^{-1}\mathbf{\Gamma}\overline{\mathbf{G}}\left(\mathbf{P}^{(n)} - \mathbf{P}^{(n-1)}\right)\right\| \le \zeta^{n+1}\left\|\mathbf{P}^{(1)} - \mathbf{P}^{(0)}\right\| \tag{16}$$

where $\kappa = \max_{\substack{1\le s\le S \\ 1\le l\le L}} \frac{\gamma_s^l\left(\sum_{j\neq s}^S \sum_{l'=1}^L V_{|l-l'|}G_{j,s}^{l'}\right)}{G_{s,s}^l}$. It follows that for $\forall n, M \ge 0$,

$$\left\|\mathbf{P}^{(n+M)} - \mathbf{P}^{(n)}\right\| = \sum_{m=1}^M \left\|\mathbf{P}^{(n+m)} - \mathbf{P}^{(n+m-1)}\right\| \overset{(b)}{\le} \frac{\zeta^n}{1-\zeta}\left\|\mathbf{P}^{(1)} - \mathbf{P}^{(0)}\right\| \tag{17}$$

(b) is verified if $\frac{\gamma_s^l\left(\sum_{j\neq s}^S \sum_{l'=1}^L V_{|l-l'|}G_{j,s}^{l'}\right)}{G_{s,s}^l} < 1$, $\forall l, \forall s$. Hence, we obtain a Cauchy sequence which is a convergent sequence. Moreover, it is straightforward to demonstrate that $\phi(\cdot)$ is a contraction function. Therefore, the power allocation scheme converges to a unique fixed point [12] $\mathbf{P}^\star = \left(\mathbf{I} - \mathbf{G}^{-1}\mathbf{\Gamma}\overline{\mathbf{G}}\right)^{-1}\mathbf{G}^{-1}\mathbf{\Gamma}\overline{\mathbf{N}}$. ∎

Remark 1. First, the criterion in Theorem 2 is a convergence condition per subcarrier. Secondly, we notice that when $l = 1$, our proposed sufficient condition (15) coincides with the convergence criterion given in [13] for the water-filling.

To be able to use (12) as a solution to the NPBG, the value of γ_s^l, $\forall s, l$ is required. From (12), we see there exists a one-to-one mapping from P_s^l to γ_s^l, $\forall s, l$. This one-to-one mapping is defined by $P_s^l = \gamma_s^l \widehat{I}_s^l$ where

$$\widehat{I}_s^l \triangleq \frac{\left(\overline{N}_s^l + \sum_{j\neq s}^S \sum_{l'=1}^L P_j^{l'} V_{|l-l'|}G_{j,s}^{l'}\right)}{G_{s,s}^l}$$

Define the following variable

$$C_s^l \triangleq \frac{G_{s,s}^l}{\sum_{j\neq s}^S \sum_{l'=1}^L V_{|l-l'|}G_{j,s}^{l'}}$$

Let $\mathbf{\Gamma}_s = (\gamma_s^1, \cdots, \gamma_s^L)^\top$ be the SINR vector for secondary user s. At the nth round, $\mathbf{\Gamma}_s^{(n)}$ can be found by solving the following convex optimization problem

$$\max_{\mathbf{\Gamma}_s \geq 0} \sum_{l=1}^{L} \gamma_s^l \hat{I}_s^{l,(n-1)} \left(1 + \sum_{k=1}^{\mathcal{K}} G_{s,k}^l \left(\sum_{l' \in \mathcal{I}_{s,k}^l} \mu_k^{l'} V_{|l-l'|} \right) \right)$$

$$\text{s.t. } \widehat{\mathcal{R}}_s \leq \sum_{l=1}^{L} B \log_2 \left(1 + \gamma_s^l \right)$$

$$\gamma_s^l \leq C_s^l - \delta_1, \ \forall l$$

(18)

The optimal solution of problem (18) is given by

$$\gamma_s^{l\star} = \left[\frac{\lambda_s \frac{B}{\ln 2}}{\hat{I}_s^{l,(n-1)} \left(1 + \left(\sum_{k=1}^{\mathcal{K}} G_{s,k}^l \left(\sum_{l' \in \mathcal{I}_{s,k}^l} \mu_k^{l'} V_{|l-l'|} \right) \right) \right)} - 1 \right]_0^{C_s^l - \delta_1}$$

(19)

Where λ_s is the dual associated with the rate constraint. In fact, it is important to notice that without the second constraint, problem (18) is equivalent to problem (6). The criterion in (15) is embedded into the optimization problem (18) as a constraint in order to assure the convergence of the algorithm to a fixed point. Notice that an infinitesimal positive constant δ_1 is deducted from the convergence criterion to relax the constraint. The game $\mathcal{G}$ given in (6) is solved by alternately solving problem (18) and substituting each γ_s^l, $\forall s, l$ in (12). The proposed new algorithm is summarized as

---

Algorithm 2. New distributed algorithm to solve (6)

1: **Input** A solution accuracy $\epsilon > 0$ and a feasible $\mathbf{P}^0$.
2: Initialize $\boldsymbol{\mu}^0$, set $n = 0$ and let $t(k, l) = 1$;
3: **repeat**
4:     $n = n + 1$;
5:     Obtain $\mathbf{\Gamma}_s^{(n)}$, $\forall s$ by solving problem (18);
6:     Calculate $P_s^{l,(n)}$, $\forall s, l$ by using (12).
7:     Assychronous update of the interferences prices μ_k^l, $\forall k, l$

$$\mu_k^{l,n} = \begin{cases} \left[\mu_k^{l,n-1} + \delta \left(I_k^{l,n} - I_k^{l,\max} \right) \right]^+, & \text{if } n+1 = n_k^{l,t(k,l)} \\ \mu_k^{l,n-1}, & \text{otherwise} \end{cases}$$

    If $n + 1 = n_k^{l,t(k,l)}$ then $t(k, l) = t(k, l) + 1$;
8: **until** $n > \mathcal{N}$ or $\left| \left(\mathcal{U}(\mathbf{P}_s^n; \boldsymbol{\mu}) - \mathcal{U}(\mathbf{P}_s^{n-1}; \boldsymbol{\mu}) \right) / \mathcal{U}(\mathbf{P}_s^{n-1}; \boldsymbol{\mu}) \right| \leq \epsilon$, $\forall s$
9: **Output** $\mathbf{P}^n$.

From the structure of the proposed Algorithm 2, we see that it always converges to a unique and fixed NE point of the game $\mathcal{G}$, the solution is given by $\mathbf{P}^\star = \left(\mathbf{I} - \mathbf{G}^{-1} \mathbf{\Gamma} \overline{\mathbf{G}} \right)^{-1} \mathbf{G}^{-1} \mathbf{\Gamma} \overline{\mathbf{N}}$. The SINR vector will also converge. Due to space constraint, we leave the detailed proof of the SINR convergence for future publication. In Sect. 5, numerical analysis of the convergence of the SINR together with the convergence of the interference prices will be provided.

To implement our proposed distributed Algorithms 1 and 2, the secondary MTs need to measure the noise-plus-interference on each subcarrier at each iteration. This value is then feeding back to the respective secondary BS. This operation is repeated until convergence or stopping criterion of both algorithms is reached. Clearly, in terms of signalling overhead, our proposed algorithms by using only local information need little signalling overhead.

5   Numerical Results

In this section, the performance of our proposed algorithm is evaluated via numerical results. All results are conducted using Monte Carlo simulation by averaging over 300 channel realizations. We consider an underlay CR network with 2 PUs and 5 SUs. The secondary BSs are randomly located at a distance varying from 0.1 km to 0.5 km away from the primary BSs. Each MT is uniformly located within a 0.5 km radius circle from its serving BS. There are $L = 32$ subcarriers having each a bandwidth of $B = 15$ KHz.

The path loss model for the channel is $LdB(d) = 128.1 + 37.6 \times \log_{10}(d)$, where d is the distance between a BS and a MT. The shadowing's standard deviation is 6 dB and $N_0 = -174$ dBm/Hz. The primary BS has a uniform power transmission $P_p^l = \frac{P_{\max}}{L}, \forall l$ with $P_{\max} = 33$ dBm. The interference threshold $I_k^{l,\max}$ is computed by assuming only 10 % of the PU k interference-free achievable rate degradation is permitted on subcarrier l, $\forall l$. Unless otherwise stated, $\widehat{\mathcal{R}}_s = 30$ Kbits/s. The maximum number of iterations is $\mathcal{N} = 40$ while $\delta_1 = 10^{-5}$ and $\epsilon = 10^{-4}$.

To evaluate the proposed Algorithms 1 and 2, we also compare with the perfect synchronization case denoted as PS. In this case, the interference weight is $V^{\mathrm{PS}} = \{1\}$ We will clearly see that asynchronism lead to a loss of performance. Our Algorithms are initialized by assuming uniform power on each subcarrier mainly $P_{\max}/L$.

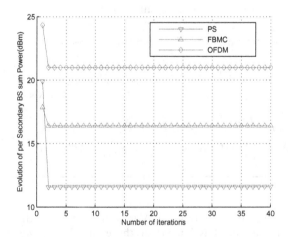

Fig. 1. Average sum secondary power versus rate constraint (Color figure online)

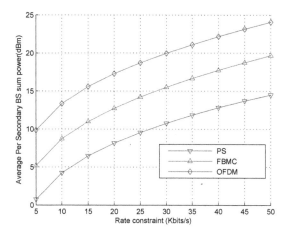

Fig. 2. Average sum secondary power versus rate constraint (Color figure online)

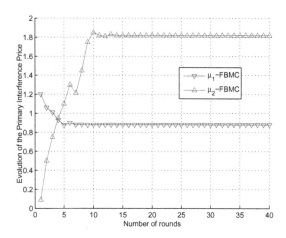

Fig. 3. Convergence behaviour of the interference price. (Color figure online)

Figures 1 and 2 portray the convergence properties and the performance of our proposed Algorithm 1 for different multi-carrier modulation scheme. Figure 1 depicts the evolution of the per secondary BS sum power. From Fig. 1, it can be clearly inferred that the proposed Algorithm 1 converges irrespective of the modulation method. It is important to observe the gap between the performance of PS and the one achieved by OFDM and FBMC. This is the consequence of inter-carrier interference induced by asynchronism and lack of cooperation.

Figure 2 depicts the performance of our proposed Algorithm 1 in terms of average sum power versus per BS power rate constraint. We can see that the sum power achieved by the proposed Algorithm 1 tends to increase as the rate constraint increases. From Fig. 2, we also observe a gain varying from 21.98 % to 22.70 % between the sum power with FBMC compared with the sum power achieved with OFDM.

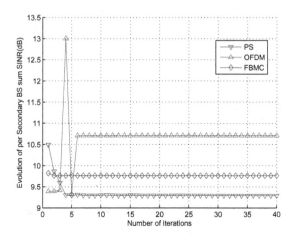

Fig. 4. Convergence behaviour of the SINR. (Color figure online)

Figure 3 demonstrates the convergence behaviour of the interference update of Algorithm 2. The convergence of interference prices is only given only for the case of FBMC. We observe that the interference prices of both primary BSs converge.

In Sect. 4, we stated that the SINR vector sequence $\{\mathbf{\Gamma}_s^{(n)}\}_{n=1}^\infty$, $\forall s$ converges. We prove our assertion by means of simulations. Indeed, Fig. 4 depicts the performance of our proposed Algorithm 2. It shows the convergence behaviour of the SINR vector. From Fig. 4, we clearly observe that the sequence of the SINR vector converges regardless of the multi-carrier modulation scheme.

6   Conclusion

In this work, we proposed two distributed algorithms to solve the problem of secondary sum power minimization for an underlay downlink asynchronous CR network with OFDM/FBMC. The problem was reformulated as a pricing-based non-cooperative game. We provide a sufficient convergence criterion to a NE point of the NPBG. Moreover, we provide a new algorithm that solves alternately power vector and SINR vector. The new algorithm always converges to a unique fixed NE point. Furthermore, we have through numerical results validated the efficiency of the proposed schemes. The simulation results highlighted the advantages of using FBMC over OFDM for asynchronous network.

References

1. Haykin, S.: Cognitive radio: brain-empowered wireless communications. IEEE J. Sel. Areas Commun. **23**(2), 201–220 (2005)
2. Zhao, Q.: Sadler, B.M: A survey of dynamic spectrum access. IEEE Signal Process. Mag. **24**(3), 79–89 (2007)
3. Tandra, R., Mishra, S.M.: Sahai, A: What is a Spectrum hole and what does it take to recognize one? Proc. IEEE **97**(5), 824–848 (2009)

4. Weiss, T.A., Jondral, F.K.: Spectrum pooling: an innovative strategy for the enhancement of spectrum efficiency. IEEE Commun. Mag. **42**(3), S8–S14 (2004)
5. Shaat, M., Bader, F.: Low complexity power loading scheme in cognitive radio networks: FBMC capability. In: IEEE 20th International Symposium on, Personal, Indoor and Mobile Radio Communications, pp. 2597–2602, September 2009
6. Zhang, H., Le Ruyet, D., Roviras, D., Sun, H.: Uplink capacity comparison of OFDM/FBMC based cognitive radio networks. In: IEEE International Conference on, Communications (ICC), pp. 1–5 (2010)
7. Shaat, M., Bader, F.: A two-step resource allocation algorithm in multicarrier based cognitive radio systems. In: IEEE, Wireless Communications and Networking Conference (WCNC), pp. 1–6, April 2010
8. Pang, J.-S., Scutari, G., Palomar, D.P., Facchinei, F.: Design of cognitive radio systems under temperature-interference constraints: a variational inequality approach. IEEE Trans. Signal Process. **58**(6), 3251–3271 (2010)
9. Hong, M., Garcia, A.: Equilibrium pricing of interference in cognitive radio networks. IEEE Trans. Signal Process. **59**(12), 6058–6072 (2011)
10. Medjahdi, Y., Terre, M., Le Ruyet, D., Roviras, D., Nossek, J.A., Baltar, L.: Intercell interference analysis for OFDM/FBMC systems. In: IEEE 10th Workshop on Signal Processing Advances in Wireless Communications (SPAWC), pp. 598–602 (2009)
11. Suraweera, H.A., Smith, P.J., Shafi, M.: Capacity limitsand performance analysis of cognitive radio with imperfect channel knowledge. IEEE Trans. Veh. Technol. **59**(4), 1811–1822 (2010)
12. Meyer, C.D.: Matrix Analysis and Applied Linear Algebra. Society for Industrial and Applied Mathematics (2000)
13. Pang, J.-S., Scutari, G., Facchinei, F., Wang, C.: Distributed power allocation with rate constraints in gaussian parallel interference channels. IEEE Transactions on Information Theory **54**(8), 3471–3489 (2008)

Invisible Hands Behind 3.5 GHz Spectrum Sharing

Liu Cui[1]([✉]) and Martin Weiss[2]

[1] Department of Computer Science, West Chester University, West Chester, USA
lcui@wcupa.edu
[2] School of Information Sciences, University of Pittsburgh, Pittsburgh, USA
mbw@pitt.edu

Abstract. There has been considerable discussion surrounding the barriers to spectrum sharing in the literature. Among those is the 'trust gap' that exists, according to the PCAST report. Trust is a complex human construct that significantly includes risk. In this paper, we examine the risks faced by the different user classes proposed by the FCC for sharing in the 3.5 GHz band. We argue that the "invisible hands" of spectrum sharing in this band is the balance between spectrum sharing gain and associated risks. We find that both gains and risks can be linked to the distance between incumbents' systems and Citizen Broadband Radio Services (CBRS)' systems. The risk portfolio is linked to spectrum rights that each tier has, since the rights they have determine risks and risk mitigation strategies. We further propose a model to calculate spectrum sharing utilities for different tiers. The optimized utility determines the distance between incumbents and CBRS systems.

Keywords: Spectrum sharing · 3.5 GHz

1 Introduction

On April 21, 2015, Federal Communications Commission (FCC) released the Report and Order (R&O) for the 3.5 GHz band [1]. In this document, FCC describes the creation of "Citizens Broadband Radio Service" (CBRS) for this band, which will be implemented by allowing non-federal users to share spectrum with incumbents. Incumbents include Department of Defense (DoD) Radar Systems in 3550–3650 MHz band, Fixed Satellite Services (FSS) and grandfathered terrestrial wireless operations in 3650–3700 MHz. The sharing arrangement between federal and non-federal usage will take place under a three-tiered sharing framework enabled by a Spectrum Access System (SAS). The highest tier, incumbent users, receives interference protection from other users. The CBRS itself contains two tiers: Priority Access Licenses (PALs) and General Authorized Access (GAA). PAL holders receive interference protection from GAA applications. GAA users receives no interference protection from other CBRS users.

© ICST Institute for Computer Sciences, Social Informatics and Telecommunications Engineering 2016
D. Noguet et al. (Eds.): CROWNCOM 2016, LNICST 172, pp. 412–423, 2016.
DOI: 10.1007/978-3-319-40352-6_34

This three tier spectrum sharing arrangement only provides a spectrum sharing framework. It does not promise a future with widely adopted spectrum sharing. Thus, promoting spectrum sharing in 3.5 GHz is an important question after FCC's rulemaking. Otherwise, 3.5 GHz may face the same situation as TVWS, which opened unlicensed access in 2008 but has not been widely utilized [2].

An underlying reason for the slow adoption of spectrum sharing is that there are risks associated with this approach. Incumbents are understandably concerned about potential interference, so they defend their rights to licensed frequencies. Potential CBRS users are uncertain about the regulation and spectrum environment (i.e., their usage rights as well as the collective action rights [3]) so both service providers and device manufacturers may be cautious about spectrum sharing in 3.5 GHz[1].

In order to promote spectrum sharing in 3.5 GHz, it is essential that risk management strategies for each usage tier in 3.5 GHz be developed. Since the rights, missions, applications, etc. vary with usage tier. Spectrum users of each tier face different risks than users of other tiers and thus need specific strategies to cope with these risks. To address this, we analyze spectrum sharing in 3.5 GHz with the specific purpose of clarifying the rights and quantifying risks for each tier as well as identifying appropriate risk management. Finally, we analyze the trade off between benefits associate spectrum sharing and costs associate with risk. This trade off is the invisible hand behind spectrum sharing that determines the size of exclusion and protection zone, and applications in CBRS.

2   Rights in 3.5 GHz

Risks and rights may be closely coupled with one another when rights are poorly defined. So we begin by introducing the users of 3.5 GHz and their de jure and de facto rights. We begin by a brief reprise of the rights framework discussed in [3,4]

In her work, the Nobel prize winning economist Elinor Ostrom attended to the rights and governance of so-called common pool resource systems. These are systems in which one user's consumption diminishes another user's consumption opportunities (i.e., the uses are "subtractible" or "rivalrous") and in which exclusion is difficult or costly. [3] argued that spectrum best fits this type of good under the current technology endowment. They also modified Ostrom's rights framework as shown in Table 1. This table describes the rights endowment of five different user types for five different types of rights. The first two rights (i.e., Reception and Transmission) can be considered *usage* rights because they relate to the operation of the how the system is used, while the remaining three are referred to as *collective action* rights and refer to the design of the rights system. In particular, management rights refer how the resource is used and managed,

---

[1] In his keynote address to IEEE DySPAN in 2015, Dr. Ranveer Chandra of Microsoft Research indicated that ASICs for TVWS had been designed, but that their manufacture was deferred until sufficient demand for TVWS devices could be demonstrated.

exclusion rights refer the determination of who has access to the resource, and alienation rights refer to the rights to transfer any of the other rights.

In spectrum sharing systems, interference is inevitable, which introduces risk. Thus, we briefly discuss interference from the perspective of this framework. In a prior paper, we examined the possibility of creating a right out of what might otherwise be considered an "externality[2]" [5]. Referring to the framework in Table 1, interference occurs when a signal from one authorized sender impinges on another authorized receiver.[3] In general, this occurs as an unexpected result of the management regime that was determined through the exercise of collective action rights, even if it is the result abnormal propagation circumstances.

Table 1. Distribution of rights by user type

	Full owner	Prop-rietor	Auth. claimant	Auth. sender	Auth. rcvr
Reception	X	X	X	X	X
Transmission	X	X	X	X	
Management	X	X	X		
Exclusion	X	X			
Alienation	X				

With this background in mind, we examine the stakeholders in the 3.5 GHz band.

2.1   Incumbents

The current spectrum allocation in 3.5 GHz band is already very complicated. Through the collective action processes of the CSMAC and FCC, the entire 3550–3700 MHz was divided into two sub-bands: 3550–3650 MHz and 3650–3700 MHz. A two tiered hierarchical usage rights scheme was developed for each sub-band (an exercise of the management rights) that are called primary usage and secondary usage. Further, both federal and non-federal usage was permitted under both primary and secondary usage. We begin with primary usage in the 3550–3650 MHz sub-band: The R&O granted primary usage rights to limited non-federal Fixed Satellite Service (FSS) if they existed prior to the effective date of the R&O. As well, federal fixed and mobile (except aeronautical mobile) radar systems were given primary usage rights. The R&O granted subordinate (i.e., secondary) usage rights to federal Radio Location Services (RLS) and certain low power non-federal applications.

---

[2] In economics, an externality is a cost or benefit that affects a party who did not choose to incur that cost or benefit (Source: http://cafehayek.com/wp-content/uploads/2011/11/Carl-Dahlman.pdf).

[3] Rogue transmissions are from unauthorized senders and so fall outside of this framework.

3650–3700 MHz is less crowded than 3550–3650 sub-band. Here, primary usage rights are granted to some federal RLS sites and ships for radar. Secondary usage rights are granted to wireless broadband services.

Superior (i.e. primary) usage rights imply the right to receive without interference from authorized suboardinate (secondary) users. Stated differently, this means that incumbents have the highest priority in spectrum access. But this classification does not address interference between different rights holders of the same class. That is, do some users have super-primary rights? Some DoD radar systems including ground-based, shipborne, and airborne platforms, which are used in conjunction with weapons control systems, may cause interference even to other primary users. In addition, incumbents with primary usage rights have the right to deploy both fixed and mobile transmitters that in line with their mission.

2.2   Citizens Broadband Radio Service

The authorization of CBRS rights is limited by inferiority to the primary users' rights. Further, the R&O envisions two further tiers: PALs, and GAA. In the framework of Table 1, PAL usage rights are superior to GAA usage rights.

Both classes of users have collective action rights through the FCC process. Of these, management rights are the most dynamic since the licensing procedures for PALs addresses exclusion rights. One important challenge ex ante in the domain of exercising management rights is the question of appropriately balancing the spectrum allocation between PALs and GAA. This must be accomplished ex ante since PALs will be auctioned (see below). PALs need to be sufficiently attractive to attract bids while simultaneously fostering a robust GAA ecosystem for innovation.

Considering the comments and suggestions from different stakeholders, the FCC concluded that a maximum of 70 MHz, 7 channels (10 MHz each), should be reserved for PALs in any given license area at any time. In addition, every PAL can aggregate up to four channels in any given license area to encourage competition. The remainder of the available frequencies is made available for GAA usage. The definitions are as follows:

PALs. Any prospective licensee who meets basic FCC qualifications is eligible for PALs. All applicants for PALs must demonstrate their qualification to hold an authorization and demonstrate how a grant of authorization would serve the public interest. Census tracts is the license size for PALs. PALs have three-year non-renewable license terms - with the ability to aggregate up to six years up-front. Finally, PALs will be assigned by competitive bidding.

GAA. FCC reasoned that a license-by-rule licensing framework would allow for rapid deployment of small cells by a wide range of users, including consumers, enterprises, and service providers, at low cost and with minimal barriers to entry.

GAA users may only use FCC certified Citizens Broadband Radio Services Devices (CBSDs) and must register with the SAS. Consistent with rules governing CBSDs, devices operating on a GAA basis must provide the SAS with all information requirement by the rules, including operator identification, device identification, and geo-location information, upon initial registration and as required by the SAS. Moreover, only fixed CBSDs are allowed at this stage.

We now translate this policy into the rights framework of Table 1. According to the R&O,

"To ensure that essential federal radiolocation systems operating in the band continue their operations without impact from the sharing arrangements, we are prohibiting CBSDs from causing harmful interference to, or claiming protection from, federal stations aboard vessels (shipborne radars) and at designated groundbased radar sites. In addition, authorized users of CBSDs must not claim protection from airborne radars and airborne radar receivers must not claim protection from CBSDs operating in the Citizens Broadband Radio Service."

The notion of "interference protection" in this exerpt from the R&O means that the transmission rights of CBSDs are subordinate to the receiving rights of the incumbent users. Furthermore, this management regime does not limit federal RLS transmission rights in order to preserve CBSDs reception rights. That is, they do not have the right of interference protection from incumbents. Transmission and reception rights are similarly organized within CBRS so that PALs' reception rights are superior to GAA's transmission rights (i.e., they have the right of interference protection from GAA users); like RLS, GAA users have no rights of interference protection from incumbent and other Citizens Broadband Radio Service users. This management and exclusion regime is encoded in a Spectrum Access System (SAS), which is a spatial database that is used to implement the rights regime described the previous paragraph in real time.

3   Invisible Hands Behind Spectrum Sharing in 3.5 GHz

In 1776, Adam Smith mentioned "the invisible hand" in his classic book "The Wealth of Nations" to describe the natural force that guides free market capitalism through competition for scare resources. He pointed out that no regulation of any type would be necessary to ensure the mutually beneficial exchange of good and services in a free market, since the "invisible hand" (the intent that each participant tried to maximize self-interest) will lead to the most mutually beneficial manner.

There is an "invisible hand" in spectrum sharing in 3.5 GHz as well. Although the exclusion and protection zone is currently determined by incumbents and regulators, a boundary can be determined voluntarily when considering both the benefit of spectrum sharing and cost associated with risks. For example, two extreme cases may exist: (1) when advanced technology can control all transmitters and monitor spectrum usage in real-time at low cost, the exclusion zone will be very small if it even exists; (2) when it is difficult to control transmitters and monitor spectrum usage in a timely manner and cost effective way, the exclusion zone will be very large.

3.1 Risks in 3.5 GHz

According to FCC's Technological Advisory Council, there are three categories to evaluate risks: (1) corporate metrics, (2) service metrics, (3) RF metrics. Coporate metrics include the ability to complete a mission, loss in revenue and profits. Service metrics measures the availability and quality of the service. RF metrics measure signal to interference and noise ratios, absolute interfering signal level, etc. [6].

In this paper, RF metric is evaluated by interference estimation described in Sect. 3.1, service metrics is evaluated by spectrum access opportunities described in Sect. 3.1, and the corporate metric is evaluated by profit that gained wireless service providers described in Sect. 3.2. Moreover, as a wireless service provider, they do not passively accept risks. Instead, users with different rights have various strategies to cope with risks. Section 3.1 describes risk mitigation strategies for each tier.

Interference Estimation. The primary RF risk for incumbents is that CBSDs may bring harmful interference to their systems, which negatively impacts their ability to carry out their mission. Similarly, CBSDs also have risks in receiving interference. Specifically, PALs may receive interference from incumbents, and GAAs may receive interference from both incumbents and PALs. We determine the potential interference that comes from a higher tier to the lower tier. For example, PALs estimate the interference that comes from incumbents, and GAAs estimate the interference that comes from both incumbents and PALs.

Although we don't have the information on transmission power level, we can follow the reverse engineer the NTIA exclusion zone calculations. First, we determine the maximum interference level that PALs and GAAs can accept. We assume that PALs can accept interference level (I^P) up to -20 dBm and GAAs can only accept interference level (I^G) up to -30 dBm. Then, we adopt the same path loss model L_p that NTIA used for calculating exclusion zones. Finally, the acceptable incumbent transmission power (X^P) is determined as $X^P = I^P + L_p$ and $X^G = I^G + L_p$ for PALs and GAAs respectively.

$$L_p = 69.55 + 26.16 \log f - 13.82 \log h_b - a(h_m) \\ + (44.0 - 6.55 \log h_b) \log d \tag{1}$$

$$a(h_m) = \begin{cases} 3.2(\log(11.75 h_m))^2 - 4.97, & \text{L City} \\ (1.1 \log(f - 0.7)h_m - 1.56 \log f + 0.8, & \text{M/S City} \end{cases} \tag{2}$$

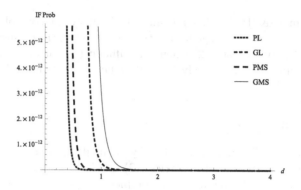

Fig. 1. Probability of receiving interference from incumbents $\sigma = 10$

Therefore, the probability of interference is determined as:

$$P(P_t > X^i) = 1 - \frac{1}{2}[1 + erf(\frac{x - \mu}{\sigma\sqrt{2\pi}})], \quad i = P, G \tag{3}$$

when we assume the transmission power level (P_t) follows a normal distribution, $P_t = \frac{1}{\sigma\sqrt{2\pi}}e^{-\frac{(x - \mu)^2}{2\sigma^2}}$, with mean μ and deviation σ.

Figures 1 and 2 show the probability of interference for situations with respect to the distance between transmitters and receivers, CBSDs' interference threshold, and path loss model in large and medium/small city. We assume that $\mu = 82$ dBm, $\sigma = 10$ and $\sigma = 100$ in Figs. 1 and 2 respectively. It is clear that when the distance (d) between transmitters and receiver stays the same, PALs in a large city (PL) have the lowest probability of interference because PALs have less sensitivity to interference than GAAs and the path loss in a large city is more severe than a medium/small city. When the path loss decreases (in the medium/small city case), the probability of interference increases in PMS. A similar pattern can be recognized for GAAs. Further, GAAs have higher probability of interference in the same geographic region than PALs, since GAA's interference threshold is higher than PALs. Moreover, the probability of interference decreases with increases of d, due to the path loss factor. Compare Figs. 1 and 2, when σ increase from 10 to 100, the transmission power is less centralized to the mean. Therefore, the probability of interference increases.

Spectrum Access Opportunities Estimation. According to the R&O, all frequency bands that are not occupied by incumbents and PALs can be utilized by GAAs. In the ideal case, GAAs have at least 80 MHz in any geographic areas outside the exclusion zones (recall that the total shareable frequency is 150 MHz, 70 MHz is allocated to PALs, so the remaining 80 MHz can be used for GAAs.) Let us calculate the spectrum access opportunities for GAAs on those 80 MHz. It is assumed that GAAs can perfectly detect each other and there is no interference from PALs and incumbents.

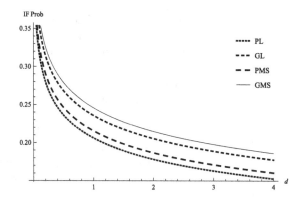

Fig. 2. Probability of receiving interference from incumbents $\sigma = 100$

Since GAAs are controlled by a SAS on a FCFS base, we adopt a queueing model to quantify the probability of the spectrum access opportunity. It is assumed that GAAs arrival process follows a Poisson distribution with mean λ, and the departure process follows exponential distribution with mean μ. It is further assumed that the capacity of the system is 8 channels with 10 MHz bandwidth. Consequently, we adopt $M/M/C$ queue with $C = 8$.

The most important metric is the probability of waiting (P_Q). Thus, the probability of spectrum is available is $(1\text{-}P_Q)$.

$$P_Q = \sum_{i=C}^{\infty} P_i = \frac{(Ca)^C}{C!} \frac{1}{1-a} P_0 \tag{4}$$

where, C is the total number of available channel, $a = \dfrac{\lambda}{\mu}$ and

$$P_0 = \left(\sum_{i=0}^{C-1} \frac{(Ca)^i}{i!} + \frac{(Ca)^C}{C!(1-a)} \right)^{-1} \tag{5}$$

In the future, we will tailor this probability according to different applications' requirement, such as elastic vs inelastic services by using metrics like mean waiting time $(\overline{W})$ and probability that waiting time is greater than threshold t $(P(W > t))$.

Coping with Risks. Under a subordinate rights structure, spectrum users with different priority have different risk metrics and risk measurement/protection methods. Spectrum users with higher priority have the right to protect their own services. However, spectrum users with lower priority can only estimate risks and make informed decision accordingly.

Protection. There are many ways to protect the system from interference: geographic separation, frequency separation, time separation are three dominate approaches. Since CBRS shares spectrum with current users, incumbents focus on geographic and time separation. Here, risk protection starts with the incumbents' own requirements, such as Signal to Noise and Interference Ratio (SNIR). Then, incumbents determine exclusion (and protection) zone where no one else can operate based on PALs and GAAs allowable transmission power limits as well as an appropriate path loss model. Even outside the exclusion zones, CBSDs can only transmit when incumbents are absent.

PALs have interference protection rights over GAAs and focus on frequency and time separation. This means that PALs have dedicated frequency bands where no other CBSDs can operate. The time seperation is implemented by the SAS through explicit grants of GAA's transmission rights, which, in turn, is based on the PAL's traffic.

Estimation. For CBRS, although they do not have the right of interference protection from incumbents, they can actively cope with risks by estimating potential risks before hand and then making informed decisions and using flexible management. CBRSs may have different goals; for example, hospitals and public safety provide life critical services that can hardly be measured by money while commercial services can be evaluated by revenue and customer satisfaction.

Accordingly, we will estimate risks for Citizens Broadband Radio Services from the QoS perspective and then link it to potential revenue. Two risk metrics are applied in analyzing QoS.

- Probability of interference. In this metric, spectrum users calculate the probability that they may be interfered by users with higher priority.
- Probability of getting spectrum without waiting. This risk metric only applies to GAAs, since they do not have spectrum reservation.

When we link QoS to potential revenue, we assume that the maximum revenue that CBSDs can earn occurs when there is no interference and spectrum is available all the time. We assume that the potential revenue that CBSDs can earn linearly decreases with probability of interference and the probability of waiting for available spectrum.

3.2 Benefit for CBRS

The invisible hands behind the spectrum sharing is to maximize the overall benefit getting from spectrum sharing in 3.5 GHz. We assume that the benefit for spectrum sharing with PALs (U^P) is calculated as the maximum benefit per user (u^p) times number of users/devices covered by PALs (N_c^p). Moreover, this maximum value can be achieve when there is no interference from incumbents $(1 - P^I(i))$ as shown in Eq. 6.

$$U^P = u^p \times N_c^p \times (1 - P^I(i));$$ (6)

GAA are more poorly situated than PALs, since they accept interference from both incumbents and PALs. Also, they cannot reserve spectrum, since the

SAS allocates frequency bands to GAA on a FCFS basis. In other words, GAAs can provide services when spectrum is available. As a result, there are three risk metrics for GAAs: the probability of getting interference from incumbents $P^I(i)$, the probability of getting interference from PALs $P^G(i)$, and the probability that a frequency band is not available right away P_Q. The benefit of spectrum sharing in GAA (U^G) equals the maximum benefit per user/device (u^g) times number of users/devices covered by GAA (N_c^g), and this benefit can only be achieved when spectrum is available without interference $(1 - P^I(i))(1 - P^P(i))(1 - P_Q)$ as shown in Eq. 7.

$$U^G = u^g \times N_c^g \times (1 - P^I(i)) \times (1 - P^P(i)) \times (1 - P_Q) \tag{7}$$

According to NTIA exclusion zone calculation, the population density decreases with the increase of exclusion zone radius (d). Therefore the N_c^p and N_c^g are a function of exclusion zone radius.

$$N_c^i = \frac{1}{d} \times \pi r_i^2, \quad i = p, g \tag{8}$$

In order to calculate benefit of spectrum sharing, we need to start by estimating risks in spectrum sharing (illustrated in Sect. 3.1).

4 Numerical Results

In this section, we discuss some representative figures that demonstrate the invisible hands, i.e., the tradeoff between benefit and cost, that determine spectrum sharing. In Fig. 3, the assumption is that user/device density decreases with the exclusion zone radius (d). Although the interference also decrease with the increase of exclusion zone radius, density in this case is the dominate factor in determining the utility for PALs. Therefore, the smaller the d is, the higher the utility.

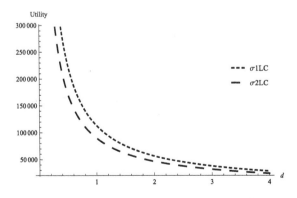

Fig. 3. PALs utility with changing user/device density

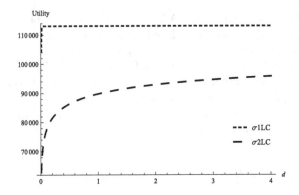

Fig. 4. PALs utility with constant user/device density

Figure 4 assumes the user/device density does not change according to the exclusion zone radius. Device to Device (D2D) communication networks can be one example of this scenario. Clearly, users have higher utility when they are away from incumbents, since the probability of interference from incumbents decreases. Further, utilities when $\sigma 1$ is applied is higher than utilities under $\sigma 2$, since the interference level in $\sigma 1$ is higher than $\sigma 2$. GAAs utility shows the same trend with lower utility value.

5   Conclusion

There is currently a great deal of interest in spectrum sharing in the 3.5 GHz band. Success in this band is likely to encourage spectrum sharing in other bands, so an important challenge is learn how to incentivize spectrum sharing under this three tier spectrum sharing prototype while protecting spectrum users' systems and investment. This is complicated by the presence of risk because the cost-benefit calculus becomes more complex. In this paper, we sought to identify and quantify the risks faced by the different classes of users in 3.5 GHz band as a first step to support users' decisionmaking. We did not address the impact of risk mitigation strategies, which change with the rights that different spectrum users hold.

For Adam Smith, the "invisble hand" guided markets in equilibrium. In the case of spectrum sharing, the "invisible hand" cautions users from investing because of the risks they face. These risks loom largest in the absence of experience, so quantifying and modelling the risks is a first step toward developing risk mitigation strategies that should cause the "invisible hand" to facilitate transactions rather than stop them.

Acknowledgement. This work was supported by the US National Science Foundation under grant 1247546.

References

1. Commission, F.C., et al.: Amendment to the commissions rules with regard to commercial operations in the band 3550–3650 MHz. FCC Gaussian noise Docket, no. 12–354 (2015)
2. Robyn, D., Jackson, C., Bazelon, C.: Unlicensed operations in the lower spectrum bands: Why is no one using the tv white space and what does that mean for the FCC's order on the 600 MHz guard bands? (2015)
3. Weiss, M.B., Lehr, W.H., Acker, A., Gomez, M.M.: Socio-technical considerations for spectrum access system (sas) design (2015)
4. Cui, L., Gomez, M.M., Weiss, M.B.: Dimensions of cooperative spectrum sharing: Rights and enforcement. In: 2014 IEEE International Symposium on Dynamic Spectrum Access Networks (DYSPAN), pp. 416–426. IEEE (2014)
5. Weiss, M.B., Cui, L.: Spectrum trading with interference rights. In: 2012 7th International ICST Conference on Cognitive Radio Oriented Wireless Networks and Communications (CROWNCOM), pp. 135–140. IEEE (2012)
6. FCC: A case study of risk-informed interference assessment: Metsat/lte coexistence in 1695–1710 MHz (2015)

Aggregate Interference in Random CSMA/CA Networks

June Hwang[1], Jinho Choi[2], Riku Jäntti[3], and Seong-Lyun Kim[2(✉)]

[1] Samsung Electronics, Mobile Communications Business, Suwon, Korea
june77.hwang@samsung.com
[2] Department of Electrical and Electronic Engineering, Yonsei University,
50 Yonsei-Ro, Seodaemun-Gu, Seoul, Korea
jhchoi@ramo.yonsei.ac.kr, slkim@yonsei.ac.kr
[3] Department of Communications and Networking, School of Electrical Engineering,
Aalto University, 00076 Aalto, Finland
riku.jantti@aalto.fi

Abstract. In this paper, we investigate the cumulative distribution function (CDF) of the aggregate interference in CSMA/CA networks measured at an arbitrary time and position. We assume that nodes are deployed in an infinite two-dimensional plane by the Poisson point process (PPP). To find the effective active node density we analyze the distributed coordinate function (DCF) dynamics in a common sensing area and obtain the steady-state power distribution. The results of a massive simulation using Network Simulator-2 (NS-2) show a high correlation with the derived CDF.

Keywords: Aggregate interference · CSMA/CA · DCF · Poisson point proce · NS-2

1 Introduction

1.1 Motivations

Due to the inherent scarcity of frequency spectrum and increasing wireless traffic demands, frequency reuse has become an essential key technological issue associated with contemporary wireless communication systems. Frequency reuse intrinsically causes interference between wireless links using the same frequency. Accordingly, the state of the aggregate interference at an arbitrary position in the random node topology has become of great importance. Currently, unlicensed spectrum is considered to be a supplementary spectrum of Long Term Evolution (LTE) in the license-assisted access (LAA) system in the 3rd Generation Partnership Project (3GPP). For the system, the main incumbent networks are wireless local area networks (WLANs) which are based on IEEE 802.11.

This work was supported by the National Research Foundation of Korea (NRF) grant funded by the Korea government (MSIP) (NRF-2014R1A2A1A11053234).

© ICST Institute for Computer Sciences, Social Informatics and Telecommunications Engineering 2016
D. Noguet et al. (Eds.): CROWNCOM 2016, LNICST 172, pp. 424–436, 2016.
DOI: 10.1007/978-3-319-40352-6_35

Because of the widespread deployment of WLAN systems, initiation of LTE operation in the unlicensed spectrum must be done carefully. In this context, it is necessary to understand the characteristics of the aggregate interference of IEEE 802.11 networks. Furthermore, the interference can be controlled using the relationships discovered among the protocol parameters. Such control is useful in optimizing the operation of densely deployed WLAN and to protect incumbent systems against LTE interference for cases in which the system shares spectrum [1].

In this paper, we are interested in the aggregate interference of random carrier sense multiple access/collision avoidance (CSMA/CA) networks that are based on the IEEE 802.11 distributed coordination function (DCF). We test and analyze the interference at the protocol level, which reflects the contention and signaling processes of DCF. Consequently, the goal is to obtain the statistical inference of the aggregate interference and to verify the results via simulations. Our analysis tool is the stochastic geometry [2].

Because of the complexity, most of the previous work has focused on ALOHA-like systems in which the aggregate interference can be analyzed by assuming that the transmitting nodes have independent locations and behaviors [3]. In a network of CSMA/CA nodes, every communication entity first senses the ongoing transmission in the channel and then determines when to start transmitting. Consequently, transmission by a node will impact on its neighbors' channel access. To reflect the effects, the authors adopt the Matérn approximation on active node density to derive the optimal carrier sense threshold in a CSMA/CA network in [4,5]. However, they directly adopt the approximation without verifying the validity of the approximation for a realistic network situation.

To utilize the characteristics of aggregate interference in practical cases, we would like to find the aggregate interference distribution in practical CSMA/CA networks, not an ideal CSMA network. Modeling a practical network requires a hybrid method that considers the both dependent and independent point processes together. To capture the collision and idle time effects caused by imperfect contention of the real-life CSMA/CA operation, we propose modeling the system using the Poisson Point Process (PPP) with the *effective active node density*.

1.2   Summary of Contributions and Organization of this Paper

Our idea is to use the PPP for calculating the aggregate interference of the CSMA/CA network, but with a new density λ', called *effective active node density* reflecting all the CSMA parameters. Section 3 is devoted to describing how we obtain λ', and its verification by massive NS-2 simulations is contained in Sect. 4. For the readers who are more interested in our results, please directly jump to Sect. 4. Our paper has the following notable results:

- **The effective node density reflecting CSMA/CA MAC layer operation is derived in Sect. 3.**

- The aggregate interference using our derived effective node density is verified using the NS-2 and MATLAB simulations in Sect. 4. The distribution of the CSMA/CA aggregate interference is neither normal nor log-normal distribution.

2   Point Process for Modeling Random CSMA/CA Networks

In this section, we focus on determining which type of point process is suitable for modeling CSMA/CA networks. In the point process, a mark can be assigned to each point independently, which is useful for modeling node-oriented properties. In particular, the case in which the number of nodes in a network is Poisson-distributed and their positions at a given time instant are independent of each other, is adequately explained by means of the PPP. The method to derive the aggregate power emitted from points at an arbitrary position under the independent marked PPP was previously studied as a *shot noise field*.

2.1   Inappropriateness of PPP and Dependent Point Process

The PPP approach as it currently exists may be insufficient to model the CSMA/CA. The reason is that it does not reflect the carrier sensing philosophy. In the carrier sensing operation, a sensing node always senses the shared medium and it delays its transmission once it senses that the medium is busy. The result is that active nodes are affected by each other, which means that the process is not independent.

Let us now consider the dependent point process as a possible alternative. Here, the dependent point process means that some initially deployed points are discarded or selected by the metric relative to the other points' marks or locations. There are two dependent point processes that are most closely related to the modeling of CSMA/CA networks [3]: the Matérn hardcore (MHC) process and the simple sequential inhibition (SSI) point process. In [6], the authors compared the aggregate power distributions of PPP, MHC, and SSI with simulations, and concluded that SSI is most appropriate for modeling CSMA networks. However, their result is not fully acceptable because they considered neither the details of practical MAC layer parameters nor the channel characteristics. A related paper [7] tries to solve the above issue by considering the backoff timer. However, there is still no consideration of the collision case, and the problem of underestimating the node intensity has not been fully solved.

2.2   Revisit of PPP with a New Density

In real situations, the concurrent transmission in an exclusion area occurs with some probability, not with deterministic patterns. This stochastic characteristic of real networks can be appropriately modeled using the independent point process. Therefore, we believe that a possible way to model a CSMA/CA network is again to use the independent PPP, but with a new effective active node density (λ') reflecting MAC layer operations.

For our analysis, we consider an infinite plane where the transmitting nodes are deployed randomly at positions specified by a Poisson distribution with intensity λ. Each node transmits with constant power p. The radio channel attenuates with the pass-loss exponent $\alpha = 4$ and Rayleigh fading.

3   Effective Active Node Density

In this section, we obtain the effective active node density, which is defined as the average number of transmitting nodes per unit area. To derive the density, we first introduce the concept of a *mutual sensing area* (Sect. 3.1). Then we derive the probability of a number of active nodes (Sect. 3.2) and the power distribution of the channel in that area (Sect. 3.3). Lastly, we obtain the effective active node density (Sect. 3.4).

3.1   Effective Carrier Sensing Range and Mutual Sensing Area

Let us introduce a CS range R such that a sensing node can sense any ongoing transmission in this range. Then within a disk of radius $\frac{R}{2}$, every node senses each other. We set this disk as the mutual sensing area.

CS is based on the threshold γ, i.e., if the sensed power level at a sensing node is greater (or lower) than γ, a sensing node regards the channel as busy (or idle). The definition of R can be also interpreted as a minimum energy detection boundary. We assume that there is a dominant interferer near the sensing node because the energy detection is most affected by the strongest interferer at large CS threshold. Then, the CS probability versus the distance to this interferer is calculated as follows:

$$\mathbb{P}[\text{Channel is busy}] = \mathbb{P}[\frac{p_i}{r^4} + \nu \geq \gamma], \tag{1}$$

where p_i is a random variable (RV) representing the product of the fading effect and the constant transmission power from a typical node i, r is the distance between the sensing node and the interferer, and ν is the receiver noise power. Considering Rayleigh fading, p_i follows $\text{Exp}(1/p)$ with a constant transmission power p.

Consequently, with the CS range R, we convert the stochastic CS to a deterministic one. First, the average residual sensing area is calculated by integrating the parts of the circumference, of which the radius and the center are r and the sensing node, respectively. The CS probability of a point on this circumference is from (1):

$$\int_0^\infty 2\pi r \cdot \mathbb{P}[\frac{p_i}{r^4} + \nu \geq \gamma] \mathrm{d}r = \int_0^\infty 2\pi r e^{-\frac{1}{p}(\gamma-\nu)r^4} \mathrm{d}r = \frac{\pi^{3/2}}{2\sqrt{\frac{\gamma-\nu}{p}}} \tag{2}$$

Assuming that the deterministic CS region should have the same average residual sensing area (sensing resolution) as the stochastic CS, we get the CS distance R as follows:

$$R = \frac{1}{\sqrt{2}} \left(\frac{\pi p}{\gamma - \nu}\right)^{1/4} \tag{3}$$

By using this deterministic CS distance, which we will call the *effective carrier sensing range*, the interference is regarded as Boolean at a given distance rather than stochastic.

Let us consider an infinite plane in which the nodes are randomly deployed. Assume that there is an arbitrary disk having a radius $R/2$ in the plane (mutual sensing area), where every node in this area can sense other nodes' transmissions according to the definition of the CS range R. By using the PPP, the number of deployed nodes in the mutual sensing area follows a Poisson distribution with the parameter $\lambda\pi(\frac{R}{2})^2$ as follows (Fig. 1):

$$\mathbb{P}[N = n] = \frac{\{\lambda\pi(\frac{R}{2})^2\}^n}{n!} \exp\left(-\lambda\pi\left(\frac{R}{2}\right)^2\right) \tag{4}$$

Once we know $\mathbb{P}[N = n]$, we derive the probability of the number of active nodes, N_a, in the mutual sensing area $\mathbb{P}[N_a = a|N = n]$ and the power distribution at an arbitrary time instant in the mutual sensing area.

3.2 Number of Active Nodes, N_a in a Mutual Sensing Area

For the explanation of this section, let us define the following probability first:

Definition 1. p_{on} is the probability that there are ongoing transmissions in a given mutual sensing area at a certain time.

Consider a given mutual sensing area H. Let the *residual sensing area* be defined as the sensing node's CS area, excluding H. The active node is defined here as the sensing node that has no ongoing transmissions in its residual sensing area. The CS results for each sensing node in H are random. Therefore, N_a is an RV that varies within $[0, N]$.

For a sensing node in H to be active, the CS result sensed from its residual sensing area must be idle, and that sensed from H must also be idle. We can find the distribution of N_a in H as in (5) when the probability p_{on} of

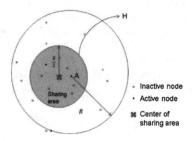

Fig. 1. If we pick up a certain mutual sensing area, every transmitting node in that area has its own residual sensing area. If there is no ongoing transmission in its residual sensing area, that node is the active node.

Definition 1 is given:

$$\mathbb{P}[N_a = a|N = n] = \sum_{\eta=0}^{8} P_{n,a,\eta} p_\eta, \quad a = 0,\ldots,n, \tag{5}$$

where

$$P_{n,a,\eta} = \binom{n}{a} \left(\frac{\eta}{8}\right)^a \left(1 - \frac{\eta}{8}\right)^{n-a}, \quad p_\eta = \sum_{D=0}^{8} O_\eta^D p_{on}^D (1 - p_{on})^{8-D}. \tag{6}$$

D and η are the discrete variables within $[0, 8]$. The value of O_η^D corresponding to D and η is given in Table 1. A description of the detailed derivation of (5), (6) and Table 1 is necessary but due to the space limitation, we only report the results.

Table 1. O_η^D

η	D									Sum(η)
	0	1	2	3	4	5	6	7	8	
0	0	0	0	8	38	48	28	8	1	131
1	0	0	0	24	24	8	0	0	0	56
2	0	0	12	16	8	0	0	0	0	36
3	0	0	8	8	0	0	0	0	0	16
4	0	0	8	0	0	0	0	0	0	8
5	0	8	0	0	0	0	0	0	0	8
6	0	0	0	0	0	0	0	0	0	0
7	0	0	0	0	0	0	0	0	0	0
8	1	0	0	0	0	0	0	0	0	1
Total	1	8	28	56	60	56	28	8	1	256

3.3 Steady-State Power Distribution in a Mutual Sensing Area

In this subsection, we derive the steady state power distribution in a mutual sensing area based on the distribution of number of active nodes in the area. We use the power distribution to obtain p_{on} in the next subsection. As shown in [8] and subsequent researches, the backoff stage of each node is random at a certain time, which can be elaborated through a two-dimensional Markov chain. We have two main quantities for addressing this: p_c is the collision probability for the transmission of each node, and τ is the transmission probability of a node at a randomly chosen time slot.

By following the notations of [5], we have the BEB dynamics with a maximum backoff stage m, a maximum retry limit K ($\geq m+2$), and an initial window size W_0. The probability τ that a node transmits in a randomly chosen time slot is:

$$\tau = \left\{ \frac{(1-p_c)W_0(1-(2p_c)^m)}{2(1-p_c^K)(1-2p_c)} + \frac{2^m W_0(p_c^m - p_c^K)}{2(1-p_c^K)} - \frac{1}{2} \right\}^{-1}. \tag{7}$$

Again, p_c is obtained as $p_c = 1 - (1-\tau)^{N_a-1}$, where N_a is the number of active nodes. We can solve the system dynamics by solving independent Equations p_c and (7), and the existence of this solution is guaranteed by the fixed point theorem [8].

Then, the probability that i nodes transmit simultaneously at an arbitrary time slot, given that N_a transmitting nodes are deployed in a mutual sensing area, is computed as follows:

$$p_a(m) = \mathbb{P}[i = m | N_a = a] = \binom{a}{m} \tau^m (1-\tau)^{a-m}, \quad m = 0, \ldots, a. \tag{8}$$

Each transmitting node's operation in a mutual sensing area is synchronized, since the medium is sensed perfectly and every node uses the inner clock. Idle time is segmented into multiple slot times (σ). All events (idle, success and collision) can be distinguished by their own time lengths. At an arbitrary time slot called the *virtual time slot*, the medium in the mutual sensing area is in one of three events, and the virtual time slot has the random duration T_v. We assume that the payload size is PAY for all nodes. In the basic mode,

$$T_v = \begin{cases} \sigma, & \text{for idle,} \\ T_s^{BAS} (= PHY + \lceil \frac{(MAC+PAY)}{R_s} \rceil T_s + SIFS + ACK + DIFS), & \text{for success,} \\ T_c^{BAS} (= PHY + \lceil \frac{(MAC+PAY)}{R_s} \rceil T_s + DIFS), & \text{for collision,} \end{cases}$$

where PHY, $SIFS$, ACK, and $DIFS$ are the durations of the PHY header, SIFS (short interframe space) time, ACK packet, and DIFS (DCF interframe space) time, respectively. MAC, R_s and T_s are the MAC header size, symbol rate and symbol duration, respectively. Besides, superscript BAS denotes the basic mode and RTS denotes the RTS-CTS mode in this subsection.

T_v has the PMF induced from (8) such as $p_a(0)$, $p_a(1)$, and $1 - p_a(0) - p_a(1)$, which are for idle, successful transmission, and collision, respectively. We derive the mean virtual time slot, $\mathbb{E}[T_v]$, using this PMF for each mode.

$$\mathbb{E}[T_v^{BAS}] = \sigma p_a(0) + T_s^{BAS} p_a(1) + T_c^{BAS}(1 - p_a(0) - p_a(1)),$$
$$\mathbb{E}[T_v^{RTS}] = \sigma p_a(0) + T_s^{RTS} p_a(1) + T_c^{RTS}(1 - p_a(0) - p_a(1)).$$

The distribution of the number of concurrent transmissions (which is also the power distribution) is based on this PMF. In each virtual time slot, the number of concurrent transmissions varies from 0 to N_a because it is possible that nobody transmits in a certain virtual time slot, even if there are some active nodes in the mutual sensing area. In the basic mode, nobody transmits during σ. During $SIFS$ and $DIFS$ in both the successful and collision slot times, nobody transmits. During the packet transmission time ($PHY + \lceil \frac{(MAC+PAY)}{R_s} \rceil T_s$ and ACK) in a successful slot, one node transmits, while multiple nodes transmit during $PHY + \lceil \frac{(MAC+PAY)}{R_s} \rceil T_s$ in a collision slot. In the RTS-CTS mode, the power density is changed in the same manner. The actual power distribution, $\mathbb{P}[j\text{nodes transmit}|N_a = a]$ is obtained as follows:

$$B_a^{BAS}(j) = \frac{1}{\mathbb{E}[T_v^{BAS}]} \cdot \begin{cases} \sigma p_a(0) + (SIFS + DIFS)p_a(1) + DIFS(1 - p_a(0) - p_a(1)), & j = 0 \\ (PHY + \lceil \frac{(MAC+PAY)}{R_s} \rceil T_s + ACK)p_a(1), & j = 1 \\ (PHY + \lceil \frac{(MAC+PAY)}{R_s} \rceil T_s)p_a(j), & 2 \leq j \leq a. \end{cases} \quad (9)$$

With (9), the probability of a busy channel in a mutual sensing area p_{on} is $\sum_{j=1}^{a} B_a^{BAS/RTS}(j) = 1 - B_a^{BAS/RTS}(0)$.

3.4 Effective Active Node Density

So far, we have introduced a mutual sensing area using the spatial boundary of the Boolean CS operation, and have derived the distribution of the number of nodes, which is $\mathbb{P}[N = n]$ as in Eq. (4) in Sect. 3.1. We derived the probability of the number of active nodes for a given number of users in the mutual sensing area, $\mathbb{P}[N_a = a|N = n]$ as in Equation (5) in Sect. 3.2. Based on the transmission probability τ, the power distributions in the mutual sensing area can be calculated as in Eq. (9) in Sect. 3.3. With the results, we can formulate Eq. (10) from the definition of p_{on}. We can get the value of p_{on} by finding the intersection of the right and left hand sides of (10) numerically, which we will call p_{on}^*. In this section, $B_a(j)$ can be either of $B_a^{BAS}(j)$ and $B_a^{RTS}(j)$ according to the mode that we consider.

$$p_{on} = \sum_{n,a} \sum_{j=1}^{a} B_a(j) = \sum_{n=0}^{\infty} \mathbb{P}[N = n] \sum_{a=0}^{n} \mathbb{P}[N_a = a|N = n] \sum_{j=1}^{a} B_a(j)$$

$$= \sum_{n=1}^{\infty} \frac{\{\lambda \pi (\frac{R}{2})^2\}^n}{n!} e^{-\lambda \pi (\frac{R}{2})^2} \sum_{a=1}^{n} \sum_{\eta} P_{n,a,\eta} \{ \sum_{D} O_\eta^D p_{on}^D (1 - p_{on})^{8-D} \} \sum_{j=1}^{a} B_a(j) \quad (10)$$

If we obtain p_{on}^*, the distribution of the number of transmitting nodes in the mutual sensing area can be derived as in (11), where the number of actual transmitting nodes (active and non-frozen) in the area is denoted by Z.

$$\mathbb{P}[Z = z] = \sum_{n=z}^{\infty} \left(\frac{\{\lambda \pi (\frac{R}{2})^2\}^n}{n!} e^{-\lambda \pi (\frac{R}{2})^2} \sum_{a=z}^{n} \mathbb{P}[N_a = a|N = n] \right) B_a(z), \text{ for } z \in \{0, 1, \dots\} \quad (11)$$

Then, the expected number of transmitting nodes is derived from this result:

$$\mathbb{E}[Z] = \sum_{z=0}^{\infty} z \cdot \mathbb{P}[Z = z] \quad (12)$$

The effective active node density is defined as the average number of transmitting nodes per unit area. Thus, we finally obtain the effective active node density as follows:

$$\lambda' = \frac{\mathbb{E}[Z]}{\pi (\frac{R}{2})^2}. \quad (13)$$

This is used in the cumulative distribution function (CDF) and the probability density function (PDF) of the aggregate interference. We plot the resulting CDF and PDF and compare these with the simulation results in Sect. 4.

4   Verification of the Analysis

In this section, we plot the effective node density (λ') of (13), and the CDF and PDF of the aggregate interference using λ'. Next, we compare the derived results with those obtained in the NS-2 and Matlab simulations. In the NS-2 simulation, the MAC/PHY parameters and channel model are given so that the CS radius R is determined to be 50, 70 and 100 (m).

4.1   Simulation Setup

MATLAB Simulation for MHC and SSI. We deployed the points using the MHC and SSI processes, as explained in [6], with MATLAB. The exclusion radius r_e is given as 70m. For a given number of nodes, the aggregated power was measured at O, which is the center of ball B with radius $R_M, R_S (= 282m)$ for MHC or SSI, respectively. The number of deployed nodes was generated using a Poisson distribution with the parameter $\lambda|B|$, where $\lambda(= 1, 2, 3, 4, 5 * 10^{-4})$ is the initial node density and $|B|$ is the area of ball B. We repeated this procedure for more than 100,000 iterations.

NS-2 Simulation for PPP. To verify the analysis results, we conducted simulations using NS-2 [9], which includes wireless PHY and MAC layer patches for the realistic IEEE 802.11 DCF standard [10]. This enabled us to realistically simulate the PHY and MAC stacks of the IEEE 802.11 DCF. The simulation parameters, which are the defaults for IEEE 802.11a PHY and MAC and are from the previous research.

The simulation conducted in this paper is full-scaled, which takes a long time to collect meaningful results for two reasons. First, each simulation per geometry scenario takes a long time. NS-2 traces all of the packet-level transactions with the received power recorded at every receiver. In the post-processing stage, the calculation of the received power from all of the ongoing transmissions at a measuring node takes computation time. Moreover, the simulation time itself (not the computation time) has to be long enough to reflect the steady-state behavior, which theoretically requires infinite investigation time. Second, to get a sound statistical inference of PPP, we repeat the per-geometry simulation 50 times since all of the resulting PDFs of the aggregate interference obtained from the simulations converge before 30 repetitions. We repeat this process for each combination of the PHY and MAC layer parameters.

Saturated traffic was assigned to all transmitters so that there was no idle time by the traffic itself during the simulations. The background grid for all of the simulation scenarios was always a 500 m by 500 m square. The transmission times for RTS, CTS, PPDU (PHY+MAC+PAY) with 500 B (or 1000 B) of payload and ACK were 52, 44, 728 (or 1396), and 44 (μs), respectively.

4.2   Discussions

Analysis results of p_{on} and λ'. p_{on} in this subsection refers to p_{on}^*, which can be found from (10), and is the final solution of p_{on} for simplicity of expression. In Fig. 2, the values of p_{on} are shown for various combinations of MAC parameters. The factor that affects p_{on} the most is the effective CS distance R, followed by λ, which is the initial node density. As λ increases, p_{on} naturally increases due to the increased congestion level. The p_{on} shows a mixture of linear and log functions. Within the same R, the combination that has the lowest p_{on} is the RTS mode and short payload. In general, RTS-CTS mode has a lower congestion level than the basic mode. With a large payload size, RTS-CTS mode is more favorable because two reasons: (1) a large payload in the basic mode makes for a higher congestion level, (2) the ratio of data transmission is large enough to compensate for RTS-CTS packet overheads.

This p_{on} was used in the new density λ' (13) and we plotted this as shown in the second figure of Fig. 2. This figure shows that the smaller R makes for a higher λ', which is the opposite of p_{on}. This is understandable, since a smaller R signifies more insensitivity to the interference. We expected the result of $R = 0$, the ALOHA system, to approach the line $\lambda' = \lambda$ in the figure. By showing the $\lambda' = \lambda$ line and the curves together, Fig. 2 also addresses the size of the gap between the original node density and effective node density, showing the effectiveness of CSMA/CA MAC. The bold curves are from the approximated node density modeled by the MHC. As shown in the figure, the variation of λ' is higher than that of MHC for varying λ. The figure shows the gap in the aggregate interference between simplified MHC and the real situation. As shown in the next section, our aggregate power distribution adopting λ' is the most accurate among the other point processes.

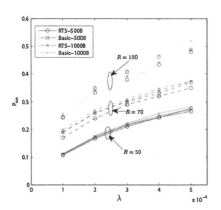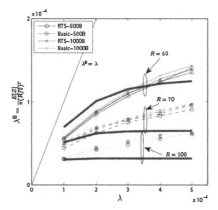

Fig. 2. P_{on} and new node density λ' versus λ for all combinations of MAC layer parameters. The thicker curves are from the Matèrn hardcore process. (Color figure online)

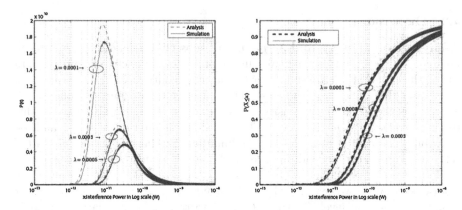

Fig. 3. Probability density and cumulative distribution of aggregate interference in the condition of RTS mode, 500B payload and $R = 70$(m).

Comparison of the Resulting Aggregate Interference with the Simulation. The PDF and CDF of the analysis in each node density showed high correlations with those of the NS-2 simulations as seen in Fig. 3.

Although at first glance they resemble a log-normal distribution, they are asymmetric based on the main lobe. They are definitely neither normal nor log-normal distributions. This is notable as some research efforts in the signal processing field assume that the aggregate interference follows normal (in dBm unit) or log-normal (in W unit) distributions. For the other features, the higher the mean of the aggregate power, the lower the probability of that mean value. Therefore, low-mean high-probability and high-mean low-probability patterns are shown in all of the results.

Compared with dependent point processes, at any given λ value, our analysis is the closest one to the simulation results, as depicted in Fig. 4. MHC and SSI do not have sufficient MAC and PHY layer parameters to reflect the real situation, while our analysis can model any combination of the system parameters, as in Fig. 5.

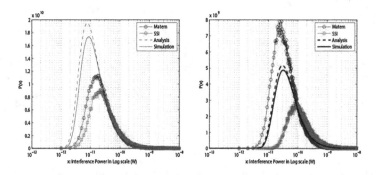

Fig. 4. Probability density of aggregate interference when λ is 0.0001 and 0.0005.

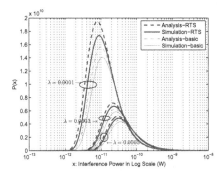

Fig. 5. Probability density of aggregate interference in the case of 500B payload and $R = 70$. (Color figure online)

As shown in the figures so far, our model of the aggregate interference has slightly lower values than that of the simulation in each case, because the simulator allows the *capture* situation. In our analysis, a collision between transmitters is regarded as a failure of transmission, and this increases each collided node's backoff stage. In contrast, there might be a successful transmission even when multiple nodes in a CS area are transmitting at the same time. This is because if the ratio of one incoming signal to the others is higher than a certain threshold, the stronger incoming signal can be decoded.

From these results, we learned the following lessons: If the network is required to maintain a lower interference than a certain level, there are multiple combinations of parameters that need to be controlled. Those controllable parameters are R, the transmission mode, payload size, etc. This can be used for the interference management in uncontrolled interference limited systems.

5   Conclusion

In this paper, we analyzed the aggregate interference from randomly deployed CSMA/CA nodes. Due to the imperfection of the CSMA/CA protocol, the transmission of each node in the network is not fully dependent, but is able to be modeled by independent point process with the new node density.

Our framework derived to find this value reveals the relation of the MAC parameters and the effective node density. Although the exact closed form expression of the interference distribution cannot be obtained, quite accurate interference distribution can be obtained by our methodology responding to the variation of MAC parameters. Furthermore, the sound simulation using NS-2 certifies that the analysis is enough to be used for optimizing the system parameters in uncontrolled WiFi hot spots or to protect incumbent systems in the case of secondary spectrum access.

References

1. Koufos, K., Ruttik, K., Jäntti, R.: 'Aggregate interference from WLAN in the TV white space by using terrain-based channel model. In: Proceedings of CROWN-COM (2012)
2. Baccelli, F., Błaszczyszyn, B.: Stochastic geometry and wireless networks volume 1: theory. Found. Trends Netw. **3**(3–4), 249–449 (2009)
3. Andrews, J., Ganti, R.K., Haenggi, M., Jindal, N., Weber, S.: A primer on spatial modeling and analysis in wireless networks. IEEE Comm. Mag. **48**(11), 156–163 (2010)
4. Kim, D.M., Kim, S.-L.: An iterative algorithm for optimal carrier sensing threshold in random CSMA/CA wireless networks. IEEE Comm. Lett. **17**(11), 2076–2079 (2013)
5. Hwang, J., Kim, S.-L.: Cross-layer optimization and network coding in CSMA/CA-based wireless multihop networks. IEEE/ACM Trans. Netw. **19**(4), 1028–1042 (2011)
6. Busson, A., Chelius, G., Gorce, J.-M.: Interference modeling in CSMA multi-hop wireless networks. Technical report INRIA, pp. 1–21, inria-00316029-ver3 (2009)
7. ElSawy, H., Hossain, E.: A modified hard core point process for analysis of random CSMA wireless networks in general fading environments. IEEE Trans. Commun. **61**(4), 1520–1534 (2013)
8. Bianchi, G.: Performance analysis of the IEEE 802.11 distributed coordination function. IEEE J. Sel. Areas Commun. **18**(3), 535–547 (2000)
9. Network Simulator-2. http://isi.edu/nsnam/ns/
10. Chen, Q., Schmidt-Eisenlohr, F., Jiang, D., Torrent-Moreno, M., Delgrossi, L., Hartenstein, H.: Overhaul of IEEE 802.11 modeling and simulation in NS-2. In: Proceedings of ACM MsWiM 2007 (2007)

Throughput Capacity Analysis of a Random Multi-user Multi-channel Network Modeled as an Occupancy Problem

Vincent Savaux[1], Apostolos Kountouris[2], Yves Louët[1(✉)], and Christophe Moy[1]

[1] Signal, Communication and Embedded Electronics (SCEE) Research Group/IETR, CentraleSupélec, Rennes, France
{vincent.savaux,yves.louet,christophe.moy}@centralesupelec.fr
[2] Orange Labs, Grenoble, France
apostolos.kountouris@orange.com

Abstract. In this paper, we model the random multi-user multi-channel access network by using the well known *occupancy problem* from probability theory. Furthermore, we combine this with a network interference model in order to derive the achievable throughput capacity of such networks. The mathematical developments and results are illustrated through various simulations results. The proposed model is particularly relevant in analyzing the performance of networks where the users are not synchronized neither in time nor in frequency as it is often the case in various Internet of Things (IoT) applications.

Keywords: Throughput capacity · Network interference · Occupancy problem

1 Introduction

The issue of transmitting asynchronous signals on a single channel has been studied for several decades [1] and it led to some protocols such as ALOHA (ALOHAnet) proposed by N. Abramson in 1985 [2]. Since then, many solutions have been proposed in the literature to overcome the distortions induced by the collisions among the transmitted signals. [3] and references therein provide an extensive overview of solutions based on packet retransmissions and [4] treats solutions based on signal coding. More recently, the model has been extended to the case of random frequency channel access besides random time channel access [5]. The authors proposed to model the interferences induced by the collisions of ultra narrow-band signals featuring a random frequency channel access in a context of the Internet-of-Things (IoT) applications.

In this paper, we propose to further analyze the issue of signal collisions in the network interference problem [6,7] by considering random multi-channel,

---

Vincent Savaux is with b<>com, Rennes, France, since January 2016.

© ICST Institute for Computer Sciences, Social Informatics and Telecommunications Engineering 2016
D. Noguet et al. (Eds.): CROWNCOM 2016, LNICST 172, pp. 437–447, 2016.
DOI: 10.1007/978-3-319-40352-6_36

multi-user access. In other words we extend the analysis to account for random behavior along the frequency dimension as well where each user of the network transmits not only at random times but also on randomly chosen channels within the band. The random frequency access of N_{ut} users into N_c channels can be modeled by the *occupancy problem* used in probability theory [13]. From this model we subsequently derive an analytical expression of the achievable throughput capacity of such a network, i.e. the probability that randomly transmitted signals are properly decoded at the receiver. To the best of our knowledge, the *occupancy problem* is an original approach for modeling a multi-channel, multi-user access network. Moreover, various simulations illustrate the obtained theoretical results.

The remaining of the paper is organized as follows: Sect. 2 presents the model of the considered network, and provides a reminder of the *occupancy problem*. The expression of the throughput capacity is derived in Sect. 3, and simulations are provided in Sect. 4. Finally, Sect. 5 concludes this paper.

2   System Model

2.1   Network Model

In this paper, it is assumed that all the transmitters (called users in this paper) of the network are distributed in the two-dimensional plane according to an homogeneous Poisson point process with a given intensity λ (in users per unit area), whose distribution is given as follows:

$$f_{poi}(k) = \frac{\lambda^k}{k!} e^{-\lambda}. \tag{1}$$

As depicted in Fig. 1, base transceiver stations (BTS) are also located in the plane, in order to pick up the signals from users. For instance considering the BTS at the center of Fig. 1, we assume it services a cell of radius R_c that contains N_{ut} users, indicated as small black circles. The users outside this cell (white circles in Fig. 1) are potential interfering users. It should be noted, however, that the users may interfere with each other as well. According to the defined parameters, the intensity λ is equal to $N_{ut}/(\pi R_c^2)$.

The role of a BTS consists in scanning and sampling the band $\mathcal{B}$ of bandwidth B which is regularly subdivided into N_c channels $\{\mathcal{B}_0, \mathcal{B}_1, .., \mathcal{B}_{N_c-1}\}$ of width B/N_c. The considered random multi-channel multi-user transmission model can be formalized as follows:

– Each user can randomly access, i.e. select, one of the N_c channels with a probability $\mathbb{P}_a = \frac{1}{N_c}$. In that way, a reuse[1] of one or many channels may occur, as shown in Fig. 2-(a). In the following, we denote by $\mathcal{R}$ the reuse factor of the channels, i.e. the number of users sharing the same channel.

---

[1] Meaning that a channel can be selected by more than one user.

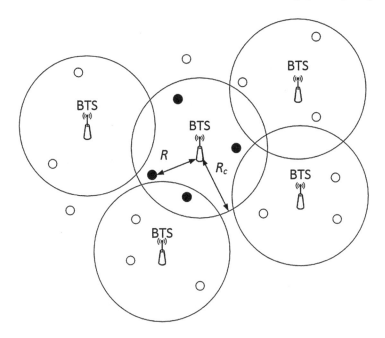

Fig. 1. Poisson point model for the spatial distribution of the users.

– A slotted traffic scheme is assumed for each user (see Fig. 2-(b)). Therefore, an asynchronous slotted traffic is considered for each channel at the BTS, due to the independence between users. The probability of transmission of a packet (or duty cycle[2]) is denoted by $q \in [0, 1]$, and for simplicity, it is supposed that q is the same for each user. As depicted in Fig. 2-(b), collisions may occur between packets when at least two users transmit at the same time on the same channel.

According to the model in [6,10], the power received at the BTS from a user at a distance R can be defined as

$$P_r = \frac{P_t G}{R^{2c}}, \tag{2}$$

where P_t denotes the transmitted power, and G is a random variable that depends of the propagation environment. Several models can be used to describe G, depending on the shadowing (mainly due to large objects), and the multipath fading (mainly due to the constructive or destructive combinations of the replicas of the transmitted signal). The term $1/R^{2c}$ is defined as the far-field path loss, where c depends on the propagation environment and is in the range of 0.8 to 4 [6]. It should be emphasized that the far-field path loss fits the considered

---

[2] Note that *duty cycle* is sometimes used to refer to the overlapping factor between to packets, as in [6].

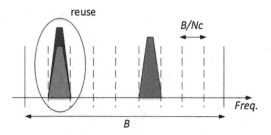

(a) Random frequency access to the channels. A reuse $\mathcal{R} > 1$ occurs when at least 2 users choose the same channel.

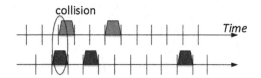

(b) Slotted-asynchronous transmission. A collision occurs when to users transmit at the same time in the same channel.

Fig. 2. Frequency and time channel access. (Color figure online)

model, in which the users are supposed to be at least several meters away from the BTS.

2.2   Occupancy Problem

The random access to N_c channels by N_{ut} users is an instance of the *occupancy problem* in probability. This theory provides useful tools to deal with the time-frequency use of the band $\mathcal{B}$ in the considered transmission model. In particular, two theorems will be used in this paper.

Theorem 1. *Let N_{ut} users randomly accessing N_c channel with a probability $\mathbb{P}_a = \frac{1}{N_c}$. Then the probability that b channels are used (at least by one user) is*

$$\mathbb{P}(b) = \binom{N_c}{N_c - b} \sum_{\nu=0}^{b} (-1)^{\nu} \binom{b}{\nu} \left(1 - \frac{N_c - b + \nu}{N_c}\right)^{N_{ut}}. \tag{3}$$

Proof. see [13], Chap. 4.

Theorem 2. *Under the same assumptions as in Theorem 1, the probability that $\mathcal{R}$ users access the same channel $(0 \leq \mathcal{R} \leq N_{ut})$ is*

$$\mathbb{P}(\mathcal{R}) = \binom{N_{ut}}{\mathcal{R}} \left(\frac{1}{N_c}\right)^{\mathcal{R}} \left(1 - \frac{1}{N_c}\right)^{N_{ut}-\mathcal{R}}. \tag{4}$$

Proof. Let $\mathbb{P}(j)$ the probability of the event "the j-th channel is chosen", and $\mathbb{P}(X_j = \mathcal{R})$ the probability that $\mathcal{R}$ users use the j-th channel. As $\mathbb{P}(j) = \frac{1}{N_c}$ and $\mathbb{P}(X_j = \mathcal{R})$ are equal for any j, then we obtain

$$\mathbb{P}(\mathcal{R}) = \sum_{j=1}^{N_c} \mathbb{P}(j)\mathbb{P}(X_j = \mathcal{R}) \tag{5}$$

$$= \sum_{j=1}^{N_c} \frac{\mathbb{P}(X_j = \mathcal{R})}{N_c} \tag{6}$$

$$= \mathbb{P}(X_j = \mathcal{R}). \tag{7}$$

The reuse factor $\mathcal{R}$ can be defined as equal to the sum $X_j = \mathcal{R} = \sum_k x_{j,k}$, where $x_{j,k}$ is a random variable which counts the number of users in the channel j. Since the users access to the channels independently of each other, $x_{j,k}$ are the results of Bernoulli trials with probability $1/N_c$ to access the channel j. Therefore, the sum $X_j = \sum_k x_{j,k}$ has the binomial distribution defined in (4).

3 Deriving the Throughput Capacity

This section deals with the analysis of the throughput capacity, whose definition is given hereafter. In the rest of the paper, we use the so-called physical interference model [7,8], in which the following condition is imposed:

– a message from a user in the cell is successfully decoded if the corresponding signal-to-interference-plus-noise ratio (SINR) exceeds a given threshold γ_p. Thus, successful decoding for a user k in a given channel n (with reuse factor $\mathcal{R}$, and $1 \leq n \leq N_C$) requires that:

$$SINR_{k,n} = \frac{P_{r,k,n}}{I_{k,n} + \sigma_n^2} \geq \gamma_p, \tag{8}$$

where $P_{r,k,n}$ is the received power of the k-th probe user in the n-th channel as defined in (2), and σ_n^2 is the noise power in the n-th channel. The term $I_{k,n}$, using the general definition in [6], is the interference power that can be written as

$$I_{k,n} = \sum_{\substack{i=1 \\ i \neq k}}^{\infty} \frac{P_t \Delta_{i,n} G_{i,n}}{R_{i,n}^{2c}}, \tag{9}$$

where $\Delta_{i,n}$ is the overlapping factor between the i-th interfering signal and the proper signal k. It is assumed that $\Delta_{i,n}$ obeys a uniform distribution in $[0,1]$.

By convention, we consider that the $\mathcal{R}-1$ first terms of the sum in (9) correspond to the users, and the indexes $i > \mathcal{R}$ point out the interfering users that are located outside the considered cell.

It has been demonstrated in [9] that the interference defined as the superposition of numerous signals from users distributed according to a homogeneous Poisson process on a plane can be modeled as a α-stable distribution [11,12], which can be seen of a generalization of the Gaussian distribution. No analytic expression of the probability density function (pdf) of the α-stable law can be derived, but its characteristic function is expressed as

$$\phi(t) = \exp(-\gamma|t|^\alpha(1 - j\beta\mathrm{sign}(t)\omega(t,\alpha))), \tag{10}$$

where

$$\omega(t,\alpha) = \begin{cases} \tan(\frac{\pi\alpha}{2}), & \text{if } \alpha \neq 1 \\ -\frac{2}{\pi}\ln(|t|), & \text{if } \alpha = 1 \end{cases}, \tag{11}$$

and

$$\mathrm{sign}(t) = \begin{cases} -1, & \text{if } t < 0 \\ 0, & \text{if } t = 0 \\ 1, & \text{if } t > 0 \end{cases}. \tag{12}$$

According to [6], the parameters α, β, and γ are defined as

$$\alpha = \frac{1}{c}$$
$$\beta = 1$$
$$\gamma = \pi\lambda_\mathcal{R}C_{1/c}^{-1}P_t^{1/c}E\{\Delta_{k,n}^{1/c}\}E\{G_{k,n}\},$$

where

$$C_\alpha = \begin{cases} \frac{1-\alpha}{\Gamma(2-\alpha)\cos(\pi\alpha/2)}, & \text{if } \alpha \neq 1 \\ \frac{2}{\pi}, & \text{if } \alpha = 1 \end{cases}, \tag{13}$$

with $\Gamma(.)$ the gamma function. It is worth noting that $\lambda_\mathcal{R}$ is now a variable function of $\mathcal{R}$, namely $\lambda_\mathcal{R} = \mathcal{R}/(\pi R_c^2)$. The throughput capacity denoted by $\mathcal{T}_{\mathcal{R},n,R_{k,n}}$ for a given reuse factor $\mathcal{R}$ in the n-th channel, and located at a distance $R_{k,n}$ from the BTS, can be defined as

$$\mathcal{T}_{\mathcal{R},n,R_{k,n}} = \mathbb{P}(\text{transmit})\mathbb{P}(\text{no outage}), \tag{14}$$

where $\mathbb{P}(\text{transmit}) = q_\mathcal{R} = 1-(1-q)^\mathcal{R}$ is the probability that a channel is occupied, and $\mathbb{P}(\text{no outage}) = \mathbb{P}(SINR_{k,n} \geq \gamma_p)$. The probability $\mathbb{P}(SINR_{k,n} \geq \gamma_p)$ can be rewritten by substituting (9) into (8), and by using the law of total probability as

$$\mathbb{P}(SINR_{k,n} \geq \gamma_p) = E_{\{G_{k,n}\}}\{\mathbb{P}_{\{I_{k,n}\}}\left(I_{k,n} \leq \frac{P_t G_{k,n}}{\gamma_p R_{k,n}^{2c}} - \sigma_n^2\right)\big|G_{k,n}\}, \tag{15}$$

and hence

$$\mathbb{P}(SINR_{k,n} \geq \gamma_p) = E_{\{G_{k,n}\}}\{F_{I_{k,n}}\left(\frac{P_t G_{k,n}}{\gamma_p R_{k,n}^{2c}} - \sigma_n^2\right)\}$$

$$= E_{\{G_{i,n}\}}\{\int_{-\infty}^{\frac{P_t G_{k,n}}{\gamma_p R_{k,n}^{2c}} - \sigma_n^2} \int_{-\infty}^{+\infty} \phi_{I_{k,n}}(t)e^{-jtx}dtdx\}, \quad (16)$$

where $F_{I_{k,n}}$ is the cumulative distribution function (cdf) of $I_{k,n}$. Several closed-form of (16) corresponding to different types of shadowing and fading have been derived in [6,10]. In particular, it should be noted that the expectation in (15) disappears if the case *path-loss only* $G_{k,n} = 1$ is considered, and

$$\mathbb{P}(SINR_{k,n} \geq \gamma_p) = F_{I_{k,n}}\left(\frac{P_t G_{k,n}}{\gamma_p R_{k,n}^{2c}} - \sigma_n^2\right). \quad (17)$$

The Rayleigh fading leads to the following expression:

$$\mathbb{P}(SINR_{k,n} \geq \gamma_p) = \exp\left(-\frac{R_{k,n}^{2c}\sigma_n^2\gamma_p}{P_t}\right)$$

$$\times \exp\left(-\frac{\lambda_{\mathcal{R}}C_{1/c}^{-1}\Gamma(1 + \frac{1}{c})E\{\Delta_{k,n}^{1/c}\}}{\cos(\frac{\pi}{2c})}\left(R_{k,n}^{2c}\gamma_p\right)^{1/c}\right). \quad (18)$$

Since the users are homogeneously distributed in the cell, then the distance $R_{k,n}$ obeys the uniform distribution denoted by $\mathcal{U}(0, R_c)$, and defined as:

$$f_u(x) = \begin{cases} \frac{1}{R_c}, & \text{if } x \in [0, R_c] \\ 0, & \text{else} \end{cases}. \quad (19)$$

The throughput capacity is then obtained by averaging $\mathbb{P}(SINR_{k,n} \geq \gamma_p)$ on the interval $[0, R_c]$ as:

$$\mathcal{T}_{\mathcal{R},n} = q_{\mathcal{R}} E_{R_{k,n}}\{\mathcal{T}_{\mathcal{R},n,R_{k,n}}\}$$

$$= q_{\mathcal{R}} \int_0^{R_c} f_u(R_{k,n})\mathbb{P}(SINR_{k,n} \geq \gamma_p)dR_{k,n}. \quad (20)$$

Note that the bound 0 of the integral in (20) should be replaced by a positive value according to the far-field model (typically ¿1 meter). Besides, this also avoids the division by zero if the *path-loss only* model in (17) is used. Since we consider a slotted-asynchronous packet transmission, then the value $E\{\Delta_{k,n}^{1/b}\}$ can be derived by following the developments in [10], which lead to:

$$E\{\Delta_{k,n}^{1/c}\} = q^2 + 2q(1 - q)\frac{c}{1 + c}. \quad (21)$$

Since the users are independent and the N_c channels have the same probability of access $\mathbb{P}_a$, the throughput capacity for the given n-th channel with reuse factor $\mathcal{R}$ is given by:

$$\mathcal{T}_\mathcal{R} = \mathbb{P}_a \mathbb{P}(\mathcal{R}) \mathcal{T}_{\mathcal{R},n}, \qquad (22)$$

and the overall achieved throughput capacity considering all the users of the cell is defined as the following weighted sum:

$$\mathcal{T} = \sum_{n=1}^{N_c} \mathbb{P}_a \sum_{\mathcal{R}=0}^{N_{ut}} \mathbb{P}(\mathcal{R}) \mathcal{T}_{\mathcal{R},n} = \sum_{\mathcal{R}=0}^{N_{ut}} \mathcal{T}_\mathcal{R}, \qquad (23)$$

where $\mathbb{P}(\mathcal{R})$ is given in (4). It can be noticed in the developments from (8) to (23) that at least five parameters have an influence on the throughput capacity value $\mathcal{T}$: the radius of the cell R_c, the number of users N_{ut}, the number of channels N_c, the duty cycle q, the threshold γ_p, and the noise level σ_n^2. Note that we assume $\sigma_n^2 = \sigma^2$ for any channel $1 \leq n \leq N_c$.

4   Simulations Results

In this section, we present the simulations results related to the throughput capacity defined as in (23). In all the simulations, the value of c has been arbitrarily set equal to 1.2, $N_c = 10$, and a Rayleigh fading is considered. Therefore the "no-outage" probability is modeled by (18). Figure 3 depicts $\mathcal{T}$ versus: (a) the radius of the cell R_c (in m), (b) the number of users N_{ut}, (c) the threshold γ_p, and (d) the duty cycle q. In this first series of simulations, the SNR defined as P_t/σ^2 has been set equal to 30 dB. In Fig. 3-(a), it can be seen that $\mathcal{T}$ values decrease where R_c increases, which reflects the fact that the average signal power $E\{P_{r,k,n}\}$ received at the BTS is lower in large cells than in small cells. Figure 3-(b) shows that the throughput capacity achieves a maximum for a given N_{ut} value that we can denote by N_{ut}^m. We can deduce that in the range $N_{ut} \leq N_{ut}^m$, $\mathcal{T}$ is limited by $\mathbb{P}(\text{transmit})$ (i.e. the number of transmitting users induces a low value of $\mathbb{P}(\text{transmit})$), while in the range $N_{ut} \geq N_{ut}^m$, $\mathcal{T}$ is limited by $\mathbb{P}(\text{no outage})$ (i.e. the users which transmit induce a large amount of interference). In Fig. 3-(c) it is verified that the throughput capacity decreases where γ_p increases, according to the network interference model in (8). Similarly to Fig. 3-(b), we observe in Fig. 3-(d) that $\mathcal{T}$ increases with q when $N_{ut} = 10$, whereas it reaches a maximum before decreasing for higher numbers of users ($N_{ut} = 100$ and $N_{ut} = 1000$). This phenomenon is due to the presence of the interferences when either or both N_{ut} and q are high-valued.

Previous results are completed by Fig. 4, which presents $\mathcal{T}$ versus SNR, in the SNR range from 0 to 60 dB. Note that very high SNR values are consistent with the system model using narrow band signals. The depicted results show the influence of the interferences which lead to an upper bound of the achievable throughput capacity in the proposed system.

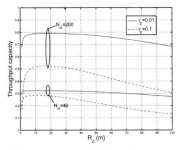

(a) $\mathcal{T}$ versus R_c, for SNR=30 dB, $q = 0.1$.

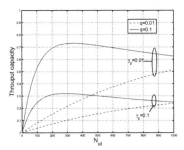

(b) $\mathcal{T}$ versus N_{ut}, for SNR=30 dB, $R_c = 100$ m.

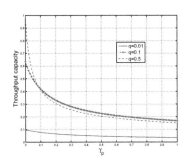

(c) $\mathcal{T}$ versus γ_p, for SNR=30 dB, $R_c = 100$ m.

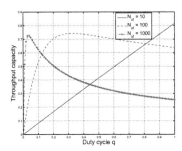

(d) $\mathcal{T}$ versus q, for SNR=30 dB, $R_c = 100$ m, and $\gamma_p = 0.01$.

Fig. 3. Throughput capacity $\mathcal{T}$ versus (a) R_c (in m), (b) N_{ut}, (c) γ_p, and (d) q.

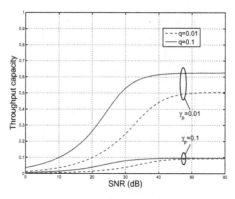

(a) T versus SNR, for $N_{ut} = 100$.

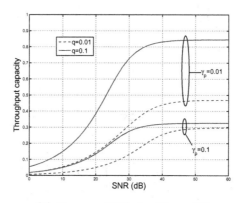

(b) T versus SNR, for $N_{ut} = 400$.

Fig. 4. Throughput capacity T versus SNR, for (a) $N_{ut} = 100$, (b) $N_{ut} = 400$, $R_c = 100\,\mathrm{m}$.

5   Conclusion

In this paper, we modeled a random multi-channel multi-user network using the, well known in probability, *occupancy problem*. We subsequently derived an analytical expression of the achievable *throughput capacity* of such a system. The mathematical developments and the simulations results revealed that the throughput capacity values largely depend on the cell size, the threshold used in the network interference model, the number of users in the network and the duty cycle of the signals. It is worth noting that the proposed approach can be used as a basis for the performance analysis of specific networks using random multi-channel multi-user access, such as in the context of the IoT applications. Further work will extend the proposed model to a more general case where no channelization of the band is imposed and the user transmissions may occupy different bandwidths within the band.

Acknowledgment. This work has been funded by Orange with grant agreement code: E06301.

References

1. Massey, J.L., Mathys, P.: The collision channel without feedback. IEEE Trans. Inf. Theory **31**(2), 192–204 (1985)
2. Abramson, N.: The development of ALOHANET. IEEE Trans. Inf. Theory **31**(2), 119–123 (1985)
3. Nardelli, P.H.J., Kaynia, M., Cardieri, P., Latva-aho, M.: Optimal transmission capacity of Ad Hoc networks with packet retransmissions. IEEE Trans. Wirel. Commun. **11**(8), 2760–2766 (2012). ISSN: 1536-1276
4. Thomas, G.: Capacity of the wireless packet collision channel without feedback. IEEE Trans. Inf. Theory **46**(3), 1141–1444 (2002)
5. Do, M.-T., Goursaud, C., Gorce, J.-M.: Interference modelling and analysis of random FDMA scheme in ultra narrowband networks. In: Proceedings of AICT 2014 (2014)
6. Win, M.Z., Pinto, P.C., Shepp, L.A.: A mathematical theory of network interference and its applications. Proc. IEEE **97**(2), 205–230 (2009)
7. Cardieri, P.: Modeling interference in wireless Ad Hoc networks. IEEE Commun. Surv. Tutor. **12**(4), 551–572 (2010)
8. Cardieri, P., Nardelli, P.H.J.: A survey on the characterization of the capacity of Ad Hoc wireless networks. In: 20th Chapter of Mobile Ad-Hoc Networks: Applicationss, pp. 453–472 (2011)
9. Ilow, J., Hatzinakos, D.: Analytic alpha-stable noise modeling in a Poisson field of interferers or scatterers. IEEE Trans. Signal Process. **46**(6), 1601–1611 (1998)
10. Pinto, P.C., Win, M.Z.: A unifying framework for local throughput in wireless networks. 10 pages (2010). arXiv:1007.2814
11. Shao, M., Chrysostomos, L.N.: Signal processing with fractional lower order moments: stable processes and their applications. Proc. IEEE **81**(7), 986–1010 (1993)
12. Samoradnitsky, G., Taqqu, M.: Stable Non-Gaussian Random Processes: Stochastic Models with Infinite Variance. Chapman and Hall, New York (1994)
13. Feller, W.: An Introduction to Probability Theory and its Applications, 3rd edn. Wiley, New York (1968)

Understanding Current Background Noise Characteristics: Frequency and Time Domain Measurements of Noise on Multiple Locations

Alexandros Palaios[(✉)], Vanya M. Miteva, Janne Riihijärvi,
and Petri Mähönen

Institute for Networked Systems, RWTH Aachen University, Aachen, Germany
apa@inets.rwth-aachen.de

Abstract. One of the important factors for the successful deployment of cognitive radio networks is keeping interference levels to minimum to primary users. The noise is usually expected to be thermal noise and non-interesting component on understanding a complex interplay between primary and secondary systems. Moreover, although usually it is assumed that excess interference is generated by secondary users it is worth of remembering that also primary systems generate harmonics and intermodulation components that can harm other primary systems. In this paper we report results from the measurement campaign that aims to find out how much there is excess noise over thermal noise floor. The observed noise levels cast questions on the widely used assumptions that we need to consider only AWGN thermal noise. We conclude the paper showing that the observed excess noise levels can have an effect on the capacity of the primary and secondary systems.

Keywords: Background noise characteristics · Frequency and time domain measurements · Noise on multiple locations · Interference levels

1 Introduction

There is a considerable body of work in the direction of how to minimize interference towards primary systems. Surprisingly though the hidden assumption is that the primary systems experience only thermal noise and any extra interference is caused by the secondary systems. However, transceiver chains are not built from ideally linear components and sharp filters. Thus many byproducts are still created and there is an open question how much primary systems themselves emit noise like components. This is important for both secondary and other primary access networks since emitted noise byproducts effectively raise the noise floor and cause reduced capacity. Moreover, the number of the wireless systems have tremendously increased, but the existing noise measurement literature is over 30 years old [1,2].

It is not only that wireless networks proliferated in the 1990 s but also there has been shift from large cells supporting wireless telephony to small cells data

© ICST Institute for Computer Sciences, Social Informatics and Telecommunications Engineering 2016
D. Noguet et al. (Eds.): CROWNCOM 2016, LNICST 172, pp. 448–461, 2016.
DOI: 10.1007/978-3-319-40352-6_37

networks. Moreover, the number of services, networks and users has increased considerably the last twenty years [3].

The current literature on noise levels is very limited and is focusing either on studying time dynamics of very narrow frequency bands and thus but failing to capture how noise can be prevalent in wide bandwidths [4]. At the same time the pulse characteristics seem to be very dependent on devices emitting [5]. Finally, the available literature is focusing exclusively on specific transceiver technologies and modulation schemes [6]. All this limits the general understanding on observable noise levels.

We have recently finalized the first measurement campaign studying noise with wideband measurements (450 MHz to 3000 MHz). We have measured both indoor and outdoor environments trying to highlight the existence of noise on different representative localities such as city center, indoor office environments, transmission towers, industrial locations, power transmission lines and streets.

The rest of this paper is structured as follows. In Sect. 2 we describe our hardware configuration and the locations we studied. In Sect. 3 we present and discuss about the fundamental methodological problems on capturing noise levels, while not assuming cooperation from license owners. In Sect. 4 the exploratory data analysis is presented and in Sect. 6 we show how the observed noise levels can influence typical assumption made in link and network analysis. Finally, in Sect. 7 we present in details a specific example how noise can reduce the assumed capacity levels of systems. We conclude in Sect. 7.

2   Hardware Configuration and Locations Studied

We start by describing in details the studied locations and the hardware components used.

2.1   Locations Studied

For understanding the variability of noise sources and the shape of radio environment one has to do measurements with high spectral and time resolution. We have focused on large band spanning from 450 MHz to 3000 MHz and for the time domain we capture time dynamics on the order of milliseconds. Finally, it is also important to understand the variability in space. One assumes that noise changes over different localities and in this paper we report results from nine different locations. This selection is a subset from our larger planned measurement campaign. The locations captured in this study are summarized below

1. Office area: At this location both indoor (office rooms) and outdoor measurements were conducted.
   (a) For the indoor measurements three different types of environments were chosen:
       – Empty office, with no electrical equipment inside it.
       – Working office, with a large number of office equipment devices (such as computers, phones and fluorescent lamps.).

- Server room, where a large number of computational servers and electric cooling systems are working constantly.
  (b) Office district: The measurement setup was located next to a street with low to average amount of vehicle traffic. The location itself comprises mainly R&D companies and university buildings.
2. Street level at a business area: A crossroad with a low traffic intensity, located nearby the office district.
3. TV tower: The measurement location is in the suburbs of the city, very close to the main radio and DVB-T tower of Aachen, which makes it a very good choice for examining the possible transmitted byproducts and out-of-band emissions in the DVB-T bands.
4. City center: A measurement spot located in the part of the city, next to a busy road with high vehicle traffic.
5. Power transmission lines: The location is right outside of the city, very close to a large number of high voltage transmission power lines. The measurement setup was positioned at around 50–70 meters from the main power lines.
6. Factory: The measurements were conducted in a factory. During the measurement different high technology lasers were used for cutting, bending, sawing, and welding.

2.2   The Measurement Platform

The most important hardware component used is the high end spectrum analyzer (SA); the Agilent/Keysight E4440A. The device was selected due to its high sensitivity levels. The device achieves sensitivities below -125 dBm per 10 kHz for all the covered frequency spans. The tradeoff here is the power consumption. As we describe below we had to pay special attention for supporting portable measurements with such a heavy device. For supporting the operation of the SA as well as deploying our measurement setup on different locations we built a mobile measurement platform as depicted in Fig. 1 that contains three compartments. At the bottom there is a large car battery compartment that operates the spectrum analyzer when AC power is not available (a power converter was used to convert the DC power from the battery to AC). At the middle there is the SA compartment and above it there is the laptop compartment. The laptop controls the SA and stores the measurement data. The spectrum analyzer was calibrated by plugging in a 50Ω terminator at the RF input frontend. The last component of our setup is the high-end antenna (Schwarzbeck SBA 9113) that features very high degree of omnidirectionality. In some of the locations we had to turn off the preamplifier to protect the device from overloading. We specify later, in which locations preamplifier was turned off, leading to slightly lower sensitivity.

For the frequency type of measurements SA operated in sweep mode. Since an increased frequency resolution was important for spotting very narrowband signals (i.e. typically byproducts of a transmitter chain before the upconversion stage) and avoiding compression of those with large bandwidths, we selected to use a very narrow resolution bandwidth (RBW) of 10 kHz. The achieved time resolution between the different sweeps were on the order of tenths of seconds.

Fig. 1. The measurement Platform, a.k.a "Blue Box". (Colour figure online)

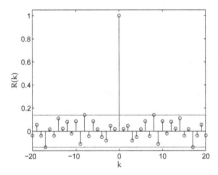

Fig. 2. Autocorrelation function of uncorrelated time series.

The 200 sweeps for the span of 450 MHz–3000 MHz last around 5 h and 30 min. The time domain measurements were conducted in zero span mode. During that time, the spectrum analyzer locks in to a fixed frequency with a given RBW at a fixed frequency and acquires the measurements faster than few microseconds. Both methods are useful but in this paper we focus mostly on frequency domain results. The frequency domain analysis has a priority as it provides us a baseline information on the seriousness of extra noise components in different frequency bands.

3 Tools for Noise Analysis

3.1 Noise Definition

We start by explaining how we define noise in our measurements. We conducted power measurements, covering wide frequency bands, a large part of which are licensed and already used for transmission. Therefore, the existence of power levels above the noise floor of the device can originate either from licensed transmitters or noisy byproducts. This means that methodologically one can not be sure if a reception is a noise byproduct or a weak licensed transmission. This is the reason we first look at the bands that are unlicensed or are used by weak licensed transmitters. We start by studying frequency bands that are unlicensed and no transmissions are allowed.

In Germany the Radio Astronomy band (1400 MHz to 1427 MHz) is a good example to monitor. Moreover, we included the cellular Guard Bands (Duplex Gap), such as the GSM, LTE and UMTS ones, as these should also be free from any intentional transmissions. For simplicity we call these bands as GSM 900, LTE 800 and UMTS but emphasize here that we are referring to their Guard Bands. Finally we observed also some licensed bands, most notably DVB-T digital-TV frequencies. In the commercial bands licenses are typically given with a specific spectrum mask (for the DVB-T system that is 8 MHz in Europe). In such a case observing very narrow band signalskHz is probably a noisy byproduct that should be considered as noise and not as part of licensed transmission.

3.2   Analysis Tools

Here we will present briefly the tools we used for processing our data and acquiring the characteristics of noise sources.

The autocorrelation analysis can be used to analyze whether subsequent values of time series are independent from each other. The normalized autocorrelation function (ACF) R for discrete time series X_T can be expressed as

$$R(k) = \frac{E[(X_T - \mu)(X_{T+k} - \mu)]}{\sigma^2},\qquad(1)$$

where the term in the numerator is the autocovariance and σ^2 is the variance. For perfect white noise the ACF is a Dirac pulse. The confidence interval (CI) of a 95 % significance level is equal to

$$CI = [\mu - 1.96\sigma/\sqrt{n}; \mu + 1.96\sigma/\sqrt{n}],\qquad(2)$$

where n is the number of samples. For uniformly distributed white noise ($\mu = 0$ and $\sigma = 1$) and $n = 200$ the confidence interval becomes CI = [-0.1386 ; 0.1386]. Figure 1 shows an example of the ACF of uncorrelated time series with $n = 200$, including the corresponding confidence interval.

The Shapiro-Wilk (SW) test can be used to show divergence of a data distribution from the Gaussian one. It is defined as

$$W = \frac{\sum_{i=1}^{n}(a_i z_i)^2}{\sum_{i=1}^{n}(z_i - \bar{z})^2},\qquad(3)$$

where z_i is the i-th order statistic from an ordered random sample $z_1 < z_2 < ... < z_k$ and $\bar{z}$ is the sample mean. The coefficients a_i are determined as

$$a_i = (a_1, a_2, ..., a_k) = \frac{m^T V^{-1}}{\sqrt{m^T V^{-1} V^{-1} m}}, \tag{4}$$

where $m = (m_1, m_2, .., m_k)^T$ is the vector of expected values of standard normal order statistics and V is the corresponding covariance matrix of those statistics. The value of the statistic W lies between zero and one, with a value of one indicating that the data has normal distribution.

In order to examine how the radio noise may affect deployed wireless networks we will have a look at channel capacity and how increased levels of radio noise could deteriorate it. As per Shannon-Hartley theorem the capacity of a certain channel can be calculated by

$$C = B \log_2(1 + S/N), \tag{5}$$

where B is the bandwidth of the channel in Hz, S is the average received signal power over the bandwidth in Watts and N is the average noise over the bandwidth also in Watts.

Firstly, we have to calculate a reference capacity for the particular technology, which we can compare with the measurements from the actual environment. The reference sensitivity needs to be set to the typical level a receiver has. For example, in the case of GSM the reference sensitivity level varies from -104 dBm to -87 dBm per 200 kHz of bandwidth for the different types of receivers [7]. The reference noise level is equal to:

$$N_{\text{ref}} = -174 + 10 \log_{10}(B) + NF \text{dBm Hz} \tag{6}$$

where NF is the noise figure of the receiver.

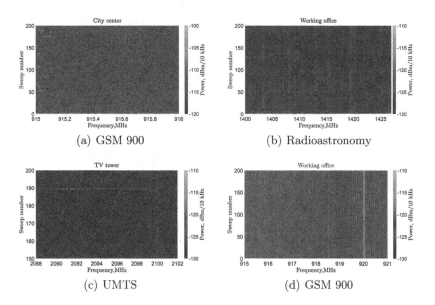

(a) GSM 900

(b) Radioastronomy

(c) UMTS

(d) GSM 900

Fig. 3. Spectograms of received power for different bands.

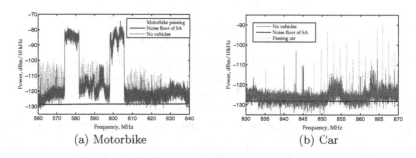

(a) Motorbike   (b) Car

Fig. 4. Shapiro-Wilk time-frequency plots. (Color figure online)

4 Exploratory Data Analysis

We start by studying the noise characteristics of our measurements. We consider a typical outdoor scenario at the street level of the Business Area close to a road intersection. We study at part of the DVB-T and LTE bands to understand how vehicle traffic can affect the overall noise contribution. The measurement setup was located right next to the stop sign at the crossroad with the antenna at around 5 m distance from the passing vehicles. We measured the frequency range before there were any cars present and after different vehicles passed by our equipment.

Figure 4(a) depicts the results from the measurement when a motorbike passed by. It can be seen from the figure that a large number of byproducts appear with power levels of up to -100 dBm/10 kHz. Figure 4(b) shows a second example of emitted byproducts when a car drove by our setup. One can see from the Figure that there are a large number of byproducts with levels of up to -90 dBm/10 kHz, the majority of which is concentrated to a frequency band from 850 MHz to 870 MHz.

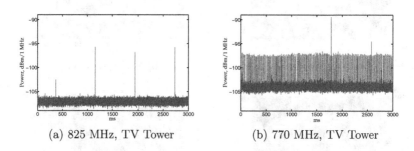

(a) 825 MHz, TV Tower   (b) 770 MHz, TV Tower

Fig. 5. Parts of the DVB-T band when different vehicles are passing by the measurement setup. (Color figure online)

We continue by analyzing the Radioastronomy and LTE 800 bands (that should be free of all man-made transmissions). The results for different location

Table 1. Mean power level and standard deviation for radioastronomy and LTE 800 guard band relative to the noise floor of the analyzer.

Location	Radioastronomy band		LTE800 Duplex Gap	
	rel. mean, dB	rel. σ, dB	rel. mean, dB	rel. σ, dB
Amplifier On:				
Empty Office	0	+0.0005	+0.078	+0.0034
Server room	+2.695	-0.0001	+7.282	+0.0276
Working Office	+1.205	+0.0635	+6.700	+0.3012
TV tower	+0.496	+0.0133	+0.047	+0.0249
Amplifier Off:				
City center	+0.328	-0.0013	+0.166	+0.0023
Office district	+0.307	-0.0009	+0.310	-0.0040
Transmission lines	0	-0.0029	+0.402	+0.0480
Working factory	+0.293	-0.0011	+0.240	+0.0038
Device Noise	mean, dBm	σ, dB	mean, dBm	σ, dB
Amplifier On	-127.299	1.7931	-128.184	1.7942
Amplifier Off	-114.590	1.7667	-114.255	1.7604

are shown in Table 1. We calculated the mean of the received power for the whole band. Regardless of the amplifier being off in some of the locations, a constant positive value was captured. One should note that indoor offices seem to have clearly persistent and strong excess noise levels. The reader should note that these results are aggregated statistics and smooth out details. Hence, they should be only used as a first step on finding the most noisy interesting bands to study in details.

5   Noise Variability

We continue by studying noise look like inside the different bands. In Fig. 3(a) the duplex gap of the GSM 900 band at the city center is depicted. The overall noise levels are slightly higher than in the other studied environments but the noise gets visible at the beginning of the band, at around 915 MHz that is concentrated around several frequency bins. It is also interesting that there is some time variation (sparks) and observerd RX-power goes up to -100 dBm. Most likely, since this noise is located at the beginning of the band it might be an adjacent band transmission leaking outside the permitted frequency mask. One more representative example is shown in Fig. 3(b), for the Radioastronomy band observed in the working office environment. This is different for, Fig. 3(c), where the UMTS guard band in the working office is depicted. In the Fig. 3(c) there is only one constant time noisy byproduct at 2100 MHz and a very wide band one that is time dependent and lasts only for the duration of the sweep (around 3 s). Finally, we consider the GSM 900 guard band. The results are shown in Fig. 3(d), where the noise seems to be existing in the whole band with different spectral characteristics. Power levels vary also considerably with the strongest one being -110 dBm at 920 MHz. Such high noise levels with such idle spectrum coverage show that the specific working environment either has lots of electronic devices or some defective component is causing this behavior.

It is clear from the presented examples that noise varies a lot in frequency and time domain. As expected we have found that the noise is highly location dependent. The spatial persistence of noise components vary from few tens of meters to up to more than hundreds of meters. The aggregated statistics can make whole bands to look less noiser than what they in reality are.

We will provide one more example to highlight the very rich nature of noisy byproducts giving an example of the time dynamics. In Fig. 6 we show noise byproducts that the measurements captured nearby TV-transmission tower. The measurements were done in zero span mode at two different frequencies, namely in the 825 MHz and 770 MHz for the TV Tower environment. In Fig. 6(a), the captured signal consists interference (noise) bursts that are somehow interleaved around 800 ms or so. In Fig. 6(b) 770 MHz band is shown. The time dynamics of this "noise channel" is much richer and the signal is composed from bursts that repeat every few ms.

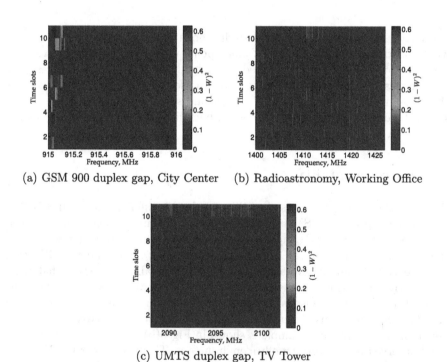

(a) GSM 900 duplex gap, City Center   (b) Radioastronomy, Working Office

(c) UMTS duplex gap, TV Tower

Fig. 6. Results from zero span measurements showing the detailed time domain structure of the noise.

6   Gaussianity Assumption

In this section we revisit AWGN channel assumptions [8–10]. Here, we would focus the Gaussianity characteristics of the AWGN channel and have a look at the time domain measurements to see if they resemble a Gaussian distribution. Finally, we study the time independency between samples that are also part of the AWGN channel.

For the Gaussianity we conducted the Shapiro-Wilk test (per frequency bin). That allows us to compare our captured data against the normal distribution. For this analysis we had to collect together multiple time domain measurements to calculate the Shapiro-Wilk statistic and thus 10 Shapiro-Wilk statistic time slots are given. Our first example is shown in Fig. 4(a) for the GSM 900 band in the City Center. The noise components around the 915 MHz that were described in the previous section are clearly seen as strong deviations from the Gaussianity assumption. We continue with Fig. 4(b) showing the Shapiro-Wilk statistics result for the Radioastronomy Band in the working office environment. The deviations from the Gaussian distribution are now well spread within the band and the problem is not just concentrated in few frequency bins as before. Finally in Fig. 4(c) we show results for the UMTS band where a strong deviation is captured that only lasted for some time before it disappeared. It is interesting though how widely the noise is spread within the band covering more than 12 MHz.

We continue by providing two more examples for the Radioastronomy and the UMTS duplex gap. In Fig. 7(a) the Radioastronomy band for the working office area is depicted. The corresponding spectogram was already presented in Fig. 4(b) and the Shapiro-Wilk statistics do a good job capturing what we have described above. The deviations are spread all over the frequency band. In Fig. 7(b) the UMTS duplex gap for the TV tower location is shown. The deviations are not as widespread as before and tend to be concentrated at the beginning of the band and even more so at the end of the gap.

We finalize this section by analyzing also at the time independency assumption of subsequent samples. We calculated the autocorrelation for the measurement samples and the results are presented in Fig. 8. It is clear that for most of the captured frequency bins the autocorrelation gives a value which shows that

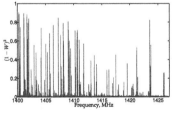
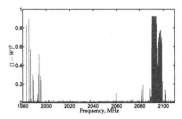

(a) Radioastronomy Working Office   (b) UMTS duplex gap, TV Tower

Fig. 7. Deviation from gaussianity assumption. Shapiro-Wilk tests.

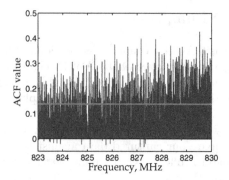

Fig. 8. Deviations from AWGN in the LTE Band in the working office. (Colour figure online)

the samples are not independent. The red line shows the limit for values to be considered independent and most of the values are larger than this threshold. Similar examples have been also found in other bands. From this set of measurements results it is clear that this large variability (frequency, time and locations) makes modeling of the excess noise rather difficult if not impossible.

7 Capacity Reduction from Increased Noise Levels

In this section we show how increased noise floor can impact the capacity of a radio systems. Digital receiver technology has progressed considerably on decoding and we are now able to exploit very weak signals. The sensitivity of receivers is superior to what they were 20 years ago. Thus, the increased noise levels have more impact on our system performance as we tend to operate in a low SNR regime, especially on the cell edges of mobile systems. We present first an example that is technology independent in the sense that we do not apply specific modulation and demodulation schemes. We employ the normalized capacity to calculate the lost capacity with excess noise. More specifically we evaluate the relative degradation of the Shannon spectral efficiency $\log(1 + S/N)$ for a signal power S and noise level N for different values of the input parameters derived from the measurements. We calculate the reference capacity C_{ref} for each band assuming first only the thermal noise floor. We then repeat the calculation ba adding the man-made interference to show how access noise components can have a similar effect as co-channel interference.

Figures 9(a) and (b) show that there is almost no reduction in the capacity for the empty office and the TV tower locations. However, for the rest of the locations the results at cell edge conditions can cause high capacity loss of up to 65 % and substantial reductions of up to 20 % even in rather high primary system power signal conditions, i.e., -80 dBm. The server room and the working office experience the highest MNP (Mean Noise Power) for the LTE800 and GSM900 duplex gaps compared to the other two locations. It is interesting to observe that the highest noise levels (that have some strong time domain constant type

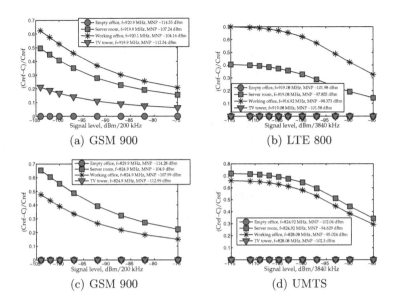

Fig. 9. Capacity reduction with the effect of measured noise.

of behavior) were spotted in the indoor environments. The reader should note that our measurements in the frequency domain were more prone in capturing long-time noise sources.

We continue estimating a possible capacity reduction for the LTE800 and the GSM900 technologies using measured values duplex gaps. In Figs. 9(c) and (d) we can see that the capacity reduction is a bit higher than in the GSM case. This is expected since the reference sensitivity for UMTS receivers is lower than the one for GSM. Situation is also similar for the empty office and the TV tower environments as before, although this does not necessarily mean that these locations are in total less noisy than others.

For the above results, we used a constant noise figure of 10 dB for the calculations of the capacity. Next, we give more examples that are based on today's specific receivers noise figures. In the Table 2 we summarize typical values for several different technologies that were used in our calculations. We produce 3-D plots that allows the interested reader to see exactly how noise can affect different type of receivers.

Figure 10(a) presents the results when the capacity is calculated with respect to the GSM technology receiver sensitivities. The capacity reduction in the empty office has a constant behavior, i.e., equals zero, down to noise figure of 7 dB, after which it starts gradually increasing. In such case the UMTS and LTE base stations, which generally have low noise figures as seen from Table 2, might experience problems. However, for receivers using Bluetooth and ZigBee technology, which have higher noise figures, as expected will not be affected by these noise levels.

Table 2. Typical noise figure for various technologies.

Technology	Noise figure, dB
GSM900	7
UMTS:	
Base station	2
User equipment	7
LTE:	
Base station	5
User equipment	9
802.11 (WLAN)	10
Bluetooth	23
ZigBee	20

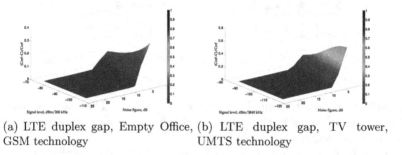

(a) LTE duplex gap, Empty Office, GSM technology

(b) LTE duplex gap, TV tower, UMTS technology

Fig. 10. Examples of the theoretical capacity reduction.

Finally, Figs. 10(a) and (b) also allow us to compare how the same noise could affect different technologies. We see from the Figures that for the same locations and frequencies the capacity reduction for the GSM case is lower compared to the UMTS one, which is due to the more strict requirements on the reference sensitivity of the receiver in UMTS.

8   Conclusions

In this paper we reported preliminary results from our high-resolution noise measurement campaign. We have shown that the noise has very different characteristics in different localities, both in frequency and time domains. Aggregated statistics at the same time can not represent exact noise levels since they heavily average out the noisy byproducts.

We were able to observe strong radio noise in all locations at different frequency bins and time scales, which shows that man-made noise level has significantly increased. It is at these levels that should be taken into account from contemporary designs of new-systems, and particularly be considered by dynamic spectrum access community. We show strong deviations from the theoretical Gaussian model. Whiteness of the noise, time independency, and the Gaussian structure of noise were found to not hold for many different frequency and time bands on different localities. Finally we would like to stress that noise measurements are not just new spectrum measurements focusing on the low level signals, but are a distinct type of characterization of the radio environment with unique challenges.

Acknowledgment. The authors would like to thank RWTH Aachen University and the German Research Foundation (Deutsche Forschungsgemeinschaft, DFG) for providing financial support.

References

1. Middleton, D.: Man-made noise in urban environments and transportation systems: models and measurements. IEEE Trans. Veh. Technol. **22**(4), 148–157 (1973)
2. Gemikonakli, O., Aghvami, A.: Impact of non-Gaussian noise on the performance of M-ary CPSK signalling transmitted through two-link nonlinear channels. IEEE Proc. Commun. Speech Vis. **136**(5), 328–332 (1989)
3. Bangerter, B., Talwar, S., Arefi, R., Stewart, K.: Networks and devices for the 5G era. IEEE Commun. Mag. **52**(2), 90–96 (2014)
4. Chandra, P.: Measurements of radio impulsive noise from various sources in an indoor environment at 900 MHz and 1800 MHz. In: The 13th IEEE International Symposium on Personal, Indoor and Mobile Radio Communications, pp. 639–643, vol. 2, September 2002
5. Shongwey, T., Vinck, A., Ferreira, H.: On impulse noise and its models. In: 18th IEEE International Symposium on Power Line Communications and its Applications, pp. 12–17, March 2014
6. Unawong, S., Miyamoto, S., Morinaga, N.: Receiver design of CDMA system for impulsive radio noise environment. In: International Symposium on Electromagnetic Compatibility Proceedings, pp. 316–319, May 1997
7. ETSI TS 145 005 V12.5.0 Digital cellular telecommunications system (Phase 2+); Radio Transmission and Reception, ITU-R, Technical report, April 2015
8. Suraweera, H., Smith, P., Surobhi, N.: Exact outage probabilityof cooperative diversity with opportunistic spectrum access. In: IEEE International Conference on Communications Workshops, pp. 79–84, May 2008
9. Atapattu, S., Tellambura, C., Jiang, H.: Energy detection based cooperative spectrum sensing in cognitive radio networks. IEEE Trans. Wirel. Commun. **10**(4), 1232–1241 (2011)
10. Ubaidulla, P., Aissa, S.: Optimal relay selection and power allocation for cognitive two-way relaying networks. IEEE Wirel. Commun. Lett. **1**(3), 225–228 (2012)

Utilization of Licensed Shared Access Resources in Indoor Small Cells Scenarios

Eva Perez[1]([✉]), Karl-Josef Friederichs[1], Andreas Lobinger[1],
Bernhard Wegmann[1], and Ingo Viering[2]

[1] Bell Labs Research, Nokia, Munich, Germany
{eva.perez,karl-josef.friederichs,andreas.lobinger,
bernhard.wegmann}@nokia.com
[2] Nomor Research, Munich, Germany
viering@nomor.de

Abstract. The increasing traffic demand will require additional spectrum to be used for mobile broadband. Licensed Shared Access (LSA) is one option for mobile network operators (MNO) to provide access to spectrum resources of other radio services, which are underutilized for specific time intervals and location areas, ensuring interference free coexistence between the sharing partners, i.e. this utilization of the spectrum requires decoupling of the resources in spatial or time domain. Indoor small cell deployments are particularly interesting for such a sharing scenario, due to the additional attenuation from the walls providing additional decoupling of the two systems. This article analyzes the network planning feasibility for LSA spectrum usage in indoor small cell scenarios. On basis of real indoor-to-outdoor propagation measurements and existing propagation models, a minimum decoupling range is determined where the LSA signal penetrating to the outdoor area falls below a certain threshold that guarantees interference free operation of the incumbent.

Keywords: Spectrum sharing · Spectrum efficiency · Licensed Shared Access (LSA) · Authorized Shared Access (ASA) · Indoor small cells

1 Introduction

The tremendous traffic increase observed during the last years, requires that the Mobile Network Operators (MNO) enhance their network capacities by improving the radio efficiency of their networks, increasing number of sites, etc. Furthermore, also additional spectrum for Mobile Broadband (MBB) provision will be required. Spectrum can be considered as the real estate for Mobile Broadband.

Exclusive access is the traditional means of making spectrum available to MNO. Currently there are around 600 MHz of total spectrum assigned to MBB. Although exclusive access is the preferred option, the amount of spectrum available is limited [1].

The original version of this chapter was revised.
An erratum to this chapter can be found at 10.1007/978-3-319-40352-6_63

© ICST Institute for Computer Sciences, Social Informatics and Telecommunications Engineering 2016
D. Noguet et al. (Eds.): CROWNCOM 2016, LNICST 172, pp. 462–470, 2016.
DOI: 10.1007/978-3-319-40352-6_38

An alternative option to have access to additional spectrum is the use of the unlicensed spectrum. However, unlicensed spectrum requires new add-ons for LTE to provide fair co-existence for LTE with existing technologies [2]. Two similar solutions considering unlicensed spectrum are being standardized: LTE-U in the LTE-U Forum and Licensed Assisted Access (LAA) in 3GPP.

On the other side, getting exclusive sharing rights from other licensed bands, even though restricted in time and location, is a very efficient mean to boost operators spectrum resources for MBB use. Mobile networks target to offer predictable quality of service; therefore it is required that sufficient control mechanisms are implemented when applying spectrum sharing. Authorized Shared Access (ASA), also known as Licensed Shared Access (LSA)[1] provides a solution for bands belonging to other radio services that cannot easily be re-farmed or totally vacated by their incumbent users but where actual spectrum usage is underutilized and infrequent [3,4].

Under LSA, a national authority can grant rights to a few LSA licensees to utilize portions of an incumbent spectrum that are unused, at a given location and time, for mobile communications provided that it is not creating harmful interference outside of that spatially and/or temporally defined area.

Due to the popularity of smart phones and tablets, most of the data usage is indoor, necessitating extensive indoor small cells deployments. Around 80 % of all mobile broadband traffic is consumed by users located indoor. Indoor small cells provide the lowest total cost of ownership (TCO) for providing coverage and capacity in indoor hot spots in enterprises and public buildings [5].

Small cell indoor capacity enhancements by additional use of LSA spectrum is rather appealing due to the lower transmit power and the attenuation of the radiation through the walls of the building which helps to minimize the interference outside the building where the incumbent requires undistorted usage of its services. On the other hand, the variety of materials, building structures, wall distributions, windows, etc. increases the complexity to detect and control the interference from the indoor cells to areas outside of the building.

The next sections of this paper provides a detailed study of interference from the indoor small cells, and their use of the LSA spectrum. In Sect. 2, the interference detection in the case of LSA spectrum used by indoor small cells is introduced, in Sect. 3, the field measurements campaign carried out for model validation is described, and in Sect. 4, based on the results from the previous section, it is described how the LSA spectrum can be used respecting the requirements from the sharing arrangements and sharing framework.

2   Interference Detection

For LSA utilization, both parties, the incumbent and the LSA licensee contractually guarantee protection against harmful interference from both the incumbent and LSA licensees, thus allowing them to provide a predictable quality of service [3].

---

[1] The Radio Spectrum Policy Group (RSPG) and the European Commission largely adopted and generalized the concept but renamed it to Licensed Shared Access where ASA is framed within LSA.

The advantage of lower interference from the indoor small cells deployments, compared to the outdoor macro cells, is the possibility to use the LSA spectrum closer to the border of the protected area used by the incumbent, where usage of the macro cells, due to their higher power and the missing shielding by walls, would not be possible.

Before the LSA spectrum can be shared among different partners, a sharing framework must be established with the regulator, and a sharing arrangement must be established with the incumbent. Once sharing framework and sharing arrangement are established, the LSA licensee is allowed to use the spectrum when the conditions defined in both the sharing framework and the sharing arrangement are fulfilled. These conditions can include technical requirements, such as interference thresholds in certain areas where the incumbent users may use the LSA spectrum (protection zones). Hence, the LSA licensee needs a mechanism to control its radiation emission, respecting the requirements defined at the sharing arrangement and sharing framework, while exploiting the maximum from the transmission.

A previous study [6], described a method to adapt the transmission characteristics of a MNO macro network such that maximum coverage for LSA spectrum usage is achieved. In this study, the MNO network using the LSA spectrum consist of an indoor small cell deployment. Such a deployment requires a more complex propagation model to estimate the interference from the indoor network towards outside of the building, considering the penetration losses through the outer wall of the building as well as walls inside the building. Due to that most of the existing propagation models are considering outdoor-to-indoor but not the opposite way from indoor-to-outdoor that is required in this context, field measurements have been carried out to validate the considered model.

3   Indoor-to-Outdoor Propagation Validation by Field Measurements

In principle user measurements can be used to detect the interference in forbidden areas. However, it cannot always guarantee that there are no areas without measurements covered by base stations operating with LSA frequencies. Therefore, in addition to the user measurements, an interference prediction based on propagation models is also needed. This interference prediction requires information known from network planning tools: location of the sites, frequency, transmit power, orientation and gain of the antennas. If available, user measurements can further enhance the prediction quality. These user measurements could be obtained, for example, by enabling the functionality Minimization of Drive Test (MDT) [7].

Most of the propagation models consider basically the outdoor use case [8–10], and in the case of having also indoor cases, the focus is normally in the propagation from outdoor to indoor [11,12]. Moreover, in the few cases where a model from indoor to outdoor is considered, the main purpose is to maximize the coverage from indoor cells also outside of the buildings, while for the LSA use case, the objective is the opposite.

In order to check the feasibility of this model outdoor field measurements have been carried out from indoor base station deployments and compared with propagation model predictions.

3.1 Propagation Model

The WINNER II project [13], provides a reasonable model for this study, where one of the scenarios analyzed is the indoor to outdoor propagation (Fig. 1).

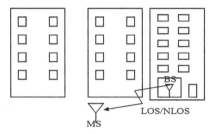

Fig. 1. Indoor to outdoor scenario

According to this model, the total path loss (PL) from indoor to outdoor is obtained from the sum of three components:

$$PL = PL_b + PL_{tw} + P_{in} \qquad (1)$$

where:

- PL_b are the propagation losses from the transmitter to the receiver location considering the outdoor model
- PL_{tw} are the propagation losses due to penetration of the outer wall, considering the angle of penetration (θ), and the losses when the signal is crossing the wall perpendicularly (W_e), or tangentially (WG_e).

$$PL_{tw} = W_e + WG_e(1 - cos(\theta))^2 \qquad (2)$$

- PL_{in} are the propagation losses inside the building, considering the number of walls (n_W), and the penetration loss per inner wall (L_W):

$$PL_{in} = n_W L_W \qquad (3)$$

Table 1 shows the penetration losses parameters considered for the simulations.

Table 1. Penetration losses

Parameter		Value
W_e	Loss through external wall for the perpendicular penetration	18 dB
WG_e	Loss through external wall for the parallel penetration	15 dB
L_w	Loss through the indoor wall	5 dB

3.2   Indoor-to-Outdoor Field Measurements

The measurements took place in the Nokia Campus, in Munich. Due to the unavailability of equipment for the 2.3 GHz band, typically used for LSA in Europe, the measurement campaign was done using Wi-Fi access points operating at 5.4 GHz, which enable easy signal strength measurements by available tools and further analysis. In principle also the 2.4 GHz Wi-Fi could have been used, but in our case that was too much interfered by a large number of active SSIDs.

The access point (AP) was located in one of the meeting rooms in the ground floor, and the signal strength from this AP, transmitting at 5.4 GHz, was measured outside the buildings, in the street immediately next to the meeting room where the AP was located, as well as in the other streets surrounding the building. The measurements were taken in winter, having snow on the streets.

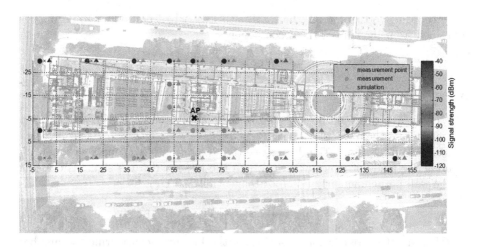

Fig. 2. Measurements and model prediction results

Figure 2 shows the signal strength measured and the propagation model predictions results at the measurement points in the campus.

First of all, both simulation and measurement results show that the signal strength of an indoor cell outside the building is still considerably high. This

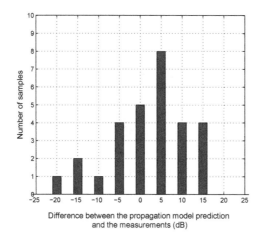

Fig. 3. Histogram of the difference between the propagation model prediction and the measured signal strength

effect is good when, the purpose of the deployments is to provide certain level of coverage outside the building. However, in the case of utilization of LSA spectrum, the purpose of the deployment of the small cells is to provide good coverage inside the building, but almost no coverage outside the building.

It is also observed that the results obtained from propagation model differ quite heavily with differences of up to 20 dB in both directions, under and over estimation.

Figure 3 shows the histogram of the difference between the results from the model and the measurements. The inaccuracy of the model requires an additional guard range of about 20 dB for confident interference prediction.

4   Indoor Deployment Options for Utilization of LSA Spectrum

Once the interference from the indoor small cells is estimated, it can be determined for which small cell base station depending on their location inside the building the LSA spectrum can be used.

At this step, the following information from the incumbent is required:

- Protection zone: Area where the incumbent users will use the spectrum
- Interference criterion: Maximum level of interference allowed by the incumbent spectrum users, that guarantees normal operation of the incumbent services.

Comparable to the LSA macro deployment study, the objective is to find optimal small cell indoor deployment and base station configuration that provides best LSA spectrum efficiency for allowed area while keeping the interference within the allowed range, which is investigated for a group of reference points

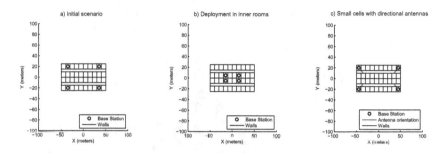

Fig. 4. Small cell indoor deployment options

in the protection zone, to compare it with the maximum interference allowed by the incumbent.

The previously described path loss model has been used to calculate the out-of-building emission for three different small cell indoor deployment options as given in Fig. 4. The deployment effecting parameters are Tx power, location of the base stations and their antenna characteristics. The received signal strength of the LSA frequency used indoor has been calculated for all reference points and is obtained summing up the signals from each indoor cell.

The three different deployment options of the small cell base stations inside the buidling can be classified as following:

(a) *Initial scenario:* The initial scenario consists of a single floor of a building, where there are four small cells placed at four different rooms.
(b) *Deployment of LSA small cells in inner rooms:* This option decreases the interference by locating the small cells in the inner rooms of the building, increasing the penetration losses due to the higher number of walls between the transmitter and the exterior of the building.
(c) *LSA small cells with directional antennas:* The small cells are located close to the external walls of the building, but oriented towards the inner part of the building. Although there will be some backward lobes of the antenna, the main lobe of the antenna will point to the inner part of the building, reducing the out-of-building emission received.

Considering the case of the wireless cameras [14,15], with an interference criterion (acceptable received interference power) of −95 dBm and the Tx power of 30 dBm of the small cell base stations the maximum interfere distance is determined, i.e. the further location measured from outer wall of the building.

Figure 5 shows the maximum interfered distance from the building for the three indoor small cell deployments, which corresponds to the further location from the outer wall where the signal exceeds the interference criterion.

The use of directional antennas provides the lowest out-of-building emission and allows the closest location of incumbent operation. For instance, if the equipment receiving the signal from wireless cameras is at least 40 m away from the building, the indoor small cells can be used without harming the wireless cameras communications.

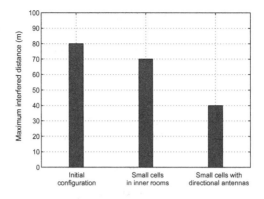

Fig. 5. Maximum interfered distance per small cell deployment

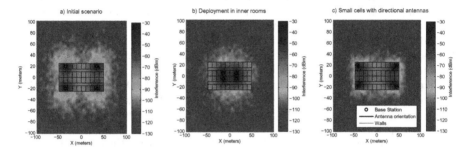

Fig. 6. Interference map per small cell deployment

The so-called heat maps visualizing the interference from all indoor sites are shown in Fig. 6.

5   Conclusions

LSA gives the mobile broadband operator the opportunity to use additional spectrum, which is currently licensed to other systems, and which is quite often unused for certain time intervals in certain locations. The LSA spectrum usage by small cells in indoor deployment outperforms the usage of outdoor macro cells since interference-free usage can be guaranteed even quite close to protection areas. With intelligent indoor small cell deployments the interference distance can be reduced to below a hundred meters, compared to couple of kilometers in case of the macro cells.

To determine that distance, network planning methods using empirical outdoor-to-indoor can be used. Although the match between the measurements and the model is not completely equivalent, the model can be used to estimate the areas where the current MNO cells may interfere.

The indoor small cell deployment option with directional antennas pointing to the inner part of the building showed the minimum out-of-building emissions. In case of LSA spectrum usage is planned is recommended to use this option.

References

1. Optimising spectrum utilisation towards 2020. White Paper, Nokia, Espoo (2014). http://networks.nokia.com/file/30301/optimising-spectrum-utilisation-towards-2020
2. LTE for unlicensed spectrum. White Paper, Nokia, Espoo (2014). http://networks.nokia.com/file/34596/lte-for-unlicensed-spectrum
3. CEPT: ECC Report 205 - Licensed Shared Access (LSA). Technical report, February 2013
4. Khun-Jush, J., Bender, P., Deschamps, B., Gundlach, M.: Licensed shared access as complementary approach to meet spectrum demands: benefits for next generation cellular systems. In: ETSI Workshop on Reconfigurable Radiosystems, Cannes (2012)
5. Indoor deployment strategies. White Paper, Nokia (2014). http://networks.nokia.com/file/32531/indoor-deployment-strategies
6. Perez, E., Friederichs, K.-J., Viering, I., Diego Naranjo, J.: Optimization of authorised/licensed shared access resources. In: 2014 9th International Conference on Cognitive Radio Oriented Wireless Networks and Communications (CROWNCOM), pp. 241–246, June 2014
7. Universal Terrestrial Radio Access (UTRA) and Evolved Universal Terrestrial Radio Access (E-UTRA); Radio measurement collection for Minimaztion of Drive Test (MDT), 3GPP. Technical Specification, Technical report TS 37.320, September 2014
8. Evolved Universal Terrestrial Radio Access (E-UTRA); Further Advancements for E-UTRA physical layer aspect, 3GPP. Technical report, TR 36.814, March 2010
9. Universal Mobile Telecommunications System (UMTS): Radio Frequency (RF) system scenarios, 3GPP Technical report, TR 25.914, October 2012
10. Hata, M.: Empirical formula for propagation loss in land mobile radio services. IEEE Trans. Veh. Technol. **29**(3), 317–325 (1980)
11. Hillery, W., Cudak, M., Ghosh, A., Vejlgaard, B.: Inside-out: can indoor femto cells satisfy outdoor coverage and capacity needs? In: IEEE Globecom Workshops (GC Wkshps), pp. 339–344, December 2013
12. Predicting coverage and interference involving the indoor-outdoor interface. White Paper, OFCOM (2007). http://stakeholders.ofcom.org.uk/binaries/research/technology-research/report1.pdf
13. Pekka, K., Juha, M., Lassi, H., et al.: WINNER II channel models part II radio channel measurement and analysis results. WINNER II, Munich, Germany, Technical report IST-4-027756 D, vol. 1 (2007)
14. CEPT: ECC Report 172 - Broadband Wireless Systems Usage in 2300–2400 MHz. Technical report, April 2012
15. CEPT: ECC Report 204 - Spectrum use and future requirements for PMSE. Technical report, February 2014

When Does Channel-Output Feedback Enlarge the Capacity Region of the Two-User Linear Deterministic Interference Channel?

Victor Quintero[1]($^{\boxtimes}$), Samir M. Perlaza[1], Iñaki Esnaola[2], and Jean-Marie Gorce[1]

[1] Institut National de Recherche en Informatique et en Automatique (INRIA), CITI Laboratory, 6 Av. des Arts, 69621 Villeurbanne, France
{victor.quintero-florez,samir.perlaza,jean-marie.gorce}@inria.fr
[2] Department of Automatic Control and Systems Engineering, The University of Sheffield, Mappin Street, Sheffield S1 3JD, UK
esnaola@sheffield.ac.uk

Abstract. The two-user linear deterministic interference channel (LD-IC) with noisy channel-output feedback is fully described by six parameters that correspond to the number of bit-pipes between each transmitter and its corresponding intended receiver, i.e., $\overrightarrow{n}_{11}$ and $\overrightarrow{n}_{22}$; between each transmitter and its corresponding non-intended receiver i.e., n_{12} and n_{21}; and between each receiver and its corresponding transmitter, i.e., $\overleftarrow{n}_{11}$ and $\overleftarrow{n}_{22}$. An LD-IC without feedback corresponds to the case in which $\overleftarrow{n}_{11} = \overleftarrow{n}_{22} = 0$ and the capacity region is denoted by $\mathcal{C}(\overrightarrow{n}_{11}, \overrightarrow{n}_{22}, n_{12}, n_{21}, 0, 0)$. In the case in which feedback is available at both transmitters, $\overleftarrow{n}_{11} > 0$ and $\overleftarrow{n}_{22} > 0$, the capacity is denoted by $\mathcal{C}(\overrightarrow{n}_{11}, \overrightarrow{n}_{22}, n_{12}, n_{21}, \overleftarrow{n}_{11}, \overleftarrow{n}_{22})$. This paper presents the exact conditions on $\overleftarrow{n}_{11}$ (resp. $\overleftarrow{n}_{22}$) for observing an improvement in the capacity region $\mathcal{C}(\overrightarrow{n}_{11}, \overrightarrow{n}_{22}, n_{12}, n_{21}, \overleftarrow{n}_{11}, 0)$ (resp. $\mathcal{C}(\overrightarrow{n}_{11}, \overrightarrow{n}_{22}, n_{12}, n_{21}, 0, \overleftarrow{n}_{22})$) with respect to $\mathcal{C}(\overrightarrow{n}_{11}, \overrightarrow{n}_{22}, n_{12}, n_{21}, 0, 0)$, for any 4-tuple $(\overrightarrow{n}_{11}, \overrightarrow{n}_{22}, n_{12}, n_{21}) \in \mathbb{N}^4$. Specifically, it is shown that there exists a threshold for the number of bit-pipes in the feedback link of transmitter-receiver pair 1 (resp. 2), denoted by $\overleftarrow{n}_{11}^\star$ (resp. $\overleftarrow{n}_{22}^\star$) for which any $\overleftarrow{n}_{11} > \overleftarrow{n}_{11}^\star$ (resp. $\overleftarrow{n}_{22} > \overleftarrow{n}_{22}^\star$) enlarges the capacity region, i.e., $\mathcal{C}(\overrightarrow{n}_{11}, \overrightarrow{n}_{22}, n_{12}, n_{21}, 0, 0) \subset \mathcal{C}(\overrightarrow{n}_{11}, \overrightarrow{n}_{22}, n_{12}, n_{21}, \overleftarrow{n}_{11}, 0)$ (resp. $\mathcal{C}(\overrightarrow{n}_{11}, \overrightarrow{n}_{22}, n_{12}, n_{21}, 0, 0) \subset \mathcal{C}(\overrightarrow{n}_{11}, \overrightarrow{n}_{22}, n_{12}, n_{21}, 0, \overleftarrow{n}_{22}))$. The exact conditions on $\overleftarrow{n}_{11}$ (resp. $\overleftarrow{n}_{22}$) to observe an improvement on a single rate or the sum-rate capacity, for any 4-tuple $(\overrightarrow{n}_{11}, \overrightarrow{n}_{22}, n_{12}, n_{21}) \in \mathbb{N}^4$ are also presented in this paper.

The CITI Laboratory is a joint lab between the Institut National de Recherche en Informatique et en Automatique (INRIA), Université de Lyon and Institut National de Sciences Apliquées (INSA) de Lyon. This research was supported in part by the European Commission under Marie Sklodowska-Curie Individual Fellowship No. 659316 (CYBERNETS), Universidad del Cauca, Popayán, Colombia, and the Administrative Department of Science, Technology and Innovation of Colombia (Colciencias), fellowship No. 617-2013.

© ICST Institute for Computer Sciences, Social Informatics and Telecommunications Engineering 2016
D. Noguet et al. (Eds.): CROWNCOM 2016, LNICST 172, pp. 471–483, 2016.
DOI: 10.1007/978-3-319-40352-6_39

Keywords: Linear deterministic interference channel · Noisy channel-output feedback · Capacity region

1   Introduction

Channel-output feedback is an interference management technique that significantly increases the number of degrees of freedom (DoF) for the two-user Gaussian interference channel (IC) in most of the interference regimes [1]. Essentially, in the very strong interference regime, the DoF gain provided by perfect-channel output feedback can be arbitrarily large when the interference to noise ratios (INRs) and signal to noise ratios (SNRs) grow to infinity. One of the reasons why feedback provides such a surprising benefit stems from the fact that it uses interference to create alternative paths to the existing point-to-point paths. For instance, in the two-user IC, feedback creates a path from transmitter 1 (resp. transmitter 2) to receiver 1 (resp. receiver 2) in which symbols that are received at receiver 2 (resp. receiver 1) are fed back to transmitter 2 (resp. transmitter 1), which decodes the messages and retransmits them to receiver 1 (resp. receiver 2). Another metric to determine the benefits of feedback is the number of generalized DoF (GDoF), see [3] for a discussion on DoF and GDoF, as well as other metrics. The GDoF gain due to feedback in the IC depends on the topology of the network and the number of transmitter-receiver pairs in the network. For instance, in the symmetric K-user cyclic Z-interference channel, the DoF gain does not increase with K [4]. In particular, in the very strong interference regime, the DoF gain is shown to be monotonically decreasing with K. Alternatively, in the fully connected symmetric K-user IC with perfect feedback, the number of GDoF per user is shown to be identical to the one in the two-user case, with an exception in a particular singularity, and totally independent of the exact number of transmitter-receiver pairs [5]. It is important to highlight that the network topology, the number of transmitter-receiver pairs and the interference regime are not the only parameters determining the effect of feedback. Indeed, the presence of noise in the feedback links turns out to be another relevant factor. As shown later in this paper, in the case in which one transmitter-receiver pair is in a high interference regime (the interfering signal is stronger than the intended signal) and the other is in a low interference regime (the interfering signal is weaker than the intended signal), the use of feedback in the former does not enlarge the capacity region, even in the case of perfect output feedback. Conversely, using feedback in the latter might enlarge the capacity region depending on the SNR of the feedback link. The exact values of the feedback SNRs beyond which the capacity region is enlarged depend on all the other channel parameters: two forward SNRs and two forward INRs. In [6], the capacity region of the two-user Gaussian IC (GIC) with noisy channel output feedback is approximated to within a constant number of bits for the symmetric case. These results are generalized in [11] for non-symmetric cases. However, from the available descriptions of the capacity regions with and without feedback, identifying whether or not the existence of a feedback link with a given SNR enlarges the capacity region is not a trivial task.

An alternative for dealing with the challenges described above is to study the GIC via its linear deterministic IC (LD-IC) approximation [7], for which the capacity region is perfectly known [8,10]. The two-user LD-IC with noisy channel output feedback (LD-IC-NOF) is fully described by six parameters: $(\overrightarrow{n}_{11}, \overrightarrow{n}_{22}, n_{12}, n_{21}, \overleftarrow{n}_{11}, \overleftarrow{n}_{22}) \in \mathbb{N}^6$. There exists a mapping between the parameters describing the two-user LD-IC and the parameters describing the GIC. More specifically, there are two forward SNRs $(\overrightarrow{\mathrm{SNR}}_i \geqslant 1)$; two forward INRs $(\mathrm{INR}_{ij} \geqslant 1)$; and two backward SNRs $(\overleftarrow{\mathrm{SNR}}_i \geqslant 1)$, with $i \in \{1,2\}$ and $j \in \{1,2\} \setminus \{i\}$. In the LD-IC, the parameters of the GIC are mapped into the number of bit-pipes between each transmitter and its corresponding intended receiver, i.e., $\overrightarrow{n}_{ii} = \lfloor \frac{1}{2} \log_2(\overrightarrow{\mathrm{SNR}}_i) \rfloor$; between transmitter j and receiver i i.e., $n_{ij} = \lfloor \frac{1}{2} \log_2(\mathrm{INR}_{ij}) \rfloor$; and between each receiver and its corresponding transmitter, i.e., $\overleftarrow{n}_{ii} = \lfloor \frac{1}{2} \log_2(\overleftarrow{\mathrm{SNR}}_i) \rfloor$. An LD-IC without feedback corresponds to the case in which $\overleftarrow{n}_{11} = \overleftarrow{n}_{22} = 0$ and the capacity region is denoted by $\mathcal{C}(\overrightarrow{n}_{11}, \overrightarrow{n}_{22}, n_{12}, n_{21}, 0, 0)$. In the case in which feedback is available at both transmitters, $\overleftarrow{n}_{11} > 0$ and $\overleftarrow{n}_{22} > 0$, the capacity is denoted by $\mathcal{C}(\overrightarrow{n}_{11}, \overrightarrow{n}_{22}, n_{12}, n_{21}, \overleftarrow{n}_{11}, \overleftarrow{n}_{22})$.

This paper presents the exact conditions on $\overleftarrow{n}_{11}$ (resp. $\overleftarrow{n}_{22}$) for observing an improvement in the capacity region $\mathcal{C}(\overrightarrow{n}_{11}, \overrightarrow{n}_{22}, n_{12}, n_{21}, \overleftarrow{n}_{11}, 0)$ (resp. $\mathcal{C}(\overrightarrow{n}_{11}, \overrightarrow{n}_{22}, n_{12}, n_{21}, 0, \overleftarrow{n}_{22})$) with respect to $\mathcal{C}(\overrightarrow{n}_{11}, \overrightarrow{n}_{22}, n_{12}, n_{21}, 0, 0)$, for any 4-tuple $(\overrightarrow{n}_{11}, \overrightarrow{n}_{22}, n_{12}, n_{21}) \in \mathbb{N}^4$. More specifically, it is shown that there exists a threshold for the number of bit-pipes in the feedback link of transmitter-receiver pair 1 (resp. 2), beyond which the capacity region of the two-user LD-IC-NOF can be enlarged, i.e., $\mathcal{C}(\overrightarrow{n}_{11}, \overrightarrow{n}_{22}, n_{12}, n_{21}, 0, 0) \subset \mathcal{C}(\overrightarrow{n}_{11}, \overrightarrow{n}_{22}, n_{12}, n_{21}, \overleftarrow{n}_{11}, 0)$ (resp. $\mathcal{C}(\overrightarrow{n}_{11}, \overrightarrow{n}_{22}, n_{12}, n_{21}, 0, 0) \subset \mathcal{C}(\overrightarrow{n}_{11}, \overrightarrow{n}_{22}, n_{12}, n_{21}, 0, \overleftarrow{n}_{22}))$, with strict inclusion. The exact conditions on $\overleftarrow{n}_{11}$ (resp. $\overleftarrow{n}_{22}$) to observe an improvement on a single rate or the sum-rate capacity, for any 4-tuple $(\overrightarrow{n}_{11}, \overrightarrow{n}_{22}, n_{12}, n_{21}) \in \mathbb{N}^4$ are also presented in this paper. Surprisingly, these values can be expressed in closed-form using relatively simple expressions that depend on some of the parameters $\overrightarrow{n}_{11}, \overrightarrow{n}_{22}, n_{12}$ and n_{21}.

Based on these results, several relevant engineering questions arise in this setting. For instance, in which of the two transmitter-receiver pairs must the feedback link be implemented if the objective is to improve: (a) the individual rate of the transmitter-receiver pair in which feedback is implemented; (b) the individual rate of the other transmitter-receiver pair; or (c) the sum-rate of both transmitter-receiver pairs. In each of these scenarios, the feedback SNR, either $\overleftarrow{n}_{11}$ or $\overleftarrow{n}_{22}$, must be bigger than a given threshold for the improvement to be observed. Interestingly, for each of these scenarios there exists a complete different answer. As a by-product of the results described above, the exact values of $\overleftarrow{n}_{11}$ or $\overleftarrow{n}_{22}$ for which feedback does not enlarge the capacity region are also identified.

2 Linear Deterministic Interference Channel with Noisy-Channel Output Feedback

Consider the two-user LD-IC-NOF, with parameters $\overrightarrow{n}_{11}$, $\overrightarrow{n}_{22}$, n_{12}, n_{21}, $\overleftarrow{n}_{11}$ and $\overleftarrow{n}_{22}$ described in Fig. 1. The parameters $\overrightarrow{n}_{ii}$, n_{ij} and $\overleftarrow{n}_{ii}$ with $i \in \{1, 2\}$ and $j \in \{1, 2\} \setminus \{i\}$, are non-negative integers. Parameter $\overrightarrow{n}_{ii}$ represents the number of bit-pipes between transmitter i and receiver i; parameter n_{ij} represents the number of bit-pipes between transmitter j and receiver i; and parameter $\overleftarrow{n}_{ii}$ represents the number of bit-pipes between receiver i and transmitter i (feedback). At transmitter i, with $i \in \{1, 2\}$, the channel-input $\boldsymbol{X}_i^{(n)}$ at channel use n, with $n \in \{1, \ldots, N\}$, is a q-dimensional binary vector $\boldsymbol{X}_i^{(n)} = \left(X_{i,1}^{(n)}, \ldots, X_{i,q}^{(n)} \right)^{\mathsf{T}}$, with

$$q = \max\left(\overrightarrow{n}_{11}, \overrightarrow{n}_{22}, n_{12}, n_{21} \right), \tag{1}$$

and N the block length. At receiver i, the channel-output $\overrightarrow{\boldsymbol{Y}}_i^{(n)}$ at channel use n is also a q-dimensional binary vector $\overrightarrow{\boldsymbol{Y}}_i^{(n)} = \left(\overrightarrow{Y}_{i,1}^{(n)}, \ldots, \overrightarrow{Y}_{i,q}^{(n)} \right)^{\mathsf{T}}$. The input-output relation during channel use n is given as follows

$$\overrightarrow{\boldsymbol{Y}}_i^{(n)} = \boldsymbol{S}^{q-\overrightarrow{n}_{ii}} \boldsymbol{X}_i^{(n)} + \boldsymbol{S}^{q-n_{ij}} \boldsymbol{X}_j^{(n)}, \tag{2}$$

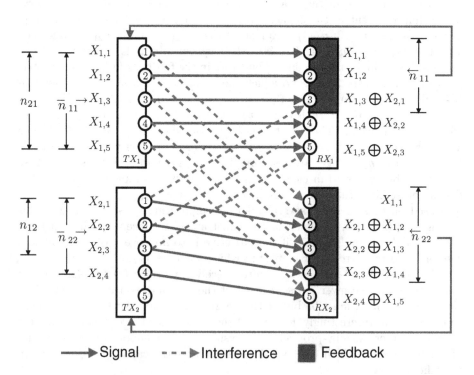

Fig. 1. Two-user linear deterministic interference channel with noisy channel-output feedback. The bit-pipe line number 1 represents the most significant bit.

and the feedback signal available at transmitter i at the end of channel use n is:

$$\overleftarrow{\boldsymbol{Y}}_{i}^{(n)} = \boldsymbol{S}^{(q-\overleftarrow{n}_{ii})^{+}} \overrightarrow{\boldsymbol{Y}}_{i}^{(n-d)}, \tag{3}$$

where d is a finite feedback delay, additions and multiplications are defined over the binary field, $\boldsymbol{S}$ is a $q \times q$ lower shift matrix, and $(\cdot)^{+}$ is the positive part operator.

Transmitter i sends M_i information bits $b_{i,1}, \ldots, b_{i,M_i}$ by sending the codeword $\left(\boldsymbol{X}_i^{(1)}, \ldots, \boldsymbol{X}_i^{(N)}\right)$. The encoder of transmitter i can be modeled as a set of deterministic mappings $f_i^{(1)}, \ldots, f_i^{(N)}$, with $f_i^{(1)} : \{0,1\}^{M_i} \rightarrow \{0,1\}^q$ and $\forall n \in \{2, \ldots, N\}$, $f_i^{(n)} : \{0,1\}^{M_i} \times \{0,1\}^{q(n-1)} \rightarrow \{0,1\}^q$, such that

$$\boldsymbol{X}_i^{(1)} = f_i^{(1)}\left(b_{i,1}, \ldots, b_{i,M_i}\right) \text{ and} \tag{4}$$

$$\boldsymbol{X}_i^{(n)} = f_i^{(n)}\left(b_{i,1}, \ldots, b_{i,M_i}, \overleftarrow{\boldsymbol{Y}}_i^{(1)}, \ldots, \overleftarrow{\boldsymbol{Y}}_i^{(n-1)}\right). \tag{5}$$

At the end of the block, receiver i uses the sequence $\boldsymbol{Y}_i^{(1)}, \ldots, \boldsymbol{Y}_i^{(N)}$ to generate the estimates $\hat{b}_{i,1}, \ldots, \hat{b}_{i,M_i}$. The average bit error probability at receiver i, denoted by p_i, is calculated as follows

$$p_i = \frac{1}{M_i} \sum_{\ell=1}^{M_i} \mathbb{1}_{\{\hat{b}_{i,\ell} \neq b_{i,\ell}\}}. \tag{6}$$

A rate pair $(R_1, R_2) \in \mathbb{R}_+^2$ is said to be achievable if it satisfies the following definition.

Definition 1 (Achievable Rate Pairs). The rate pair $(R_1, R_2) \in \mathbb{R}_+^2$ is achievable if there exists at least one pair of codebooks $\mathcal{X}_1^N$ and $\mathcal{X}_2^N$ with codewords of length N, and the corresponding encoding functions $f_1^{(1)}, \ldots, f_1^{(N)}$ and $f_2^{(1)}, \ldots, f_2^{(N)}$ such that the average bit error probability can be made arbitrarily small by letting the block length N grow to infinity.

Denote by $\mathcal{C}(\overrightarrow{n}_{11}, \overrightarrow{n}_{22}, n_{12}, n_{21}, \overleftarrow{n}_{11}, \overleftarrow{n}_{22})$ the capacity region of the two-user LD-IC-NOF with parameters $\overrightarrow{n}_{11}$, $\overrightarrow{n}_{22}$, n_{12}, n_{21}, $\overleftarrow{n}_{11}$, and $\overleftarrow{n}_{22}$. Lemma 1 fully characterizes the set $\mathcal{C}(\overrightarrow{n}_{11}, \overrightarrow{n}_{22}, n_{12}, n_{21}, \overleftarrow{n}_{11}, \overleftarrow{n}_{22})$.

Lemma 1 (Lemma 6 in [10]). The capacity region $\mathcal{C}(\overrightarrow{n}_{11}, \overrightarrow{n}_{22}, n_{12}, n_{21}, \overleftarrow{n}_{11}, \overleftarrow{n}_{22})$ of the two-user LD-IC-NOF is the set of non-negative rate pairs (R_1, R_2) that satisfy $\forall i \in \{1, 2\}$ and $j \in \{1, 2\} \setminus \{i\}$:

$$R_i \leqslant \min\left(\max\left(\overrightarrow{n}_{ii}, n_{ji}\right), \max\left(\overrightarrow{n}_{ii}, n_{ij}\right)\right), \tag{7a}$$

$$R_i \leqslant \min\left(\max\left(\overrightarrow{n}_{ii}, n_{ji}\right), \max\left(\overrightarrow{n}_{ii}, \overleftarrow{n}_{jj} - \left(\overrightarrow{n}_{jj} - n_{ji}\right)^{+}\right)\right), \tag{7b}$$

$$R_1 + R_2 \leqslant \min\left(\max\left(\overrightarrow{n}_{22}, n_{12}\right) + \left(\overrightarrow{n}_{11} - n_{12}\right)^{+}, \max\left(\overrightarrow{n}_{11}, n_{21}\right)\right.$$

$$\left. + \left(\overrightarrow{n}_{22} - n_{21}\right)^{+}\right), \tag{7c}$$

$$R_1 + R_2 \leqslant \max\left((\overrightarrow{n}_{11} - n_{12})^+, n_{21}, \overrightarrow{n}_{11} - (\max(\overrightarrow{n}_{11}, n_{12}) - \overleftarrow{n}_{11})^+\right)$$
$$+ \max\left((\overrightarrow{n}_{22} - n_{21})^+, n_{12}, \overrightarrow{n}_{22} - (\max(\overrightarrow{n}_{22}, n_{21}) - \overleftarrow{n}_{22})^+\right), \quad (7d)$$
$$2R_i + R_j \leqslant \max(\overrightarrow{n}_{ii}, n_{ji}) + (\overrightarrow{n}_{ii} - n_{ij})^+$$
$$+ \max\left((\overrightarrow{n}_{jj} - n_{ji})^+, n_{ij}, \overrightarrow{n}_{jj} - (\max(\overrightarrow{n}_{jj}, n_{ji}) - \overleftarrow{n}_{jj})^+\right). \quad (7e)$$

3 Preliminaries

3.1 Definitions

Let $\alpha_i \in \mathbb{Q}$ be the interference regime of transmitter-receiver pair i, with $i \in \{1,2\}$ and $j \in \{1,2\} \setminus \{i\}$,

$$\alpha_i = \frac{n_{ij}}{\overrightarrow{n}_{ii}}. \quad (8)$$

For each transmitter-receiver pair i, there exist five possible interference regimes (IRs): very weak IR (VWIR), i.e., $\alpha_i \leqslant \frac{1}{2}$, weak IR (WIR), i.e., $\frac{1}{2} < \alpha_i \leqslant \frac{2}{3}$, moderate IR (MIR), i.e., $\frac{2}{3} < \alpha_i \leqslant 1$, strong IR (SIR), i.e., $1 < \alpha_i \leqslant 2$ and very strong IR (VSIR), i.e., $\alpha_i > 2$ [9]. The scenarios in which the desired signal is stronger than or equal to the interference ($\alpha_i \leqslant 1$), namely VWIR, WIR and MIR, are referred to as low-interference regimes (LIRs). Conversely, the scenarios in which the desired signal is weaker than the interference ($\alpha_i > 1$), namely SIR and VSIR, are referred to as high-interference regimes (HIRs). In the two-user LD-IC, it is possible to observe up to twenty-five possible interference regimes, given α_1 and α_2. However, only twelve cases are of real interest. This is because the transmitter-receiver pairs can be indifferently labeled and thus, for instance, studying the case in which $\alpha_1 \leqslant \frac{1}{2}$ and $\alpha_2 > 2$ is the same as studying the case in which $\alpha_1 > 2$ and $\alpha_2 \leqslant \frac{1}{2}$.

The main results of this paper are presented using a list of events (Boolean variables) that are fully determined by the parameters $\overrightarrow{n}_{11}$, $\overrightarrow{n}_{22}$, n_{12}, and n_{21}. For instance, given the parameters $(\overrightarrow{n}_{11}, \overrightarrow{n}_{22}, n_{12}, n_{21})$, the events (9)–(25) describe some combinations of interference regimes that are particularly interesting. Let $i \in \{1,2\}$ and $j \in \{1,2\} \setminus \{i\}$ and define the following events:

$$E_1: \quad \alpha_1 \leqslant 1 \quad \wedge \quad \alpha_2 \leqslant 1, \ (i \text{ and } j \text{ in LIR}) \quad (9)$$

$$E_{2,i}: \quad \alpha_i \leqslant \frac{1}{2} \wedge 1 < \alpha_j \leqslant 2, \ (i \text{ in VWIR and } j \text{ in SIR}) \quad (10)$$

$$E_{3,i}: \quad \alpha_i \leqslant \frac{1}{2} \wedge \quad \alpha_j > 2, \ (i \text{ in VWIR and } j \text{ in VSIR}) \quad (11)$$

$$E_{4,i}: \frac{1}{2} < \alpha_i \leqslant \frac{2}{3} \wedge \quad \alpha_j > 1, \ (i \text{ in WIR and } j \text{ in HIR}) \quad (12)$$

$$E_{5,i}: \frac{2}{3} < \alpha_i \leqslant 1 \wedge 1 < \alpha_j \leqslant 2, \ (i \text{ in MIR and } j \text{ in SIR}) \quad (13)$$

$$E_{6,i}: \frac{2}{3} < \alpha_i \leqslant 1 \wedge \quad \alpha_j > 2, \ (i \text{ in MIR and } j \text{ in VSIR}) \quad (14)$$

$$E_{7,i} : \frac{1}{2} < \alpha_i \leqslant 1 \wedge \quad \alpha_j > 1, \; (i \text{ in WIR or MIR and } j \text{ in HIR}) \tag{15}$$

$$E_{8,i} : \quad \alpha_i > 1 \wedge \quad \alpha_j \leqslant 1, \; (i \text{ in HIR and } j \text{ in LIR}) \tag{16}$$

$$E_9 : \quad \alpha_1 \leqslant \frac{2}{3} \wedge \quad \alpha_2 \leqslant \frac{2}{3}, \; (i \text{ and } j \text{ in VWIR or WIR}) \tag{17}$$

$$E_{10,i} : \quad \alpha_i \leqslant \frac{2}{3} \wedge \frac{2}{3} < \alpha_j \leqslant 1, \; (i \text{ in VWIR or WIR and } j \text{ in MIR}) \tag{18}$$

$$E_{11,i} : \frac{2}{3} < \alpha_i \leqslant 1 \wedge \quad \alpha_j \leqslant \frac{2}{3}, \; (i \text{ in MIR and } j \text{ in VWIR or WIR}) \tag{19}$$

$$E_{12} : \quad \alpha_1 > 2 \wedge \quad \alpha_2 > 2, \; (i \text{ and } j \text{ in VSIR}). \tag{20}$$

Some other auxiliary events are considered. The event in which the signal from transmitter i is stronger (resp. weaker) in its intended receiver than in its non-intended receiver is denoted by $E_{13,i}$ (resp. $\widetilde{E}_{13,i}$), i.e.,

$$E_{13,i} : \overrightarrow{n}_{ii} > n_{ji}, \tag{21}$$

$$\widetilde{E}_{13,i} : \overrightarrow{n}_{ii} < n_{ji}. \tag{22}$$

The event in which the sum of the number of bit-pipes in the direct links is bigger (resp. smaller) than the sum of the number of bit-pipes in the cross-interference links is denoted by $E_{14,i}$ (resp. $\widetilde{E}_{14,i}$), i.e.,

$$E_{14} : \overrightarrow{n}_{11} + \overrightarrow{n}_{22} > n_{12} + n_{21}, \tag{23}$$

$$\widetilde{E}_{14} : \overrightarrow{n}_{11} + \overrightarrow{n}_{22} < n_{12} + n_{21}. \tag{24}$$

The event in which the number of bit-pipes in the direct link j is bigger than the sum of bit-pipes in both cross-interference links is denoted by

$$E_{15,i} : \overrightarrow{n}_{jj} > n_{ij} + n_{ji}. \tag{25}$$

The event in which the sum of the number of bit-pipes in the direct links is bigger than the sum of the number of bit-pipes in one cross-interference link and twice the number of the bit-pipes in the other cross-interference link is denoted by

$$E_{16,i} : \overrightarrow{n}_{ii} + \overrightarrow{n}_{jj} > n_{ij} + 2n_{ji}. \tag{26}$$

Finally, the event in which the sum of the number of bit-pipes in the direct links is bigger than the number of bit-pipes in one cross-interference link is denoted by

$$E_{17,i} : \overrightarrow{n}_{ii} + \overrightarrow{n}_{jj} < n_{ij}. \tag{27}$$

Combining the events (9)–(27), five main events are identified:

$$S_{1,i} : (E_1 \wedge E_{13,i}) \vee (E_{2,i} \wedge E_{13,i}) \vee (E_{3,i} \wedge E_{13,i} \wedge E_{14}) \vee (E_{4,i} \wedge E_{13,i} \wedge E_{14})$$
$$\vee (E_{5,i} \wedge E_{13,i} \wedge E_{14}) \vee \left(E_{6,i} \wedge \widetilde{E}_{13,j} \wedge E_{14}\right), \tag{28}$$

$$S_{2,i} : \left(E_{3,i} \wedge \widetilde{E}_{13,j} \wedge \widetilde{E}_{14}\right) \vee \left(E_{7,i} \wedge \widetilde{E}_{13,j} \wedge \widetilde{E}_{14}\right) \vee \left(\overline{E}_1 \wedge \widetilde{E}_{13,j}\right), \tag{29}$$

$$S_{3,i} : (E_1 \wedge \overline{E}_{13,i}) \vee (E_{2,i} \wedge \overline{E}_{13,i}) \vee (E_{3,i} \wedge E_{13,j} \wedge \overline{E}_{13,i}) \vee (E_{4,i} \wedge E_{13,j} \wedge \overline{E}_{13,i})$$
$$\vee (E_{5,i} \wedge E_{13,j} \wedge \overline{E}_{13,i}) \vee (E_{6,i} \wedge E_{13,j}) \vee (\overline{E}_1 \wedge E_{13,j}) \vee (E_{8,i}), \tag{30}$$
$$S_{4,i} : (E_9 \wedge E_{13,i} \wedge E_{13,j}) \vee (E_{10,i} \wedge E_{13,i} \wedge E_{13,j} \wedge E_{16,i} \wedge E_{16,j})$$
$$\vee (E_{11,i} \wedge E_{13,i} \wedge E_{13,j} \wedge E_{16,i} \wedge E_{16,j}) \tag{31}$$
$$S_{5,i} : E_{12} \wedge E_{17,i} \wedge E_{17,j}. \tag{32}$$

For all $i \in \{1, 2\}$ the events $S_{1,i}$, $S_{2,i}$ and $S_{3,i}$ exhibit the property stated by the following lemma.

Lemma 2. For all $i \in \{1, 2\}$ and for all $(\overrightarrow{n}_{11}, \overrightarrow{n}_{22}, n_{12}, n_{21}) \in \mathbb{N}^4$, only one of the events $S_{1,i}$, $S_{2,i}$ and $S_{3,i}$ is true.

Proof. The proof follows from verifying that for all $i \in \{1, 2\}$ and $j \in \{1, 2\} \setminus \{i\}$, the events (28)–(30) are mutually exclusive. For instance, consider that the event $(E_1 \wedge E_{13,i})$ in (28) is true. Then, $S_{1,i}$ is true and $E_{2,i}$, $E_{3,i}$, $E_{4,i}$, $E_{5,i}$, $E_{6,i}$, $E_{7,i}$ and $E_{8,i}$ hold false, which implies that $S_{2,i}$ and $S_{3,i}$ hold false as well, since all events in (29) and (30) hold false. The same verification can be made for all the remaining events in (28). This proves that if $S_{1,i}$ is true then $S_{2,i}$ and $S_{3,i}$ hold simultaneously false. The same verification can be done for showing that when $S_{2,i}$ holds true (resp. $S_{3,i}$), both events $S_{1,i}$ and $S_{3,i}$ (resp. $S_{1,i}$ and $S_{2,i}$) hold simultaneously false. Finally following the same reasoning it can be verified that if any pair of the events $\{S_{1,i}, S_{2,i}, S_{3,i}\}$ is false, the remaining event is necessarily true. This completes the proof.   ∎

For all $i \in \{1, 2\}$ the events $S_{4,i}$ and $S_{5,i}$ exhibit the property stated by the following lemma.

Lemma 3. For all $i \in \{1, 2\}$ if one of the events $S_{4,i}$ or $S_{5,i}$ holds true, then the other necessarily holds false.

Proof. The proof of Lemma 3 follows along the same lines of the proof of Lemma 2.   ∎

3.2   Rate Improvement Metrics

The rate improvements are given in terms of the following metrics [8, 10]: (a) maximum individual rate improvements Δ_1 and Δ_2; and (b) maximum sum-rate improvement Σ, with $\Delta_i \in \mathbb{R}_+$ and $\Sigma \in \mathbb{R}_+$ for $i \in \{1, 2\}$.

Let $\mathcal{C}_1 = \mathcal{C}(\overrightarrow{n}_{11}, \overrightarrow{n}_{22}, n_{12}, n_{21}, \overleftarrow{n}_{11}, \overleftarrow{n}_{22})$ and $\mathcal{C}_2 = \mathcal{C}(\overrightarrow{n}_{11}, \overrightarrow{n}_{22}, n_{12}, n_{21}, 0, 0)$ be the capacity region with noisy channel-output feedback and without feedback, respectively. In order to formally define Δ_1, Δ_2 and Σ, consider a two-user LD-IC-NOF with parameters $\overrightarrow{n}_{11}$, $\overrightarrow{n}_{22}$, n_{12}, n_{21}, $\overleftarrow{n}_{11}$, and $\overleftarrow{n}_{22}$. The maximum improvement of the individual rate R_i, $\Delta_i(\overrightarrow{n}_{11}, \overrightarrow{n}_{22}, n_{12}, n_{21}, \overleftarrow{n}_{11}, \overleftarrow{n}_{22})$, due to the effect of channel-output feedback with respect to the case without feedback is

$$\Delta_i(\overrightarrow{n}_{11}, \overrightarrow{n}_{22}, n_{12}, n_{21}, \overleftarrow{n}_{11}, \overleftarrow{n}_{22}) = \max_{R_j > 0} \sup_{\substack{(R_i, R_j) \in \mathcal{C}_1 \\ (R_i^\dagger, R_j) \in \mathcal{C}_2}} R_i - R_i^\dagger, \tag{33}$$

and the maximum improvement of the sum rate $\Sigma(\overrightarrow{n}_{11}, \overrightarrow{n}_{22}, n_{12}, n_{21}, \overleftarrow{n}_{11}, \overleftarrow{n}_{22})$ with respect to the case without feedback is

$$\Sigma(\overrightarrow{n}_{11}, \overrightarrow{n}_{22}, n_{12}, n_{21}, \overleftarrow{n}_{11}, \overleftarrow{n}_{22}) = \sup_{\substack{(R_1, R_2) \in \mathcal{C}_1 \\ (R_1^\dagger, R_2^\dagger) \in \mathcal{C}_2}} R_1 + R_2 - (R_1^\dagger + R_2^\dagger). \quad (34)$$

In the following, when feedback is exclusively used by transmitter-receiver pair i, i.e., $\overleftarrow{n}_{ii} > 0$ and $\overleftarrow{n}_{jj} = 0$, then the maximum improvement of the individual rate of transmitter-receiver k, with $k \in \{1, 2\}$, and the maximum improvement of the sum rate are denoted by $\Delta_k(\overrightarrow{n}_{11}, \overrightarrow{n}_{22}, n_{12}, n_{21}, \overleftarrow{n}_{ii})$ and $\Sigma(\overrightarrow{n}_{11}, \overrightarrow{n}_{22}, n_{12}, n_{21}, \overleftarrow{n}_{ii})$, respectively. Hence, this notation $\Delta_k(\overrightarrow{n}_{11}, \overrightarrow{n}_{22}, n_{12}, n_{21}, \overleftarrow{n}_{ii})$ replaces either $\Delta_k(\overrightarrow{n}_{11}, \overrightarrow{n}_{22}, n_{12}, n_{21}, \overleftarrow{n}_{11}, 0)$ or $\Delta_k(\overrightarrow{n}_{11}, \overrightarrow{n}_{22}, n_{12}, n_{21}, 0, \overleftarrow{n}_{22})$, when $i = 1$ or $i = 2$, respectively. The same holds for the notation $\Sigma(\overrightarrow{n}_{11}, \overrightarrow{n}_{22}, n_{12}, n_{21}, \overleftarrow{n}_{ii})$ that replaces $\Sigma(\overrightarrow{n}_{11}, \overrightarrow{n}_{22}, n_{12}, n_{21}, \overleftarrow{n}_{11}, 0)$ or $\Sigma(\overrightarrow{n}_{11}, \overrightarrow{n}_{22}, n_{12}, n_{21}, 0, \overleftarrow{n}_{22})$, when $i = 1$ or $i = 2$, respectively.

4   Main Results

4.1   Enlargement of the Capacity Region

In this subsection, the capacity region of a two-user LD-IC-NOF with parameters $(\overrightarrow{n}_{11}, \overrightarrow{n}_{22}, n_{12}, n_{21})$, when feedback is available only at transmitter-receiver pair i, i.e., $\overleftarrow{n}_{ii} > 0$ and $\overrightarrow{n}_{jj} = 0$, is denoted by $\mathcal{C}(\overrightarrow{n}_{11}, \overrightarrow{n}_{22}, n_{12}, n_{21}, \overleftarrow{n}_{ii})$ instead of $\mathcal{C}(\overrightarrow{n}_{11}, \overrightarrow{n}_{22}, n_{12}, n_{21}, \overleftarrow{n}_{11}, 0)$ or $\mathcal{C}(\overrightarrow{n}_{11}, \overrightarrow{n}_{22}, n_{12}, n_{21}, 0, \overleftarrow{n}_{22})$, when $i = 1$ or $i = 2$, respectively. Following this notation, Theorem 1 identifies the exact values of $\overleftarrow{n}_{ii}$ for which the strict inclusion $\mathcal{C}(\overrightarrow{n}_{11}, \overrightarrow{n}_{22}, n_{12}, n_{21}, 0, 0) \subset \mathcal{C}(\overrightarrow{n}_{11}, \overrightarrow{n}_{22}, n_{12}, n_{21}, \overleftarrow{n}_{ii})$ holds, with $i \in \{1, 2\}$.

Theorem 1. Let $i \in \{1, 2\}$, $j \in \{1, 2\} \setminus \{i\}$ and $\overleftarrow{n}_{ii}^* \in \mathbb{N}$ be

$$\overleftarrow{n}_{ii}^* = \begin{cases} \max\left(n_{ji}, (\overrightarrow{n}_{ii} - n_{ij})^+\right) & \text{if } S_{1,i} = \text{True} \\ \overrightarrow{n}_{jj} + (\overrightarrow{n}_{ii} - n_{ij})^+ & \text{if } S_{2,i} = \text{True}. \end{cases} \quad (35)$$

Assume that $S_{3,i} = \text{True}$. Then, for all $\overleftarrow{n}_{ii} \in \mathbb{N}$, $\mathcal{C}(\overrightarrow{n}_{11}, \overrightarrow{n}_{22}, n_{12}, n_{21}, 0, 0) = \mathcal{C}(\overrightarrow{n}_{11}, \overrightarrow{n}_{22}, n_{12}, n_{21}, \overleftarrow{n}_{ii})$. Assume that either $S_{1,i} = \text{True}$ or $S_{2,i} = \text{True}$. Then, for all $\overleftarrow{n}_{ii} \leqslant \overleftarrow{n}_{ii}^*$, $\mathcal{C}(\overrightarrow{n}_{11}, \overrightarrow{n}_{22}, n_{12}, n_{21}, 0, 0) = \mathcal{C}(\overrightarrow{n}_{11}, \overrightarrow{n}_{22}, n_{12}, n_{21}, \overleftarrow{n}_{ii})$ and for all $\overleftarrow{n}_{ii} > \overleftarrow{n}_{ii}^*$, $\mathcal{C}(\overrightarrow{n}_{11}, \overrightarrow{n}_{22}, n_{12}, n_{21}, 0, 0) \subset \mathcal{C}(\overrightarrow{n}_{11}, \overrightarrow{n}_{22}, n_{12}, n_{21}, \overleftarrow{n}_{ii})$.

Proof. The proof of Theorem 1 is presented in [12]. ∎

Theorem 1 shows that under event $S_{3,i}$ in (30), implementing feedback in transmitter-receiver pair i does not bring any capacity region enlargement. Alternatively, under events $S_{1,i}$ in (28) and $S_{2,i}$ in (29), the capacity region can be enlarged whenever $\overleftarrow{n}_{ii} > \overleftarrow{n}_{ii}^*$. That is, there exists a threshold on the SNR of the feedback link beyond which it is possible to observe a capacity region enlargement.

An interesting observation is that the threshold $\overleftarrow{n}_{ii}^*$ beyond which feedback is useful is different under event $S_{1,i}$ in (28) and event $S_{2,i}$ in (29). In general when $S_{1,i}$ holds true, the enlargement of the capacity region is due to the fact that feedback allows using *interference as side information* [13]. More specifically, when feedback is used at transmitter-receiver pair i and $\overleftarrow{n}_{ii} > \max\left(n_{ji}, (\overrightarrow{n}_{ii} - n_{ij})^+\right)$, transmitter i obtains part of the information sent by transmitter j. This information can be re-transmitted by transmitter i to cancel the interference it produced at receiver i when it was first transmitted by transmitter j. Interestingly, the interference perceived at receiver j due to this re-transmission can be cancelled given that this information was reliably decoded when it was first sent by transmitter j. This allows transmitter-receiver pair i or j to improve its individual rate. Alternatively, when $S_{2,i}$ in (29) holds true, the enlargement of the capacity region occurs thanks to the fact that some of the bits that cannot be transmitted directly from transmitter j to receiver j, that is, those transmitted via the bit-pipes $\overrightarrow{n}_{jj} + 1, \ldots, \max(\overrightarrow{n}_{jj}, n_{ij})$, can arrive to receiver j via an alternative path: transmitter j - receiver i - transmitter i - receiver j. For this to be possible at least the $\left(\overrightarrow{n}_{jj} + (\overrightarrow{n}_{ii} - n_{ij})^+ + 1\right)$-th (feedback) bit-pipe from receiver i to transmitter i must be noise-free, i.e., $\overleftarrow{n}_{ii} > \overrightarrow{n}_{jj} + (\overrightarrow{n}_{ii} - n_{ij})^+$.

4.2   Improvement of the Individual Rate R_i by Using Feedback in Link i

Implementing channel output feedback in transmitter-receiver pair i might allow increasing the individual rate of either transmitter-receiver pair i or j. Theorem 2 identifies the exact values of $\overleftarrow{n}_{ii}$ for which the individual rate R_i can be improved, given the parameters $(\overrightarrow{n}_{11}, \overrightarrow{n}_{22}, n_{12}, n_{21})$ and $\overleftarrow{n}_{jj} = 0$ in the two-user LD-IC-NOF.

Theorem 2. *Let $i \in \{1,2\}$, $j \in \{1,2\} \setminus \{i\}$ and $\overleftarrow{n}_{ii}^{\dagger} \in \mathbb{N}$ be*

$$\overleftarrow{n}_{ii}^{\dagger} = \max\left(n_{ji}, (\overrightarrow{n}_{ii} - n_{ij})^+\right). \tag{36}$$

Assume that either $S_{2,i} = \text{True}$ or $S_{3,i} = \text{True}$. Then, for all $\overleftarrow{n}_{ii} \in \mathbb{N}$, $\Delta_i(\overrightarrow{n}_{11}, \overrightarrow{n}_{22}, n_{12}, n_{21}, \overleftarrow{n}_{ii}) = 0$. Assume that $S_{1,i} = \text{True}$. Then, when $\overleftarrow{n}_{ii} \leqslant \overleftarrow{n}_{ii}^{\dagger}$, it holds that $\Delta_i(\overrightarrow{n}_{11}, \overrightarrow{n}_{22}, n_{12}, n_{21}, \overleftarrow{n}_{ii}) = 0$; and when $\overleftarrow{n}_{ii} > \overleftarrow{n}_{ii}^{\dagger}$, it holds that $\Delta_i(\overrightarrow{n}_{11}, \overrightarrow{n}_{22}, n_{12}, n_{21}, \overleftarrow{n}_{ii}) > 0$.

Proof. The proof of Theorem 2 is presented in [12].   ∎

Theorem 2 highlights that under events $S_{2,i}$ in (29) and $S_{3,i}$ in (30), the individual rate R_i cannot be improved by using feedback in transmitter-receiver pair i, i.e., $\Delta_i(\overrightarrow{n}_{11}, \overrightarrow{n}_{22}, n_{12}, n_{21}, \overleftarrow{n}_{ii}) = 0$. Alternatively, under event $S_{1,i}$ in (28), the individual rate R_i can be improved, i.e., $\Delta_i(\overrightarrow{n}_{11}, \overrightarrow{n}_{22}, n_{12}, n_{21}, \overleftarrow{n}_{ii}) > 0$, whenever $\overleftarrow{n}_{ii} > \max\left(n_{ji}, (\overrightarrow{n}_{ii} - n_{ij})^+\right)$. It is worth noting that under event $S_{3,i}$

in (30), the capacity region cannot be improved via feedback (Theorem 1) and thus, none of the individual rates can be improved as suggested by Theorem 2. Alternatively, under event $S_{2,i}$ in (29), the capacity region can be enlarged (Theorem 1) but the individual rate R_i cannot be improved (Theorem 2). This implies that the capacity improvement occurs due to the fact that R_j can be improved. More specifically, in this case: $\Delta_i(\overrightarrow{n}_{11}, \overrightarrow{n}_{22}, n_{12}, n_{21}, \overleftarrow{n}_{ii}) = 0$ and $\Delta_j(\overrightarrow{n}_{11}, \overrightarrow{n}_{22}, n_{12}, n_{21}, \overleftarrow{n}_{ii}) > 0$. This implies that using feedback in transmitter-receiver pair i is exclusively beneficial for transmitter-receiver pair j, as shown in the following section.

4.3 Improvement of the Individual Rate R_j by Using Feedback in Link i

Implementing channel output feedback in transmitter-receiver pair i might allow increasing the individual rate of transmitter-receiver pair i or j; or both individual rates. This reveals the altruistic nature of implementing feedback as suggested in [2]. Theorem 3 identifies the exact values of $\overleftarrow{n}_{ii}$ for which the individual rate R_j can be improved by using feedback in transmitter-receiver pair i, given the parameters $(\overrightarrow{n}_{11}, \overrightarrow{n}_{22}, n_{12}, n_{21})$ and $\overleftarrow{n}_{jj} = 0$ in the two-user LD-IC-NOF.

Theorem 3. Let $i \in \{1, 2\}$, $j \in \{1, 2\} \setminus \{i\}$ and $\overleftarrow{n}_{ii}^{\dagger} \in \mathbb{N}$ be

$$\overleftarrow{n}_{ii}^{\dagger} = \begin{cases} \max\left(n_{ji}, (\overrightarrow{n}_{ii} - n_{ij})^+\right) & \text{if } S_{1,i} = \text{True} \\ \overrightarrow{n}_{jj} + (\overrightarrow{n}_{ii} - n_{ij})^+ & \text{if } S_{2,i} = \text{True.} \end{cases} \qquad (37)$$

Assume that $S_{3,i} = \text{True}$. Then, for all $\overleftarrow{n}_{ii} \in \mathbb{N}$, $\Delta_j(\overrightarrow{n}_{11}, \overrightarrow{n}_{22}, n_{12}, n_{21}, \overleftarrow{n}_{ii}) = 0$. Assume that either $S_{1,i} = \text{True}$ or $S_{2,i} = \text{True}$. Then, when $\overleftarrow{n}_{ii} \leqslant \overleftarrow{n}_{ii}^{\dagger}$, it holds that $\Delta_j(\overrightarrow{n}_{11}, \overrightarrow{n}_{22}, n_{12}, n_{21}, \overleftarrow{n}_{ii}) = 0$; and when $\overleftarrow{n}_{ii} > \overleftarrow{n}_{ii}^{\dagger}$, it holds that $\Delta_j(\overrightarrow{n}_{11}, \overrightarrow{n}_{22}, n_{12}, n_{21}, \overleftarrow{n}_{ii}) > 0$.

Proof. The proof of Theorem 3 is presented in [12]. ∎

Theorem 3 shows that under event $S_{3,i}$ in (30), implementing feedback in transmitter-receiver pair i does not bring any improvement on the rate R_j. This is in line with the results of Theorem 1 that states that under event $S_{3,i}$ in (30), implementing feedback in transmitter-receiver pair i does not enlarge the capacity region.

In contrast, under events $S_{1,i}$ in (28) and $S_{2,i}$ in (29), the individual rate R_j can be improved ($\Delta_j(\overrightarrow{n}_{11}, \overrightarrow{n}_{22}, n_{12}, n_{21}, \overleftarrow{n}_{ii}) > 0$) whenever $\overleftarrow{n}_{ii} > \overleftarrow{n}_{ii}^{\dagger}$. It is important to highlight that under event $S_{1,i}$, the threshold on $\overleftarrow{n}_{ii}$ for increasing the individual rate R_i i.e., $(\overleftarrow{n}_{ii}^{\dagger})$, and R_j i.e., $(\overleftarrow{n}_{ii}^{\dagger})$, are identical, see Theorems 2 and 3. This shows that in this case, the use of feedback in transmitter-receiver pair i, with $\overleftarrow{n}_{ii} > \overleftarrow{n}_{ii}^{\dagger} = \overleftarrow{n}_{ii}^{\dagger}$, simultaneously improves both individual rates. Under event $S_{2,i}$, using feedback in transmitter-receiver pair i, with $\overleftarrow{n}_{ii} > \overleftarrow{n}_{ii}^{\dagger}$, exclusively benefits transmitter-receiver pair j, which can improve its own individual rate.

4.4   Improvement of the Sum-Capacity

Implementing channel output feedback in transmitter-receiver pair i might allow increasing the sum-capacity. Theorem 4 identifies the exact values of $\overleftarrow{n}_{ii}$ for which the sum-capacity can be improved, for parameters $(\overrightarrow{n}_{11}, \overrightarrow{n}_{22}, n_{12}, n_{21})$ and $\overleftarrow{n}_{jj} = 0$ in the two-user LD-IC-NOF.

Theorem 4. Let $i \in \{1,2\}$, $j \in \{1,2\} \setminus \{i\}$ and $\overleftarrow{n}_{ii}^{+} \in \mathbb{N}$ be

$$\overleftarrow{n}_{ii}^{+} = \begin{cases} \max\left(n_{ji}, (\overrightarrow{n}_{ii} - n_{ij})^{+}\right) & \text{if } S_{4,i} = \text{True} \\ \overrightarrow{n}_{jj} + (\overrightarrow{n}_{ii} - n_{ij})^{+} & \text{if } S_{5,i} = \text{True}. \end{cases} \tag{38}$$

Assume that $S_{4,i} = \text{False}$ and $S_{5,i} = \text{False}$. Then, for all $\overleftarrow{n}_{ii} \in \mathbb{N}$, $\Sigma(\overrightarrow{n}_{11}, \overrightarrow{n}_{22}, n_{12}, n_{21}, \overleftarrow{n}_{ii}) = 0$. Assume that $S_{4,i} = \text{True}$ or $S_{5,i} = \text{True}$. Then, when $\overleftarrow{n}_{ii} \leqslant \overleftarrow{n}_{ii}^{+}$, it holds that $\Sigma(\overrightarrow{n}_{11}, \overrightarrow{n}_{22}, n_{12}, n_{21}, \overleftarrow{n}_{ii}) = 0$; and when $\overleftarrow{n}_{ii} > \overleftarrow{n}_{ii}^{+}$, it holds that $\Sigma(\overrightarrow{n}_{11}, \overrightarrow{n}_{22}, n_{12}, n_{21}, \overleftarrow{n}_{ii}) > 0$.

Proof. The proof of Theorem 4 is presented in [12].   ∎

Theorem 4 identifies the conditions under which implementing feedback in transmitter-receiver pair i improves the sum-capacity whenever $\overleftarrow{n}_{ii} > \overleftarrow{n}_{ii}^{+}$, that is, $\Sigma(\overrightarrow{n}_{11}, \overrightarrow{n}_{22}, n_{12}, n_{21}, \overleftarrow{n}_{ii}) > 0$. Theorem 4 highlights that one of the necessary but not sufficient conditions for improving the sum-capacity by implementing feedback in transmitter-receiver pair i is that either (a) at least one transmitter-receiver pair must be in VWIR or WIR; or (b) both transmitter-receiver pairs must be in VSIR. This follows immediately from observing that for $S_{4,i}$ or $S_{5,i}$ to hold true, at least one of the events E_9, $E_{10,i}$, $E_{11,i}$ or E_{12} must hold true. Interestingly, Theorem 4 shows that if at least one transmitter-receiver pair is in SIR, then the sum-capacity cannot be improved. Finally, note that the thresholds $\overleftarrow{n}_{ii}^{+}$ in the events $S_{4,i}$ and $S_{5,i}$ coincide with those observed in Theorem 1.

5   Conclusions

This paper presented the exact conditions on the feedback parameters $\overleftarrow{n}_{11}$ and $\overleftarrow{n}_{22}$, beyond which the capacity region of the two-user LD-IC-NOF can be enlarged for any 4-tuple $(\overrightarrow{n}_{11}, \overrightarrow{n}_{22}, n_{12}, n_{21}) \in \mathbb{N}^4$. More specifically, the exact values of $\overleftarrow{n}_{11}$ (resp. $\overleftarrow{n}_{22}$) for which $\mathcal{C}(\overrightarrow{n}_{11}, \overrightarrow{n}_{22}, n_{12}, n_{21}, 0, 0) \subset \mathcal{C}(\overrightarrow{n}_{11}, \overrightarrow{n}_{22}, n_{12}, n_{21}, \overleftarrow{n}_{11}, 0)$ (resp. $\mathcal{C}(\overrightarrow{n}_{11}, \overrightarrow{n}_{22}, n_{12}, n_{21}, 0, 0) \subset \mathcal{C}(\overrightarrow{n}_{11}, \overrightarrow{n}_{22}, n_{12}, n_{21}, 0, \overleftarrow{n}_{22}))$, with strict inclusion. The exact conditions on $\overleftarrow{n}_{11}$ (resp. $\overleftarrow{n}_{22}$) to observe an improvement on a single rate or the sum-rate capacity, for any 4-tuple $(\overrightarrow{n}_{11}, \overrightarrow{n}_{22}, n_{12}, n_{21}) \in \mathbb{N}^4$ were also presented. Interestingly, there exist conditions in the two-user LD-IC-NOF in which the use of feedback does not enlarge the capacity region.

References

1. Suh, C., Tse, D.N.C.: Feedback capacity of the Gaussian interference channel to within 2 bits. IEEE Trans. Inf. Theory **57**(5), 2667–2685 (2011)
2. Perlaza, S.M., Tandon, R., Poor, H.V., Han, Z.: Perfect output feedback in the two-user decentralized interference channel. IEEE Trans. Inf. Theory **61**(10), 5441–5462 (2015)
3. Jafar, S.A.: Interference alignment: a new look at signal dimensions in a communication network. Found. Trends Commun. Inf. Theory **7**(1), 1–134 (2010)
4. Tandon, R., Mohajer, S., Poor, H.V.: On the symmetric feedback capacity of the K-user cyclic Z-interference channel. IEEE Trans. Inf. Theory **59**(5), 2713–2734 (2013)
5. Mohajer, S., Tandon, R., Poor, H.V.: On the feedback capacity of the fully connected K-user interference channel. IEEE Trans. Inf. Theory **59**(5), 2863–2881 (2013)
6. Le, S.-Q., Tandon, R., Motani, M., Poor, H.V.: Approximate capacity region for the symmetric Gaussian interference channel with noisy feedback. IEEE Trans. Inf. Theory **61**(7), 3737–3762 (2015)
7. Avestimehr, S., Diggavi, S., Tse, D.N.C.: Wireless network information flow: a deterministic approach. IEEE Trans. Inf. Theory **57**(4), 1872–1905 (2011)
8. Quintero, V., Perlaza, S.M., Gorce, J.-M.: Noisy channel-output feedback capacity of the linear deterministic interference channel. In: IEEE Information Theory Workshop. Jeju Island, Korea (2015)
9. Etkin, R.H., Tse, D.N.C., Hua, W.: Gaussian interference channel capacity to within one bit. IEEE Trans. Inf. Theory **54**(12), 5534–5562 (2008)
10. Quintero, V., Perlaza, S.M., Gorce, J.-M.: Noisy channel-output feedback capacity of the linear deterministic interference channel. Technical report 456, INRIA Grenoble - Rhône- Alpes, Lyon, France, January 2015
11. Quintero, V., Perlaza, S.M., Esnaola, I., Gorce, J.-M.: Approximated capacity of the two-user Gaussian interference channel with noisy channel-output feedback. Technical report 8861, INRIA Grenoble - Rhône- Alpes, Lyon, France, March 2016
12. Quintero, V., Perlaza, S.M., Esnaola, I., Gorce, J.-M.: When does channel-output feedback enlarge the capacity region of the two-user linear deterministic interference channel?. Technical report 8862, INRIA Grenoble - Rhône- Alpes, Lyon, France, March 2016
13. Shannon, C.E.: Channels with side information at the transmitter. IBM J. Res. Dev. **2**(4), 289–293 (1958)

Hardware Architecture and Implementation

A Flexible 5G Receiver Architecture Adapted to VLSI Implementation

Vincent Berg[(✉)] and Jean-Baptiste Doré

CEA-LETI, Minatec Campus, Grenoble, France
{vincent.berg,jean-baptiste.dore}@cea.fr

Abstract. A flexible data frame structure adapted to 5G operations and designed to support high bandwidth pipes or sporadic traffic is described. The frame structure imposes to consider receiver architectures that are adapted to orthogonal frequency division mutliplexing (OFDM) for structured synchronous traffic and alternative flexible asynchronous waveforms such as filterbank multicarrier (FBMC) for sporadic traffic. OFDM and FBMC receivers are reviewed and a new flexible receiver architecture is then proposed and described. The design of the new architecture is centered on a memory unit complemented with co-processor units improving the flexibility of the digital signal processing operations of the receiver. The architecture is particularly adapted to application specific integrated circuit. The throughput imposed on the memory and the associated data receiver bus has been evaluated. The evaluation concluded that the throughput is suitable for very large scale integration implementations.

Keywords: FBMC · OFDM · LTE · 5G · Architecture

1 Introduction

So far, the appetite for broadband service has fueled the development of mobile cellular networks. Mobile communications started with wireless real time voice communications in the first and second generations of cellular systems (1G and 2G) to provide reliable voice connectivity everywhere. It was then followed by internet data connectivity in the third generation (3G) when the adoption of laptop computers became widespread to bring internet on-the-go. Finally, the advent of the smart-phone accelerated the demand for high bandwidth with the world information accessed at the tip of everyone's finger everywhere at anytime. Therefore, the aim to deliver high-bandwidth pipes has logically been the main driver for the current fourth generation (4G) also called Long Term Evolution (LTE) and LTE-Advanced (LTE-A).

In order to maximize spectral efficiency, strict synchronization and orthogonality between users within a single cell is imposed by LTE and LTE-A standards. However, sporadic traffic has emerged as an important service for future generations of cellular networks (5G). Machine Type Communications (MTC)

© ICST Institute for Computer Sciences, Social Informatics and Telecommunications Engineering 2016
D. Noguet et al. (Eds.): CROWNCOM 2016, LNICST 172, pp. 487–497, 2016.
DOI: 10.1007/978-3-319-40352-6_40

devices of the Internet of Things are expected to inherently generate sporadic data traffic to the network and should not be forced to be integrated into the constrained synchronization procedure of LTE-A in order to limit signaling overhead. Furthermore, a previously unforeseen mechanism designed to save battery usage of the handset also called fast dormancy has resulted in significant control signaling growth. This mechanism causes the user equipment (UE) to go into a deep sleep mode and break any connection to the network. When the UE changes back to an active state the mobile has to go through a complete synchronization procedure again. This phenomenon is another significant source of sporadic traffic on the network [1].

Furthermore, because spectrum is scarce and expensive, its utilization should be as optimal as possible. However, the nature of the sporadic traffic causes significant fragmentation. Therefore carrier aggregation will be implemented to achieve much higher rates by dynamically aggregating non-contiguous frequency bands [2]. However, legacy LTE-A imposes generous guard bands to other legacy networks to satisfy spectral mask requirements because of the poor frequency localization of OFDM.

Therefore relaxed synchronization and access to fragmented spectrum have been considered as key parameters for future generations of wireless networks [1,2]. This requirement of spectrum agility has encouraged the study of alternative multicarrier waveforms such as filter bank multicarrier (FBMC)to provide better adjacent channel leakage performance without compromising spectral efficiency [3].

So far, few studies have been realized to evaluate the architecture trade-offs of hardware implementations of FBMC transceivers. An implementation of a software defined radio platform has been described in [4]. In [5], a complete design and prototyping flow from algorithm specifications to on-board validation and demonstration has been evaluated in the context of 5G using FBMC waveforms. Finally, The authors of [6] demonstrated one of the most achieved concepts with a real time non-synchronous mulituser FBMC transceiver operating over-the-air on fragmented spectrum. All these results have demonstrated the feasibility of prototyping FBMC transceivers with a reasonable level of complexity and are adapted to today hardware platforms. However, coexistence of new asynchronous waveforms with legacy systems (e.g. LTE) should be investigated and architectures adapted to new scenarios should be optimised. The objective of this paper is to propose and evaluate an architecture of implementation suitable to very large scale integration (VLSI) targets (FPGA, ASIC) that could support both OFDM and FBMC receivers.

The paper is organized as follows. First, a review of the so called unified frame structure for 5G is described and the FBMC system model introduced. A current architecture of an FBMC receiver is then analyzed. Finally, a new flexible architecture is proposed and evaluated.

2   Scenario and Model

2.1   Unified Frame Structure for 5G

In order to provide a uniform service experience to users with the premises of heterogeneous networking but also higher data rates, the authors of [1] introduced the concept of the unified frame structure for 5G. The idea is to provide a flexible multi-service solution in an integrated air interface. A frame, divided into different areas of services has been proposed. Four types of traffic have been devised to allow for flexible operation. An example of the proposed frame is shown in Fig. 1: type I and II represent high data rate traffic for video or other high bandwidth services, type I possibly also carries real-time traffic. Type III and IV are dedicated to sporadic asynchronous MTC traffic. Different levels of traffic scheduling have thus been considered: strictly scheduled and organized traffic as already in place in LTE and LTE-A is dedicated to high bandwidth data pipes while sporadic traffic uses contention-like based approaches with random access designed to efficiently enable MTC type payloads (Type III and IV) and bring an efficient solution to the fast dormancy issue.

In order to be efficient, this structure clearly demands to revisit the strict synchronism and orthogonality that prevails in current LTE-A systems. This new requirement leads to rethink the transmission technique and consequently the transceiver structure of the 5th generation of cellular networks.

Alternative waveforms such as UFMC [7], GFDM [8] and FBMC have thus been considered. The motivation of these new waveforms is to keep the flexibility of multicarrier modulations while the frequency response of each carrier is controlled by introducing a filterbank centered on every active carrier and based on the same prototype response. This prototype filter can be selected to minimize adjacent channel interference. As the filtering is embedded in the digital modulation no additional filter is required and more flexibility is obtained.

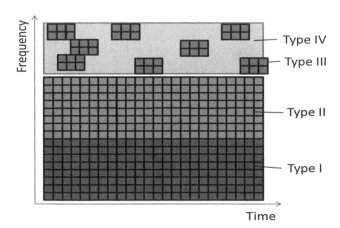

Fig. 1. 5G unified frame structure proposed by [1] (Color figure online)

When considering the unified frame structure, a mix of synchronous (high data rate pipe) and asynchronous traffic should coexist. In this 5G scenario, considering legacy waveforms (i.e. OFDM) for structured synchronous traffic and flexible asynchronous waveforms (i.e. FBMC) for sporadic traffic is very likely. Therefore, an architecture of implementation suitable to VLSI targets (FPGA, ASIC) adapted to both OFDM and FBMC reception should be evaluated.

2.2 OFDM Receiver Architecture

The definition of architectures of OFDM has been widely investigated in the literature [9]. A typical architecture of an OFDM receiver is depicted in Fig. 2. A time domain (TD) synchronization module estimates the start of the multicarrier symbol. The information is used to align a N-point FFT that is processed on the received data every $N+N_{GI}$ samples, where N_{GI} is the size of the guard interval of the OFDM. The N points generated by the FFT are then simultaneously stored to a memory unit for later processing and used by a frequency domain synchronization detector to estimate the carrier frequency offset (CFO).

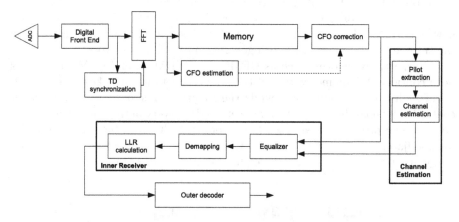

Fig. 2. Typical OFDM receiver block diagram

On the channel estimation datapath, CFO compensation is first performed in the frequency domain using a feed-forward approach. Then, the channel coefficients are estimated on the pilot subcarriers before interpolation for every active subcarrier. Once the channel is estimated on all the active subcarriers the response is stored in a dedicated channel response memory. Depending on the pilot carrier distribution within the time frequency grid, a time interpolation may also be performed. The data buffered in the memory unit are then processed through a one-tap per subcarrier equalizer. Demapping and Log-Likelihood Ratio (LLR) computation complete the inner receiver architecture. A soft-input Forward Error Correction (FEC) decoder finally recovers the originally sent messages.

2.3 FBMC Receiver Architecture

FBMC Review. A multicarrier system can be described by a synthesis/analysis filter bank, i.e. a transmultiplexer structure. The synthesis filter bank is composed of a set of parallel transmit filters. FBMC waveforms utilize a prototype filter designed to give a good frequency localization of the subcarriers. The prototype filter considered in this paper is based on the frequency sampling technique [10]. This technique gives the advantage of using a closed-form representation that includes only a few adjustable design parameters.

The most significant parameter is the duration of the impulse response of the prototype filter also called overlapping factor, K. The impulse response of the prototype filter is given by [10]:

$$h(t) = G_P(0) + 2 \sum_{k=1}^{K-1} (-1)^k G_P(k) \ cos \left(\frac{2\pi k}{KN} (t+1) \right) \qquad (1)$$

where $G_P(0..3) = \left[1, 0.97195983, \frac{1}{\sqrt{2}}, 1 - G_P(1)^2 \right]$ for an overlapping factor of $K = 4$ and N is the number of carriers. The larger the overlapping factor K, the more localized the signal will be in frequency. Adjacent carriers significantly overlap with this kind of filtering. In order to keep adjacent carriers orthogonal, real and pure imaginary values alternate on successive carrier frequencies and on successive transmitted symbols (Offset-QAM modulation is used) for a given carrier at the transmitter side. The well-adjusted frequency localization of the prototype filter guarantees that only adjacent carriers interfere with each other. This allows for a more flexible operation than OFDM for Frequency Division Multiple Access (FDMA), i.e. non synchronous flexible frequency division multiple access.

Most of the published receiver architectures are based on PolyPhase Network (PPN) receivers [10]. In this scheme, the filterbank process is applied in the time domain before the FFT using a polyphase filter. It reduces the size of the FFT but makes the receiver less tolerant to large channel delay spread or synchronization mismatch of the FFT. Therefore, this strategy is not well adapted to the reception of non synchronous users. In [3], the authors describe a high performance receiver architecture denoted FS-FBMC (Frequency Spreading FBMC). One advantage of this architecture comes from the fact that time synchronization may be performed in the frequency domain independently of the position of the FFT [3]. This is realized by combining time synchronization with channel equalization. Moreover, good performance for channel exhibiting large delay spread is achieved [3]. This asynchronous frequency domain processing of the receiver provides a receiver architecture that allows for multiuser asynchronous reception particularly adapted to the envisaged scenarios.

FS Based Receiver. FBMC waveforms are expected to be spectrally more efficient than OFDM when relaxed synchronization between users is considered. Therefore, a preferred architecture for FBMC receivers should be able

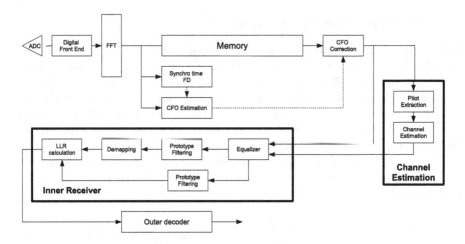

Fig. 3. Typical FBMC receiver block diagram [6]

to efficiently demodulate the signal in the frequency domain without a priori knowledge of the FFT timing alignment (i.e. the location of the FFT block) [6]. A FBMC receiver architecture based on this criteria is given in Fig. 3. A free-running FFT of size KN is processed every blocks of $N/2$ samples generating KN points that are stored in a memory unit for later processing. In parallel a frequency domain synchronization detector detects the start of burst and directly estimates the CFO at the output of the FFT. On the channel estimation datapath, CFO compensation is first performed in the frequency domain using a feed-forward approach. Then, as for OFDM, channel coefficients are estimated on the pilot subcarriers before being interpolated on every active subcarrier. Once the channel is estimated on all the active subcarriers the response is stored for each user in a dedicated channel response memory. The data buffered in the memory unit is then processed through a one-tap per subcarrier equalizer before filtering by the FBMC prototype filter. Demapping and Log-Likelihood Ratio (LLR) computation complete the inner receiver architecture. As far as the LLR computation is concerned, processing is slightly different for FBMC than OFDM. Indeed, in case of a FS-FBMC architecture based receiver, the computation of the LLR associated to a bit from an observation symbol is a function of $2K - 1$ channel coefficients [3].

3   Hardware Architecture of a Flexible FBMC Multicarrier Receiver

3.1   Description

Considering the envisaged scenario for 5G described in Sect. 2, a flexible receiver architecture should be able to dynamically receive legacy LTE signal and asynchronous 5G waveforms. An architecture adapted to the aforementioned unified

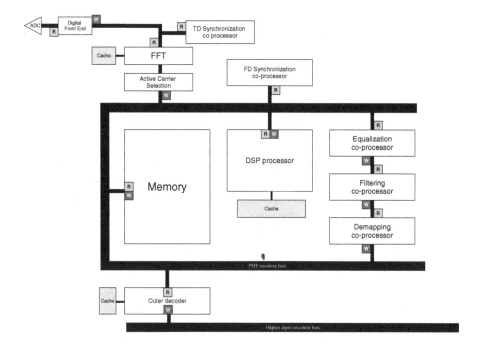

Fig. 4. Flexible architecture adapted to 5G unified frame structure

frame structure has been depicted in Fig. 4. The proposed architecture considers dynamic support for both OFDM and FBMC.

A central memory unit dedicated to the physical layer (PHY) is at the core of the flexible receiver. A set of co-processor units are able to access this central memory through a high speed PHY receiver data bus. These modules include a FD synchronization co-processor (Fig. 4), a FFT co-processor (FFT and Active Carrier selection in Fig. 4), a DSP processor, an Equalization/demapping co-processor (Equalization, filtering and demapping co-processor in Fig. 4) and an Outer decoder processor. A control plane dedicated to transfer control information has been omitted on purpose in Fig. 4 to improve clarity of the figure. The information transiting through the control plane is of relatively low throughput.

The sampled signal received at the analog-to-digital converter is first conditioned by the digital front end to the appropriate sampling frequency into a baseband signal. A TD synchronization processor is at the output of the digital front end and determines the beginning of the burst when in OFDM mode. The module runs in parallel to the FFT module that can either be controlled by the TD synchronization module (when in OFDM mode) or be in free running mode. Appropriate control sets the size of the FFT (N when in OFDM mode or KN when in FBMC mode). The FFT module is followed by an active carrier selection module that selects the active carriers and can write the result to the Memory unit through the PHY receiver bus. In the case of the 20MHz LTE mode, 1201 carriers are typically selected out of the 2048 points at the output of the FFT.

A FD synchronization co-processor can then read the blocks of data samples at the output of the FFT through the PHY receiver bus in order to either estimate CFO when in OFDM mode or estimate CFO and detect start of burst when in FBMC mode. FD synchronization output control signals are then shared through the control plane bus.

Equalization, demapping and LLR computation are hard-wired functions assuming a data-flow architecture. Compared to classical OFDM processing, FBMC includes an extra frequency domain filtering module. The module is therefore bypassed when in OFDM mode. Once the demapping is done, LLR values are written back to the shared memory for further processing by the Outer decoder. As mentioned in the previous section although slightly different demapping follows is very similar process between FBMC and OFDM.

A dedicated digital signal processor (DSP) that can access to its dedicated cache memory has been considered for processing operations such as deframing, pilot extraction and channel estimation. This choice has been driven by the amount of control that these operations required which are therefore more adapted for implementation by an embedded software processing unit. Finally, a dedicated outer module with its internal (cache) memory completes the receiver. The output of the outer decoder is connected to the higher layer bus.

The design of such an architecture where the processing data path is centered on a memory unit has been driven by mainly two motivations. First, the architecture gives more flexibility to the sequencing of the different co-processing units, second and foremost it avoids unnecessary duplication of memory banks. This latter advantage is particularly beneficial for ASIC implementation as large memory banks scale well when using submicron technologies. The main drawback however comes from the constraints that are imposed on the memory and high speed data bus throughput as all the samples written to or read from the memory have to go through the same interface (i.e. the PHY receiver bus). The memory bus throughput has therefore been estimated in the following section to evaluate the relevance of the proposed architecture.

3.2   Memory Bus Throughput Estimation

In order to estimate the constraints that have been put on the PHY receiver bus, an analysis of the throughput has been realized for this architecture. Throughput has been first evaluated analytically for the key modules of the receiver (FFT and the equalizer) for OFDM and for FBMC reception. The results have been summarized in Table 1 in samples per second. The following parameters have been introduced: F_{CS} refers to the frequency spacing between the carrier of the multicarrier modulation, N_a, the number of active carriers, m the modulation order, γ is the ratio of reference over data signal. In the case of the LTE 20MHz mode, γ is equal to 4.76 % or 0.0476.

At the input of the FFT, the data throughput is the same when the receiver receives OFDM or FBMC signals and is equal to the carrier spacing times the number of carriers in samples per second. Typically, for the worst case of LTE, the 20 MHz mode, carrier spacing is equal to 15kHz and N to 2048 points.

Table 1. Analytic sample throughput at the memory bus

		OFDM	FBMC
FFT	Input	$N_{FFT}F_{CS}$	NF_{CS}
	Output	$\frac{N}{N+N_{GI}}NF_{CS}$	$2KNF_{CS}$
	Active carriers selection	$\frac{N}{N+N_{GI}}N_aF_{CS}$	$2KN_aF_{CS}$
Equalizer	Input	$\frac{N}{N+N_{GI}}N_aF_{CS}$	$2KN_aF_{CS}$
	Output filtering	$\frac{N}{N+N_{GI}}N_aF_{CS}$	N_aF_{CS}
	Output demapper (2^m-QAM)	$mN_a\frac{N}{N+N_{GI}}F_{CS}$	mN_aF_{CS}
DSP	Input	$\gamma\frac{N}{N+N_{GI}}N_aF_{CS}$	$\gamma 2KN_aF_{CS}$
	Output	$N_a\frac{N}{N+N_{GI}}F_{CS}$	$2KN_aF_{CS}$

In this case, the input throughput at the FFT is equal to $30.72M\,samples/s$. For OFDM, the FFT output average throughput is then divided by the ratio between N and $N + N_{GI}$ as the guard interval is removed. Then active selection further reduces the throughput by N_a/N. The output of the FFT module is then used as an input to the equalizer, LLR calculation increases the throughput by the modulation order m. For FBMC, when FS-FBMC is considered, the FFT output throughput is however multiplied by $2K$ and therefore significantly increased. Similarly to OFDM, active carrier selection reduces the throughput by N_a/N. The output of the FFT module is also used as an input to the equalizer, where the throughput is divided by $2K$ once the prototype matched filtering is applied. Throughput is then increased as per OFDM by the modulation order m after LLR calculation. The throughput of data processed through the DSP module essentially consists of the channel estimation and interpolation. The output of the FFT is read on the pilot tones only and input throughput is therefore equal to the throughput of the FFT output (after active carriers selection) scaled by γ for both OFDM and FBMC. Then channel state information is interpolated and output for every active carriers. Output throughput is thus equal to the equalizer output throughput of the FFT after active carrier selection.

A numerical application for the LTE $20MHz$ mode has then been derived to evaluate the throughput of data at the bus and summarized in Table 2. As throughput is also dependent on the finite precision of the registers implemented in the receiver function, the following assumptions have been made: the input of the FFT is assumed to be a complex 12-bit input signal. Its output is assumed to be on 16 bits because of the FFT bit growth. Then, input to the equalizer includes both FFT output and Channel state information on 16 bit registers. Finally, LLR values are estimated to be sufficient on 6-bit registers. With these assumptions, For OFDM, the PHY receiver bus should sustain an overall write throughput of $1682Mb/s$ and read throughput of $1707Mb/s$. Assuming a 32-bit (resp. 64 bit) transfer bus, this is equivalent to a data bus throughput of $53Mw/s$ (resp. $26Mw/s$) for write operations and $53Mw/s$ (resp. $27Mb/s$) for read operations. This is relatively low when ASIC submicron implementations are considered.

Table 2. Constraint on bus throughput

		System				Quantization		
		OFDM LTE (20MHz)	FBMC (20MHz)			OFDM LTE (20MHz)	FBMC (20MHz)	
FFT	Input	30,72	30,72	Msamp/s	2 x 12	737,28	737,28	Mb/s
	Ouput	28,70	245,76	Msamp/s	2 x 16	918,46	7864,32	Mb/s
	Active Carrier Selection	16,82	144,00	Msamp/s	2 x 16	538,16	4608,00	Mb/s
Equalizer	Input	16,82	144,00	Msamp/s	4 x 16	1076,32	9216,00	Mb/s
	Output Filtering	16,82	18,00	Msamp/s	/	/	/	
	Ouptut (64QAM)	100,91	108,00	MLLR/s	1 x 6	605,43	648,00	Mb/s
DSP	Input	0,80	6,86	Msamp/s	2 x 16	25,63	219,43	Mb/s
	Output	16,82	144,00	Msamp/s	2 x 16	538,16	4608,00	Mb/s
Outer decoder	Input	100,91	108,00	MLLR/s	1 x 6	605,43	648,00	Mb/s
			Total W			1681,75	9864,00	Mb/s
			Total R			1707,38	10083,43	Mb/s
			Total W+R			3389,13	19947,43	Mb/s
		32 bits BUS	Total W			52,55	308,25	Mw/s
			Total R			53,36	315,11	Mw/s
			Total W+R			105,91	623,36	Mw/s
		64 bits BUS	Total W			26,28	154,13	Mw/s
			Total R			26,68	157,55	Mw/s
			Total W+R			52,96	311,68	Mw/s

However, for FBMC implementation and when similar quantization levels as the levels considered for OFDM reception are assumed, the architecture gives an overall aggregated throughput on the PHY receiver bus that is 5.9 times larger for write operations and for read operations. When a 32 bit (resp. 64 bit) data bus is considered, FBMC receptions requires an aggregated throughput of $308Mw/s$ (resp. $154Mw/s$) for read operations and $315Mw/s$ (resp. $158Mw/s$) write operations. The aggregated throughput assuming a 32-bit data bus seems to be in the upper limit of the possibilities available for ASIC submicron technologies, while 64-bit data bus is acceptable. Furthermore, throughput constraints are well balanced between write and read operations.

4 Conclusion

A flexible data frame structure designed to support high bandwidth pipes and sporadic traffic introduced by MTC and fast dormancy has been described. This new requirement imposed by 5G applications has led to consider legacy waveforms such as OFDM for structured synchronous traffic and alternative flexible asynchronous waveforms such as FBMC for sporadic traffic. In order to evaluate an architecture for a flexible receiver adapted to both OFDM and FBMC reception for VLSI targets, OFDM and FBMC receivers have been reviewed. A new type of flexible architecture receiver has then been proposed and described. A memory block combined with a high speed data bus is at the core of the newly proposed architecture. The design of an architecture centered on a memory unit combined with co-processor units. It improves the flexibility of the digital signal processing operations of the receiver as the sequencing of the processing operations is more flexible. Furthermore, the architecture limits the amount of memory blocks in the design which is particularly beneficial for ASIC implementation. Finally, the throughput on the memory and the associated PHY receiver bus that this new architecture imposes has been evaluated. The paper concluded that the throughput imposed by FBMC reception is almost 6 times bigger for write operations and for read operations in comparison to OFDM reception. Although the difference is significantly high, the maximum estimated throughput is adapted to the constraints imposed by VLSI implementations when a 64-bit bus is implemented even when $20MHz$ bandwidth scenarios are considered.

Acknowledgments. The research leading to these results was supported by the European Commission under grant agreement 671563 (H2020-ICT-2014-2 / ICT-14-2014), Flexible and efficient hardware/software platforms for 5G network elements and devices (Flex5GWare).

References

1. Wunder, G., et al.: 5GNOW: non-orthogonal, asynchronous waveforms for future mobile applications. In: IEEE Communications Magazine, 5G special issue, February 2014
2. Wunder, G., et al.: 5GNOW: Challenging the LTE design paradigms of orthogonality and synchronicity. In: 2013 IEEE 77th Vehicular Technology Conference (VTC Spring) (2013)
3. Doré, J.-B., Berg, V., Cassiau, N., Kténas, D.: FBMC receiver for multi-user asynchronous transmission on fragmented spectrum. In: EURASIP Journal, Special Issue on Advances in Flexible Multicarrier Waveform for Future Wireless Communications, vol. 2014 (2014)
4. Dziri, A., Alexandre, C., Zakaria, R., Le Ruyet, D.: SDR based prototype for filter bank based multi-carrier transmission. In: 2014 11th International Symposium on Wireless Communications Systems (ISWCS), pp. 878–882, August 2014
5. Nadal, J., Nour, C., Baghdadi, A., Lin, H.: Hardware prototyping of FBMC/OQAM baseband for 5g mobile communication. In: 2014 25th IEEE International Symposium on Rapid System Prototyping (RSP), pp. 72–77, October 2014
6. Berg, V., Dore, J.-B., Noguet, D.: A multiuser FBMC receiver implementation for asynchronous frequency division multiple access. In: 2014 17th Euromicro Conference on Digital System Design (DSD), pp. 16–21, August 2014
7. Vakilian, V., Wild, T., Schaich, F., ten Brink, S., Frigon, J.-F.: Universal-filtered multi-carrier technique for wireless systems beyond LTE. In: 2013 IEEE Globecom Workshops (GC Wkshps), pp. 223–228, December 2013
8. Fettweis, G., Krondorf, M., Bittner, S.: GFDM - generalized frequency division multiplexing. In: 2009 IEEE 69th Vehicular Technology Conference. VTC Spring 2009, pp. 1–4, April 2009
9. Speth, M., Fechtel, S., Fock, G., Meyr, H.: Optimum receiver design for wireless broad-band systems using OFDM. i 47(11), pp. 1668–1677 (1999)
10. FP7 european project - phydas: physical layer for dynamic spectrum access and cognitive radio. http://www.phydyas-ict.org

Evolutionary Multiobjective Optimization for Digital Predistortion Architectures

Lin Li[1(✉)], Amanullah Ghazi[2], Jani Boutellier[2], Lauri Anttila[3],
Mikko Valkama[3], and Shuvra S. Bhattacharyya[1,3]

[1] ECE Department, University of Maryland, College Park, MD 20742, USA
{lli12311,ssb}@umd.edu
[2] Department of Computer Science and Engineering, University of Oulu,
Oulu, Finland
{amanullah.ghazi,jani.boutellier}@ee.oulu.fi
[3] Department of Electronics and Communications Engineering,
Tampere University of Technology, Tampere, Finland
{lauri.anttila,mikko.e.valkama}@tut.fi

Abstract. In wireless communication systems, high-power transmitters suffer from nonlinearities due to power amplifier (PA) characteristics, I/Q imbalance, and local oscillator (LO) leakage. *Digital Predistortion (DPD)* is an effective technique to counteract these impairments. To help maximize agility in cognitive radio systems, it is important to investigate dynamically reconfigurable DPD systems that are adaptive to changes in the employed modulation schemes and operational constraints. To help maximize effectiveness, such reconfiguration should be performed based on multidimensional operational criteria. With this motivation, we develop in this paper a novel evolutionary algorithm framework for multiobjective optimization of DPD systems. We demonstrate our framework by applying it to develop an adaptive DPD architecture, called the *adaptive, dataflow-based DPD architecture (ADDA)*, where Pareto-optimized DPD parameters are derived subject to multidimensional constraints to support efficient predistortion across time-varying operational requirements and modulation schemes. Through extensive simulation results, we demonstrate the effectiveness of our proposed multiobjective optimization framework in deriving efficient DPD configurations for run-time adaptation.

Keywords: Digital predistortion · Multiobjective optimization · Evolutionary algorithms

1 Introduction

In wireless communication systems, I/Q mismatch, power amplifier (PA) nonlinearities, and signal leakage in the local oscillator (LO) are implementation-related problems that must be addressed before the direct-conversion principal can be deployed. In the frequency domain of the transmitted signal, the effects of

© ICST Institute for Computer Sciences, Social Informatics and Telecommunications Engineering 2016
D. Noguet et al. (Eds.): CROWNCOM 2016, LNICST 172, pp. 498–510, 2016.
DOI: 10.1007/978-3-319-40352-6_41

these impairments are translated as power leakage into adjacent channels. Digital predistortion (DPD) is a widely investigated technique (e.g., see [2–4, 7, 9]) to counteract such impairments by applying carefully-calculated distortion to the signal prior to transmission.

A major challenge in deploying DPD architectures for cognitive radio systems is the dynamic optimization of key DPD parameters subject to time-varying and multidimensional constraints on system performance. A general approach to such optimization is to perform efficient search at design time (i.e., off-line) across alternative DPD configurations, and to then select from the search results a set of configurations that are Pareto-optimal, and that effectively cover the targeted range of operational scenarios and their trade-offs. These selected, "Pareto-optimized" configurations can then be stored in memory, and switched across during system operation based on time-varying changes in communication system requirements. Here, "Pareto-optimized" configurations refer to configurations that are Pareto-optimal with respect to the applied search process, while "Pareto-optimal" configurations refer to configurations that are globally optimal in a Pareto sense.

In this paper, we develop a novel framework for systematic derivation of Pareto-optimized DPD system configurations that can be applied to adaptive DPD implementations. Our framework builds on the methodology of multiobjective evolutionary algorithms (e.g., see [12]), and incorporates adaptations of this methodology to efficiently handle distinguishing characteristics of DPD system optimization. We refer to our framework for DPD system optimization as the framework for *Evolutionary Adaptive DPD Implementation (EADI)* or ("EADI Framework").

We demonstrate the EADI Framework in this paper by applying it to develop an adaptive DPD architecture, called the *adaptive, dataflow-based DPD architecture (ADDA)*, where Pareto-optimized DPD parameters are derived subject to multidimensional constraints to support efficient predistortion across time-varying operational requirements and modulation schemes. While the ADDA architecture is used to concretely demonstrate the capabilities of the EADI Framework, the EADI Framework is not specific to any particular DPD architecture, and can readily be adapted to work across a variety of parameterized DPD architectures. Exploring such adaptations is a useful direction for future work that emerges from the developments of this paper.

The design evaluation metrics (optimization objectives) targeted in our development of the EADI Framework and ADDA architecture in this paper are system energy consumption, adjacent channel power ratio (ACPR), and system accuracy. We abbreviate this set of metrics as *EAA*.

The ADDA is a parameterized architecture that can be configured dynamically to achieve a range of EAA trade-offs. The DPD design space that we consider consists of three design parameters: the polynomial order, bit-width, and filter order. This design space is modeled in the EADI Framework, and optimization results from the framework are used to extract a subset of generated Pareto-optimized configurations (settings of the DPD parameter values).

This subset of configurations provides the set of DPD system modes that will be implemented in the ADDA architecture. The set of DPD modes provided in the ADDA configuration set is made available during operation such that predistortion trade-offs can be reconfigured among the different options in the configuration set based on dynamically changing operational requirements.

To demonstrate and experiment with the ADDA, we apply the *lightweight dataflow environment* (*LIDE*), which is a design tool for dataflow-based design and implementation of signal processing systems [8]. Dataflow graphs provide a useful form of model-based design in many areas of signal processing, and wireless communications (e.g., see [11]). We map the signal flow structure of the ADDA into actors (dataflow-based signal processing components) in LIDE, and implement the internal functionality of these actors using the Verilog hardware description language (HDL).

We demonstrate the effectiveness of the EADI Framework through extensive simulations, and validate the capabilities of the ADDA through hardware synthesis.

2   Related Work

Unlike earlier DPD architectures (e.g., see [3,5]), the DPD algorithm proposed in [1] is one of the first DPD techniques that jointly compensates for PA non-linearities and I/Q modulator impairments. This DPD architecture employs an extended parallel Hammerstein structure, which decomposes DPD operation into direct and conjugate predistortion subsystems. Such a decomposed structure provides additional degrees of freedom in system design. In this paper, we exploit the decomposed, parallel structure of the DPD method introduced in [1] and we present new methods to search the design space, and derive Pareto-optimized realizations for this form of DPD architecture.

In architectures for cognitive radios, adaptive DPD systems that operate under Pareto-optimized configurations are highly desirable due to the multidimensional space of relevant implementation metrics. However, prior work on system-level DPD optimization has emphasized single-objective optimization of ACPR [2,9]. These works employ a form of search technique called genetic algorithms, which are closely related to evolutionary algorithms, to optimize DPD ACPR performance. However, the resulting solutions may not be efficient in terms of energy consumption or accuracy. Furthermore, the underlying design methodology does not produce multiple alternative configurations that may be employed for dynamic reconfiguration based on time-varying changes in operational requirements. The methods that we develop in this paper address these limitations, respectively, through development of the (1) EADI Framework for multidimensional, Pareto-optimized DPD configuration, and (2) ADDA for reconfigurable DPD architecture implementation based on configurations that are derived by the EADI Framework.

3   Adaptive Dataflow-Based DPD Architecture

The ADDA architecture developed in this paper is based on the algorithm presented in [1]. This DPD algorithm operates in two stages. In the *coefficient estimation* stage, the DPD filtering coefficients are estimated. The estimated coefficients are then employed in the *DPD filtering* stage for actual predistortion of the input signal. Since the first stage is intended for off-line computation, the ADDA architecture and EADI optimization process are focused only on the second (filtering) stage.

Figure 1 illustrates the dataflow model of the DPD filtering subsystem that is employed in the ADDA. Here, the mode selection actor dynamically selects the DPD operational mode based on the current application scenario (i.e., based on the current modulation and requirements on EAA) and finds the corresponding parameter settings for that mode in its local memory, and distributes these DPD parameter values to the polynomial computation actor and all of the filter actors. Following [1], we decompose the signal processing for the applied DPD algorithm into separate direct and conjugate parts.

With the parameters obtained from the mode selection actor, the polynomial computation actor computes the polynomial basis function defined in [1] for both the direct and conjugate branches. The computed polynomials are then sent to their corresponding branches and filtered by the filter actors in those branches. These filter actors are implemented with integrated use of LIDE and Verilog, as described in Sect. 1. As shown in Fig. 1, the filtered samples (one output sample from each filter) are summed to produce a single sample as the final predistorted output. Based on the analysis in [4], where a similar dataflow model is constructed for the DPD algorithm in [1], most of the computation and energy consumption is concentrated in the filter actors. Thus, in this paper, we map only the filter actors to hardware, and focus our design optimization processes on the filter actors.

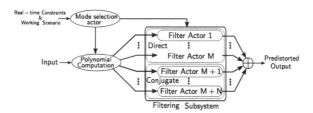

Fig. 1. Dataflow graph model of the predistortion filter.

4   Optimization Metrics and Design Space

4.1   Optimization Metrics

In this subsection, we elaborate on the three objectives in our targeted design optimization problem. As defined in Sect. 1, we refer to these metrics collectively as *EAA*.

Energy Measurement. As explained in Sect. 3, we focus our energy measurement on the energy consumed by the filtering subsystem, and the figure of merit that we employ is the filtering energy expended to producing a single output sample, which is denoted by the *energy per sample* (*eps*). To calculate eps, we use the total power consumption of all FIR filters used in the predistortion subsystem, which we denote as P_{FIR}. The eps metric is then defined as eps $= P_{FIR} \times C/F$, where C represents the average number of clock cycles required by the filter actors to process a single new input sample, and F represents the clock frequency. In our design, both F and C are fixed for each configuration. Thus, eps is proportional to P_{FIR}, and we can therefore use P_{FIR} as optimization objective for our evolutionary algorithm process. Also, we report results for P_{FIR} in Sect. 6 (instead of eps) as our assessment of the energy efficiency of each configuration.

We implement the DPD filtering subsystem using the Altera EP2C35F672C6 FPGA from the Cyclone II family. To facilitate efficient design space exploration within the EADI optimization process, we model the power consumption as a function of the design vector $[P\,Q\,\mathbf{BW}^T\,\mathbf{FO}^T]^T$. The definitions of the quantities P, Q, $\mathbf{BW}$ and $\mathbf{FO}$ are given in Sect. 5.

Our approach to system-level DPD power estimation starts by first measuring the total power consumption of a single branch under all valid filter order and bit-width values using Altera PowerPlay Analyzer. The power consumption for a specific DPD configuration is then estimated as

$$Power_{est} = \sum_{p \in I_P} Power_p(bw_p, fo_p) + \sum_{q \in I_Q} Power_q(bw_q, fo_q), \qquad (1)$$

where I_P and I_Q are the set of direct branches and conjugate branches, respectively; bw_x and fo_x are the bit-width and filter order for branch x, respectively, and $Power_x(bw_x, fo_x)$, the power consumed by branch x with bit-width bw_x and filter order fo_x, is obtained from the aforementioned power measurement process.

ACPR Measurement. ACPR is a metric that is commonly used to assess the extent of out-of-band energy leakage [7]. ACPR is defined as the ratio of the mean power centered on the adjacent channel to the mean power centered on the desired channel, as shown in (2).

$$\text{ACPR} = 10 \log_{10} \frac{\int_{\omega_A} S(\omega) d\omega}{\int_{\omega_D} S(\omega) d\omega}. \qquad (2)$$

Here, $S(\omega)$ denotes the power spectral density of the postdistorter input signal s_n, and ω_A and ω_D denote the frequency bands of the adjacent channel and desired channel, respectively.

Accuracy Measurement. We measure the accuracy of candidate DPD designs by the *error vector magnitude* (*EVM*) and *symbol error rate* (*SER*). The former is considered as an optimization objective and the latter as a constraint on the derived configurations. The EVM measures the distortion of original symbols under the influence of non-linearities introduced by the PA and DPD. This distortion is calculated as

$$\text{EVM(Pf)} = 10 \log_{10} \left(\frac{1}{K} \sum_{k=1}^{K} |X_0(k) - \hat{X}^{\text{Pf}}(k)|^2 \right), \tag{3}$$

where Pf represents a certain profile (finite sequence) $X_0(1), X_0(2), \ldots, X_0(K)$ of symbols to be transmitted, and $\hat{X}^{\text{Pf}}(k)$ is the kth actual transmitted symbol under Pf.

SER is measured as the average rate of erroneous symbol transmissions. This rate is determined as

$$\text{SER(Pf)} = \frac{1}{K} \sum_{k=1}^{K} I(X_0(k) - \hat{X}^{\text{Pf}}(k)), \tag{4}$$

where $I(x)$ (the *indicator function*), has value 1 if $x \neq 0$ and 0 otherwise. We require that all of the configurations extracted for mapping into the ADDA must have zero SER.

4.2 Design Space

In this section, we elaborate on the selected DPD parameters that define the predistorter design space associated with the ADDA.

Polynomial Orders. As mentioned in Sect. 3, the DPD algorithm proposed in [1] splits its signal processing into a direct part and a conjugate part, which enables use of different polynomial orders for direct and conjugate signal terms. For example, a DPD system can be realized with fifth-order for the direct signal and only third-order for the conjugate signal. We denote the polynomial order for the direct signal and conjugate signal by P and Q, respectively. Following [1], only odd values for P and Q are considered. Thus, the number of branches (or filter actors) that is employed in a specific DPD configuration is given by $N_{branch} = (P+1)/2 + (Q+1)/2$. In our experiments, we set the domain D of valid values for both P and Q as $D = \{1, 3, 5, 7, 9\}$. Thus, there are in total 25 $P - Q$ combinations in our targeted design space.

Bit-Widths. Intuitively, smaller bit-widths for data storage and computation lead to less energy consumption. However, signal processing accuracy may be traded off as a consequence. To incorporate this trade-off between energy efficiency and accuracy, we incorporate bit-width as a parameter of ADDA, and as a design space component of EADI. Considering requirements on system accuracy and constraints on hardware resources, we set the range of allowable bit-widths in our experiments as $\{5, 6, \ldots, 15\}$. Additionally, we allow different branches to be configured with different bit-widths in the same design. This leads to great flexibility in design optimization, and a correspondingly large design space — if there are m branches used in a specific design, then the total number of valid bit-width combinations is 11^m.

Filter Orders. Similar to the bit-width design, the filter used in each branch may also have different number of coefficients. We denote this parameter as *filter order*. The filter order parameters would also significantly affect the trade-offs among EAA. The range of filter order in this work is set to be $\{1, 2, 3, 4, 5\}$.

According to the above description, our design space is too huge for exhaustive search. As a numerical example, given the aforementioned ranges for the system parameters, the design space would contain more than 55^{10} configurations.

5 Multiobjective Optimization Using Evolutionary Algorithm

As motivated in Sect. 4, the DPD design space addressed in this work is a complex multidimensional space that is too large to be evaluated using exhaustive search techniques. Therefore, we apply a heuristic search strategy called *evolutionary algorithms (EAs)*, including a particular form of EA, called *strength Pareto EA (SPEA)*, that is suited for multiobjective optimization [12]. We select the SPEA approach due to its efficiency and scalability in addressing complex optimization problems, and its customizability to different kinds of design spaces and optimization criteria. This latter feature makes the EADI Framework readily adaptable across different kinds of DPD architectures and communication system constraints.

5.1 Problem Encoding

The parameters involved in the DPD design optimization problem are polynomial orders, bit-widths, and filter orders. Each configuration can be represented throughout the EA process by a vector, specified as $[P\ Q\ \mathbf{BW}^T\ \mathbf{FO}^T]^T$. Here, P and Q are the direct and conjugate polynomial order, respectively. As described in Sect. 4, the maximum number of branches considered in the design space is 10 (at most 5 branches for both the direct signals and the conjugate signals). Thus, $\mathbf{BW}$ is a vector with 10 dimensions representing bit-width settings for up to 10 branches, where each dimension represents the bit-width associated with the corresponding branch. For the branches that are not used, the corresponding vector elements are set to zero. Similar conventions are applied to generate the 10-dimensional vector $\mathbf{FO}$ of filter order settings.

As discussed in Sect. 1, the objective space of the EADI Framework encompasses average power consumption, ACPR and EVM. Thus, the objective vector can be formulated as $[P_{FIR}\ ACPR\ EVM]$ with units (mW, dBc, dB). Here, P_{FIR} is the power consumption, as estimated by the method discussed in Sect. 4, and ACPR and EVM are calculated according to (2) and (3), respectively.

5.2 Optimization Process

The EADI optimization process is executed separately for each modulation type that is to be supported in the targeted ADDA platform. The resulting Pareto-optimized configurations for the different modulation types are then collected and stored in the ADDA memory. This enables the ADDA to dynamically to select among different modulation types, and among different operational trade-offs for each modulation type.

As mentioned previously, the work flow of the EADI optimization process is based on the SPEA methodology for multidimensional search. For details on SPEA, we refer the reader to [12].

According to SPEA, the population set (set of candidate solutions or *individuals*) ρ contains the individuals generated during each SPEA iteration, and the external set $\bar{\rho}$ maintains selected non-dominated individuals among all individuals generated so far up through the current iteration. Here, we say that an individual x dominates another individual y if x is superior to y in terms of at least one design evaluation metric, and x is not inferior to y in terms of any metric. A *non-dominated* individual is one that is not dominated by any individual.

We initialize ρ with a well-distributed population across the design space. For each possible $P - Q$ combination, we generate two design vectors by selecting the corresponding bit-width and filter order values randomly from their valid ranges. Thus, the size of ρ, denoted by **N**, is 50 individuals.

During each iteration, each individual in ρ is evaluated to generate the objective vector $[P_{FIR} \text{ ACPR EVM}]$. The individuals that do not satisfy certain modulation-specific constraints (defined in Sect. 6) are ignored. Only the remaining non-dominated individuals are copied to $\bar{\rho}$. If the size of $\bar{\rho}$ exceeds a predefined maximum population size $\bar{N}_{max}$, a k-means clustering algorithm is used to classify the members in $\bar{\rho}$ into $\bar{N}_{max}$ groups. This allows us to limit the size of $\bar{\rho}$ while maintaining a diverse population in $\bar{\rho}$ by retaining a "representative" individual of each group in $\bar{\rho}$ [12].

After updating of $\bar{\rho}$ during an optimization iteration (*generation*), individuals from both ρ and $\bar{\rho}$ are selected to generate a "mating pool" ρ'. This selection process is performed randomly in a manner such that the probability of an individual's selection for the mating pool is larger for individuals with smaller fitness values. Here, "fitness" is a measure of the quality of an individual; smaller fitness values imply higher quality solutions. The *recombination operator* selects pairs of individuals ("parents") in ρ', and for each selected pair, two new individuals ("children") are generated with probability p_r.

Each generated child (from recombination) undergoes a process of random modification by a *mutation operator* with probability p_m. After all recombination and mutation operations are completed on the mating pool ρ', the resulting new population is assigned as the current population ρ for the next generation. The individuals that comprise the set $\bar{\rho}$ after **T** generations are the Pareto-Optimized solutions obtained by the EADI Framework. Here, **T** is a pre-defined number of optimization iterations that is to be executed by the SPEA.

The values p_r, p_m, and **T** are design parameters of the optimization process that can be set through experimentation or by selecting commonly-used values from the literature.

These general concepts of fitness measures, recombination operators, and mutation operators are standard components of EAs. They are applied to form an optimization process that has analogies to processes by which living species evolve. However, these three operators need to be designed specifically for each optimization context. In the remainder of this section, we discuss how these operators have been designed in the EADI Framework.

5.3   Fitness Measure

Based on the SPEA approach, each individual $\mathbf{i} \in \bar{\rho}$ is assigned a real value $S(\mathbf{i}) \in [0, 1)$, which is referred to as the *strength* of $\mathbf{i}$. If N represents the number of individuals in the set ρ, then $S(\mathbf{i})$ is calculated as the ratio of (a) the number of individuals in ρ that are dominated by $\mathbf{i}$ to (b) $(N + 1)$. The fitness of $\mathbf{i}$ is equal to $S(\mathbf{i})$. The fitness of an individual $\mathbf{i} \in \rho$ is calculated by summing the strengths of all individuals $\mathbf{j} \in \bar{\rho}$ that dominate $\mathbf{i}$, and then adding one to this sum. We add one to the sum here in order to guarantee that members in $\bar{\rho}$ have better fitness than members in ρ (since fitness is to be minimized).

5.4   Recombination Operator

Recombination is a process of selecting parent solutions and producing child solutions from them that integrate properties of the corresponding parent solutions. The inputs of the recombination operation are the configuration vectors of the two selected parents $\mathbf{Y_1}$ and $\mathbf{Y_2}$, and the outputs are either (a) the same two parents $\mathbf{Y_1}$ and $\mathbf{Y_2}$ (with probability $(1 - p_r)$) or (b) the configuration vectors of two generated children (with probability p_r), denoted by $\mathbf{C_1}$ and $\mathbf{C_2}$.

In the latter case (when children are generated), the process of generating each child individual $\mathbf{C_k}$, $k = 1, 2$ from the two parents is summarized as follows: (i) assign P, Q values (polynomial orders) from $\mathbf{Y_1}$ or $\mathbf{Y_2}$ to $\mathbf{C_k}$ with equal probability subject to the requirement that the generated pair of P and Q values for C_1 and C_2 cannot be identical to each other; (ii) set the bit-width and filter order values of each child $\mathbf{C_k}$ to the corresponding values of an average vector Y_{avg}: $Y_{avg} = \gamma(Y_1, Y_2)$, where $\gamma(Y_1, Y_2)$ first computes the average $(\mathbf{Y_1} + \mathbf{Y_2})/2$, and for each component in this average vector that is not integer-valued, the operator replaces the component by its floor or ceiling with equal probability; and (iii) set the bit-widths and filter orders of the unused branches in the children to be zero.

5.5   Mutation Operator

In EAs, mutation operators are employed to help promote diversity from one generation of a population to the next by randomly modifying selected solution components ("genes") within individuals. In the EADI Framework for ADDA implementation, the genes for potential mutation are taken to be the vector-valued settings of **BW** and **FO**. The specific gene (**BW** or **FO**) to which modification is to be applied is selected randomly with equal probability, and then a single component of the selected vector that is to be modified is selected randomly (with equal probability among all vector components). The mutation operator replaces the value of the selected vector component with a uniform random value drawn between the given upper and lower bounds for that component.

6   Experimental Setup and Simulation Results

To validate the EADI Framework and ADDA platform, and to demonstrate their capabilities, we experiment with three LTE modulation schemes — QPSK, 16–QAM, and 64–QAM. The multiobjective optimization process is performed separately for each of the three modulation schemes, and then the resulting Pareto-optimized solution sets are integrated into the ADDA as discussed in Sect. 5. For all three modulation schemes, we employ the following SPEA parameter settings: (i) $T = 100$ (number of generations); (ii) $N = 50$ (population size); (iii) $\bar{N}_{\max} = 20$ (maximum size of external set); (iv) $p_r = 0.8$ (recombination rate); (v) $p_m = 0.2$ (mutation rate). These values for generic SPEA settings are values that are commonly used in the literature (e.g., see [10,12]).

The constraint on ACPR used in the EADI Framework for all three modulations is -45.0 dBc. The constraints on EVM are -15.1 dB, -18.1 dB, and -22.0 dB for QPSK, 16–QAM, and 64–QAM, respectively. The constraint on SER is that it should be zero.

To help validate the effectiveness of the EADI Framework in deriving high quality DPD configurations, we apply a *partial search* (*PS*) method to solve the same multiobjective optimization problem. PS involves performing a complete search on a reduced design space. PS is also a widely-applied method for obtaining Pareto fronts in multiobjective optimization problems (e.g., see [6]).

In our PS approach, we reduce the search space by equalizing the bit-widths and filter orders of all the filters used in all branches and apply the same valid parameter value ranges as used in the SPEA process. Thus, the reduced design space contains $5 \times 5 \times 11 \times 5 = 1375$ configurations. We evaluate these 1375 configurations exhaustively with the P_{FIR}, ACPR, SER and EVM computations, as described in Sect. 4. We then remove the undesirable solutions based on the same SER, ACPR and EVM constraints as applied in the SPEA. Finally, we collect all of the non-dominated configurations from the resulting design space as the Pareto front obtained by the PS.

In the PS process, we estimate P_{FIR} using relevant FPGA design tools (Altera PowerPlay Analyzer), while in the EADI process, we estimate P_{FIR} using the power estimator introduced in Sect. 4. The estimator of Sect. 4 enables faster power estimation (at some expense in accuracy), which is important because very large numbers of candidate solutions are evaluated during the EADI process.

For the Pareto-optimized configurations achieved by EADI, we also estimate P_{FIR} using FPGA tools to obtain more accurate power estimation results for the derived Pareto front. In the results that we report in the remainder of this section, the comparison between the quality of the two solution sets (PS and EADI) is based on the same (more accurate) power estimation method — i.e., using FPGA tools.

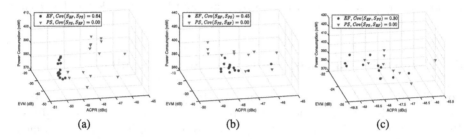

(a)   (b)   (c)

Fig. 2. Pareto-optimized solutions obtained from the EADI Framework and PS for (a) QPSK, (b) 16–QAM, (c) 64–QAM.

The Pareto fronts derived by the EADI Framework and PS for the three selected modulations are shown in Fig. 2(a) to (c). We use *coverage of two sets* (*Cov*) measurements [12] to evaluate the quality of the solution sets produced by the EADI Framework and PS, which we denote by S_{EF} and S_{PS}, respectively. Given a multiobjective design space, and two sets α and β of candidate solutions in this space, $Cov(\alpha, \beta) = dom(\alpha, \beta)/size(\beta)$, where $dom(\alpha, \beta)$ is the number of solutions in β that are dominated by at least one solution in α. Coverage results for each of the three modulation schemes are given in Fig. 2(a) to (c) along with plots of S_{EF} and S_{PS}. Here, we see that $Cov(S_{PS}, S_{EF})$ is uniformly zero over all three modulations, while the values for $Cov(S_{EF}, S_{PS})$ indicate that significant proportions of the PS solutions are dominated by results from the EADI Framework.

We also measured that the PS method requires approximately 91 hours to evaluate the three optimization metrics for the 1375 given configurations, and extract the Pareto front, while the evaluation and Pareto front extraction by the EADI Framework takes only about 1 hour. We conclude from these results involving *Cov* and optimization time that the EADI Framework significantly outperforms the PS method in terms of both the quality of the obtained Pareto fronts and run-time efficiency.

To concretely demonstrate DPD performance trade-offs realized in the proposed ADDA architecture, we first classify the individuals in the Pareto front obtained by EADI into three groups according to their power consumption levels. Then we select one representative individual in each group and store it in ADDA as a DPD working mode. The selected design vectors and their corresponding P_{FIR}-ACPR-EVM measurements under three modulations in LTE are listed in Table 1. From this table, we see that for the Pareto-optimized parameter settings

Table 1. Selected pareto-optimized parameter settings for LTE under different modulations. The design evaluation metrics are shown in the format $(P_{FIR}, ACPR, EVM)$ with units (mW, dBc, dB).

	Power Level	P,Q	BW		FO		Performance
			Direct	Conj	Direct	Conj	
QPSK	Low	3, 1	12, 9	12	5, 3	4	380.47, −49.99, −43.41
	Medium	3, 1	13, 11	12	3, 4	4	387.57, −50.15, −43.75
	High	3, 3	15, 10	13, 5	4, 5	4, 1	397.52, −50.14, −44.07
16–QAM	Low	3, 1	11, 12	12	4, 4	5	380.96, −48.38, −24.29
	Medium	3, 3	12, 12	12, 5	3, 4	5, 3	385.34, −48.15, −30.90
	High	5, 3	12, 12, 10	11, 5	5, 4, 2	4, 4	395.11, −46.09, −31.69
64–QAM	Low	3, 1	11, 11	10	4, 3	3	382.97, −48.28, −22.89
	Medium	3, 1	12, 12	15	4, 3	4	395.76, −48.08, −25.21
	High	5, 3	12, 12, 11	14, 11	3, 3, 5	4, 4	409.62, −48.34, −24.71

obtained by EADI, P is always greater than or equal to Q, which validates the argument in [1] that the higher orders of the conjugate predistorters are weak, and a smaller Q value is therefore preferred. Also, in general, the branches corresponding to the lower polynomial orders are configured with higher bit-widths and filter orders compared to the branches corresponding to higher polynomial orders. This results from the higher order signals being relatively weak for both direct and conjugate parts.

7   Conclusions

In this paper, we have presented a novel framework, called the Evolutionary Adaptive DPD Implementation (EADI) Framework, for multiobjective optimization of digital predistortion (DPD) systems. The targeted optimization objectives include system energy consumption, adjacent channel power ratio (ACPR), and system accuracy. We apply the EADI Framework to develop an architecture, called the adaptive, dataflow-based DPD architecture (ADDA), where Pareto-optimized DPD parameter settings are derived to support efficient, adaptive predistorter operation. Simulation results demonstrate the effectiveness of the EADI Framework in deriving efficient DPD configurations across time-varying modulation schemes subject to multidimensional constraints. The extracted Pareto-optimized configurations also help to validate assumptions in the DPD literature about preferred DPD parameter settings. Finally, the EADI Framework is shown to significantly outperform a partial search method in terms of both optimization time efficiency and the quality of the derived Pareto fronts.

Acknowledgements. This research was supported in part by Tekes, the Finnish Funding Agency for Innovation; and the U.S. National Science Foundation.

References

1. Anttila, L., Händel, P., Valkama, M.: Joint mitigation of power amplifier and I/Q modulator impairments in broadband direct-conversion transmitters. IEEE Transactions on Microwave Theory and Techniques **58**(4), 730–739 (2010)
2. Çiflikli, C., Yapící, A.: Genetic algorithm optimization of a hybrid analog/digital predistorter for RF power amplifiers. Analog Integrated Circuits and Signal Processing **52**(1), 25–30 (2007)
3. Ding, L., et al.: Compensation of frequency-dependent gain/phase imbalance in predistortion linearization systems. IEEE Transactions on Circuits and Systems I: Regular Papers **55**(1), 390–397 (2008)
4. Ghazi, A., et al.: Low power implementation of digital predistortion filter ona heterogeneous application specific multiprocessor.In: Proceedings of the International Conference on Acoustics, Speech,and Signal Processing, pp. 8391–8395. Florence, Italy (2014)
5. Hilborn, D.S., Stapleton, S.P., Cavers, J.K.: An adaptive direct conversion transmitter. IEEE Transactions on Vehicular Technology **43**(2), 223–233 (1994)
6. Llamocca, D., Pattichis, M.: Dynamic energy, performance, and accuracy optimization and management using automatically generated constraints for separable 2D FIR filtering for digital video processing. Transactions on Reconfigurable Technology and Systems 7(4). Article No. 31(2015)
7. Nizamuddin, M.: Predistortion for nonlinear power amplifiers with memory. Ph.D. thesis, Virginia Polytechnic Institute and State University (2002)
8. Shen, C., Plishker, W., Wu, H., Bhattacharyya, S.S.: A lightweight dataflow approach for design and implementation of SDR systems. In: Proceedings of the Wireless Innovation Conference and Product Exposition (2010)
9. Sills, J.A., Sperlich, R.: Adaptive power amplifier linearization by digital predistortion using genetic algorithms. In: Proceedings of the Radio and Wireless Conference, pp. 229–232 (2002)
10. Sindhya, K., Miettinen, K., Deb, K.: A hybrid framework for evolutionary multiobjective optimization. IEEE Transactions on Evolutionary Computation **17**(4), 495–511 (2013)
11. Wang, L.H., et al.: Dataflow modeling and design for cognitive radio networks. In: Proceedings of the International Conference on Cognitive Radio Oriented Wireless Networks, pp. 196–201 (2013)
12. Zitzler, E.: Evolutionary algorithms for multiobjective optimization: Methods and applications. Ph.D. thesis, Swiss Federal Institute of Technology (ETH) Zurich (1999)

Experimental Study of an Underlay Cognitive Radio System: Model Validation and Demonstration

Hanna Becker[1]([✉]), Ankit Kaushik[1], Shree Krishna Sharma[2],
Symeon Chatzinotas[2], and Friedrich Jondral[1]

[1] Communications Engineering Lab, Karlsruhe Institute of Technology (KIT),
Karlsruhe, Germany
hanna.e.becker@gmail.com, {ankit.kaushik,friedrich.jondral}@kit.edu
[2] SnT - Securityandtrust.lu, University of Luxembourg,
Luxembourg City, Luxembourg
{shree.sharma,symeon.chatzinotas}@uni.lu

Abstract. Cognitive radio is one of the potential contenders that address the problem of spectrum scarcity by making efficient use of the currently allocated spectrum below 6 GHz. A secondary access to the licensed spectrum is only possible, if the cognitive radio systems restrict the interference to the primary systems. However, the performance analysis of such a cognitive radio system is a challenging task. Currently, performance evaluation of underlay systems is limited to theoretical analysis. Most of the existing theoretical investigations make certain assumptions in order to sustain analytical tractability, which could be unrealistic from the deployment perspective. Motivated by this fact, in this work, we validate the performance of an underlay system by means of laboratory measurements, and consequently propose a hardware demonstrator of such a system. Moreover, we present a graphical user interface to provide insights to the working of the proposed demonstrator and highlight the main issues faced during this experimental study. (This work was partially supported by the National Research Fund, Luxembourg under the CORE projects "SeMIGod" and "SATSENT".)

Keywords: Cognitive radio · Underlay system · Power control · Dynamic access · Empirical validation · Demonstrator

1 Introduction

The amount of data transmitted over wireless channels is constantly increasing. However, the available spectrum is scarce and expensive, with more and more operators competing for their share of it. Therefore, ways have to be found to use the available spectrum more efficiently. Cognitive radio networks do so by enabling dynamic spectrum access to multiple systems. Secondary access to the licensed spectrum has been extensively investigated in the literature and is mainly categorized in terms of three cognitive radio paradigms [1]:

© ICST Institute for Computer Sciences, Social Informatics and Telecommunications Engineering 2016
D. Noguet et al. (Eds.): CROWNCOM 2016, LNICST 172, pp. 511–523, 2016.
DOI: 10.1007/978-3-319-40352-6_42

1. An interweave system exploits time gaps in the spectrum of primary users for data transmission.
2. An overlay system involves higher network layers to employ advanced coding algorithms to transmit data simultaneously with other systems.
3. In an underlay system, spectrum access is enabled only if the interference power received at primary users is below a certain amount. This can be achieved, for instance, by employing a power control mechanism at the secondary transmitter.

The existing investigations in [2–4] depicted the performance limits in terms of throughput achieved at the secondary receiver for the underlay system. However, the performance evaluation has been limited to theoretical analysis, which tends to make certain assumptions (for instance, perfect knowledge of channel), that are not applicable in hardware implementations [5]. Recently, hardware implementations in context to cognitive radio systems have started to receive significant attention [6–8], however these deployments are mainly concerned with the interweave system. In this regard, we provide insights for the deployment of underlay systems, in this paper. More specifically, we extend the mathematical framework derived in [9] to validate the performance of underlay systems by means of experimental analysis. To complement the analysis presented in [9], the main contributions of this paper are as follows:

1. Empirical validation: We set up a suitable hardware environment, perform measurements and evaluate their results by comparing them with the theoretical expressions.
2. Upon validating the mathematical model, we propose to deploy a hardware demonstrator of the underlay system. We present a graphical user interface to provide further insights to the working of the demonstrator.

This paper is organized as follows: Sect. 2 introduces the system model. Section 3 describes the experimental setup and the validation of the mathematical model. Section 4 portrays the implementation of the underlay system's hardware demonstrator. Finally, Sect. 5 concludes the paper.

2   System Model

The analysis done is this paper is based on the signal model illustrated in [9].

2.1   Underlay Scenario

Cognitive Relay (CR) is a cognitive radio small cell deployment that facilitates secondary access to indoor devices (IDs) [10]. Figure 1 shows such a scenario, where the CR acts as a secondary transmitter (ST), transmitting data to a secondary receiver (SR) represented by an ID. The channels between the primary receiver (PR) and ST and between the ST and SR are modeled in terms of path loss factors (α_p, α_s) and small-scale fading gains (g_p, g_s). A power control mechanism is employed at the ST to ensure that interference received at the PR is below a certain level. For this mechanism, it is necessary to acquire the knowledge of the channel between the ST and the PR. As proposed in [9], the ST can retrieve this information by listening to a pilot or beacon signal transmitted by the PR.

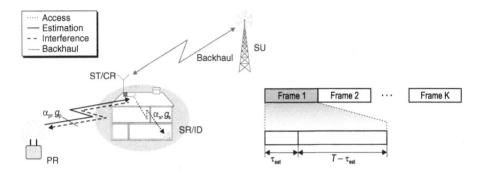

Fig. 1. Underlay scenario and frame structure [9]

We consider slotted medium access for the secondary system with a frame duration of T. For the ST to be able to satisfy the interference constraints at the PR, we consider channel reciprocity of the primary link. T is designed such that the channel can be assumed to remain constant within it. Based on this premise, g_p and g_s are constant within one frame and included in α_p and α_s for further analysis.

In order to implement a power control mechanism, we have divided the frame interval in two phases, refer to Fig. 1. During the first phase of duration τ_{est} (estimation time), the ST measures the received power of the pilot signal transmitted by the PR. Based on this received-power, the ST estimates α_p by relating it to the known PR transmit power (P_{tran}) and adapts its own transmit power for the secondary link (P_{cont}) accordingly. During the second phase duration i.e., $T - \tau_{est}$, the ST transmits data to the ID with the controlled power P_{cont}.

The sequence of events portrayed by the underlay scenario from Fig. 1 can be summarized as:

1. The PR sends a pilot signal with power P_{tran} to the ST.
2. The ST measures the power received (P_{rcvd}) from this signal.
3. From P_{rcvd}, the ST estimates α_p. We assume that the ST has the knowledge of P_{tran}.
4. From α_p, the ST calculates P_{cont}. It is scaled such that, in case of perfect channel reciprocity and the absence of noise on the primary link, the interference power arriving at the PR (P_p) has the value of the interference temperature (θ_I). In control theory terms, θ_I is the setpoint for P_p.

5. The ST transmits data to the SR with P_{cont}. In the context of this work, we send an unmodulated sinusoidal signal. This is mathematically equivalent to the constant power signal sent by the PR (refer to [9] and the references therein).
6. The SR receives the data signal with power P_s. It provides this value back over a feedback channel to the ST, where it is used to estimate the expected throughput of the secondary link (R_s).
7. Due to the presence of noise at the ST, the ST encounters variations in P_{rcvd}, which further affects P_{cont} and, in addition with noise at the PR, finally translates to variations in P_p around θ_I. This may severely degrade the performance of the cognitive radio system. In order to control these variations, an interference constraint in terms of probability of confidence (P_c) has been proposed in [9].

2.2   Stochastic Model

According to [9], P_{rcvd} can be modeled as a non-central chi-squared distribution with the following probability density function (pdf) [11]:

$$f_{P_{\text{rcvd}}}(x) = \frac{N}{2\sigma_p^2}\left(\frac{Nx}{\lambda}\right)^{\frac{N-2}{4}}\exp\left(-\frac{Nx+\lambda}{2\sigma_p^2}\right)I_{\frac{N}{2}-1}\left(\frac{\sqrt{Nx\lambda}}{\sigma_p^2}\right),\tag{1}$$

where N is the degree of freedom, i.e. the number of samples used for determining P_{rcvd}, σ_p^2 is the noise variance of the in-phase or quadrature-phase component of the received pilot signal (y_{rcvd}, refer to [9]), and $I_{\frac{N}{2}-1}(\cdot)$ is the modified Bessel function of the first kind of order $\frac{N}{2}-1$ [12]. Furthermore,

$$\lambda = \sum_{n=1}^{N}|\mathbb{E}\left[y_{\text{rcvd}}[n]\right]|^2 = N \times A^2\tag{2}$$

is the non-centrality parameter, where $y_{\text{rcvd}}[n]$ represents the discrete sample at the ST [9]. As our pilot signal is a sinusoid with a constant amplitude which is down-converted by an IQ demodulator at the ST, the complex samples have a constant envelope of value A, which explains the simplification in (2).

The system variables P_{cont}, P_p, and R_s are derived from P_{rcvd} in [9], where the respective pdfs $f_{P_{\text{cont}}}(\cdot)$ and $f_{P_p}(\cdot)$ are also provided. In [9], $f_{R_s}(\cdot)$ represented a pdf of the capacity. Here, we modify this expression to determine the pdf of the secondary throughput

$$f_{R_s}(x) = \frac{T}{T-\tau_{\text{est}}}\frac{NK\theta_I\alpha_s\ln 2}{2\sigma_p^2\sigma_s^2}\left(\frac{p(x)+1}{[p(x)]^2}\right)e^{-\frac{N}{2\sigma_p^2}\left(\frac{K\theta_I\alpha_s}{p(x)\sigma_s^2}+\alpha_pP_{\text{tran}}\right)}\tag{3}$$

$$\times\left(\frac{K\theta_I\alpha_s}{p(x)\alpha_pP_{\text{tran}}\sigma_s^2}\right)^{\frac{N}{4}-\frac{1}{2}}I_{\frac{N}{2}-1}\left(\frac{N}{\sigma_p^2}\sqrt{\frac{K\theta_I\alpha_pP_{\text{tran}}\alpha_s}{p(x)\sigma_s^2}}\right),$$

$$\text{with } p(x) = 2^{\frac{Tx}{T-\tau_{\text{est}}}} - 1 .$$

The definition of P_c can be retrieved from [9]. It is based on the cumulative distribution function (cdf) of P_p[1]

$$F_{P_p}(x) = Q_{\frac{N}{2}}\left(\sqrt{\frac{NP_{\text{tran}}\alpha_p}{\sigma_p^2}}, \sqrt{\frac{N\alpha_p\theta_I K}{\sigma_p^2 x}}\right), \tag{4}$$

with parameters defined in [9]. $Q_{\frac{N}{2}}(\cdot)$ is the Marcum Q-function [12].

It is challenging to determine the parameter σ_p^2 utilized in most of the theoretical expressions, accurately. We decided to approximate σ_p^2 by setting it equal to the variance of the envelope of y_{rcvd}, as this provided the best fit of the model function to the measurement values.

3 Validation

3.1 Experimental Setup

Figure 2 illustrates the experimental setup used for validation. The primary link is implemented via a cable and attenuators. By doing so, we were able to acquire a large number of system variable realizations measured under similar conditions, which we needed for validating the stochastic model.

The CR/ST is implemented in a Universal Software Radio Peripherals (USRP) B210 from Ettus Research [13]. There, upon arrival, the pilot signal is down-converted to an intermediate frequency, band-pass filtered, down-converted to baseband and decimated. The first two steps were carried out to avoid I/Q imbalance and remove the receiver's DC offset and the flicker noise (1/f) around the DC. Due to the small bandwidth of the pilot signal, these effects were the bottleneck of our validation and had to be accounted for. The decimation is performed to reduce the effect of correlation between the samples due to oversampling, since the model function $f_{P_{\text{rcvd}}}(\cdot)$ required independent and identically distributed energy samples [9]. Finally, the measurement data is analyzed offline using Matlab.

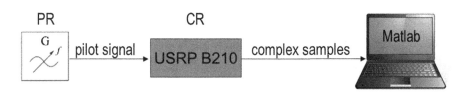

Fig. 2. Measurement setup for the validation of the stochastic model, laptop image from [14]

---

[1] In [9], we discovered a small typing error in the cdf of P_p, in this paper, we present the exact version of it.

3.2   Validation of System Variables

Since the stochastic model is the basis of the further performance analysis that will be carried out over P_c and R_s, as a first step, we validate the pdfs of the system variables P_{rcvd}, P_{cont}, P_p, and R_s, from Sect. 2.2 and [9]. To this end, measurements with the setup in Fig. 2 have been performed for different values of received signal-to-noise ratio at the ST over the primary link ($\mathrm{SNR}_{rcvd}{}^2$). The measurement data was plotted in terms of histograms and scaled such that it represented the relative frequency (f_{rel}). Figure 3 compares the histograms from the measurements and plotted pdfs using the analytical expressions for different system parameters. The plots show that the theoretical expressions very accurately capture the performance of real world cognitive radio systems.

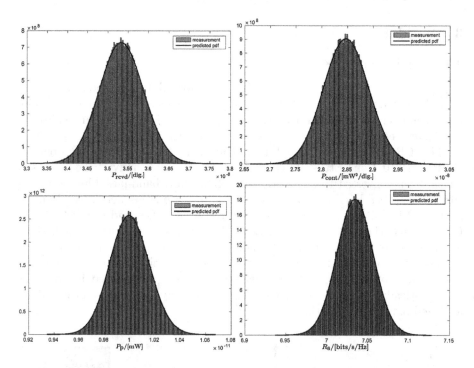

Fig. 3. Theoretical expressions of the pdf and experimental results of different system variables (parameters from Table 1)

We repeated the experiment for different values of SNR_{rcvd}. It was observed that for a considerable range of $\mathrm{SNR}_{rcvd} \in (4, 30)$ dB, the theoretical expressions depicted a significant accuracy to the experimental data, refer to Table 2. The accuracy was quantified in terms of relative error (e_{rel}) defined as

---

2 As noise power, we used the measured receiver noise floor.

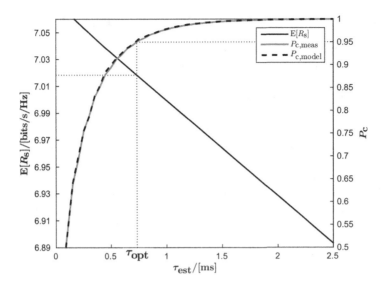

Fig. 4. Estimation-throughput tradeoff (parameters from Table 1), $P_{c,meas}$: empirical values of P_c, $P_{c,model}$: analytical values of P_c.

$$e_{rel} = \frac{1}{n_{bins}} \times \sum_{n=1}^{n_{bins}} \frac{f_{P_{rcvd}}[n] - f_{rel}[n]}{f_{rel}[n]}, \tag{5}$$

where n_{bins} is the number of histogram bins with $f_{rel}[n] \neq 0$.

Table 1. Values of the parameters used for the performing experiments.

Parameter	SNR_{rcvd}	N/τ_{est}	θ_I	T	α_s	σ_s^2
Value	22 dB	100/0.5ms	-110 dBm	100 ms	1[a]	2.1355×10^{-10} [b]

[a] The channel gain of the ST-SR link $\alpha_s \in (0, 1)$ was set to its maximum theoretical value for this analysis.
[b] The value represents the measured receiver noise floor (digital value) of the in-phase or quadrature-phase components.

Table 2. e_{rel} from 5 for various SNR_{rcvd} (parameters from Table 1)

$SNR_{rcvd}/[dB]$	4.08	9.10	14.11	19.12	24.09	29.09	34.03	39.38	45.08
e_{rel}		0.0568	0.0601	0.0522	0.0437	0.0506	0.0634	0.1179	0.0800	0.1695

3.3   Validation of the Estimation-Throughput Tradeoff

Finally, we validate the performance in terms of estimation-throughput trade-off to yield a suitable estimation time that satisfies the interference constraint on P_c and maximizes the achievable throughput. In contrast to the theoretical analysis presented in [9], in Fig. 4 we provide an empirical validation to the performance of the underlay system. Clearly, this tradeoff considers that a large τ_{est} will improve the performance of the primary system by reducing the variations in P_p. This improvement is depicted in terms of an increase in P_c. On the other hand, the increase in τ_{est} reduces the achievable secondary throughput. Figure 4 also includes a validation of P_c. This is achieved by comparing its empirical values with its analytical expressions for different τ_{est}. In contrast to the analytical model [9], the empirical values of P_c were determined using a numerical integration in the region within the confidence interval $(1 \pm \mu) \times \theta_I$, where μ is the accuracy as defined in [9]. Hence, with this verification, we conclude that the estimation-throughput tradeoff proposed in [9] is suitable for hardware implementation.

4   Implementation of a Demonstrator

In this section, we provide the details on the implementation of a demonstrator for the underlay system.

4.1   Estimation Time

As we already verified the dependence of P_c and R_s on τ_{est} (refer to Fig. 4), it is challenging to select τ_{est} such that the system adheres to the interference constraints at the PR and still achieves the highest possible secondary throughput. To analyze this problem, we introduce a new parameter called the optimized estimation time (τ_{opt}). It is the τ_{est} that maximizes the secondary throughput according to equation (11) in [9] for a certain value of SNR_{rcvd}, μ and a target value of P_c defined as $\bar{P}_c$. In Fig. 4, this optimization process is indicated graphically by the dotted lines, where, from a fixed $\bar{P}_c = 0.95$, we acquire $\tau_{opt} \approx 0.75$ ms, which corresponds to $\mathbb{E}[R_s] \approx 7.02$ bits/s/Hz.

However, this analysis is carried out for a fixed value of SNR_{rcvd}. Under real conditions, due to channel fading, SNR_{rcvd} is not known. In this sense, it is not possible to determine τ_{opt}. To resolve this issue, we propose a procedure, whereby we analyze the variations of τ_{opt} for different values of SNR_{rcvd}, refer to Fig. 5, and select τ_{opt}'s maximum value. By doing this, we are able to satisfy the interference constraint for all realizations of the channel. In addition, we consider different values of $\bar{P}_c$. It is observed that τ_{opt} increases with the decrease in SNR_{rcvd} and attains saturation below a certain SNR_{rcvd}.[3]

---

[3] For varying θ_I, while the shape of the curves changed slightly, the upper limits for τ_{opt} remained constant.

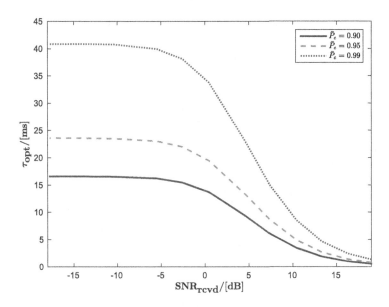

Fig. 5. τ_{opt} over SNR_{rcvd}, $\theta_{\mathrm{I}} = $ -110 dBm, $\mu = 0.05$

The explanation of this behavior is given in the following: For large values of SNR_{rcvd}, P_{cont} is low, hence the variations of P_{p} around θ_{I} are low and consequently a lower value of τ_{opt} is needed to maintain these variations within the confidence interval. Very weak received signals, on the other hand, cannot be distinguished from noise by the USRP, due to the quantization limit of the analog-to-digital converter in the receiver chain. This is why, below a certain SNR_{rcvd}, all received signals yield the same value of τ_{opt}.

We use this analysis for determining the τ_{est} in the implementation of our demonstrator. Since we target $\bar{P}_{\mathrm{c}} = 0.95$, we choose a fixed τ_{est} of 24 ms, which is the maximum value determined from Fig. 5. By doing so, we seek to satisfy the interference constraints at the PR, at the cost of a decreased performance in R_{s}, particularly at higher SNR_{rcvd}, where τ_{opt} achieves a low value.

4.2   Simplifications

The main objective of this paper is to demonstrate the basic principle of an underlay scenario, in view of this, we will consider the following reasonable simplifications in the proposed analytical framework:

1. We do not consider the hardware implementation of the SR, that is, it is regarded virtual in the system (refer to Fig. 6).
2. According to the model, the path loss is determined using [9]

$$\alpha_{\mathrm{p}} = \frac{\mathbb{E}\left[P_{\mathrm{rcvd}}\right] - \sigma_{\mathrm{p}}^2}{P_{\mathrm{tran}}} .$$ (6)

This is not possible in practical situations, where only a single realization of P_{rcvd} is available. Hence, we determine the path loss based on this realization. As σ_{p}^2 is negligible compared to P_{rcvd}, it can be further simplified

$$\alpha_{\mathrm{p}} = \frac{\mathbb{E}\left[P_{\mathrm{rcvd}}\right] - \sigma_{\mathrm{p}}^2}{P_{\mathrm{tran}}} \approx \frac{P_{\mathrm{rcvd}} - \sigma_{\mathrm{p}}^2}{P_{\mathrm{tran}}} \approx \frac{P_{\mathrm{rcvd}}}{P_{\mathrm{tran}}} = \tilde{\alpha}_{\mathrm{p}} \; . \tag{7}$$

By not averaging over multiple realizations of P_{rcvd}, we expect a higher variance in the resulting powers P_{cont} and P_{p}.

3. The model involves a frame synchronization (in case of Time Division Duplexing) between PR and ST, which is complicated. To simplify this matter, we propose Frequency Division Duplexing between the PR and the ST: We transmit and receive the signals using two different frequencies (2.422 GHz and 2.423 GHz) over two separate antennas, as illustrated in Fig. 6. With this technique, the channel reciprocity may be compromised.

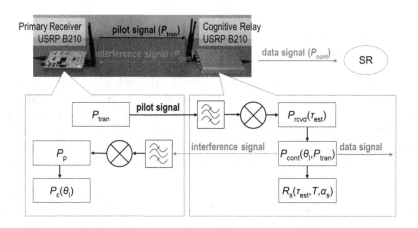

Fig. 6. Setup and block diagram of demonstrator

Mapping the steps described in Sect. 2.1 onto hardware and applying the above-mentioned simplifications, we acquire the signal flow illustrated in Fig. 6, which we have implemented in GNU Radio using the available blocks therein.

4.3   User Interaction and Observations

Figure 7 shows the user interfaces of the demonstrator, providing insights to the parameters evaluated at the PR (for instance, P_{p} and P_{c}) and the CR/ST (for instance, P_{rcvd}, P_{cont}, and R_{s}). We have performed hardware calibration in the demonstrator to provide physical significance to the digital values obtained from the USRPs, hence the displayed units. As the SR has not been implemented in the hardware, to incorporate the effect of α_s on the performance of the system, we employ a slider to modify its value.

As expected, changing the value of θ_I at the CR/ST changes the measured value of P_p at the PR to approximately the same value. This phenomenon is highlighted in Fig. 7. At the same time, the values of R_s and P_{cont} adapt accordingly. This demonstrates that the received power estimation done at the ST by listening to the pilot based channel, thereby acquiring the channel knowledge and performing the power control, is working in accordance to the underlay principle.

The response to the dynamic conditions can be verified by changing the distance between the PR and ST, the effect can be captured by observing the changes in P_{rcvd} and other parameters depending on it. As the distance is increased beyond a certain value, the ST operates at its maximum transmit power. This event is indicated in the user interface.

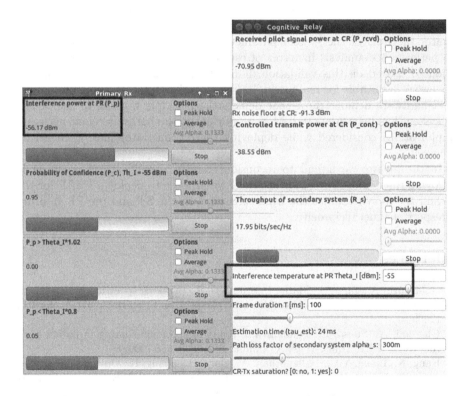

Fig. 7. A snapshot of the performance parameters displayed in the user interfaces

With $\mu = 0.05$, the demonstrator does not provide the target value of 0.95 for P_c, as the variations in P_p are higher as expected. Certainly, this issue is partly caused by the simplifications undertaken in 7, which have to be accounted for in future implementations. Another possible reason for this observation is that we used a pilot signal produced by a signal generator in the previous analysis, which offers a higher signal quality than the one produced by a USRP in the demonstrator. Moreover, because of the separate links for sensing and transmission and the frequency separation of 1 MHz, the channel reciprocity in our demonstrator may be compromised compared with the theoretical model. To resolve this issue, we increase the tolerance limit to $\mu = 0.20$, which leads to the desired P_c of 0.95. On this account, we will consider the signals being transmitted by a USRP for validation, in the future. Despite this, we have been able to demonstrate the principle working of an underlay system that employs a power control mechanism at the ST to limit the excessive interference at the PR.

5   Conclusion

In this paper, we have analyzed the performance of an underlay system from a deployment perspective. To this end, an existing analytical framework [9] has been validated. In this regard, the validation of a stochastic model that incorporates the pdfs of the system parameters has been considered. In addition, the performance analysis in terms of estimation-throughput tradeoff has been validated. Based on this validation, it has been illustrated that the proposed framework is suitable for real world deployments. Upon the experimental analysis, a hardware demonstrator that depicts the principle working of the underlay system has been proposed. More importantly, the hardware challenges and simplifications considered while deploying the demonstrator have been briefly discussed.

In the future, we intend to reconsider certain simplifications made while deploying the demonstrator, for instance, we propose to deploy a USRP for the SR and try to synchronize the frame structure at the ST and the PR in order to respect channel reciprocity.

References

1. Goldsmith, A., Jafar, S., Maric, I., Srinivasa, S.: Breaking spectrum gridlock with cognitive radios: an information theoretic perspective. Proc. IEEE **97**(5), 894–914 (2009)
2. Ghasemi, A., Sousa, E.: Fundamental limits of spectrum-sharing in fading environments. IEEE Trans. Wirel. Commun. **6**(2), 649–658 (2007)
3. Kang, X., Liang, Y.C., Nallanathan, A., Garg, H., Zhang, R.: Optimal power allocation for fading channels in cognitive radio networks: ergodic capacity and outage capacity. ieee trans. wirel. commun. **8**(2), 940–950 (2009)
4. Musavian, L., Aissa, S.: Fundamental capacity limits of cognitive radio in fading environments with imperfect channel information. IEEE Trans. Commun. **57**(11), 3472–3480 (2009)
5. Sharma, S., Bogale, T., Chatzinotas, S., Ottersten, B., Le, L., Wang, X.: Cognitive radio techniques under practical imperfections: a survey. IEEE Commun. Surv. Tutorials **17**(4), 1858–1884 (2015)

6. Nguyen, T., Nguyen, T., Nguyen, H., Dang, K.: Hardware implementation of reception diversity techniques for spectrum sensing efficiency enhancement in cognitive radio network. In: 2013 Third World Congress on Information and Communication Technologies (WICT), pp. 69–73 December 2013

7. Anas, N., Mohamad, H., Tahir, M.: Cognitive radio test bed experimentation using USRP and Matlab/Simulink. In: 2012 IEEE Symposium on Computer Applications and Industrial Electronics (ISCAIE), pp. 229–232, December 2012

8. Combes, R., Proutiere, A.: Dynamic rate and channel selection in cognitive radio systems. IEEE J. Sel. Areas Commun. **33**(5), 910–921 (2015)

9. Kaushik, A., Sharma, S., Chatzinotas, S., Ottersten, B., Jondral, F.: Estimation-throughput tradeoff for underlay cognitive radio systems. In: 2015 IEEE International Conference on Communications (ICC), pp. 7701–7706, June 2015

10. Kaushik, A., Raza, M., Jondral, F.: On the deployment of cognitive relay as underlay systems. In: 2014 9th International Conference on Cognitive Radio Oriented Wireless Networks and Communications (CROWNCOM), pp. 329–334 (2014)

11. Charalambous, C., Menemenlis, N.: Stochastic models for short-term multipath fading channels: chi-square and Ornstein-Uhlenbeck processes. In: Proceedings of the 38th IEEE Conference on Decision and Control, vol. 5, pp. 4959–4964 (1999)

12. Jeffrey, A., Zwillinger, D.: Table of Integrals, Series, and Products. Elsevier Science, Boston (2000)

13. Ettus Research, 2 December 2015. http://www.ettus.com/

14. Sedykh, I.: laptop-icon_500x500, 2 December 2015; Copyright information: Creative Commons Attribution 2.0 Generic (CC BY 2.0) (2013). https://www.flickr.com/photos/shmectorcom/8616375606

Flexible In-Band Full-Duplex Transceivers Based on a Modified MIMO RF Architecture

Alexandre Debard$^{(\boxtimes)}$, Patrick Rosson, David Dassonville,
and Vincent Berg

CEA-LETI, Minatec Campus, 17 rue des martyrs, 38054 Grenoble, France
{alexandre.debard,patrick.rosson,david.dassonville,
vincent.berg}@cea.fr

Abstract. In-band full-duplex transceivers are considered for future generations of cellular network systems. This paper proposes to evaluate the performance of in-band full-duplex transceivers using a modified architecture based on
hardware available for multiple-input multiple-output transceivers. A hybrid
self-interference cancellation technique using an auxiliary transmitter is therefore introduced. Performance is evaluated using simulation models and is
confirmed by hardware experimentation. The main limiting factors of the proposed architecture are analyzed and improvements to the architecture are then
suggested.

Keywords: In-Band Full-Duplex · Transceiver · MIMO architecture · RF
impairments

1 Introduction

The massive adoption of smartphones together with the need to always be connected has
accelerated the demand for broadband mobile connectivity. Therefore, the capacity of
cellular networks should continue to increase to meet end-user requirements. A possible
solution considered by cellular operators is to further improve spectral efficiency. Multiple Input Multiple Output (MIMO) techniques [1, 2] and relay techniques [3, 4] which
have been studied for the last twenty years allow cellular operators to improve bit rate and
capacity. Another alternative consists in increasing the useful bandwidth or in deploying
new locations for base stations. All options increase the cost and/or the power consumption of the base station or the mobile equipment. Recently In-Band Full-Duplex
(IBFD) solution has been studied and demonstrated in the Wireless Local Area Network
(WLAN) context [5–8]. It seems to be a promising approach to resolve the asymmetric
data flows introduced by Frequency Division Duplex (FDD) communication or to reduce
latency in Time Division Duplex (TDD) communication. It may optimize time-frequency
resources and can under certain conditions increase the overall capacity [9].

Basically, the main problem of IBFD transceivers is self-interference. In a full-duplex
point-to-point communication, both transmitters and receivers simultaneously transmit
and receive in the same frequency band. Therefore each transmitter generates a powerful
signal which simultaneously creates a well-known self-interference at the receiver. To
receive the useful low power signal, the transceiver must cancel this self-interference.

© ICST Institute for Computer Sciences, Social Informatics and Telecommunications Engineering 2016
D. Noguet et al. (Eds.): CROWNCOM 2016, LNICST 172, pp. 524–535, 2016.
DOI: 10.1007/978-3-319-40352-6_43

Several methods can be used to cancel the self-interference but all request a cancellation in the analog domain. This can be done at the antenna level and at RF level. The important level of cancellation that is required cannot be only reached in the digital domain mainly because of the analog-to-digital converter which does not support such a dynamic range.

In 2013, Stanford proposed a new IBFD architecture based on a single antenna with circulator, a digitally controlled broadband analog filter and a digital canceller [8]. Stanford demonstrated an IBFD single input single output (SISO) transmission at 2.4 GHz. The full-duplex transceiver was able to reject self-interference level of around 110 dB over 80 MHz of bandwidth. A first RF cancellation step of 62 dB is provided by a circulator and a broadband analog filter. Then, a digital canceller adds 48 dB of cancellation. The overall transmitted signal of 20 dBm was reduced to the thermal noise level of −90 dBm. The main contributions of Stanford are the broadband analog filter and the digital canceller that takes into account the non-linear effect.

The solution proposed by Stanford [8] is very performant but very complex especially the digitally controlled analog filter. This analog filter consists in 16 delay lines with 16 attenuators controlled by 7 bits command for a range of 31.75 dB and successive steps of 0.25 dB. The Peregrine Semiconductor attenuator components used in this analog broadband filter are not perfect. The RF attenuation is not exactly the expected attenuation, the relative phase changes when the attenuation or frequency increases [10]. This supposes a calibration process. Moreover the delay lines need to be adapted to the frequency band and the antenna structure.

The next section introduces the proposed architecture. Part III presents the self-interference cancellation technique and its performance. In part IV, the canceller is evaluated through a simple implementation on real signal.

2 Flexible Architecture

In this paper, an alternative architecture is studied in order to reduce the complexity of the analog broadband filter proposed by the Stanford solution. The envisaged solution is based on current RF MIMO transceiver architectures. As previously mentioned, current generations of cellular systems support MIMO communication, meaning that most of the transceiver supports multiple transmitters. In this paper we will try to evaluate a full-duplex architecture using these multiple transmitters. The second transmitter called auxiliary transmitter will replace the broadband analog filter. It is called Hybrid Self-Interference Cancellation (HSIC). In our context, the target requirements are less ambitious than the Stanford requirements but our solution is simpler to integrate and updates from current device architectures. The MIMO FDD transceiver becomes a single ended IBFD transceiver. This proposal allows the Radio Resource Management (RRM) to select between classical FDD MIMO or IBFD single-end transmissions. The transceiver becomes flexible and can switch from a configuration to another according to the scenario. If the channel conditions as well as the interference environment are favourable for MIMO transmission, MIMO operation is selected. Otherwise IBFD can be considered. Impact of coexistence between MIMO and IBFD will not be studied in this paper. Few components are added compared to a classical MIMO transmit architecture. The rest of the transceiver does not change. Only the baseband takes the self-interference cancellation into account. This evolution corresponds to a smaller technology gap than previously thought architectures and should make the migration to IBFD more acceptable. The overall flexible architecture is presented in Fig. 1.

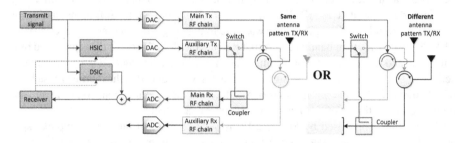

Fig. 1. In-band Full-duplex transceiver using MIMO architecture

3   Theoretical Study

3.1   Overview

This section presents the simulation model. It is used for performance evaluation of the interference cancellation technique. This simulation chain consists of several blocks. Each block models component of the transceiver taking into account most of the imperfections. We chose models for converters (i.e. DAC and ADC), phase noise of local oscillators, amplifiers non-linearities, circulator and antenna matching and isolation. Most of the models can be found in reference [11]. In this theoretical study, we focus our effort on the impact of the transmitter noise and the transmitter non-linearity on the performances. Baseband equivalent models have been considered.

3.2   Models

The model used for quantization, phase noise, Low Noise Amplifier (LNA) and converters are presented in [11]. Time varying effects have not been considered. For Power Amplifier (PA), we use the same model as [11] regarding the amplitude non-linearity and we chose the Saleh model [12] for the phase distortion:

$$\varphi_{out} = \varphi_{in} + \frac{\alpha_\varphi |x_{in}|}{1 + \beta_\varphi |x_{in}|^2} \tag{1}$$

Where φ_{out} is the phase of the output signal of the PA and φ_{in} its phase at the input. α_φ and β_φ are the parameters of the model and $|x_{in}|$ is the input signal's magnitude. To model the complex transfer function of the circulator, the S-parameters of a 2.4 GHz circulator connected to an antenna have been measured. The measured response is then

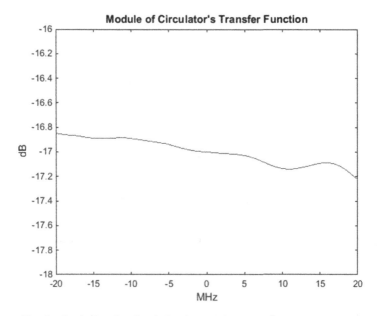

Fig. 2. Equivalent baseband circulator and antenna frequency response

translated to baseband for simulation and given in Fig. 2 for reference. Note that this response includes the reflection coefficient of the antenna.

The filters used in the simulation are 3rd-order Butterworth-type Infinite Impulse Response (IIR) filters. The developed simulation can be applied in baseband or with a transposition to an Intermediate Frequency (IF).

3.3 Hybrid Analog and Digital Self-interference Cancellation HSIC

The analog canceller aims at reducing the linear contribution of the self-interference and avoids the ADC saturation. It is in fact a hybrid analog-digital solution, as defined in [13]. It is based on an auxiliary synchronous transmitter. From the digital baseband domain, the auxiliary RF transmitter generates an analog signal which is subtracted from the main reception path. Please note that the RF components of the first and second transmitters are identical nevertheless the circulator on the main transmitter and the coupler introduces important mismatches. In practice, these mismatches must be estimated and compensated to optimize the analog cancellation performances. The filters to be estimated correspond to the h_{t1} and h_{t2} on Fig. 3.

If linearity is assumed, the signal at the output of the analog canceller is given by:

$$y_a(t) = (x_1 * h_{t1})(t) - (x_2 * h_{t2})(t) + r(t) \tag{2}$$

Where x_1 and x_2 are the outputs of the digital transmitters, r is the received signal coming from the antenna and $*$ being the convolution product.

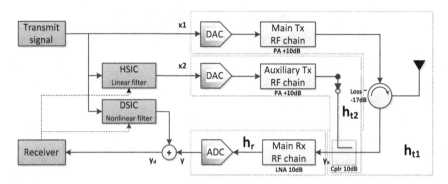

Fig. 3. Schematized system with transmission filters

The linear contribution of the self-interference at the output of the analog canceller is given by:

$$self\_interf(t) = (x_1 * h_{t1})(t) - (x_2 * h_{t2})(t) \tag{3}$$

Therefore to minimize the self-interference, the following equality should be reached:

$$\frac{X_2(f)}{X_1(f)_2} = \frac{H_{t1}(f)}{H_{t2}(f)} \tag{4}$$

Where $X_1(f)$, $X_2(f)$, $H_{t1}(f)$ and $H_{t2}(f)$ are respectively the Fourier transforms of x_1, x_2, h_{t1} and h_{t2}. Only the discrepancy between h_{t1} and h_{t2} is important. This difference is estimated in a two-step calibration measurement. Firstly, the filter between x_1 and y (denoted h_1) is estimated when nothing is transmitted on the auxiliary path. Secondly, the filter between x_2 and y (denoted h_2) is estimated when nothing is transmitted on the main path. The signal x_2 is derived from h_1, h_2 and x_2 as follows:

$$x_2(k) = x_1(k) * TF^{-1}\left(\frac{H_{t1}(f)}{H_{t2}(f)}\right) = x_1(t) * TF^{-1}\frac{H_1(f)}{H_2(f)} \tag{5}$$

As

$$H_i = H_{ti}H_r \text{ for } i = 1, 2 \tag{6}$$

Where TF^{-1} is the inverse Fourier transform. Filters h_1 and h_2 are estimated thanks to a reference sequence in the time domain as mentioned in the reference [8].

3.4   Digital Non-linear Self-interference Cancellation DSIC

The digital cancellation objective is to remove the remaining self-interference. As previously mentioned, the analog cancellation does not consider non-linear terms, so the digital cancellation is needed to take care of the nonlinear contribution [8]. Only non-linear terms of odd orders can interfere in the useful frequency band [11]. Considering these non-linear terms, we have:

$$y(k) = \sum_{m\,odd,k=-n,...,n} x(k)(|x(k)|)^{m-1} * h_m(k) \tag{7}$$

The considered filter estimation technique is similar to the one of the hybrid analog part, is realized for each order m.

The studies consider lower non linearity effect on the auxiliary RF transmitter. This assumption is supposed to be right if the circulator isolation is important compared to the coupling factor. In this case, the hybrid canceller asks for an output power of the auxiliary chain lower than the main transmitter. The nonlinear contribution comes from the main RF transmitter, the circulator and the RF receiver.

3.5   Performances

The performances of the system are characterized by the self-interference after the HSIC and finally after the DSIC. In the simulation there is no useful signal received by the antenna. The transmitted signal is made of random OFDM symbols. The OFDM signal has a PAPR of about 10 dB. The output power of the amplifier is set to 20dBm. The parameters used for the models described in Sect. 3.2 are:

- The DACs and ADC have a resolution of 16 bits. Their maximum Integral Non Linearity (INL) is set to a level of 2 Least Significant Bit (LSB), their maximum Differential Non Linearity (DNL) is 0.3 LSB. The maximum output peak-to-peak voltage of the DACs is 5.6 V and the maximum input peak-to-peak voltage of the ADC is 0.3 V.
- The phase noise is characterized by a -110 dB noise floor and a 2-order $1/f^3$ filter.
- The gain of the PA is 10 dB, its 1 dB compression point is 32 dBm and its third order interception point is 40 dBm. The parameters chosen for the Saleh model are $\alpha = 4.5$ and $\beta = 1.1$.
- The gain of the LNA is 10 dB, its 1 dB compression point is -10 dBm and its third order interception point is 0 dBm. Its noise output is at a power of -90 dBm. This noise is taken as the receiver's noise floor.

Simulations highlight the influence of the transmitter noise on performances. Indeed, an all-analog cancellation system as the one presented by [8] is designed to be able to cancel the transmitter's noise included in the self-interference. A hybrid system like ours cannot remove any noise as transmit (x1) and mirrored (x2) paths are uncorrelated. This result was then expected. It also appeared that in our case, the phase noise could be cancelled as we included in the simulation the fact that the phase noises brought by all our mixers was the same but only with different delays. This hypothesis is justified by the fact that mixers use the same local oscillator.

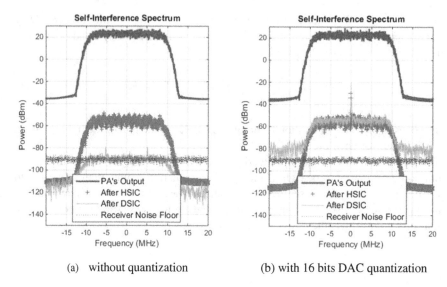

(a)   without quantization   (b) with 16 bits DAC quantization

Fig. 4. Simulation results

The simulations show that the transmitter noise is by far the most important limiting factor in this architecture, as the system will easily exceed the 100 dB of cancellation when no quantization effects is considered. However, the cancellation is clearly limited by quantization when it is activated. Indeed we can see on Fig. 4 that the signal will not decrease under around −50 dBm after DSIC.

When assuming uniform distribution of quantization error between $\left[-\frac{q}{2},\frac{q}{2}\right]$, the formula of quantization noise power is given by:

$$QN = \frac{q^2}{12R} \tag{8}$$

Where q is the quantization step. R is the impedance which is 50 ohms. With 16 bit resolution DAC and 5.6 V peak-to-peak voltage, the formula gives an output noise power in all the DAC bandwidth equal to −75 dBm. In the simulations and with the error distribution, the output power is greater and equals to −70 dBm. If we had INL and DNL imperfections, this power would further increase up to −62 dBm.

Then we have to consider the fact that this noise power evolves along the RF chain. On the main transmitter, it is amplified by the PA, then attenuated by the circulator isolation (which is of about 17 dB, Fig. 2), and then amplified again by the LNA. On the auxiliary transmitter chain, the link budget is different as the circulator is replaced by the coupler (10 dB instead of −17 dB). Figure 5 shows both transmitter noise levels along the transmitter receiver path. Both transmitter signals are also included.

We can also notice in Fig. 4. (a) that the HSIC would not reduce the self-interference below about −55 dBm either, even without quantization. This limitation must be due to

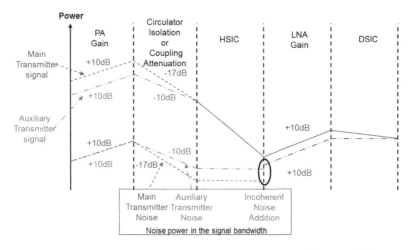

Fig. 5. Evolution of the power versus different RF components (without variable RF gain)

the non-linear distortions of the signal that are not canceled by the HSIC. Indeed, when calculating the power of the difference between the output of the PA and the linearly amplified signal, we get −38 dBm. That power is then attenuated by the circulator of 17 dB which gives −55 dBm and explains the results. Note that the auxiliary transmitter is used at a lower power level so generates less non-linear terms.

4 Simplified Hardware Implementation

In order to understand the effect of the transmitter noise and non-linearity, a simplified hardware implementation has been done using generated signals. It can be considered as a first step to evaluate HSIC performances using real life implementation.

4.1 Overview

The implementation is based on a Red Pitaya [14] board supporting two transmitters and one receiver all synchronized. This board is connected to a personal computer for the transmitted data flow and off-line evaluation. Additional connected RF components complete the demonstration. The useful transmitted baseband signal is digitally transposed on an IF and then converted in the analog domain. An amplifier used in a non-linear zone creates non-linearities. The experimentation is realized in a controlled environment. The antenna and the circulator are emulated by a loss which is approximately equal to −20 dB. After that the auxiliary transmitted signal is added thanks to a 10 dB coupler. For the demonstration, a 3 dB splitter is added to measure the performance and the HSIC. Then, a LNA with variable attenuator is inserted to adapt the input power level to the next amplifier (Fig. 6).

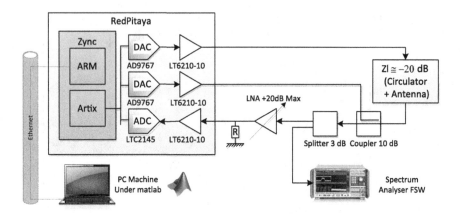

Fig. 6. Testbed overview

4.2 Red Pitaya Board

The Red Pitaya unit is an embedded oscilloscope and signal generator running on Linux operating system. It includes Radio Frequency signal acquisition and generation technologies, FPGA, Digital Signal Processing and CPU processing. The original Red Pitaya board is designed on a Zynq-7010 component and supports two transmitters and two receivers.

The board has been customized with a Zynq7020 instead of the Zynq7010. It contains a dual core ARM Cortex A9 and a Xilinx Artix-7 FPGA with 74 K programmable logic cells.

This board has also two analog outputs and two analog inputs with sampling converters working at 125 MHz. DAC and ADC sample signals at 125 MHz over 14 bits. Table 1 summarizes the main parameters of the hardware elements.

Table 1. Main parameters of the Red Pitaya board.

Functions	Components	Main characteristic
Digital	Zynq 7020	ARM dual core Cortex A9, ARTIX 7
2 DAC	AD9767	14 bits 125 MHz over ±1 V
ADC	LTC2145	14 bits 125 MHz over ±1 V
Amplifier	LT6210-10	Low Noise Op 0.95nV/$\sqrt{Hz}$ Amp Family 1.6 GHz

4.3 Waveform Definition

An Orthogonal Frequency Division Multiplexing (OFDM) waveform has been used at the transmitter. It spans over 20 MHz of bandwidth and it is transmitted on an IF of around 20 MHz.

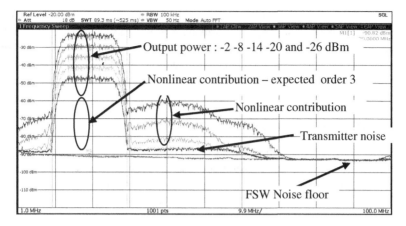

Fig. 7. Transmitted spectrum for different output powers

4.4 Transmitter Noise and Non-linearity

Figure 7 shows the spectrum at the output of the amplifier of the hardware demonstrator. Different output powers are transmitted from −2 dBm down to −26 dBm by 6 dB steps to estimate the nonlinear contributions and transmitter noise.

4.5 Performances

The following figure shows the spectrum at the output of the amplifier and after the HSIC more exactly after the splitter (Fig. 8).

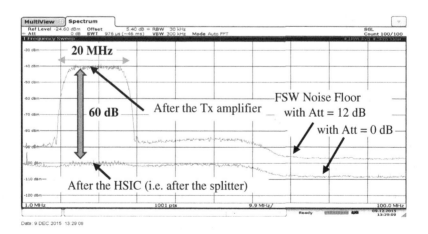

Fig. 8. HSIC performances

The isolation and the HSIC canceller reduce the self-interference by about 60 dB over the 20 MHz band. The DSIC performances is not showed as it cannot reduce self-interference lower than the transmitter noise received after the ADC. The transmitter is the critical parameter for high performances. The next section proposes an evolution of the architecture to reduce the effect of the transmitter noise.

4.6   RF Architecture Evolution

Several issues have been identified thanks to the theoretical study and the practical measurements. First of all, the transmitter noise must be as low as possible. Three solutions could be proposed:

(a) To add a direct RF path from the main transmitter as in the classical approach,
(b) To increase the oversampling to spread the DAC noise over the overall bandwidth and
(c) To use a variable gain on both transmitters to get same level of noise power at the output of the analog canceller.

Moreover, it is useful to add a controlled attenuator before the LNA. This avoids saturation of the LNA and ADC during the calibration phase when transmitted signals are very strong (compared to the received signal coming from the antenna) (Fig. 9).

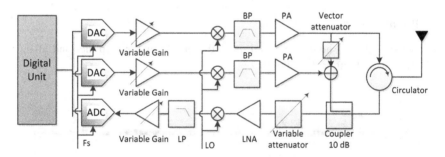

Fig. 9. Architecture evolution

5   Conclusion and Perspective

Based on a flexible RF architecture which is able to switch between FDD MIMO transceiver and IBFD single ended transceiver, we evaluate self-interference cancellation techniques taking into account major RF impairments. As expected, the performance of the transmitters is critical in the proposed architecture as both transmitters have a strong impact on the first analog canceller. The transmitter noise and non-linearity effects are critical parameters. With the proposed architecture, the transmitter noise should be reduced as much as possible. Oversampling is a possibility along with a reduction of the transmitted bandwidth. A trade-off needs to be found between cancellation performance and transmission bandwidth related to full-duplex bit rate. The non-linearity effect is also important but its impact can be compensated. Nevertheless simulations show that non-linear cancellation is very complex if we suppose that both RF transmitters work in their nonlinear zone. This work has been completed by practical measurements to validate the model and increase our understanding. To conclude, direct analog cancellation path from the main transmitter to the receiver seems to be unavoidable to provide high level of interference cancellation. Future work should investigate an architecture compromise that is based on multiple transmitters (a main transmitter for signal output and a mirror transmitter for signal cancellation) combined with a very simple RF canceller. This new proposed architecture seems to be a good trade-off between the complex solution of Stanford [8] and our first architecture.

Acknowledgements. The research leading to these results was supported by the European Commission under grant agreement 671563 (H2020-ICT-2014-2/ICT-14-2014), Flexible and efficient hardware/software platforms for 5G network elements and devices (Flex5GWare).

References

1. Telatar, I.E.: Capacity of multi-antenna Gaussian channels. Europ. Trans. Telecommu. **10**, 585–595 (1999)
2. Goldsmith, A., Jafar, S.A., Jindal, N., Vishwanath, S.: Capacity limits of MIMO channels Selected Areas in Communications. IEEE J. **21**, 684–702 (2003)
3. Cover, T., Gamal, A.E.: Capacity theorems for the relay channel. IEEE Trans. Inf. Theory **25**, 572–584 (1979)
4. Kramer, G., Gastpar, M., Gupta, P.: Cooperative strategies and capacity theorems for relay. IEEE Trans. Netw. Inf. Theory **51**, 3037–3063 (2005)
5. Choi, J., Kim, T., Bharadia, D., Seth, S., Srinivasan, K., Levis, P., Katti, S., Sinha, P.: Practical, real-time, full duplex wireless. In: International Conference on Mobile Computing and Networking (2011)
6. Duarte, M., Dick, C., Sabharwal, A.: Experiment-driven characterization of full-duplex wireless systems. IEEE Trans. Wirel. Commun. **11**, 4296–4307 (2012)
7. Everett, E., Sahai, A., Sabharwal, A.: Passive self-interference suppression for full-duplex infrastructure nodes. IEEE Trans. Wirel. Commun. **13**, 680–694 (2014)
8. Bharadia, D., McMilin, E., Katti, S.: Full duplex radios. In: SIGCOMM 2013, Hong Kong (2013)
9. Goyal, S. et al.: Improving small cell capacity with common-carrier full duplex radios. In: IEEE International Conference on Communications (ICC), pp. 4987– 4993 (2014)
10. Perigrine Semiconductor, Product specification 50 ohms RF Digital Attenuator 7bits 31.75 dB 9 KHz–8 GHz
11. He, J., Yang, J., Kim, Y., Kim, A.S.: System level time domain behavioral modeling for a mobile WiMax transceiver. In: Behavioral Modeling and Simulation Workshop, Proceedings of the 2006 IEEE International, pp. 138–143, San Jose (2006)
12. Saleh, A.A.M.: Frequency-independent and frequency-dependent nonlinear models of TWT amplifiers. IEEE Trans. Commun. **29**, 1715–1720 (1981)
13. Hua, Y., Ma, Y., Liang, P., Cirik, A.: Breaking the barrier of transmission noise in full-duplex radio. In: IEEE Military Communications Conference, pp. 1558—1563 (2013)
14. Red pitaya hardware specifications v1.1.1. Red Pitaya (2014)

Large-Signal Analysis and Characterization of a RF SOI-Based Tunable Notch Antenna for LTE in TV White Space Frequency Spectrum

Essia Ben Abdallah[✉], Serge Bories, Dominique Nicolas, Alexandre Giry, and Christophe Delaveaud

CEA, LETI, MINATEC Campus, Univ. Grenoble-Alpes, 38054 Grenoble, France
{essia.benabdallah,serge.bories,dominique.nicolas,
alexandre.giry,christophe.delaveaud}@cea.fr

Abstract. In this paper, a compact frequency-agile notch antenna for LTE low-band using TV white space frequencies is designed and fabricated. The antenna aperture tuning is provided by a SOI CMOS tunable capacitor. The tunable capacitor RF non linearity analysis was studied and measured. The simulated and experimental performances are presented and demonstrate antenna tuning operation from 800 MHz down to 500 MHz. High linearity is validated with measured ACLR levels lower than -30 dBc up to 22 dBm input power in the considered frequency range.

Keywords: Tunable notch antenna · TV white space · Tunable capacitor · LTE · Linearity

1 Introduction

Due to the rapid demand for data over cellular networks, it has became difficult to cover all the traffic with current frequency bands. Thus, more frequency will be required to attend the increasing demand. However, it is becoming more and more challenging to allocate parts of spectrum for the data demand since most of the frequency bands which are suitable to mobile communications, are already assigned to the existing wireless systems.

The concept of Cognitive Radio (CR) proposed by [1] presents a solution to lower the usage of these widely used bands. The CR principles enable the unlicensed users to dynamically locate the unused spectrum segments and to communicate via these unused spectrum segments. CR used with TV White Space (TVWS) is one of solutions to excel spectrum resources shortage. TVWS are frequencies available for unlicensed use at locations where the spectrum is not being used by licensed services, such as television broadcasting. This spectrum is located from 470 MHz to 790 MHz in Europe [2].

The antennas usually used for this band are geometrically large to cover the entire frequency band. Several wideband antennas have been proposed in these bands [3]. In [4], an asymmetric fork-like printed monopole antenna is presented for DVB-T application, which achieves a -10 dB bandwidth of 451–912 MHz but with a size of

D. Noguet et al. (Eds.): CROWNCOM 2016, LNICST 172, pp. 536–544, 2016.
DOI: 10.1007/978-3-319-40352-6_44

247×35 mm^2. In [5], an UHF wideband printed monopole antenna has been introduced with dimension of 120×240 mm^2. This antenna achieves a gain higher than -6 dBi and an efficiency better than 14 % over the bandwidth 470–862 GHz. Reconfigurable antennas, which are much more compact and have a low profile structures, are the most compelling for TVWS where instantaneous bandwidth is only a narrow part of the overall band.

Therefore, in this paper a tunable miniaturized notch antenna working in the TVWS bands is designed using a tunable capacitor. Since transmit RF circuits generally operate at high power level, linearity is a critical parameter to prevent distortions or inter-channel interferences. As a result some typical reconfigurable RF component as varactor cannot be implemented. A large signal analysis and characterization of the tunable antenna has been performed by realizing adjacent channel leakage Ratio (ACLR) measurements on the antenna prototype using a LTE signal.

In this paper, a frequency-agile notch antenna using a SOI CMOS tunable capacitor and addressing TVWS from 510 MHz to 900 MHz has been studied, implemented and measured. First, TC tunable capacitor specifications are provided. Second, design, analysis and antenna measurements (impedance and efficiency) are presented. Finally, the linearity analysis is discussed.

2 Tunable Capacitor

The Tunable Capacitor (TC) used in this study is a SOI CMOS integrated circuit based on a network of binary-weighted switched capacitors. Floating body FET transistors are used as low-loss switches to select the appropriate capacitance value. Multiple FET transistors are stacked in series in order to prevent breakdown in OFF-state by providing voltage division of the high-power RF signal [6].

The 5-bit tunable capacitor has been modeled and designed under ADS simulation tool. The capacitance is digitally controlled through a SPI interface and it can be tuned from 1.3 pF to 7.1 pF at 1 GHz, Fig. 1.

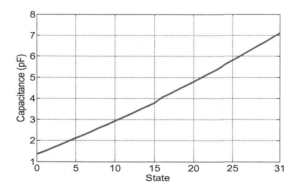

Fig. 1. Simulated capacitance (@1 GHz) vs TC state in shunt configuration

One of the most commonly emphasized electrical specifications for the tunable component is the Quality Factor (Q), which is determined by the resistive (dissipative) losses of the component. Tuning states versus obtained capacitance quality factor at 1 GHz is given in Fig. 2. It can be observed that a minimum quality factor of 40 is achieved for the highest TC state, whereas a maximum Q of 103 is obtained for the lowest state. This, demonstrate that the equivalent series resistance (ESR) is inversely proportional to the tuning state.

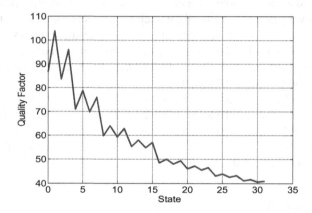

Fig. 2.   Simulated quality factor (@1 GHz) vs TC state in shunt configuration

3   Frequency-Reconfigurable Antenna

3.1   Antenna Design

The proposed antenna is a notch antenna designed on FR4 substrate ($\varepsilon r = 4.3$, tg $\delta = 0.025$) with a thickness of 0.8 mm. The slot size is 18 mm by 3 mm (which correspond to $\lambda/32.6$ by $\lambda/196.7$ at 510 MHz) etched on a ground plane size of 103 mm by 50 mm ($\lambda/5.7 * \lambda/11.8$ at 510 MHz), typical smartphone size [7]. Microstrip feeding line and control lines for the SOI tunable capacitor are located on the bottom layer. An open stub is set at the end of the microstrip line to match the antenna impedance (Fig. 3a). Orthogonal orientation and central positioning of the coaxial connector on the PCB helps to minimize interactions with measurement cable and disturbance of the radiated EM field as the surface currents are mainly located at the edges of the PCB, which contribute to the radiation. A small circuit powered by a miniature battery is used to maintain the tunable capacitor value during the anechoic chamber characterization (Fig. 3b). The tunable capacitor is positioned at the open end of the notch, where high electric fields are concentrated, to allow the antenna tuning at TVWS bands.

The input impedance is shown in Fig. 4. Good agreement is observed between simulated and measured results. The unloaded antenna resonance frequency is 2.3 GHz which does not correspond to $\lambda_g/4$ due to the capacitive effect of the TC footprint at the open end of the slot.

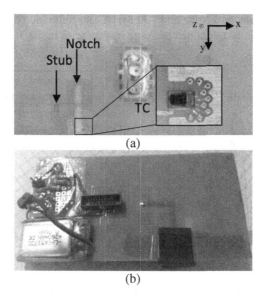

Fig. 3. Antenna structure (a) top view (b) bottom view

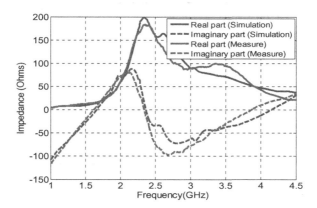

Fig. 4. Comparison of input impedance for the unloaded antenna

3.2   Antenna Tuning

The tunable capacitor has been modeled under ADS and each state is imported into the EM simulator. Figure 5 gives a comparison between simulation and measurement of the antenna response for a fixed capacitor connected at the end of the stub ($C_{stub} = 1$ pF) and different settings of the tunable capacitor. Simulated resonance frequency can be tuned from 550 MHz up to 960 MHz. The achievable bandwidth at −6 dB reflection coefficient ranges from 12 MHz to 70 MHz (Fig. 6) when operating frequency increases. The measured frequency response of the notch antenna for the different tunable capacitor settings ranges between 510 MHz and 900 MHz. The difference between simulations and measurements can be explained by some inaccuracies in the TC model.

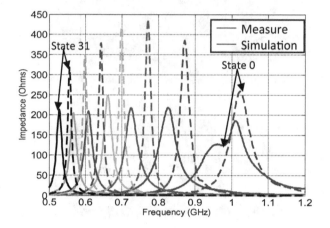

Fig. 5. Comparison of simulated and measured antenna input impedance (real part)

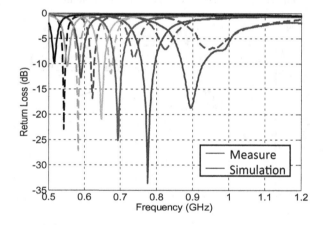

Fig. 6. Comparison of simulated and measured antenna response

To address TVWS, a limited set of capacitor states (from 5 to 31) will be considered. Frequencies from 470 MHz to 510 MHz are not addressed with actual devise. It is so easily to cover them, just to use another tunable capacitor with a higher tuning ratio or to connect a fixed capacitor in parallel configuration. The time estimating to finish transition between states has been measured to about 5 μs. Details will be published in other works.

The measured antenna radiation patterns including the co polarization and the cross polarization in both (xz) and (xy) planes at 510 MHz and 770 MHz are presented in Fig. 7. The gain was measured in the CEA-Leti anechoic chamber. It is important to note that radiation measurements are carried on without any metallic coaxial cable nor

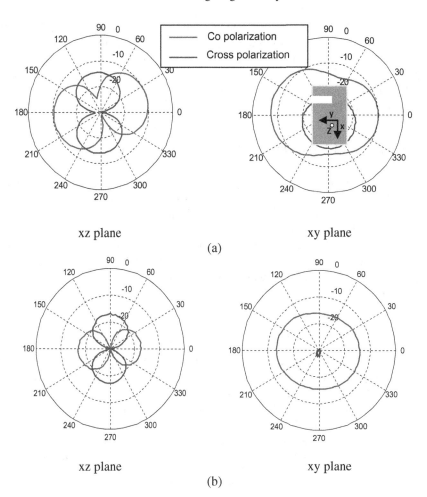

Fig. 7. Measured radiation pattern (realized gain) of the antenna at (a) 770 MHz and (b) 510 MHz (Color figure online)

control interface not to perturb the antenna radiation and consequently the efficiency estimation. The gain decreases as expected when the antenna is tuned to lower frequencies, at 770 MHz the notch length is $\lambda/21$ and at 510 MHz the notch is $\lambda/33$. In fact, as the antenna is tuned away from its natural resonance, higher currents run on the miniaturized surface and the series resistance (ESR) of the tuning component causes higher insertion losses. The ohmic losses in the TC ESR which are proportional to the square of these high currents, increase and cause higher efficiency degradation. Thus, it is a trade-off between antenna size, radiation performances and ESR. This ESR value results of both the IC design and the IC technology.

As a result of the reduction in realized gain, a total efficiency degradation is observed (-10 dB at 770 MHz). In order to demonstrate losses introduced by the ESR, an ideal TC (ESR = 0 Ω) was simulated and results indicate that for state 5 the antenna resonance frequency is 770 MHz and the corresponding total efficiency is −2.5 dB and for state 31, the total efficiency is −5.8 dB. Equally at low frequency, the antenna is miniaturized and consequently, bandwidth and efficiency decrease as predicted by the fundamental limits of electrically small antennas [8], these are the losses of the finite conductivity. The TC ESR decreases the efficiency with a shift of 7.5 dB for state 5. In next works, methods to reduce ESR and improve the antenna efficiency will be addressed.

4 Analysis and Characterization of Tunable Antenna Linearity

One critical specification for digital communication systems is the Adjacent Channel Leakage Ratio (ACLR) corresponding to signal distortion leaking in neighboring channels. Leakage power influences the system capacity as it interferes with the reception in adjacent channels. Therefore it must be rigorously controlled to guarantee correct communication for all subscribers in a network. ACLR is the ratio of the power in the adjacent channels to the power in the transmit channel. ACLR limits are given in the standard for the whole system, however up to now there are no specific constraints for the antenna.

4.1 ACLR Measurement Setup

ACLR measurements are made using a spectrum analyzer and the required test signals are built using a signal generator. In the following setup, a Vector Signal Generator with an internal baseband generator is connected to the antenna in transmission to allow generation of a LTE signal and the received signal from the measurement (Horn) antenna is connected to a Spectrum Analyzer (Fig. 8). For LTE, depending on the considered signal bandwidth, the adjacent channels are located at ± 1.4 MHz, ± 3 MHz, ± 5 MHz, ± 10 MHz, ± 15 MHz and ± 20 MHz offsets [9].

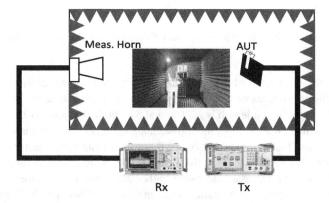

Fig. 8. Measurement setup around the anechoic chamber

4.2 ACLR Measurement Results

Figure 9 shows the measured ACLR characteristic of the tunable antenna for TC state 5 and a 16 QAM LTE 10 MHz uplink signal at 770 MHz for different antenna input power (Pin) ranging from 5 dBm to 23 dBm. Dotted line corresponds to the noise floor. The tunable antenna respects the standard ACLR specification of −30 dBc up to 22 dBm input power at 770 MHz.

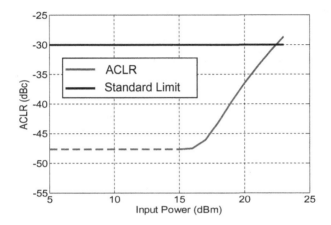

Fig. 9. Measured ACLR versus input power with a LTE 16 QAM 10 MHz signal at 770 MHz (Color figure online)

ACLR was measured for different TC states and frequencies using an LTE 16 QAM 10 MHz uplink signal.

Table 1 summarizes the linearity performance of the tunable antenna.

Table 1. Measured power at the input of the tunable antenna for ACLR = −30 dBc

State	State 5	State 10	State 15	State 31
Frequency	770 MHz	685 MHz	630 MHz	510 MHz
Pin (ACLR = −30 dBc)	22 dBm	22 dBm	23 dBm	22 dBm

The tunable antenna respect the standard specification up to 22 dBm input power which corresponds to the LTE UE transmit power as specified in 3GPP TS36.101 (UE core specification) [9].

5 Conclusion

In this paper, a miniature tunable notch antenna is proposed to address LTE low bands in TVWS spectrum. The notch antenna incorporates a SOI CMOS tunable capacitor at its open-end to operate in the different communication bands ranging from 510 MHz to 900 MHz. Measured results demonstrate trade-off between compact size, limited instantaneous operating bandwidth and efficiency. The RF linearity characterization of the tunable antenna has been performed through ACLR measurements on the antenna prototype and obtained results demonstrate that the proposed tunable antenna respects the standard specifications. Due to the dynamic frequency allocation within TVWS, fast transition times between frequency bands are expected to satisfy high quality communications. At last notice the proposed antenna solution is not restricted to LTE but is also compatible with other waveforms.

References

1. Mitola III, J.: Cognitive Radio, Ph.D. thesis, KTH, Stockholm, Sweden (2000)
2. Mueck, M., Noguet, D.: TV White Space standardization and regulation in Europe. In: 2011 2nd International Conference on Wireless Communication, Vehicular Technology, Information Theory and Aerospace & Electronic Systems Technology (Wireless VITAE), pp. 1–5, 28 February 2011–3 March 2011
3. John, M., Ammann, M.J.: A compact shorted printed monopole antenna for TV white space trials. In: 2013 7th European Conference on Antennas and Propagation (EuCAP), pp. 3713–3715, 8–12 April 2013
4. Huang, C., Jeng, B., Yang, C.: Wideband monopole antenna for DVB-T applications. Electron. Lett. **44**, 1448–1450 (2008)
5. Liang, X., Jin, R., Zhao, Y., Geng, J.: Compact DVB-T printed monopole antenna. In: 2010 International Workshop on Antenna Technology (iWAT), pp. 1–4, 1–3 March 2010
6. Nicolas, D., Giry, A., Ben Abdallah, E., Bories, S., Tant, G., Parra, T., Delaveaud, C., Vincent, P., Po, F.C.W.: SOI CMOS tunable capacitors for RF antenna aperture tuning. In: 2014 21st IEEE International Conference on Electronics, Circuits and Systems (ICECS), pp. 383–386, 7–10 December 2014
7. Le Fur, G., Lach, C., Rudant, L., Delaveaud, C.: Miniature reconfigurable multi-antenna system for IMT-advanced band. In: IEEE International Symposium on Antennas and Propagation (APSURSI), pp. 1207–1210, 3–8 July 2011
8. Chu, L.J.: Physical limitations on omni-directional antennas. J. Appl. Phys. **19**, 1163–1175 (1948)
9. http://www.etsi.org/deliver/etsi_ts/136199/136/101/10.03.00/ts_136101v100300.pdf

On the FPGA-Based Implementation of a Flexible Waveform from a High-Level Description: Application to LTE FFT Case Study

Mai-Thanh Tran[✉], Matthieu Gautier, and Emmanuel Casseau

IRISA, University of Rennes 1, Lannion, France
{mai-thanh.tran,matthieu.gautier,emmanuel.casseau}@irisa.fr

Abstract. The Field Programmable Gate Array (FPGA) technology is expected to play a key role in the development of Software Defined Radio (SDR) platforms. To this aim, leveraging the nascent High-Level Synthesis (HLS) tools, a design flow from high-level specifications to Register-Transfer Level (RTL) description can be thought to generate processing blocks that can be reconfigured at run-time. Based on such a flow, this paper describes the architectural exploration of a Fast Fourier Transform (FFT) for Long Term Evolution (LTE) standard. Synthesis results show the tradeoff between reconfiguration time and area that can be achieved with such an approach.

Keywords: High-level synthesis · Software defined radio · FPGA · LTE · Hardware implementation · Design flow

1 Introduction

Advanced wireless communication standards are designed with various requirements in terms of data transmission rate, spectral efficiency and multiple channel bandwidths. To fulfil these requirements, many configurations of the waveform (PHY layer) features are allowed such as the number of antennas, the coding rate, the modulation scheme or the number of subcarriers in the case of Orthogonal Frequency Division Multiplexing (OFDM) modulation. In such a context, new needs of PHY layer implementation appear while the hardware implementation has to change from one configuration to one another in a short time, refereed to as run-time flexibility in this document.

An emergent technology that answers these new needs is Software Defined Radio (SDR) that allows both flexibility and fast prototyping capabilities from a high-level description [13]. However, when implementing the processing on Digital Signal Processors (DSP), SDR suffers from important power consumption and limited performance as compared to dedicated hardware fabrics. FPGA-based SDR is an old paradigm [8] offering a good tradeoff between reconfiguration capability and processing power. Fast prototyping capability of an FPGA-based

© ICST Institute for Computer Sciences, Social Informatics and Telecommunications Engineering 2016
D. Noguet et al. (Eds.): CROWNCOM 2016, LNICST 172, pp. 545–557, 2016.
DOI: 10.1007/978-3-319-40352-6_45

SDR is achieved by leveraging High-Level Synthesis (HLS) principles and tools to generate Register-Transfer Level (RTL) descriptions from high-level specifications [18]. However, the issue of run-time flexibility is still opened. This paper discusses the FPGA-based implementation of a run-time hardware reconfiguration of a flexible waveform from its high-level description. The proposed methodology mainly aims at analyzing the performance of using a multi-mode processing block with control signals or Dynamic Partial Reconfiguration (DPR) to provide flexibility.

In this paper, the example of LTE standard is addressed. Among many configurations, this standard specifies that the computation of an OFDM symbol can be performed over several numbers of subcarriers among $\{128, 256, 512, 1024, 1536, 2048\}$ and one symbol has a duration of $66.67\,\mu s$ [22]. Because Fast Fourier Transforms (FFTs) are used to compute the OFDM symbols, this paper discusses the FPGA implementation of a flexible FFT function that can operate with the different configurations of the LTE standard.

The main contributions of this paper are:

- To implement run-time reconfiguration from a high-level description of a processing block,
- To propose a flexible FFT implementation that covers the LTE configuration modes,
- To perform Design Space Exploration (DSE) of the proposed implementations using HLS capabilities.

The paper is organized as follows. A discussion over related works is given in Sect. 2. The flow to design a reconfigurable component from its high-level description is introduced in Sect. 3. Section 4 discusses the implementation of the flexible FFT for LTE purpose. RTL synthesis results and DPR performance for different reconfigurations are given in Sect. 5. Finally, conclusions and perspectives are drawn in Sect. 6.

2   Related Works

Several proposals attempted to meet the flexibility requirements of an SDR by using software-based approaches. Indeed, software gives an abstraction level that enables more control over the hardware-based approaches. Two complementary approaches have been proposed namely, the SDR-specific languages to design the waveform [1,16,21] and the SDR middleware to provide the building environment [10,12]. They both take advantage of the abstraction level given by the software to achieve both compile-time and run-time flexibility.

Our proposal aims at keeping a high specification level while addressing FPGA platforms. To this end, HLS turns out to be a good candidate to achieve such a high abstraction level. The recent development of *HLS tools* allows the consideration of components described in *C/C++* languages. It raises the abstraction level compared to hardware languages like *VHDL* and *Verilog* dedicated to RTL-based architectures. HLS fast prototyping capability enables the compile-time flexibility of an FPGA-based SDR [17].

There are two kinds of works that address the run-time flexibility of a FPGA-based SDR. The firsts propose to design multi-mode RTL components with control signals to switch between the different modes [5]. The others are based on DPR. Run-time DPR, refereed to as *Hardware reconfiguration* in the following, is the ability to reconfigure part of the FPGA (e.g. a functionality at the hardware level) while the rest of the FPGA continues to work. It is a research topic since the 90s [15] and it is now commonly used in FPGAs, since Xilinx and Altera provide such circuits [2,4]. The main advantages of the *Hardware reconfiguration* are to add hardware flexibility and to reuse hardware area, allowing power consumption and production cost reductions.

Based on the proposed methodology, a flexible FFT is proposed in this paper for LTE specifications. Hundreds of architectures for 128- to 2048-point FFT has been proposed by varying the degree of parallelism and the radix factorization [6,19]. These implementations are optimized in terms of speed, memory used and hardware logic requirements. A reconfigurable FFT for which algorithmic modifications allow the reuse of the resources while switching form one FFT to one another can be found in [22]. While all of these components are described at the RTL level so require very good skills in hardware design, our approach aims at providing a FFT design methodology from a high-level description to jointly achieve fast DSE and run-time reconfiguration.

3   SDreconf: Design Flow for Software Defined Reconfiguration

As mentioned in Sect. 2, there are different ways to achieve a flexible processing block while implementing it onto a FPGA. The first one is to design a multi-mode processing block and the second one is to use DPR (*Hardware reconfiguration*). In our approach, a multi-mode processing block can be described using dedicated algorithmic modifications of the processing block (*Algorithmic reconfiguration*) or with an automatic generation using a HLS encapsulation (*Software reconfiguration*). The goal of our design flow is to choose or combine these reconfigurations while describing the processing block at a high-level of description. This work is based on one commercially available HLS tool: VivadoHLS from Xilinx. It produces a RTL description of an application from its C-like specification. This section details the ways towards the generation of a flexible block.

Software reconfiguration: This reconfiguration uses HLS encapsulation to generate a MULTI_MODE_BLOCK. The method uses the different modes of a BLOCK and generates a MULTI_MODE_BLOCK with a control input to switch between the modes. Algorithm 1 describes this encapsulation in the case of two modes BLOCK_A and BLOCK_B.

The advantages of this method are its simplicity, the rapid prototyping capability provided by HLS and the short reconfiguration time (one clock cycle). However, the resources can be important in that case. Actually the HLS tool does not share the resources efficiently although the modes are timewize mutually exclusive.

Algorithm 1. Software reconfiguration for the automatic generation of a multi-mode processing block.

---

function MULTI_MODE_BLOCK(inputs, outputs, control)
    switch control **do**
        case A
            BLOCK_A(inputs, outputs)
        case B
            BLOCK_B(inputs, outputs)
end function

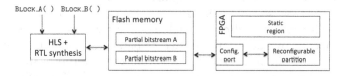

Fig. 1. Design approach based on hardware reconfiguration.

Hardware reconfiguration: In dynamic partial reconfiguration, the FPGA is divided into several regions being static (the areas that are not modified) or reconfigurable (the reconfigurable partitions). Each mode has its own partial bitstream. The partial bitstreams are stored in a memory and a software processor controls which partial bitstream is loaded into the reconfigurable partition at a particular time. The reconfigurable partition size must cover the area of the largest mode. Figure 1 shows the example of the *Hardware reconfiguration* of two modes BLOCK_A and BLOCK_B. The two modes are processed separately using first HLS and then RTL synthesis to generate two partial bitstreams. Part of this flow can be automated, at least to have an estimation of the performance. The main advantage is that the modes share the same area. The drawback is the reconfiguration time that depends on the size of the partial bitstream.

Algorithmic reconfiguration: For this kind of reconfiguration, the designer has to hand-code a dedicated processing block being intrinsically flexible. Signals are used to control the modes. Algorithmic optimizations should be done so that the HLS tool can share the resources between the modes.

Figure 2 shows the design tradeoff between the resources and the reconfiguration time when the three kinds of reconfiguration are considered. *Algorithmic reconfiguration* is used to decrease the resources compared to *Software reconfiguration* and to decrease the reconfiguration time compared to *Hardware reconfiguration*. It provides the best performance in term of resources/reconfiguration time tradeoff. However, depending on the processing blocks, time to code the algorithmic reconfiguration can be important compared to *Software reconfiguration*.

Based on these three kinds of reconfiguration, the design flow used in this work is shown in Fig. 3. The different modes of a processing block can be provided by hand-coding or using a *HLS tool* to generate different versions of a processing

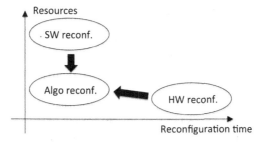

Fig. 2. Tradeoff between resources and reconfiguration time for the different reconfigurations.

block by modifying synthesis constraints like throughput, latency, data size, etc. Such tools make it easy to explore a set of solutions via DSE [9, 14, 20] considering a given architecture.

In Fig. 3, *Performance constraints* are user-defined constraints such as resources/area, reconfiguration time, throughput or latency. The *Performance analysis* compares the performance of the three paths to the user-defined constraints. The basic idea is to first analyze the performance of the *Software reconfiguration* and *Hardware reconfiguration* paths and to use the *Algorithmic reconfiguration* if the *Performance constraints* are not met.

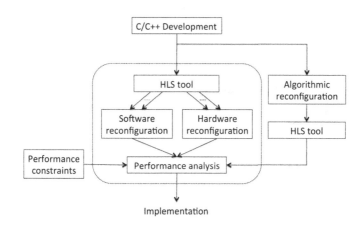

Fig. 3. Design flow for Software Defined Reconfiguration.

We have experimented this design flow with the rapid prototyping of a flexible FFT. An architecture exploration was performed allowing the comparison of the three kinds of reconfiguration.

4   Building a Flexible FFT

In this section, a flexible FFT is designed using the proposed approach. Addressing LTE standard, the resulting FFT component should have six modes corresponding to these FFT sizes: $\{128, 256, 512, 1024, 1536, 2048\}$. To introduce the different kinds of reconfiguration, preliminary results are first given in Sects. 4.1 and 4.2 with the design of a flexible FFT with two sizes: $\{128, 2048\}$. This design is based on two hand-coded FFT functions using the radix-2 Decimation-In-Time (DIT) algorithm which is the simplest and most common form of the Cooley-Tukey algorithm [7].

Experimental tools and setup: Vivado HLS 2013.3 is used for the high-level synthesis. DPR is setup with PlanAhead 14.6. The xc6vlx240tff1156 FPGA is targeted from Virtex 6 family as ML605 evaluation board will be used for future demonstration. Syntheses are based on a 100 MHz clock frequency.

Table 1. Performance of *Software reconfiguration* for a FFT with 2 modes (128/2048).

Processing block	Block_FFT128()	Block_FFT2048()	Two_Mode_Block()
BRAM	6	12	12
DSP	17	65	82
LUT	1017	2522	3459
FF	862	2443	3241
Latency	5362	72410	5491/72411

4.1   Software Reconfiguration

First, two functions for 128- and 2048-point FFTs have been hand-coded and synthesized separately (Block_FFT128() and Block_FFT2048() respectively). Then the FFT with 2 modes (128/2048) has been designed using *Software reconfiguration*: using Algorithm 1, a function Two_Mode_Block() is generated from Block_FFT128() and Block_FFT2048(). Table 1 shows the synthesis results and the latency (in number of clock cycles) of the three processing blocks. The resources are given with logical components such as the number of BRAM, DSP slices, LUT and FF. As expected, Block_FFT2048() requires more resources than Block_FFT128(): the number of BRAM to store cos and sin coefficients and input data is twofold and the number of DSP slices is 4 times more.

Except for the number of BRAM, Table 1 shows that the resources used by Two_Mode_Block() are (a little bit less than) the sum of the resources used by the two FFT blocks when synthesized separately. The HLS tool does not share the resources between the two functions although they are not executed at the same time. In this case, from the resources point of view, *Software reconfiguration* appears not to be an efficient solution to implement a flexible block. Latency is similar to mono-mode blocks.

Table 2. Performance for *Hardware reconfiguration* of a FFT with 2 modes (128/2048).

	Processing block: Resources needed		Partition: Resources used		
	FFT 128	FFT 2048	FFT 128	FFT 2048	FFT 128/2048
BRAM	6	12	6	17	17
DSP	17	65	24	68	68
LUT	1017	2522	1440	4080	4080
FF	862	2443	2880	8160	8160
Bitstream size	n/a	n/a	138672 Bytes	416016 Bytes	2 × 416016 Bytes
Reconf. time	n/a	n/a	10.98 ms	32.9 ms	32.9 ms
Latency	5362	72410	5362	72410	5362/72410

4.2 Hardware Reconfiguration

The two functions BLOCK_FFT128() and BLOCK_FFT2048() are now used for *Hardware reconfiguration*. In Xilinx's FPGAs [4], the functions to be dynamically placed are mapped into an area called a reconfigurable partition. Generally speaking, a partition has a rectangle shape and uses resources according to this area even if they are not needed.

Two partitions have been first generated using PlanAhead tool: one for the FFT 128 only and one for the FFT 2048 only. Table 2 shows the resources used by these partitions. Bitstream sizes are also given. Because FFT 128's partition uses less logical components than FFT 2048's one, it is smaller than FFT 2048's one. The reconfiguration time depends on the bitstream size so the reconfiguration time for FFT 128 is smaller than FFT 2048's one. Reconfiguration time is computed from PRCC tool (Partial Reconfiguration Cost Calculator) from Technical University of Crete [3] assuming that the reconfiguration controller is an on-chip MicroBlaze processor[1].

For comparison, Table 2 also shows the resources needed by FFT 128 and FFT 2048 when they are placed as static logic (*i.e.* not as a reconfigurable module, so resources needed are same as in Table 1).

To perform the DPR of the 2 functions, a third partition called FFT 128/2048 in Table 2 has been defined. In this case, the two functions are placed on the same partition, *i.e.* on the same area of the FPGA. For each type of logical component, the resources used by this partition are based on the more costly case. In our case, FFT 2048 partition always needs the largest number of resources whatever the kind of logical component. Thus the resulting FFT 128/2048 partition is based on the FFT 2048 partition.

Reconfigurable partition's latency is equal to the latency of the function when synthesized alone onto a static region. The hardware reconfiguration needs 32.9 ms to switch from one mode to the other one whereas only one clock cycle

---

[1] Higher throughput up to 400 MBytes/s may be reached using a dedicated controller so that reconfiguration time can be reduced.

is required with *Software reconfiguration*. It means many OFDM symbols will be lost in practice when changing the mode with hardware reconfiguration.

4.3   Algorithmic Reconfiguration

The FFT for LTE standard should have six modes: $\{128, 256, 512, 1024, 1536, 2048\}$. Based on the previous results, a *Software reconfiguration* will generate a huge component as the resources are not shared. *Hardware reconfiguration* makes resource sharing possible but may require a long reconfiguration time. In this section, we first present a power-of-two point FFT for algorithmic reconfiguration to share the resources between its different modes. Then, a FFT 1536 function is presented. Sharing the resources of these 2 functions is discussed.

Power-of-two point FFT for algorithmic reconfiguration: The power-of-two point FFT has 5 modes $\{128, 256, 512, 1024, 2048\}$. A dedicated control signal is used to decide the mode and the HLS tool deals with the FFT_size as a variable. Indeed, as presenting in Algorithm 2, FFT_size and FFT_stages are calculated based on the control signal value. When FFT_size and FFT_stages are determined, a standard three-loop structure for the FFT based on radix-2 is computed. The first loop determines the stage. The second loop chooses butterflies with the same twiddle factor at each stage. Last loop computes all the chosen butterflies.

This algorithmic reconfiguration generates a BLOCK_FFTPOW2() function with only one main FFT core for the five different modes. With this function, the resources should be approximately the ones used by the largest FFT (*i.e.* 2048).

1536-point FFT: By applying the Cooley-Tukey algorithm [7] for a FFT size of 1536, the BLOCK_FFT1536() function can be generated using three BLOCK_FFT512() functions and one radix-3 function [11] as shown in Fig. 4. First,

---

Algorithm 2. Algorithmic reconfiguration for the power-of-two point FFT.

---

   function BLOCK_FFTPOW2(inputs, outputs, control)   ▷ $0 \leq$ control ≤ 4
      FFT_size_max $= 2048$
      FFT_stages_max $= 11$
      FFT_size $=$ FFT_size_max $>>$ control   ▷ FFT_size $= \frac{\text{FFT\_size\_max}}{2^{\text{control}}}$
      FFT_stages$=$ FFT_stages_max - control
      Bit_reverse()   ▷ re-range the order of bits before calculating
      for i $= 0$ **to** FFT_stages **do**
          Calculate_index()   ▷ choose the stage, prepare for possible twiddles
          for j $= 0$ **to** FFT_size/2 **do**
              Computing_twiddles()   ▷ determine coefficients for radix 2
              for k $= 0$ **to** FFT_size/2 **do**
                 Radix_2(inputs, outputs)
   end function

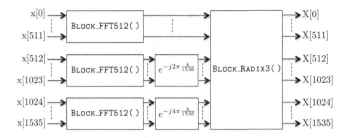

Fig. 4. Functional description of Block_FFT1536() function.

the 1536 inputs of the FFT are split into three parts. Those parts are computed as three 512-point FFTs independently. Then, while the first part is kept as it is, the second and the third ones are multiplied by twiddles factors. Last, the radix-3 function is applied to compute the final results.

Resource sharing may theoretically be done between Block_FFT512() functions of the 1536-point FFT and the power-of-two point FFT because they are both based on the radix-2 algorithm. On the contrary, the Block_Radix3() function is based on radix-3 thus it can not share resources.

5 Performance Results

This section presents the architectural exploration of a flexible FFT for LTE standard. First, design space will be explored for the function Block_FFTpow2(). Then, the combination between Block_FFTpow2() and Block_FFT1536() will be addressed using *Software* and *Hardware* reconfigurations.

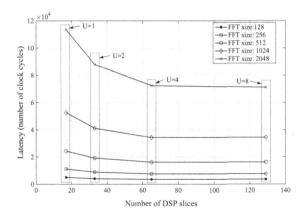

Fig. 5. DSE of Block_FFTpow2() function.

5.1 Design Space Exploration of the Power-of-two Point FFT

In order to generate a flexible FFT that respects the design constraints (area, latency, throughput, ...), HLS allows the DSE of a processing block by using compilation directives. This part presents the DSE of the function BLOCK_FFTPOW2(). Several directives are made available on a typical HLS tool (e.g. memory mapping, pipeline, loop unrolling, inlining, ...). They make it possible to optimize the design for area or latency. In this study, because of the loop structures and the data dependencies of the FFT, loop unrolling is used. Loop unrolling reduces the total loop iterations by duplicating (with a factor U) the loop body so that we can tradeoff between area and latency.

Figure 5 shows the latency of BLOCK_FFTPOW2() processing block as a function of the number of DSP slices. In practice, four multi-mode components have been generated by varying the unroll factor U. Each component is characterized by its numbers of DSP slices. The number of DSP slices increases with the unroll factor while the latency decreases with the unroll factor. One's can see a floor effect appears in Fig. 5. Due to the BRAM accesses (read and write data from/to memory) that reach their bounds for $U = 4$, increasing the number of DSP resources is not useful in practice.

The same behavior is also obtained for the number of LUT and FF. Thus $U = 4$ seems to be a good tradeoff between number of resources and latency.

5.2 Proposed Flexible FFT Implementations for LTE Standard

Performance of the multi-mode FFT with software reconfiguration: Software reconfiguration is applied first to design the FFT with 6 modes for LTE. Using Algorithm 1, a MULTI_MODE_BLOCK_LTE() function is generated from the two functions BLOCK_FFTPOW2() and BLOCK_FFT1536(). Table 3 shows the synthesis results and latency (in number of clock cycles) of the three processing blocks. $U = 4$ is used for BLOCK_FFTPOW2(). The resources used by MULTI_MODE_BLOCK_LTE() are almost the sum of the resources used by BLOCK_FFTPOW2() and BLOCK_FFT1536() when synthesized separately. As observed in Sect. 4.1, the HLS tool does not share the resources between the two functions even if they are not executed at the same time.

Table 3. Performance of a FFT for LTE standard with *Software reconfiguration.*

Processing block	BLOCK_FFTPOW2()	BLOCK_FFT1536()	MULTI_MODE_BLOCK_LTE()
BRAM	12	14	26
DSP	65	40	103
LUT	2553	3054	5256
FF	2497	2010	4299
Latency	cf. Fig. 5 - U=4	52198	cf. Fig. 5/52198

Table 4. Performance of a FFT for LTE standard with *Hardware reconfiguration*.

	Processing block: Resources needed		Partition: Resources used		
	Pow.-of-two FFT	FFT 1536	Pow.-of-two FFT	FFT 1536	FFT for LTE
BRAM	12	14	17	14	17
DSP	65	40	68	56	68
LUT	2553	3054	4080	3360	4080
FF	2497	2010	8160	6720	8160
Bitstream size	n/a	n/a	416016 Bytes	277344 Bytes	2 × 416016 Bytes
Reconf. time	n/a	n/a	32.9 ms	21.96 ms	32.9 ms
Latency	cf. Fig. 5 - U=4	52198	cf. Fig. 5 - U=4	52198	cf. Fig. 5/52198

Performance of the multi-mode FFT with hardware reconfiguration: Hardware reconfiguration is now applied on the two functions BLOCK_FFTPOW2() and BLOCK_FFT1536(). Two partitions are first generated: one for the power-of-two point FFT only and one for the FFT 1536 only. Then, a partition is finally created for the DPR of the 2 FFTs.

Table 4 shows the synthesis results. The partition for FFT 1536 is smaller than the power-of-two point FFT's one. Actually, number of BRAMs is greater but function BLOCK_FFT1536() uses less DSP slices than BLOCK_FFTPOW2() so that its area is smaller. Thus, when combining the 2 FFTs into one partition, the resulting partition is based on power-of-two point FFT's partition.

Compared with *Software reconfiguration*, the multi-mode FFT based on hardware reconfiguration uses less resources (BRAM and DSP are the more area costly logical components). When the FFT size has to be modified but is still a power of two, in both cases only one clock cycle is required to reconfigure. However, 32.9 ms are required to reconfigure when switching from a 1536-point FFT and a power-of-two point FFT (or vice versa) with hardware reconguration whereas only one clock cycle is required with *Software reconfiguration*.

6   Conclusion

This paper presents a methodology for the implementation of run-time reconfiguration in the context of FPGA-based SDR. The proposed design flow allows the exploration between dynamic partial reconfiguration and control signal based multi-mode design. This architectural tradeoff relies upon HLS and its associated design optimizations.

A flexible FFT for LTE standard is implemented as a case study. The proposed component combines both DPR (to deal with FFT size of 1536) and algorithmic reconfiguration (when FFT size is a power of two). Synthesis results show the tradeoff that could be achieved between the reconfiguration time and the FPGA resource utilization. Future work is to explore the implementation of other processing functions and the automation of the design flow.

References

1. GNU Radio: The free and open software radio ecosystem. www.gnuradio.org
2. Increasing Design Functionality with Partial and Dynamic Reconfiguration in 28-nm FPGAs. Altera White Paper, WP-01137-1.0. www.altera.com
3. Kyprianos Papadimitriou, Microprocessor and Hardware Laboratory, Technical University of Crete. Partial Reconfiguration Cost Calculator. http://users.isc.tuc.gr/~kpapadimitriou/prcc.html
4. Partial Reconfiguration User Guide, UG702 (v14.1) (2012). www.xilinx.com
5. Casseau, E., Le Gal, B.: Design of multi-mode application-specific cores based on high-level synthesis. Integr. VLSI J. **45**(1), 9–21 (2012). Elsevier
6. Chen, J., Hu, J., Lee, S., Sobelman, G.E.: Hardware efficient mixed Radix-25/16/9 FFT for LTE systems. IEEE Trans. Very Large Scale Integr. (VLSI) Syst. **23**(2), 221–229 (2015)
7. Cooley, J.W., Tukey, J.W.: An algorithm for the machine calculation of complex Fourier series. Math. Comput. **19**, 297–301 (1965). American Mathematical Society
8. Cummings, M., Haruyama, S.: FPGA in the software radio. IEEE Commun. Mag. **37**(2), 108–112 (1999)
9. Gautier, M., Ouedraogo, G.S., Sentieys, O.: Design space exploration in an FPGA-based software defined radio. In: Euromicro Conference on Digital System Design (DSD), Verona, Italy, pp. 22–27, August 2014
10. Gelonch, A., Revs, X., Marojevik, V., Ferrús, R.: P-HAL: a middleware for SDR applications. In: SDR Forum Technical Conference, November 2005
11. Freescale Semiconductor Incorporated. Software Optimization of DFTs and IDFTs Using the StarCore SC3850 DSP Core. Application Note AN3980 (2009)
12. Jianxin, G., Xiaohui, Y., Jun, G., Quan, L.: The software communication architecture: evolutions and trends. In: IEEE Conference on Computational Intelligenceand Industrial Applications (PACIIA), November 2009
13. Jondral, J.F.: Software-defined radio: basics and evolution to cognitive radio. EURASIP J. Wirel. Commun. Netw. **3**, 275–283 (2005)
14. Le Moullec, Y., Diguet, J.-P., Ben Amor, N., Gourdeaux, T., Philippe, J.-L.: Algorithmic-level specification, characterization of embedded multimedia applications with design trotter. J. VLSI Signal Process. Syst. Signal Image Video Technol. **42**(2), 185–208 (2006)
15. Lemoine, E., Merceron, D.: Run time reconfiguration of FPGA for scanning genomic databases. In: IEEE Symposium on FPGAs for Custom Computing Machines, pp. 90–98, April 1995
16. Lin, Y., Mullenix, R., Woh, M., Mahlke, S., Mudge, T., Reid, A., Flautner, K.: SPEX: a programming language for software defined radio. In: Software Defined Radio Technical Conference and Product Exposition (SDR-Forum), November 2006
17. Ouedraogo, G.-S., Gautier, M., Sentieys, O.: A frame-based domain-specific language for rapid prototyping of FPGA-based software-defined radios. EURASIP J. Adv. Signal Process. **2014**(1), 164 (2014)
18. Ouedraogo, G.S., Gautier, M., Sentieys, O.: Frame-based modeling for automatic synthesis of FPGA-software defined radio. In: IEEE International Conference on Cognitive Radio Oriented Wireless Networks and Communications (CROWN-COM), June 2014
19. Pitkanen, T., Takala, J.: Low-power application-specific processor for FFT computations. In: IEEE International Conference on Acoustics, Speech and Signal Processing (ICASSP), pp. 593–596, April 2009

20. So, B., Hall, M.W., Diniz, P.C.: A compiler approach to fast hardware design space exploration in FPGA-based systems. In: Proceedings of the ACM SIGPLAN Conference on Programming language design and implementation (PLDI), pp. 165–176, New York, USA (2002)
21. Willink, E.D.: The waveform description language: moving from implementation to specification. In: IEEE Military Communications Conference (MILCOM) (2001)
22. Yang, C.-H., Tsung-Han, Y., Markovic, D.: Power and area minimization of reconfigurable FFT processors: a 3GPP-LTE example. IEEE J. Solid State Circ. **47**(3), 757–768 (2012)

Performance of Fractional Delay Estimation in Joint Estimation Algorithm Dedicated to Digital Tx Leakage Compensation in FDD Transceivers

Robin Gerzaguet[2(✉)], Laurent Ros[1], Fabrice Belvéze[3], and Jean-Marc Brossier[1]

[1] Image and Signal Department,GIPSA-Lab, Saint Martin d'Hères, France
[2] CEA-Leti, Grenoble, France
robin.gerzaguet@cea.fr
[3] ST-Microelectronics, Grenoble, France

Abstract. This paper deals with the performance of the fractional delay estimator in the joint complex amplitude / delay estimation algorithm dedicated to digital Tx leakage compensation in FDD transceivers. Such transceivers are affected from transmitter-receiver signal leakage. Combined with non linearity of components in the received path, it leads to a pollution in the baseband signal. The baseband polluting term depends on the equivalent Tx leakage channel, modeling leakages and the received path. We have proposed in [7,8] a joint estimation of the complex gain and the fractional delay and derived asymptotic performance of the complex gain estimator, that showed the necessity of the fractional delay estimation. In this paper, we propose a comprehensive study of the fractional delay estimation algorithm and its analytic performance. The study is based on the analysis of the S-curve and loop noise variance of the timing error detector, from which an approximation of the asymptotic performance of the joint estimation algorithm is derived.

Keywords: Tx Leakage · FDD transceiver · Digital compensation · Least-Mean-Square algorithm · Joint estimation · S-curve

1 Introduction

Cognitive radios offer the possibility to improve the spectrum uses and to adapt the transmission scheme to optimize the sharing of the available bandwidth, which leads to important constraints on the hardware components, located on the physical layer [1]. These constraints thus lead to performance limitations [21], and we focus on this paper on a hardware impairments that occurs when the radio follows a Frequency Division Duplexing (FDD) scheme.

Compact wireless transceivers can be based on Time Division Duplexing or on FDD modes [15] to multiplex transmission and reception links. For a frequency division duplexing framework, which is the case of the study presented here, the transmission and the reception are done simultaneously, using two different carrier frequencies.

© ICST Institute for Computer Sciences, Social Informatics and Telecommunications Engineering 2016
D. Noguet et al. (Eds.): CROWNCOM 2016, LNICST 172, pp. 558–568, 2016.
DOI: 10.1007/978-3-319-40352-6_46

A surface acoustic wave (SAW) duplexer is often used to connect the received (Rx) and the transmitted (Tx) path to a common antenna [11] (see Fig. 1). As the duplexer does not provide infinite attenuation between the Rx path and the Tx path, the transmitted signal can leak into the Rx path [6] leading to the so-called Tx Leakage (TxL) phenomenon. As the uplink and the downlink bands are spectrally separated, the received signal will not be directly impaired by the leakage of the transmitted signal which is filtered by the Low Pass Filter (LPF) after the demodulation.

However, due to the non linearity and imperfections of components in the analog Rx stage, especially the Low Noise Amplifier (LNA) [18] and the demodulator [14], intermodulation products can shift downward to baseband square component of the Tx Leakage signal [13]. As a consequence, this polluting signal will impair the received signal and can severely degrade the performance. The pollution is potentially detrimental in the cell edge context (i.e. when the receiver is far from the base station), where the power of the received signal is low, and the power of the transmitted signal is strong [10] leading to a strong polluting signal and thus a low signal to interference ratio.

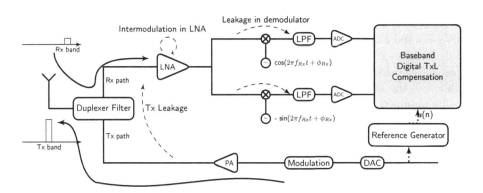

Fig. 1. Classical frequency division duplexing chain in a RF Transceiver with Tx Leakage baseband pollution. The PA denotes the power Amplifier, the ADC is the Analog to Digital Converter, and the DAC denotes the Digital to Analog Converter

To avoid this pollution, passive methods based on analog filtering can be implemented. Such mitigation methods consist in adding a band pass filter to attenuate the leaked transmitted signal in the Rx stage before or after the LNA [2]. As radio frequency (RF) transceivers contain more and more digital parts, and as signal-processing techniques are becoming an area of interest for RF impairments problematic [5], several digital compensation methods have been investigated in the past few years for the Tx Leakage (TxL) compensation [6,7,10].

In this paper, we complete the performance analysis of our previously proposed joint estimation (JE) algorithm dedicated to Tx Leakage compensation [7].

This algorithm both estimates the TxL channel approximated by a time varying complex gain and the fractional delay (FD). It works with two reference based least-mean-square (LMS) algorithms. As [7] mainly focuses on the performance of the complex gain estimator (with and without *a priori* knowledge of the polluting FD) and not the FD nor the JE estimation algorithm itself, we focus in this paper on the FD estimation part. The two main objectives, in addition to being a comprehensive study, are to prove that the FD estimation algorithm can lock around the desired FD value and to derive an approximation of the asymptotic performance of the JE algorithm. The FD estimation algorithm has similitudes with data-aided algorithm that were designed for phase and timing synchronization as well as for automatic gain control [9,16,19], however in the actual estimation process, the RX signal is considered as noise, and an image of the Tx signal is considered as pilot datas (used as reference). In this paper, as we focus on the FD estimator, we assume a constant complex channel gain and we derive analytic formulae of the S-curve of the FD detector [16]. The S-curve represents the characteristic of the detector. For a timing detector, it describes the output of the detector with respect to the delay error between the two input components. Mathematically, it is computed as the conditional expectation of the so-called error signal that updates the FD estimation. The shape of the S-curve is an important point to understand the tracking performance of a Delay-Locked Loop [20]. Using then the general framework of the tracking loop analysis, a linear approximation of the estimation error variance of the FD stage is established from the S-curve. This result is finally used to derive an approximation of the asymptotic performance of the JE algorithm, using the primary results obtained in [7].

This paper is organized as follows. We give the baseband polluting model in Sect. 2. We recall the joint estimation of the complex gain and the fractional delay in Sect. 3. We derive the performance of the fractional delay estimation algorithm in Sect. 4. Section 5 validates our method and theoretical results through simulations.

2   Baseband Model and Issues

The discrete time observation model sampled at $T_{\mathrm{Rx}} = 1/F_{\mathrm{Rx}}$ is expressed as

$$d(n) = x(n) + b(n) + s_{TxL}(n) \tag{1}$$

where $x(n)$ is the desired signal, which is assumed to be uncorrelated and zero-mean, of variance σ_x^2, $b(n)$ is the white additive Gaussian noise, of variance σ_b^2, and $s_{TxL}(n)$ is the TxL polluting signal. The baseband polluting model of the TxL phenomenon is due to the cascade of several impairments, and more precisely to the combination of duplexer finite isolation, low-noise amplifier non linearity and coupling on oscillator [6,18]. As a consequence, the baseband polluting term can be expressed as

$$s_{TxL}(n) = \left[h_{\mathrm{Rx}}(t) * |h_{\mathrm{D}}(t) * \tilde{s}_{Tx}(t)|^2 \right]_{t=nT_{Rx}} \tag{2}$$

where h_D and h_{Rx} denotes respectively the duplexer equivalent channel and the Rx chain equivalent channel, which models the impact of the Duplexer and the Rx chain, and $*$ denotes the convolution. In this paper, we suppose that the duplexer is frequency flat as this assumption is widely used in the literature [4,6,12], and that the Rx filter is known (or estimated) within a complex gain introduced by the coupling at the oscillator. Besides, due to both digital and analogs blocks and propagation, the baseband polluting model is affected by a delay Δ that can be separated between an integer part D of the Rx sampling time and a fractional part δ. Thus, the model can be expressed as

$$s_{TxL}(n;\Delta) = \beta_{\text{TxL}}(n) \left[|h_D(t) * \tilde{s}_{Tx}(t)|^2 \right]_{t=(n-D)T_{Rx}-\delta} \tag{3}$$

where $\beta_{\text{TxL}}(n)$ models the global TxL complex channel gain at time index n. The JE algorthm is an adaptive method based on two the LMS algorithms that can then handle time-varying parameters estimation, as presented in [7] with the tracking of the complex gain with appropriate step-size. However, as we focus on the static performance of the FD estimator in this paper, we assume a constant TxL channel for the theoretical analyis We have shown in [7] that the fractional delay cannot be neglected and we have proposed a joint estimation algorithm that is recalled in the next part. This structure is piloted by a reference signal $u(n)$, synthesized in the reference generator (see Fig. 1 and (4)) from the baseband Tx samples, with the known or estimated Rx filter h_{Rx} :

$$u(n) = \left[h_{Rx}(t) * |\tilde{s}_{Tx}(t)|^2 \right]_{t=nT_{Rx}} \tag{4}$$

It is to note that in practise, it may need additional processing such as an upsampler, a low pass filter, and a resampler.

3   Joint Estimation Algorithm

The structure of the joint estimation algorithm is described on Fig. 2. It is composed of 2 blocks, one dedicated to the complex gain estimation, with a classic one tap LMS approach and the fractional delay is estimated with another LMS where a steepest descend is applied on the FD estimation. At each iteration, the reference signal $u(n)$ is delayed of the estimated FD $\delta(n)$ with an interpolated structure that can be for example a Farrow structure [3].

$$u_{\delta_n}(n) = \sum_{j=0}^{L} u(n-j) \left(\prod_{\substack{i=0 \\ i \neq j}}^{L} \frac{\delta - i}{j - i} \right) \tag{5}$$

The fractional delay estimation algorithm is based on the minimisation of the instantaneous square error $|e(n)|^2$, to which a gradient is applied. The final joint estimation algorithm can be expressed as:

$$e(n) = d(n) - \hat{\beta}_{\text{TxL}}(n)u_{\delta_n}(n) \tag{6}$$

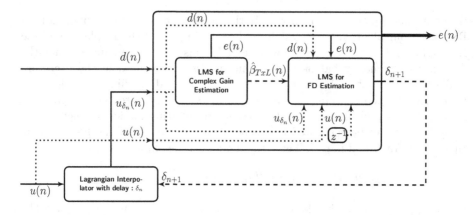

Fig. 2. Joint estimation framework for the TxL compensation

$$\hat{\beta}_{\text{TxL}}(n+1) = \hat{\beta}_{\text{TxL}}(n) + \mu u_{\delta_n}(n)e(n) \tag{7}$$

$$\Delta_u(n) = u(n-1) - u(n) \tag{8}$$

$$\delta_{n+1} = \delta_n + \nu\Re\left\{\left[\hat{\beta}_{TxL}(n)\Delta_u(n) - u_{\delta_n}(n)L(n)\right]e^*(n)\right\} \tag{9}$$

$$L(n+1) = \left(1 - \mu u_{\delta_n}(n)^2\right)L(n) + \mu d(n)\Delta_u(n) + 2\mu\hat{\beta}_{\text{TxL}}(n)u_{\delta_n}(n)\Delta_u(n) , \tag{10}$$

where $e(n)$ is the compensated output, $\hat{\beta}_{\text{TxL}}(n)$ the complex gain estimation, μ the step-size of the complex gain estimator, $L(n) = \partial\hat{\beta}_{\text{TxL}}(n)/\partial\delta$ and ν, the constant step size of the fractional delay estimator. It can be seen that this algorithm is recursive, online (as it provides a compensated output $e(n)$ at each iteration), and with low complexity.

4 Performance of the FD Estimation Process

In this section, we focus on the FD estimation algorithm. To perform the analytical study of this algorithm, we assume a perfectly known channel ($\hat{\beta}_{\text{TxL}}(n) = \beta_{\text{TxL}}$). This algorithm can be considered as a delay-locked loop synchronization algorithm and its performance can be studied with the same approach as described in [16,20]. Thus, the FD algorithm is equivalent to a loop algorithm (see Fig. 3) piloted by its error signal denoted $e_\delta(n)$. The stability and the performance of the proposed FD estimation algorithm can be studied using classical error detector open-loop analysis tools.

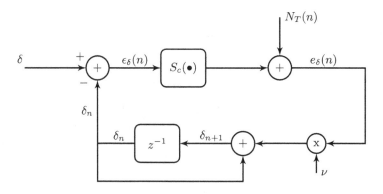

Fig. 3. Equivalent scheme of the FD estimation algorithm

The error signal $e_\delta(n)$ can be decomposed to the sum of the conditional expectation $E[e_\delta(n)|\delta(n)]$ and an additional random zero mean term. The first term is denoted S-curve and is function of the FD estimation error $\epsilon_\delta(n) = \delta - \delta(n)$ and the second term is called loop noise and is denoted $N_T(n)$.

$$S_c(\epsilon_\delta(n)) = E[e_\delta(n)|\delta(n)] \tag{11}$$

$$N_T(n) = e_\delta(n) - S_c(\epsilon_\delta(n)) \tag{12}$$

Based on the proposed FD algorithm (6)–(10), the S-curve can be expressed

$$S_c(\delta, \delta(n)) = P_{\beta_{\text{TxL}}}[-\text{sinc}(\delta) + \text{sinc}(1 - \delta)$$
$$+\text{sinc}(\delta_n) - \text{sinc}(1 - \delta_n)]\sigma_{Tx}^4 \tag{13}$$

with $\text{sinc}(x) = \sin(\pi x)/(\pi x)$, $P_{\beta_{\text{TxL}}} = |\beta_{\text{TxL}}|^2$ the power of the TxL channel, that is assumed to be known. In the context of small error of the FD estimation process, it is possible to linearize the S-curve around its stable equilibrium point [17]:

$$S_c(\epsilon_\delta(n)) \approx D \times \epsilon_\delta(n) \tag{14}$$

where $D = P_{\beta_{\text{TxL}}} \times \sigma_{Tx}^4$ (with σ_{Tx}^2 the variance of the transmitted samples) is the slope of the S-curve at the stable equilibrium point. (14) has been obtained using the approximation $\text{sinc}(x) \approx 1 - 2x$ for small x. As $D > 0$, it shows that the FD estimator is stable, and converges to δ (*i.e.* the loop will lock within the range of FD). Besides, as $S_c(0) = 0$, the FD estimator is unbiased. Assuming a white loop noise for small FD error, the variance of the FD estimation stage can be derived as [16]:

$$\sigma_{\epsilon_\delta}^2 = \frac{\nu}{D(2 - \nu D)}\Gamma_{N_T}[0] \tag{15}$$

with $\Gamma_{N_T}[0]$ the loop noise auto-correlation at zero delay, that can be expressed as

$$\Gamma_{N_T}[0] = \sigma_{Tx}^2 P_{\beta_{\text{TxL}}}(\sigma_x^2 + \sigma_b^2), \tag{16}$$

From (14), (15) and (16) and under the aforementioned assumptions, the variance of the FD estimation process can finally be approximated as

$$\sigma_{\epsilon_\delta}^2 \approx \frac{\nu(\sigma_x^2 + \sigma_b^2)}{4\left[1 - \nu\sigma_{\text{Tx}}^4 P_{\beta_{\text{TxL}}}\right]} \tag{17}$$

Using the performance in terms of Signal to Interference Ratio (SIR) of the complex gain estimator (assuming a constant TxL channel) derived in presence of a non compensated FD are described in [7], we can now express the performance of the JE algorithm, using the variance of the FD estimator expressed in (17) :

$$\text{SIR}_{\text{comp}} \approx -10\log_{10} \left[\frac{\mu\sigma_u^2}{2 - \mu\sigma_u^2} + +|\beta_{0\text{TxL}}|^2 \frac{\sigma_{\epsilon_\delta}^2}{\sigma_{x_b}^2} \right. \tag{18}$$
$$\left. + \frac{\mu\sigma_u^2 \left(|\beta_{0\text{TxL}}|^2\right) \sigma_{\epsilon_\delta}^2}{\sigma_{x_b}^2 \left(2 - \mu\sigma_u^2\right)} \right]$$

5   Simulations

The performance of the proposed algorithm, and more precisely the performance of the FD algorithm are further analyzed by simulations. We first plot on Fig. 4 the theoretical, the linearized theoretical and simulated S-curve versus the estimation error of the FD estimator in open loop. It is shown that the theory is corroborated and shows that the S-curve can be linearized when the estimation error is low (between -10 and 10 percent of the sampling time). It is also shown that the FD estimation loop will lock and is unbiased as $S_c(0) = 0$.

On Fig. 5, we represent the theoretical and simulated loop noise zero-delay auto-correlation $\Gamma_{N_T}[0]$ versus the FD estimation error. It is shown that when the FD estimation error is low, the assumption of a white loop noise with variance expressed in (16) leads to accurate results.

We finally consider the performance of the JE algorithm on Fig. 6. We consider a white unitary variance desired noisy signal polluted by a TxL signal that follows (3). The interference level is set to 0 dB and the power of the desired noisy signal is set to -80 dB. The complex gain algorithm is piloted by a non interpolated reference $u(n)$ defined with (4) and the JE algorithm uses a Farrow structure to apply the estimated FD to $u(n)$. We also assume a constant TxL channel, and we represent the performance of the complex gain estimator and the JE algorithm for several values of FD versus the step-sizes μ and $\nu(\sigma_x^2 + \sigma_b^2)$ (as ν is normalized with the power of the entry signal). It is shown that the performance can be dramatically reduced if the FD estimation part is not activated, and secondly that the JE algorithm greatly improves the asymptotic performance. Is is also shown that the asymptotic performance of the structure can be approximated with (18) leading to a direct link between a desired asymptotic level and step-size value for the FD estimator ν.

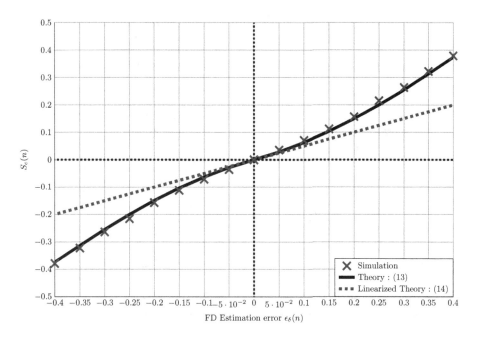

Fig. 4. Theoretical, simulated and linearized S-curve versus the FD estimation error

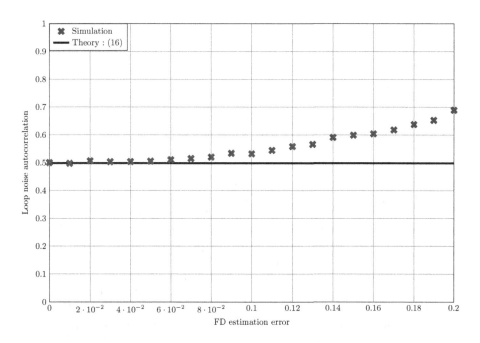

Fig. 5. Theoretical and simulated zero-delay autocorrelation of loop noise versus estimation error of the FD estimator

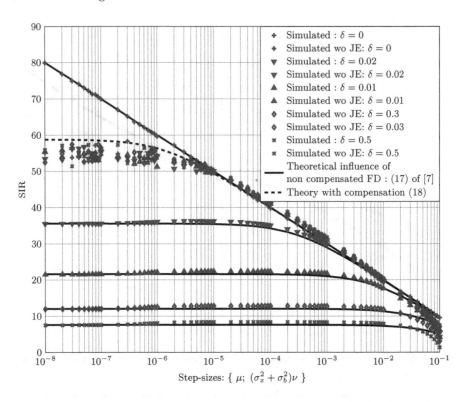

Fig. 6. Asymptotic performance of the complex gain estimator and the JE algorithm for several values of the step-sizes and for different values of fractional delays.

6   Conclusion

This paper deals with the performance of a joint estimation algorithm dedicated to the compensation of the digital Tx Leakage in RF transceivers. The strong constraints that apply in cognitive radios lead to a potential detrimental loss of performance due to hardware impairments and we focus here on a pollution that occurs in FDD transceivers, related to the pollution of an image of the transmitter stage on the receiver stage. Based on the initial algorithm proposed in [7], we focus in this paper on the fractional delay estimation algorithm that can be considered as a synchronisation algorithm piloted by an interpolated reference signal. We first have derived the analytical S-curve, defined as the conditional expectation of the error signal that controls the loop, and we have deduced an approximation of the asymptotic estimation error variance of the FD algorithm. This result can be used to properly tune the JE algorithm step-size as the proposed approximated theoretical performance formula of the whole JE algorithm shows good accordance with simulations. As a perspective to this work, an implementation on a FDD device (or software defined radio) would be useful in order to validate the method and the theoretical results on experimental measurements.

References

1. Cabric, D., Brodersen, R.: Physical layer design issues unique to cognitive radio systems. In: Proceedings of the 16th International Symposium on Personal, Indoor and Mobile Radio Communications (PIMRC), vol. 2, pp. 759–763, September 2005
2. Dufrene, K., Weigel, R.: Highly linear IQ downconverter for reconfigurable wireless receivers. In: The European Conference on Wireless Technology, pp. 19–22, October 2005
3. Farrow, C.W.: A continuously variable digital delay element. In: IEEE International Symposium on Circuits and Systems, vol. 3, pp. 2641–2645 (1988)
4. Faulkner, M.: DC offset and IM2 removal in direct conversion receivers. IEE Proc. Commun. 149(3), 179–184 (2002)
5. Fettweis, G., Lohning, M., Petrovic, D., Windisch, M., Zillmann, P., Rave, W.: Dirty RF: a new paradigm. In: Proceedings of the IEEE 16th International Symposium on Personal, Indoor and Mobile Radio Communications (PIMRC), vol. 4, pp. 2347–2355 (2005)
6. Frotzscher, A., Fettweis, G.: A stochastic gradient LMS algorithm for digital compensation of Tx leakage in Zero-IF-Receivers. In: Proceedings of the IEEE Vehicular Technology Conference (VTC), pp. 1067–1071 (2008)
7. Gerzaguet, R., Ros, L., Belveze, F., Brossier, J.-M.: Joint estimation of complex gain and fractional delay for Tx leakage compensation in FDD transceivers. In: Proceeding of the 21st International Conference on Electronics Circuits and Systems (ICECS) (2014)
8. Gerzaguet, R., Ros, L., Belveze, F., Brossier, J.-M.: Performance of a digital transmitter leakage LMS-based cancellation algorithm for multi-standard radio-frequency transceivers. Digit. Sig. Proc. 51, 35–46 (2016)
9. Gerzaguet, R., Ros, L., Brossier, J.-M., Belveze, F., Ghandour-Haidar, S.: Self-adaptive stochastic rayleigh flat fading channel estimation. In: Proceedings of the 18th International Conference on Digital Signal Processing (DSP 2013), pp. 1–6, July 2013
10. Kiayani, A., Anttila, L., Valkama, M.: Modeling and dynamic cancellation of Tx-Rx leakage in FDD transceivers. In: Proceedings of IEEE 56th International Midwest Symposium on Circuits and Systems (MWSCAS), pp. 1089–1094, August 2013
11. Knox, M.: Single antenna full duplex communications using a common carrier. In: Proceedings of IEEE 13th Annual Wireless and Microwave Technology Conference (WAMICON), pp. 1–6, April 2012
12. Lederer, C., Huemer, M.: LMS based digital cancellation of second-order Tx intermodulation products in homodyne receivers. In: Proceeding of IEEE Radio and Wireless Symposium (RWS), pp. 207–210 (2011)
13. Lederer, C., Huemer, M.: The influence of DC offsets on the digital cancellation of second-order Tx intermodulation distortions in homodyne receivers. In: Proceedings of IEEE International Conference on Wireless Information Technology and Systems (ICWITS), pp. 1–4 (2012)
14. Manstretta, D., Brandolini, M., Svelto, F.: Second-order intermodulation mechanisms in CMOS downconverters. IEEE J. Solid-State Circuits 38, 394–406 (2003)
15. Martikainen, H.: Analysis of duplexing modes in the IEEE 802.16 wireless system. In: Proceedings of European Wireless Conference (EW), pp. 849–856, April 2010
16. Mengali, U., D'Andrea, A.: Synchronization Techniques for Digital Receivers. Applications of Communications Theory. Springer, New York (1997)

17. Meyr, H., Moeneclaey, M., Fechtel, S.A.: Digital Communication Receivers. Proakis (1998)
18. Razavi, B.: Design considerations for direct-conversion receivers. IEEE Trans. Circuits Syst. (Analog Digit. Sig. Process.) **44**, 428–435 (1997)
19. Ros, L., Hijazi, H., Simon, E.P.: Complex Amplitudes tracking loop for multipath channel estimation in OFDM systems over slow to moderate fading. Sig. Process. **97**, 134–145 (2014)
20. Simon, E., Raoof, K., Ros, L.: Optimization of symbol timing recovery for multi-user DS-CDMA receivers. In: Proceedings of IEEE of Acoustics, Speech, and Signal Processing (ICASSP), vol. 4, pages IV-604-7, April 2003
21. Tang, H.: Some physical layer issues of wide-band cognitive radio systems. In: First IEEE International Symposium on New Frontiers in Dynamic Spectrum Access Networks (DySPAN), pp. 151–159, November 2005

Predictive Channel Selection for over-the-Air Video Transmission Using Software-Defined Radio Platforms

Marko Höyhtyä$^{(\boxtimes)}$, Juha Korpi, and Mikko Hiivala

VTT Technical Research Centre of Finland Ltd.,
P.O. Box 1100, 90571 Oulu, Finland
{marko.hoyhtya,juha.korpi,mikko.hiivala}@vtt.fi

Abstract. This paper demonstrates a predictive channel selection method by implementing it in software-defined radio (SDR) platforms and measuring the performance using over-the-air video transmissions. The method uses both long term and short term history information in selecting the best channel for data transmission. Controlled interference is generated in the used channels and the proposed method is compared to reference methods. The achieved results show that the predictive method is a practical one, able to increase the throughput and reduce number of collisions and channel switches by using history information intelligently.

Keywords: Cognitive radio · Spectrum databases · Dynamic spectrum access

1 Introduction

Cognitive radio (CR) techniques have been studied intensively for over a decade, focusing mainly on dynamic spectrum access oriented operation. Numerous techniques have been developed and analyzed, including spectrum sensing, power and frequency allocations, beacon signaling, and spectrum databases. Only a subset of the proposed techniques have been implemented and tested in real systems to see their practicality. This paper focuses on channel selection problem in a changing radio environment and demonstration of the proposed method in a practical system.

Importance of history information and knowledge on primary traffic patterns in channel selection was shortly discussed already in [1]. Later, the problem has been studied intensively and prediction methods for both stochastic and deterministic traffic have been developed [2–8]. For example, a deterministic long-term component can be seen in several bands such as cellular mobile communication systems due to daily rhythm of the users [3]. Traffic pattern estimation method for exponential traffic has been proposed in [4]. A more general method able to classify traffic patterns and select the prediction method based on this information is proposed in [5]. Switching delay has been included in the channel selection to decide whether to switch a channel or not based on channel prediction and switching overhead in [6]. The method is developed further in [7] where an adaptive sensing policy is developed to detect the primary user appearance as fast as possible. Sequential channel sensing policy is studied also in [8].

© ICST Institute for Computer Sciences, Social Informatics and Telecommunications Engineering 2016
D. Noguet et al. (Eds.): CROWNCOM 2016, LNICST 172, pp. 569–579, 2016.
DOI: 10.1007/978-3-319-40352-6_47

The sensing procedure and channel selection can be made faster by reducing the number of channels to sense in the first place. Both short term and long term information can be used to guide the process. A channel selection method that was described in [9, 10] uses long term information on the use of primary channels to select the most promising ones to be sensed and exploited by cognitive radios at the requesting time. These channels are investigated in more detail over short term to find the best channels for data transmission. Both long term and short term data are stored in databases to be able to predict which channels look most promising for secondary use.

The proposed hybrid method that uses both sensing and databases is a promising approach to be used e.g., in different spectrum sharing scenarios of future fifth generation (5G) systems, and also in military environments. A hybrid method is the most probable step in-between pure database access and sensing based access to the spectrum. This paper demonstrates the described method by implementing it in a software-defined radio system. Verification is performed by transmitting video over a cognitive link and measuring the performance regarding error rates, channel switches, and throughput. Achieved results are compared to reference methods that are not using prediction in the channel selection.

The paper is organized as follows. Section 2 describes used channel selection methods starting from the intelligent hybrid one. Section 3 defines the demonstration environment and measurement results are presented in Sect. 4. Time domain analysis and discussions about possible improvements to the demonstrator are given in Sect. 5. Conclusions are drawn in Sect. 6.

2   Description of the Channel Selection Methods

2.1   The Smart Channel Selection Method

Simplistic view of the method is shown in Fig. 1. In the first phase a CR sends query to the long term database to receive a set of promising channels among M possible ones. The set is selected e.g., based on the long term spectrum occupancy data. Time and capacity estimations can be used to define channels that are suitable, offering needed time for the requested transmission. Given N channels are sensed to know whether they are free or not and the sensing information is stored in the ST database.

The short term database classifies the type of traffic in different channels which enables use of specific prediction methods for each traffic type, making prediction results accurate. Then, future idle times are predicted using the classification result and the history data. The P channels with the longest idle times are selected into use and the rest $N–P$ channels are returned to be offered to other users requesting access to spectrum. After channel selection is made, the CR can send data for predefined period of time, sensing periodically the channel to be sure that it is still free for transmission. Thus, use of long term database shortens the sensing time by reducing the channels to be sensed. The use of short term database reduces the channel switching rate and collisions with primary users. Therefore, more time is left for data transmission and consequently, capacity of the system is increased.

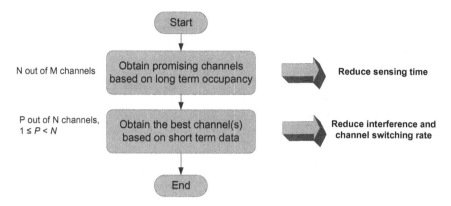

Fig. 1. Simplistic view of the proposed method.

2.2 Methods to Compare

No Channel Switching at all. The simplest way to operate in the spectrum is to stay always in the same channel. Thus, the first method to compare is no channel switching at all–method. If there is interference in the channel, the system suffers and there is degraded quality of service during that period of time. The system may also be required to stop transmitting totally and wait until the channel is free again.

Change to the Next Predefined Frequency. An improved step to the previous method is to change frequency when there is interference in the current channel. This can be done in many different ways. The simplest one is to predefine the next frequency to switch into. The advantage of this method is to be able to find a good channel to operate. A disadvantage is that it may take several switches since the channel to switch into may also be under interference.

Change to the Free Frequency. It is wise to switch into a channel that is available for transmission even though it requires more resources in sensing and finding those potential channels. This method may randomly select any of the free frequencies or jump into next free frequency whenever interference occurs at the current operational channel. This kind of reactive channel switching is proposed e.g., in [11, 12]. Since only instantaneous information about the availability of channels is used, switching may need to be performed quite often, depending on the primary user spectrum use.

3 Demonstration Environment

Figure 2 presents a block diagram of the measurement set-up. We are using SDR platforms for a data link, five interfering transmitters, and a spectrum sensor. A photograph of the environment is shown in the Fig. 3. The measurement environment is physically located at VTT premises in Oulu, Finland.

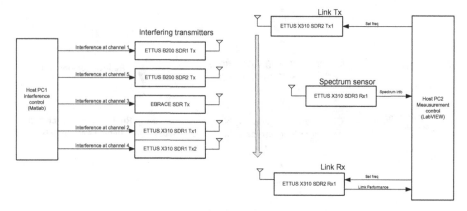

Fig. 2. Block diagram of the measurement setup.

Fig. 3. Demonstration setup.

Interference generation is made using Matlab controlled SDR-platforms. We use USRP B200 and X310 platforms from Ettus Research [13] together with the EBRACE SDR platform which, like the USRPs, is also a field-programmable gate array (FPGA) based SDR platform. The SDR platforms are used to generate continuous data to five different frequency bands. The type of the platforms is not important for the measurements. In fact, many other controllable interference sources could be used with the same effect. The transmission power of the interfering transmitters has been set high enough to cause strong interference to the selected band. The lengths of the continuous busy and idle times are both parametrized for each frequency separately.

In general, data traffic transmitted in a network can be characterized by traffic patterns. These patterns can be classified as [1]: (1) deterministic patterns, where the transmission is ON, then OFF during the fixed time slot, and (2) stochastic patterns, where traffic can be described only in statistical terms. Thus, values for busy and idle times in the demonstrator can be set either with fixed values or e.g. exponentially distributed random values.

Suppose we have a vector of n samples of idle times from the channel i, $\mathbf{X}^i = (x_1^i, x_2^i, \ldots, x_n^i)$. Assuming exponentially distributed idle times with traffic parameter $\lambda_{\text{OFF}} > 0$ the probability density function of the exponential distribution is

$$f(x) = \begin{cases} \lambda_{\text{OFF}} e^{-\lambda_{\text{OFF}} x}, & x \geq 0 \\ 0, & x < 0 \end{cases} \tag{1}$$

The maximum likelihood (ML) estimate for the idle time is $\hat{T}_{\text{OFF}} = \bar{x}$, where $\bar{x} = (1/n) \sum_{j=1}^{n} x_j$ is the sample mean [5]. Thus, the best prediction of the next idle time is the average of the previous ones. The same model applies also for busy times. In practice, traffic patterns of different channels might vary over time. Thus, the observation interval for average calculation should be restricted.

Interference detection for the used channel is done by measuring the block error rates (BLER) in the receiver. BLER is defined as the ratio of the number of erroneous blocks received to the total number of blocks sent, expressed as a percentage. It is used in 3GPP Long Term Evolution (LTE) systems during link radio monitoring, typically aiming to have the BLER below 10 %. It can be improved e.g., by adaptive modulation and coding or by changing to a new frequency. In our proof-of-concept implementation, once the block error rate exceeds a threshold value we decide to change the channel. The next channel is selected according to methods presented in Sect. 2.

Spectrum measurement for other channels is done in 100 MHz bandwidth, and by also selecting the system bandwidth to be 100 MHz, all of the used channels can be monitored simultaneously. This keeps the spectrum sensing simple in our measurement set-up. For each used frequency, after averaging over a few measurements, we use a simple threshold to decide whether the channel is interfered or not. If the measured power level is above the threshold, the channel is considered interfered. Measurement control reads this binary (busy/free) information and decides the next free frequency where to jump to if a channel change is needed.

Actual data link is in our case a modified real-time LTE link based on National Instruments LabVIEW Communications LTE Application Framework version 1.0 [14] where we have added the frequency switching algorithms. With the data link we can measure the throughput, error rate and number of frequency changes. Measurements were done with LTE Modulation and Coding Scheme (MCS) number 24, i.e. with 64QAM and code rate 3/4, and using 20 MHz bandwidth. Throughput was recorded both for all of the physical downlink shared channel (PDSCH) data and for the user data, which in our case was video data streamed over the link. The graphical user-interface (GUI) of the measurement control is presented in Fig. 4. User interface shows in real-time which channel is currently used, which channels are under interference, and

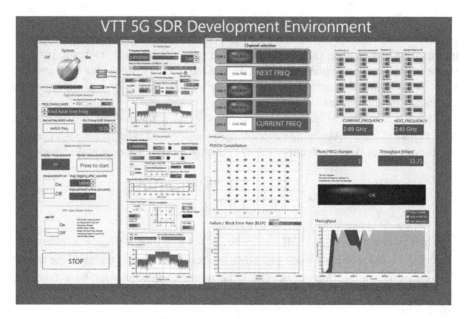

Fig. 4. Graphical user interface.

what is the next channel to switch into when interference occurs. The GUI shows also spectrum information and enables selecting the different channel selection methods on the fly.

Measurements presented in this paper were made in the 2.4 GHz industrial, scientific, and medical (ISM) band which is not fully controlled environment. We noticed that there was also other traffic present at the band during the measurements. The total 100 MHz bandwidth was divided into five equal size channels. Especially, channels 1, 2 and 3 were slightly interfered but the channels 4 and 5 were mostly free of other traffic. The other traffic was mostly general WiFi traffic in the office environment. The measurements were made during the night time and during the weekends when the amount of this other, uncontrolled traffic was very small. Our radio link used a turbo code with 3/4 code rate, and frame error rate (FER) was low on all channels if we ourselves were not generating any controlled interference with our interference generators.

4   Results

Measurement results for video transmission with different channel selection methods are given in Figs. 5 and 6. Measurements were conducted for each method over a 13500 s period. The presented results are average results over four consecutive measurement periods. For brevity, the used methods are named as:

Mode0: No channel changes

Mode1: Change to next (predefined) frequency

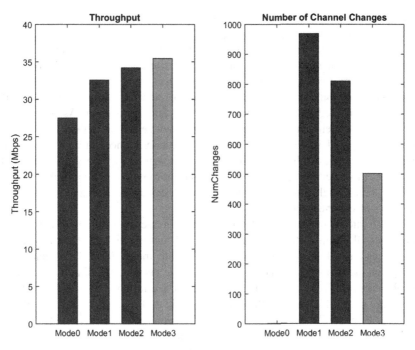

Fig. 5. Measured throughput and number of channel changes with all the methods, random interference traffic.

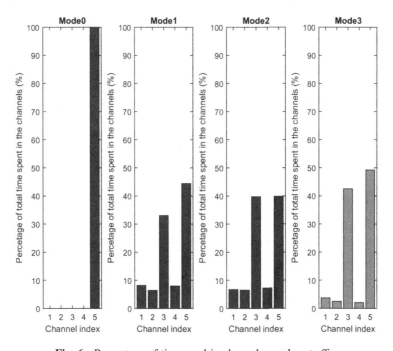

Fig. 6. Percentage of time used in channels, random traffic.

Mode2: Change to next free frequency

Mode3: Change to the best free frequency

Figure 5 shows the measured throughput and number of channel changes for all the channel selection methods with the random interference. The busy and idle periods for used channels are given in Table 1. The given values represent both mean values for exponentially distributed interference traffic and fixed values for deterministic traffic. Same values are used with each mode to have a fair comparison. As is seen in the figure, the more intelligence is added to the channel selection method the higher the achieved throughput is. Mode0 suffers during the interference since it is not able to change the channel. Ability to switch improves right away the performance. The proposed method, i.e., Mode3 achieves the highest throughput since it is able to predict and select the channels offering longest idle times for transmission and thus, minimize the number of channel changes over the experiment.

Figure 6 shows in more detail how different modes use different channels. The Mode0 uses all the time the best channel. Mode1 and Mode2 select the next channel randomly and use bad channels 1, 2, and 4 quite much. Mode3 concentrates the operation on the two best channels with longest idle periods, avoiding the use of other channels whenever it is possible. Only when there are no good channels free, the bad channels are used. The results with the deterministic traffic shown in Fig. 7 confirm the same conclusions. The advantage of the intelligent method is roughly the same regardless the type of the interference traffic in the channels.

The previous results were achieved with two good and three bad channels. We made also experiments with one, three, and four good channels to see the impact. When there are three good channels the trend looks still the same. However, the advantage is not that large anymore since the random methods also tend to select good channels more often. With four good channels the intelligent method still concentrates on three best ones since almost all the time some of them is available. When there is only one good channel to be used the performance is heavily dependent on the quality difference between channels. Purely from the throughput perspective the Mode0 can be better than other random methods since waiting in the good channel can be better than switching all the time among bad ones. Also in this case the intelligent method provided the best performance in measurements. From the quality of experience point of view, it is often better to change the channel since the waiting times and related video stoppage can be quite long. This is especially true if the interference is strong and continuously occurring in periods of several seconds.

Table 1. Idle and busy periods for used channels.

	Channel 1	Channel 2	Channel 3	Channel 4	Channel 5
Idle time	11 s	5 s	37 s	8 s	56 s
Busy time	17 s	8 s	10 s	5 s	21 s

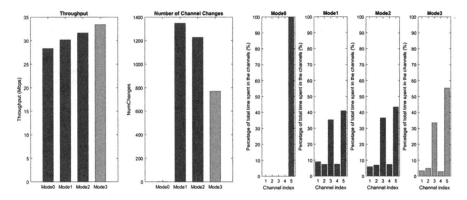

Fig. 7. Performance results with deterministic interference traffic.

5 Time Domain Analysis

The impact of the intelligent method in time domain can be seen in Fig. 8. The use of the long term database shortens the sensing time due to reduced number of sensed channels. The short term data reduces the number of channel switching by concentrating the operation on the channels with longest idle times. Thus, more time is left for data transmission.

This combination of databases is the main advantage of the proposed method when compared to other predictive approaches such as ones in [2–8]. The operation in the demonstration setup is presented in more detail in Fig. 9, showing the steps needed from the occurrence of the interference to synchronized data transmission in a new channel. The performance of the system could be improved through optimization of the sensing and switching times which could be achieved e.g., with a fully FPGA based decision making since FPGA processing speed is much faster than software based processing. This would speed up the procedures T1-T3 in Fig. 9 also by eliminating the

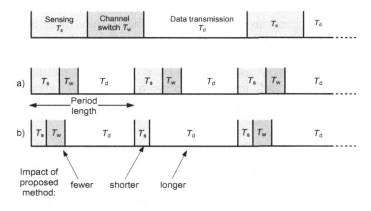

Fig. 8. Impact of the method: a) original frame b) with the intelligent method.

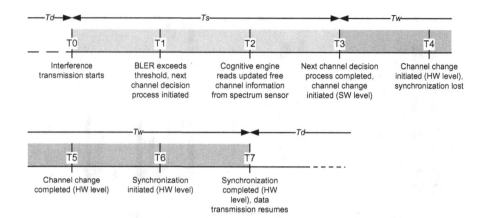

Fig. 9. Details of the time domain operation during a channel switch.

need to exchange control data through the interface between software and hardware layers in the LabVIEW framework. In the current setup the switching decisions are made with software which makes the total decision cycle clearly longer.

6 Conclusions

Use of history information enables a radio system to operate efficiently in a spectrum sharing radio environment. This paper has studied the channel selection problem by implementing a predictive method in a software-defined radio demonstrator and comparing its performance in the same system to several reference methods. Achieved results show that the proposed method increases the throughput and decreases interference towards other sharing systems. The quality of experience was clearly better for video streaming studied in the demonstration setup when the proposed predictive method was used.

As a possible future step, the setup can be developed further by inclusion of steerable antenna techniques such as the method proposed in [15] to improve the sharing in spatial domain. In addition, current implementation did not include the classification algorithm to be able to recognize the traffic pattern in the channels and select the optimal prediction method accordingly. An improvement to the operation would be also the speed-up of frequency switching by implementing a fully FPGA based decision making. Finally, the measurements could be made in totally interference-controlled environments with a channel emulator and/or in an isolated chamber.

References

1. Haykin, S.: Cognitive radio: brain-empowered wireless communications. IEEE J. Sel. Area Comm. **23**, 201–220 (2005)
2. Clancy, T.C., Walker, B.D.: Predictive dynamic spectrum access. In: SDR Forum Technical Conference, Orlando (2006)

3. López-Benitez, M., Casadevall, F.: An overview of spectrum occupancy models for cognitive radio networks. In: Casares-Giner, V., Manzoni, P., Pont, A. (eds.) Networking Workshops 2011. LNCS, vol. 6827, pp. 32–41. Springer, Heidelberg (2011)
4. Gabran, W., Liu, C.H., Pawelczak, P., Cabric, D.: Primary user traffic estimation for dynamic spectrum access. IEEE J. Sel. Area Comm. **31**, 544–558 (2013)
5. Höyhtyä, M., Pollin, S., Mämmelä, A.: Improving the performance of cognitive radios through classification, learning, and predictive channel selection. Adv. Electron. Telecommun. **2**, 28–38 (2011)
6. Kahraman, B., Buzluka, F.: A novel channel handover strategy to improve the throughput in cognitive radio networks. In: International Wireless Communications and Mobile Computing Conference, pp. 107–112 (2011)
7. Zhang, C., Shin, K.G.: What should secondary users do upon incumbents return? IEEE J. Sel. Area Comm. **31**, 417–428 (2013)
8. Shokri-Ghadikolaei, H., Fischione, C.: Analysis and optimization of random sensing order in cognitive radio networks. IEEE J. Sel. Area Comm. **33**, 803–819 (2015)
9. Höyhtyä, M., Vartiainen, J., Sarvanko, H., Mämmelä, A.: Combination of short term and long term database for cognitive radio resource management. In: 3rd International Symposium on Applied Sciences and Communication Technologies, Rome (2010)
10. Höyhtyä, M., Sarvanko, H., Vartiainen, J.: Method and device for selecting one or more resources for use from among a set of resources. U. S. Pat. Appl. US20130203427 A1 (2013)
11. Jing, X., Mau, S.-C., Raychaudri, D., Matyas, R.: Reactive cognitive algorithms for co-existence between 802.11b and 802.16a networks. In: IEEE Global Telecommunications Conference, St. Louis, pp. 2465–2649 (2005)
12. Feng, S., Zhao, D.: Supporting real-time CBR traffic in a cognitive radio sensor network. In: IEEE Wireless Communications and Networking Conference, Sydney, (2010)
13. Ettus Research. http://www.ettus.com/
14. National Instruments. http://www.ni.com/
15. Paaso, H., Mämmelä, A., Patron, D., Dandekar, K.R.: DoA estimation through modified unitary MUSIC algorithm for CRLH leaky-wave antennas. In: 24th International Symposium on Personal Indoor and Mobile Radio Communications, London, pp. 311–315 (2013)

Next Generation of Cognitive Networks

Uplink Traffic in Future Mobile Networks: Pulling the Alarm

Jessica Oueis$^{(\boxtimes)}$ and Emilio Calvanese Strinati

CEA, LETI, Minatec, 17 rue de Martyrs, 38054 Grenoble, France
{jessica.oueis,emilio.calvanese-strinati}@cea.fr

Abstract. Mobile wireless networks are designed and dimensioned according to mobile users downlink traffic. Downlink has been dominating wireless traffic since early 3G systems. Nowadays, we are witnessing a massive integration of novel applications and services into wireless networks, mostly through cloud architecture and technologies. The accessibility of cloud services from mobile devices increases demand over uplink traffic, which increase in uplink traffic enlarges the uplink to downlink ratio. Are current wireless networks capable of serving the increasing uplink traffic? Are DL-based network dimensioning and mechanisms adapted for such a change in traffic patterns? We launch a call for a future challenge to overcome in wireless networks: the uplink. In this paper, we set the issue of an inevitable uplink traffic explosion in future mobile networks. We discuss contributing factors in uplink traffic changes, as well as some research solutions for increasing networks uplink capacity.

Keywords: Uplink traffic · Cloud · Future networks · Mobile networks

1 Introduction

Cellular networks have always been designed, dimensioned, and deployed based on the downlink (DL) mobile users' demand and traffic patterns. The reason for leveraging downlink traffic was the asymmetry — then true — between uplink (UL) and DL traffic. In other terms, the capacity required in the downlink was much higher than the one required in the uplink. Therefore, designing networks with higher data rates to offer in downlink than in uplink was trivial. More precisely, within early 2G based cellular networks, the traffic load for both UL and DL have been roughly the same. This has also been the case for the very early 3G systems. It is not until the 3.5G and 4G systems that downlink traffic load greatly surpassed uplink requirements [1]. In these systems, with the eruption of IP based networks and high speed access to the Internet through cellular networks, traffic is dominated by downlink. The data explosion in downlink and uplink was asymmetrical. While downlink traffic grew exponentially, uplink traffic was also subject to an increase, however, the traffic demands in both directions were not equal. Mobile users downloaded more than they uploaded. The estimated

© ICST Institute for Computer Sciences, Social Informatics and Telecommunications Engineering 2016
D. Noguet et al. (Eds.): CROWNCOM 2016, LNICST 172, pp. 583–593, 2016.
DOI: 10.1007/978-3-319-40352-6_48

ratio of uploaded to downloaded data is about 1:7 [2]. Thus, current mobile networks are dimensioned based on the amount of data mobile users are downloading according to downlink traffic models. As the fastest growing segment of the communication industry, wireless communication, and especially cellular systems, have experienced, and are still experiencing, exponential growth over the last decade. Many new applications, services, and technologies have and will integrate the wireless network. The way mobile users see, use, and exploit mobile networks have changed. Mobile networks are nowadays the provider of unlimited number of heterogeneous services that differ in data requirements. As some are mainly downlink based, others have equal requirements of uplink and downlink traffic, or depend on large amount of data upload, like cloud storage for example. Today, the asymmetry between UL and DL is reduced, and sometimes inverted. These changes evoke a set of questions: What is the impact of network evolution on uplink traffic? Have networks started experiencing uplink traffic explosion? Should networks continue to be designed, planned and dimensioned according to downlink traffic only? What has been done to increase uplink network capacities? These questions are of great importance, especially under the fact that very low attention has been given to UL traffic models comparing to DL. Indeed, uplink traffic lacks of tractable models since it depends on users actions and unplanned interventions that are often less easily accessible and predictable. In contrast to downlink traffic that has been given significant attention, attempts to model the uplink have been limited [3]. With an increasing number of connected devices and mobile subscribers, the integration of cloud enabled technologies in wireless networks, the convergence of IoT systems, the development of M2M and MTC platforms, and many other factors, it is important to understand if and how new communication networks will cope with challenging uplink traffic loads. The idea is not about uplink rising over downlink traffic. We do not assume or consider that uplink traffic overtakes the downlink — although this might be the case in specific scenarios. We only present the uplink as a new important player that should be considered when setting network design and dimensions. Even though there are no precise forecasts on uplink, the traffic pattern change is inevitable. A study by NSN in 2013 showed that the overall UL to DL usage ratio reaches approximatively 1:2.4 [4]. In addition, the Ericsson mobility report of 2012 shows that UL to DL ratio reaches 1:1 for bi-directional applications such as P2P TV, email, and P2P sharing [5]. With the availability of high data rate services, new applications are enabled, and mobile devices energy consumption increases. Cloud technologies, sensor networks, device to device communications and social networking are all growing trends that increase uplink traffic and do not rely solely on downlink traffic. All of these trends introduce applications where mobile users create content and launch actions on the network, which changes the classic UL-based traffic pattern adopted in wireless networks. The research community, aware of the upcoming uplink traffic volume change, is already proposing some solutions in the network for improving uplink capacity.

In the remainder of this paper, we present the major factors that contribute to the uplink traffic explosion in the current/future mobile networks. Then we discuss some of the efforts that have already started by the wireless community to improve current networks uplink capacity in order to cope with mobile users' increasing uplink demand.

2   Contributing Factors in Uplink Data Traffic Increase

2.1   Increase in Number of Mobile Subscribers and Devices

The number of mobile subscriptions and mobile devices has been constantly growing since the first deployments of cellular mobile networks. From 6.4 billion mobile subscriptions in 2012 to 7.2 billion in early 2015, the ever-increasing index is to reach 9.2 billion by 2020 according to latest mobility reports [6]. Mobile broadband that was accounted for 2.9 billion out of the 7.1 billion subscriptions will grow its share to occupy 7.7 billion out of the 9.2 billion subscriptions in 2020, which is around 85 % of all subscriptions. As the number of fixed broadband and the number of related devices such as mobile PCs and mobile routers will have very low growth, and the number of total mobile subscribers and subscriptions will increase linearly, the number of mobile subscriptions will increase exponentially. Smartphones, which already are the main mobile equipment (2.6 out of 2.9 billion), are expected to double in number by 2020. Mobile broadband will be accessible to everyone and mobile devices will continue to outnumber the earth population. By 2020, mobile phones will be in possession of 90 % of humans over 6 years old. The growth of mobile devices and users showed in numbers gives an idea of how data traffic (in both uplink and downlink) could increase.

2.2   Evolution of Cellular Networks

Since wireless Internet, wireless generations adopting new technologies for increasing system capacity have been designed and deployed. The increasing users' traffic demand required a network evolution to cope with constant changes. However, for all consecutive technologies and wireless generations, downlink data rate far exceeded uplink. Due to possible technical challenge and asymmetry in traffic demand, mobile networks were always dimensioned to assure higher DL capacity. Table 1 shows the difference in up and downstream data rates among technologies. Note that the table shows advertised peak data rates, which are usually higher than nominal achieved rates. Evolution of wireless networks and users' traffic demand are in perpetual evolution and growth, one implying the other. Indeed, wireless network evolves to *"give more"* for mobile users and cope with their increasing traffic. At the same time, when offered more capacity, mobile users would like to *"do more"* with their mobile equipment through the wireless network. Numbers show that the proliferation of new wireless generations offering higher service quality attracts mobile users. Since the introduction of HSPA and then LTE, the number of mobile users continues to grow strongly. In the third quarter of 2012 HSPA and LTE subscriptions increased by 13 and

65 million respectively. As for GSM/EDGE it attracted then 20 million new subscriptions. With LTE proliferation in the market, the numbers in the first quarter of 2015 are as follows: 105 million additions for LTE, 60 million for HSPA, and a decline of 30 million for GSM/EDGE. These numbers and Fig. 1 show how the market follows the offer of new technologies and increasing service quality. LTE will have, alone, 3.7 billion subscriptions by 2020. In conclusion, the number of mobile users and the evolution of cellular networks are joint in an escalating increase relationship; where the increase of the first requires improvement in cellular networks, which re-attracts more mobile users to subscribe.

Table 1. Cloud architecture evolutions comparison

Technology	Generation	Downstream (Mbits/sec)	Upstream (Mbits/sec)
EDGE	2.5G	1.6	0.5
EVDO (Rev A)	3G	2.45–3.1	0.15–1.8
HSPA	3G	0.384–14.4	0.384–5.76
HSPA+	3.5G	21–678	5.8–168
LTE	4G	100–300	50–75
LTE-A	4G	1000	500

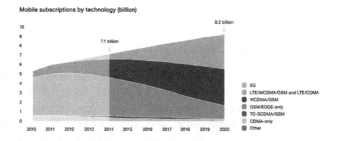

Fig. 1. Mobile subscriptions by technology [6]

2.3   Emergence of Cloud Technologies and Dense Heterogeneous Networks

Cloud technologies are progressively but rapidly integrating wireless networks. Cloud radio access network, remote cloud computing, cloudlets, and mobile edge cloud, are all new technologies and architectures in which the cloud concept integrated wireless networks. Cloud technologies in mobile networks consist on delegating computing, storage, data processing, and other resources consuming functionalities to a computing entity instead of performing the tasks on the mobile devices. Cloud in wireless networks can take different forms. It can be a centralized remote server pool, a nearby cloudlet, or an edge computing platform. Aside from cloud computing, cloud can be used as a remote storage location. As mobile equipments in general suffer from lack of resources of computing and storage, mobile users are more and more relying on outsourcing required storage and computation capacities. With cloud storage, mobile users can take photos or record videos with their mobile devices and directly upload them for saving on the cloud instead of their devices. In such applications, uplink is as important as downlink and thus should be taken into account in network dimensioning.

Another emerging deployment technology in wireless networks is Heterogeneous networks (HetNets). All mobile users are not served by the same type of base stations. Along with classic large coverage macro-cells, cellular networks are being intensively deployed with pico-cells, relays, and femtocells. One of the main motivations and interests of heterogeneous networks is offloading heavily loaded macro-cells. Users in reach of a femtocell, for example, will communicate with the latter instead of a congested macro-cell. As femtocells are deployed at closer distances from mobile users, communication channels between femtocells and mobile users are very often characterized by a better signal to noise ratio. Due to the lack of tractable models, the impact of such offloading on the uplink performance is not well understood [3].

Uplink traffic modeling has not gained the same attention as downlink. Both directions differ fundamentally in access modes, heterogeneity of transmitters, and resources management. The invasion of wireless networks by cloud enabled heterogeneous network certainly has an effect on traffic patterns especially in uplink, since new offloading opportunities are available to consumers. With the adoption of offloading computation and the concept of virtual machines (VM) and enable applications such as videoconferencing in enterprises and improved network mobility support, upload speeds become critical against users' experience quality and content efficient delivery to the cloud. With the development of cloud technologies, upload speed and capacity will continue to have an important impact.

2.4   New Applications and Services Ecosystem

Cloud based wireless networks are the next breakout of the wireless communication. Cloud is integrating many functionalities of the wireless networks and increasing their capabilities. Whether a remote cloud or an edge cloud, the cloud unlocks a whole new ecosystem of services and applications. Application developers have now the door open to new types of applications that can be run on the cloud and that were not adapted before to the mobile concept due to heavy resources requirements. Furthermore, cloud and services providers work on increasing their infrastructure ability and performance through improving availability and reliability: An evolving ecosystem that will push forward the cloud based offer and demand, and thus create higher cloud related traffic requirements. Among the applications that are now compatible with the mobile network, we distinguish different types of traffic requirements. Some applications require very high downlink and/or uplink traffic with varying latency constraints. Applications that comply with downlink based networks include streaming basic video and music and web browsing, where upload requirements are relatively low. Streaming relies basically on high downlink traffic, as for web browsing it has in general lower traffic requirements. However, numerous applications do not comply with that model. Many applications require roughly the same amount of upload and download such as web conferencing (cloud-based), video conferencing, tele-medicine, virtual office and connected vehicles safety applications [7]. Others, on the contrary, require more traffic upload than download such as web electronic health records, virus scanning, face recognition, cloud storage, and aggregated data analysis. Hence, the heterogeneity in new services and applications has non-negligible impact on traffic patterns and on the importance of uplink. The diversity of services offered through the Internet requires a management of network capacities in order to avoid both functional and economical harm to wireless communication infrastructure and businesses and their customers.

2.5 Crowded Networks Scenarios

Mobile networks are designed based on peak network traffic and the ability to serve in peak hours. This has led into excessive energy consumption. Several solutions were proposed for this problem such as base station sleeping. Furthermore heterogeneous networks deployment helps by offloading traffic from congested macro-cells onto smaller base stations. Now that solutions exist, the network should be dimensioned to keep its efficiency in peak data traffic scenarios. Peak traffic does not only concern downlink, uplink traffic is also subject to peak demands. Crowded scenarios are the best example for such situations. We take the example of a football stadium were thousands of people are gathered to watch a game. In such situations, mobile users share their experience through social networks, texting or talking. They post photos and videos during matches. A study by Ericsson [6] about the FIFA 2014 football games showed that social networking and texting were used during the matches and traffic peaked at half time. Ensuring a good user experience in such scenarios is a challenge to operators. Network planning and optimization are necessary. What is important to notice in crowded scenarios is the footprint of uplink traffic. According to the same study, the ratio of uplink in total data traffic was as high as 50 % during the final game of the world cup. The normal ratio in the same location is between 12 and 17 %. The increase in uplink traffic is clearly non negligible and should be taken into account during network planning and dimensioning. The study showed that 61 % posted or sent pictures via the Internet, and only 25 % used the Internet to find and download content related to the world cup. The numbers also showed that more users posted videos (33 %) than watched videos (18 %) through the Internet. Video uploading data usage is quite high especially that smartphones and tablets camera technology is quite advanced and 4 K video enabled. Furthermore, in many sports events, uplink traffic surpasses downlink in some time windows. With the accessibility of high data rate services such as 3G and 4G, mobile users will be more active uploading data and using social networking. 4G users, which have higher data rates, are more active than 3G users. This proves what was stated in Sect. 2.2 that users with higher services will want to do more. Even though 30 % of data traffic was handled by 4G networks, 4G users consumed 70 % more data than 3G users.

2.6   Sensor and MTC Networks

The evolution of sensor networks and Machine Type Communications (MTC) has been more than evident in the last decade. Mobile devices, sensors, and all types of objects already or soon will be equipped with sensors and RF circuits in order to integrate the wireless communication network. The Internet of Things is a well-known application of such networks that is creating a new trend and imposing a breakthrough in network management. In most wireless sensor networks scenarios, data is aggregated from end equipment (sensors) into a gateway that communicates with the network infrastructure. As for machine to machine (M2M) communications that allow devices of the same type to intercommunicate is also requiring an increasing role in the wireless network. Any device to device (D2D) communications can be established through different scenarios, where control link can be managed by end devices or the network. In D2D communications, especially scenarios where control is done by the network, uplink traffic is at least equal to downlink. Figure 2 shows a simplified level architecture of how the M2M system will be connected to wireless networks. The capillary network, which represents the set of communicating machines, aggregates data in a M2M gateway that is connected to the network and may use cellular communication. The convergence of sensor networks and cellular has been studied in literature. The European FP7 SENSEI project [8] focuses on the integration of WSN (Wireless Sensor Networks) and actuator networks. With the expected

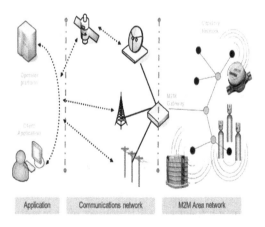

Fig. 2. High level simplified M2M architecture

increasing deployment of IoT devices and services, the traffic generated by these devices may change current traffic patterns. In fact, regarding the number of connected IoT devices, an estimated 50 billion "things" will be connected to the Internet by 2020 according to Cisco [9]. As for traffic patterns, current traffic models do not take into account traffic generated by smart devices. And since mobile networks are designed according to those patterns, they may not be adapted to such applications that are mostly uplink traffic generating. It is then necessary to understand how smart devices will affect network traffic and include it in network optimization and dimensioning.

3   Uplink Improvement Related Work

3.1   Range Extension in Heterogeneous Networks

Coverage Range Extension (CRE) in heterogeneous networks is a technique that can help increasing the uplink/downlink fairness. In an area covered by both macro-cell and a small-cell, the MUE/ Base Station (BS) association is based on the downlink received signal power only. And since small cells are characterized by a smaller transmit power than macro-cells, and are randomly deployed, they are expected to have large areas with low signal to interference (SIR) conditions [10]. In the uplink, the strength of the signal does not depend on the BS transmit power. It depends on the mobile device transmit power and the received signal power at base stations depends on the channel gain. This results in boundaries mismatch between uplink and downlink handover in heterogeneous networks. And since small cell coverage ranges are smaller than those of macro-cells, we notice unfair distributions of data rates between macro and small cells due to different loadings of connected users. The proposed solution is to balance the load between macro and small base stations by expanding the range of small cells (see Fig. 3). This is achieved by associating users to base stations based on path loss instead of received signal power. This will be in favor of uplink network performance since minimum path loss association maximizes uplink coverage rate [3]. Nevertheless, range expansion lead to high interference levels in the downlink which imposes using interference coordination techniques.

(a) Base stations coverage limits without CRE   (b) Base stations coverage limits with CRE

Fig. 3. Cell Range Expansion (CRE) impact of base stations footprint

3.2 Downlink and Uplink Decoupling

From the first generation to 4G, downlink and uplink of cellular networks have been coupled. Indeed, mobile users' equipments have been connected with the same base station in both uplink and downlink directions. As mentioned earlier, the best base station and user equipment couple is not necessarily the same for both directions. While for uplink it is best to connect UEs with the base station with the highest received signal power, for downlink, the best association is the one that minimizes path loss. Adopting a downlink centric association negatively affects load balancing in heterogeneous networks as well as uplink overall performance. Nevertheless, adopting an uplink centric association through cell range expansion creates interference problems for uplink users. As a solution, uplink and downlink association decoupling has been proposed in order to optimize communications in both directions [1,11]. Association decoupling is expected to increase uplink SNR and reduce transmit power, improve uplink interference conditions, improve uplink data rate, allow distribution of users among macro and small cells, and achieve more efficient resources utilization and uplink rates. This technique indeed proved to achieve up to 200 % improvement in the 5th percentile uplink throughput in a simulation based on a live Vodafone LTE test network deployment in London [11]. Nonetheless, the concept of uplink and downlink decoupling is considered as one of the components of future cellular networks [12,13]. However, this technique requires changes in system design since it needs mechanisms to allow acknowledgment process between serving base stations for uplink and downlink, strong synchronization, and data connectivity between base stations.

3.3 Uplink CoMP Techniques

Uplink Coordinated Multi Point (CoMP) is a new technology introduced with the LTE systems, which consists on jointly processing signals that are received at different antennas and/or base stations. It is the uplink analogy of CoMP where a single user is served by more than one base station (see Fig. 4). In uplink CoMP, users' signals are captured by more than a base station and processed jointly. Uplink CoMP can be deployed through three different scenarios: Intra-cell CoMP, Inter-cell CoMP, and between macro and small cells in heterogeneous networks. Inter-site CoMP is easily deployed since all signals information are inside one cell. Intra-site and heterogeneous CoMP require however low delay high capacity backhaul support between base stations. We note that uplink CoMP is transparent to mobile users in the sense that mobile equipment do not need to be aware of the base stations receiving their signal. Therefore, uplink CoMP does not change the association complexity on the mobile equipment side. By jointly processing received signals at different base stations, uplink CoMP results in uplink improvement. Uplink CoMP achieves uplink gain from both macro diversity reception and from enabling uplink/downlink decoupling in heterogeneous networks. Uplink perceived capacity is improved in high interference and poor coverage conditions. Important gains can be achieved especially in locations where uplink and downlink optimal associations are not the same, i.e. in locations where the most powerful received signals and the minimum channel path loss are not of the same base station. In a full scale field trial in LTE network [14] uplink CoMP proved to achieve 3 Mbps improvement in uplink throughput, and 100 % throughput gain if coupled with downlink/uplink decoupling.

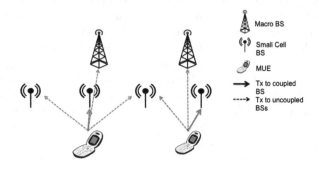

Fig. 4. Uplink CoMP usecase example

4   Conclusion

In this paper, we set the challenge facing uplink communication in future wireless networks. We focused on the upcoming increase in uplink traffic demand. We discussed several factors that contribute in validating changes in traffic uplink bottleneck and help prove that the upcoming uplink traffic explosion is a fact. Inscriptions number increase, wireless networks evolution, emerging cloud technologies, cloud enabled ecosystem, and convergence of sensor and actuators networks are among the factors with influence on uplink traffic. We then discussed some research work and studies that are proposed by the communication society that can help improve uplink network resources management, and increase uplink capacity in current and future networks. However, there is still no clear uplink traffic patterns that can validate if the existing efforts are enough for coping with the upcoming challenge. Until the emergence of a clear vision of how future networks will encompass uplink traffic explosion, we keep the alarm on.

References

1. Boccardi, F., Andrews, J.G., Elshaer, H., Dohler, M., Parkvall, S., Popovski, P., Singh, S.: Why to Decouple the Uplink and Downlink in Cellular Networks and How To Do It. CoRR, abs/1503.06746 (2015)
2. Electronic Communications Committee (ECC) CEPT, Asymmetry of Mobile Backhaul Networks (2012). http://www.cept.org/documents/se-19/6560/se19

3. Singh, S., Zhang, X., Andrews, J.G.: Uplink rate distribution in heterogeneous cellular networks with power control and load balancing. In: 2015 IEEE International Conference on Communication Workshop (ICCW), pp. 1275–1280, June 2015
4. Nokia Solutions and Networks, Nokia Solutions and Networks TD-LTE Frame Configuration Primer, November 2013
5. Ericsson, Ericsson Mobility Report, On the Pulse of the Networked Society, November 2012. http://www.ericsson.com/res/docs/2012/ericsson-mobility-report-november-2012.pdf
6. Ericsson, Ericsson Mobility Report, on the Pulse of the Networked Society, June 2015.   http://www.ericsson.com/res/docs/2015/ericsson-mobility-report-june-2015.pdf
7. Cisco, Cisco Global Cloud Index: Forecast and Methodology, 2013–2018 (2014). http://www.digital4.biz/upload/images/10_2013/131028130134.pdf
8. SENSEI, SENSEI EU project (2010). www.sensei-project.eu
9. Cisco, Fog Computing and the Internet of Things: Extened the Cloud to Where the Things Are (2015). https://www.cisco.com/web/solutions/trends/iot/docs/computing-overview.pdf
10. Qualcomm, LTE Advanced: Heterogeneous Networks, January 2011. https://www.qualcomm.com/media/documents/files/lte-heterogeneous-networks.pdf
11. Elshaer, H., Boccardi, F., Dohler, M., Irmer, R.: Downlink and uplink decoupling: a disruptive architectural design for 5G networks. In: 2014 IEEE Global Communications Conference (GLOBECOM), pp. 1798–1803, December 2014
12. Andrews, J.G.: Seven ways that hetnets are a cellular paradigm shift. IEEE Commun. Mag. **51**(3), 136–144 (2013)
13. Boccardi, F., Heath, R.W., Lozano, A., Marzetta, T.L., Popovski, P.: Five disruptive technology directions for 5G. IEEE Commun. Mag. **52**(2), 74–80 (2014)
14. Li, L., Butovitsch, P.: Uplink CoMP, the applications in LTE heterogeneous networks: principles and the field trial. In: 2013 15th IEEE International Conference on Communication Technology (ICCT), pp. 309–314, November 2013

Adaptive Channel Selection among Autonomous Cognitive Radios with Imperfect Private Monitoring

Zaheer Khan[(⊠)] and Janne Lehtomäki

CWC, University of Oulu, Oulu, Finland
zaheer@ee.oulu.fi

Abstract. We analyze the problem of autonomous cognitive radios (CRs) competing for multiple potentially available channels that may offer different rewards due to their non-homogeneity. The non-homogeneity in channels may lead to payoff distribution conflict among CRs, as each CR would prefer to select the more desirable channels. In our model, CRs are not able to observe the channel selections of other competing CRs. Rather, they get an imperfect signal from which the channel selections must be inferred. We study an adaptive *win-shift lose-randomize (WSLR)* strategy that (without centralized coordination) enables the CRs to autonomously reach an efficient and fair payoff distribution outcome. We study the autonomous channel selection problem under different primary user (PU) occupancy models; analyze the proposed strategy under imperfect signals. We also investigate the impact of deviations by a selfish CR on the performance of the proposed strategy.

Keywords: Adaptation · Autonomous cognitive radios · Heterogeneous channels · Opportunistic spectrum access

1 Introduction

Both the FCC and a recent EU report have recommended the adoption of spectrum sharing technologies, including cognitive radio (CR), to address the rising demand for high-bandwidth wireless service [1,2]. Regulatory bodies are currently defining how cognitive systems with dynamic spectrum access capabilities will be allowed to operate [3].

CR wireless systems are a collection of wireless network entities that are able to adapt intelligently to the environment through observation, exploration and learning. In sensing-based OSA, cognitive radios (CRs) monitor the environment to reliably detect the primary user (PU) signals and operate whenever the band is empty. In practice, detection of PUs may rely on a combination of sensing and the use of geolocation spectrum occupancy databases [4].

When autonomous CRs have to search multiple potentially available channels for spectrum opportunities, they face competition from one another to access a channel. The end result of this competition is reduced CR throughput due to collisions among CRs that transmit simultaneously in the same channel.

© ICST Institute for Computer Sciences, Social Informatics and Telecommunications Engineering 2016
D. Noguet et al. (Eds.): CROWNCOM 2016, LNICST 172, pp. 594–604, 2016.
DOI: 10.1007/978-3-319-40352-6_49

Moreover, further payoff distribution conflict among multiple CRs may arise when potentially available channels offer different rewards due to their non-homogeneity.

The following example illustrates this payoff distribution conflict. Consider a communication system with two autonomous CRs and two potentially available channels. Suppose that the reward that a CR receives from the use of a given channel is proportional to the probability that the channel is free from PU activity. Consider the case where channel A is available with greater probability than channel B. When the two CRs choose their actions autonomously and without coordination, then if a particular channel is simultaneously sensed free by the two CRs and both of them decide to transmit on the channel, a collision occurs. To maximize the total system reward would require the CRs to autonomously arrive at orthogonal channel utilizations, where one CR opportunistically utilizes channel A and the other CR opportunistically utilizes channel B. Clearly, there is a source of conflict for the CRs over such an orthogonal channel utilization, as each CR prefers the orthogonal outcome in which it selects channel A and the other CR selects channel B.

In this paper, the question we seek to answer is how CRs can autonomously arrive at an outcome that maximizes the total average reward (the total average number of successful transmissions) in the distributed CR network in a way that also minimizes the payoff distribution conflict among autonomous CRs.

In this paper, we evaluate an adaptive WSLR strategy that maximizes the total average reward in the network by leading to reduced likelihood of collisions among CRs. The proposed strategy also leads the autonomous CRs to engage in intertemporal sharing of the rewards from cooperation. The concept of fairness we focus on is envy-freeness [5]. Using simulation results we also compare the performance of our proposed strategy against other existing strategies. We study channel selection among autonomous CRs under imperfect private monitoring which implies that CRs are not able to directly see the actions of their opponents. Rather, they get an imperfect signal from which the channel selection must be inferred. We explore the impact of false alarms, channel errors, and co-channel interference tolerance on the behavior of autonomous CRs and on the performance of the proposed adaptive strategy. By co-channel interference tolerance, we mean that there is a small but non-zero probability that, even when two CRs transmit simultaneously on the same channel, one of the transmissions is correctly decoded by its intended receiver. We also evaluate the effect of varying the number of channels that a CR can sense on the performance of the proposed scheme. Moreover, we also show that the performance of the proposed strategy is not strongly affected by different primary user (PU) occupancy models.

The works in [6–8] have also studied the problem of multichannel cognitive medium access and proposed learning and allocation strategies for distributed CRs. However, the works in [6–8] assumed that each CR cooperatively follows the same strategy (protocol). Unfortunately, in the presence of non-homogeneous spectrum resources this assumption is not valid, as non-homogeneity of spectrum resources may induce some CRs to deviate from the protocol to maximize their own usage at the expense of the total CR system throughput. It is useful to study these scenarios as a model in which competition and conflict among autonomous CRs searching multiple channels for spectrum opportunities is analyzed. In this paper, we build on our previous research presented in [9]. However, different from this work, in [9] the CRs are assumed to have perfect monitoring, i.e., they get a perfect signal from which the action of other CRs must be inferred. Different from our work in [10], in this paper we evaluate the impact of correlation in channel occupancy by a PU in consecutive time slots.

The paper is organized as follows. The system model is presented in Sect. 2, while in Sect. 3 we present, analyze and compare the WSLR strategy to related strategies proposed in other works. Finally, Sect. 4 summarizes our main conclusions.

2   System Model

We examine a multichannel CR network in which N autonomous CRs have M potentially available channels. Let $\mathcal{M} = \{1, 2, ..., M\}$ represent the set of (potentially available) channels and $\mathcal{N} = \{1, 2, ..., N\}$ represent the set of autonomous CRs. We investigate the proposed method under: (1) The (i.i.d.) model of PU channel occupancy (also adopted by [6]), in which for each channel, the PU activity in a time slot is independent of the PU activity in other time slots and is also independent of the PU activity in other channels; (2) The second model which considers correlation in channel occupancy by a PU in consecutive time slots. In this model, the state of each channel is described by a two-state Markov chain, with α_i indicating the transition probability for the ith channel from PU-occupied to PU-free and β_i indicating the transition probability from PU-free to PU-occupied. This PU activity model is also adopted in [11]. We show (see numerical results in Sect. 3.2, Table 1) that the proposed method is not strongly affected by the stochastic model of the PU behavior. In this work, for simplicity, we utilize the (i.i.d.) model of PU channel occupancy for theoretical analysis. The primary user duty cycle statistics are known to the autonomous CRs. In practice, the autonomous CRs may obtain the primary user duty cycle statistics through the use of geolocation databases [4]. Each CR can sense only one channel at a time and, due to hardware constraints, at any given time each CR can either sense or transmit, but not both. To protect transmissions by the incumbent, the detection probability $(P_{d,i})$ of an autonomous CR i is fixed at a desired target value, $P_{d,i} = P_d$, for all $i \in \mathcal{N}$. In practice, P_d is required to be close to 1. In the literature this is defined as the constant detection rate (CDR) requirement [12]. For a fixed target detection probability, the false alarm probability of a CR is a variable. In this paper, we consider the effect of varying probabilities of false alarm.

Different reward values may be associated with potentially available channels. The difference in reward values of potentially available channels can be due to a variety of reasons, such as channels with different bandwidth and channels with different availability for opportunistic use. In sensing-based OSA, when multiple channels are potentially available for transmissions, time-slotted multiple access is often adopted [6,13]. In such systems, the primary users and CRs are both

Table 1. Total average reward per time slot in the CR network and highest average envy ratio between a pair of CRs in the network as a function of $N = M$ for different strategies. With and without false alarms.

$P_{fa} = 0,$	$N = M = 4$		$N = M = 6$		$N = M = 8$	
	$\sum_{i=1}^N G_i$	Υ	$\sum_{i=1}^N G_i$	Υ	$\sum_{i=1}^N G_i$	Υ
$rand - C$	3.1803	$\frac{0.8902}{0.6804} = 1.3$	4.37	$\frac{0.8902}{0.6804} = 1.3$	5.3997	$\frac{0.8998}{0.4999} = 1.8$
$WSLR$ i.i.d	3.1856	$\frac{0.7965}{0.7962} = 1$	4.3374	$\frac{0.7226}{0.7221} = 1$	5.3910	$\frac{0.6617}{0.6612} = 1$
$WSLR$ Markov	3.2	$\frac{0.807}{0.804} = 1$	4.31	$\frac{0.7186}{0.7140} = 1$	5.21	$\frac{0.6516}{0.6510} = 1$
$Rand$	1.5	1	2.1996	1	2.8384	1
EWD	1.5697	$\frac{0.4245}{0.2967} = 1.43$	2.3501	$\frac{0.4125}{0.2885} = 1.43$	3.0367	$\frac{0.3926}{0.2917} = 1.34$
$P_{fa} = 0.1,$	$N = M = 4$		$N = M = 6$		$N = M = 8$	
	$\sum_{i=1}^N G_i$	Υ	$\sum_{i=1}^N G_i$	Υ	$\sum_{i=1}^N G_i$	Υ
$rand - C$	2.9741	$\frac{0.8402}{0.6982} = 1.2$	4.2908	$\frac{0.8468}{0.5773} = 1.47$	5.1744	$\frac{0.8484}{0.5200} = 1.63$
$WSLR$ i.i.d	2.9949	$\frac{0.7348}{0.7329} = 1$	4.2882	$\frac{0.7222}{0.7218} = 1$	5.1679	$\frac{0.6464}{0.6441} = 1$
$WSLR$ Markov	3.03	$\frac{0.7597}{0.7595} = 1$	4.208	$\frac{0.7023}{0.7013} = 1$	5.107	$\frac{0.6388}{0.6385} = 1$
$Rand$	1.7640	1	2.5032	1	3.1552	1
EWD	1.8076	$\frac{0.4798}{0.3693} = 1.3$	2.6736	$\frac{0.4638}{0.3557} = 1.3$	3.4284	$\frac{0.4394}{0.3554} = 1.24$

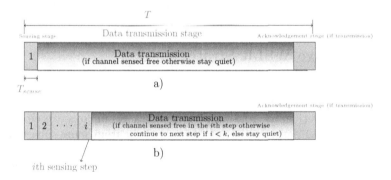

Fig. 1. Time slot structure: a) Single-channel sensing policy: When in a given time slot only one sensing step is available. b) Sequential-channel sensing policy: When in a given time slot more than one sensing step is available.

assumed to use a time-slotted system, and each primary user is either present in a channel for the entire time slot, or absent for the entire time slot [6,7].

The sensing stage in each time slot is divided into a number of sensing steps. Each sensing step is used by a CR to sense a different channel (see Fig. 1). In practice, improving the sensing accuracy implies increasing the sensing duration, whereby CRs may not be able to sense all the channels within the duration of a slot. We evaluate our proposed strategy for the scenario where the number of sensing steps k that a CR can utilize in a given time slot varies from 1 to M. If a CR finds a channel free in its ith sensing step, it transmits in that channel. However, if in all sensing steps channels are found to be busy, then the CR stays silent for the remaining duration of that time slot (see Fig. 1). When a free channel is found in the ith sensing step, the durations of the sensing stage and data transmission plus acknowledgement stage are iT_{sense} and $T - iT_{sense}$, respectively, where T_{sense} is the time required to sense each channel, T is the total duration of each slot and $T >> T_{sense}$. When multiple autonomous CRs search multiple potentially available channels, then from an individual CR perspective one of the following three events will happen in each sensing step: (1) It is the only one to find a channel free and transmit; the CR then has the channel for itself for the remainder of the time slot; (2) It finds that the channel is occupied by the PU or by another CR, then it continues looking in the next sensing step; (3) It visits a given channel, finds it free and transmits, but so does at least one other CR; a collision occurs. A CR infers that a collision has occurred whenever it fails to receive an acknowledgement (ACK) for a transmitted data frame. It should be noted that in this context, the channel sensing order $\mathbb{P}$, i.e., the order in which radios competing for the channels visit those channels, will affect their probability of successful access.

The vector of observation error probabilities is given by $\mathbf{e} = (P_{fa}, \sigma, \pi)$, where P_{fa} represents the probability of a false alarm, σ represents the probability that co-channel interference can be tolerated, i.e., when two CRs transmit simultaneously on the same channel, one of the transmissions is correctly decoded by its intended receiver with probability σ, and π represents the probability of a channel error.

3   An Adaptive WSLR Strategy

In this section, we propose an adaptive Win-shift, lose-randomize (WSLR) strategy, where adaptations are in the autonomous choice, by CRs, of the channel sensing order. Note that the sensing order that a CR employs can either come from the space of all permutations of M channels or from some subset thereof. However, the work in [14] shows that CRs can increase their average number of successful transmissions by adaptively selecting sensing orders from a predefined Latin Square of M channel indices. Note that when CRs select sensing orders from a Latin Square, $|\mathcal{S}| = M$, and two or more CRs can collide only if they select the same sensing order. We consider the case where each CR employs a common pre-defined sequence matrix (a Latin Square) Φ to select a sensing order in which k potential channels are to be visited in a given time slot, where k takes integer values between 1 to M.

For a given number of channels M there can be many Latin Squares. To select a sensing order from a common predefined Latin Square, CRs can employ any of the many Latin Squares. However, to make the analysis tractable, we assume that each CR employs a circulant matrix (which is an example of a Latin Square). For example, with $M = 3$, the circulant matrix Φ is given as:

$$\Phi = \begin{matrix} s_1 \\ s_2 \\ s_3 \end{matrix} \begin{pmatrix} 1 & 2 & 3 \\ 2 & 3 & 1 \\ 3 & 2 & 1 \end{pmatrix}$$

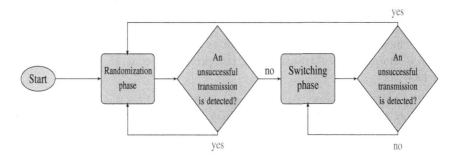

Fig. 2. The Win-shift, lose-randomize (WSLR) strategy.

The WSLR strategy is described in Fig. 2. The randomization and switching phases of the WSLR strategy (see Fig. 2) are explained as follows. In the *randomization phase*, each CR utilizes $\mathbf{p} = [\frac{1}{|\mathcal{S}|}, \frac{1}{|\mathcal{S}|}, \cdots, \frac{1}{|\mathcal{S}|}]$ ($|\mathcal{S}|$-element probability vector) for the selection of a sensing order. In the *switching phase*, the CR updates its current sensing order s_i to s_j where $j = (i \bmod |\mathcal{S}|) + 1$, i.e., in the next time slot, it shifts to the next sensing order to visit the channels.

The WSLR strategy is meant to address three aims:

1) Convergence: Note that when N CRs independently and randomly (with equal probability) select a sensing order (among $|\mathcal{S}| = N$ sensing orders) in each time slot, then the probability of arriving at orthogonal sensing orders in a time slot is $(1/N)^N (N!)$, and consequently the expected time required to arrive at orthogonal sensing orders is $N^N (1/N!)$. Clearly, this random strategy is inefficient as even when a CR attains a singleton status, i.e., the sensing order it has selected was not selected by another CR, it randomizes and with high probability it may lose the singleton status in the next time slot. In contrast to that, the WSLR strategy requires that singleton CRs should shift and non-singleton CRs should randomize. This reduces the number of CRs that randomly select a sensing order in the next time slot and hence increases the probability of arriving at orthogonal sensing orders. Hence with perfect private monitoring the WSLR strategy converges to conflict-free (orthogonal) sensing orders (provided that the number of CRs is less than or equal to M). For detailed proof of convergence of adaptation based sensing order selection strategies, we have results in [19, Theorem 4.1 and 4.2].

Unfortunately, when $\pi > 0$ (channel errors), in a given time slot if all the CRs have orthogonal sensing order selections, then an unsuccessful transmission (due to channel error) in any later time slot will lead a CR to erroneously move from shifting to randomization phase, in which case orthogonality may be lost, leading to reduced reward values. However, in Sect. 3.2 we show that for imperfect private monitoring our proposed strategy enables the CRs to increase their expected reward as compared to the random selection of sensing orders and other strategies.

2) Intertemporal sharing of rewards: Since different sensing orders may result in different rewards, intertemporal sharing of the sensing orders among autonomous CRs is achieved by allowing a CR to shift to the next sensing order if it has not observed an unsuccessful transmission, i.e., private outcome (U), in the previous time slot.

3) Discourage deviations: To discourage deviations, i.e., the CRs that select the sensing orders with higher rewards may prefer to again select those sensing orders in the next rounds, some punishment mechanism must be devised. This is achieved by triggering a switch to the randomization phase when an unsuccessful transmission is observed. Section 3.1 will further describe how the proposed mechanism discourages deviations by any of the autonomous CRs.

3.1   Analysis of the Adaptive WSLR Strategy

Without loss of generality, we assume that the channels are ordered by increasing probability of the PU being present, i.e., $\theta_1 \leq \theta_2 \leq, \cdots, \leq \theta_M$. For efficient channel utilization, we consider the scenarios where the N CRs utilize the N top rows of Φ for the selection of sensing orders. This is reasonable as the channel indices $1, 2, ..., M$ are ordered by increasing probability of the PU being present, hence the top N rows of Φ dominate in terms of having channels (in their initial columns) where PU's are less likely to be present. Note that for $N = M$, the entire matrix Φ is utilized by a CR for the selection of sensing orders. Let $\mathbf{S}_N$ represent the matrix of the top N rows of Φ.

Next, through extensive simulations, we analyze and compare the performance of the WSLR strategy for perfect and imperfect private monitoring.

3.2   Simulation Results and Comparison with Other Strategies

Using simulation our aim is to compare the performance (e.g., in terms of total average reward per time slot in the CR network, expected reward of a CR per time slot, and the maximum envy ratio between a pair of CRs) of the WSLR strategy against: (1) When all CRs utilize random selection of sensing orders, *Rand strategy*; (2) the randomize after every collision (rand-C) strategy proposed in [6]. In the rand-C strategy, initially each CR independently and randomly (with equal probability) selects a sensing order. In the next time slots, a CR randomly (with equal probability) selects a new sensing order only if it has experienced a collision in the previous slot; otherwise, it retains the previously selected sensing order; and (3) An autonomous CR i considers deviating from the WSLR strategy while all other CRs follow the strategy. The studied deviations by the CR i are: (a) Always select the preferred sensing order $s_1 = (1, 2, ..., M)$, fixed deviation (FD); (b) Always select s_1 with probability $q = 0.75$ and s_2 with probability $(1 - q)$, weighted deviation (WD); and (c) Always select s_1 with probability $q = 0.75$ and $s_2, s_3, ..., s_N$ with probabilities $[\frac{(1-q)}{(N-1)}, \frac{(1-q)}{(N-1)}, ..., \frac{(1-q)}{(N-1)}]$, extended weighted deviation (EWD). Moreover, we also evaluate the effect of varying the number of sensing steps on the performance of the proposed scheme. Note that calculations for expected reward per time slot of the CR i (G_i) are performed by Monte Carlo method using 15,000 Monte Carlo runs for dynamic channel selection process using different scenarios. $\Theta = (0.1, 0.2, 0.2, 0.3, 0.3, 0.5, 0.5, 0.5, 0.5, 0.5)$ is utilized PU duty cycle statistics vector (i.i.d PU occupancy model) for channels 1 to M respectively. For Markov Occupancy model, $(0.6000, 0.6000, 0.6000, 0.6000, 0.6000, 0.5, 0.5, 0.5, 0.5, 0.5)$ is the vector of α_i's (see Sect. 2 for details) and $(0.0660, 0.1500, 0.1500, 0.2600, 0.2600, 0.5, 0.5, 0.5, 0.5, 0.5)$ is the vector of β_i's for 1 to M channels respectively.

In Table 1, we evaluate different strategies in terms of the total average reward per time slot in the CR network and the highest envy ratio between a pair of CRs. Table 1 shows that the proposed method is not strongly affected by the stochastic model of the PU behavior. It also shows that the WSLR strategy performs equally well as the rand-C strategy in terms of maximizing the total average reward per time slot and performs significantly better in terms of ensuring envy-freeness among the competing CRs. It can also be seen from Table 1 that the random selection of sensing orders (rand) significantly reduces the total average reward per time slot in the CR network. When a CR considers deviating, while all other CRs follow the WSLR strategy, (see EWD in Table 1), it can be seen that there is no incentive in deviation from the WSLR strategy. Moreover, a deviating CR significantly reduces the total expected reward per time slot as compared to when all the CRs adopt the WSLR strategy. Figure 3a and b evaluate the expected reward per time slot achieved by a CR i using the different strategies under different scenarios, when CRs have perfect private monitoring and when their private monitoring is imperfect. From the two figures we can see that the WSLR strategy achieves the highest expected reward per time slot for the CR i as compared to when it considers deviating (while all the other CRs follow the WSLR strategy) and when all N CRs in the network utilize random selection of sensing orders. Note that in Fig. 3a and b, the loss in the expected reward of the CR i (when it adopts the WSLR strategy or random selection) is due to the non-homogeneity in channel availability statistics. The availability probabilities of the first 5 channels are at least 70 %, and the availability probabilities of the last five channels are 50 %. Hence, as $N = M$ increases, the expected reward of the CR i decreases, as with the increasing number of CRs the number of potentially available channels also increases but with high probability of a PU being present. Figure 3c evaluates the effect of varying the number of sensing steps on the performance of the different strategies in terms of expected reward of the CR i per time slot. It can be seen in Fig. 3c that when all the CRs with utilize the WSLR strategy then the expected reward per time slot of a CR increases as the number of sensing steps increase. However, it can also be seen in Fig. 3c that for the deviating CR (when all other CRs follow the WSLR strategy) there is little or no gain when more sensing steps are utilized for sensing. This is because for a given N and M, as k increases and when all the CRs utilize the WSLR strategy, it becomes more likely for a CR to find free channels in the later sensing steps as the WSLR strategy allows them to arrive at conflict-free allocations. When a CR deviates (while all other CRs stay on the WSLR strategy), the CR can either find

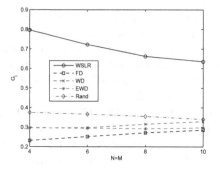

(a) Vector **e** is set to **e** $= (P_{fa}, \sigma, \pi) = (0, 0, 0)$.

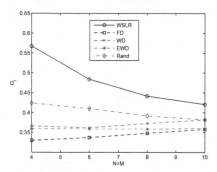

(b) Vector **e** is set to **e** $= (0.1, 0.05, 0.05)$.

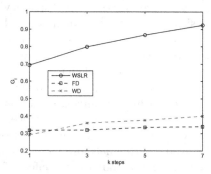

(c) Vector **e** is set to **e** $= (P_{fa}, \sigma, \pi) = (0, 0, 0)$.

Fig. 3. a) and b) Expected reward per time slot of the CR i (G_i) as a function of $N = M$ CRs for different scenarios. c) Expected reward per time slot of the CR i as a function of the number of sensing steps k, with $N = 5$ CRs, $M = 9$ channels. PU occupancy model is i.i.d.

a free channel in its initial sensing steps (when the CR is the sole radio following this sensing order) as it selects with high probability the sensing order s_1 in which the availability probabilities of the first 2 channels are at least 90 %, or else it may collide with the other CRs (when some other CR also selects the same sensing order) and fail to find a free channel during that time slot. Hence, increasing the number of sensing steps has either little or no gain for the deviating CR i.

4   Concluding Remarks

We have studied the problem of coexistence and competition among multiple autonomous CRs for a shared pool of non-homogeneous spectrum resources. For efficient and fair co-existence, we present an adaptive WSLR strategy that does not require coordination from a centralized entity and utilizes noisy feedback (imperfect signals) to infer the actions of other CRs. To analyze the impact of imperfect signals of a CR on the performance of the proposed WSLR strategy, using simulations, we have explored the effects of false alarms, channel errors and co-channel interference tolerance on the performance of our proposed strategy. We have shown that the proposed strategy increases the average number of successful transmissions in the network and minimizes the payoff distribution conflict among competing CRs.

Acknowledgement. This work was funded by Academy of Finland under the Academy Post Doctoral funding grant of Dr. Zaheer Khan.

References

1. Schmidt, E.E., Mundie, C.: Realizing the full potential of government-held spectrum to spur economic growth (2012). http://www.whitehouse.gov/administration/eop/ostp/pcast
2. Bataller, H., Commission, E.: Promoting the shared use of radio spectrum resources in the internal market (2012). http://www.eesc.europa.eu
3. Melvin, J.: US regulators ok T-mobile testing of shared use of airwaves, 15 Aug 2012. http://www.reuters.com/article/2012/08/15
4. Ghasemi, A., Sousa, E.S.: Spectrum sensing in cognitive radio networks: requirements, challenges and design trade-offs. IEEE Commun. Mag. **46**(4), 32–39 (2008)
5. Lipton, R. J., Markakis, E., Mossel, E., Saberi, A.: On approximately fair allocations of indivisible goods. In: Proceedings of the 5th ACM Conference on Electronic Commerce, EC 2004, pp. 125–131 (2004)
6. Anandkumar, A., Michael, N., Tang, A.: Opportunistic spectrum access with multiple users: learning under competition. In: Proceedings of the IEEE International Conference on Computer Communications (INFOCOM), pp. 1–9 (2010)
7. Liu, K., Zhao, Q.: Distributed learning in cognitive radio networks: multi-armed bandit with distributed multiple players. In: Proceedings of the IEEE International Conference on Acoustics, Speech, and Signal Processing (ICASSP), pp. 3010–3013 (2010)
8. Gai, Y., Krishnamachari, B., Jain, R.: Learning multiuser channel allocations in cognitive radio networks: a combinatorial multi-armed bandit formulation. In: Proceedings of the IEEE International Dynamic Spectrum Access Networks (DySPAN), pp. 1–9 (2010)

9. Khan, Z., Lehtomaki, J. J., DaSilva, L. A., Latva-aho, M., Juntti, M.: Adaptation in a channel access game with private monitoring. In: Proceedings of the Global Communications Conference, IEEE GLOBECOM 2013 (2013). Accepted for publication. http://www.ee.oulu.fi/~zaheer/Globecom2013.pdf
10. Khan, Z., Lehtomaki, J., DaSilva, L., Hossain, E., Latva-aho, M.: Opportunistic channel selection by cognitive wireless nodes under imperfect observations, limited memory: a repeated game model. IEEE Trans. Mob. Comput. **15**, 173 (2015)
11. Su, H., Zhang, X.: Cross-layer based opportunistic MAC protocols for QoS provisionings over cognitive radio wireless networks. IEEE J. Sel. Areas Commun. **26**, 118–129 (2008)
12. Peh, E., Liang, Y.-C.: Optimization for cooperative sensing in cognitive radio networks. In: Proceedings of the IEEE International Wireless Communication Networking Conference (WCNC), pp. 27–32 (2007)
13. Liu, K., Zhao, Q., Krishnamachari, B.: Distributed learning under imperfect sensing in cognitive radio networks. In: Proceedings of IEEE Asilomar Conference on Signals, Systems, and Computers (2010)
14. Khan, Z., Lehtomaki, J., DaSilva, L., Latva-aho, M.: Autonomous sensing order selection strategies exploiting channel access information. IEEE Trans. Mob. Comput. **12**(2), 274 (2013)

An Analysis of WiFi Cochannel Interference at LTE Subcarriers and Its Application for Sensing

Prasanth Karunakaran[(⊠)] and Wolfgang Gerstacker

Institute for Digital Communications, Friedrich-Alexander Universität
Erlangen-Nürnberg, Erlangen 91058, Germany
{prasanth.karunakaran,wolfgang.gerstacker}@fau.de

Abstract. LTE-Unlicensed (LTE-U) involving the deployment of LTE in unlicensed bands has been gaining significant interest lately. The standardization of LTE-U has been proposed for Release 13 of 3GPP LTE. The two main requirements mandated for such an operation are the coexistence mechanisms with the WiFi systems and the sensing before transmission by an LTE-U device. Often in literature the interference in the LTE-U scenario is modeled as white Gaussian noise. There is a need to understand the properties of interference in such scenarios more accurately by taking into account the physical layer specifications of the LTE and the WiFi standards. To this end, we analyze the interference generated by a WiFi transmitter at an LTE receiver and characterize its correlation properties. We show that the interference powers across the LTE subcarriers exhibit a periodic behavior and that this can be exploited to develop sensing schemes selective to WiFi signals.

Keywords: LTE-U · WiFi · Interference · Statistics · Sensing · LBT

1 Introduction

The deployment of LTE systems [1] in the unlicensed spectrum, referred to as LTE-Unlicensed (LTE-U), has been gaining interest recently [2,3]. Even though an unlicensed spectrum is less reliable than a licensed spectrum, the LTE base stations (BSs) with carrier aggregation (CA) capabilities can potentially exploit the unlicensed spectrum to their advantage, improving their data rates [2]. Another possibility of interest is to consider LTE small cells as an alternative to WiFi small cells as illustrated in Fig. 1. One of the major issues in this context is the coexistence with other systems operating in the unlicensed band such as WiFi [2–9]. LTE, being developed originally for the licensed spectrum, lacks the required coexistence mechanisms for the unlicensed band. Due to these reasons, researchers from academia as well as industry have been studying this problem recently. Introducing a duty cycle in the LTE transmission is often considered as a suitable modification for enabling coexistence with the WiFi systems. In the throughput analysis of an LTE system for such situations, the WiFi interference when present is often treated as white noise [4–6]. However, it has been pointed

© ICST Institute for Computer Sciences, Social Informatics and Telecommunications Engineering 2016
D. Noguet et al. (Eds.): CROWNCOM 2016, LNICST 172, pp. 605–617, 2016.
DOI: 10.1007/978-3-319-40352-6_50

out that the WiFi and the LTE systems have different physical layer characteristics and there is a need to study the impact of these differences [7] which we
attempt to address in this work.

Most of the common WiFi standards such as 802.11a, 802.11n, 802.11ac
etc. use an orthogonal frequency division multiplexing (OFDM) transmission
with a subcarrier spacing of 312.5 kHz whereas the subcarrier spacing in LTE
is 15 kHz. In the first part of this work, we analyze the properties of the co-
channel interference generated by an OFDM WiFi transmitter at an LTE device.
Starting from the continuous-time formulation of a WiFi signal, we derive the
expression for the resulting co-channel interference due to the WiFi signal at
the LTE subcarriers under a multipath propagation channel. Furthermore, the
cross-correlation properties of the interference at the different subcarriers are
derived and the accuracy of the analysis has been confirmed through simulations.
Since the LTE subcarriers are significantly narrower than the WiFi subcarriers,
the correlators corresponding to the subcarriers in an LTE receiver essentially
perform a spectral analysis fine enough to measure the spectrum between the
WiFi subcarriers. Each subcarrier in WiFi has a Sinc shaped spectrum which
causes the power spectral density (PSD) of a WiFi signal to fall and rise from
one WiFi subcarrier to the other. Because an LTE receiver is able to observe
these variations, the WiFi interference power decreases and increases across the
LTE subcarriers with a period corresponding to the WiFi subcarrier spacing. It
is also shown that the magnitudes of the off-diagonal elements in the covariance
matrix of the WiFi interference contributions at the different subcarriers are
significantly smaller than that of the diagonal elements in the covariance matrix.
We also provide a brief discussion of some of the implications for an LTE system.

Another requirement for LTE-U is the inclusion of sensing before transmission (SBT), also known as listen before talk (LBT) [2,8,9], to check the availability of a channel. The regulations governing the operation in the unlicensed
spectrum in certain countries make this feature mandatory. The sensing functionality may be introduced at both the LTE transmitters and the receivers.
Since multiple devices may be present within an LTE small cell, enabling the
sensing at the devices can help to overcome the hidden node problem as illustrated in Fig. 1, where the LTE transmitter's sensing is rendered ineffective by

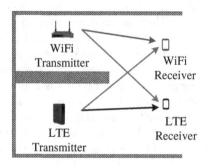

Fig. 1. WiFi-LTE interference scenario.

the blocking wall. Numerous algorithms for SBT have been developed in the context of cognitive radio (CR) systems such as energy detection (ED), cyclostationarity detection (CD) etc. [10]. Among them, ED has the least complexity. However, its operating point on the receiver operating characteristic (ROC) is highly sensitive to noise variance estimation errors. Algorithms based on CD are more complex but they do not require the knowledge of the noise variance making them constant false alarm detectors (CFADs). Though these algorithms may be useful, our interest is to find CFADs that can be easily incorporated into LTE devices. Therefore, in the second part of this work, we propose a sensing technique for WiFi signals, namely autocorrelation feature detection (AFD), that makes use of the periodic variation of the WiFi signal power across the LTE subcarriers and does not require the knowledge of the noise variance. This scheme can be easily implemented in LTE devices because it utilizes the signals received at the LTE subcarriers. The signal powers across the subcarriers can be interpreted as a PSD. As the PSD and the autocorrelation function (ACF) are Fourier transform pairs, the periodic fluctuations in the PSD indicate strong peaks in the ACF. The period of these fluctuations is equal to the WiFi subcarrier spacing (312.5 kHz) and the corresponding peaks in the ACF occur at $\pm\frac{1}{312.5\,\text{kHz}} = \pm 3.2\,\mu\text{s}$. Since this location is determined by the subcarrier spacing of the WiFi system, it is unlikely that signals other than a WiFi signal generate significant contributions at the feature position. In addition, we also obtain an approximate analytical expression for the false alarm probability of the AFD. It turns out that the analytical result agrees well with the false alarm probability achieved in the simulations. Our results show that the detector achieves a good detection performance even at low signal-to-noise ratios (SNRs) and that AFD can successfully distinguish a WiFi signal from an LTE signal. Furthermore, provided that the signal is contained in the sensing bandwidth, the AFD's performance is not affected by frequency offsets. Even for a low sensing bandwidth, the algorithm yields a decent performance which might turn out to be useful for LTE-M devices with low bandwidths designed for machine-type communications [13]. Excluding the OFDM discrete Fourier transform, which already is implemented in LTE devices, the complexity of this method is comparable to that of a time-domain ED with the same sensing duration.

The paper is organized as follows. Section 2 describes the system model and provides an analysis of the WiFi signal at the LTE subcarriers. In Sect. 3, the problem of SBT is introduced, the AFD is developed and its performance is evaluated via simulations. Section 4 concludes the paper.

2   System Model

The continuous-time signal transmitted by a single antenna WiFi transmitter can be written as

$$s(t) = \sqrt{P} \sum_{k \in \mathcal{F}} \sum_{l=-\infty}^{\infty} a_{k,l} \text{rect}\left(\frac{t + T_{cp} - lT}{T}\right) e^{j2\pi f_k(t-lT)}, \tag{1}$$

where $a_{k,l}$ is the zero mean transmit symbol with unit variance at subcarrier k of frequency f_k, which is an integer multiple of the subcarrier width Δf, and OFDM symbol l. $\mathcal{F}$ is the set of active WiFi subcarrier indices, $\mathrm{rect}(\frac{t}{T})$ represents a rectangular pulse with an amplitude of 1 situated at the interval $[0, T]$, $T_u = \Delta f^{-1}$ denotes the duration of the useful part of the OFDM symbol, T_{cp} stands for the duration of the cyclic prefix (CP) used in WiFi, and $T = T_u + T_{cp}$ is the duration of an OFDM symbol in WiFi. P refers to the transmit power per subcarrier. For typical WiFi systems, the values of these parameters are $T_u = 3.2\,\mu s$, $\Delta f = 312.5\,kHz$, $T_{cp} = 0.8\,\mu s$ and $T = 4.0\,\mu s$. For an M_τ-tap discrete multipath channel with the weight function $h(t, \tau) = \sum_{m=0}^{M_\tau - 1} h_m(t)\delta(\tau - \tau_m)$ ($h_m(t)$: channel coefficient of the mth path at time t, τ_m: delay of the mth path, M_τ: number of paths), the received signal contribution from the WiFi signal can be expressed as

$$r(t) = \int_{-\infty}^{\infty} s(t - \tau)h(t, \tau)d\tau \tag{2}$$

$$= \sqrt{P} \sum_{k \in \mathcal{F}} \sum_{m=0}^{M_\tau - 1} \phi_{mk} \sum_{l=-\infty}^{\infty} h_m(lT)a_{k,l}\theta_{kl}\mathrm{rect}\left(\frac{t - \tau_m + T_{cp} - lT}{T}\right) e^{j2\pi f_k t}, \tag{3}$$

where $\phi_{mk} = \exp(-j2\pi f_k \tau_m)$ and $\theta_{kl} = \exp(-j2\pi f_k lT)$. Here, we assume that $h_m(t)$ remains constant within symbol intervals $[lT, (l+1)T]$, $\forall l$. The contribution from $r(t)$ at a given subcarrier of frequency ν_i of an LTE receiver for the pth OFDM symbol, g_{ip}, can be obtained as follows

$$g_{ip} = \frac{1}{T_{u1}} \int_{pT_1}^{pT_1 + T_{u1}} r(t)e^{-j2\pi \nu_i t} dt \tag{4}$$

$$= \frac{\sqrt{P}}{T_{u1}} \sum_{k \in \mathcal{F}} \sum_{m=0}^{M_\tau - 1} \phi_{mk} \sum_{l=-\infty}^{\infty} h_m(lT)a_{k,l}\theta_{kl}$$

$$\times \int_{pT_1}^{pT_1 + T_{u1}} \mathrm{rect}\left(\frac{t - \tau_m + T_{cp} - lT}{T}\right) e^{j2\pi(f_k - \nu_i)t} dt \tag{5}$$

$$= \frac{\sqrt{P}}{T_{u1}} \sum_{m=0}^{M_\tau - 1} \sum_{k \in \mathcal{F}} \sum_{l \in \mathcal{L}(p, \tau_m)} \phi_{mk} h_m(pT_1)a_{k,l}\theta_{kl} \underbrace{\int_{t_e(l, \tau_m, p)}^{t_n(l, \tau_m, p)} e^{j2\pi(f_k - \nu_i)t} dt}_{\beta_{ki}(l, \tau_m, p)}, \tag{6}$$

where $T_{u1} \approx 66.7\,\mu s$ is the duration of the useful part of an LTE OFDM symbol which is equal to the reciprocal of the subcarrier width in LTE $\Delta f_1 = 15\,kHz$, $T_1 \approx 71.4\,\mu s$ represents the duration of an LTE OFDM symbol including its CP, and $\mathcal{L}(p, \tau_m)$ is the set of symbols which fall within the integration interval. $t_e(l, \tau_m, p)$ and $t_n(l, \tau_m, p)$ are the earliest and the latest time positions, respectively, of the lth WiFi symbol received over the mth path if this contribution lies in the integration interval of the pth LTE symbol. Here, we have also assumed that the channel remains constant within an LTE symbol interval. $\beta_{ki}(l, \tau_m, p)$ can be simplified as follows

$$
\beta_{ki}(l, \tau_m, p) = \begin{cases} t_n(l, \tau_m, p) - t_e(l, \tau_m, p), & \text{if } f_k = \nu_i, \\[2mm] \dfrac{e^{j2\pi(f_k - \nu_i)t_n(l)} - e^{j2\pi(f_k - \nu_i)t_e(l)}}{j2\pi(f_k - \nu_i)}, & \text{otherwise.} \end{cases}
\tag{7}
$$

Now, assuming uncorrelated transmit symbols and uncorrelated path gains, the correlation between the WiFi interference at subcarriers i and $i - o$ for the pth LTE symbol can be calculated as

$$
r_{f,p}(i, o) = \mathbb{E}[g_{ip} g^*_{(i-o)p}] = \frac{P}{T^2_{u1}} \sum_{m=0}^{M_\tau - 1} P_m \sum_{k \in \mathcal{F}} \sum_{l \in \mathcal{L}(p, \tau_m)} \beta_{ki}(l, \tau_m, p) \beta^*_{k(i-o)}(l, \tau_m, p),
\tag{8}
$$

where $P_m = \mathbb{E}[|h_m(pT_1)|^2]$, assuming weak-sense stationary channel coefficients. For typical channels with maximum excess delay less than the cyclic prefix length used in LTE systems, the time correlation $r_{t,i}(p, o) = \mathbb{E}[g_{ip} g^*_{i(p-o)}]$ is zero for $|o| > 1$ as the contributing symbols are uncorrelated. As g_{ip} and $g_{i(p\pm 1)}$ may share the same OFDM symbol at the integration boundaries, $r_{t,i}(p, o)$ may be non-zero for $|o| = 1$. However, this effect can be considered to be weak as the number of unshared symbols is about 15. It is also of interest to compute the correlation between interference at different subcarriers for a given channel state $\mathbf{h} = [h_0(pT_1), \cdots, h_{M_\tau - 1}(pT_1), \tau_0, \cdots, \tau_{M_\tau - 1}]^T$, denoted as $r_{f,p}(i, o, \mathbf{h})$. This parameter is of relevance to channel coding/decoding, link adaptation etc. which are performed over short intervals over which the channel is essentially static, and can be expressed as follows

$$
r_{f,p}(i, o, \mathbf{h}) = \mathbb{E}[g_{ip} g^*_{(i-o)p} | \mathbf{h}],
$$

$$
= \frac{P}{T^2_{u1}} \sum_{k \in \mathcal{F}} \sum_{m=0}^{M_\tau - 1} \sum_{n=0}^{M_\tau - 1} h_m(pT_1) h^*_n(pT_1) \phi_{mk} \phi^*_{nk}
\tag{9}
$$

$$
\times \sum_{l \in \mathcal{L}(p, \tau_m) \cap \mathcal{L}(p, \tau_n)} \beta_{ki}(l, \tau_m, p) \beta^*_{k(i-o)}(l, \tau_n, p).
\tag{10}
$$

Simulations have been performed in order to confirm the anaytical results. In our simulations, random 16-QAM symbols have been selected as transmit symbols. The WiFi subcarrier frequencies used are $\Delta f \times [-26, \cdots, -1, 1, \cdots, 26]$ and an oversampling of 8 times the sampling rate of a 20 MHz LTE receiver is used to avoid aliasing effects. The analytical and simulation results for the correlations are presented in Figs. 2 and 3 which shows $r_{f,p}(i, 0)$ and $r_{f,p}(i, 0, \mathbf{h})$. The simulation results agree well with the analytical expressions. In both figures, it can be seen that the power fluctuates with a period of $\frac{\Delta f}{\Delta f_1} \approx 20$ subcarriers. This is because the magnitude of $\beta_{ki}(l, \tau_m, p)$ is inversely proportional to the

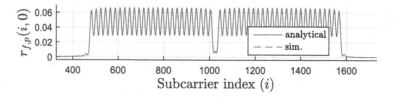

Fig. 2. Variance $r_{f,p}(i,0)$ of g_{i0} for a single tap channel with $P_0 = 1$.

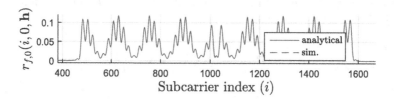

Fig. 3. Variance $r_{f,p}(i,0,\mathbf{h})$ of g_{i0} for a two tap channel with $h_0(t) = 1$, $h_1(t) = 0.5$, $\tau_0 = 0.0\,\mu\text{s}$ and $\tau_1 = 0.5\,\mu\text{s}$.

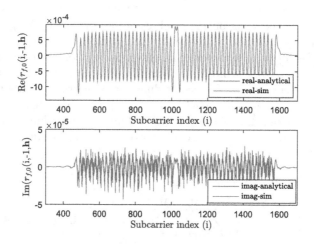

Fig. 4. Cross-correlation $r_{f,0}(i,-1,\mathbf{h})$ for a single path channel with $h_0(t) = 1$, $\tau_0 = 0.0\,\mu\text{s}$.

frequency separation $(f_k - \nu_i)$ and hence, the closer the LTE subcarrier is located to an active WiFi subcarrier the larger the power collected from it. Results for the cross-correlation $r_{f,0}(i,-1,\mathbf{h})$ at an offset of -1 are presented in Fig. 4. The analytical results again agree well with the simulations. It is interesting to note that the magnitude of the cross-correlation is significantly smaller than that of the variance in Fig. 3. These results indicate that the interference may be considered as uncorrelated across subcarriers. However, the significant difference in interference powers at different subcarriers must be taken into account in performance evaluations when WiFi interference is assumed to be present. The knowledge of these power variations may be exploited to improve the channel estimation, channel quality computation, resource allocation etc. when a WiFi signal is interfering at an LTE receiver. Furthermore, since the periodic power variation is a characteristic of an LTE system with WiFi interference, such a fluctuation may be interpreted as a particular feature of a WiFi signal and can be used to distinguish between a WiFi signal and a non-WiFi signal in sensing, which is further explored in the next section.

3   WiFi Signal Sensing

The sensing before transmission in case of a WiFi signal potentially beginning at $t = 0$ by an LTE device is equivalent to a binary hypothesis testing problem given by

$$H_0 : y_{ip} = n_{ip}; \ i = 1, 2, \cdots, N; \ p = 0, 1, \cdots, M - 1, \tag{11}$$
$$H_1 : y_{ip} = g_{ip} + n_{ip}; \ i = 1, 2, \cdots, N; \ p = 0, 1, \cdots, M - 1,$$

where H_1 and H_0 denote the hypotheses where a WiFi transmitter is active and inactive, respectively. y_{ip} is the received sample at the ith subcarrier of the pth OFDM symbol, n_{ip} denotes the circularly symmetric additive white Gaussian noise (AWGN) at the receiver ($n_{ip} \sim \mathcal{CN}(0, \sigma_n^2)$, σ_n^2 : noise variance) and g_{ip} represents the WiFi signal contribution defined according to (4). The noise is assumed to be independent and identically distributed (i.i.d.) over the subcarriers as well as the OFDM symbols with a variance given by $\sigma_n^2 = \frac{N_{0,PSD}}{T_{u1}}$ where $N_{0,PSD}$ is the noise power spectral density. N and M stand for the maximum number of subcarriers (LTE DFT size) and the number of LTE OFDM symbols used in the sensing process, respectively. We restrict our attention to sensing schemes that can be implemented in the LTE subcarrier domain and also without the knowledge of the noise variance, thereby making them suitable for existing LTE receivers.

3.1   Autocorrelation Feature Detection (AFD)

In this approach, we make use of the periodic fluctuations of the variance of g_{ip} ($i = N_0, \cdots, N_1$) resulting from the difference in subcarrier widths of LTE and WiFi systems as illustrated in Figs. 2 and 3. N_0 and N_1 are adjusted to select the bandwidth used in sensing. The number of subcarriers used for sensing is $N_{tot} = N_1 - N_0 + 1$ ($1 \leq N_0 < N_1 \leq N$). According to the LTE standard, a 20 MHz receiver uses $N = 2048$ and its sampling frequency is $N \Delta f_1 = 30.72$ MHz. For this receiver, setting $N_0 = 1$ and $N_1 = N$ selects the whole 30.72 MHz bandwidth for the sensing process. The time average of the powers at the different subcarriers can be interpreted as an estimate of the power spectral density of a WiFi signal after passing through the correlators corresponding to the LTE subcarriers. Periodic fluctuations in the power spectral density indicate a strong peak in its DFT coefficient sequence, which can

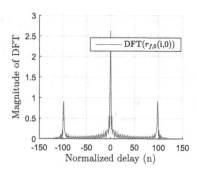

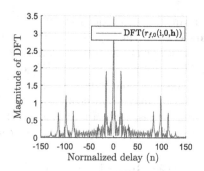

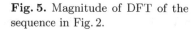

Fig. 5. Magnitude of DFT of the sequence in Fig. 2.

Fig. 6. Magnitude of DFT of the sequence in Fig. 3.

be interpreted as the complex conjugate of an autocorrelation function. If the averaging is performed over a duration smaller than the coherence time of the channel, additional peaks may be present whose locations depend on the path delays. The channel independent peaks in the N_{tot}-point DFT of the vector $\mathbf{r}_{f,p}(0) = [r_{f,p}(N_0, 0), \cdots, r_{f,p}(N_1, 0)]^T$ are fixed by the subcarrier widths and located at the integer closest to $\frac{\pm N_{tot} \Delta f_1}{\Delta f}$. They can be considered as a particular feature corresponding to a WiFi signal.

For $N_{tot} = 2048$ ($N_0 = 1$, $N_1 = 2048$), the peaks occur at the normalized delays ± 98 (i.e., $\pm 3.2 \, \mu s$). This is illustrated in Figs. 5 and 6 where the magnitudes of the autocorrelation function computed from the 2048-point DFT of the sequences $r_{f,p}(i, 0)$ (long-term averaging) and $r_{f,p}(i, 0, \mathbf{h})$ (short-term averaging) are shown. The additional peaks in Fig. 6 are due to the static two path channel. The results imply that when a channel delay exists at $3.2 \, \mu s$, and the averaging length is short, this peak is spurious for the detection of WiFi signals. However, since the maximum delays encountered in indoor small cells environments are typically much smaller than $3.2 \, \mu s$ [11], we can safely assume that the peaks at $\pm 3.2 \, \mu s$ are from a WiFi signal. To develop a detector exploiting this feature, we define $r_i = \frac{1}{M} \sum_{p=1}^{M} |y_{ip}|^2$, $\mathbf{r}_y = [r_{N_0}, \cdots, r_{N_1}]^T$ and $\mathbf{v}_y = \mathbf{F} \mathbf{r}_y$, where $\mathbf{F}$ is the unitary N_{tot}-point DFT matrix. $\mathbf{r}_y$ is a vector of time averaged subcarrier powers and therefore, its DFT $\mathbf{v}_y$ is conjugate symmetric. Without loss of generality, we assume N_{tot} is even. Let $v_y(n)$ indicate the element of $\mathbf{v}_y$ corresponding to the normalized delay n. Then, our feature detector is expressed as follows

$$T_2 = \frac{|\mathcal{D}_1| \sum_{n \in \mathcal{D}_0} |v_y(n)|^2}{|\mathcal{D}_0| \sum_{n \in \mathcal{D}_1} |v_y(n)|^2} > t_2, \tag{12}$$

where $\mathcal{D}_0 = \{n_0\}$ (n_0: the positive normalized delay corresponding to the feature position), $\mathcal{D}_1 = \{1, \cdots, \frac{N_{tot}}{2} - 1\} \sim \mathcal{D}_0$ ($\sim$: set difference, $|\mathcal{D}_k|$: cardinality of $\mathcal{D}_k$) contains all the positive delays except n_0 and t_2 is the threshold for a desired false alarm probability which can be computed analytically. In the following, we assume that H_0 holds unless otherwise specified. In this case, r_i is the sum of the squared magnitudes of M i.i.d. complex Gaussian random variables. For large M, using the central limit theorem, approximation $r_i \sim \mathcal{N}(a, b^2 - a^2)$ becomes accurate where $a = \mathbb{E}[r_i] = \sigma_n^2$ and $b^2 = \mathbb{E}[|r_i|^2] = \sigma_n^4(1 + \frac{1}{M})$. Also, for $i \neq j$, $\mathbb{E}[r_i r_j^*] = \mathbb{E}[r_i]\mathbb{E}[r_j^*] = \sigma_n^4$. Therefore, $\mathbf{r}_y \sim \mathcal{N}(\boldsymbol{\mu}_r, \mathbf{C}_r - \boldsymbol{\mu}_r \boldsymbol{\mu}_r^H)$, where $\boldsymbol{\mu}_r = \sigma_n^2 \mathbf{1}$ (**1**:vector with all elements equal to one) and $\mathbf{C}_r$ is a circulant matrix whose first column is given by $\mathbf{c}_{r1} = \sigma_n^4 \mathbf{1} + [\frac{\sigma_n^4}{M}, 0, \cdots, 0]^T$. Now, $\mathbb{E}[\mathbf{v}_y] = \sigma_n^2 \mathbf{F}\mathbf{1} = \sigma_n^2 [\sqrt{N_{tot}}, 0, \cdots, 0]^T$ and using the diagonalization property of circulant matrices, we obtain $\mathbb{E}[\mathbf{v}_y \mathbf{v}_y^H] = \mathbf{F}\mathbf{C}_r\mathbf{F}^H = \text{diag}(\sqrt{N_{tot}}\mathbf{F}\mathbf{c}_{r1}) = \sigma_n^4 \cdot \text{diag}(N_{tot} + \frac{1}{M}, \frac{1}{M}, \cdots, \frac{1}{M})$. This shows that the denominator of T_2 is the sum of squared magnitudes of $(\frac{N_{tot}}{2} - 2)$ i.i.d. complex Gaussian random variables which has a central Chi-squared distribution with $2(\frac{N_{tot}}{2} - 2)$ degrees of freedom [12], and the numerator is Chi-squared distributed with 2 degrees of freedom. Therefore, for large M, T_2 under H_0 follows a central F-distribution with 2 numerator degrees of freedom and $2(\frac{N_{tot}}{2} - 2)$ denominator degrees of freedom [12]. The threshold for a desired false alarm probability can be obtained using the corresponding tail probability function. Figure 7 shows the analytical and simulation results for the false alarm probability for $M = 20$ and $N_{tot} = 2048$. The results indicate that the approximations adopted provide a sufficient accuracy.

An interesting aspect of this method is that it is unaffected by frequency offsets that translate the WiFi spectrum to another part of the slice of subcarriers under inspection. This is because the magnitude of the DFT coefficients is not affected by translations. Furthermore, this method is quite insensitive to non-WiFi signals because it targets a WiFi specific feature. It can also be applied in LTE devices designed for bandwidths smaller than 20 MHz (for e.g., LTE-M [13]).

3.2   Autocorrelation Detection (AD)

If our interest is to detect the presence of any signal (not only WiFi), all the normalized delays where the signal is expected to have a significant contribution and the average autocorrelation contribution from noise is zero can be considered. This is accomplished by setting $\mathcal{D}_0 = \{1, \cdots, L_+\}$, where L_+ is chosen such that all the DFT coefficients where the contributions of the signals to be detected have appreciable strengths are included in $\mathcal{D}_0$. The threshold for a desired false

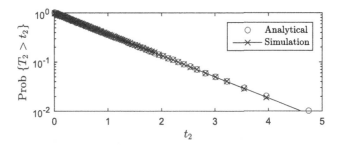

Fig. 7. Tail probability of T_2 for $M = 20$ and $N = 2048$.

alarm probability can be obtained by performing an analysis similar to the one used for the AFD.

3.3 Performance Comparison

Simulations have been conducted to compare the performance of the AD and the AFD. In indoor small cell scenarios, the user mobility is expected to be quite limited. For example, at a carrier frequency of 5.5 GHz and for a Jakes' Doppler power spectrum, the coherence time, defined as the time separation at which the magnitude of the time autocorrelation function of the channel becomes half of its peak value, for a speed of 7 km/hr is given by 6.7 ms which spans about 93 LTE OFDM symbols. Since the value of $M = 20$ used in the simulations is much smaller than 93, the channel is assumed to be quasi-static

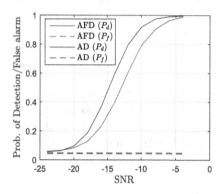

Fig. 8. WiFi signal sensing with $M = 20$ (1.4 ms) and $N_{tot} = 2048$ (Bandwidth $= 30.72$ MHz).

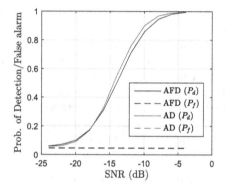

Fig. 9. WiFi signal sensing with $M = 20$ (1.4 ms), $N_{tot} = 1332$ (Bandwidth $= 20$ MHz) and a frequency offset of $150\Delta f_1$.

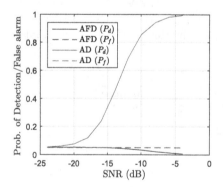

Fig. 10. LTE signal sensing with $M = 20$ (1.4 ms) and $N_{tot} = 1332$ (Bandwidth $= 20$ MHz).

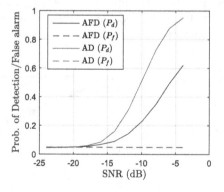

Fig. 11. WiFi signal sensing with $M = 20$ (1.4 ms) and $N_{tot} = 128$ (Bandwidth $= 1.92$ MHz).

over the sensing interval. However, this is not a requirement for the proposed algorithms. A three tap channel is used with the path delays located at $\tau_0 = 0$ µs, $\tau_1 = 1$ µs and $\tau_2 = 2$ µs, and relative powers of 0 dB, -3 dB and -6 dB, respectively. The number of channel realizations has been set to 10000. The noise PSD is calculated as $N_{0,PSD} = F_{rx}K_BT_k$ where $F_{rx} = 8$ dB is the receiver noise figure, $K_B = 1.38 \times 10^{-23}$ Ws/K is the Boltzmann's constant and $T_k = 293$ K is the room temperature. The WiFi signal generated in simulations occupies the central 52 subcarriers other than the DC subcarrier and the LTE signal generated corresponds to a 20 MHz signal where the central 1200 carriers excluding the DC carrier are loaded. Three subcarrier slices, corresponding to a spectral width of 30.72 MHz ($N_0 = 1, N_1 = 2048$), 20 MHz ($N_0 = 359, N_1 = 1690$) and 1.92 MHz ($N_0 = 961, N_1 = 1088$), are chosen to realize different sensing bandwidths. The SNR is always referred to the total signal power of the original 20 MHz WiFi signal and a receiver bandwidth of 30.72 MHz for which the noise power is -91 dBm. The total receive signal power is equally divided between the used subcarriers and is increased from -115 dBm to -95 dBm, changing the SNR from -24 dB to -4 dB. The desired false alarm probability is chosen as 0.05 and the analytical results are used to set the threshold for both the schemes. The simulation results for the probability of detection (P_d) and the probability of false alarm (P_f) are presented in Figs. 8, 9, 10 and 11. AD uses $L_+ = n_0 + 10$. The simulation results show that the achieved false alarm probability is quite close to the target value, confirming the accuracy of the analysis. AD outperforms AFD because the former collects the contribution from all significant taps including the feature. Figure 9 depicts the performance under a frequency offset of 150 LTE subcarrier spacings and demonstrates the immunity of the autocorrelation based methods to frequency offsets, as discussed in Sect. 3.1. For the very low sensing bandwidth case of Fig. 11, both the schemes suffer due to the reduction in number of samples. The impact on the AFD is more significant than on the AD. However, as shown Sect. 3.1, the noise contributions affecting the test statistic can be reduced by increasing M and the performance can be improved even for the low bandwidth cases. Therefore, this method is suitable for LTE devices with low system bandwidths which might turn out to be useful in LTE-M scenarios [13]. Finally, the detection performance of the algorithms for a 20 MHz system when an LTE signal is used instead of a WiFi signal is shown in Fig. 10. As expected, AD is also suitable for the LTE signal. AFD is not triggered because it targets a specific feature that results when a WiFi signal passes through an LTE receiver. Since the maximum channel delay used in the simulation is 2 µs, there is hardly any contribution from the LTE signal at the feature position. Due to the way T_2 is defined, the contribution of the LTE signal in the denominator of T_2 scales down the numerator term more heavily as the LTE signal strength is raised. As a result, when the signal power is increased, the PDF of T_2 under H_1 shifts to the left side compared to the PDF of T_2 under H_0 and the detection probability falls below the false alarm probability. This does not occur for a WiFi signal due to the strength of the feature position in the numerator of T_2. Regarding complexity of the proposed schemes, compared to a time-domain energy detector the main increase in complexity stems from the additional DFT that is needed for each symbol. However, an LTE device already includes a DFT module as a part of its OFDM receiver. Besides, DFTs can be implemented efficiently via the fast Fourier transform.

4   Conclusions

In this work, we have analyzed the correlation properties of the interference introduced by a WiFi transmitter at an LTE receiver. It is shown that the interference powers across the LTE subcarriers from a WiFI system exhibit a periodic behavior. The knowledge of this behavior might turn out to be useful in the design of algorithms for LTE-U scenarios. Furthermore, it is demonstrated that the periodic variation of the interference can be exploited in sensing, even by low bandwidth LTE devices, to detect WiFi signals as well as to distinguish WiFi signals from other signals.

References

1. 3GPP TS 36.211: Evolved Universal Terrestrial Radio Access (E-UTRA): Physical Channels and Modulation. Version 12.1.0 (2014). http://www.3gpp.org/dynareport/36211.htm
2. QUALCOMM: Extending LTE Advanced to Unlicensed Spectrum. White paper (2014). http://www.qualcomm.com/media/documents/white-paper-extending-lte-advanced-unlicensed-spectrum
3. Abinader, F.M., et al.: Enabling the coexistence of LTE and Wi-Fi in unlicensed bands. IEEE Commun. Mag. **52**(11), 54–61 (2014)
4. Sagari, S., Seskar, I., Raychaudhuri, D.: Modeling the coexistence of LTE and WiFi heterogeneous networks in dense deployment scenarios. In: Proceedings of the IEEE International Conference on in Communication (ICC), pp. 2301–2306 (2015)
5. Bhorkar, A., Ibars, C., Zong, P.: On the throughput analysis of LTE and WiFi in unlicensed band. In: Proceedings of the 48th Asilomar Conference on in Signals, Systems and Computers, pp. 1309–1313 (2014)
6. Rupasinghe, N., Güvenc, I.: Licensed-assisted access for WiFi-LTE coexistence in the unlicensed spectrum. In: Proceedings of the IEEE Globecom Workshops, pp. 894–899 (2014)
7. Zhang, R., Wang, M., Cai, L.X., Zheng, Z., Shen, X.: LTE-unlicensed: the future of spectrum aggregation for cellular networks. IEEE Wirel. Commun. **22**, 150–159 (2015)
8. Chen, C., Ratasuk, R., Ghosh, A.: Downlink performance analysis of LTE and WiFi coexistence in unlicensed bands with a simple listen-before-talk scheme. In: Proceedings of the 81st IEEE Vehicular Technology Conference (VTC Spring), pp. 1–5 (2015)
9. Bhorkar, A., Ibars, C., Papathanassiou, A., Zong, P.: Medium access design for LTE in unlicensed band. In: Proceedings of the IEEE Wireless Communications and Networking Conference (WCNC), pp. 369–373 (2015)
10. Yucek, T., Arslan, H.: A survey of spectrum sensing algorithms for cognitive radio applications. IEEE Commun. Surv. Tutorials **11**, 116–130 (2009)

11. Durgin, G., Kukshya, V., Rappaport, T.: Wideband measurements of angle and delay dispersion for outdoor and indoor peer-to-peer radio channels at 1920 MHz. IEEE Trans. Antennas Propag. **51**, 936–944 (2003)
12. Kay, S.M.: Fundamentals of Statistical Signal Processing: Detection Theory. Prentice Hall, New York (1998)
13. Nokia Networks: LTE-M - Optimizing LTE for the Internet of Things. White paper (2015). http://networks.nokia.com/innovation/futureworks-publications

Dynamic Sleep Mode for Minimizing a Femtocell Power Consumption

Rémi Bonnefoi[✉], Christophe Moy, and Jacques Palicot

CentraleSupélec/IETR, CentraleSupélec Campus de Rennes,
35510 Cesson-sévigné, France
{remi.bonnefoi,christophe.moy,jacques.palicot}@centralesupelec.fr

Abstract. The use of power control (PC) and discontinuous transmission (DTx) can reduce the average power consumption of mobile base stations (BS). In this paper, the power consumption of a picocell or a femtocell are analyzed in a time division multiple access (TDMA) scenario. The minimization of the power consumption is viewed as a constrained optimization problem and a closed form of the compromise between transmit power and transmission time which minimizes the average power consumption is found. Moreover, we show that for a base station with a low transmit power and a sufficient power saving during sleep mode, such as picocells and femtocells, the average power consumption is minimum when the service time is minimized. Finally, numerical results show that discontinuous transmission greatly decreases the average power consumption of the base station.

Keywords: Sleep mode · Discontinuous transmission · Green communications

1 Introduction

Decreasing the power consumption of a mobile network allows to reduce both the operating cost for mobile operator and the ecological footprint of the network. It has been shown that deploying more and more pico and femto base stations can reduce the global carbon footprint of mobile communication technologies [1]. Moreover, power control and cell DTx allow to curtail power consumption without reducing users quality of service (QoS).

With cell DTx, the base station can switch between active and sleep mode to reduce the average power consumption. Two kinds of DTx are possible, dynamic and quasi-static DTx [2]. With quasi-static DTx cells are switched off or into deep sleep mode during off peak hours (for some minutes or hours) if there is no users in the cell coverage. Quasi-static DTx is often used with cell shaping to reduce the energy consumption of wireless networks [3]. On the contrary, with dynamic DTx, the base station is switched to sleep mode for a very short time (a few microseconds). During those short sleeping periods, some of the most consuming elements of the radio frequency chain such as the power amplifier are switched off. These two type of DTx were studies during the EARTH project [4].

© ICST Institute for Computer Sciences, Social Informatics and Telecommunications Engineering 2016
D. Noguet et al. (Eds.): CROWNCOM 2016, LNICST 172, pp. 618–629, 2016.
DOI: 10.1007/978-3-319-40352-6_51

An example of use of dynamic DTx can be found in the LTE standard. Indeed, in this standard, a 10 ms frame is composed by ten 1 ms sub-frames and up to six of those sub-frames can be MBSFN (multicast broadcast single frequency network). During a MBSFN sub-frame, the base station is switched to sleep mode and the power consumption reduced [5]. In [6], the reduction of power consumption induced by MBSFN sub-frames is studied. In [7], the power amplifier is switched on and off to save energy while maintaining the cell coverage and quality of service. In those three papers, power control is not used and user are always served using the whole available power. Contrary to papers previously mentioned which only use DTx, in this paper we analyze the joint use of PC and dynamic DTx to save energy.

In this paper, we suppose that a picocell or a femtocell serves several users and knows the channel gain and the throughput demand (capacity demand) of each of them. Time is divided in short time frames (which, for example, can last 1, 5, 10 or 100 ms as we do not restrict to LTE context). In each frame, users are served successively using time division multiple access (TDMA) over the whole available band and when all users have been served, the base station switches to sleep mode for the rest of the frame. During each frame, each user is served during a specific service time (duration of data transmission to this user). This service time must be adjusted so that each user has a certain QoS (capacity demand) while minimizing the average power consumption of the base station. The optimization problem solved in this paper has already been raised in [8] with simulation results but we propose here a closed form solution.

We are in a Cognitive Green Radio (GR) context [9] and the system studies can be modeled as a simplified cognitive cycle in three steps (sensing, decision making and adaptation). During the sensing period, the base station collects the propagation conditions and the capacity constraint of all users. Then, the service time and transmit power are computed during the decision making period and applied during the adaptation period to reduce the power consumption. The simplified cognitive cycle is shown in Fig. 1. In this paper we focus on the decision making algorithm.

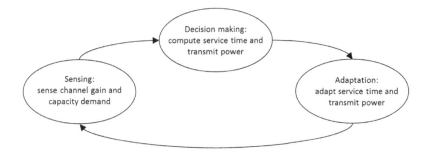

Fig. 1. Simplified cognitive cycle for considered cognitive Green radio scenario

In this paper, the problem of power minimization is first reformulated before proposing a resolution scheme. Then, this problem is solved in a closed form. We show that, in some cases, this closed form solution can be expressed using the W-Lambert function [10] which is the reciprocal bijection of $x \mapsto xe^x$. Finally, we show that in a realistic context, the average power consumption of a picocell or a femtocell is minimized by minimizing the service time.

The rest of this paper is organized as follows. In Sect. 2 the system model is presented and the optimization problem is introduced. The optimization problem is reformulated in Sect. 3 and solved in Sect. 4. Finally, in Sect. 5 numerical simulations illustrate the reduction of power consumption.

2   System Model

The power consumption of a base station can be accurately modeled by a linear model [8]:

$$\begin{cases} P_{supply} = P_0 + mP_{TX} & if \ 0 < P_{TX} \le P_{max} \\ P_{supply} = \qquad P_s & else \end{cases} \tag{1}$$

where P_0 is the static power consumption, P_s the power consumed when the base station is in sleep mode. P_{TX} is the transmit power which is below the maximum transmit power of the base station denoted P_{max}. m is the slope of the load dependence. Figure 2 illustrates the power consumption of the base station during a TDMA frame.

In [11] the authors list some realistic values for P_0, P_s, P_{max}, and m. Those values are mentioned in Table 1.

In this paper, we denote N_u the number of users served by the base station during a frame, T the frame duration and t_i the time used to serve user i.

We denote μ_i the proportion of time used to serve user i:

$$\mu_i = \frac{t_i}{T} \tag{2}$$

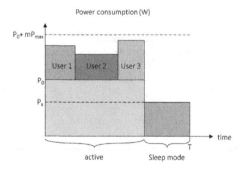

Fig. 2. Variation of the power consumption of a base station during one frame, in this scheme, three users are served by the base station and the base station switches to sleep mode until the end of the frame after having served all

Table 1. Power values for different base stations types from [11]

BS type	P_{max}	P_0	m	P_s
Macro	20.0	130.0	4.7	75.0
Micro	6.3	56.0	2.6	39.0
Pico	0.13	6.8	4.0	4.3
Femto	0.05	4.8	8.0	2.9

In a TDMA scheme, all users have the same bandwidth denoted B. Moreover, for each user, the expression of the channel capacity is:

$$C_i = B\mu_i \log_2(1 + \rho_i) \tag{3}$$

where ρ_i is the Signal to Noise Ratio (SNR) of this user. This last is expressed as:

$$\rho_i = \frac{P_{Tx}^i |h_i|^2}{N} = \frac{P_{Tx}^i |h_i|^2}{kTB} = \frac{P_{Tx}^i G_i}{B} \tag{4}$$

where P_{Tx}^i the transmit power used for user i, h_i is the channel gain, N denotes the noise power, k is the Boltzmann constant, T the temperature in Kelvin and $G_i = \frac{|h_i|^2}{kT}$ is a gain proportional to the channel gain of user i. For a given value of μ_i, there is only one value of the transmit power which allows to exactly meet the capacity constraint C_i and its expression is:

$$P_{Tx}^i = \frac{B}{G_i} \left(2^{\frac{C_i}{B\mu_i}} - 1 \right) \tag{5}$$

Using Eqs. 1 and 5 we can derive the mean power consumption of the base station during one frame:

$$P_{mean} = \left(1 - \sum_{i=1}^{N_u} \mu_i\right) P_s + \sum_{i=1}^{N_u} \mu_i \left(P_0 + m \frac{B}{G_i} \left(2^{\frac{C_i}{B\mu_i}} - 1 \right) \right) \tag{6}$$

The objective is to minimize P_{mean} under constraints:

$$P_{Tx}^i \leq P_{max} \qquad \forall i \in [1; N_u] \tag{7}$$
$$\mu_i \geq 0 \qquad \forall i \in [1; N_u] \tag{8}$$

and

$$\sum_{i=1}^{N_u} \mu_i \leq 1 \tag{9}$$

3   Problem Reformulation

3.1   Existence of Solutions

Before minimizing the average power consumption, we have to analyze the existence of solutions. This problem has a solution if and only if the sum of minimum portions of time required to serve each user is smaller than 1. We denote $\mu_{i\,min}$ the shortest portion of time which ensures the required QoS of user i. The expression of $\mu_{i\,min}$ is:

$$\mu_{i\,min} = \frac{C_i}{B \log_2\left(1 + \frac{P_{max} G_i}{B}\right)} \tag{10}$$

If $\mu_i < \mu_{i\,min}$ the transmit power P^i_{Tx} computed with 5 will be higher than P_{max} and thus the constraint of Eq. 7 will not be met. This value is computed by using the maximum transmit power P_{max} for user i. Thus, the smallest portion of time necessary to serve all users is:

$$\sum_{i=1}^{N_u} \mu_{i\,min} = \sum_{i=1}^{N_u} \frac{C_i}{B \log_2\left(1 + \frac{P_{max} G_i}{B}\right)} \tag{11}$$

And, the ability to serve users can be written as $\sum_{i=1}^{N_u} \mu_{i\,min} \leq 1$, i.e.:

$$\sum_{i=1}^{N_u} \frac{C_i}{B \log_2\left(1 + \frac{P_{max} G_i}{B}\right)} \leq 1 \tag{12}$$

3.2   New Formulation of the Optimization Problem

To better understand the problem, we can rewrite Eq. 6 as:

$$P_{mean} = \underbrace{\left(\sum_{i=1}^{N_u} \mu_i\right) P_0 + \left(1 - \sum_{i=1}^{N_u} \mu_i\right) P_s}_{\text{Average static power consumption } P_m^{stat}} + \underbrace{mB \sum_{i=1}^{N_u} \mu_i \frac{1}{G_i}\left(2^{\frac{C_i}{B\mu_i}} - 1\right)}_{\text{Average dynamic power consumption } P_m^{dyn}} \tag{13}$$

The average static power consumption P_m^{stat} is minimized when $\sum_{i=1}^{N_u} \mu_i$ is minimum (when the values of μ_i are set to $\mu_{i\,min}$). In opposite, the dynamic power consumption P_m^{dyn} is minimized when the power is spread over time. Resolving this optimization problem is equivalent to find, for each user, the best compromise between these extremes.

P_m^{stat} and P_m^{dyn} have an upper and a lower bound.

$$0 \leq P_m^{dyn} \leq mP_{max} \quad and \quad P_s \leq P_m^{stat} \leq P_0 \tag{14}$$

Those bounds can be used to see the relative importance of P_m^{stat} and P_m^{dyn}. We can note that for femtocells and picocells, $mP_{max} \ll P_s$. Consequently, the average static power consumption is much higher than the average dynamic power consumption.

Constraints of Eqs. 7 and 8 can be rewritten in a single constraint on μ_i using $\mu_{i\,min}$. These constraints are equivalent to:

$$\mu_i \geq \mu_{i\,min} \qquad \forall i \in [1; N_u] \tag{15}$$

The last constraint remains unchanged as in Eq. 9.

4 Minimizing Power Consumption

In this section, we denote $\mu_{i\,opt}$ the value of μ_i which minimizes the power consumption.

4.1 Resolution Scheme

We have,

$$\frac{\partial^2 P_{mean}}{\partial \mu_i \partial \mu_j} = 0 \qquad \forall i \neq j \tag{16}$$

And,

$$\frac{\partial^2 P_{mean}}{\partial^2 \mu_i} = \frac{mB}{G_i} 2^{\frac{C_i}{B\mu_i}} \left(\frac{C_i}{B}\right)^2 \frac{\ln(2)^2}{\mu_i^3} \geq 0 \quad \mu_i \in \,]\mu_{i\,min}; 1] \tag{17}$$

Thus, the Hessian matrix of the function of Eq. 13 is diagonal and the ith diagonal element of this matrix is positive. As a consequence, the function studied is convex. To solve this non-linear constrained optimization problem with inequality constraints in a closed form, we use the Karush-Kunh-Tucker (KKT) conditions. According to these conditions, each inequality constraint must be considered either as an inequality or as an equality. In other words, among the 2^{N_u+1} possible Lagrangian, only one leads to the solution. In order to find this Lagrangian and to solve the optimization problem, the following methodology is employed:

- Find the optimum of the problem without constraints. After this resolution, if the result of this calculation satisfies Eqs. 9 and 15, this solution is optimal and all constraints are strict inequalities constraints;
- If, for user i, the value of $\mu_{i\,opt}$ found in the preceding step does not satisfies Eq. 15, we must set $\mu_{i\,opt} = \mu_{i\,min}$;
- If those new values of $\mu_{i\,opt}$, verify the last constraint (Eq. 9) then the problem is solved, else, the average power consumption is optimal without employing DTx and is a problem of power allocation. This power allocation problem has already been formulated in [12] but not solved.

4.2 Users Considered Independently

The values of μ_i which potentially minimize the mean power consumption are:

$$\mu_{i\,opt} = \frac{C_i \ln(2)}{B} \frac{1}{\mathcal{W}\left(e^{-1}\left[\frac{G_i}{B} \frac{P_0 - P_s}{m} - 1\right]\right) + 1} \tag{18}$$

In this equation, the function $\mathcal{W}(.)$ is the W-Lambert function. This expression of $\mu_{i\,opt}$ is always positive and $\mu_{i\,opt}$ decreases when G_i increases.

If $\mu_{i\,opt} \leq \mu_{i\,min}$, $\mu_{i\,opt}$ must be set to $\mu_{i\,min}$. For each user, the value of $P_0 - P_s$ from which $\mu_{i\,opt} \leq \mu_{i\,min}$ is:

$$P_0 - P_s = \left[\frac{mB}{G_i} + mP_{max}\right] \ln\left(1 + \frac{P_{max}G_i}{B}\right) - mP_{max} \qquad (19)$$

Using the inequality:

$$\ln(1+x) \geq \frac{x}{1+x} \qquad \forall x \in \mathbb{R}^+ \qquad (20)$$

We prove that this value is positive. As a consequence, whatever the value of the G_i, there is a value of $P_0 - P_s$ from which it is preferable to serve the user with P_{max} during $\mu_{i\,min}$. Moreover, the derivative of this expression with respect to G_i is:

$$\frac{\partial(P_0 - P_s)}{\partial G_i} = \frac{m}{G_i}\left[P_{max} - \frac{B}{G_i}\ln\left(1 + \frac{P_{max}G_i}{B}\right)\right] \qquad (21)$$

Using 20, we can see that this derivative is positive. Consequently, the value of $P_0 - P_s$ from which $\mu_{i\,opt}$ is equal to $\mu_{i\,min}$ increases with G_i. Thus, the lower is G_i, the more we should serve the user during $\mu_{i\,min}$.

Moreover, there is a value of $P_0 - P_s$ from which the mean power consumption is minimum when all users are served during $\mu_{i\,min}$. To compute this value we assume that the SNR is bounded and we denote ρ_{max} its maximum. As a consequence, G_i can be upperbounded by :

$$G_i \leq \frac{\rho_{max}B}{P_{max}} \qquad (22)$$

We denote $(P_0 - P_s)_l$ the value of $P_0 - P_s$ from which all users must be served using $\mu_{i\,min}$. This quantity can be computed using Eqs. 19 and 22:

$$(P_0 - P_s)_l = mP_{max}\left(\left(1 + \frac{1}{\rho_{max}}\right)\ln(1 + \rho_{max}) - 1\right)$$
$$\approx mP_{max}(\ln(\rho_{max}) - 1) \qquad (23)$$

In a cellular network, the SNR is generally below $\rho_{max} = 1000$ $(30dB)$. Table 2 lists the value of $(P_0 - P_s)_l$ for each base station type.

For pico and femto base stations, $(P_0 - P_s)_l$ is slightly higher than current values. Consequently, almost all users will be served during $\mu_{i\,min}$ and only a few users will be served during a longer time. Thus, the average power consumption is very close to its minimum when all users are served during $\mu_{i\,min}$. In other word, a very simple and efficient way to serve all users optimally is to serve them during $\mu_{i\,min}$ with all the available transmit power.

On the contrary, for macro and micro base stations, $(P_0 - P_s)_l$ is greater than P_0 and much greater than the current value of $P_0 - P_s$.

Table 2. Comparison between current and limit value of $P_0 - P_s$

BS type	$(P_0 - P_s)_l$	Realistic value of $P_0 - P_s$
Macro	556.1	55
Micro	96.9	17
Pico	3.1	2.5
Femto	2.4	1.9

4.3   Constraint on the Sum

A consequence of the results of the previous section is that constraint of Eq. 9 will almost always be met for picocells and femtocells. Moreover, if all users are served during $\mu_{i\,min}$ as previously envisaged the constraint on the sum is always met.

For macro and micro base stations, $(P_0 - P_s)_l$ is greater than P_0 and much greater than the current value of $P_0 - P_s$. In this case, the constraint of Eq. 9 will not always be met, and more attention should be paid to the management of this case. This case will be studied in a future paper.

5   Numerical Results

The model proposed in this paper is new and thus it is difficult to compare it with the litterature. We use simulations to compare three different strategies:

- A simple strategy with MBSFN suframes: for each user, the number of subframes is chosen so that the channel capacity exceeds its capacity demand. Only one user can be served in each subframe and up to six subframes can be switched to sleep mode when they are not used to serve one user.
- The use of DTx only, all users are served during $\mu_{i\,min}$ with P_{max}.
- The optimal strategy, all users are served during $\mu_{i\,opt}$ with the corresponding transmit power.

The base station is a femtocell and its power characteristics are those of Table 1. The number of users served by the base station is defined by a Poisson process with parameter $\lambda = 2$. The distance between users and the base station is uniformly distributed between 10 and 200 meters. TDMA is used to manage multiple users access. Central frequency is $f = 2GHz$ and a 10 MHz band is used for all users. The pathloss is computed using the ITU pathloss model for indoor propagation [13]. With this model, the expression of the pathloss in a small area is:

$$L = 20\log_{10}(f) + 33\log_{10}(d) - 28 + P_f(n_{floor}) \tag{24}$$

where f is the central frequency in MHz, d the distance between the user and the base station and $P_f(n_{floor})$ the floor penetration loss factor:

$$P_f(n_{floor}) = 15 + 4(n_{floor} - 1) \qquad n_{floor} > 0 \tag{25}$$

where n_{floor} is the number of floors between the base station and the user, this number is equidistributed between 0 and 5. A log-normal shadowing with a standard deviation of 10dB is added. Moreover, the antenna gain is zero and the noise figure is 9 dB for all users. We suppose that the SNR is below $\rho_{max} = 30dB$ and that all users have the same capacity constraint.

To compute the minimum power consumption, the following algorithm is used:

1: Compute $\mu_{i\,min}$
2: Compute $\mu_{i\,opt}$, the optimum without constraint using Eq. 18
3: **if** $\mu_{i\,opt} \leq \mu_{i\,min}$ **then**
4:     $\mu_{i\,opt} = \mu_{i\,min}$
5: **end if**
6: **if** $\sum_{i=1}^{N_u} > 1$ **then**
7:     use an iterative algorithm to compute the optimum (for example the interior point method)
8: **end if**

Figure 3 shows the evolution of the average power consumption as a function of the capacity constraint.

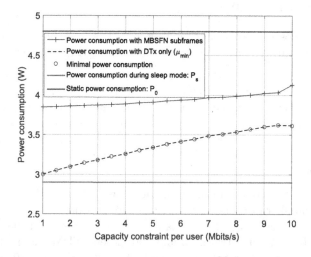

Fig. 3. Comparison between MBSFN subframes, DTx only and the joint use of DTx and power control

If DTx and PC were not used, the average power consumption would be $P_0 + mP_{max} = 5.2W$. Thus, with MBSFN subframes, the power reduction is between 20 and 26 % whereas, with the proposed DTx, this reduction is between 31 and 42 %.

Moreover, we can see that on Fig. 3 that the circles are on the dashed line. Thus, for picocells and femtocells, it is optimal to use DTx only. Two reasons explain this result:

– For femtocells, the average dynamic power consumption is much lower than the average static power consumption;
– The value of $P_0 - P_s$ is close to the limit value $(P_0 - P_s)_l$ from which all users must be served during $\mu_{i\,min}$.

This two reasons can be studied independently of one another. We will first analyze the effect of the reduction of $P_0 - P_s$ before showing the effect of an increase in P_{max} (increase of the dynamic power consumption).

On Fig. 3, the evolution of the mean power consumption is nearly linear. To understand this, we can write the expression of the expectation numerically computed. If all users are served during $\mu_{i\,min}$ the expression of this expectation is:

$$\mathbb{E}\left(P_{mean}\right) = P_s + C\frac{\left[(P_0 - P_s) + mP_{max}\right]}{B} \times \mathbb{E}\left(\sum_{i=1}^{N_u} \frac{1}{\log_2\left(1 + \frac{P_{max}G_i}{B}\right)}\right) \quad (26)$$

where C is the capacity constraint. As long as this capacity constraint is not too high, its increase doesn't change the expectation of the total service time.Thus, the evolution is linear. For higher capacity constraint, an increase in capacity changes the number of users served and thus the shape of the curve.

Figure 4 shows the evolution of the mean power consumption versus P_s.

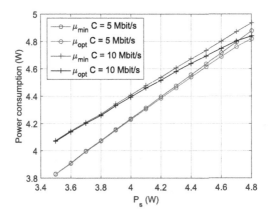

Fig. 4. Evolution of P_{mean} versus P_s, an increase in P_s slightly influes on the gap between the power consumption with DTx only and the minimal power consumption

For this simulation, all propagation conditions and the characteristics of the base station but P_s remains unchanged. The increase of P_s implies a decrease of $P_0 - P_s$ from 1.2 W to 0W.

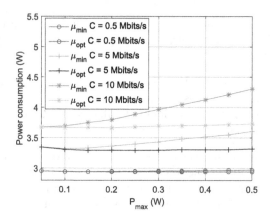

Fig. 5. Evolution of P_{mean} versus P_{max}, an increase in P_{max} greatly increases the gap between the power consumption with DTx only and the minimal power consumption

We can see on Fig. 4 that when $P_0 - P_s$ decreases, the mean power obtained by serving all users during $\mu_{i\,min}$ becomes slightly higher than the theoretical limit.

Figure 5 shows the evolution of the mean power consumption versus P_{max} for different values of C.

On Fig. 5 the mean power consumption obtained using $\mu_{i\,min}$ is compared with the theoretical minimal power consumption. When P_{max} increases, the gap between the two curves increases. For example, for a capacity demand of 10 Mbit/s, the use of $\mu_{i\,min}$ instead of $\mu_{i\,opt}$ increases the power consumption by 5 % when $P_{max} = 0.25W$ and by 15 % when $P_{max} = 0.5W$. Thus, if P_{max} is higher than the values indicate in Table 1, it is no longer optimal to use $\mu_{i\,min}$ for all users.

6   Conclusion

In this paper we studied the compromise between power and time which minimize the average power consumption of a base station in TDMA. We saw that for current picocells and femtocells, the power is minimized when the service time is minimized.

This conclusion is no longer true when the power consumed during sleep mode (P_s) and the maximum transmit power (P_{max}) increase. In particular, this conclusion is not true for macrocells and microcells. For these base station types, it is more difficult to find the compromise sought. This case will be studied in a future work.

Acknowledgments. Part of this work is supported by the project SOGREEN (Smart pOwer Grid for Energy Efficient small cell Networks), which is funded by the French national research agency, under the grant agreement coded: N ANR-14-CE28-0025-02 and by Région Bretagne, France.

References

1. Fehske, A., Fettweis, G., Malmodin, J., Biczok, G.: The global footprint of mobile communications: the ecological and economic perspective. IEEE Commun. Mag. **49**, 55–62 (2011)
2. Saker, L., Elayoubi, S.-E., Chahed, T.: Minimizing energy consumption via sleep mode in green base station. In: 2010 IEEE Wireless Communications and Networking Conference (WCNC), pp. 1–6, April 2010
3. Naoues, M., Noureddine, H., Bodinier, Q., Zhang, H., Palicot, J.: Wifi-based platform for energy saving in wireless networks. In: 2014 IEEE Online Conference on Green Communications (GreenCom), November 2014
4. EARTH project deliverable 4.2, Green Radio Technologies, January 2012
5. Migliorini, D., Stea, G., Caretti, M., Sabella, D.: Power-aware allocation of mbsfn subframes using discontinuous cell transmission in LTE systems. In: IEEE 78th Vehicular Technology Conference (VTC Fall), pp. 1–5, September 2013
6. Frenger, P., Moberg, P., Malmodin, J., Jading, Y., Godor, I.: Reducing energy consumption in LTE with cell DTX. In: 2011 IEEE 73rd Vehicular Technology Conference (VTC Spring), pp. 1–5, May 2011
7. Chatzipapas, A., Alouf, S., Mancuso, V.: On the minimization of power consumption in base stations using on/off power amplifiers. In: Online Conference on Green Communications (GreenCom), IEEE, pp. 18–23, September 2011
8. Holtkamp, H., Auer, G., Haas, H.: On minimizing base station power consumption. In: IEEE Vehicular Technology Conference (VTC Fall), pp. 1–5, September 2011
9. Palicot, J., Zhang, H., Moy, C.: On the road towards green radio. Radio Sci. Bull. **347**, 40–56 (2013)
10. Corless, R., Gonnet, G., Hare, D., Jeffrey, D., Knuth, D.: On the Lambert W function. Adv. Comput. Math. **5**(1), 329–359 (1996)
11. Auer, G., Giannini, V., Desset, C., Godor, I., Skillermark, P., Olsson, M., Imran, M., Sabella, D., Gonzalez, M., Blume, O., Fehske, A.: How much energy is needed to run a wireless network? IEEE Wirel. Commun. **18**, 40–49 (2011)
12. Holtkamp, H., Auer, G., Haas, H.: Minimal average consumption downlink base station power control strategy. In: IEEE 22nd International Symposium on Personal Indoor and Mobile Radio Communications (PIMRC), pp. 2430–2434, September 2011
13. ITU Recommandation, Propagation data and prediction methods for the planning of indoor radiocommunication systems and radio local area networks in the frequency range 300 MHz to 100 GHz, July 2015

Energy Detection Performance with Massive Arrays for Personal Radars Applications

Francesco Guidi[1,2]([⊠]), Anna Guerra[3], Antonio Clemente[1,2], Davide Dardari[3], and Raffaele D'Errico[1,2]

[1] CEA, LETI, MINATEC Campus, 38054 Grenoble, France
{francesco.guidi,antonio.clemente,raffaele.derrico}@cea.fr
[2] Univ. Grenoble-Alpes, 38000 Grenoble, France
[3] DEI, University of Bologna, Via Venezia 52, 47521 Cesena, Italy
{anna.guerra3,davide.dardari}@unibo.it

Abstract. The idea to adopt massive arrays for personal radars applications is facing a rapid growth, thanks to the high scanning resolution achievable with the large number of antennas employed. In fact, such multi-antenna systems enable the possibility to detect and localize surrounding objects through an accurate beamforming procedure. In this paper we show a classical energy-detection approach for target ranging and localization, where the threshold is designed according to the receiver noise only, since an ideal laser-beam antenna is considered. Successively, we show the ambiguities that could arise when the presence of side-lobes cannot be neglected (e.g., when considering real massive arrays instead of ideal pencil-beam like radiation patterns) and we propose a set of guidelines that can be followed from a system design point-of-view to overcome this issue.

Keywords: Massive arrays · Personal radar · Target detection · Side-lobes

1 Introduction

The adoption of massive arrays is facing a rapid growth in several ranging and localization applications, such as personal radars [1], thanks to the possibility to achieve a precise and high-scanning resolution given by the large number of adopted antennas [2].

The concept of personal radar has been recently proposed in [1,3] where it has been shown the possibility to jointly use millimeter-waves (mmW) and wideband massive arrays technologies for indoor environment mapping. Thanks to this technology, it is possible to avoid the adoption of a dedicated very high-directional antenna with mechanical steering, as proposed in [4–6], which can not be easily integrated into portable radar devices. The consequent near-pencil beam of massive arrays returns a precise angle and range information thus making the modeling and characterization of the environment with personal radars very similar to that based on laser.

© ICST Institute for Computer Sciences, Social Informatics and Telecommunications Engineering 2016
D. Noguet et al. (Eds.): CROWNCOM 2016, LNICST 172, pp. 630–641, 2016.
DOI: 10.1007/978-3-319-40352-6_52

Different literature has been produced for the analysis of the localization performance of wideband large antenna arrays. In fact, wideband signals are the best candidate to achieve high ranging performance [7], but a strict phase control in beamforming, i.e. the adoption of precise and costly phase shifters and delay lines, becomes necessary to assure a perfect signal alignment. A cheaper and alternative solution is to adopt digitally controlled phase shifters implementing a discrete set of phase shifts at the price of a reduced signal alignment and an increased level of side-lobes [8, 9]. Despite the high-ranging accuracy which can be achieved by the adoption of such systems, all these effects have to be accounted for when target detection is performed for different steering directions.

In this paper we propose a low-complexity non-coherent detection scheme, where the detection of objects is performed by a massive array which steers its beam in different directions in order to detect and localize objects. Thanks to the near-pencil beam array considered, all the measured contributions are associated to the considered steering direction. According to the receiver performance, we describe a set of guidelines to be followed when energy detection with massive arrays is performed.

The rest of the paper is organized as follows. In Sect. 2 we first show the threshold design, and successively we evaluate the side-lobe effects when the previously defined threshold based on receiver noise is adopted. Finally, in Sect. 3 we report a case study where real massive arrays are considered, and we discuss a possible solution to overcome the issue when energy detection is performed.

2   Target Detection Scheme

The personal radar concept is based on the idea that the surrounding objects are detected and localized thanks to the beamforming procedure enabled by massive arrays.

The system herein considered exploits monostatic scattering, i.e. the transmitter and receiver are co-located. For each steering direction θ_b, the massive array steers its main beam towards that direction, and collects the overall backscattered response in order to detect and localize objects. To our purpose, first we account for a laser-like antenna, with a radiation pattern that permits to neglect the side-lobes effect, and thus, we adopt a classical constant false alarm rate (CFAR) approach accounting for the receiver noise only. Second, we focus our analysis to real antennas in which side lobes might cause false target detection. In fact, the presence of a target could be detected and assigned in a certain direction even if it is not effectively located in that part, as shown in Fig. 1. In addition, they could even cause errors in the ranging procedure, i.e. the distance of target 1 and 2 of Fig. 1 can be confused. Thus, once the threshold has been set, we theoretically evaluate the impact of real massive arrays on the detection performance. For our specific case, we consider massive arrays at 60 GHz which can be considered a natural candidate for personal radars applications due to their radiation characteristics.

In the following we describe the receiver scheme considered, and the threshold design by accounting for an ideal laser-beam antenna.

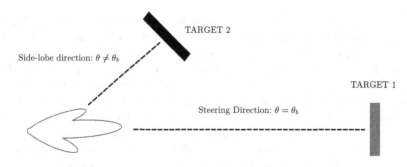

Fig. 1. Considered scenario, where for a steering direction θ_b, the signal reflected from a target in the side-lobe direction is collected.

2.1 Receiver Scheme

The detection scheme we propose is a non-coherent approach based on energy detection to account for the complete uncertainty we have on the received waveform shape.

For each steering direction θ_b, define the received signal as

$$r(t, \theta_b) = s(t, \theta_b) + n(t) \tag{1}$$

with $n(t)$ indicating the noise term and $s(t, \theta_b)$ the received waveform including all the signals coming from the steering direction θ_b.

The received signal is first passed through an ideal bandpass filter with center frequency f_c to eliminate out-of-band noise.[1] The filtered signal is denoted by

$$y(t, \theta_b) = x(t, \theta_b) + z(t) \tag{2}$$

where $x(t, \theta_b) = s(t, \theta_b) \otimes h_F(t)$ and $z(t) = n(t) \otimes h_F(t)$ with $h_F(t)$ being the impulse response of the filter.

Energy evaluations are performed over time interval T_{ED}, with $N_{bin} = \lfloor T_f / T_{ED} \rfloor$ representing the number of integration bins each time frame T_f is divided in, that is

$$e_m(\theta_b) = \int_{(m-1)\,T_{ED}}^{m\,T_{ED}} [y(t, \theta_b)]^2 \, dt \tag{3}$$

with $m = 1, \ldots, N_{bin}$. The detection strategy consists in comparing each element $e_m(\theta_b)$ with a threshold ξ_m. If the energy value of at least one bin is above the threshold, then the target is detected and it is assumed present in the steering direction $\theta = \theta_b$.

We define the following two figures of merit: (i) the probability of false alarm (PFA) P_{FA} as the probability of deciding that a target is detected, when it is not effectively within the considered scenario, due to presence of receiver noise;

---

[1] This operation is necessary since the receiver is energy-based.

(ii) the crossing probability P_c as the probability that the threshold is overcome due to the signal backscattered from a target placed in the scenario.

If the threshold is exceeded for $m = \hat{m}$, the coordinate $\hat{m}$ leads to an estimate of the target time-of-arrival (TOA) and, jointly with the steering direction θ_b, it provides the spatial position of the target in surrounding environment. Consider now the elements $e_m(\theta_b)$ of the decision energy bins. The presented decision rule consists in

$$\text{Decide}: \begin{cases} \widehat{\mathcal{H}}_0, & \text{if } e_m(\theta_b) < \xi_m \quad \forall\{m\}, \\ \widehat{\mathcal{H}}_1, & \text{if } \exists\{m\} \quad \text{s.t.} \quad e_m(\theta_b) \geq \xi_m. \end{cases} \tag{4}$$

Define now, for each energy bin, the normalized energy detector test

$$\Lambda_m(\theta_b) = \frac{2}{N_0} e_m(\theta_b) \underset{\widehat{\mathcal{H}}_0}{\overset{\widehat{\mathcal{H}}_1}{\gtrless}} \tilde{\xi}_m \tag{5}$$

where $\tilde{\xi}_m = \frac{N_0}{2}\xi_m$. According to [10] we have

$$\Lambda_m(\theta_b) = \frac{2}{N_0} \int_{(m-1)T_{\text{ED}}}^{m\,T_{\text{ED}}} [y(t,\theta_b)]^2 \, dt \simeq \frac{1}{\sigma^2} \sum_{i=(m-1)N}^{mN} [y_i(\theta_b)]^2 \tag{6}$$

where $N - 2WT_{\text{ED}}$, $\sigma^2 = N_0 W$ is the noise variance, and y_i are the sampling expansion coefficient of the equivalent low-pass (ELP) of $y(t)$ [10], taken at Nyquist rate W in each interval T_{ED}.

2.2   Threshold Evaluation Criteria with Ideal Pencil-Beam Pattern

When an ideal pencil-beam antenna is considered, we aim to preserve that the probability of false alarm (PFA) due to the receiver noise does not exceed a certain value. Thus, in the presence of only the noise, i.e. $y(t, \theta_b) = z(t)$, Eq. (5) can be written as

$$\Lambda_m(\theta_b) = \frac{2}{N_0} \int_{(m-1)T_{\text{ED}}}^{m\,T_{\text{ED}}} [z(t)]^2 \, dt \simeq \frac{1}{\sigma^2} \sum_{i=(m-1)N}^{mN} (z_i)^2. \tag{7}$$

In order to set the threshold, it is well known that the output of the energy detector is distributed according to a central Chi-square distribution, with probability density function (PDF)

$$f_C(\alpha, \beta) = \frac{\alpha^{(\frac{\beta}{2}-1)}}{2^{\frac{\beta}{2}} \Gamma\left(\frac{\beta}{2}\right)} e^{-\frac{\alpha}{2}}, \quad \alpha \geq 0 \tag{8}$$

where $\Gamma(\cdot)$ is the gamma function [11, p. 255] and β is the number of degrees of freedom.

Considering (8), a threshold-crossing event at the mth bin, that is, $\Lambda_m(\theta_b) > \tilde{\xi}_m$, results in a single-bin $p_{\mathrm{FA}}^{(m,b)}$ given by [12]

$$p_{\mathrm{FA}}^{(m,b)} = \tilde{\Gamma}\left(\frac{N}{2}, \frac{\tilde{\xi}_m}{2}\right) \tag{9}$$

with $\tilde{\Gamma}$ denoting the regularized Gamma function [13].

To properly set the threshold, the joint false alarms for all bins have to be taken into account. Thus, for a considered steering direction θ_b, the overall desired false alarm probability is given by

$$P_{\mathrm{FA}}^\star = 1 - \prod_{m=1}^{N_{\mathrm{bins}}} \left(1 - p_{\mathrm{FA}}^{(m,b)}\right) \approx N_{\mathrm{bins}} \cdot p_{\mathrm{FA}} \tag{10}$$

where we have assumed that all bins are statistically independent and $p_{\mathrm{FA}}^{(m,b)} = p_{\mathrm{FA}} \ \forall\, m$, and consequently we can set a desired $p_{\mathrm{FA}}^\star$ per bin as

$$p_{\mathrm{FA}}^\star \approx \frac{P_{\mathrm{FA}}^\star}{N_{\mathrm{bins}}}. \tag{11}$$

Finally we can write

$$\tilde{\xi} = 2\left[\mathrm{Inv}\tilde{\Gamma}\left(\frac{N}{2}, \frac{P_{\mathrm{FA}}^\star}{N_{\mathrm{bins}}}\right)\right] \tag{12}$$

where $\mathrm{Inv}\tilde{\Gamma}(\cdot, \cdot)$ is the inverse gamma regularized function. Note that with such approach, the threshold does not depend on the bin index, and it is set to keep the probability of false alarm (PFA) due to the receiver noise to a desired value $P_{\mathrm{FA}}^\star$.

2.3 Side-Lobes Effects in Energy Detection Schemes

The previous threshold has been set according to a central Chi-square distribution, having accounted for the presence of the noise receiver only. On the contrary, during the steering procedure, in real scenarios we associate all the contributions deriving from the antenna pattern to that of the steering direction θ_b.

This approximation is often incorrect, especially when real arrays are adopted, as shown in Fig. 2. Thus, in order to evaluate the impact of the side-lobes in the target detection performance, we evaluate the probability that the threshold is crossed due to the presence of a target in the side-lobe direction. In particular, define $x(t, \theta_b) = x_{\mathrm{sl}}(t, \theta_b)$ the received backscattering response under the assumption that no target is in the steering direction θ_b, i.e. the target 1 of Fig. 1 is not present. The new random variable (RV) is now distributed as a non-central Chi-square distribution

$$f_{\mathrm{NC}}(\alpha, \lambda, \beta) = \frac{1}{2}e^{-\frac{\alpha+\lambda}{2}}\left(\frac{\alpha}{\lambda}\right)^{\frac{\beta-2}{4}} I_{\frac{\beta}{2}-1}(\sqrt{\alpha\lambda}), \quad \alpha \geq 0 \tag{13}$$

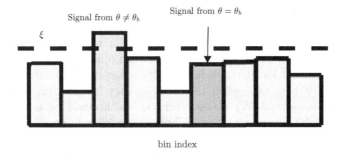

Fig. 2. Example of energy detector output, where energy-bins are compared with the threshold and might cause errors in the localization procedure.

where $I_\kappa(\cdot)$ denotes the κth order modified Bessel function of the first kind [11, p. 374] and probability density function (PDF) $f_{\mathrm{NC}}(\alpha, \lambda, \beta)$ [10], with β being the number of degrees of freedom and λ the non-centrality parameter (NCP).

Due to the presence of signals coming from side-lobes direction, the normalized decision variable results into

$$\Lambda_m(\theta_b) = \frac{2}{N_0} \int_{(m-1)\,T_{\mathrm{ED}}}^{m\,T_{\mathrm{ED}}} [x_{\mathrm{sl}}(t, \theta_b) + z(t)]^2 \, dt \simeq \frac{1}{\sigma^2} \sum_{i=(m-1)N}^{mN} (x_{\mathrm{sl}i}(\theta_b) + z_i)^2 \tag{14}$$

where $x_{\mathrm{sl}i}(\theta_b)$ are the sampling expansion coefficients of the equivalent low-pass (ELP) of $x_{\mathrm{sl}}(t, \theta_b)$ [10]. In particular, the presence of $x_{\mathrm{sl}i}(\theta_b)$ leads to the non-centrality parameter (NCP) $\lambda_m(\theta_b) = 2\gamma_m(\theta_b)$ [10,14], where $\gamma_m(\theta_b)$ is the side-lobe-level-to-noise ratio (SLLNR) per bin, given by

$$\gamma_m(\theta_b) = \frac{1}{N_0} \int_{(m-1)\,T_{\mathrm{ED}}}^{m\,T_{\mathrm{ED}}} x_{\mathrm{sl}}(t, \theta_b)^2 \, dt \simeq \frac{1}{2\sigma^2} \sum_{i=(m-1)N}^{mN} [x_{\mathrm{sl}i}(\theta_b)]^2 \ . \tag{15}$$

A threshold-crossing event at the mth bin, that is, $\Lambda_m(\theta_b) > \tilde{\xi}_m$, results in a single-bin $p_c^{(m,b)}$ given by [12]

$$p_c^{(m,b)} = Q_h \left(\sqrt{\lambda_m(\theta_b)}, \sqrt{\tilde{\xi}_m} \right) \tag{16}$$

with $Q_h(\alpha, \beta) = \int_\beta^\infty x \left(\frac{x}{\alpha}\right)^{k-1} \exp\left(-\frac{x^2 + \alpha^2}{2}\right) I_{k-1}(\alpha x) \, dx$ denoting the generalized Marcum's Q function of order $h = \beta/2$ [13].

Since the signals components deriving from side-lobes direction are undesired, we aim that the threshold is not exceeded due to such signals, i.e. $p_c^{(m,b)} \le p_{\mathrm{FA}}^\star$. If it is not the case, target detection and ranging could be wrongly performed. In the following, the impact of real antenna patterns on the detection performance is investigated, and possible solutions to counteract such effect are reported.

3 Case Study

We now consider the previously described system in order to evaluate what happens when real antennas are employed instead of ideal laser-beam antennas, which are accounted for the threshold design. Despite the analysis conducted is general, i.e. it can be applied to any frequency bandwidth, here we focus on millimeter-waves (mmW) massive arrays, which could be one of the next fifth generation (5G) key technologies. This choice leads to an effective radiated isotropic power (EIRP) constrained by Federal Communications Commission (FCC) regulations as described in [15]. Thus, we first report threshold values in order to achieve a desired $P_{FA}^{\star}$ based on the receiver noise only. Once the threshold has been defined, we evaluate the bin-crossing probability for different values of the non-centrality parameter (NCP), and finally we evaluate possible values in practical scenarios.

3.1 Threshold Setting

According to the analysis of Sect. 2, we now evaluate the threshold considering the receiver noise level. In particular, if otherwise indicated, we consider a time frame $T_f = 100$ ns, a bandwidth $W = 1$ GHz and a time integration interval $T_{ED} = 1$ ns.[2] In this way, by setting an overall $P_{FA}^{\star} = 10^{-3}$, it is $p_{FA}^{\star} = 10^{-5}$ which gives the threshold $\tilde{\xi}^{\star}$ reported in Fig. 3. What it is important to remark is that such desired normalized threshold has been set according to the noise receiver only. Consequently, the impact of signals deriving from side-lobe directions has to be evaluated.

3.2 Side-Lobes Effects

When realistic antennas are adopted, and targets outside the steering direction are present, the threshold might be overcome. Consequently, such targets are wrongly associated with θ_b, which translates into a possible detection and localization error. Thus, by considering (16), it is possible to estimate such effects when $\lambda_m(\theta_b)$ is greater than 0.

In particular, we considered in (16), the $\xi^{\star}$ obtained in Fig. 3 in order to preserve an overall $P_{FA}^{\star} = 10^{-3}$. The obtained results are reported in Fig. 4, where it is evicenced that for $\lambda_m(\theta_b) < 0.2$, the single bin $p_c^{(m,b)}$ is still close to the desired value of 10^{-5}. On the contrary, if we account for $\lambda_m(\theta_b) \approx 2.25$, it is even $p_c = 10^{-3} = P_{FA}^{\star}$. Obviously, for such values of $\lambda_m(\theta_b)$, the system is not robust for target detection in θ_b, as it is extremely sensitive to the presence of a target outside the desired direction.

In the following, we try to map generic values of the non-centrality parameter (NCP) $\lambda_m(\theta_b)$ to those which can be obtained when massive arrays, as the ones described in [3], are adopted in practical applications.

---

[2] From [16], the threshold is accurate for large values of N, whereas for low $W T_{ED}$ values, approximations could improve the accuracy of the threshold. Here we kept $N = 2 W T_{ED}$ since the effects do not affect the validity of the analysis.

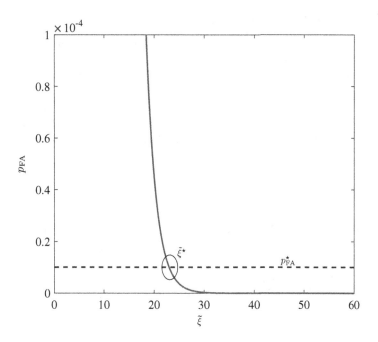

Fig. 3. Threshold choice in order to guarantee the target p^*_{FA}.

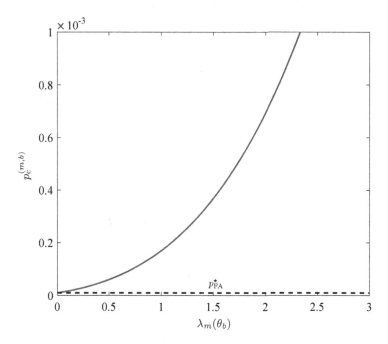

Fig. 4. Bin-crossing probability when the threshold $\tilde{\xi}^\star$ of Fig. 3 is adopted.

3.3   Numerical Evaluation of the NCP

Considering the previous results, we now map the obtained non-centrality parameter (NCP) $\lambda_m(\theta_b)$ into possible real values. In particular, we considered a receiver noise figure $F = 4\,\text{dB}$ and $T_0 = 290\,\text{K}$. A simple and practical solution is to consider free-space propagation from the target to the radar section, and to assume the entire backscattered energy contained into one bin, which represents a worst case scenario. Successively, we dimension $\lambda_m(\theta_b)$ according to the expected path-loss of the signal in each bin from a side-lobe direction. We obtained

$$\lambda_m(\theta_b) = 2\gamma_m(\theta_b) = \frac{1}{\sigma^2} \int_W S(f) \cdot G_{\text{sl}}^2(f, \theta_b) M \frac{c^2}{f^2 (4\pi)^3 d_m^4} df \qquad (17)$$

where M is the target radar cross-section in the side-lobe direction,[3] $G_{\text{sl}}(f, \theta_b)$ is the maximum side-lobe gain in the steering direction θ_b, $S(f)$ is the transmitted power spectral density (PSD) and d_m is the target-array distance. Note that the power spectral density (PSD) has been set so that EIRP, evaluated according to G_{max}, is compliant with the Federal Communications Commission (FCC) regulations. In our scenario we fixed EIRP$= 30\,\text{dBm}$.

As existing antennas, we consider 15×15 massive arrays, which are a possible candidate for this kind of applications thanks to their narrow beam [3], by accounting for a different number of quantization bits which impact in the array pattern. As an example, Table 1 reports different values of G_{max} and G_{sl} at the central frequency $f_c = 60\,\text{GHz}$.[4] The side lobe level (SLL) represents the difference (in [dB]) between the maximum gain and the peak of the main side lobe G_{sl}.

In Fig. 5, $\lambda_m(\theta_b)$ values are reported according to different quantization bits, the bin index (i.e. the target distance from the TX/RX) and different values of M. Consequently, according to Fig. 4, we found that $\lambda_m(\theta_b)$ is often above 0.2, which was found as a limit value in order to preserve $p_c^{(m,b)} = p_{\text{FA}}^\star$ in the

Table 1. SLL and maximum gain at $f_c = 60\,\text{GHz}$ for different phase compensation conditions.

$\theta_b = 0°$			$\theta_b = 20°$		
Quantization	G_{max} [dBi]	SLL [dB]	Quantization	G_{max} [dBi]	SLL [dB]
Perfect	27.1	23.3	Perfect	26.8	22.8
3 bits	26.9	24.6	3 bits	26.5	20.7
2 bits	26.4	20.6	2 bits	25.9	19.4
1 bit	23.5	16.8	1 bit	22.2	13.6

---

[3] Note that here we neglected the dependency of M with the frequency.

[4] The values account also for the spillover loss when massive arrays, such as transmitarrays [2], are excited with an external source.

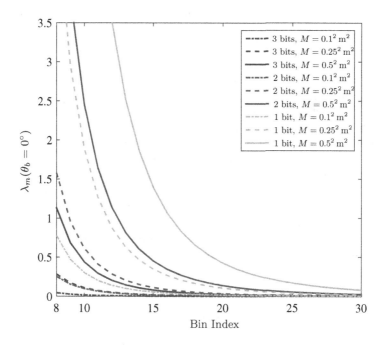

Fig. 5. NCP values for different values of M and for $\theta_b = 0°$.

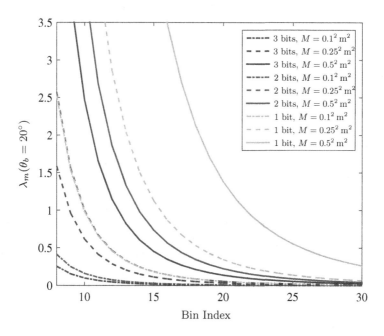

Fig. 6. NCP values for different values of M and for $\theta_b = 20°$.

presence of side-lobes. Note that the values of $\lambda_m(\theta_b)$ are also strictly related to the steering direction. In fact, when electronically beamsteering is performed, the side lobe level (SLL) might increase, as also reported in Table 1. Indeed, also the $\lambda_m(\theta_b)$ values change, as clearly evidenced in Fig. 6, especially when a low number of quantization bits is adopted. This effect suggests that the side lobe level (SLL) should be treated differently for each steering direction, as it is a key design parameter to be taken into account both for the massive array choice and for the threshold evaluation criterion. From one side, in order to reduce the interfering signal coming from directions different from θ_b, it is important to reduce at most the side lobe level (SLL): for example, in our case study, at least 3 quantization bits are required to preserve reliable performance. On the other side, the choice of the massive array could be jointly performed with other operations in order to improve the performance. In fact, different techniques can be adopted in order to counteract side lobes effects on target detection. A possibility could be the adoption of a side-lobe blanker. In particular, a guard channel, which can be omni-directional or adaptive according to the direction, can be implemented to eliminate impulsive interference (hostile or from other neighboring radars) [17,18]. Analogously, in [19], a technique to mitigate the image artifacts due to the sidelobes of the random array is reported. All these solutions are appealing, but they can not be adopted due to their computational complexity and the use of coherent receivers.

A simple and effective solution could be the conception of new threshold design strategies, which accounts for both the receiver noise and the impact of the non-central parameters $\lambda_m(\theta_b)$. Future works will consider a constant false alarm rate (CFAR) approach where the $P_{FA}^\star$ and the different level of interferers per bin are used to properly set the threshold.

4   Conclusions

In this paper we analyzed the impact of massive arrays side-lobes into detection performance for personal radars applications. In particular, in order to keep both the antenna array complexity and the cost low, a discrete set of phase shifts are often adopted for beamforming at the expense of an increased side-lobe level. In these situations, the design of a threshold accounting only for the receiver noise is not sufficient to guarantee the correct functioning of the system in terms of detection performance. In fact, as demonstrated by simulation results, the presence of side-lobes could drastically increase the crossing probability even when there is no target in the steering direction. This effect poses several attentions in the massive array choice according to its maximum side lobe level (SLL). Future studies will investigate the design of a threshold which accounts for the side-lobes effect in each steering direction.

Acknowledgments. This research was supported in part by the IF-EF Marie-Curie project MAPS (Grant 659067) and by the European H2020 project XCycle (Grant 635975).

References

1. Guidi, F., Guerra, A., Dardari, D.: Personal mobile radars with millimeter-wave massive arrays for indoor mapping. IEEE Trans. Mobile Comput. **15**(6), 1471–1484 (2016)
2. Clemente, A., Dussopt, L., Sauleau, R., Potier, P., Pouliguen, P.: Wideband 400-element electronically reconfigurable transmitarray in X band. IEEE Trans. Antennas Propag. **61**(10), 5017–5027 (2013)
3. Guerra, A., Guidi, F., Clemente, A., D'Errico, R., Dussopt, L., Dardari, D.: Application of transmitarray antennas for indoor mapping at millimeter-waves. In: Proceedings of the IEEE European Conference on Networks and Communationa (EUCNC) (2015)
4. Dissanayake, M.W.M.G., Newman, P., Clark, S., Durrant-Whyte, H., Csorba, M.: A solution to the simultaneous localization and map building (SLAM) problem. IEEE Trans. Robot. Autom. **17**(3), 229–241 (2001)
5. Jose, E., Adams, M.: An augmented state SLAM formulation for multiple line-of-sight features with millimetre wave RADAR. In: Proceedings of the IEEE/RSJ International Conference on Intelligent Robots and System, pp. 3087–3092, Aug 2005
6. Jose, E., Adams, M., Mullane, J., Patrikalakis, N.: Predicting millimeter wave radar spectra for autonomous navigation. IEEE Sens. J. **10**(5), 960–971 (2010)
7. Shen, Y., Win, M.: Fundamental limits of wideband localization; part I: A general framework. IEEE Trans. Inf. Theory **56**(10), 4956–4980 (2010)
8. Rondinelli, L.: Effects of random errors on the performance of antenna arrays of many elements. In: IRE International Convention Record, vol. 7, pp. 174–189 (1959)
9. Ruze, J.: Antenna tolerance theory - A review. Proc. IEEE **54**(4), 633–640 (1966)
10. Urkowitz, H.: Energy detection of unknown deterministic signals. Proc. IEEE **55**(4), 523–531 (1967)
11. Abramowitz, M., Stegun, I.A.: Handbook of Mathematical Functions wih Formulas, Graphs, and Mathematical Tables. United States Department of Commerce, Washington, D.C. (1970)
12. Mariani, A., Giorgetti, A., Chiani, M.: Effects of noise power estimation on energy detection for cognitive radio applications. IEEE Trans. Commun. **59**(12), 3410–3420 (2011)
13. Chiani, M.: Integral representation and bounds for marcum Q-function. IEEE Electr. Lett. **35**(6), 445–446 (1999)
14. Guidi, F., Decarli, N., Bartoletti, S., Conti, A., Dardari, D.: Detection of multiple tags based on impulsive backscattered signals. IEEE Trans. Commun. **62**(11), 3918–3930 (2014)
15. FCC: Federal Communications Commission (FCC), 18–07-0082-01-0000d1, Amendment of Part 15 Rules for License-Exemppt 57–64 GHz Band
16. Slepian, D., Sonnenblick, E.: Eigenvalues associated with prolate spheroidal wave functions of zero order. Bell Sys. Tech. J. **44**, 1745–1758 (1965)
17. Nickel, U.: Detection with adaptive arrays with irregular digital subarrays. In: Proceedings of the IEEE Radar Conference, pp. 635–640, April 2007
18. Richmond, C.: Performance of the adaptive sidelobe blanker detection algorithm in homogeneous environments. IEEE Trans. Sig. Process. **48**(5), 1235–1247 (2000)
19. Tsao, J., Steinberg, B.: Reduction of sidelobe and speckle artifacts in microwave imaging: the CLEAN technique. IEEE Trans. Antennas Propag. **36**(4), 543–556 (1988)

Energy Management of Green Small Cells Powered by the Smart Grid

Mouhcine Mendil[1,2]([✉]), Antonio De Domenico[1], Vincent Heiries[1],
Raphaël Caire[2], and Nouredine Hadj-said[2]

[1] CEA, LETI, MINATEC, 38054 Grenoble, France
mouhcine.mendil@grenoble-inp.org
[2] G2Elab, Univ. Grenoble Alpes, 38054 Grenoble, France

Abstract. In this paper, we investigate energy management strategies
for a small cell base station powered by local renewable energy, local
storage, and the smart grid to simultaneously minimize electricity expenditures of the mobile network operators and enhance the life span of the
storage device. Simulation results in different cases show that important
cost reductions can be achieved by properly using the battery.

Keywords: Green communication · Small cell · Battery · Smart grid ·
Renewable energy · Energy controller

1 Introduction

The increasing growth of data traffic has led to a massive deployment of Small
cell Base Stations (SBSs) to offer improved capacity and coverage [1]. As a consequence, the energy demand of cellular networks is growing, essentially because
of the power consumption done at the level of base stations [2]. Based on this,
managing the energy usage is primordial for Mobile Network Operators (MNOs)
to ensure the economic and environmental sustainability of the future heterogeneous cellular networks.

Several concepts have been proposed to improve the energy consumption
in mobile networks addressing network planning, protocols, and equipment [3].
Additionally, a lot of interest has been shown towards Renewable Energy (RE)
usage in cellular networks as it provides the ability to lower the carbon emissions
(by reducing dependence on fossil fuels) and realize long term cost savings thanks
to reduced operating expenditure (OPEX) [4]. Moreover, local harvested energy
enables off-grid base station deployment, where the connection to the electricity
grid is expensive or impossible [5].

The difficulty associated with integrating RE sources is due to their inherent
intermittence. In fact, their power fluctuates over the day and does not always
correspond to the imminent energy demand. Energy storage is then introduced

The research leading to these results has received funding from the French Agence
Nationale de la Recherche in the framework of the SOGREEN project (ANR-14-
CE29-0025-01).

© ICST Institute for Computer Sciences, Social Informatics and Telecommunications Engineering 2016
D. Noguet et al. (Eds.): CROWNCOM 2016, LNICST 172, pp. 642–653, 2016.
DOI: 10.1007/978-3-319-40352-6_53

to ensure the reliability of RE and maintain the balance between energy supply and demand. In addition, the Smart Grid (SG) brings new opportunities to enable a better utilization of RE sources by allowing a two-way flow connection of decentralized production to the power grid.

In this paper, we are interested in the energy cost minimization of a SBS powered by the SG, the RE, and a local battery. Existing works in the literature have adopted two different approaches to address this issue. The first consists on formulating a cost minimization problem assuming that all the characteristics of the model are known or perfectly predictable. In this category, Leithon et al. [6] have studied an offline energy management for green base stations connected to the SG and equipped with a battery. He showed the impact of different energy pricing profiles and battery setups on the energy cost by solving the optimization problem using the Karmarkar algorithm. The second category uses online (or adaptive) methods to take into account the uncertainty of power price, production, and consumption. In particular, Niyato et al. [7] have investigated an online stochastic approach based on multi-period recourse to minimize the energy cost of a SG-powered green micro BS. This study has been extended in [8] by allowing a two way energy flow between the BS and the power grid. Additionally, by using the Kalman filter to forecast the power consumption and the RE generation profiles, the benefit of estimation-based models has been discussed in [9].

It is important to realize that the battery is an expensive investment of the system, and enhancing its life span is vital for an efficient return on investment. However, to the best of our knowledge, none of the proposed strategies have taken into consideration the battery life maximization. This motivated us to investigate the design of an energy controller and model its stochastic environment to jointly optimize the energy cost while operating the system in the most effective way to improve the battery life span.

The rest of the paper is organized as follows. System architecture is provided in Sect. 2. In Sect. 3, we propose a formulation of the cost minimization problem while extending the battery life span. Results are presented in Sect. 4. Finally, conclusions and perspectives are discussed in Sect. 5.

2   System Architecture

In the proposed architecture (illustrated in Fig. 1), the SBS, deployed to offer high data rate services to local mobile users, is powered by two sources of energy: the SG and RE. RE usage provides several benefits compared with a classic grid-powered SBS such as long-term cost savings and reduced carbon emissions. Moreover, a battery is used as a local storage device to offer flexibility in the energy utilization. The system is interconnected to the SG in a two-way energy flow, i.e., energy can be sold or bought from it. Finally, an energy supervision system (ESS) is in charge of scheduling the energy flow between each component of the system to reduce the cost of energy transactions with the SG and improve the battery life span. In the following, we present the chosen model for each component of the system.

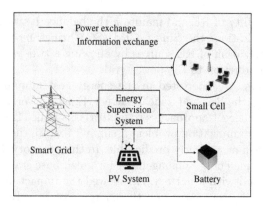

Fig. 1. System architecture of small cell power supervision.

2.1   Small Cell Base Station Power Consumption Model

We assume that the SBS load ρ follows a non-homogeneous Poisson process, which intensity depends on time. Additionally, the SBS can be either in the active state ($\rho > 0$) or sleep state ($\rho = 0$). The following equation gives the overall SBS power consumption P_{BS} [W] as a function of the traffic load [10]:

$$P_{BS}(t) = \begin{cases} P_0 + \Delta_p \cdot \rho(t) \cdot P_{max}, & \text{if } 0 < \rho(t) \le 1 \\ P_{sleep}, & \text{if } \rho(t) = 0 \end{cases} \tag{1}$$

where P_0 is the power consumption at the minimum non-zero output power, Δ_p is the slope of the input-output power consumption, P_{max} is the maximum output power, and P_{sleep} is the power consumed in sleep mode.

2.2   Energy Storage Model

We use a Lithium-ion battery as the power storage device in our architecture. It can be charged by the locally produced energy or from the SG, and discharged to cater the SBS or sell energy to the SG. The battery is described by two parameters: its power and its State Of Charge (SOC), which describes the present battery capacity as a percentage of the nominal capacity C_N [Ah] (a SOC of 100 % means fully-charged and 0 % means fully-discharged). In the following, we present the selected models for the battery SOC and power.

State of Charge Model. The SOC variation is generally calculated using current integration. The rate at which the battery is charged or discharged, noted C_{rate} [s^{-1}], corresponds to the charge or discharge current intensity $i(t)$ [A] relative to the battery nominal capacity:

$$C_{rate}(t) = \frac{i(t)}{3600 \cdot C_N} \tag{2}$$

Periodically, for a given C_{rate}, we use the Ampere-Hour integral model to estimate the SOC variation [11]:

$$z(t + \delta t) = z(t) + \eta \int_t^{t+\delta t} C_{\text{rate}}(u)\mathrm{d}u \tag{3}$$

where δt represents the time between two SOC estimations and η is the battery Coulombic efficiency, equals to η_{dis} when discharging and η_{chg} when charging.

Battery Power Model. The battery is a pack that consists of individual modules, which are composed of cells organized in series and parallel. For simplicity of description, we assume that the battery pack comprises n_s cell modules connected in series, where each cell module comprises an individual cell. The following equation gives a simplified relation between charge or discharge current $i(t)$ and the voltage of cell k V_k [V] [12]:

$$V_k(t) = OCV(z(t)) + R_k \cdot i(t) \tag{4}$$

where OCV [V] (Open Circuit Voltage) is the cell voltage when the cell is disconnected from any circuit, z is the battery SOC, and R_k [Ω] is the internal resistance of the cell k, which depends on several parameters (SOC, current intensity, temperature, and State Of Health (SOH)) [13]. The OCV relationship with a given SOC can be measured experimentally by allowing the battery to reach equilibrium after being disconnected from any load for a long period of time [14]. Reiterating this method for different SOCs, the obtained OCV-SOC look-up table can then be used to elaborate an analytical OCV model. In this paper, we consider an n-order polynomial approximation model such that [15]:

$$OCV(z(t)) = \sum_{j=0}^n a_j \cdot z^j(t) \tag{5}$$

where $(a_j)_{j=1..n}$ are the polynomial coefficients calculated from the experimental OCV-SOC dependency function.

As a sign convention, we assume that the charge (resp. discharge) current and power have a positive (resp. negative) sign. Consequently, the power P_{batt} [W] of the battery can be calculated using the sum of all cell powers :

$$P_{\text{batt}}(t) = \sum_{k=1}^{n_s} i(t) \cdot V_k(t) \tag{6}$$

From (Eq. 2) to (Eq. 6), and by supposing for simplicity that C_{rate} is constant during the period δt, we can rewrite the battery power formula as a function of two consecutive SOC values:

$$P_{\text{batt}}(z(t), z(t+\delta t)) = \sum_{k=1}^{n_s} \sum_{j=0}^n A_{j,k} z^j(t) z(t+\delta t)$$
$$- B_{j,k} z^{j+1}(t) + \alpha^2 \cdot R_k \cdot z^2(t+\delta t) \tag{7}$$

Such that,

$$A_{j,k} = \alpha \cdot (a_j - 2\alpha \cdot R_k \cdot \delta_{1,j}),$$
$$B_{j,k} = \alpha \cdot (a_j - \alpha \cdot R_k \cdot \delta_{1,j}),$$
$$\alpha = \frac{3600 \cdot C_N}{\eta \cdot \delta t}$$

and $\delta_{1,j}$ is the Kronecker symbol, equals to 1 when $j = 1$ or 0 otherwise.

Battery Ageing. In general, batteries must operate within a safe operating area restricted by temperature, current, and voltage windows [11]. Not respecting these restrictions leads to a rapid attenuation of the battery performance (capacity loss and decrease of charge and discharge efficiencies) and even results in safety problem. The voltage restrictions can be translated into recommendations for the operating range of the battery SOC. In this paper, we restrict the battery usage on the specific range of the SOC $\Delta_{\mathrm{soc}} = [20\%, 90\%]$ (see Fig. 2). Additionally, by using (Eq. 3), the current restriction can be reformulated as a limitation of the SOC variation in each decision period:

$$\forall t, \ \Delta SOC_{\min} \leq z(t+1) - z(t) \leq \Delta SOC_{\max} \tag{8}$$

2.3   Harvested Energy Model

A solar panel is used in the proposed architecture to collect solar energy and transform it into electricity via photo-voltaic (PV) effect. We assume that the solar radiation I_g [W/m^2] varies on a hourly basis and depends on several factors such as geographical location, time of the day, and local weather.

Let I_t be the random variable corresponding to the solar radiation at hour t. The vector $(I_1, ..., I_{24})$ of daily radiation is supposed to follow a multivariate Gaussian distribution (or Gaussian Process) $GP(\mu_{\mathrm{irrad}}, \Sigma_{\mathrm{irrad}})$, where μ_{irrad} is

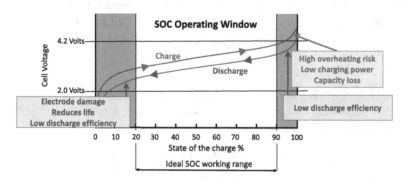

Fig. 2. Recommendations for the operating range of SOC of lithium ion battery [16].

a vector of size 1×24 composed of the hourly average radiations of the day, and Σ_{irrad} is the covariance matrix 24×24. These parameters are inferred from historical measures of solar radiations during one year [17]. Given the utilization of real historical data, the obtained stochastic process can capture all the phenomena that influence the solar radiation. Then, the hourly photo-voltaic output power P_{PV} [W] is given by the following relation [18]:

$$P_{\mathrm{PV}}(t) = \eta_{\mathrm{PV}} \cdot S \cdot I_{\mathrm{g}}(t) \tag{9}$$

where η_{PV} is the energy conversion efficiency of the solar panels and S [m^2] is the total panels surface.

2.4 Price Signal Model

Maintaining a permanent balance between the power consumption and production is a major requirement for power grid operators to guarantee the security of energy supply. Dynamic pricing, which consists in varying the energy price with time, is a promising mechanism to adapt consumption profiles to the energy availability. In this study, we consider a stochastic dynamic energy price: the buying price $p_{\mathrm{buy}}(t)$ [$/kWh], i.e., the price at which energy is bought from the SG, is modeled by the Gaussian process $GP(\mu_{\mathrm{price}}, \Sigma_{\mathrm{price}})$, where μ_{price} is a vector of size 1×24 composed of the hourly average buying prices of the day, and Σ_{price} is the covariance matrix. These parameters are inferred from historical data of electricity pricing for residential customers during one year [19]. Moreover, the price at which energy is sold back to the SG is set proportional to the buying electricity price such that $p_{\mathrm{sell}} = \kappa \cdot p_{\mathrm{buy}}$ [6].

3 Energy System Supervisor

We aim at minimizing the energy expenditures of a SBS powered by the SG, the RE, and a local battery under the constraint of the battery operating range Δ_{soc}, $\Delta SOC_{\mathrm{min}}$, and $\Delta SOC_{\mathrm{max}}$ defined in Sect. 2.2. This optimization consists in managing the energy exchange between the SBS and the power grid over a horizon divided into T decision periods. We consider the length of a period to be Δ_t, in which the SBS load, the PV power, and the energy price are fixed. The ESS, in charge of the energy management, is composed of two layers:

1. The High Level Controller (HLC) minimizes the energy cost by imposing an objective value of SOC to the battery at each decision period.
2. The Low Level Controller (LLC) manages the energy flow between each subsystem in real time to realize the HLC objective while respecting the energy supply-demand balance.

Therefore, the ESS can schedule the amount of energy to exchange with the SG by selecting a succession of SOCs (the SOC variation means that the battery is being charged or discharged, see Sect. 2.2). In fact, for a given SOC value, if the energy locally produced is not sufficient to power the SBS and the battery, the LLC can evaluate the missing energy and notify the HLC to buy it from the SG. Similarly, if the energy produced or offered by the battery is excessive compared to the consumption, the surplus is sold. In this paper, we focus on the objective and constraints of the optimization problem at the HLC level to find the optimal SOC strategy $\mathbf{z}^* = (z^*(1), ..., z^*(T+1))$, which are defined as follows:

$$\mathbf{z}^* = \underset{(z(1),...,z(T+1))\in\mathbb{R}^{T+1}}{\mathrm{argmin}} \sum_{t=1}^{T} p(t) \cdot E(t, z(t), z(t+1)) \tag{10}$$

Subject to

$$\frac{E(t, z(t), z(t+1))}{\Delta t} = P_{\mathrm{BS}}(t) + P_{\mathrm{Batt}}(z(t), z(t+1)) - P_{\mathrm{PV}}(t) \tag{11}$$

$$z(t) \in \Delta_{\mathrm{soc}}, t = 1, ..., T+1 \tag{12}$$

$$\Delta SOC_{\min} \leq z(t+1) - z(t) \leq \Delta SOC_{\max}, t = 1,...,T \tag{13}$$

where $(z(1), ..., z(T+1))$ is the multivariable decision vector that represents the battery SOCs over the optimization horizon, E is the amount of energy exchanged with the SG, and p is the buying energy price when $E > 0$ or the selling price when $E < 0$. (Eq. 10) is the objective function to minimize, which corresponds to the long term cost due to power transaction with the electrical grid. At all times, the balance between the power supply and demand is illustrated by the constraint (Eq. 11). In addition, during all the decision periods, the constraints (Eqs. 12 and 13) on the SOC have to be respected to improve the battery life span.

4   Results and Discussions

We consider a finite horizon of 24 h, i.e., $T = 24$ and $\Delta_t = 1$ h. The profiles illustrated in Fig. 3 describe the average hourly SBS load, solar radiation, and energy buying price used in our simulations to model the intensity of the SBS load, the average vector of the solar radiation, and the average vector of the energy price, respectively. Concerning the SBS, the traffic load grows progressively and reaches the maximum around 21:00-22:00. In addition, we assume that the traffic between 2:00 and 9:00 is handled by the under-layer macro base station, such that the SBS load in this period is zero. For the solar radiation, the profile is characterized by a peak around midday and positive values during daytime. Finally, the energy price is marked by an increasing trend from low prices late at night to high values attained during the afternoon.

The battery OCV (Eq. 5) is modeled by a 2nd order polynomial such that $OCV(z(t)) = 2.9 + 0.13 \cdot z(t) - 0.008 \cdot z^2(t)$. Other simulation settings for each component of the system are summarized in Table 1. Without loss of generality, we consider that the battery parameters (nominal capacity, cell resistance, and

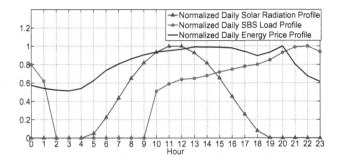

Fig. 3. Normalized average solar radiation [17], SBS load (based on [21]), and energy price profiles [19].

Table 1. Simulation Parameters.

	Parameter	Value	Parameter	Value
SBS	P_0	13.6 W	Δ_p	4
	P_{max}	0.13 W	P_{sleep}	8.6 W
Battery	n_s	5	C_N	12 Ah
	$\forall k\ R_k$	50 mΩ	η_{chg}	96 %
	η_{dis}	100 %	z_0	30 %
	ΔSOC_{min}	−70 %	ΔSOC_{max}	70 %
Solar panel	η_{PV}	14 %	S	0.25 m^2
Energy price	κ	90 %		

charge/discharge efficiency) are independent of the SOH, current intensity, and temperature.

Energy management of the SBS is a challenging issue due to the uncertainty in the environment. To address this issue, we first solve the non-linear constrained problem of Sect. 3 in the ideal case, i.e. the variations of the SBS load, solar radiation and energy price over all the optimization period are perfectly predicted by the ESS. The objective is to obtain the maximum performance of the ESS in term of the energy cost that can serve as an upper-bound in the realistic case, where the stochastic variables cannot be totally forecast. To converge to the optimal solution, we perform multiple runs of the interior-point algorithm implemented in Matlab [20]. Then we compare the performance of the above described *ideal* strategy with three other strategies averaged over five years:

1. The *reference strategy* systematically buys energy from the SG, in which the battery and the solar panel are not used.
2. The *naive strategy* seeks to reduce the immediate energy cost. At each decision period, either the PV production is sufficient to cover the SBS consumption, in which case the energy surplus is sold or the SBS consumes more than the energy produced, in which case the missing energy is purchased from the SG. Consequently the battery is never used.

3. The *optimized strategy* utilizes the solution of the optimization problem of Sect. 3, obtained in the case where the SBS load, the solar radiation, and the energy price variations correspond exactly to their respective average profiles (Fig. 3). Consequently, this strategy exploits the a priori knowledge of the environment to implement the same policy every day, for five years, without further adaptation to the actual variation of the stochastic environment variables.

Figure 4 represents the average SBS energy consumption, RE production, energy transactions with the SG, and energy stored in the battery with the ideal strategy. Notice that the energy transaction can be positive or negative, which means that the energy is purchased or sold to the SG, respectively. In general, the ESS buys electricity at night, when the PV system can not produce any energy, to power the SBS or/and store it into the battery. Additionally, we can observe that the amount of energy purchased from the SG depends closely on the energy price. Once the PV production becomes available or when the price is high, the ESS prioritizes the use of the energy produced by the PV panels and the energy already stored in the battery to feed the SBS, and sells a quantity of the surplus to the SG. Notice that all the decisions made by the ESS are consistent with

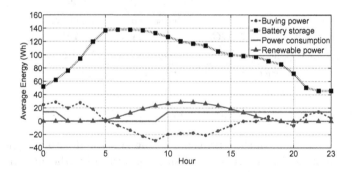

Fig. 4. Energy transaction, consumption, production, and storage with the ideal strategy, averaged over 5 years.

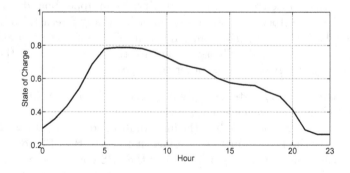

Fig. 5. Battery SOC with the ideal strategy, averaged over 5 years.

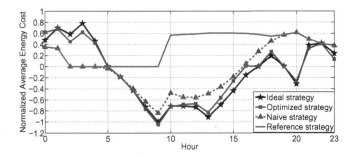

Fig. 6. Normalized daily energy cost for different strategies, averaged over 5 years.

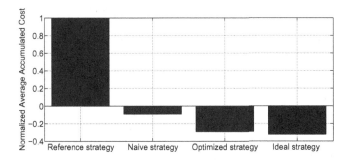

Fig. 7. Normalized strategies accumulated daily cost.

the recommended operating SOC range Δ_{soc} and SOC variations ΔSOC_{min} and ΔSOC_{max} as shown in Fig. 5.

Next we analyze the behavior of the *ideal* strategy in the light of the three strategies presented earlier. Figure 6 illustrates how the decisions in each case impact the hourly energy cost: although powering the SBS is a priority for all strategies, the energy surplus (when existing) is not similarly handled. On the one hand, the absence of local storage leads the *naive strategy* to always sell back the energy as soon as there is production excess; on the other hand, the *ideal* and *optimized* schemes buy electricity when the price is low and store energy for later consumption or transaction with the SG.

In Fig. 7, we compare the average cost over a day for each strategy normalized with respect to the *reference strategy* cost. The largest cost saving of 132 % compared to the *reference* is naturally achieved in the *ideal* case. The information about the energy consumption, production, and price trends carried in the average profiles allow the *optimized strategy*, even though not being totally adapted to its environment, to perform only 10 % less compared to the *ideal strategy* cost. Finally, the naive strategy achieves only one third of the *ideal strategy* cost savings. To finalize, we can observe that the supervised usage of the battery plays an important role for two reasons: 1. it allows more flexibility in energy purchase such that the system does not buy electricity only to match the energy consumption, but also to feed the SBS later when buying energy becomes expensive and 2. it creates opportunities to increase the RE value by offering it to sell when the prices are high.

5   Conclusion

In this paper, we have investigated the impact of using local renewable production and local storage to reduce small cell energy expenditures. We have proposed a controller that can jointly optimize the energy cost and maximize the battery lifespan. Simulation results have shown that the solution achieves very large cost reduction compared to basic strategies while respecting the battery constraints. As a future work, we aim at modeling the battery SOH and evaluating the quantitative impact of the proposed solutions on the battery life span improvement. We will also consider an online approach to minimize the SBS energy cost in stochastic environment.

References

1. Small Cells Deployment Market Status 2015 Report. http://scf.io/en/documents/050_-_Market_status_statistics_Q1_2014_-_Mobile_Experts.php
2. Hassan, H.A.H., Nuaymi, L., Pelov, A.: Renewable energy in cellular networks: a survey. In: IEEE GreenCom, pp. 1–7 (2013)
3. Koutitas, G., Demestichas, P.: A review of energy efficiency in telecommunication networks. Telfor J. **2**(1), 2–7 (2010)
4. Piro, G., Miozzo, M., Forte, G., Baldo, N., Grieco, L.A., Boggia, G., Dini, P.: Hetnets powered by renewable energy sources: sustainable next-generation cellular networks. IEEE Internet Comput. **17**(1), 32–39 (2013)
5. Wang, H., Li, H., Tang, C., Ye, L., Chen, X., Tang, H., Ci, S.: Modeling, metrics, and optimal design for solar energy-powered base station system. EURASIP J. Wireless Commun. Networking **2015**(1), 1–17 (2015)
6. Leithon, J., Sun, S., Lim, T. J.: Energy management strategies for base stations powered by the smart grid. In: IEEE GLOBECOM (2013)
7. Niyato, D., Lu, X., Wang, P.: Adaptive power management for wireless base stations in a smart grid environment. IEEE Wireless Commun. **19**(6), 44–51 (2012)
8. Kaewpuang, R., Niyato, D., Wang, P.: Decomposition of stochastic power management for wireless base station in smart grid. IEEE Wireless Commun. Lett. **1**(2), 97–100 (2012)
9. Leithon, J., Lim, T.J., Sun, S.: Online energy management strategies for base stations powered by the smart grid. In: IEEE SmartGridComm, pp. 199–204 (2013)
10. Auer, G., Giannini, V., Desset, C., Godor, I., et al.: How much energy is needed to run a wireless network? IEEE Wireless Commun. **18**(5), 40–49 (2011)
11. Lu, L., Han, X., Li, J., Hua, J., Ouyang, M.: A review on the key issues for lithium-ion battery management in electric vehicles. J. power sources **226**, 272–288 (2013)
12. Restaino, R., Zamboni, W.: Comparing particle filter and extended kalman filter for battery state-of-charge estimation. In: IEEE IECON, pp. 4018–4023 (2012)
13. Barre, A.: Statistical analysis for understanding and predicting battery degradations in real-life electric vehicle use. J. Power Sources **245**, 846–856 (2014)

14. Chang, W.-Y.: The state of charge estimating methods for battery: a review. ISRN Appl. Math. **1–7**, 2013 (2013)
15. Weng, C., Sun, J., Peng, H.: An Open-circuit-voltage model of lithium-ion batteries for effective incremental capacity analysis. In: ASME Dynamic Systems and Control Conference. American Society of Mechanical Engineers (2013)
16. Lithium Battery Failures. http://www.mpoweruk.com/lithium_failures.htm#soc
17. GeoModel SOLAR. Solar Radiation Time Series. http://geomodelsolar.eu/data/full-time-series
18. Rami, G., Tran-Quoc, T., Hadjsaid, N., Mertz, J. L.: Energy supply for remote base transceiver stations of telecommunication. In: IEEE PESGM, pp. 1916–1921 (2004)
19. Ameren. Ameren Price Database. https://www2.ameren.com/RetailEnergy/rtpDownload
20. Forsgren, A., Gill, P.E., Margaret, H.: Interior methods for nonlinear optimization. SIAM Rev. **44**(4), 525–597 (2002)
21. EARTH. D2.3, Energy Efficiency Analysis of the Reference Systems, Areas of Improvements and Target Breakdown (2010)

Min-max BER Based Power Control for OFDM-Based Cognitive Cooperative Networks with Imperfect Spectrum Sensing

Hangqi Li[(✉)], Xiaohui Zhao, and Yongjun Xu

College of Communication Engineering, Jilin University, Changchun 130012, China
{lhq14,xuyj10}@mails.jlu.edu.cn, xhzhao@jlu.edu.cn

Abstract. In this paper, a power control (PC) algorithm for multi-user Orthogonal Frequency Division Multiplexing (OFDM)-based cognitive cooperative networks under the imperfect spectrum sensing is studied to minimize total Bit Error Rate (BER) of secondary users (SUs) under the consideration of maximum transmit power budgets, signal-to-interference-and-noise ratio (SINR) constraints and interference requirements to guarantee quality of service (QoS) of primary user (PU). And a cooperative spectrum sensing (CSS) strategy is considered to optimize sensing performance. The worst-channel-state-information (worst-CSI) PC algorithm is introduced to limit the BER of SUs, which only needs to operate the algorithm in one link that CSI is worst, while the interference model is formulated under the consideration of spectrum sensing errors. In order to obtain optimal solution, the original min-max BER optimization problem is converted into a max-min SINR problem solved by Lagrange dual decomposition method. Simulation results demonstrate that the proposed scheme can achieve good BER performance and the protection for PU.

Keywords: Cooperative transmission · Imperfect spectrum sensing · OFDM-based cognitive radio networks · The worst-CSI PC algorithm

1 Introduction

Cognitive radio (CR), as an efficient technology for next generation of wireless communication, can significantly improve spectrum utilization by dynamically detecting spectrum usage and opportunistically accessing the free frequency band [1]. Generally, power control (or resource allocation) techniques are used in CR networks (CRNs), which depends on perfect channel state information (CSI) and spectrum sensing results. In practical CRNs, there are several network types, such as traditional CRNs, cognitive relay networks, OFDM-based CRNs and multi-antenna CRNs [2].

In order to expand communication scope, cooperative technology is introduced to help primary users (PUs) or secondary users (SUs) for their communications in CRNs [3]. The earliest emergence of relay networks can be traced back to the late 1970s, proposed by Dr. Cover in [4,5]. Since cognitive relay networks

© ICST Institute for Computer Sciences, Social Informatics and Telecommunications Engineering 2016
D. Noguet et al. (Eds.): CROWNCOM 2016, LNICST 172, pp. 654–667, 2016.
DOI: 10.1007/978-3-319-40352-6_54

have more advantages than traditional CRNs (i.e., non-relay network), and are suitable for actual communication scenarios (i.e., heterogeneous networks, 5G communications), in this paper, we focus on the study of power allocation problem in multiuser cognitive relay networks.

Currently, power control (PC) as an important role in the performance of CRNs can provide protection for PU and allow SUs opportunistically transmit data. In [6], a distributed PC algorithm for a multiuser CRN with multicell environments is given to address uplink interference management problem. Due to the advantage of flexible scheduling spectrum of the orthogonal frequency division multiplexing (OFDM) technology, it has been widely introduced to CRNs [7].

Obviously, the literatures mentioned above only consider PC problem under perfect spectrum sensing information, which may not be valid in practice due to time-varying channels, inevitable errors and uncertainties as results of imperfect spectrum sensing. In order to obtain good system performance and ensure quality of service (QoS) of SUs and PUs, it is necessary to take errors of spectrum sensing into consideration. Based on different optimization functions (e.g., minimization of total power allocation, maximization of capacity of SUs, maximization of energy efficiency, etc.), PC problems with the imperfect spectrum sensing have been studied from various network structures (e.g., traditional CRNs, OFDM-based CRNs, micro CRNs, etc.). Considering a traditional CRN with the imperfect spectrum sensing, PC problem is studied in [8]. In [9], for an OFDM-based CRN, the resource allocation is studied to maximize the overall capacity of SUs. Considering the imperfect spectrum sensing in CRNs with one primary network (PN) and many micro CRNs [10], a hybrid spectrum access strategy is proposed, where the capacity of secondary link is maximized. However, research of PC problem in cognitive relay networks under the imperfect spectrum sensing is quite few.

In this paper, a PC algorithm is proposed to minimize total bit error rate (BER) of SUs in OFDM-based cognitive relay networks under the imperfect spectrum sensing. Multiple PUs, multiple SUs and multiple relays are considered in our model. The min-max criteria is used to minimize total BER of SUs under practical constraints. We convert the original min-max BER optimization problem into an equivalent max-min signal-to-interference-and-noise ratio (SINR) problem solved by Lagrange dual decomposition.

The reminder of this paper is organized as follows. In Sect. 2, system model is described. Section 3 introduces cooperative spectrum sensing (CSS) scheme and describes the interference model. Next, PC problem with the imperfect spectrum sensing is formulated and the algorithm is given in Sect. 4. Section 5 presents some numerical results and analysis of the system performance. Finally, Sect. 6 provides the conclusion of the paper.

2   System Model

In this paper, we consider an overlay cognitive amplify-and-forward (AF) relay network with P PUs and L SUs as shown in Fig. 1.(a). The related explanation

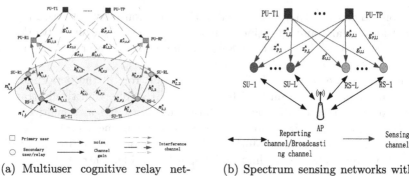

(a) Multiuser cognitive relay networks

(b) Spectrum sensing networks with an AP

Fig. 1. System model and spectrum sensing networks (Color figure online)

is given in Table 1. The set $\mathbf{L} = \{1, 2, \cdots, L\}$ denotes the number of SUs, and $\mathbf{P} = \{1, 2, \cdots, P\}$ denotes the number of PUs, and $\forall l, j \in \mathbf{L}$, $\forall p \in \mathbf{P}$. Let SU-T and SU-R (PU-T and PU-R) denote secondary (primary) transmitter and receiver, and RS denote relay node, respectively. We assume both SUs and PUs use OFDM modulation mode, in which the total bandwidth is divided into $\mathbf{N} = \{1, 2, \cdots, N\}$ orthogonal subcarriers, and $\forall n \in \mathbf{N}$. This model is a dual-hop relay network in which time-division half duplex relays are used to help communication of SUs. The direct communications from the secondary source nodes to secondary destination nodes are not considered. Under overlay spectrum sharing scenario, multiple source nodes and relays are available to obtain spectrum information in spectrum sensing phase. Relays first assist SUs to detect vacant bands via cooperative spectrum sensing, then an access point (AP) collects local detection results reported by SUs and relays. AP takes fusion criterion and makes a global decision for data transmission as shown in Fig. 1.(b). Let V_p^n and O_p^n represent the licensed spectrum unoccupied and occupied over the subcarrier n by the pth PU, respectively. $\hat{V}_p^n$ and $\hat{O}_p^n$ are used to indicate the status of the licensed spectrum estimated by secondary network.

3   Spectrum Sensing Process

3.1   Cooperative Spectrum Sensing (CSS)

Energy detector (ED) [3] is used by sensing nodes in spectrum sensing phase in order to make a decision about the spectrum occupied or unoccupied by PUs, through comparing the energy of received signal with a detection threshold. We assume that observation time spent by each subcarrier is τ/N, where τ is the observation time window on the whole licensed spectrum. And each sensing node that performs ED in a fixed bandwidth for each subcarrier is f. Therefore,

Table 1. Symbol introduction

Symbol	Specification
P_p^n	Transmit power of the pth PU-T on subcarrier n
$P_{l,1}^n$	Transmit power of the lth SU-T on subcarrier n
$P_{l,2}^n$	Transmit power of the lth relay on subcarrier n
$h_{l,1}^n$	Channel gain of the first-hop of the lth link on subcarrier n
$h_{l,2}^n$	Channel gain of the second-hop of the lth link on subcarrier n
$h_{l,p,1}^n$	Channel gain of the lth SU-T to the pth PU-R on subcarrier n
$h_{l,p,2}^n$	Channel gain of the lth relay to the pth PU-R on subcarrier n
$g_{p,l,1}^n$	Channel gain of the pth PU-T to the lth relay on subcarrier n
$g_{p,l,2}^n$	Channel gain of the pth PU-T to the lth SU-R on subcarrier n
$z_{p,l}^n$	Sensing channel gain of the pth PU-T to the lth SU-T on subcarrier n

the time bandwidth product on each subcarrier is $f\tau/N$ [3]. Let $x_p^n(i)$ be the transmit signal from the pth PU on the subcarrier n, and $\forall i \in \{1, 2, \cdots, 2f\tau/N\}$. The received signal from the pth PU on the subcarrier n at the lth SU-T and relay is given by

$$\begin{cases} y_{p,l,1}^n(i) = \sqrt{\alpha P_p^n} z_{p,l}^n x_p^n(i) + n_{p,l,1}^n(i) \\ y_{p,l,2}^n(i) = \sqrt{\alpha P_p^n} g_{p,l,1}^n x_p^n(i) + n_{p,l,2}^n(i) \end{cases} \tag{1}$$

where $y_{p,l,1}^n(i)$ and $y_{p,l,2}^n(i)$ are the received signal from the pth PU on the subcarrier n at the lth SU-T and the lth relay. P_p^n is the transmit power of the pth PU-T on the subcarrier n. $n_{p,l,1}^n(i)$ and $n_{p,l,2}^n(i)$ are the additive noise on the subcarrier n which are the independent zero-mean white Gaussian noise (AWGN) with power density N_0. α represents the state of the pth PU on the subcarrier n, which is given by

$$\alpha = \begin{cases} 1, & O_p^n \\ 0, & V_p^n \end{cases} \tag{2}$$

When the subcarrier n is unoccupied by the pth PU (i.e., V_p^n), $\alpha = 0$, otherwise $\alpha = 1$. According to energy calculation formula [11], the expressions of the received signal energy from the pth PU on the subcarrier n at the lth SU-T (i.e., $E_{p,l,1}^n$) and the lth relay (i.e., $E_{p,l,2}^n$) are

$$\begin{cases} E_{p,l,1}^n = \sum_{i=1}^{2f\tau/N} \left| y_{p,l,1}^n(i) \right|^2 \\ E_{p,l,2}^n = \sum_{i=1}^{2f\tau/N} \left| y_{p,l,2}^n(i) \right|^2 \end{cases} \tag{3}$$

We assume that channel gains are time-invariant during the sensing phase, and suppose the decision threshold of energy detector as ε at the lth SU-T and the lth relay on the subcarrier n. For $\forall k \in \{1, 2, \cdots, 2L\}$, $a_{p,k}^n$ is a binary number denoting the status of comparative results. The decision criterion is

$$\begin{cases} \hat{O}_p^n, & E_{p,l,1}^n \geq \varepsilon \\ \hat{O}_p^n, & E_{p,l,2}^n \geq \varepsilon \end{cases} \tag{4}$$

$$a_{p,k}^n = \begin{cases} 1, & \hat{O}_p^n \\ 0, & \hat{V}_p^n \end{cases} \tag{5}$$

where $\hat{V}_p^n$ and $\hat{O}_p^n$ denote the sensing result of sensing node on the subcarrier n unoccupied and occupied by the pth PU, respectively.

If $E_{p,l,1}^n > \varepsilon$ and $E_{p,l,2}^n > \varepsilon$, it indicates that the lth SU-T and the lth relay have successfully detected the presence of the pth PU on the subcarrier n that satisfies the hypothesis O_p^n (the result of sensing is $\hat{O}_p^n$). Energy collected in the process of detecting status of the pth PU on the subcarrier n at the sensing node in the frequency domain is denoted by $E_{p,k}^n$ which serves as a decision with the following distribution [11]

$$E_{p,k}^n \sim \begin{cases} \chi_{2u}^2 & , & V_p^n \\ \chi_{2u}^2 \left(2\gamma_{p,k}^n \right), & O_p^n \end{cases} \tag{6}$$

where u is equal to $f\tau/N$. χ_{2u}^2 follows a central chi-square distribution with $2u$ degrees of freedom, and $\chi_{2u}^2 \left(2\gamma_{p,k}^n \right)$ follows a non-central chi-square distribution with $2u$ degrees of freedom and a non centrality parameter $2\gamma_{p,k}^n$ [3]. And $\gamma_{p,k}^n$ is the instantaneous signal-noise ratio (SNR) of the received signal from the pth PU at the kth sensing node on the subcarrier n.

In order to insure the generality of the sensing, we take the spectrum sensing uncertainties into consideration so that we can derive the expressions of average detection probability, false-alarm probability, and miss-detection probability. In order to simplify the calculations, we assume that the decision threshold ε is a constant parameter.

$$P_{d,p,k}^n = \mathrm{E}[Pr(E_{p,k}^n > \varepsilon | O_p^n)] = Pr(\chi_{2u}^2(2\gamma_{p,k}^n) > \varepsilon) \tag{7}$$

$$P_{fa,p,k}^n = \mathrm{E}[Pr(E_{p,k}^n > \varepsilon | V_p^n)] = \frac{\Gamma(u, \frac{\varepsilon}{2})}{\Gamma(u)} \tag{8}$$

$$P_{md,p,k}^n = 1 - P_{d,p,k}^n \tag{9}$$

where $\mathrm{E}[\cdot]$ denotes the expectation and $Pr(\cdot)$ is the probability. $\Gamma(m, \tilde{x})$ is the incomplete gamma function given by $\Gamma(m, \tilde{x}) = \int_{\tilde{x}}^{\infty} v^{m-1} e^{-v} dv$, and $\Gamma(m)$ is the gamma function. $P_{d,p,k}^n$ and $P_{fa,p,k}^n$ denote the detection probability and the false-alarm probability. And $P_{md,p,k}^n$ denotes the probability of the miss-detection.

In the next sub-phase, sensing nodes report detection results to AP, which makes the global decision follow the OR fusion rule [3]

$$S_p^n = \sum_{k=1}^{2L} a_{p,k}^n = \begin{cases} \geq 1, & \hat{O}_p^n \\ 0, & \hat{V}_p^n \end{cases} \tag{10}$$

The decision at the kth sensing node is reported to AP and expressed by $a_{p,k}^n \in \{0,1\}$ for binary phase shift keying (BPSK) modulation. S_p^n denotes a parameter that clearly identifies the state of the subcarrier n (unoccupied or occupied by the pth PU). We assume that the distance between any two sensing nodes (i.e., SUs and relays) is much smaller than the distance from any sensing nodes to the primary transmitters, so that the received signal at every sensing node experiences almost identical path loss. Therefore, we can assume that we have independent and identically distributed (i.i.d.) Rayleigh fading with the instantaneous SNRs of the received signal from PUs at sensing nodes on the subcarrier n. Based on the above, we can take false-alarm probabilities $P_{fa,p,k}^n$ as identical since $P_{fa,p,k}^n$ is independent of k, and the global decision of false-alarm probability can be denoted by P_{fa}^n (i.e., $Pr(\hat{O}_p^n|V_p^n)$). In the case of the AWGN channel, the detection probabilities at the sensing nodes are independent of k, so that the detection probabilities are identical and the global decision is expressed by P_d^n (i.e., $Pr(\hat{O}_p^n|O_p^n)$). Similarly, taking the global decision of the missing-detection probability as P_{md}^n (i.e., $Pr(\hat{V}_p^n|O_p^n)$).

$$P_{fa}^n = 1 - \prod_{k=1}^{2L} (1 - P_{fa,p,k}^n) \approx 1 - (1 - P_{fa,p,k}^n)^{2L} \tag{11}$$

$$P_{md}^n = \prod_{k=1}^{2L} P_{md,p,k}^n \tag{12}$$

$$P_d^n = 1 - P_{md}^n \tag{13}$$

Considering the error probability (i.e., P_e^n) of the reporting channel on the subcarrier n, we change the expression of the miss-detection probability as

$$P_{md}^n = \prod_{k=1}^{2L} [P_{md,p,k}^n(1 - P_e^n) + (1 - P_{md,p,k}^n)P_e^n] \tag{14}$$

3.2   SINR Expression (AF Protocol)

A dual-hop communication link is considered. The first hop instantaneous SINR on the subcarrier n is denoted by $SINR_{l,1}^n$, the second hop is $SINR_{l,2}^n$. For AF protocol [12], the expression of equivalent SINR of SU link is

$$\begin{aligned} SINR_{l,eq}^n &= T(SINR_{l,1}^n, SINR_{l,2}^n) \\ &= \frac{SINR_{l,1}^n SINR_{l,2}^n}{SINR_{l,1}^n + SINR_{l,2}^n + 1} \end{aligned} \tag{15}$$

where

$$T(x,y) = \frac{xy}{x+y+1} \tag{16}$$

where $x = SINR_{l,1}^n$, $y = SINR_{l,2}^n$.

3.3   Interference Constraint

In order to guarantee the QoS of PUs, the transmit power of SUs and relays should be probably controlled. Since there is half-duplex scheme at relay nodes, the interference to the pth PU in each hop can be written as

$$I_{SP_p} = \sum_{l=1}^{L}\sum_{n=1}^{N} Pr\left(O_p^n\right) P_{md}^n P_{l,1}^n |h_{l,p,1}^n|^2 \tag{17}$$

$$I_{RP_p} = \sum_{l=1}^{L}\sum_{n=1}^{N} Pr\left(O_p^n\right) P_{md}^n P_{l,2}^n |h_{l,p,2}^n|^2 \tag{18}$$

where $Pr(O_p^n)$ is a probability that the subcarrier n is occupied by the pth PU. I_{SP_p} and I_{RP_p} are the interference produced by all SU-Ts and all relay transmitters, which must be limited by the interference temperature (IT) constraints.

4   Proposed Algorithm

The BER expressions at SU-R for multiple quadrature amplitude modulation (MQAM) (19) or multiple phase shift keying (MPSK) modulation (20) [13] over the AWGN channel are written as

$$BER_{l,MQAM}^n = \frac{4}{b}\left(1 - \frac{1}{\sqrt{M}}\right) Q\left(\sqrt{\frac{3b(SINR_{l,eq}^n/b)}{M-1}}\right) \tag{19}$$

$$BER_{l,MPSK}^n = \frac{2}{b} Q\left(\sqrt{2b \times (SINR_{l,eq}^n/b)\sin^2(\frac{\pi}{M})}\right) \tag{20}$$

where $Q\left(\bar{x}\right) = \frac{1}{\sqrt{2\pi}} \int_{\bar{x}}^{\infty} e^{-\frac{w^2}{2}} dw$ is a Gaussian Q-function. $b = \log_2 M$, M is the number of bits of the modulation symbols.

In this paper, a worst-channel-state-information (worst-CSI) PC algorithm is presented to limit total BER of SUs, which only needs to operate algorithm in one link that CSI is worst, while keeping the interference leakage to PUs below the IT level, and the maximum transmit power of SU and the relay below certain thresholds. Here we introduce the SINRs at SU-R and relay in order to guarantee the requirement for each hop. Thus, the optimization problem is formulated as **OP1**

OP1 $\quad \min\limits_{P_{l,1}^n, P_{l,2}^n} \quad \max\limits_{\forall l} \quad BER_l^n$

$$s.t. \quad C1: 0 \leq \sum_{n=1}^{N} P_{l,1}^n \leq P_{l,1}^{max}, \quad \forall l$$

$$C2: 0 \leq \sum_{n=1}^{N} P_{l,2}^n \leq P_{l,2}^{max}, \quad \forall l$$

$$C3: SINR_{l,1}^n \geq SINR_{l,1,th}^n, \quad \forall l, \forall n \tag{21}$$

$$C4: SINR_{l,2}^n \geq SINR_{l,2,th}^n, \quad \forall l, \forall n$$

$$C5: \sum_{l=1}^{L} \sum_{n=1}^{N} Pr(O_p^n) P_{md}^n P_{l,1}^n |h_{l,p,1}^n|^2 \leq I_{p,th}, \quad \forall p$$

$$C6: \sum_{l=1}^{L} \sum_{n=1}^{N} Pr(O_p^n) P_{md}^n P_{l,2}^n |h_{l,p,2}^n|^2 \leq I_{p,th}, \quad \forall p$$

where $P_{l,1}^{max}$ and $P_{l,2}^{max}$ are the maximum power budgets of SU-T and relay. $SINR_{l,1,th}^n$ and $SINR_{l,2,th}^n$ are the SINR thresholds at the relay and SU-R. $I_{p,th}$ is the interference threshold prescribed by the pth PU receiver. C1 and C2 represent the transmit power constraints of the secondary system. C3 and C4 are the SINR constraints to keep basic communication requirements of SUs. C5 and C6 denote the IT constraints at the source and relay nodes. Since the objection of **OP1** is a monotonic function about the equivalent SINR (i.e., $SINR_{l,eq}^n$), so **OP1** can be converted into

$$\textbf{OP2} \quad \max\limits_{P_{l,1}^n, P_{l,2}^n} \quad \min\limits_{\forall l} \quad SINR_{l,eq}^n \tag{22}$$
$$s.t. \quad C1 \sim C6$$

therefore, the original optimization problem (i.e., **OP1**) becomes a worst-CSI SINR maximization problem (i.e., **OP2**). The criterion about selecting the worst-CSI user is given by

$$\left|h_{l,1}^n\right|^2 \left|h_{l,2}^n\right|^2 \leq \left|h_{j,1}^n\right|^2 \left|h_{j,2}^n\right|^2 \tag{23}$$

If the channel gain of two hops can satisfy (23), we regard the lth SU as the worst-CSI user. **OP2** is not convex due to the constraints C3 and C4. In order to simplify theoretical analysis, we take C3 and C4 on reciprocal, such as

$$C3: \frac{1}{SINR_{l,1}^n} \leq \frac{1}{SINR_{l,1,th}^n} \tag{24}$$

$$C4: \frac{1}{SINR_{l,2}^n} \leq \frac{1}{SINR_{l,2,th}^n} \tag{25}$$

i.e.,

$$\frac{\frac{N_{l,1}^n}{|h_{l,1}^n|^2} + \sum_{p=1}^{P} P_p^n \frac{|g_{p,l,1}^n|^2}{|h_{l,1}^n|^2}}{P_{l,1}^n} \leq \frac{1}{SINR_{l,1,th}^n} \tag{26}$$

$$\frac{\frac{N_{l,2}^n}{|h_{l,2}^n|^2} + \sum_{p=1}^{P} P_p^n \frac{|g_{p,l,2}^n|^2}{|h_{l,2}^n|^2}}{P_{l,2}^n} \leq \frac{1}{SINR_{l,2,th}^n} \tag{27}$$

where $N_{l,1}^n$ and $N_{l,1}^n$ denote the additive noise power at the lth relay and SU-R. Define

$$\begin{cases} F_{l,1}^n = \frac{N_{l,1}^n}{|h_{l,1}^n|^2} \\ F_{l,2}^n = \frac{N_{l,2}^n}{|h_{l,2}^n|^2} \end{cases} \tag{28}$$

$$\begin{cases} G_{p,l,1}^n = \frac{|g_{p,l,1}^n|^2}{|h_{l,1}^n|^2} \\ G_{p,l,2}^n = \frac{|g_{p,l,2}^n|^2}{|h_{l,2}^n|^2} \end{cases} \tag{29}$$

Then the equivalent SINR is

$$SINR_{l,eq}^n = \frac{a_l^n P_{l,1}^n b_l^n P_{l,2}^n}{a_l^n P_{l,1}^n + b_l^n P_{l,2}^n + 1} \tag{30}$$

where a_l^n and b_l^n are given by

$$\begin{cases} a_l^n = \dfrac{1}{F_{l,1}^n + \sum\limits_{p=1}^{P} P_p^n G_{p,l,1}^n} \\ b_l^n = \dfrac{1}{F_{l,2}^n + \sum\limits_{p=1}^{P} P_p^n G_{p,l,2}^n} \end{cases} \tag{31}$$

Further more, to make the equivalent SINR tractable, we adopt the following approximation [14]

$$SINR_{l,eq}^n \approx \frac{a_l^n P_{l,1}^n b_l^n P_{l,2}^n}{a_l^n P_{l,1}^n + b_l^n P_{l,2}^n} \tag{32}$$

Define $P_{l,1}^n = x_1$, $P_{l,2}^n = x_2$, $t = \frac{1}{SINR_{l,eq}^n}$, then

$$t = \frac{1}{SINR_{l,eq}^n} = \frac{1}{b_l^n}\frac{1}{x_2} + \frac{1}{a_l^n}\frac{1}{x_1} \tag{33}$$

Therefore, **OP2** can be rewritten as

$$\begin{aligned}
&\textbf{OP3} \quad \min_{x_1,x_2} \quad \max_{\forall l} \quad t \\
&s.t. \quad C1: 0 \le \sum_{n=1}^{N} x_1 \le P_{l,1}^{max}, \quad \forall l \\
&\qquad C2: 0 \le \sum_{n=1}^{N} x_2 \le P_{l,2}^{max}, \quad \forall l \\
&\qquad C3: \frac{1}{a_l^n}\frac{1}{x_1} \le \frac{1}{SINR_{l,1,th}^n}, \quad \forall l, \forall n \\
&\qquad C4: \frac{1}{b_l^n}\frac{1}{x_2} \le \frac{1}{SINR_{l,2,th}^n}, \quad \forall l, \forall n \\
&\qquad C5: \sum_{l=1}^{L}\sum_{n=1}^{N} Pr(O_p^n) P_{md}^n x_1 |h_{l,p,1}^n|^2 \le I_{p,th}, \quad \forall p \\
&\qquad C6: \sum_{l=1}^{L}\sum_{n=1}^{N} Pr(O_p^n) P_{md}^n x_2 |h_{l,p,2}^n|^2 \le I_{p,th}, \quad \forall p
\end{aligned} \tag{34}$$

Now **OP3** is a convex problem which can be solved by the dual decomposition method [15].

First, we give a Lagrange function with the Lagrange multipliers $\lambda_{l,1}, \lambda_{l,2}, \lambda_{l,3}^n, \lambda_{l,4}^n, \lambda_{p,5}, \lambda_{p,6} \geq 0$ as follows

$$
\begin{aligned}
L(t, &\{\lambda_{l,1}\}, \{\lambda_{l,2}\}, \{\lambda_{l,3}^n\}, \{\lambda_{l,4}^n\}, \{\lambda_{p,5}\}, \{\lambda_{p,6}\}) \\
&= t + \sum_{l=1}^{L} (\lambda_{l,1}(\sum_{n=1}^{N} x_1 - P_{l,1}^{max})) \\
&+ \sum_{l=1}^{L} (\lambda_{l,2}(\sum_{n=1}^{N} x_2 - P_{l,2}^{max})) \\
&+ \sum_{l=1}^{L} (\sum_{n=1}^{N} \lambda_{l,3}^n(\frac{1}{a_l^n}\frac{1}{x_1} - \frac{1}{SINR_{l,1,th}^n})) \\
&+ \sum_{l=1}^{L} (\sum_{n=1}^{N} \lambda_{l,4}^n(\frac{1}{b_l^n}\frac{1}{x_2} - \frac{1}{SINR_{l,2,th}^n})) \\
&+ \sum_{p=1}^{P} (\lambda_{p,5}(\sum_{l=1}^{L}\sum_{n=1}^{N} Pr(O_p^n)P_{md}^n x_1 |h_{l,p,1}^n|^2 - I_{p,th})) \\
&+ \sum_{p=1}^{P} (\lambda_{p,6}(\sum_{l=1}^{L}\sum_{n=1}^{N} Pr(O_p^n)P_{md}^n x_2 |h_{l,p,2}^n|^2 - I_{p,th}))
\end{aligned}
\tag{35}
$$

The dual problem of the Lagrange function (35) is

$$
\begin{aligned}
D(t, &\{\lambda_{l,1}\}, \{\lambda_{l,2}\}, \{\lambda_{l,3}^n\}, \{\lambda_{l,4}^n\}, \{\lambda_{p,5}\}, \{\lambda_{p,6}\}) \\
&= \sum_{l=1}^{L} (\sum_{n=1}^{N} \min_{x_1,x_2} L_l^n(t, \lambda_{l,1}, \lambda_{l,2}, \lambda_{l,3}^n, \lambda_{l,4}^n, \{\lambda_{p,5}\}, \{\lambda_{p,6}\})) \\
&- \sum_{l=1}^{L} (\lambda_{l,1} P_{l,1}^{max}) - \sum_{l=1}^{L} (\lambda_{l,2} P_{l,2}^{max}) \\
&- \sum_{l=1}^{L} (\sum_{n=1}^{N} \lambda_{l,3}^n \frac{1}{SINR_{l,1,th}^n}) - \sum_{l=1}^{L} (\sum_{n=1}^{N} \lambda_{l,4}^n \frac{1}{SINR_{l,2,th}^n}) \\
&- \sum_{p=1}^{P} (\lambda_{p,5} I_{p,th}) - \sum_{p=1}^{P} (\lambda_{p,6} I_{p,th})
\end{aligned}
\tag{36}
$$

Define L_l^n as a function of x_1 and x_2

$$
\begin{aligned}
L_l^n(t, &\lambda_{l,1}, \lambda_{l,2}, \lambda_{l,3}^n, \lambda_{l,4}^n, \{\lambda_{p,5}\}, \{\lambda_{p,6}\}) \\
&= t + \lambda_{l,1} x_1 + \lambda_{l,2} x_2 + \lambda_{l,3}^n \frac{1}{a_l^n}\frac{1}{x_1} + \lambda_{l,4}^n \frac{1}{b_l^n}\frac{1}{x_2} \\
&+ x_1 \sum_{p=1}^{P} \lambda_{p,5} Pr(O_p^n) P_{md}^n |h_{l,p,1}^n|^2 \\
&+ x_2 \sum_{p=1}^{P} \lambda_{p,6} Pr(O_p^n) P_{md}^n |h_{l,p,2}^n|^2
\end{aligned}
\tag{37}
$$

Since the primal problem in (34) is convex, strong duality holds, the dual problems can be solved by an iterative manner using the gradient projection method [15]. The Lagrange multipliers in (35) can be updated by the sub-gradient method [15].

By the Karush-Kuhn-Tucker (KKT) conditions, the optimal transmit power $P_{l,1}^n$ and $P_{l,2}^n$ at SU-T and relay can be calculated by $\frac{\partial L_l^n}{\partial x_1} = 0$ and $\frac{\partial L_l^n}{\partial x_2} = 0$, such as

$$P_{l,1}^n{}^* = x_1^* = \sqrt{\frac{\frac{1}{a_l^n}(1 + \lambda_{l,3}^n)}{\lambda_{l,1} + \sum_{p=1}^{P} \lambda_{p,5} Pr(O_p^n) P_{md}^n |h_{l,p,1}^n|^2}} \tag{38}$$

$$P_{l,2}^n{}^* = x_2^* = \sqrt{\frac{\frac{1}{b_l^n}(1 + \lambda_{l,4}^n)}{\lambda_{l,2} + \sum_{p=1}^{P} \lambda_{p,6} Pr(O_p^n) P_{md}^n |h_{l,p,2}^n|^2}} \tag{39}$$

Finally, taking the optimal solutions $P_{l,1}^n{}^*$ and $P_{l,2}^n{}^*$ into (19) and (20) respectively, the optimal BER can be calculated.

The computational complexity can be roughly analyzed as follows. The optimal solutions of power allocation in an OFDMA network requires exhaustive search to find an optimal subcarrier allocation scheme for SUs. Since the number of subcarriers is N, the computational complexity at subcarrier allocation phase is $\mathcal{O}(N)$. Since there are P pairs PUs, the computational complexity of outer loop requires a complexity of $\mathcal{O}(P)$. In order to calculate the Lagrange multipliers $\lambda_{p,5}$ and $\lambda_{p,6}$, we should evaluate whether the interference power at PU receiver is below the interference threshold, i.e., $\sum_{l=1}^{L} \sum_{n=1}^{N} Pr(O_p^n) P_{md}^n x_1 |h_{l,p,1}^n|^2 \leq I_{p,th}$

and $\sum_{l=1}^{L} \sum_{n=1}^{N} Pr(O_p^n) P_{md}^n x_2 |h_{l,p,2}^n|^2 \leq I_{p,th}$, which introduce $\mathcal{O}(I \cdot D)$, where $\mathcal{O}(D)$ is complexity of finding $P_{l,1}^n{}^*$ and $P_{l,2}^n{}^*$ under the conditions of convergence respectively. Therefore, the total computational complexity of the proposed algorithm is the sum of complexities of the aforementioned steps as $\mathcal{O}(N)\mathcal{O}(P)\mathcal{O}(I \cdot D) = \mathcal{O}(NPID)$, where I is the number of iterations in algorithm.

5   Numerical Results

In this section, we present numerical results to show the effectiveness of the proposed algorithm. We assume that there are four SUs and relays (i.e., $L = 4$), one PU (i.e., $P = 1$), and four subcarriers (i.e., $N = 4$), and each SU occupies one subcarrier. Similar to [16], the normal values of the interference channel gains $h_{l,p,1}^n$, $h_{l,p,2}^n$, $g_{p,l,1}^n$ and $g_{p,l,2}^n$ are selected from the interval $(0, 0.3)$ respectively. The normal values of the channel gains $h_{l,1}^n$ and $h_{l,2}^n$ are randomly chosen from the interval $(0, 1)$ respectively. We set the target SINR on each subcarrier at SU-R and relay is $SINR_{l,1,th}^n / SINR_{l,2,th}^n = 3\,\mathrm{dB}$. The maximum transmit power of each SU-T and relay is $P_{l,1}^{max} / P_{l,2}^{max} = 1.5\,\mathrm{mW}$. We also assume that $Pr(O_p^n)$ is same for every subcarrier, e.g., $Pr(O_p^n) = 0.1$. The background noise power on each subcarrier is assumed to be identical and equal to $0.01\,\mathrm{mW}$, i.e.,

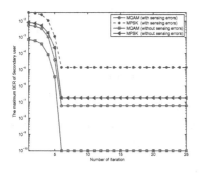

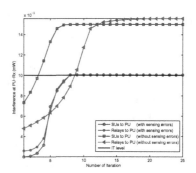

Fig. 2. Convergence of maximum BER under $P_{md}^n = 0.1$ and $Pr(O_p^n) = 0.1$.

Fig. 3. Convergence of interference under $P_{md}^n = 0.1$ and $Pr(O_p^n) = 0.1$.

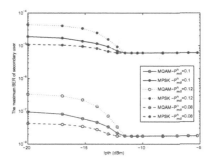

Fig. 4. Maximum BER under different $I_{p,th}$ and P_{md}^n under $Pr(O_p^n) = 0.1$.

Table 2. Maximum BER at SU-R for different P_{md}^n

Modulation form	$P_{md}^n = 0.08$	$P_{md}^n = 0.10$	$P_{md}^n = 0.12$
BPSK	1.102e-5	1.898e-5	4.412e-5
QPSK	9.401e-4	1.247e-3	1.936e-3
16PSK	9.777e-2	1.010e-1	1.064e-1
2QAM	4.264e-8	9.523e-8	3.309e-7
4QAM	9.401e-4	1.247e-3	1.936e-3
16QAM	6.618e-2	6.610e-2	7.367e-2

$N_{l,1}^n = N_{l,2}^n = 0.01$ mW [7]. The simulation results are presented from Figs. 2, 3 and 4 and Table 2.

Figure 2 shows that the maximum BER performance of selected SU link. The BER of the proposed algorithm for the given IT level $I_{p,th} = 0.01$ mW is higher than that of the PC algorithm without sensing errors, while providing the protection of PU when SUs share spectrum opportunistically. From Fig. 2, we can see that the maximum BER under the proposed algorithm for both MPSK and MQAM modulation quickly converges to the stable point, and the optimization goal is achieved by minimizing the maximum BER of the worst-CSI channel to limit the total BER of SUs. Briefly, the purpose of minimizing the BER of the system is obtained by adjusting transmit power of SU-T and relay, which improves the performance and insure the QoS of SUs.

From Fig. 3, we find that our PC algorithm under the imperfect spectrum sensing can guarantee the interference power is always below the IT level, whereas the PC algorithm without the sensing errors results in the actual received interference power exceeds the allowable region. Combining Fig. 2 with Fig. 3, we get a conclusion that the proposed algorithm can well provide the protection for PU at the cost of its BER increases.

In Fig. 4, we depict the maximum BER versus the IT level from $I_{p,th}=-20$ dBm to $I_{p,th}=-5$ dBm of our proposed algorithm for different P_{md}^n. Figure 4 shows that the maximum BER performance against $I_{p,th}$ and P_{md}^n for MPSK and MQAM ($M=2$) modulation. For a given P_{md}^n, for example, $P_{md}^n=0.1$, the maximum BER of SUs first decreases as the increasing interference power constraint and then keep flat because of the maximum transmit power constraint. What's more, we find that the BER performance under different P_{md}^n of our proposed algorithm is same when $I_{p,th}$ is large, for example, larger than -12 dBm, and the BER performance for $P_{md}^n=0.08$ is the best of three when $I_{p,th}$ is low. In fact, from another perspective, the interference power constraint stands for the distance, with the increasing distance between SU and PU, more transmit power is allocated to achieve lower BER.

Table 2 shows that the maximum BER versus different P_{md}^n, for the given $I_{p,th}$, for example, $I_{p,th}=0.01$ mW, and the transmission data for MPSK and MQAM ($M=2$, 4 and 16) modulation. From Table 2, we know that the spectrum sensing requirement is improved from $P_{md}^n=0.12$ to $P_{md}^n=0.08$ for the given modulation methods (i.e., MPSK and MQAM), and the maximum BER of the system decreases accordingly. The reason is that, with an improved spectrum sensing requirement, a spectrum hole would be detected more accurately thus less interference occurs between the primary network and the secondary network, resulting in decreased BER for the secondary transmission. Furthermore, we also find that the maximum BER of the system increases with the increase of the number of bits of the modulation symbols.

6 Conclusion

In this paper, we have investigated the issues on BER problem under the imperfect spectrum sensing in cognitive relay networks. A PC algorithm under maximum transmit power constraints, SINR constraints and interference constraints to guarantee the QoS of PUs is proposed to minimize total BER for all SUs. We find that the maximum BER of the secondary system decreases as the decreasing miss-detection probability. Besides, the proposed algorithm can well protect the communication of PU though there is a little BER increase of the secondary system at the price. In our future research, the PC optimization problem with the introduction of more complicated channels in the underlay cognitive relay networks will be conducted.

Acknowledgment. The work of this paper is supported by National Natural Science Foundation of China under grant No. 61571209.

References

1. Haykin, S.: Cognitive radio: brain-empowered wireless communications. IEEE J. Sel. Areas Commun. **23**(2), 201–220 (2005)
2. Xu, Y.J., Zhao, X.H., Liang, Y.-C.: Robust power control and beamforming in cognitive radio networks: A survey. IEEE Commun. Surveys Tuts. **17**(4), 1834–1857 (2015)
3. Letaief, K.B., Zhang, W.: Cooperative communications for cognitive radio networks. Proc. IEEE **97**(5), 878–893 (2009)
4. Cover, T.M., Gamal, A.E.: Capacity theorems for the relay channel. IEEE Trans. Inf. Theory **25**(5), 572–584 (1979)
5. El, G.A., Cover, T.M.: Multiple user information theory. Proc. IEEE **68**(12), 1466–1483 (1980)
6. Rasti, M., Hasan, M., Le, L.B., Hossain, E.: Distributed uplink powercontrol for multi-cell cognitive radio networks. IEEETrans. Commun. **63**(3), 628–642 (2015)
7. Xu, Y.J., Zhao, X.H.: Robust power control for underlay cognitive radio networks under probabilistic quality of service and interference constraints. IET Commun. **8**(18), 3333–3340 (2014)
8. Wu, Y., Tsang, D.H.: Joint bandwidth and power allocations for cognitive radio networks with imperfect spectrum sensing. Wireless Per. Commun. **57**(1), 19–31 (2011)
9. Wang, S., Zhou, Z.H., Ge, M., Wang, C.: Resource allocation for heterogeneous multiuser ofdm-based cognitive radio networks with imperfect spectrum sensing. In: IEEE INFOCOM, pp. 2264–2272 (2012)
10. Tan, X., Zhang, H., Hu, J.: Capacity maximisation of the secondary link in cognitive radio networks with hybrid spectrum access strategy. IET Commun. **8**(5), 689–696 (2014)
11. Digham, F.F., Alouini, M.-S., Simon, M.K.: On the energy detection of unknown signals over fading channels. IEEE Trans. Commun. **55**(1), 21–24 (2007)
12. Liu, Z., Yuan, H., Li, H., Guan, X., Yang, H.: Robust power control for amplify-and-forward relaying scheme. IEEE Commun. Lett. **19**(2), 263–266 (2015)
13. Xu, X., Bao, J., Cao, H., Yao, Y., Hu, S.: Energy efficiency based optimalrelay selection scheme with a ber constraint in cooperative cognitive radionetworks. IEEE Trans. Veh. Technol., January 2015. doi:10.1109/TVT.2015.2389810
14. Ge, M., Wang, S.: Energy-efficient power allocation for cooperative relaying cognitive radio networks. In: IEEE Wireless Communications and Networking Conference (WCNC), pp. 691–696, April 2013
15. Boyd, S., Vandenberghe, L.: Convex Optimization. Cambridge Univ. Press, Mar, Cambridge (2004)
16. Setoodeh, P., Haykin, S.: Robust transmit power for cognitive radio. Proc. IEEE **97**(5), 915–939 (2009)

TOA Based Localization Under NLOS in Cognitive Radio Network

Dazhi Bao[✉], Hao Zhou, Hao Chen, Shaojie Liu, Yifan Zhang,
and Zhiyong Feng

Key Laboratory of Universal Wireless Communications,
Ministry of Education, Wireless Technology Innovation Institute (WTI),
Beijing University of Posts and Telecommunications, Beijing 100876,
People's Republic of China
baodazhi@bupt.edu.cn

Abstract. In this paper, we consider cooperative localization of primary users (PU) in a cognitive radio network (CRN) using time-of-arrival (TOA). A two-step none-line-of-sight (NLOS) identification algorithm is proposed for the situation where both NLOS error distribution and channel model are not available. In the first step the TOA measurements are clustered into groups. The groups with a dispersion higher than a predefined threshold are identified as NLOS and discarded. In order to make the threshold more reasonable, Ostu's method, a threshold selection method for image processing is utilized. The second step is introduced to correct the error of possible surviving NLOS. To increase the accuracy of estimated position when line-of-sight (LOS) paths are limited, we proposed a result reconstruction method. Simulation results show that our algorithm can effectively identify NLOS paths and improve positioning accuracy compared to existing works.

Keywords: Cognitive radio network · LOS identify · Time of arrival · Location estimation · Least square method

1 Introduction

The available spectrums are very limited due to the character of electromagnetic wave itself. Cognitive radio has emerged as a promising technology to improve the spectrum utilization dramatically. One of the most important tasks for CRNs is to detect the presence or absence of primary users (PUs), which is called the spectrum sensing technique. In cognitive radio technology, there are two types of users-the primary (licensed) user and the secondary (unlicensed) user (SU). The PUs have the right of priority in using a certain frequency spectrum. The SUs on the other hand have restricted access to the available unused frequency spectrum. The SUs are allowed to use the frequency spectrum only if they do not interfere with the PUs. Information about the locations of PUs can allow cognitive networks to identify spectrum holes in space more reliably and accurately and perform location-aware intelligent routing and power control mechanisms in

© ICST Institute for Computer Sciences, Social Informatics and Telecommunications Engineering 2016
D. Noguet et al. (Eds.): CROWNCOM 2016, LNICST 172, pp. 668–679, 2016.
DOI: 10.1007/978-3-319-40352-6_55

a CRN [1,2]. Hence, locating the PUs position in a CRN is an important but challenging task.

Networks similar with the wireless sensor networks (WSN) are composed when SUs proceed cooperative spectrum sensing, which SUs are similar with the wireless sensor nodes, PUs are similar with the unknown source node (USN). Therefore, wireless sensor networks localization algorithms can be used in acquiring position of PUs of CRNs.

TOA, signal strength, and angle-of-arrival (AOA) legacy location estimation techniques can be considered as candidates for localization of PUs. AOA techniques are mostly implemented by means of antenna arrays which not suitable for rich multipath environments. On the other hand, signal strength based methods provide high accuracy only for the short ranges. Moreover, the performance of the estimator for signal strength techniques depends on the channel parameters that CRNs cannot control. Since the accuracy of TOA techniques mainly depends on the parameter that transceiver can control, it is the most suitable location estimation technique [3]. Time-of-arrival (TOA) based positioning system is also widely used for positioning in WSNs. Accurate synchronization and none-line-of-sight (NLOS) are two significant problems of TOA. Accuracy of synchronization mostly depends on bandwidth.

NLOS is another notorious factor in positioning system for degrading the accuracy of estimated results, benefiting from a large number of anchor nodes (ANs) in WSNs distributing around the detection region, much measurement information can be obtained. Four LOS paths are enough to achieve accurate TOA localization. As a result, it is generally assumed that four LOS paths exist. Therefore, we can simply identify and discard the NLOS paths and utilize the LOS for TOA positioning to improve accuracy.

A lot of study has been undertaken to deal with the problem of NLOS identification. In [4] an NLOS identification method for mobile location estimation is proposed. NLOS identification is achieved by comparing the standard deviation of TOA measurement with a threshold calculated from historical measurement noise. However, in order to obtain reliable result, the threshold needs to be determined by field experiment. In [5–8], a class of channel estimation based NLOS identification algorithms is proposed. In these methods, NLOS identification can be accomplished by examining the statistics of the multipath channel coefficients. The problem of these methods is that both signal model and the channel model are needed. Some localization algorithms require a-priori knowledge of the probability density function (PDF) of NLOS noise. In [9], a distribution test model for NLOS identification is formulated, where the positioning error is modeled as Gaussian zero mean. [10] proposed a residual weighting(RW) algorithm, in which the weight of every sensor is calculated by summing up the weighted residuals of all possible combinations and the one with the heaviest weight is identified as NLOS.

In this paper, we proposed a two-step NLOS identification algorithm which none priori information is needed. In the first step, based on the fact that NLOS errors often result in high dispersion of estimated results, we cluster all the measurements into groups. And the groups with dispersion higher than a threshold are abandoned. Considering the similarity of application scene, A threshold selection method for image processing is utilized to determine the value of threshold. We find some NLOS may survive if they are grouped with certain measurements. So in the second step, a reliability factor is defined for each remaining measurement. Then the top four measurements with the largest reliability factors are utilized for position estimating. Considering that the assumption of 4 LOS measurements may be rejected, we proposed an estimated result reconstruction method to increase the robustness under NLOS environment. Finally, simulation combined with actual measurement data is conducted to verify the performance of our algorithm and compare it with existing works.

The rest of the paper is organized as follows. In Sect. 2, we analysis the impact of NLOS errors on the estimated results. In Sect. 3, our proposed two-step algorithm for NLOS identification is detailed. In Sect. 4, the estimated result reconstruction method is introduced. Simulation results are presented to demonstrate the reliability of the method in Sect. 5 and conclusion is presented in Sect. 6.

2   System Model

2.1   LS Method for Location Estimation in TOA System

The system model under consideration is for TOA-based location estimation. There are N sensors and one USN to be localized. We define $X = (x, y)$ as the real position of the USN, $\hat{X} = (\hat{x}, \hat{y})$ the estimation of the USN location, $X_i = (x_i, y_i)$ the position of the ith sensor, $\hat{d}_i$ is the measured distance between the USN and the ith sensor. The simplified assumption of $\hat{d}_i$, which can be expressed as

$$\hat{d}_i = ct_i = d_i + v_i, \tag{1}$$

where t_i is the measured transmission time, c is the speed of light, and $d_i = \sqrt{(x_i - x)^2 + (y_i - y)^2}$ is the actual distance between the USN and the ith sensor.

$$v_i = \begin{cases} e_i, & if\ ith\ path\ is\ LOS \\ e_i + n_i, & if\ ith\ path\ is\ NLOS \end{cases} \tag{2}$$

is the total error, e_i and n_i are the TOA measurement noise and the NLOS error respectively. We assume that e_i is a Gaussian random variable with zero mean and variance σ_i. The PDF of n_i is assumed to be unknown in this paper.

Based on the above signal model, LS method in [12] can be utilized to estimate the location of USN.

Let $R = \sqrt{x^2 + y^2}, R_i = \sqrt{x_i^2 + y_i^2}$. A simplified equation incorporating all the information for localization is

$$h = G\theta + v, \tag{3}$$

where

$$h = \begin{bmatrix} \hat{d}_1^2 - R_1^2 \\ \hat{d}_2^2 - R_2^2 \\ \vdots \\ \hat{d}_N^2 - R_N^2 \end{bmatrix}, G = \begin{bmatrix} -2x_1 & -2y_1 & 1 \\ -2x_2 & -2y_2 & 1 \\ \vdots & \vdots & 1 \\ -2x_N & -2y_N & 1 \end{bmatrix} \qquad (4)$$

are the constant vector and coefficient matrix respectively.

$\theta = \begin{bmatrix} x & y & R^2 \end{bmatrix}^T$ is the vector we are to estimate eventually with $v = \begin{bmatrix} v_1^2 - 2\hat{d}_1 v_1 & v_2^2 - 2\hat{d}_2 v_2 & \cdots & v_N^2 - 2\hat{d}_N v_N \end{bmatrix}^T$ being the estimation error. The least square solution can be expressed as

$$\hat{\theta} = \left(G^T G\right)^{-1} G^T h. \qquad (5)$$

2.2 Deviation Caused by NLOS Noise

A minimum TOA system is illustrated in Fig. 1. Three sensors are involved to estimate the location of USN in 2-dimentional location system. If all the sensors are LOS paths, the three circles can almost intersect at the same point, which will be the estimated position of the USN. However, if one sensor (AN_3) suffers from NLOS noise, it will lead to a large fuzzy localization area. Therefore, the final estimated result will be inaccurate. So the size of fuzzy area can be used to determine the existence of the NLOS paths in a certain group. As a result, the relationship between the fuzzy area and NLOS errors should be analyzed.

Assume that one of the three sensors used for estimating the location of USN suffers from NLOS error, and the other two are LOS paths. Thus the estimation

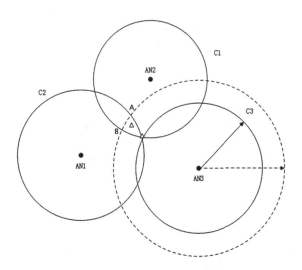

Fig. 1. NLOS effect on the TOA location estimation

error will be introduced into positioning result. According to (5), the equation which contains NLOS error can be

$$\begin{bmatrix} x + n_x \\ y + n_y \\ R^2 + n_{R^2} \end{bmatrix} = \left(G^T G \right)^{-1} G^T \begin{bmatrix} (d_1 + n_{NLOS})^2 - R_1^2 \\ d_2^2 - R_2^2 \\ d_3^2 - R_3^2 \end{bmatrix}, \tag{6}$$

where n_{NLOS} is the measurement error caused by NLOS noise. n_x and n_y are the estimation errors of X-coordinate and Y-coordinate caused by NLOS noise respectively. n_{R^2} is the deviation of R^2. The deviation between the estimated location and the actual location is

$$\bar{d} = \sqrt{n_x^2 + n_y^2}. \tag{7}$$

The value of $\bar{d}$ only has concern with the value of n_x and n_y, which can be calculated by

$$\begin{bmatrix} n_x \\ n_y \\ n_{R^2} \end{bmatrix} = \left(G^T G \right)^{-1} \begin{bmatrix} -2x_1 \\ -2y_1 \\ 1 \end{bmatrix} \left(2d_1 n_{NLOS} + n_{NLOS}^2 \right). \tag{8}$$

According to (8), we can see that the effect of NLOS noise on estimated results not only depends on the value of NLOS errors themselves, but also concerns with point coordinates of all the sensors. That is to say, when a certain NLOS point is introduced into different ANs combinations, the deviations of results are different. So it is difficult to make quantitative analysis, but we can summarize two qualitative conclusions to provide the theoretical support for our work.

- NLOS errors will surely cause fuzzy localization area, which can lead to inaccurate results.
- NLOS measurements may not cause large positioning errors when the certain NLOS measurements are introduced into specific groups. So the information obtained from some of the NLOS paths can be used to improve the positioning results when LOS paths are limited.

3   NLOS Identification

In view of the above-mentioned fact, a reasonable two-step method is proposed to determine and discard NLOS measurements in the situation without any previously known knowledge of NLOS error distribution or channel model. Then the coordinate of the USN can be calculated by LS method with the information of LOS measurements.

3.1   The First Step: Group Decision

We define a deviation factor Δd to reflects the deviation between the estimated position and the actual position.

$$\Delta d = \sqrt{(x - \hat{x})^2 + (y - \hat{y})^2}. \tag{9}$$

However, Δd is unavailable for (x, y) is unknown. So we need another approximate parameter to replace Δd.

We assume that there are N sensors available for location estimation. All the N sensors are divided into $M = C_N^K$ groups, where K is the number of sensors in each group. As was mentioned above, the minimum number of sensors required for 2-D TOA location estimation is 3. So C_K^3 estimated point coordinates of USN can be obtained in each group. After calculating the distances between all two different estimated points combinations, the distance with the maximum value $\Delta\hat{d}_k (1 \leq k \leq C_N^K)$ is defined as the dispersion degree of location results in the kth group G_k.

$$\Delta\hat{d}_k = \max_{1 \leq i,j \leq C_K^3, i \neq j} \sqrt{(\hat{x}_{ki} - \hat{x}_{kj})^2 + (\hat{y}_{ki} - \hat{y}_{kj})^2}, \tag{10}$$

where $(\hat{x}_{ki}, \hat{y}_{ki})$ and $(\hat{x}_{kj}, \hat{y}_{kj})$ are the two different estimated point coordinates in group G_k.

Then a threshold γ is set to make a distinction between the two extreme dispersion degrees cause by measurement errors and NLOS errors separately. If $\Delta\hat{d}$ is smaller than γ, that means all the paths in the group are LOS, or the deviation caused by NLOS errors is not large enough. On the contrary, at least one NLOS path must exist.

Ostu's method [11] was used in [13] to determine a threshold to divide two peaks in gray images. Considering the similarity of the application scenario, it is also utilized to obtain a reasonable threshold γ.

In the previous work, totally M dispersion degrees were obtained. With a interval (e.g. 10 m), we divide all these values of dispersion degrees into l intervals. Let $t = \lceil \gamma/10 \rceil$, the value level ranges within $G = \{1, 2, \cdots, l\}$ can be divided into two classes, as $C_0 = \{1, 2, \cdots, t\}$ and $C_1 = \{t+1, t+2, \cdots, l\}$. We define the between-class variance σ_B^2 and total-variance σ_T^2 as

$$\sigma_T^2 = \sum_{i=0}^{l-1} (i - \mu_T)^2 P_i, \quad \sigma_B^2 = \omega_0\omega_1(\mu_1 - \mu_0)^2, \tag{11}$$

where

$$
\begin{aligned}
P_i &= n_i/n, \quad \omega_0 = \sum_{i=1}^{t} P_i, \quad \omega_1 = 1 - \omega_0 \\
\mu_T &= \sum_{i=1}^{l} iP_i, \quad \mu_t = \sum_{i=1}^{TH} iP_i, \quad \mu_0 = \frac{\mu_t}{\omega_0} \\
\mu_1 &= \frac{\mu_T - \mu_t}{1 - \omega_0}.
\end{aligned} \tag{12}
$$

Here, n_i indicates the number of dispersion degrees in ith interval. $n = \sum_{i=1}^{l} n_i$ is the total number of dispersion degrees. For a selected threshold t, the class probabilities ω_0 and ω_1 represent the portions of areas occupied by object and NLOS classes respectively.

The optimal threshold can be determined by maximizing the following criterion function against the threshold.

$$\eta = \frac{\sigma_B^2}{\sigma_T^2}. \tag{13}$$

After calculating both dispersion degrees $\Delta\hat{d}_k$ and the threshold γ, we can determine whether there exist NLOS measurements in a certain group. But unfortunately, according to the conclusion we draw from (8), a small part of NLOS measurements may survive even when the groups meet the threshold condition. As a result, we propose a method to further find 4 LOS sensors from the result of the first step.

3.2   The Second Step: Weighted Ranking

Assume that there are L groups and S sensors remained in the result of group decision. Use $G_k, k = 1, 2, \cdots L$ to denote the kth group and $A_i, i = 1, 2, \cdots S$ to denote the ith sensors. For each group G_k, every AN in it will be assigned a weight of W_i, which is calculated by

$$W_k = 1/\Delta\hat{d}_k. \tag{14}$$

After evaluating all the L groups, A_i will have a total weight w_i by summing the weights of all the groups it belongs to.

$$w_i = \sum_{k=1}^{L} \lambda W_k \begin{cases} \lambda = 0, & if \quad A_{i^\cdot} \notin G_k \\ \lambda = 1, & if \quad A_{i^\cdot} \in G_k \end{cases}. \tag{15}$$

Rank these sensors according to their total weights. The 4 sensors with the heaviest weights are determined to be LOS sensors. And the final estimated position of USN can be obtained by solving the equation established only by the measurement information of these 4 sensors.

3.3   Geometric Dilution of Precision

Position results will deviate right positions a lot even if all the WSNs are LOS when all WSNs are in the similar direction of the USN. So, we define Geometric Dilution of Precision(GDOP) to determine if the group is effective.

$$GDOP = \sqrt{trace(G^T G)^{-1}} \tag{16}$$

G can be obtained from (4). The value of GDOP reflects the reversibility of coefficient matrix. The smaller of the value of GDOP, more accurate of the position results. Threshold of GDOP can be set up to determine if the WSNs group can be used to judge NLOS. From this step, some effective groups could be eliminated, complexity values can be reduced the reasonable.

Algorithm 1. Two-step Algorithm for NLOS Identification

Input: The TOA measurements $\hat{d}_i$ of the signals received by all the available sensors and the position coordinates (x_i, y_i) of all sensors;

Output: The estimated coordinate of USN;

1: Group all the N available sensors into all possible groups with different K sensors; Assume that a network with 10 sensors, there will be C_{10}^K groups.

2: Separate K sensors in a certain group into all possible combinations with different 3 sensors. And there will be C_K^3 combinations. Using all the 3 sensors combinations to estimate the position of USN;

3: Calculating $\Delta\hat{d}_k$ for group G_k;

4: Comparing $\Delta\hat{d}_k$ with the threshold TH. If $\Delta\hat{d}_k$ is smaller than TH, every AN in group G_k should be assigned a weight of W_k. Calculating the total weight of each AN by summing the weights of all the groups it belongs to.;

5: Ranking all the sensors according to their weights. Using the measurements of 4 sensors with heaviest weights to estimate the position of USN.

---

4 Estimated Result Reconstruction

We have assumed that there are at least four LOS paths exist. So that the best accuracy can be achieved by identifying and discarding NLOS measurements, and only use the LOS measurements. However, under practical situation, this assumption may be rejected. Hence, we should use as much information as possible to obtain a relatively accurate estimated result, such as the NLOS measurements with small NLOS errors. So the estimated result reconstruction(ERR) method is proposed.

In the phase of group decision, we have acquired L groups meet the threshold condition. As mentioned above, $M = C_K^3$ estimated coordinates can be obtained in each group. We assume that $\hat{X}_{k,i} = (x_{k,i}, y_{k,i})$, $1 \leq k \leq L$ and $1 \leq i \leq M$ is the ith estimated coordinate in kth group G_k. We define the variance of estimated results in G_k as

$$V_k = \frac{1}{M} \sum_{i=1}^{M} \sqrt{(\hat{x}_{k,i} - \bar{x}_k)^2 + (\hat{y}_{k,i} - \bar{y}_k)^2}, \tag{17}$$

where

$$(\bar{x}_k, \bar{y}_k) = \left(\frac{1}{M} \sum_{i=1}^{M} \hat{x}_{k,i}, \frac{1}{M} \sum_{i=1}^{M} \hat{y}_{k,i} \right). \tag{18}$$

Moreover, an estimated coordinate $(\hat{x}_k, \hat{y}_k)$ that include the information of all the measurements in G_k can be retrieved with the least square method.

For the kth group G_k, we define a weight as

$$\lambda_k = \frac{\zeta_k}{\sum_{j=1}^{L} \zeta_j}, \tag{19}$$

where $\zeta_k = 1/V_k$ is the reciprocal of G_k's variance. And the final estimated position can be calculated by

$$(\hat{x}, \hat{y}) = \left(\sum_{k=1}^{L} \lambda_k \hat{x}_k, \sum_{k=1}^{L} \lambda_k \hat{y}_k \right). \tag{20}$$

Although the accuracy of estimated result obtained by the proposed ERR method will be lower than that obtained by using only the LOS measurements, the ERR method is practical when the number of LOS measurements is small.

5   Performance Analysis

In this section, simulation experiments are carried out to show the performance of the proposed NLOS identification method and ERR method. The measurement data was obtained by a filed measurement conducted around one of the television signal transmission towers in Beijing, China. Our purpose is to estimated the position of the transmission tower. We collected the digital television signal to estimate the TOA between the transmitter and the receiver. The instruments used include Agilent N6841A RF sensor and a laptop computer.

Totally 40 measurements are obtained, consisting of 20 obvious LOS paths and 20 NLOS paths. Each time we randomly select $M(0 \leq M \leq 10)$ NLOS measurements and $10 - M$ LOS measurements for simulation test. Every result is obtained from the average of 10,000 independent simulation experiences.

5.1   Success Rate of NLOS Identification

Figure 2 compares the success rate of finding 4 LOS measurements by using only the group decision algorithm and the two-step algorithm. In the former case, the 4 sensors being selected are the measurements appearing the most in the result of group decision. And the size of each group is set to be 4.

As shown in Fig. 2, compared with group decision method, the two-step method can obtain more accurate estimated results. At the same time, it also has better performance than the RW algorithm proposed in [7].

Further simulation is made to study the effects of groups' size K on the accuracy of the results.

As can be seen in Fig. 3, if there are enough LOS measurements available for NLOS identification, it can be sure to find 4 measurements with LOS paths successfully. The result of this simulation proves that the size of groups in the first step of this algorithm will affect the accuracy of estimated results. When the number of LOS measurements is fixed, the group with larger size will obtain a more accurate result. Figure 3 also shows that when the size of groups K is larger than the number of existent LOS measurements, it will fail to identify LOS measurements for the reason that at least one NLOS AN exists in the result.

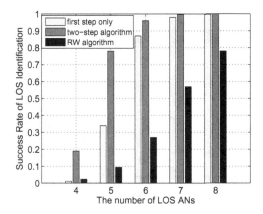

Fig. 2. Success rate of identification with different algorithm

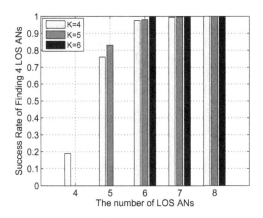

Fig. 3. Success rate of identification with different size of groups

5.2   The Accuracy of Position Estimation

Figure 4 shows the root-mean-square error (RMSE) of estimated results obtained by utilizing three different methods. The accuracy is low when all the TOA measurements are used to estimate the position of USN directly. By using the two-step algorithm, 4 optimal measurements are selected to estimate the final result. The simulation shows that it really helps improve the accuracy of estimated result. But it performs poorly when the number of LOS paths is fewer than four for the measurements with large NLOS errors are introduced into LS equations. ERR method is not particularly accurate in LOS environments, but it is robust to NLOS measurements. So it can be used when the number of LOS connections is limited.

Finally, it needs to be explained that why we didn't analyze the false alarm rate and the effects of a false detection in the positioning precision. As mentioned

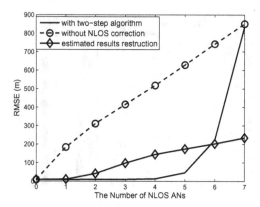

Fig. 4. Comparison of location error before and after using proposed algorithm

above, in the situation where the LOS paths are limited, both LOS measurements and the NLOS measurements with small NLOS errors are utilized to estimated the position of UNS in our proposed ERR method and the estimated results with certain precision can be obtained. For this reason, the accuracy of estimated results is more significant than alarm rate.

6 Conclusion

In this paper, we have proposed a two-step algorithm for NLOS identification in CRNs similar with WSNs through studying the effect of NLOS noise on estimated results. Compared with other methods, the proposed method can achieve NLOS error identification without any priori knowledge of NLOS environment. We divide the available measurements into all possible groups with the same size. Then the estimated results are used to determine the existence of NLOS paths in a certain group. At the same time, the weight calculated by the maximum distance between the estimated results is added to each node in the group. The ranking of the total weight for each AN determines the LOS ANs. The groups with larger size can obtain more accurate estimated results. But this will increase the requirements of both computational complexity and the number of LOS paths. So the second step of the proposed method is proposed to balance the computation complexity and the recognition accuracy. Simulation results demonstrate that the two-step algorithm performs well in determining LOS measurements, especially when the number of LOS paths is larger than 4. Accurate estimated position of USN(PU) can be obtained through solving the LS equation established by the LOS measurements information. Considering that the two-step algorithm performances poorly when the number of LOS paths is limited, we propose the ERR method to make full use of the information extracted from available measurements. Simulation shows that the ERR method is robust to NLOS environment.

Acknowledgement. This work was supported by the National Natural Science Foundation of China (61227801), the National Key Technology R&D Program of China (2014ZX03001027-003).

References

1. Saeed, N., Nam, H.: Robust multidimensional scaling for cognitive radio network localization. IEEE Trans. Veh. Technol. **64**(9), 4056–4062 (2015)
2. Tandra, R., Mishra, S.M., Sahai, A.: What is a spectrum hole and what does it take to recognize one? Proc. IEEE **97**(5), 824–848 (2009)
3. Celebi, H., Arslan, H.: Cognitive positioning systems. IEEE Trans. Wireless Commun. **6**(12), 4475–4483 (2007)
4. Wylie, M.P., Holtzman, J.: The non-line of sight problem in mobile location estimation. In: IEEE International Conference Universal Personal Communications, pp. 827–831, September 1996
5. Guvenc, I., Chong, C.-C., Watanabe, F.: NLOS identification and mitigation for UWB localization systems. In: IEEE Wireless Communications and Networking Conference, pp. 1571–1576, March 2007
6. Venkatesh, S., Buehrer, R.M.: Non-line-of-sight identification in ultra-wideband systems based on received signal statistics. IET Microwaves Antennas Propag. **1**(6), 1120–1130 (2007)
7. İsmail, G., et al.: NLOS identification and weighted least-squares localization for UWB systems using multipath channel statistics. EURASIP J. Adv. Signal Process. **2008**(1), 1–14 (2007)
8. Mucchi, L., Marcocci, P.: A new UWB indoor channel identification method. In: International Conference on Cognitive Radio Oriented Wireless Networks and Communications, pp. 58–62, August 2007
9. Chan, Y.T., Tsui, W.Y., So, H.C., Ching, P.-C.: Time-of-arrival based localization under NLOS conditions. IEEE Trans. Veh. Technol. **55**(1), 17–24 (2006)
10. Cong, L., Zhuang, W.: Non-line-of-sight error mitigation in TDOA mobile location. In: Global Telecommunications Conference, pp. 680–684, November 2001
11. Ostu, N.: A threshold selection method from gray-scale histogram. IEEE Trans. Syst. Man Cybern. SMC **8**, 62–66 (1979)
12. Cheung, K.W., So, H.C., Ma, W.K., Chan, Y.T.: Least squares algorithms for time-of-arrival-based mobile location. IEEE Trans. Signal Process. **52**(4), 1121–1130 (2004)
13. Ye, X., Cheriet, M., Suen, C.Y.: TStroke-model-based character extraction from gray-level document images. IEEE Trans. Image Process. **10**(8), 1152–1161 (2001)

Standards, Policies and Business Models

Business Models for Mobile Network Operators Utilizing the Hybrid Use Concept of the UHF Broadcasting Spectrum

Seppo Yrjölä[1(✉)], Petri Ahokangas[2], and Pekka Talmola[3]

[1] Nokia Networks, Oulu, Finland
seppo.yrjola@nokia.com
[2] Oulu Business School, Oulu, Finland
Petri.Ahokangas@oulu.fi
[3] Nokia Technologies, Salo, Finland
Pekka.hk.Talmola@nokia.com

Abstract. This paper explores and presents business models for mobile network operators (MNOs) in the novel hybrid use spectrum sharing concept of the Ultra High Frequency broadcasting spectrum (470–790 MHz) used for Digital Terrestrial TV and Mobile Broadband. The created business models indicate that MNOs would benefit significantly from the more flexible use of the UHF bands. New business models would enable them to gain faster access to new potentially lower cost, licensed, below 1 GHz spectrum in order to cope with increasing data traffic asymmetry, and to offer differentiation through personalized broadcasting and new media services. As a collaborative benefit with the broadcasting domain, the concept opens up new converging business opportunities in delivering TV and media content using MBB network with means to introduce this flexibly. Moreover, it will significantly re-shape the business ecosystem around both broadcasting and mobile broadband by introducing new co-opetitive business opportunities in business and technology towards 5G.

Keywords: Broadcasting · Business model · Mobile broadband · Mobile network operator · Spectrum sharing · UHF · 5G

1 Introduction

The mobile broadband (MBB) usage is growing at an increasing pace [1], placing growing needs for the scarce radio spectrum resource. As mobile data traffic is increasingly consisting of downstream video [2], asymmetry in mobile broadband traffic is increasing: the average downlink traffic in Europe is eight times the uplink. Changing consumer usage habits and high capacity demand anytime and anywhere put Mobile Network Operators against a disruptive change. At the same time, Digital Terrestrial Television (DTT) as the main delivery vehicle for the TV media content has been challenged by the alternative content delivery mechanisms, Over the Top (OTT) services and higher spectrum fees for all using the UHF band. Even if consumers' interest in TV content remains and even increases, the ways how TV content is delivered and consumed

© ICST Institute for Computer Sciences, Social Informatics and Telecommunications Engineering 2016
D. Noguet et al. (Eds.): CROWNCOM 2016, LNICST 172, pp. 683–694, 2016.
DOI: 10.1007/978-3-319-40352-6_56

will, and has, already started to change. Users are increasingly receiving TV content via cable, satellite, fixed broadband and, especially, via MBB, and at the same time, changing their consumption habits from linear real time to non-linear usage with the growing demand for interactivity [3].

With these sights to the future, spectrum regulators are, on one hand, considering gradually compressing and withdrawing some DTT licenses of lower demand and repurposing these frequencies for MBB. On the other hand, in order to continue fulfilling the national Public Media Service (PSM) obligations, the most used and, in particular national broadcasters', DTT licenses will continue to the foreseeable future as long as required. The traditional spectrum auctioning & re-farming process is becoming increasingly difficult in the future due to high costs, time needed, and difficulties in finding unused exclusive spectrum needed for the re-allocation process. Spectrum sharing where systems operate in the same spectrum band, has lately received growing interest among regulators considering new ways of fulfilling the different spectrum demands and to meet the mobile traffic growth while maintaining the rights of the original incumbent systems operating in the bands [4]. This business environment transformation influences the broadcaster spectrum holders, and opens up new business opportunities, as well as risks due to increasing pressure for innovative flexibility and sharing in the spectrum usage. To date, broadcasting (BC) community has not been offered incentives to change their spectrum usage. On the contrary, we have seen unilateral acts from regulators and MNOs towards further compressing DTT bands to give room for new MBB spectrum. The UHF broadcast spectrum was originally from 470 to 862 MHz, and 800 MHz band (790–862) is now been deployed for MBB use throughout the Europe. The World Radiocommunication Conference (WRC) already in 2012 made a decision on the 700 MHz band to be used for the MBB after the WRC-15 [5]. As a part of new IMT spectrum identification point of discussion, the WRC'15 addressed the co-primary allocation with mobile of the lower UHF band (470–694 MHz) that currently has a primary allocation to broadcasting [6]. Further, the FCC in the USA has lately made a decision on 600 MHz incentive auctions [7].

The co-existence between MBB and DTT on Digital Dividend (DD) spectrum has been widely addressed in regulation and standardization forums and supported by extensive research. The DD1 at 800 MHz and interleaved UHF spectrum concept has widely been studied by the Federal Office of Communications (OFCOM), e.g., [8, 9], focusing on the performance of the DVB-T receiver in the presence of interference from real LTE signals. In [10], co-existence of the DTT and the LTE in the 700 MHz band was analyzed based on system level simulations, and in [11] extended through laboratory measurements and link budget analysis. In [12], the analysis of the interference between the digital terrestrial multimedia broadcast (DTMB) and the LTE below 698 MHz was discussed. In the reference [13], generic requirements for the co-existence between DVB-T/T2 and LTE for fixed outdoor and portable indoor DTT reception is summarized based on the system level Monte Carlo simulations. In [14], Antonopoulos proposed additional physical infrastructure sharing deployments and architecture scenarios. Regulatory and system architecture scenarios towards 5G are discussed, e.g., in [15, 16].

In the recent European spectrum debate, the European Commission (EC) set up a High Level Group consisting of mobile and broadcast sectors to deliver strategic advice on the future use of the UHF spectrum. Accordingly, The European Conference of Postal and Telecommunications Administrations (CEPT) set up Task Group 6 (TG6) "Long term vision for the UHF broadcasting band" [17], to identify and analyze possible scenarios for the development of the band, taking into account technology and service development. In this paper, we focus on analyzing the scenario of *hybrid usage of the band by DTT and MBB*. In support of this scenario, the EC released a decision proposal in February 2016 to limit the terrestrial use other than BC on this band to downlink-only [18]. We considered the hybrid use of UHF, and its key enabling technologies in general, to represent one of these new emerging concepts that are expected to reshape business models and whole business ecosystems within the BC and MBB sectors [19]. This reshape is expected to provide new opportunities for value creation and capture with innovative business models for the key stakeholders. Previous works on business models for shared DTT spectrum use are limited as focus has been on TV White Spaces (TVWS) concept, e.g., [20, 21]. The general business drivers, enablers and potential impacts of the spectrum sharing on the MBB market were described in [19] and incentives and strategic dynamic capabilities for the key stakeholders in the hybrid use of the UHF were discussed in [22]. Furthermore, there are several studies on the optimal contract design, e.g., [23]. However, in earlier research there is no complete MNO business model related to the hybrid use of UHF discussed, as the focus has been on identifying the opportunities and discussing the business model only regarding a limited amount of business elements. The purpose of this paper is to explore and discuss MNO's business model transformation when they are doing business based on hybrid shared used of the UHF spectrum. Particularly, we are focusing on the European regulatory regime. This paper seeks to answer the following research questions:

(1) What are the business opportunities the hybrid usage concepts could open for MNOs?
(2) What are the key changes it may bring to MNOs´ current business models?
(3) What kind of business models MNOs may build on the identified business opportunities?

The research methodology applied in this paper is the anticipatory action learning in a future-oriented mode [24]. The business models presented are developed by utilizing the capacity and expertise of the policy, business and technology research communities.

The rest of this paper is organized as follows. First, the hybrid DTT MBB usage concept is presented in Sect. 2. Theoretical background for business models is introduced in Sect. 3. The research methodology applied and the business models for MNOs in using hybrid concept are derived in Sect. 4. And finally, conclusions are drawn in Sect. 5.

2   The Hybrid Use of UHF Broadcasting Spectrum Concept

In their vision work, the CEPT Task Group 6 created the following scenarios how the UHF band 470–694 MHz can accommodate the delivery of the TV content as well as provide additional capacity for the MBB [25]:

- Class A: Primary usage of the band by existing and future DVB terrestrial networks.
- *Class B: Hybrid usage of the band by DVB and/or downlink LTE terrestrial networks.*
- Class C: Hybrid usage of the band by DVB and/or LTE (including uplink) terrestrial networks.
- Class D: Usage of the band by future communication technologies.

In particular, the spectrum sharing scenario in the class B introduces a flexible way of transferring TV channels to mobile use while maintaining capability to deliver TV content both in conventional living room large screen use cases as well as in new mobile use cases on smart phones and tablets. In the following analysis, TV media content, consumption and delivery mechanisms are considered as different matters, and they need to be separated. Although the users interested in the TV media content remain at the same level or even increases, the ways how TV content is delivered and received will be and have already been changed. Increasingly, users are receiving TV programs via cable, satellite, fixed broadband and, especially, via MBB. In addition, non-linear usage is greatly increasing as well as the demand for interactivity.

Recent studies show that the demand and the value for DTT as the main delivery mechanism of TV content will decrease [3]. Based on this, it could be assumed that some 'underutilized' and lower valued TV frequencies will be reassigned and or shared with mobile use. As the freed TV channels can be different in different geographical areas and countries, we propose them to be assigned first for the MBB supplemental downlink (SDL) use only. The SDL is more compatible, compared with the traditional Frequency Division Duplex (FDD) or the Time Division Duplex (TDD) use, with the remaining DTT to be used in the country or across the national borders. The freed TV channels could be taken into mobile use in a flexible way by using functionalities that are already developed for shared spectrum access like, e.g., recently widely discussed Licensed Shared Access (LSA) concept [25], allowing different time schedules in different regions and countries, if needed. The SDL Carrier Aggregation (CA) technology [26] allows both the 'traditional' MBB DL and the LTE evolved Multimedia Broadcast Multicast Service Broadcast (eMBMS) [27] flexibly used for optimizing the capacity on demand. The SDL use would also support the trend that the future MBB traffic is strongly asymmetric towards downlink direction.

The evolution scenario of the hybrid use concept of the UHF spectrum can follow the market demand. Potential evolutionary scenarios for Europe are illustrated in the Fig. 1. Already in the first phase, the hybrid SDL CA concept could speed up the take-off of the 700 MHz through better co-existence characteristics with across-the-border TV transmitters. The spectrum usage can evolve so that the DTT use could be moved towards the lower end of the 470–694 MHz band, as more spectrum is freed from DTT. It should also be noted that as the SDL base stations start replacing the DTT frequencies one by one locally, there is no change in the availability of interleaved spectrum used for example for Program Making and Special Events (PMSE). Depending on the national regulation and market demand, it should also be possible in the long term vision to fully migrate to the LTE using either the SDL and/or the eMBMS to deliver TV content and hence completely replace current DTT technologies with converged delivery platform [28].

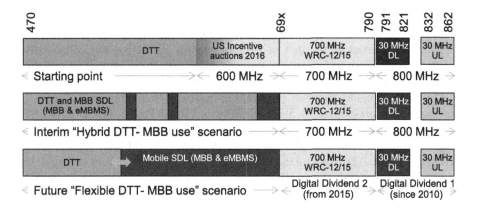

Fig. 1. Evolution of the UHF band usage with the novel hybrid use scenario.

3   Business Model Elements

Business research provides us with numerous examples of business model concepts and elements utilized. Often referred to [29] defines the business model as consisting of nine elements: key partners, key activities, key resources, value proposition, customer relationships, customer segments, channels, cost structure and revenue streams. The other view widely used in analyzing new ventures is to refer business model to comprise of the elements: strategic choices, value proposition, value creation and delivery system and value capture [30]. Traditional approaches, however, include several limitations; they do not build around the business opportunity, have only weak linkages to the systemic complexities of the business context, depict structures rather than activities, and lack the element locations prevalent in current businesses. In this paper, for the MNO we adopt the approach and conceptualization presented in [31], that helps to answer to the concerns discussed above in business modeling, and consists of the following elements:

(1) *What*: Offer, value proposition, customer segmentation, unique differentiation
(2) *How*: Key operations, basis of competitive advantage, mode of delivery, selling, marketing
(3) *Why*: Base of pricing, way of charging, cost elements and cost drivers
(4) *Where*: Location of activities/elements perspective (i.e., are activities carried out internally of by external partners) of all the preceding items.

4   Analysis of the MNO Business Models

The research methodology applied, business models created and their analysis are summarized in this section.

The Business model scenarios presented in this paper were created utilizing the Anticipatory Action Learning (AAL) approach that is a particular action research method conducted in a future-oriented mode [24]. In developing foresight, the method represents a unique style of questioning the future from transformational point of view, using business model as the unit of analysis. In this interactive and collaborative approach, conversation and dialog among cross-disciplinary participants, from multiple domains concerned with the research project is essential.

The business model elements presented in this paper were created in a series of future-oriented workshops in 2015, organized by the Finnish Future of the UHF (FUHF) research project utilizing the capacity and expertise of the policy, business and technology research communities. The research process comprised (1) identifying the critical change factors, (2) assessing their impact and possible consequences on key stakeholders, (3) building and selecting the scenario axes, (4) creating the business scenarios, and (5) evaluating them. Foresight, by definition, is future focused and its reliability and validity cannot be controlled. Instead, the qualitative focus of research is in how probable, plausible, and preferable the outcomes appear. Also, the collaborative and conversation based method how the futures were created was regarded as way to ensure the quality of the research. [32]

4.1 Business Models

Using the above summarized, future-oriented action research method. We created business models for MNOs deploying the hybrid UHF concept applying the business model framework from [31]. Potential changes in the business models caused by the deployment of the novel spectrum sharing concept were analyzed by creating foresighted business models. In the traditional model, the MNOs are using exclusively licensed IMT spectrum possibly including upper DD spectrum bands (e.g. 800 MHz). The future hybrid UHF business scenario is based on the additional flexible shared access to lower UHF spectrum. The business models were created applying the above discussed format where elements responding Why, What and How questions, are presented in the form of rectangles inside the sector in question. Internal and external operations, the Where question, is depicted in the locations of the rectangles as shown in the Fig. 2.

MNO business model in the traditional exclusive UHF licensing case.
We started with sketching business models for the MNO in the traditional case, where exclusive spectrum bands without additional downlink UHF bands are utilized. The developed business model is shown in Fig. 2. In general, an MNO wishes to maintain and grow its current market position. The overall opportunity for the MNO is to serve as a "Mobile data pipe" or a "Mobile smart data pipe" corresponding to acting merely as an access channel or providing services on top of the access, respectively.

In the *What* element of the business model, the offering is MBB services that guarantee mobility, high data rates and services to customers. Both in consumer and enterprise customer segments customers at large may be treated as a mass. MNOs' offering mainly consists of voice, messaging and data services. Traditionally, customer lock-in has been achieved via subscription. Lately, Quality of Service (QoS), in particular data speed, as well as the bundling of subscriptions and services has become important elements of value proposition and in achieving customer retention.

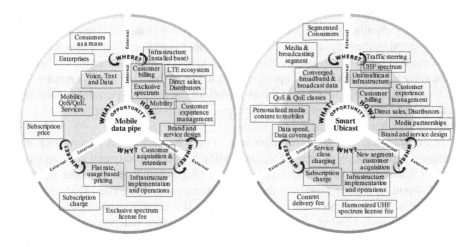

Fig. 2. MNO's "Mobile data pipe" business model in the traditional exclusive license case and the Smart ubicast business model scenario for the hybrid UHF usage case.

The key operations in the *How* element that the MNO wants to keep under its own control are customer interface and billing. In this case, competitive advantages are based on infrastructure and exclusive spectrum licenses, which guarantee QoS for end users and operational long-term certainty needed for the large infrastructure investments for the MNO. MNOs seize the harmonized and scaled 3GPP LTE ecosystem within their infrastructure. Exclusive long-term licenses and installed base infrastructure provide a strong position against new entrants who would need big investments to support full services (and coverage) to customers. The traditional sales mode include direct sales, own shops and distributors such as retail chains. MNOs exploit their existing customer base and related customer data in marketing, sales, and service design and are in unique position to leverage customer big data analytics, which offers them an opportunity to strengthen their position against new entrants.

Considering the *Why* sector, pricing is based on flat or usage based prices with regular subscription charges. The key technical cost drivers include expenditures related to implementing and operating the infrastructure with the real estate of the mobile broadband spectrum license fee. Customer acquisition and retention play essential role in the heavily competitive MBB industry.

MNO Business Model in the Hybrid UHF Case.
After the common insight on the present business model state-of-the-art, we developed a new business model for the situation where new spectrum bands based on the hybrid shared use with DTT on the UHF spectrum becomes available as presented in Fig. 2. The basic opportunity for the MNO with UHF is to gain access to lower cost spectrum and seek growth when courting to meet the growing mobile broadband data traffic needs. Gaining faster access to new licensed low frequency UHF spectrum presents an opportunity for the MNO to build data coverage with favorable propagation, and build market penetration more cost-efficiently than what would be the case of densifying the existing

infrastructure in the current exclusive higher spectrum bands. Additional downlink only spectrum in particular helps an operator to cope firstly with rising downlink asymmetry mainly caused by the video content. Secondly, it helps to open up a real option to deliver media and traditional broadcasting content in their MBB networks. An additional benefit of the early introduction of the co-primary hybrid use of the UHF bands, compared to traditional exclusive licensed, is in the avoidance of the lengthy spectrum re-farming, clearing and cross-border optimization process, which provides faster access to new spectrum on a harmonized basis.

In the *What* element, the MNO continues to offer MBB services to consumer and enterprise customers. With new additional UHF downlink capacity and coverage, the operator would differentiate in the competition by offering enhanced data rates and tailored different QoS level classes to different customer segments. The MNO can take advantage of the new spectrum so that to balance the capacity demand and service supply. As discussed earlier, content is increasingly provided independent of the platform and today's consumers have a choice of DTT and MBB platforms to be used as a delivery vehicle for the linear and non-linear media content. As a collaborative benefit, the concept opens up new business opportunities in delivering TV and media content using MBB network with the means to introduce this flexibly. A combination of the broadcast eMBMS and unicast with the SDL CA technology could generate a very efficient and flexibly integrated platform for delivering personalized media content as well as traditional broadband services to mobile users. The availability of user equipment that support the new spectrum bands and enabling SDL CA and eMBMS technologies is essential in introducing the new services. However, the intended technologies are globally harmonized, standardized and under deployment in other spectrum bands.

The basis for the offering of new services and service level in the *How* element to segmented customers lies in the dynamic load and traffic management based on both network parameters as well as the customer experience data. This combination of existing and new flexible downlink unicasting/broadcasting resources enables traffic steering between different radio access technologies and spectrum bands to offer personalized and enhanced Quality of Experience (QoE) to segmented customers. Service level differentiation can be based on strong existing spectrum and infrastructure assets to realize the full benefits of the additional spectrum. The subscriber data management and customer experience management will be unique assets in the design of new services and service levels. In order to expand offering to media distribution with collaborative benefits with media content, providers such as national TV broadcasters and content aggregators distribution channels should be expanded from still valid direct sales and distributors to broadcasters and content providers.

In the business model *Why* element, the service level differentiation could lead to new service level based pricing models charged via subscription fees. Converged media distribution services will introduce new opportunities for revenue sharing e.g., with venue owners, event organizers, content and service providers and advertisement partners. These distribution services could be further expanded to applications, firmware software and Internet of Things updates and content deliveries. The business model cost drivers and elements continue to include infrastructure, implementation and operational expenditures. However, spectrum license costs resulting from the hybrid licenses may differ from those of today's auctions. In fact, the license costs of the hybrid UHF bands will be lower due to restrictions in the conditions of using the band, but at the same time, they could be free from, e.g., coverage obligations.

4.2 Discussion

The developed future MNO business models are summarized in Table 1. The transition in the business model with additional flexible downlink data capacity is not only about avoiding costs, or scouting further growth within the mobile data, but to expand business towards ubiquitous customer experience in the merging media and ICT era. Faster access to QoS licensed below 1 GHz spectrum without mandatory coverage obligations could allow the operators to strengthen their existing market position and enabling new more personalized service level based offerings with enhanced QoS and QoE to different customer segments. New downlink bands complement the current spectrum assets, and offer improved QoE by allowing load balancing and traffic steering to match best the personalized user demand with the network capacity supply.

Table 1. Summary of developed business models.

Case	MNO business model
Traditional exclusive spectrum	Be cost effective mobile "data pipe" or mobile "smart data pipe"
Access to hybrid UHF spectrum	Seek growth through "Smart Ubicast"

The following service opportunities enabled by the flexible UHF business model were identified: (1) Extra *Mobile Broadband capacity and coverage* to cope with asymmetric data traffic benefit, (2) *Public Service Media service* to broadcast aggregated TV channels flexibly via broadcast or unicast, as demand can be met most efficiently, (3) *Live TV/Radio Broadcast*, similar to PSM, with different coverage, content, content protection and funding models, (4) *Event & Venue Casting* delivers premium content services at key events, high density locations like sports stadiums or local service businesses, (5) *Media on Demand* allows numerous users to subscribe to relevant content, e.g., news, sports, stock, weather, and a variety of user generated content either through live broadcasts or device caching, to view them at their convenience, (6) *Off-Peak Media & Software* delivers high demand pre-recorded content, e.g., TV shows, movies, YouTube, subscription content, e.g., eNewspapers, eMagazines and music, applications and firmware updates at off-peak times, and (7) *Internet of Things (IoT)* connects to the clouds to provide ease of management, location-based media services, updates and content deliveries, e.g., smart meters, public TV terminals, connected cars).

5   Conclusion

In this paper, we have developed business models for mobile broadband network operators utilizing UHF spectrum bands with the hybrid shared concept with broadcasting DTT. The concept allows MNOs to access new supplemental downlink UHF licensed

QoS spectrum bands to respond to the growing asymmetric video and new media driven downlink data traffic. In this paper, opportunity driven business models were developed to address the basic questions of a business model: What, How, Why and Where to act regarding the business. The concept discussed could open up new business opportunities for MNOs through dynamic load and traffic management of the significantly increased downlink capacity that can be used to provide different service levels to different customer segments. The lower frequency hybrid UHF bands are a cost-efficient solution for the MNO to respond to the growing data traffic demand in a flexible and timely fashion.

Moreover, the concept could help operators to win over new customers by offering personalized mobile broadband data and "ubicast" media delivery services to selected customer segments. With MBB broadcast concept on hybrid UHF spectrum, e.g., linear, traditional TV broadcast can be extended to mobile devices providing the flexibility to combine linear and non-linear TV, on-demand and interactive TV. This can significantly re-shape the business ecosystem around the mobile broadband and media, and open up new converging and co-operative business opportunities with transforming media and TV industry towards 5G. MNOs are optimally positioned to explore new business model opportunities in parallel with traditional business model.

In the future, hybrid UHF usage concept business modeling studies will need to be expanded to cover also other key stakeholders. In particular, co-operative business model with broadcast domain will be an important aspect to scout.

Acknowledgments. This work has been performed as a part of Future of UHF (FUHF) project. The authors would like to acknowledge the project consortium: Digita Networks, Elisa, Finnish Communications Regulatory Authority, Nokia, RF-tuote, Schneider Finland, Telia Sonera, Turku University of Applied Sciences, University of Turku, VTT Technical Research Centre of Finland, YLE, Åbo Akademi University and Tekes Finnish Innovation Fund.

References

1. ITU-R M.2243: Assessment of the global mobile broadband deployments and forecasts for International Mobile Telecommunications (2011)
2. Cisco white paper: Cisco Visual Networking Index: Global Mobile Data Traffic Forecast Update, 2014–2019 (2015). http://www.cisco.com/c/en/us/solutions/collateral/service-provider/visual-networking-index-vni/white_paper_c11-520862.html
3. Lewin, D., Marks, P., Nicoletti, S.: Valuing the use of spectrum in the EU. GSMA (2013)
4. The White House: Realizing the Full Potential of Government-Held Spectrum to Spur Economic Growth. PCAST Report (2012)
5. ITU-R: Final Acts - WRC-12, Geneva (2012)
6. ITU-R: Provisional Final Acts - WRC-15, Geneva (2015)
7. FCC report 12-118: Broadcast Television Spectrum Incentive Auction NPRM (2012). http://www.fcc.gov/document/broadcast-television-spectrum-incentive-auction-nprm
8. Federal Office of Communications (OFCOM): Coexistence of New Services in the 800 MHz Band with Digital Terrestrial Television-Further Modelling (2012). http://stakeholders.ofcom.org.uk/binaries/consultations/949731/annexes/DTTCo-existence.pdf

9. OFCOM report 2221/PCFT/R/1.2: The co-existence of LTE and DTT services at UHF: a field trial (2011). http://www.ofcom.org.uk/static/research/co-existenceLTEDTTservices atUHF.pdf

10. Kim, D.-H., Oh, S.-J., Woo, J.S.: Coexistence analysis between IMT system and DTV system in the 700 MHz band. In: International Conference on ICT Convergence (2012)

11. Ribadeneira-Ramírez, J., Martínez, G., Gómez-Barquero, D., Cardona, N.: Interference analysis between digital terrestrial television (DTT) and 4G LTE mobile networks in the digital dividend bands. IEEE Trans. Broadcasting, to be published

12. Li, W., et al.: Performance and analysis on LTE system under adjacent channel interference of broadcasting system. In: IEEE 12th International Conference on Information Technology, pp. 290–294 (2012)

13. ITU-R WRC-15: Agenda Item 1.2: Lower Edge of Mobile Allocation and Adjacent Band Compatibility, CPG-PTD (13)010 (2013)

14. Antonopoulos, A., Kartsakli, E., Bousia, A., Alonso, L., Verikoukis, C.: Energy-efficient infrastructure sharing in multi-operator mobile networks. IEEE Commun. Mag. **53**(5), 242–249 (2015)

15. Razzac, A.A., et al.: Dimensioning and profit sharing in hybrid LTE/DVB systems to offer mobile TV services. IEEE Trans. Wireless Commun. **21**(12), 6314–6327 (2013)

16. Calabuig, J., Monserrat, J.F., Gomez-Barquero, D.: Fifth Generation mobile networks: A new opportunity for the convergence of mobile broadband and broadcast services. IEEE Commun. Mag. **53**, 198–205 (2015)

17. ECC Report 224: Long Term Vision for the UHF broadcasting band. On long-term vision for the UHF broadcasting band out for public consultation (2014)

18. EC COM/2016/043: Proposal for a Decision of the European Parliament and of the Council on the use of the 470-790 MHz frequency band in the Union (2016)

19. Delaere, S., Ballon, P.: The business model impact of flexible spectrum management and cognitive networks. Info **9**(5), 57–69 (2007)

20. Mwangoka, J., Marques, P., Rodriguez, J.: Exploiting TV White Spaces in Europe: The COGEU Approach (2011). http://www.ict-cogeu.eu/pdf/publications/Y2/IEEE%20 DySPAN2011_COGEU_paper.pdf

21. Luo, Y., Gao, L., Huang, J.: Business modeling for TV white space networks. IEEE Commun. Mag. **53**, 82–88 (2015)

22. Chapin, J., Lehr, W.: Cognitive radios for dynamic spectrum access – The path to market success for dynamic spectrum access technology. IEEE Commun. Mag. **45**(5), 96–103 (2007)

23. Duan, L., Gao, L., Huang, J.: Contract-based cooperative spectrum sharing. In: Dynamic Spectrum Access Networks (DySPAN) IEEE Symposium, pp. 399–407 (2011)

24. Inayatullah, S.: Anticipatory action learning: Theory and practice. Futures **38**, 656–666 (2006)

25. ECC Report 205: Licensed Shared Access (2013)

26. 3GPP technical report TR 36.808: Evolved Universal Terrestrial Radio Access (E-UTRA); Carrier Aggregation; Base Station (BS) radio transmission and reception (2012)

27. 3GPP TS 25.346: Multimedia Broadcast/Multicast Service (MBMS); Protocols and Codecs

28. Yrjölä, S., Ahokangas, P., Matinmikko, M., Talmola, P.: Incentives for the key stakeholders in the hybrid use of the UHF broadcasting spectrum utilizing Supplemental Downlink: A dynamic capabilities view. In: International Conference on 5G for Ubiquitous Connectivity (5GU) (2014)

29. Osterwalder, A., Pigneur, Y.: Business Model Generation. John Wiley and Sons, Hoboken (2010)

30. Richardson, J.: The business model: an integrative framework for strategy execution. Strateg. Change **17**, 133–144 (2008)
31. Ahokangas, P., Juntunen, M., Myllykoski, J.: Cloud computing and transformation of international e-Business models. In: Sanchez, R., Heene, A.: Building Competences in Dynamic Environments, in Research in Competence-Based Management, vol 7, pp. 3–28. Emerald Group, London (2014)
32. Floyd, J.: Action research and integral futures studies: A path to embodied foresight. Foresight **44**, 870–882 (2012)

Co-primary Spectrum Sharing and Its Impact on MNOs' Business Model Scalability

Petri Ahokangas[✉], Kari Horneman, Marja Matinmikko, Seppo Yrjölä, Harri Posti, and Hanna Okkonen

Oulu Business School, University of Oulu, P.O. Box 4600, 90014 Oulu, Finland
{petri.ahokangas,marja.matinmikko,harri.posti,
hanna.okkonen}@Oulu.fi,
{kari.horneman,seppo.yrjola}@nokia.com

Abstract. This paper focuses on inter-operator spectrum sharing, specifically co-primary spectrum sharing (CoPSS), that denotes the case where two or more MNOs (mobile network operators) operate in the same frequency band. Specifically, we discuss the concept and its impact on the mobile network operators' (MNO) business model scalability potential. CoPSS has several technical and business advantages in volatile demand conditions. It highlights predefined policies and rules for sharing, utilization of subscriber and usage profiles for spectrum resource allocation, hybrid business models, value differentiation between exclusive and shared spectrum licenses, utilization of customer experience management systems (CEM) for value differentiation, and utilization of the LTE ecosystem.

Keywords: Co-primary · Spectrum sharing · Business models · MNOs

1 Introduction

Inter-MNO (mobile network operator) spectrum sharing has raised research interests under a variety of terms, concepts and settings. Inter-MNO spectrum sharing denotes the case where two or more MNOs to operate in the same frequency band. In practice, MNOs have been reluctant to consider inter-operator spectrum sharing within currently used spectrum bands. However, research indicates inter-MNO spectrum sharing to be beneficial in bursty and fluctuating traffic/spectrum demand conditions [1]. In addition, spectrum sharing between MNOs is particularly beneficial in small cells where interference can easily be controlled [2]. Also, the gains from sharing differ depending on the user locations within the cells [3].

For MNOs, starting a spectrum sharing based business is a disruptive innovation that changes and challenges their traditional strategy and business models [4]. Co-primary spectrum sharing (CoPSS) is one of the new emerging conceptions enabling inter-MNO spectrum sharing. In the CoPSS concept, licenses are issued for at least two MNOs which agree on the conditions for operating in the given band. The CoPSS concept brings new value creation and capture possibilities, i.e., business model opportunities for MNOs both from technological and business perspective. Typically, MNOs' business foci

© ICST Institute for Computer Sciences, Social Informatics and Telecommunications Engineering 2016
D. Noguet et al. (Eds.): CROWNCOM 2016, LNICST 172, pp. 695–702, 2016.
DOI: 10.1007/978-3-319-40352-6_57

comprise either an aggressive approach where the aim is to generate new revenue from new business opportunities, or a defensive approach where the aim is to increase cost efficiency within existing businesses and operations [5]. However, as CoPSS does not work in isolation but as an addition to MNOs' other business operations, attention needs to be paid to the scalability potential of the CoPSS concept, especially in small cell contexts.

This paper focuses on the aggressive approach, where the possibilities of CoPSS could open up MNOs new business opportunities. Therefore, the purpose of this paper is to *explore the CoPSS elements required for business model scalability among MNOs*. To this aim, the structure of the paper is as follows: after introduction we present the research domain, i.e., the concept and business domain for CoPSS, discuss the business model concept and scalability, and present the business domain for CoPSS. After that, we will present our key findings, the CoPSS elements contributing to business model scalability among MNOs. We end the paper with conclusions.

2   Understanding CoPSS as a Research Domain

This chapter discusses the CoPSS concept, business model concept and scalability, and CoPSS as business domain for MNOs.

2.1   Co-primary Spectrum Sharing, CoPSS

The definition of CoPSS as a dynamic spectrum-sharing concept is currently emerging [6]. The three different elements that help to define CoPSS as a concept comprise the following. First, the *type of the spectrum* authorized for sharing (licensed or license exempt). Second, the *dimensions of shared resources* (temporal, spatial, or spectral, where the first dimension refers to the length of time scale of spectrum sharing-related decisions, the second to the geographical resolution (size of area) of the decisions, and the final one to the resolution in the frequency domain (size of spectrum chunks)). Third, *degree and type of information sharing* (proactive, reactive or enhance intra-operator sharing schemes).

Based on these elements, paper [6] defines CoPSS as follows:

1. CoPSS concerns a specific spectrum band for which licenses are issued for at least two MNOs,
2. These MNOs enter into an agreement regarding the conditions of sharing,
3. CoPSS requires real-time information sharing between the MNOs, information about the type and level of sharing resolution, which is agreed between the MNOs so as to guarantee efficient spectrum sharing,
4. The dynamics of spectrum sharing in CoPSS is considerably high, approaching the level of intra-operator resource allocation, and
5. Sufficient guaranteed QoS is part of CoPSS.

Figure 1 illustrates the CoPSS concept. In this example, two MNOs originally have licensed spectrum bands. Through CoPSS they maintain a part of their licensed band for their own use while offering the remaining part to a shared spectrum pool that can be accessed by both MNOs under agreed terms.

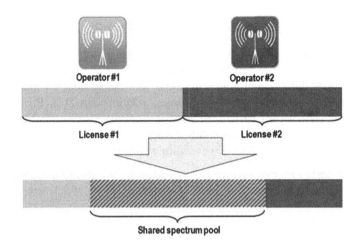

Fig. 1. CoPSS framework.

In the MNO business regarding CoPSS, the spectrum is no longer necessarily owned, but borrowed or co-owned with actors that still view each other as competitors. In this situation, we need to understand the horizontal collaborative aspects of a business relationship in parallel with the vertical, competitive aspects of a business relationship. The main theories around business relationships, such as Network Theory, Industrial Organization Theory or Strategic Alliances Theory, do not describe the dyad situation where companies compete and cooperate with each other's simultaneously [7]. Instead, coopetition is shown to emerge in horizontal relationships. Coopetition [7, p. 412] is "the dyadic and paradoxical relationship that emerges when two firms cooperate in some activities, such as in strategic alliance, and at the same time compete with each other in other activities... And (coopetition) must be regarded as the most advantageous one, when companies in some respect help each other and to some extent force each other toward, for example, more innovative performance." Especially, this can be regarded to hold when the MNOs in coopetition have differing customer profiles, the servicing of which becomes feasible for using the CoPPS band resources.

In general, sharing may take several forms from no sharing to complete sharing (commonly operated infrastructure) [8–11]. Compared to CoPSS, the performance of no sharing is lower, whereas complete sharing might bring about considerable performance improvement. Considering the costs, CoPSS might mean savings in license costs, albeit with own infrastructure. In complete sharing, the problem is the access to sharing arrangements and division of costs across the stakeholders. In no sharing the all stakeholders cover their own costs.

Coopetition has a strategic role [12] as it captures the benefits from both competition and cooperation but it can create a tension between the players. In short, the coopetition is about creating value and capturing value together between two or more players, so called value co-creation and value co-capture. Coopetition can also be seen as an enabler for developing scalable business models.

2.2 Business Model Concept and Scalability

Business model has established its position as conceptual tool that helps managers to translate abstract strategies practical, especially by looking the exploration and exploitation of opportunities and advantages of firms. This can be done by looking at how value is co-created and co-captured in business relationships [13]. Business model concept [13] revolves around business opportunity and answers four questions: what the company is offering to their customers, how it is doing it, where the activities are located, and why the company thinks it can do everything profitably:

1. *What?* Including offering, value proposition, customer segments, and differentiation,
2. *How?* Including key operations, basis of advantage, mode of delivery, selling, and marketing,
3. *Why?* Including basis of pricing, way of charging, cost elements and cost drivers, and
4. *Where?* Including the location of activities, either internally or externally to the firm.

Scalability, one unique characteristic of business, is an important concept when innovating a business model. The scalability of the business model fundamentally stems from the synchronization of a business model to the respective business opportunity. In addition, a scalable company can maintain or improve its profit margins when sales volume increases. There are two approaches related to scalability concept [14]. *Scale-up* is the vertical approach to scale a system (i.e. only one node of the system will be modified by adding more resource), and *scale-out* is the horizontal approach to scale a system (i.e. takes an effect on the whole system by adding more nodes to the system). In this paper, we consider four categories of elements that affect the scalability of CoPSS based business model among MNOs:

1. Mission criticality and uniqueness which shows the value of the concept for the customer,
2. Superior value proposition across value chain describes the strength of the business model compared to other in the value chain,
3. Potential for sustainable, continuous revenue stream, and
4. Location of the business activities.

3 CoPSS and MNO Business Model Scalability

We have evaluated the proposed CoPSS concept according to the presented business model scalability elements by arranging workshops that gathered experts from industry and academia. The identified elements are summarized in Table 1.

Table 1. Elements contributing to MNOs' business model scalability in CoPSS.

CoPSS as mission critical and unique for MNOs	• Licensed spectrum shared among competing MNOs by predefined policies and rules • Spectrum licenses for specific locations or specific universal services • Utilization of subscriber profiles and usage profiles for spectrum resource allocation
CoPSS as providing MNOs a sustainable and continuous revenue stream	• Using by MNO and other stakeholders a hybrid business model that links commerce, content, context and connection based business models, also between MNOs • Shared license may include dedicated share for base operation for each MNO and other stakeholders
CoPSS as providing MNOs a superior value proposition	• Value differentiation between exclusive license and shared license enabled • Opportunity to utilize the whole shared pool as a resource asset by each MNO • Utilization of customer experience management as a tool for value differentiation
Location of the MNOs CoPSS based business	• Business both for location-dependent services and location-independent specific universal services • Utilizing LTE ecosystem for all services to ease the standardization efforts • Virtual network operation for MNOs by local operator

For the category *mission critical and uniqueness* we have identified several features of CoPSS which might make it unique. An essential unique feature of CoPSS is that the licensed spectrum resources are being shared among competing MNOs by predefined policies and rules. The sharing could happen in certain locations like public premises or public places, where each MNO shall offer services its subscribers. The MNOs could have different subscriber profiles with different usage behaviors, and the sharing could be based on the utilization of the differences of the profiles. I.e., at a time one MNO operator might have high need for extra resources to serve its subscribers, the others could serve their customers with their basic resources. Compared to the traditional situation where the extra resources are coming from over Wi-Fi, the licensed spectrum offers continued connectivity without breaks and guaranteed quality of service with quality of service differentiation according to the used pricing model. Additionally, security is inbuilt in MNOs' services. The local nature of the sharing may include offering local services (e.g. multimedia broadcasting, advertisement etc.), which might be provided by (specific) companies located in public premises. The spectrum to be shared could be shared also with venue owners. One specific location for shared spectrum could be roadsides, and there especially for the purpose of vehicle-to-vehicle communications [15].

For the *superior value proposition* category we have also identified several features. A national regulator authority influences the CoPSS type of sharing as it will define indirectly the constraints for sensible pricing of the spectrum. This could be done e.g., by comparing the value of the spectrum sold as exclusive licenses to that of shared licenses. The MNOs will make the same comparison in their side. The value of CoPSS spectrum could be lower than that of the licensed spectrum of the same blot, but an MNO may consider that the part it has bought includes the possibility to utilize other MNOs' spectrum. This extra utilization may give opportunities to extra value proposition to subscribers. Getting this extra value depends heavily on how well MNOs could predict and utilize their subscribers' profiles and usage behaviors in time and across locations. Customer experience management (CEM) has an essential role in this, as it could be the implementation platform for all information gathering and decision making influencing the user experience. There is a need of information what content a user is requesting and what is the most suitable delivery channel for the content, considering user location and time of the delivery (e.g., certain content may allow delayed delivery in a predictable location if user movement is known). The suitability consideration of the delivery channel may include also different pricing models, which, in turn, may take into account the specific features of CoPPS spectrum.

In the *sustainable, continuous revenue stream* category MNOs may consider that in addition subscribers, their customers include also content providers, venue owners and local companies, which form a business network with hybrid business models. The hybrid business models link commerce (e.g. shops, restaurants, etc.) to the content provided (e.g. advertising, info searching, etc.) and to the context information available (e.g. shopping mall) through various connections (both wireless and wire-line). CoPSS spectrum could be considered as a common band for all MNOs and other stakeholders for delivering local content. MNOs and other stakeholders could have a basic share for the common band, enabling the generation of a base revenue to operate, and extra revenue could come from the underutilized shares of the other MNOs and stakeholders in that location.

Considering the *location* category, as mentioned earlier in text, the locations for CoPSS business are assumed to be limited places, for example such as public premises or roadsides, but there could be also specific services or applications (e.g. machine-to-machine communication, device-to-device, proximity services [16, 17]), that are not location specific but for which there could be reserved a common spectrum. The latter cases are in nature as multi-tenant services where the communicating entities may belong to different MNOs' customers. In case of both location-dependent and location-independent businesses, the scaling potential is stemming from the utilization of the LTE ecosystem. The LTE ecosystem provides solutions for small cells and proximity services, as solutions for the vehicle-to-vehicle communication build on proximity services [15]. The LTE ecosystem provides easy standardization approaches as the standardization takes place within the ecosystem. The location-dependent business might utilize the benefits of urban areas like dense population and easy implementation of infrastructure for communication. A general infrastructure provider may also provide the whole communication system and so enable the MNOs to act as local virtual operators. The spectrum licenses could also be location-specific for both local network operator and

MNOs. Although it is recognized that CoPSS can be utilized in wide and local areas, it can be acknowledged that the investments needed may vary considerably across the cases.

4 Conclusions

Co-primary spectrum sharing (CoPSS) is one of the emerging concepts to increase spectral efficiency among MNOs and other actors. In this paper, we have examined the elements of business model scalability for CoPSS, especially from the MNO perspective. We define business model scalability as a possibility to maintain or improve profit margins when sales volume increases. To reach its full potential, CoPSS as a concept for inter-MNO spectrum sharing should utilize the LTE ecosystem and technology platforms. We argue that the scalability potential of the concept stems from the mission-critical and unique features of the concept, especially from the utilization of pre-defined policies and rules for sharing and from the utilization of subscriber and usage profiles.

From location perspective, CoPSS enables both location-dependent and location-independent but specific services. CoPSS enables sustainable and continuous revenue streams through hybrid business models that allow for various combinations of connectivity, content, context and commerce services. It is also possible to grant MNOs sharing spectrum resources a dedicated share of the shared resource that guarantees them an opportunity to generate revenue. From the value differentiation perspective, CoPSS enables the differentiation between exclusive and shared spectrum resources, and if combined with customer experience management, this differentiation can be made even clearer. In conclusion, we see CoPSS showing a great deal of potential for MNOs' business operations, also in the near future.

Acknowledgments. This research is part of the CORE++ project. The authors would like to acknowledge the project consortium: VTT Technical Research Centre of Finland, University of Oulu, Centria University of Applied Sciences, Turku University of Applied Sciences, Nokia, PehuTec, Bittium, Anite, FairSpectrum, Finnish Defence Forces, Finnish Communications Regulatory Authority, and Tekes – the Finnish Funding Agency for Innovation.

References

1. Bennis, M.: Spectrum sharing for future mobile cellular systems. Ph.D. thesis, University of Oulu (2009)
2. Sousa, E.S., Alsohaily, A.: Spectrum sharing LTE-advanced small cell systems. In: Proceedings of the International Symposium on Wireless Personal Multimedia Communications (WPMC) (2013)
3. Gangula, R, Gesbert, D., Lindblom, J., Larsson, E.G.: On the value of spectrum sharing among operators in multicell networks. In: Proceedings of the IEEE VTC-Spring 2013 (2013)
4. Ahokangas, P., Matinmikko, M., Yrjölä, S., Okkonen, H., Casey, T.: "Simple rules" for mobile network operators' strategic choices in future spectrum sharing networks. IEEE Wireless Commun. **20**(2), 20–26 (2013)

5. Ahokangas, P., Matinmikko, M., Atkova, I., Yrjölä, S., Minervini, LF., Mustonen, M.: Co-opetitive business models in mobile broadband. Paper presented at the 6th Workshop on Coopetition Strategy, Umeå, Sweden (2014)

6. Ahokangas, P., Horneman, K., Posti, H., Matinmikko, M. Hänninen, T., Gonçalves, V.: Defining "co-primary spectrum sharing" – a new business opportunity for MNOs? Invited paper presented at 9th International Conference on Cognitive Radio Oriented Wireless Networks (CrownCom) June 2–4, 2014, Oulu, Finland (2014)

7. Bengtsson, L., Kock, S.: Coopetition in business networks: to cooperate and compete simultaneously. Ind. Mark. Manage. **29**, 411–426 (2000)

8. Hultell, J., Johansson, K., Markendahl, J.: Business models and resource management for shared wireless networks. In: IEEE Vehicular Technology Conference (2004)

9. Beckman, C., Smith, G.: Shared networks: making wireless communication affordable. IEEE Wirel. Commun. **12**, 78–85 (2005)

10. Frisanco, T., Tafertshofer, P., Lurin, P., Ang, R.: Infrastructure sharing and shared operations for mobile network operators: From a deployment and operations view. In: Proceedings of the IEEE ICC, Beijing, China, May 2008

11. Markendahl, J., Nilson, M.: Business models for deployment and operation of femtocell networks: Are new cooperation strategies needed for mobile operators? In: 21st European Regional ITS Conference, Copenhagen, September 2010

12. Gnyawali, D., Park, B.: Coopetition between giants: collaboration with competitors for technological innovation. Res. Policy **40**, 650–663 (2011)

13. Ahokangas, P., Juntunen, M., Myllykoski, J.: Cloud computing and transformation of international e-business models, In: Sanchez, R., Heene, A. (eds.) Building Competences in Dynamic Environments. Research in Competence-Based Management, vol. 7, pp. 3–28. Emerald Group, London (2014)

14. Agrawal, D., El Abbadi, A., Das, S., Elmore, A.J.: Database scalability, elasticity, and autonomy in the cloud. In: Yu, J.X., Kim, M.H., Unland, R. (eds.) DASFAA 2011, Part I. LNCS, vol. 6587, pp. 2–15. Springer, Heidelberg (2011)

15. TR 36.885, 3rd Generation Partnership Project, Technical Specification Group Radio Access Network, Study on LTE-based V2X Services (2015)

16. Cellular IoT Whitepaper: A Choice of Future m2 m Access Technologies for Mobile Network Operators, 28 March 2014

17. The Tactile Internet, ITU-T Technology Watch Report, August 2014

Spectrum Toolbox Survey: Evolution Towards 5G

Michal Szydelko[1(✉)] and Marcin Dryjanski[2]

[1] Movilo Ltd., Milostowska 8/9, 51-315 Wroclaw, Poland
michal.szydelko@movilo.net
[2] Grandmetric Ltd., Malta Office Park, Baraniaka 88E/F, 61-131 Poznan, Poland
marcin.dryjanski@grandmetric.com

Abstract. The ever increasing needs for more spectrum resources, and the new Radio Access Technologies (RAT) to serve Mobile Broadband (MBB) services add to the complexity of the Spectrum Toolbox in mobile networks landscape. This paper briefly describes a collection of available frequency bands, spectrum aggregation mechanisms, licensing and duplexing schemes, as well as spectrum sharing and refarming techniques. With such a classification, Spectrum Toolbox is defined and its evolution directions are discussed. It covers 3GPP LTE evolution from its first version in Release 8 up to the Release 14, which deals with LTE-A Pro enhancements. Studies on the new non-backwards compatible RAT are also covered. Finally, the potential evolution towards emerging 5G ecosystem, in the context of future Spectrum Toolbox enhancements, is presented.

Keywordss: 3GPP evolution · 5G · LTE-A pro · Spectrum management · Spectrum toolbox · WRC-15

1 Introduction

The ever increasing mobile data demand in cellular networks calls for more spectrum resources and for novel spectrum access schemes. This in turn increases the overall complexity of mobile networks. Recent 3GPP Rel-13 standardization in the Radio Access Networks (RAN) group, as well as discussions covered during 3GPP on "5G RAN" and Licensed-Assisted Access (LAA) workshops, gave clear indication on the requirement of further spectrum allocation flexibility improvements.

This paper presents Spectrum Toolbox, which may serve as a simple guide through spectrum-related solutions in mobile system landscape. Spectrum Toolbox covers such aspects as frequency bands overview, spectrum aggregation mechanisms, licensing and duplexing schemes, as well as spectrum sharing and refarming techniques. Based on 3GPP RAN standardization status[1], a brief introduction to the available solutions is presented, focusing on LTE and WLAN spectrum resources and their classification, with the aim to provide indications on further system developments towards 5G.

The remainder of this paper is organized as follows: after a brief summary on recent radio access related standardization and regulatory events, Sect. 3 covers overview of

---

[1] As of 2015/12.

© ICST Institute for Computer Sciences, Social Informatics and Telecommunications Engineering 2016
D. Noguet et al. (Eds.): CROWNCOM 2016, LNICST 172, pp. 703–714, 2016.
DOI: 10.1007/978-3-319-40352-6_58

3GPP solutions for spectrum access techniques classifying Spectrum Toolbox elements. In Sect. 4, discussion on potential future development directions is presented, followed by the final conclusions in Sect. 5.

2 Setting up the Scene for Spectrum Toolbox Discussion

This chapter presents the outcomes of recent standardization and regulatory events (i.e. late 2015 timeframe), including LAA workshop, "5G RAN" workshop, as well as World Radio Conference (WRC-15). These discussions set up the scene for the evaluation of Spectrum Toolbox, covered in the following sections.

3GPP RAN workshop on LAA[2] collected inputs from various unlicensed spectrum stakeholders, including IEEE802 committee, Wi-Fi Alliance (WFA), Wireless Broadband Alliance (WBA), as well as from the regulatory bodies representatives [1]. The goal of the workshop was to strengthen technical collaboration, especially in the areas of coexistence evaluation for unlicensed Industrial, Scientific and Medical (ISM) bands among different actors, and to follow up with finalization of the LAA feature in 3GPP specifications. The main conclusions of the workshop were related to the Listen Before Talk (LBT) mechanism design to provide fair coexistence between LTE and WLAN users within 5 GHz ISM band. Coexistence testing and performance requirements definition discussion was also started, focusing on DL-only operation of LAA within Rel-13 timeframe (while UL LAA is expected to be covered in Rel-14).

3GPP workshop on "5G RAN" development directions collected ideas and requirements on the next generation mobile networks, including 5G timeline feasibility discussion [2]. In terms of spectrum allocation for 5G, the below 6 GHz spectrum, as well as millimeter Wave (mmW) spectrum bands were discussed for new, non-backwards compatible 5G RAT, which is expected to be developed in parallel to the LTE-Advanced Pro[3] in coming 3GPP releases. As a facilitator for the technical feasibility studies and performance benchmarks, work on mmW channel models will first have to be concluded, covering spectrum bands ranging from 6 GHz up to 100 GHz[4] [3].

WRC-15 conference purpose was to allocate new frequency bands for various mobile services, ranging from road safety to global flight tracking use cases [4]. New spectrum allocation was agreed for Mobile Broadband (MBB) communication services within L-band (i.e. 1427–1518 MHz), and within lower part of the C-band (i.e. 3.4–3.6 GHz). More detailed spectrum allocation breakdown is covered in Table 1.

During WRC-15, decision on spectrum agenda studies for the next WRC-19 conference was taken, aiming at identification of 5G frequency bands above 6 GHz. In the studies the following bands are supposed to be considered: 24.25–27.5 GHz, 31.8–33.4 GHz, 37–43.5 GHz, 45.5–50.2 GHz, 50.4–52.6 GHz, 66–76 GHz, 81–86 GHz. Furthermore, it was decided that the broadcasting and mobile industry players in

---

[2] More details on LAA covered in Sect. 3.2. LAA is also called LTE-Unlicensed (LTE-U).

[3] LTE evolution in Rel-13 and beyond.

[4] However, the 5G RAT is to be defined for both below and above 6 GHz spectrum.

Table 1. WRC-15 decisions on spectrum allocation for 4G mobile services [4]

Frequency band	Geographical distribution	Spectrum availability
470–694/698 MHz	Some APAC and American countries	Auction in the USA
694–790 MHz	Global band, now including EMEA	60 MHz
1427–1518 MHz	Global band, in most countries	91 MHz
3300–3400 MHz	Global band, not Europe/North America	100 MHz
3400–3600 MHz	Global band, in most countries	200 MHz
3600–3700 MHz	Global band, not Africa/some APAC	100 MHz
4800–4990 MHz	Some APAC and American countries	190 MHz

Europe, have to conclude on the opportunity of mobile broadband technologies adaptation for future terrestrial TV requirements, while using TV UHF band (470–694 MHz). The consensus is to be reached until WRC-23 conference.

3 Spectrum Toolbox

Increasing amount of the available spectrum bands, equipped with the range of spectrum access technologies, provides highly complex system to operate and coordinate the resource usage among radio access nodes. Figure 1 presents an example of Heterogeneous Network (HetNet) comprising of macro- and Small Cell (SC) layers, accompanied with various spectrum access techniques, including e.g. Carrier Aggregation (CA) and Dual Connectivity (DC), covering licensed and unlicensed frequency bands.

Spectrum Toolbox covers the available frequency bands, spectrum aggregation mechanisms, licensing and duplexing schemes, as well as spectrum sharing and refarming techniques. Table 2 presents an overview of the Spectrum Toolbox evolution over the LTE releases, while the following sections elaborate on individual areas presented therein.

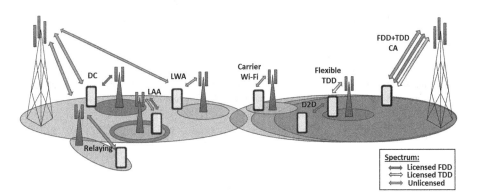

Fig. 1. Spectrum toolbox landscape in HetNet

Table 2. Spectrum Toolbox evolution across LTE releases: standardized solutions, ongoing Study Items, and future concepts for 5G discussions[a]

3GPP release	LTE: Rel-8, 9	LTE-A: Rel-10, 11, 12	LTE-A Pro: Rel-13, 14	5G phase I: Rel-15 5G phase II: Rel-16
Frequency bands [GHz][b]	0.7, 0.8, 1.8, 2.1, 2.3–2.4, 2.5–2.6 GHz	0.45 (Brasil), Digital Dividend, 1.5, 3.4–3.8 GHz	5 GHz ISM; WRC-15 bands	New bands below 6 GHz for 5G RAT; mmW: 6-100 GHz; WRC-15/19 bands
Spectrum aggregation	Single Carrier (1.4–20 MHz), symmetric DL/UL	Dual Connectivity, CA variants: -up to 5CC, -intra-/inter-band, -(non)-continuous, -FDD and/or TDD -Co-located, RRH; -asymmetric DL/UL	Massive CA (32CC), LAA (5 GHz), LWA, SDL for CA: 2.3–2.4 GHz	Multi-Connectivity with asymmetric DL/UL, SDL for CA: 700 MHz, 2.5–2.6 GHz, Lean carrier
Spectrum licensing schemes	Licensed spectrum only	Licensed, Carrier Wi-Fi	Licensed, Unlicensed, DL LAA, LWA, LSA	Co-existence of: exclusive licensed, shared license-exempt spectrum, enhanced LAA (DL + UL)
Duplexing schemes	Separate FDD, TDD	FDD and TDD (CA-based), eIMTA	FDD Flexible Duplex	Full Duplex, Additional DL-only TDD configurations
Sharing schemes (network, spectrum)	Static schemes (MOCN, MORAN)	Static schemes (MOCN, MORAN)	RSE, LSA	LSA (new bands), SC sharing, SCaaS, spectrum trading, Cognitive Radio
Spectrum refarming	Static	Static	Dynamic, DSA, MRAT Joint Coordination	Fully dynamic, opportunistic, Cognitive Radio

[a]Abbreviations in Table 2: CC – Component Carrier, DSA – Dynamic Spectrum Access, eIMTA – enhanced Interference Mitigation & Traffic Adaptation, LSA – License-Shared Access, LWA – LTE-WiFi Aggregation, MOCN – Multi-Operator Core Network, MORAN – Multi-Operator RAN, RRH – Remote Radio Head, RSE – RAN Sharing Enhancements, SCaaS – Small Cells as a Service, SDL – Supplemental Downlink
[b]3GPP introduces frequency bands and CA band combinations in a release independent manner. Per-release breakdown presented only to describe standard's evolution.

3.1 Frequency Bands

Below, the list of current 3GPP spectrum bands for E-UTRA is presented [5]. Due to the complex nature of the country-, and market-specific spectrum bands allocation, the following compilation is limited only to the presentation of general spectrum bands without detailed distinction:

- FDD [GHz]: 0.45, 0.7, 0.8, 0.9, 1.5, 1.8, 1.9, 2.1, 2.3, 2.6, 3.5
- TDD [GHz]: 0.7, 1.8, 1.9, 2.0, 2.6, 2.3-2.4, 2.5-2.7, 3.4-3.8

In case of spectrum bands for IEEE 802.11 access technologies, unlicensed frequency bands are identified and summarized in Table 3.

Table 3. IEEE802.11 spectrum bands breakdown for 802.11 protocol variants

Protocol	Release date	Frequency band [GHz]	Channel bandwidth [MHz]
802.11	1997.06	2.4	22
802.11 a	1999.09	3.7[a]; 5	20
802.11 b	1999.09	2.4	22
802.11 g	2003.06	2.4	20
802.11 n	2009.10	2.4/5	20, 40
802.11 ac	2013.12	5	20, 40, 80, 160
802.11 ad	2012.12	60	2160
802.11 ah	~2016	0.9	1, 2, 4, 8, 16
802.11 aj	~2016	45/60	540, 1080
802.11 ax	~2016	2.4/5	80, 160
802.11 ay	~2017	60	8000

[a]Licensed 3.7 GHz band; allowed by FCC in the USA.

3.2 Spectrum Aggregation Techniques

Baseline LTE standard (i.e. 3GPP Rel-8) was defined as an OFDMA[5] system supporting single carrier with various channel bandwidths (BW), defined in the range of 1.4–20 MHz. In Rel-10 LTE-Advanced, CA was introduced to aggregate multiple Component Carriers (CC) with the use of MAC layer scheduling. Up to 5CC were standardized and each individual CC was reusing Rel-8 numerology for the BW size to allow backward compatibility. Thus, the overall theoretical maximum aggregated bandwidth summed up to 100 MHz with intra-band consecutive, non-consecutive, or with inter-band spectrum aggregation options. Different component carrier allocation for UL and DL could reflect the expected traffic demand by the use of non-symmetrical configurations (e.g. 3DL CC, 1UL CC). CA introduced concepts of Primary Cell (PCell) and Secondary Cell (SCell), where the former is used for signalling and user data purposes, while the latter serves for user data only to increase the overall user's throughput.

---

[5] OFDMA used for LTE DL transmission. For LTE UL, Single Carrier FDMA is used.

Unpaired spectrum variant further improving inter-band FDD CA-based operation was provided with the SDL concept defined for L-band (i.e. 1452–1496 MHz, previously used for broadcasting services) [16]. SDL was also introduced for 717–728 MHz band, and the discussion continues to introduce the harmonized European SDL band in 738–758 MHz range, as well as to enable SDL in 2.6 GHz band [8, 9].

CA was standardized for either intra-site or inter-site[6] scenarios, based on ideal backhaul due to CA timing requirements on scheduling. Furthermore, in later stages CA allowed aggregation of TDD and FDD based component carriers.

Even though 5CC configuration was standardized in Rel-10 already, the highest CA combination being standardized for specific spectrum bands so far considers "only" up to 4 frequency bands and 4 CCs [1], e.g.:

- FDD only: 1900, 2100, 2300 + 700 SDL
- FDD + TDD: (2100, 1800, 800) FDD + 3500 TDD

Lately, work on 5CC Carrier Aggregation has been started within Rel-13 for 5DL/1UL configuration [24].

Significant change in the spectrum aggregation management was possible through the aggregation of different carriers with the use of Rel-12 Dual Connectivity feature. With DC, spectrum is aggregated in inter-site scenario, where a macro-cell serves as a mobility anchor (using so called Primary Cell Group, PCG) whereas the other radio link provided by Small Cell acts as local capacity booster (i.e. Secondary Cell Group, SCG). This feature implements Control Plane/User Plane (CP/UP) split, in order to reduce the signaling overhead, reduce the number of handovers, and to improve user experience for mobile users. CP/UP split operates by switching User Plane (UP) links among available SCs, whereas the user's CP context is maintained by the overlay macro-cell. In contrary to CA, DC scheme uses concept of the Split Bearer, where instead of aggregating MAC layer transport blocks, the PDCP Packet Data Units (PDUs) are combined, thus omitting the requirement for low latency and allowing non-ideal backhaul for SC connectivity.

Rel-13 extends spectrum aggregation mechanisms towards higher number of aggregated bands and towards the use of unlicensed spectrum for mobile networking. Massive CA enables up to 32CCs and thus theoretically provides up to 640 MHz of aggregated bandwidth for a single device, while still fulfilling backwards compatibility with LTE Rel-8 channel bandwidths. Furthermore, LAA and LTE-WiFi Aggregation (LWA)[7] are provided as features to utilize the unlicensed spectrum. LAA aggregates the licensed LTE carrier (serving as a mobility and signaling anchor - PCell) with SCell using the new LTE frame format over the unlicensed 5 GHz ISM band[8]. Similarly, in case of LWA scheme, the Carrier Wi-Fi is serving as capacity booster's counterpart, using radio level integration for uniform user experience provision over the Wi-Fi

---

[6] Based on RRH deployments using fiber for ideal backhaul.

[7] All spectrum access and spectrum aggregation schemes have deployment related limitations, which are not discussed in this paper, e.g. LWA and LAA applicable mostly for Small Cells.

[8] Other LTE-based access schemes using ISM spectrum: LTE-U – downlink-only radio access with Carrier Sensing Adaptive Transmission (CSAT) for fairness assurance; MuLTEfire – LTE-based technology without licensed PCell anchor proposed by Nokia and Qualcomm.

radio. In LWA, UE is configured by the eNB to utilize radio resources of LTE and WLAN. Another LTE-WLAN interworking mode is RAN Controlled LTE WLAN Interworking (RCLWI), defining LTE-controlled bidirectional traffic steering between LTE and WLAN. In RCLWI, LTE may send steering command to UE in order to perform traffic offloading to WLAN. LWA and RCLWI are supported in collocated, as well as in non-collocated scenarios (ideal backhaul, or non-ideal backhaul between eNB and WLAN access point, respectively).

3.3   Spectrum Licensing Schemes

Licensed spectrum allocation per Mobile Network Operator (MNO) is the basic principle for mobile networks operation, requiring acquisition of the spectrum license. However, an unlicensed spectrum usage was considered already in 3GPP Rel-8, where the Access Network Discovery and Selection Function (ANDSF) was introduced for the traffic offload to e.g. Carrier Wi-Fi access nodes. This technique was not very popular, but traffic offloading techniques evolution was continued in the subsequent 3GPP releases.

As already mentioned in the previous section, 3GPP has introduced LAA unlicensed spectrum access schemes, which became hot topic among 3GPP, WFA and IEEE. ISM spectrum usage was also considered under the LWA or RCLWI scheme. To enhance the integration of unlicensed spectrum into mobile networks, the Unlicensed Spectrum Offloading System (USOS) work was agreed for Rel-14, to define service requirements allowing MNOs to empower the clarification for: spectrum usage over unlicensed access networks, network planning and charging purposes.

Furthermore, the new spectrum sharing scheme for 2.3–2.4 GHz band was introduced, based on ETSI RRS work [10], called License Shared Access (LSA, or Authorized Shared Access, ASA). LSA allows spectrum owners (i.e. incumbents) to share their radio resources with other market players (LSA Licensees), enabling QoS support within a shared band. QoS support is achieved by the use of protection measures such as geographical exclusion, or restriction zones, within which the incumbent's receivers will not be subject to interference caused by LSA Licensees. 3GPP studied LSA access in Rel-13 [12], looking into architectural aspects of ETSI's concept for global solution provisioning, covering required information flows for static and semi-static spectrum sharing scenarios. Two alternatives for OAM based spectrum usage reconfigurations are available within LSA architecture model to handle LSA Spectrum Resource Availability Information (LSRAI) exchange over LSA$_1$ interface (Fig. 2). LSA Repository (LR) entity stores information describing Incumbent's usage and protection requirements. LR allows National Regulatory Authority (NRA) to monitor LSA operation. LSA Controller (LC) is located within licensee's domain, allowing the licensee to obtain LSA spectrum availability information from the Repository. LC controller interacts with the licensee's mobile network to support mapping of the spectrum availability information into appropriate radio nodes configurations.

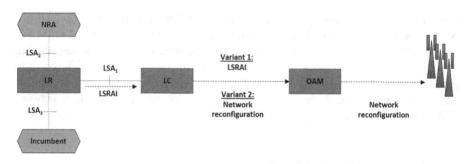

Fig. 2. Two variants for LTE RAN network elements reconfiguration under LSA scheme [11]

3.4 Duplexing Schemes

Duplexing schemes for LTE system correspond to either FDD or TDD, with pre-defined duplex-specific spectrum bands [1]. Initial Rel-8 TDD frame configurations included multiple settings for different UL and DL traffic ratios (i.e. ranging from UL heavy 2:2:6[9] up to DL heavy 8:1:1) within 10 ms radio frame, which were meant to be configured in a semi-static manner. However, TD-LTE frame setup modification requires careful inter-site coordination to avoid major interference problems, e.g. in case of inter-site configuration, where UL and DL transmission is present in the same subframe from the neighboring access nodes.

Based on TDD frame configurations described above, their dynamic adaptation for a HetNet scenario was introduced in Rel-12 with the introduction of enhanced Interference Management and Traffic Adaptation (eIMTA) feature. eIMTA enables on-the-fly changes of the TDD configuration based on instantaneous DL:UL traffic demand ratio in the particular cell. This feature was mainly foreseen for Small Cell deployments in 3.5 GHz band. In such scenario, DL-UL interference should not be an issue, since a SC with low transmit power, and operating on high frequency band is considered to be well isolated from the potential neighboring interferers (unless, dense deployment of SCs is considered), and it is not expected to deploy macro cells on such high frequency bands[10].

Furthermore, TDD frame configurations were discussed within Rel-13 to be further extended with additional DL-heavy and DL-only frames formats to support 9:1:0 and 10:0:0 options [25], but this study was left for standardization in future releases.

With the introduction of CA, TDD or FDD spectrum aggregation was possible. In case of FDD bands aggregation, it was possible to have a different number of the DL and UL CCs (e.g. refer to the SDL concept); in the case of TDD spectrum aggregation, number of carriers (and their bandwidths) obviously has to be the same for DL and UL due to the nature of TDD. Further evolution of CA configurations came with the aggregation of FDD and TDD component carriers, enabling greater flexibility in the

---

[9] DL:Sp:UL; Sp – Special subframe.

[10] However, 3.5 GHz based rooftop macro sites were considered in Japan, based on the inter-site coordination for interference management purposes.

spectrum arrangements for the operators having both, FDD and TDD spectrum licenses. In such scenario, PCell could be either FDD, or TDD (with the limit for the cross-carrier scheduling originating from the TDD PCell). One particular use case considers FDD based PCell, with the RRH based SCell using TDD.

Flexible duplex feature for FDD E-UTRA bands was also proposed during Rel-13 discussions [13]. It aimed to provide more efficient handling of the asymmetric traffic load between DL and UL, going beyond static resource allocation for both link directions, by permitting DL transmission to originate from the underutilized FDD UL frequency bands. However, based on negative feedback received from regulatory bodies, it seems that the option of flexible duplex for FDD bands will not be further considered at this stage of standardization.

3.5   Spectrum Sharing Schemes

According to the basic definition of spectrum sharing[11], some of the techniques described in the previous chapters could be considered as belonging to the spectrum sharing mechanisms, e.g. LAA spectrum access method, or LSA spectrum licensing method.

This section focuses on recent developments in the area of shared spectrum access improvements, in the form of Rel-13 RAN Sharing Enhancements (RSE) work [14]. RSE treats on RAN sharing, however it is closely related to the spectrum sharing. It improves the legacy inter-operator sharing schemes, where previously there was no knowledge on how much capacity the other participating operator is using. RSE improvements are addressing the following aspects:

- Allocation of the shared RAN resources based on the proportion of the assigned resource usage for each participating operator;
- Ability to monitor the usage of shared resources;
- Allows on-demand capacity negotiations;
- Load balancing while respecting the agreed shares of resources;
- Selective OAM access for the participating operators.

RSE is under the development for each of the following radio technologies: GSM, UTMS [15] and LTE. In terms of GSM, it is becoming an interesting improvement for some markets, where it is already expected, that there will be single shared GSM network for legacy devices support for all competing operators. At the same time, each MNO will own its individual 4G network using extra 2G spectrum resources by the means of dynamic refarming.

RSE functionality is expected to be particularly important in case of HetNets, including dense deployment of SCs for hotspots. In the authors' opinion, SC sharing will become an enabler for fruitful deployments of future networks in mature markets, where shortage of the access nodes' sites in dense environments will become a bottleneck.

---

[11] Spectrum sharing: simultaneous usage of the particular frequency band by the number of independent systems, or users.

3.6   Spectrum Refarming

Along the mobile networks evolution from 1G analogue systems up to today's LTE-A Pro cellular networks underlying spectrum regulations and standardization has been evolving to provide sufficient amount of market-, and RAT-specific frequency bands for the mobile operators. With new radio technologies becoming more spectrally efficient than the previous RAT generations and with ever increasing demand for the more licensed bands, the spectrum refarming comes as a natural solution to enable release of the legacy RAT bands. Static approach to the spectrum refarming, where reuse of 2G spectrum for the LTE services is allowed after 2G license expiration, is not too attractive, as market-specific spectrum licenses might have been granted for long periods (e.g. 20 years). Moreover, the static spectrum refarming has another drawback, i.e. the requirement of legacy RAT's devices lead-out consideration. Thus, the evolution in spectrum refarming domain already includes dynamic methods, and ultimately shifts towards Cognitive Radio based mechanisms.

So far, one promising method for dynamic spectrum refarming is being standardized under MRAT Joint Coordination framework [6, 17]. This concept utilizes Dynamic Spectrum Access (DSA) scheme[12], where the collocated LTE and GSM systems use dedicated bandwidth part whose size depends on the actual traffic demand in each radio network. This method is based on the temporal traffic statistics: e.g. when GSM load on Traffic Channel is low, LTE is allowed to use shared GSM/LTE part of the spectrum.

4   Future Evolution Directions

Considering the above solutions and the emerging standardization of 5G, the presented Spectrum Toolbox is expected to be further enhanced with new technology elements. Spectrum Toolbox will become even more complex, but it will also provide more flexibility in the spectrum allocation. Therefore, Spectrum Toolbox will have to be more adaptive and automated, evolving towards CR mechanisms, equipped with self-learning and self-optimization solutions. Some of the 3GPP study and work items considered for Rel-14 discussion are indicating future evolution directions, briefly covered below:

- Enhanced LAA (eLAA) proposal extends LAA scheme with UL consideration to enable full DC-like capabilities for unlicensed spectrum [19].
- SDL spectrum is extended by 2570–2620 MHz band for TDD or for unpaired DL within SDL framework.
- Flexible bandwidth study [20] targets the possibility to fully exploit non-standard channel bandwidths for LTE (i.e. extension of Rel-8 channel BWs) in spectrum, which is currently underutilized, e.g. in case of spectrum refarming, where the released 2G spectrum doesn't fit the Rel-8 channel bandwidths.

---

[12] Also studied in FP7 SEMAFOUR project [18].

- Lean carrier concept, after being initially rejected in Rel-12, is again brought to attention during Rel-14 discussion [21]. It is supposed to inherit from LAA radio frame, where most of the legacy PHY-layer information (including cell specific reference signals) is sent only when needed. By doing so, in the low/medium load conditions, the interference is decreased, and the eNB's power amplifiers can be switched to deep sleep mode to save energy. Furthermore, to enable easy integration with the existing system, it is proposed to provide the lean carrier with the possibility to be toggled to the legacy LTE frame when needed, e.g. serving as a legacy SCell when the PCell is overused by legacy users.
- Multi-connectivity [22] is expected to enhance DC, providing multiple links for a UE in the following two options:
  - Configuration of multiple radio links for UE, while only limited, selected set of them are active at any given moment;
  - All of the configured multiple links are active.

More advanced and novel areas of the "Towards 5G" RAN aspects include: adoption of high frequency bands, support for full duplex, and definition of the Unified Air Interface (UAI):

- New frequency bands in the range of mmWave bands have been considered for local capacity boosting and as dynamic backhaul/fronthaul solutions, where the standardization starts by studying the appropriate channel models.
- Full duplex provides the possibility to improve spectral efficiency by utilizing a single, un-paired band for simultaneous transmission and reception (i.e. without splitting the time slots onto DL and UL) possible through the use of advanced receivers.
- UAI should handle multiple different traffic types and incorporate their requirements into a single radio frame design utilizing different frame parts (e.g. in the frequency domain) with different waveform parameters and access schemes, e.g. small data packets transmitted by MTC devices can be sent in a contention-based manner, whereas MBB data is sent in synchronized scheduled manner [23].

All the above techniques provide a Spectrum Toolbox that includes a wide set of bands ranging from 450 MHz up to 100 GHz with licensed and unlicensed spectrum, covering different licensing options, as well as different access schemes, BW aggregation mechanisms, duplexing and RATs. On top of that, 5G requirements target tight integration of all these elements to unify the operation of the next generation mobile systems and to provide the possibility to adapt to different use cases and scenarios that will further complicate the overall landscape.

5   Summary and Conclusions

The aim of this paper was to present set of the available and currently discussed spectrum access techniques, considered as the enabler for the 5G pre-study discussions within 3GPP Rel-14. Based on the performed analysis, the authors are convinced that the ongoing 5G-related discussions should already address all the possible spectrum resources, while considering various licensing schemes. Unified approach to this complex problem of Radio Resources Management is an enabler for the spectrum-efficient future networks. This is due to the fact, that the unified approach to RRM is still considered to be a missing piece in the current multi-RAT mobile networks landscape. In order to enable efficient usage of new spectrum licensing schemes (like LSA), considering scenarios which are evolving towards Ultra Dense Networks (UDN), the radio resources coordination shall be addressed on multiple levels, namely inter-MNO, inter-RAT, inter-site, inter-layer, inter-band dimensions. To achieve this, it is already obvious that the high level flexibility in the RF domain will be required. Additionally, high level of "programmability" of the baseband units will be needed, relying on Software Defined Networking (SDN) techniques, leading to dynamic spectrum access. On top of that, the overall design of future networks should natively incorporate SON engines, to manage the network towards unified user experience provided across multiple converged radio access technologies.

References

1. 3GPP, Report of 3GPP RAN workshop on Licensed-Assisted Access (LAA) (2015)
2. Flore, D.: RAN workshop on 5G: Chairman Summary, 3GPP RAN, RWS-150073 (2015)
3. 3GPP, RP-151606, Study on channel model for frequency spectrum above 6 GHz (2015)
4. Ericsson, ITU WRC-15 summary (2015)
5. 3GPP, TS36.104, Base Station (BS) radio transmission and reception, v13.1.0 (2015)
6. 3GPP, RP-150417, Study on Multi-RAT Joint Coordination (2015)
7. 3GPP, RP-121414, Introduction of LTE 450 MHz band in Brasil (2012)
8. 3GPP, RP-152042, TR 36.895 v1.0.0 European 700 Supplemental Downlink band (738–758 MHz) in E-UTRA (2015)
9. 3GPP, RP-152004, LTE FDD 2.6 GHz SDL band (2570–2620 MHz) (2015)
10. ETSI 103 154, System requirements for operation of Mobile Broadband Systems in the 2 300 MHz - 2 400 MHz band under LSA, v1.1.1 (2014)
11. 3GPP, TR32.855, Study on OAM support for LSA, v0.4.0 (2015)
12. 3GPP, SP-150056, Study on OAM support for LSA (2015)
13. 3GPP, RP-150470, Study on regulatory aspects for flexible duplex for E-UTRAN (2015)
14. 3GPP, RP-130330, RAN Sharing Enhancements (2015)
15. 3GPP, SP-140637, GERAN UTRAN Sharing Enhancements (2014)
16. 3GPP, TR37.814, L-band for Supplemental Downlink (2014)
17. 3GPP, TR37.870, Study on Multi-RAT joint coordination (2015)
18. FP-7 SEMAFOUR, D6.6 Final report on a unified self-management system for heterogeneous radio access networks (2015)
19. 3GPP, RP-151978 New Work Item on enhanced LAA for LTE (2015)
20. 3GPP, RP-151889, LTE bandwidth flexibility enhancements (2015)
21. 3GPP, RP-151982, New Work Item on Lean Carrier for LTE (2015)
22. 3GPP, RP-151791, Enhancements for densely deployed small cells in LTE (2015)
23. Wunder, G., et al.: 5GNOW: non-orthogonal, asynchronous waveforms for future mobile applications. Commun. Mag. 52, 97–105 (2014)
24. 3GPP, RP-151671, LTE Advanced inter-band CA Rel-14 for 5DL/1UL (2015)
25. 3GPP, RP-142248, Study on possible additional configuration for LTE TDD (2015)

Workshop Papers

A Reconfigurable Dual Band LTE Small Cell RF Front-end/Antenna System to Support Carrier Aggregation

Cyril Jouanlanne[1,2], Christophe Delaveaud[1,2(✉)],
Yolanda Fernández[3], and Adrián Sánchez[3]

[1] Univ. Grenoble-Alpes, Grenoble, France
{cyril.jouanlanne, christophe.delaveaud}@cea.fr
[2] CEA, LETI, MINATEC Campus, 17 rue des Martyrs, 38054 Grenoble, France
[3] TTI, PCTCAN, C/Albert Einstein 14, 39011 Santander, Spain
{yfernandez, asanchez}@ttinorte.es

Abstract. In this contribution, the prototype of an energy efficient RF front-end combined to a miniaturized highly efficient antenna system is presented. This demonstrator is designed for LTE small cell base station applications and can address both LTE band 7 & 20. It supports carrier aggregation (CA) for capacity network improvement and is compatible with inter-band CA as well as contiguous and non-contiguous intra-band CA. The measurement results show an energy consumption saving of around 30 % and a size reduction factor of 3 for the antenna system compared to regular base station antennas. The proposed design demonstrates the possibility of increasing the energy efficiency as well as the level of integration of a small cell base station despite the use of new technology enablers such as CA technique.

Keywords: Carrier aggregation · Multi-band antennas · Reconfigurable solutions · Energy efficiency

1 Introduction

One of the key aspects in future network deployments is the increasing bandwidth demand in confrontation with spectrum availability. ITU has proposed the technique of carrier aggregation (CA) supporting up to 100 MHz system bandwidth for LTE-Advanced (LTE-A) [1]. This technique enables aggregation of multiple component carriers (CCs) providing operators the maximum flexibility for using their available spectrum. Therefore LTE-A provides much higher throughputs than otherwise possible with the potential of achieving more than 1 Gb/s throughput for downlink (DL) and 500 Mb/s throughput for uplink (UL).

Three different CA modes are defined: intra-band contiguous CA (CCs are contiguous within the same band), intra-band non-contiguous CA (CCs are non-contiguous in the same band) and inter-band CA (CCs are in different bands).

To handle bandwidths up to 100 MHz in different frequency bands, innovative solutions are required for antenna and RF front-end. Concerning the antenna solution, the focus is on the bandwidth optimization enabling antenna miniaturization. Thus, the

© ICST Institute for Computer Sciences, Social Informatics and Telecommunications Engineering 2016
D. Noguet et al. (Eds.): CROWNCOM 2016, LNICST 172, pp. 717–728, 2016.
DOI: 10.1007/978-3-319-40352-6_59

antenna system instantaneous bandwidth has been reduced to the minimum in order to miniaturize the antenna size without impacting its performance. Frequency agility technique has been used to compensate this bandwidth reduction by giving the ability to the antenna to reconfigure its frequency bandwidth according to the active CA configuration. This technique has been so far mainly used in handset antenna design for CA purposes as demonstrated in [2]. On the topic of RF front-ends, reconfigurable solutions in terms of bandwidth and frequency of operation are demanded. However, peak-to-average power ratio (PAPR) is an important issue in CA, because CC aggregation causes an increase in PAPR. For the DL, the eNB should afford the PAPR increase caused by multiple CCs.

This paper addresses the issue of two innovative reconfigurable solutions to support CA. One solution is a multi-band frequency agile antenna which is able to adjust its bandwidth according to the active CA configuration. As well as a reconfigurable RF front-end capable to work at different operating points to improve energy efficiency depending on CA configuration.

For experimental validation, the proposed scenario is a small cell base station supporting intra-band CA in LTE band 7 (2620–2690 MHz) and inter-band CA in LTE band 20 (791–821 MHz) & LTE band 7 (2620–2690 MHz).

This work is structured as follows: Sect. 2 describes the multi-band frequency agile antenna. Section 3 presents the reconfigurable RF front-end. Section 4 depicts the small cell reconfigurable carrier aggregation demonstrator. Finally, Sect. 5 concludes the paper, highlighting the main results.

2 Multi-band Frequency Agile Antenna

A dual band (LTE band 7 & band 20) dual access frequency agile antenna system capable to adapt its frequency bandwidth to the active CA configuration has been developed. Frequency agility is used to reduce the antenna instantaneous bandwidth which enables miniaturization without performance reduction.

2.1 Design

Due to its low operating frequency bandwidth, only band 20 antenna requires a miniaturization. Based on the fact that the size of an antenna is proportional to its bandwidth [3, 4], the main idea with this antenna design is to do the miniaturization by limiting the frequency bandwidth to a single band 20 sub-channel instead of the whole band 20, reducing the instantaneous bandwidth from 70 MHz to twice 10 MHz. Thus, the antenna has been made dual resonant with two very narrow closely spaced resonances and the frequency agility technique has been used to tune the antenna resonances at the right frequency according to the active CA configuration.

The antenna system (Fig. 1) is based on two microstrip patch antennas, one covering LTE band 7 and the other one LTE band 20. Band 20 antenna is miniaturized thanks to classical loading and folding techniques. Both antennas are co-located on a square 100 × 100 mm² ground plane and the antenna system only takes up a small

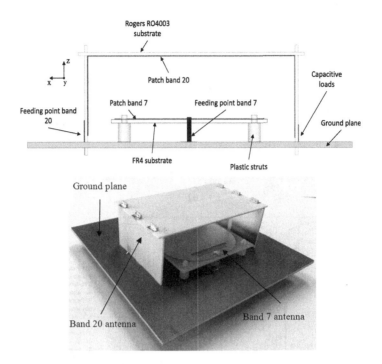

Fig. 1. Drawing and picture of the antenna prototype.

volume of $66 \times 54.5 \times 23$ mm^3 ($\lambda_0/5.7 \times \lambda_0/7 \times \lambda_0/16.5$ with $\lambda_0 = 379$ mm @791 MHz).

This mechanical arrangement with band 20 antenna right above band 7 antenna is possible due to their orthogonal polarizations as well as the fact that their respective strong near field spots are distinct and do not interfere significantly. This aspect is developed and illustrated in [5].

Two digitally tunable capacitors (DTCs) are used as frequency agility active components in this design. DTCs have been selected due to their superior linearity performance compared to other tuning components such as varactor diodes. They are also very easy to use and only require a SPI bus to control their effective capacitance.

2.2 Performance Evaluation

A measurement campaign was carried out to evaluate the antenna system electrical and radiated characteristics.

Figure 2 shows the reflection coefficient (S11) at the antenna system band 20 and band 7 ports. A reflection coefficient below -10 dB (worst case −7.7 dB) for band 20 and −12 dB for band 7 is achieved whichever the antenna state. A very good agreement is obtained between simulated and measured reflection coefficient in band 7. The comparison is equally good in band 20. The more visible discrepancies reflect very little difference due to low values shown with a logarithmic scale.

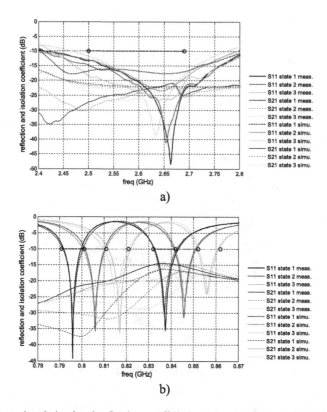

Fig. 2. Measured and simulated reflection coefficients at band 20 (a) and band 7 (b) RF ports and isolation between both RF ports for different DTC states.

Similar remark can be made with the isolation parameters (S21) that are close in simulation and measurement. An isolation level higher than 14.5 dB within band 20 and 20 dB within band 7 is demonstrated between both RF ports for the three states. This level of isolation will ensure a good electromagnetic compatibility between both antennas.

A peak realized gain (computed over two orthogonal elevation planes) better than 3 dBi for band 20 and 8 dBi for band 7 (Fig. 3) has been measured. These gain levels will ensure the good connectivity of the small cell base station.

The antenna system frequency agility is also demonstrated in Figs. 2 and 3, where the band 20 antenna ability to tune its frequency bandwidth in order to cover the three band 20 sub-channels is shown.

Realized gain patterns of band 20 antenna (Fig. 4) show that the main beam is oriented towards the Z + axis ($\pm 10°$) at both frequencies. The size of the ground plane being about $\lambda_0/4$, it is too small to ensure its reflector role. Therefore, the back radiation is relatively high (front to back ratio of about 3 dBi) and the directivity is consequently limited. Realized gain patterns of band 7 antenna (Fig. 5) show that the main beam is oriented towards the Z + axis ($\pm 10°$). Due to the sufficient size of the ground plane,

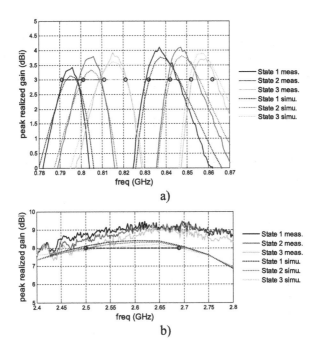

a)

b)

Fig. 3. Measured and simulated peak realized gain of band 20 (a) and band 7 (b) antennas for different DTC states.

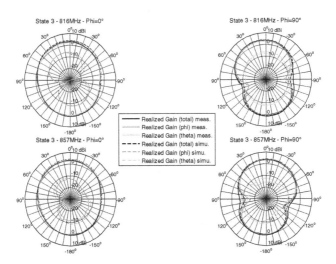

Fig. 4. Measured and simulated realized gain pattern of band 20 antenna.

there is almost no radiation towards the Z- hemisphere meaning a good front to back ratio (greater than 15 dB). This kind of radiation pattern was targeted in order to have the best performance once the base station located against a wall for instance.

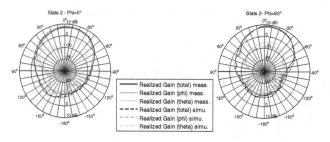

Fig. 5. Measured and simulated realized gain pattern of band 7 antenna.

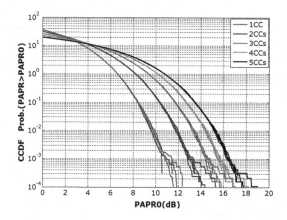

Fig. 6. CCDF simulation results from 1CC to 5CCs using E-TM1.1.

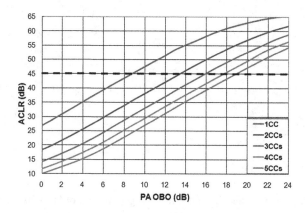

Fig. 7. ACLR simulated results vs PA OBO from 1CC to 5CCs using E-TM1.1 supporting intra-band contiguous CA.

3 Reconfigurable RF Front-end

A reconfigurable RF front-end capable to adapt to different CA configurations with different number of aggregated CCs providing energy savings has been developed. The solution is based on a reconfigurable power amplifier (PA) working at different operating points which provides optimized output power levels for CA.

3.1 Design

To support different CA configurations with different number of aggregated CCs, base station transmitter specifications [6] were considered in order to define mainly PA requirements. Depending on the number of aggregated CCs and CA configuration, different output back-off (OBO) levels at PA are required related to PAPR requirements. To calculate OBO specification, it is normally sufficient to apply 0.01 % complementary cumulative distribution function (CCDF) of the waveform [7]. CCDF characterizes the weighted probability of the signal excursions that lead to distortions, by indicating the number of samples where the signal peak power exceeds its average power by a certain value. CC aggregation causes an increase in PAPR due to the influence of aggregated CCs on the unwanted emissions and particularly on adjacent channel leakage power ratio (ACLR), which should be at least 45 dB.

To evaluate technical specifications, E-UTRA test models [8] are defined. E-UTRA test model 1.1 (E-TM1.1) verifies most PA specifications.

Firstly, simulations merging from 2CCs to 5CCs were carried out to evaluate LTE-A waveform and characterize its CCDF curves (Fig. 6). These simulations were performed with bandwidths from 3 MHz to 20 MHz, showing slight variations.

Furthermore, the OBO impact over ACLR specification was evaluated. Several simulations were performed modifying the OBO at the PA in order to determine the 45 dB ACLR threshold. Figure 7 presents the ACLR simulated results depending on the PA OBO from 1CC to 5CCs using E-TM1.1 supporting intra-band contiguous CA.

Table 1 summarizes the simulated results by combining the 0.01 % waveform CCDF evaluation and the ACLR evaluation to support intra-band contiguous CA from 1CC to 5CCs. Total OBO requirement must be the most restrictive value between both simulations.

Table 1. OBO requirements to support intra-band contiguous CA from 1CC to 5CCs.

Intra-band contiguous CA	OBO requirement		
	0.01 % CCDF evaluation	45 dB ACLR evaluation	Global evaluation
1 CC	9.6 dB	8.7 dB	9.6 dB
2 CCs	11.8 dB	13.4 dB	13.4 dB
3 CCs	13.3 dB	15.9 dB	15.9 dB
4 CCs	14.3 dB	17.7 dB	17.7 dB
5 CCs	15.2 dB	18.9 dB	18.9 dB

As shown in Table 1, the higher the number of CCs is, the higher is the impact of ACLR specification over OBO requirement.

The study was also extended to intra-band non-contiguous CA. It was checked that increasing the gap among CCs, OBO requirement reduces mainly owing to a lower impact of ACLR specification. For instance, in case of having 2CCs with a gap between CCs equal to the bandwidth, OBO requirement decreases to 12.6 dB. With a gap of twice the bandwidth, it reduces up to 11.8 dB.

3.2   Prototype

To validate the PA OBO requirement analysis for different CA modes, a reconfigurable RF front-end was developed (Fig. 8). It supports intra-band CA allowing up to 3CCs in LTE band 7 and inter-band CA in LTE band 20 & band 7. The hardware prototype is composed of three modules: the signal generation module, the RF gain block module and the reconfigurable PA capable to work at different operating points to optimize CA energy efficiency performance.

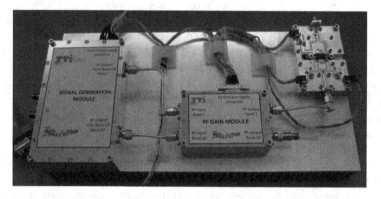

Fig. 8. Picture of the reconfigurable RF front-end prototype.

The signal generation module translates LTE baseband signals (E-UTRA test models) into LTE band 20 & band 7 RF signal, acting as an up-converter/modulator. Three RF signals are generated at tuning frequencies and appropriately combined using two RF splitters and one RF switch to configure different CA configurations. The RF gain module amplifies the RF signals from the signal generation module, also includes a digital attenuator to adjust the output power level to ease the characterization of the reconfigurable PA. Finally, the reconfigurable PA for LTE band 7 is integrated.

A console has been designed to configure different CA configurations defining the operating frequencies and the number of aggregated CCs.

The commercial PA, AFT20S015 N from Freescale, was implemented as the reconfigurable PA to validate the proposed solution. Its 1 dB compression output power (P1 dB) is about 38 dBm at 28 V drain voltage. Moreover, it was tested at drain voltages from 28 V to 14 V (Fig. 9).

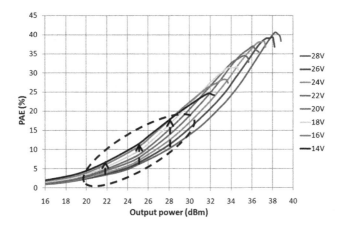

Fig. 9. PAE results in AFT20S015 N for different operating points.

The higher the drain voltage is, the higher is the possible delivered output power level. Nevertheless, depending on CA configuration there is an operating point optimized to deliver the required output power improving EE compared to single operating point as shown Fig. 9.

3.3 Performance Evaluation

For proof-of-concept validation, the most restrictive CA mode evaluated with the prototype is 3CCs in intra-band contiguous CA configuration, which requires 15.9 dB OBO. Considering 38 dBm P1 dB for 28 V drain voltage and aggregating 3CCs there is an increase in 4.7 dB in the average power due to the grouping. Therefore it can assume about 17 dBm as the average power per CC. Table 2 presents the associated operating points and the power-added efficiency (PAE) enhancement for different CA configurations. The evaluation was completed for intra-band contiguous CA up to 3CCs and for intra-band non-contiguous CA merging 2CCs with a gap equal to the bandwidth and twice the bandwidth.

For 1CC of 17 dBm average power and 9.6 dB OBO requirement, 16 V operating point is enough, providing 50 % PAE enhancement as shown Table 2. For 2CCs, the average output power is 20 dBm and different OBO requirements are associated to different CA configurations which can be fulfilled using different operating points, providing between 25 % and 39 % PAE enhancement.

4 Reconfigurable CA Demonstrator

A demonstrator combining the antenna system and the RF front-end was realized showing the compatibility between both subsystems.

The interfaces between the multi-band frequency agile antenna and the reconfigurable RF front-end are SMA connectors, one for each frequency band. Therefore RF

Table 2. Energy efficiency evaluation for different CA configurations.

CA configuration	PA OBO (dB)	Operating point (V)	Output power (dBm)	PAE (%)	PAE enhancement (%)
1 CC	9.6 dB	28 V	27.3 dBm	9.4 %	
		16 V	17.9 dBm	2.8 %	50 %
		28 V		1.4 %	
2 CCs non-contiguous (Gap = 2*Bandwidth)	11.8 dB	28 V	22.7 dBm	3.9 %	39 %
		20 V	17.1 dBm	1.8 %	
		28 V		1.1 %	
2 CCs non-contiguous (Gap = Bandwidth)	12.6 dB	28 V	21.6 dBm	3.3 %	
		21 V	16.9 dBm	1.7 %	35 %
		28 V		1.1 %	
2 CCs contiguous	13.4 dB	28 V	21.2 dBm	3 %	
		22 V	17.4 dBm	1.6 %	25 %
		28 V		1.2 %	
3 CCs contiguous	15.9 dB	28 V	16.7 dBm	1 %	–

cables are required to connect both prototypes. A future version of the demonstrator would integrate the RF components on the back side of the antenna ground plane.

A console was developed in order to manage both prototypes simultaneously. Depending on the active CA configuration, the reconfigurable RF front-end is programmed with the appropriate frequencies. Furthermore, the LTE band 20 PA is only

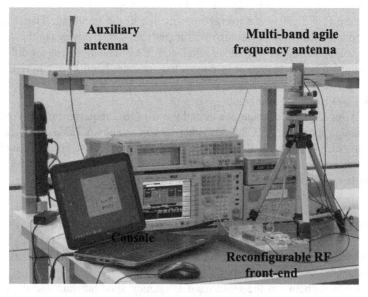

Fig. 10. Reconfigurable CA demonstrator.

activated in inter-band CA configuration to save energy during intra-band CA. In the multi-band frequency agile antenna, the DTCs are also programmed to tune the antenna resonances properly.

To validate both prototypes, a second antenna emulating the user equipment antenna was used to perform a radio link as shown Fig. 10.

Different CA configurations were performed combining both prototypes. Figure 11 shows one example of the test results at the auxiliary antenna with 2CCs in intra-band contiguous CA configuration. The demonstrator proves the ability of both prototypes to support CA.

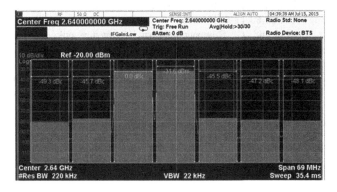

Fig. 11. Test results from the reconfigurable CA demonstrator at the auxiliary antenna for 2CCs in intra-band contiguous CA.

5 Conclusions

In this contribution, energy efficiency miniaturized dual band LTE small cell base station using CA for capacity network improvement is demonstrated.

The dual band reconfigurable RF front-end supports CA up to 3 CCs and provides energy saving up to 50 % using different operating points depending on the active CA configuration. It supports intra-band contiguous and non-contiguous CA as well as inter-band CA. The dual band antenna system, developed in parallel with the RF front-end, is capable to adapt its frequency bandwidth according to the RF front-end configuration (CA mode). This frequency bandwidth optimization has led to a miniaturization of the antenna system by a factor 3 without impacting its electrical performance.

The measured performances of both sub-systems are analyzed and a reconfigurable CA demonstrator combining the RF front-end and the antenna system is presented.

Acknowledgments. The present work was carried out within the framework of Celtic-Plus SHARING project.

References

1. Zukang, S., Papasakellariou, A., Montojo, J., Gerstenberger, D., Fangli, Z.: Overview of 3GPP LTE-advanced carrier aggregation for 4G wireless communications. IEEE Commun. Mag. **50**, 122–130 (2012)
2. Avser, B., Rebeiz, G.M.: Low-profile 700–970 MHz and 1600–2200 MHz Dual-band tunable antenna for carrier aggregation systems. In: IEEE Antennas and Propagation Society International Symposium (APSURSI) (2014)
3. Best, S.: Optimization of the bandwidth of electrically small planar antennas. In: Antenna Applications Symposium, Monticelo, IL (2009)
4. Sievenpiper, D.F., Dawson, D.C., Jacob, M.M., Kanar, T., Kim, S., Long, J., Quarfoth, R.G.: Experimental validation of performance limits and design guidelines for small antennas. IEEE Trans. Antennas Propag. **60**(1), 8–19 (2012)
5. Jouanlanne, C., Delaveaud, C.: Compact dual-band frequency agile antenna designed for carrier aggregation LTE small cell. In: 20th International Symposium on Wireless Communication Systems (2015)
6. 3GPP TR36.104, Release 12: Base Station Radio Transmission and Reception (2015)
7. Kowlgi, S. Berland, C.: Linearity considerations for multi-standard cellular base station transmitters. In: European Microwave Conference, pp. 226–229 (2011)
8. 3GPP TR36.141 Release 12: Base Station Conformance Testing (2015)

Energy Efficient Target Coverage
in Partially Deployed Software Defined
Wireless Sensor Network

Slavica Tomovic$^{(\boxtimes)}$ and Igor Radusinovic

University of Montenegro, Džordža Vašingtona bb,
81000 Podgorica, Montenegro
{slavicat,igorr}@ac.me

Abstract. Limited energy resources of sensor nodes are one of the main weaknesses of wireless sensor networks (WSNs). It has long been recognized that conventional methods of data transmission in WSNs are energy inefficient. However, implementation of coordinated, energy-aware routing and power control strategies among sensor nodes is difficult due to distributed network control. Software defined networking (SDN) is a new networking paradigm which overcomes this issue by decoupling the network control and data planes. As an emerging technology, originally envisioned for wired networks, SDN cannot be expected to completely replace traditional WSNs in near future. Therefore, in this paper, we investigate how to save energy in partially deployed software-defined WSN (SD-WSN). In particular, the paper considers the scenario of WSN deployed for monitoring set of targets with known locations, and analyses how the incremental SDN deployment and various power- mode switching policies could affect the WSN lifetime.

Keywords: SDN · WSN · Target coverage · Routing · Energy efficiency

1 Introduction

Wireless sensor networks (WSNs) consist of small, usually low-powered devices (sensor nodes), that measure specific parameters of the environment (e.g. temperature, pressure, motion, etc.). Each sensor node (SN) has the wireless communication capability, which enables it to send report messages to the network gateway. The gateway further delivers the gathered information to more powerful Internet-connected device that can process it. In this way, WSN can significantly reduce, or even completely eliminate the need for human involvement in many civilian, industrial, agricultural, and military applications [1].

Energy is the main resource constraint of SNs because their power sources are usually batteries with limited capacity. WSN lifetime is one of the key factors that determines the functionality and the accuracy of the sensing applications. Thus, methods that optimize the energy utilization of SNs are of great importance.

This paper focuses on the *target coverage problem* [2] in WSNs with large number of SNs randomly deployed to monitor (cover) set of targets with known locations. The main

© ICST Institute for Computer Sciences, Social Informatics and Telecommunications Engineering 2016
D. Noguet et al. (Eds.): CROWNCOM 2016, LNICST 172, pp. 729–740, 2016.
DOI: 10.1007/978-3-319-40352-6_60

application requirement is to cover all the targets and regularly deliver sensed data to the gateway as long as possible. One way to increase the lifetime of such a WSN is to schedule sensor nodes' activities. Since all SNs in the network share the common sensing task, if a target is monitored by multiple SNs, turning off some nodes does not affect the overall system function as long as target coverage is guaranteed. Although scheduling of sensor nodes's activities has been studied in literature, most of the proposed solutions are focused on minimizing the number of active SNs used for target coverage [2–7]. They assume that energy is consumed only for sensing, i.e. that each SN consumes a same amount of energy in the active power mode. However, in practice, energy consumed for data transmission predominantly determines the WSN lifetime. Energy is not only consumed by a SN which generates data, but also by all SNs along its route to the gateway. Thus, optimal scheduling decisions cannot be made without global view of the network state, which is lacking in traditional WSNs with a distributed control plane. Also, it should be noted that efficient energy utilization is usually of more importance than overall energy consumption, because unbalanced energy utilization can cause network partition even when a large number of SNs have a maximum residual energy [7]. Thus, small energy consumption does not always lead to increased WSN lifetime. Distributed routing protocols are one of the main causes of unbalanced energy utilization. For example, conventional multi-hop communication scheme based on Minimum Transmission Energy (MTE) routing, often leads to equally short WSN lifetime as direct communication with the gateway [8]. In MTE network, all nodes serve as routers for other nodes. Because SNs close to the gateway are most engaged in transmission of data, they have a tendency to drain their energy resources soon compared to the more distant nodes. This results in network partition and degradation of the network coverage.

Considering that distributed control plane of traditional WSNs prevents global resource optimization and smart traffic management, we propose the use of software defined networking (SDN). SDN is technology initially proposed for wired networks, which separates control logic from the forwarding hardware [9]. In SDN networks, the control plane is placed on a logically centralized controller, which maintains a global view of the network, interacts with simple forwarding devices and provides a programming interface for network management applications. Leveraging centralized intelligence of the SDN controller, it is possible to dynamically alter the network behaviour and deploy new applications in real time [10, 11].

The target coverage problem that we consider is motivated by the scenario where SDN sensor nodes (SDNSNs) are incrementally deployed in traditional ad-hoc WSN. Particularly, we have focused on MTE-based WSNs, with only a small percentage of SDNSNs deployed. The key questions we are trying to answer are:

1. Whether it is possible to do effective traffic engineering and prolong the network lifetime with only a small percentage of SDNSNs?
2. Which features SDNSNs and regular sensor nodes must have in order to be able to cooperate in the same WSN?
3. How various activity scheduling algorithms affect the lifetime of the considered WSN model?

The rest of the paper is organized as follows. In Sect. 2 we outline the network model assumed in the analysis and the proposed routing scheme. The analysed activity scheduling algorithms are described in Sect. 3. Simulation results are presented in Sect. 4. Concluding remarks are given in Sect. 5.

2   System Architecture

We consider WSN (Fig. 1) comprising of three main components: regular SNs, SDNSNs and SDN controller, which is integrated at externally supplied gateway (e.g. base station) and makes routing decisions for SDNSNs.

SNs are randomly deployed close to the targets with known locations that need to be continuously observed. At regular intervals, they send report messages regarding the

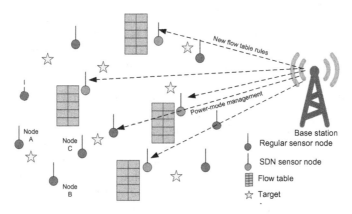

Fig. 1. The proposed network model.

observed status of the targets to the base station (BS). SNs can be in active or low-energy sleep mode. The network activity is organized into rounds, which consists of configuration and sensing phase. In the configuration phase, SDN controller makes routing decisions and decisions regarding the power mode of SNs in the sensing phase. The set of active sensor nodes in each round should be chosen by taking into account two constraints: (i) each target must covered by at least one node at any time; (ii) there must be a feasible route from each SN that participates in target monitoring towards the gateway. Thus, the set of active nodes includes nodes which are supposed to monitor the targets and generate report messages, and nodes which are needed only for relaying the messages towards the BS. The routing and scheduling algorithms are centralized and designed with the objective of maximizing the WSN lifetime, which is measured by the time that elapses from the network initialization to the moment when one or both of the mentioned constraints could not be met due to lack of energy.

Our initial assumption is that no changes are made to regular SNs, i.e. they are completely unaware of the existence of SDN elements in the network. The SDN controller is able to control regular SNs only by sending configuration commands that alternate the power mode. It is not able to control their forwarding behaviour. We have assumed that regular SNs run MTE routing protocol, which is a conventional protocol in today's WSNs. The routing principle of MTE protocol is to choose the next-hop on the route such that energy consumed for transmission is minimized. For example, node A in Fig. 1 would transmit to node C through node B if and only if:

$$d_{AB}^2 + d_{BC}^2 \leq d_{AC}^2 \qquad (1)$$

In the above formula, d denotes distance between the specified points. To make such forwarding decision, MTE node needs to know all functional SNs within its coverage area and their position. These information may be obtained via periodic exchange of "keep-alive" messages between SNs. Since SDNSNs have to participate in this exchange, we assumed that MTE routing daemon is running on them as well. Note that MTE routing logic makes sense only if nodes can vary the amount of transmit power. Technological advances in radio hardware make this assumption reasonable. The power level at which data will be transmitted is determined based on the location of the next-hop neighbour. For regular nodes that is always the neighbour on the way to the BS which requires the minimum power consumption, while for SDNSNs that could be any node within their radio range.

SDNSNs have a role similar to OpenFlow switches [12] in wired SDN networks. They perform data forwarding according to the controller's instructions stored in flow tables. Due to the specificity of WSN environment, format of flow table is modified compared to the OpenFlow specification. We assumed the table format proposed in [13]. Each flow table entry consists of: (i) matching rule - which defines the charac- teristics of packets that belong to the same traffic flow; (ii) action - which defines the processing steps; and (iii) counters - which serve for statistical purposes. As shown in Table 1, matching rules contain several window blocks which refer to blocks of bytes that will be matched against the packets. Each *window* consists of four fields, which define: the number and location of bytes that are supposed to be analysed, relational operator that is used during the block analysis, and value which is matched against the specified block of bytes. For our study, only two *actions* are of interest: forwarding and deactivation of the radio interface. Thus, the *action value* field indicates either the next hop on the route or time interval during which the radio interface should be turned off. For example, the first entry from Table 1 specifies that all packets that have in bytes 2 and 3 values 172 and 24 must be forwarded to node 170.16. The value in the last column indicates that 17 packets of this flow have been processed by SDNSN up to now. SDNSNs perform MTE routing only when no route to SDN controller is known. This could happen when control communication is lost due to bad conditions on the wireless channel.

Table 1. Format of flow table proposed in [13].

Window 1				...	Action		Stats
Size	Operator	Address	Value		Type	Value	
2	=	2	172.24	...	Forward	170.21	17
2	=	2	170.16	...	Drop	1	3
2	≠	2	170.25	...	Forward	170.22	3

SDN control communication relies on three types of messages: beacon packets, packet-in requests and flow-mod responses [13]. Beacon packets are periodically broadcasted by the BS. These packets contain one-byte field which indicates the number of hops required to reach the BS/controller from the transmitting SN. BS initially sets this byte to zero. Upon receiving a beacon packet, each SN checks whether the incremented byte value is less than the current estimate of the distance to the BS (initially ∞). If yes, the current estimate and estimate in the beacon packet are updated. Only if the received beacon packet is updated, it will be broadcasted further. In this way, SDNSNs learn a path towards the controller. SDNSN sends packet-in message to SDN controller when a received data packet does not match any of the rules in the flow table. This message carries the packet's header, based on which the controller can reactively make routing decisions. To inform SDN nodes on routing decisions, the controller generates flow-mod messages, which contain elements of flow table entries.

In order to be able to make smart routing and activity management decisions, SDN controller needs information about the network topology and SNs' characteristics such as SDN capability, location, radio range and energy consumption model. As will be discussed in the next chapters, node's energy consumption mostly depends on radio characteristics, distance to the next-hop node and amount of data transmitted. We assume that the first factor is known by controller in advance. The second factor is calculated based on the locations of the node and its next-hop neighbour. Information about the number of bits transmitted by SN during some period of time may be derived from the flow table counters. Because regular nodes do not have flow tables, we assume that BS has a flow table with separate entry for each SN. These entries perform matching on the packet source address, such that the counters indicate number of bytes originated at each node which successfully reached the BS. SNs also consume energy when forward data generated by other nodes in the network. However, SDN controller knows routes from each node to the BS that were used during the analysed time interval. Therefore, by knowing the routes and the flow table counters it can estimate a total amount of traffic carried by each SN. Computation of residual energy is performed at the beginning of each configuration phase by considering the last estimation, time when the last estimation was made, route collection, current and previous state of the corresponding counter in the flow table of the BS. Although SDN controller may not be able to predict precisely when SN runs out of energy (e.g. it is possible that some transmitted data are lost), estimations of residual energy could help in making efficient control decisions.

We have proposed the routing algorithm for the considered heterogenous WSN model in [11]. The key step of the algorithm is to create the reduced network graph. The reduced network graph differs from the connectivity graph by the number of links. It includes all links that originate at SDNSNs, and links that connect MTE nodes with their next-hop neighbours. The link cost function is defined as:

$$Cost(l) = \frac{EC(l)^{\alpha}}{RE(l_{src})^{\beta}} \tag{2}$$

In the above formula $EC(l)$ denotes energy needed to transmit and receive a packet on the link l, while $RE(l_{src})$ denotes the residual energy of the link source node. Parameters α

and β define the relative impact of these two factors on total link cost. They can be chosen in way to prefer the minimum energy paths or paths with nodes having the most energy, or a combination of the above. Once the link costs are determined, the least cost paths are calculated by the Dijkstra algorithm [14].

3 Node Activity Scheduling

In this section we present three algorithms for WSN activity management, which have been used in our simulation analysis.

Energy-Aware CPNS algorithm (EACPNS) is centralized version of the Coverage-Preserving Node Schedule (CPNS) algorithm [5]. CPNS algorithm aims to minimize energy consumption by minimizing the number of SNs that are used for target monitoring during each round of the WSN operation. It also provides guarantees that its decisions will not jeopardize the target coverage. The algorithm is distributed, and runs on each SN during the configuration phase. A node uses local neighbour information to decide whether to turn off itself or not. If the whole sensing area (in our case just targets within the sensing area) is covered by neighbouring nodes, this node can be turned off without reducing the network coverage [5]. In order to avoid "blind points", which may occur when two neighbouring nodes make decisions in the same time (because each believes that the other one will be in active state during the sensing phase), each node in the network delays its decision for a small, random period of time, and then notifies the neighbours. EACPNS is centralized, slightly modified version of CPNS algorithm. This version of the algorithm firstly sorts SNs in increasing order in terms of residual energy, and then determines the activity status of each node one by one. In this way, low-energy nodes have priority to be deactivated first if their targets could be monitored by neighbouring nodes with more energy. We have constrained selection of active nodes only on those that are able to reach the BS, either directly or by multi-hop communication.

Minimum Energy Consumption Algorithm (MECA) aims to minimize overall energy consumption during the sensing phase. To achieve this, besides energy consumption of SNs that are used for target monitoring, it takes into account energy consumption of relay nodes as well. Since energy consumption depends on several factors, such as: energy needed for communication with the BS and amount of generated data, we have used simulated annealing algorithm [15] to find an acceptable solution. The input argument of the algorithm are the coordinates of nodes that have targets within their sensing areas. More precisely, for each target, a set of candidate nodes is determined. As was the case with EACPNS algorithm, only nodes that have sufficient energy and which can communicate with the BS are taken into account. Also, in order to assure balanced energy consumption among the "monitoring nodes", at the beginning of the configuration phase SDN controller computes average energy level in the network, and removes nodes whose energy level is below the average value from the list of the candidates. The output of the algorithm is a subset of candidate nodes that is considered the most optimal from all solutions analysed during the algorithm's runtime. Optimality of the solution is evaluated based on total energy consumption that the solution requires for one round of WSN operation. The initial solution is chosen randomly (one candidate node for each target), and used as an input argument of the algorithm. At each iteration, the next state, which is in our case a new subset of the candidate nodes, is derived from previous one by perturbing coordinates of the corresponding nodes. The candidate nodes that have locations closest to the newly obtained coordinates become the new subset of candidate nodes that is going to be analysed. Given the current state at iteration k, represented by subset of candidate nodes C with energy cost Ec, the new state, represented by subset of candidate nodes C' with energy cost Ec', will become the current state with the probability:

$$p_k = \begin{cases} e^{-(Ec'-Ec)/\alpha_k}, \ Ec' > Ec \\ 1 \qquad\qquad , \ Ec' < Ec \end{cases} \tag{3}$$

where α_k is control parameter which increases with increasing k. We have configured this parameter according to the suggestions given in [16].

The third algorithm that we have analysed is based on CWGC algorithm [17]. CWGC algorithm is organized in three phases. In phase 1, a "communications tree" is constructed, which connects all SNs with the BS over the path with the least weight. We have adjusted this phase to the routing algorithm presented in Sect. 2. In phase 2, CWGC uses a greedy algorithm to determine a set of SNs that will monitor the targets and generate data in the following round of the WSN operation. This is done iteratively, by choosing node with the largest profit in each iteration. The node's profit is calculated as the ratio of the number of uncovered targets in the sensing range of the node, and weight of the path which connects the node with the BS in the communication tree. After selecting a monitoring node, the path weights of the upstream nodes in the communication tree are updated according to formula:

$$w(p_s) = w(p_s) + (e_{TX} + e_{RX}) \cdot B_r \cdot w(p_s)/E_r(s) \tag{4}$$

where B_r is the number of bits that the selected node is expected to generate in the round r, e_{TX} and e_{RX} are energies consumed by upstream node s for receiving and transmitting bit to the next-hop neighbour, and $E_r(s)$ is residual energy of the node s. Note that if a path weight of a SN changes, its profit value changes as well. At the end of each iteration, the selected SN is marked as active node and all targets within its sensing area are considered already covered in the following iterations. The process repeats until all targets are covered. In the final phase, the algorithm outputs a sub-tree of the *communication tree* on the basis of the selected monitoring nodes.

4 Performance Evaluation

In order to verify the effectiveness of the partially deployed SD-WSN and different algorithms for node activity scheduling in such an environment, we carried out set of simulations in MATLAB. Throughout the simulations, we have considered several random network configurations distributed in an 200 m $\times$ 200 m area, where each node is assigned an initial energy of 0.25 J. In the considered network scenarios, 20 target points are uniformly distributed over the area, while BS is positioned in the centre of the area. All SNs are assumed to have the same sensing range $Rs = 40$ m and the same maximum communication range $Rc = 80$ m. Values of α and β parameters in link cost definition have been set to 1 and 4 respectively.

WSN operation is organized into rounds. At the beginning of each round, SDN controller makes decisions regarding the activity status of SNs. We assumed that SNs send data at fixed rate. Each active SN which has one or more targets in its sensing area, generates ten 2000b data packets in regular intervals during the sensing phase of the round. Energy consumption in sleep mode has been neglected. By default, the round duration is set to be long enough to embrace ten regular reportings of the monitoring nodes. However, the SDN controller is able to define a smaller round duration when a selected set of active nodes does not have sufficient energy.

In simulations, we assumed the same radio model discussed in [8, 11, 16]. Energy consumed for the transfer of a k bit message between two SNs separated by a distance of r meters has been calculated as follows:

$$
E_T = E_{Tx}k + E_{amp}k = \begin{cases} E_{Tx}k + \varepsilon_{FS}r^2k, & r \leq r_o \\ E_{Tx}k + \varepsilon_{TW}r^4k, & r \geq r_o \end{cases} \tag{5}
$$

$$
E_R = E_{Rx}k \tag{6}
$$

where E_T denotes the total energy dissipated in the transmitter and E_R represents the total energy dissipated in the receiver electronics. Parameters E_{Tx} and E_{Rx} are per-bit energy dissipations for transmission and reception respectively. In transmission, additional energy is dissipated to amplify the signal (E_{amp}). Parameters ε_{FS} and ε_{TW} denote amplifier parameters corresponding to the free-space and two-ray propagation models respectively, while r_o is threshold distance given by the expression:

$$
r_o = \sqrt{\varepsilon_{FS}/\varepsilon_{TW}} \tag{7}
$$

All the radio model parameters have been configured to values used in [8, 11, 16] ($E_{Tx} = E_{Rx} = 50nJ/bit$, $\varepsilon_{FS} = 10pJ/bit/m^2$, and $\varepsilon_{TW} = 0.0013pJ/b/m^4$).

In the first experiment, we have studied the impact of node density on the lifetime of WSN in scenarios where MTE routing protocol is used by all SNs, and where 25 % of SNs have SDN capability and perform routing according to the centralized algorithm described in [11]. In the first scenario (MTE-CPNS), CPNS algorithm is used for managing activity of SNs, because we assume that all control decisions in MTE network are made in a distributed manner. The performance of SDN-based WSN is evaluated for each of the activity scheduling algorithms described in Sect. 3 (SDN-MECA, SDN-EACPNS, SDN-CWGS scenarios).

The Fig. 2 shows the WSN lifetime expressed in number of the reporting cycles as a function of the number of SNs in the network. The shown results are an average of multiple runs, each with a random generated topology. It should be noted that depending on the network topology, the obtained results sometimes varied substantially, but performance ordering of the considered network configurations remained the same in all simulations performed. From Fig. 2 we can see that the network lifetime increases as the

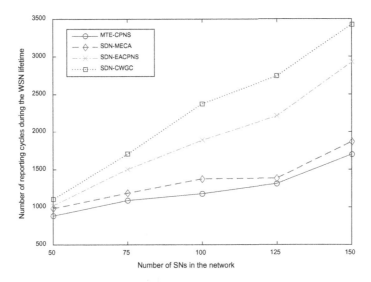

Fig. 2. The WSN lifetime vs. node density.

network density increases. This is because more SNs can be used to sense the targets and forward data to the BS. The results show that the hybrid WSN, consisted of 75 % of regular and 25 % of SDNSNS, for the same initial energy always produces a longer WSN lifetime than traditional WSN with MTE routing scheme and CPNS activity scheduling algorithm. The short WSN lifetime in MTE-CPNS scenario is dominantly a consequence of distributed routing. In traditional MTE WSNs, nodes which are close to the gateway forward the largest amount of data and suffer higher energy losses. These nodes "die" very fast, causing the network partitioning. Thus, even when there are nodes with sufficient energy that could be used for target coverage, it happens that the generated data cannot reach the BS because the upstream relaying nodes are out of energy. If we take WSN with 100 SNs as a representative example, in traditional MTE network the first node "dies" in 270th reporting cycle, in average. On the other side, in all the analysed SDN scenarios the first dead node occurs much later (Fig. 3). This could be attributed to a more balanced energy consumption among SNs.

In simulations of partially deployed SDWSNs, the MECA algorithm for node activity scheduling has shown the worst performance. This suggests that energy consumption is not a good criterion for selecting active nodes. As discussed earlier, when the objective is to maximize the WSN lifetime, overall energy consumption is not that important as to assure that each node consumes energy evenly. MECA algorithm fails to achieve a balanced energy consumption. The results from Fig. 3 support this claim, since we can see that the increase in WSN density does not always prolongs the lifetime of all SNs. However, the main reason of the poor performance of MECA algorithm is a large amount of generated data within the network. Figure 4 shows an average number of active SNs during one round of WSN operation. The smaller this number is, the smaller amount of data is generated. The shown results indicate the advantages of greedy algorithms for node activity scheduling, which in each iteration

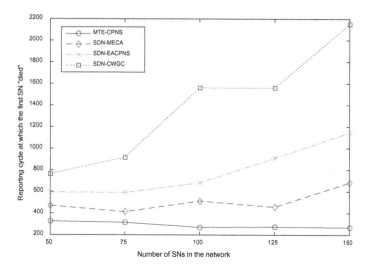

Fig. 3. The reporting cycle at which first SN runs out of energy vs. network density.

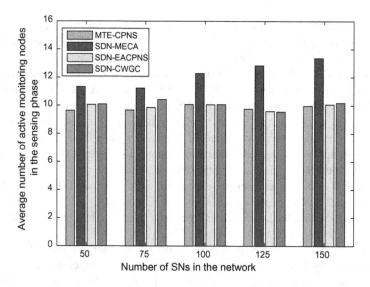

Fig. 4. The average number of active "monitoring" SNs during one round of WSN operation.

choose one active node according to specific criteria (CWGC, EACPNS, CPNS). When a SN is selected to be in active state, all of its targets are considered covered in the following iterations of the greedy algorithms. Thus, there is a lower probability for redundant data to be generated, which happens in cases when multiple SNs monitor the same target. On the other side, MECA algorithm rather examines the optimality of the complete solutions, consisted of one candidate node for each of the targets. Although the MECA's optimization function tends to decrease the number of active nodes, when the running time of the algorithm is limited to a reasonable number of iterations (200 in our simulation setup), the final solution may result in a high level of data redundancy. However, we can see that the introduction of even a small percentage of SDNSNs has been sufficient to compensate the inefficient scheduling decisions and prolong the WSN lifetime. Apparently, CWGC algorithm most efficiently exploits the benefits of centralized control among all the analysed activity scheduling algorithms. This is because CWGC uses the knowledge of the routing paths to select monitoring SNs which will not cause the "transmission bottlenecks" [17]. On the other side, EACPNS algorithm efficiently balances energy consumption among the monitoring nodes, but it neglects the fact that some of them might use the same relay SNs to deliver data to the gateway. Since the routing paths remain valid during the whole sensing phase, energy losses of the "bottleneck" SNs could be significant. In particular, the proposed WSN architecture with CWGC activity scheduling algorithm achieves 22.3 % longer lifetime than with the EACPNS algorithm in topology with 100 SNs.

5 Conclusion

When WSN is densely deployed for the purpose of monitoring a set of targets with known locations, the energy savings could be achieved by scheduling SNs to work alternatively. However, due to distributed control plane, in traditional WSNs it is hard to achieve balanced energy consumption. Routing and scheduling decisions are made by individual SNs, based on incomplete, local view of the network state, which prevents global resource optimization and smart traffic management. In this paper, we have analysed how the mentioned problems could be alleviated with incremental deployment of SDN-enabled SNs. The hybrid WSN model is presented, which assumes cooperation between traditional SNs, that run distributed MTE routing protocol, and SDN-enabled SNs. Through simulations we have shown that if information regarding the positions and capabilities of SNs are available, the proposed WSN model promises a significant increase in the WSN lifetime even when a small percentage of SDN-enabled SNs is deployed. Further on, we have pointed out the importance of a centralized view of the routing paths for activity scheduling algorithms. In particular, CWGC algorithm [17] has been identified as very beneficial for the considered WSN scenario.

The results presented in this paper only indicate potential benefits of incremental SDN deployment into traditional WSNs. In the next phase of our research, we will evaluate performance of the proposed architecture more accurately, by taking into account the influence of proportion of SDNSNs on WSN lifetime as well as energy waste due to control overhead and other more accurate wireless channel models.

Acknowledgments. This work has been supported by the EU FP7 project Fore-Mont (Grant Agreement No. 315970 FP7-REGPOT-CT-2013) and the BIO-ICT Centre of Excellence (Contract No. 01-1001) funded by the Ministry of Science of Montenegro and the HERIC project.

References

1. He, T., et al.: Energy-efficient surveillance system using wireless sensor networks. In: ACM International Conference on Mobile Systems, Applications and Services (2004)
2. Cardei, M., Thai, M.T., Li, Y., Wu, W.: Energy-efficient target coverage in wireless sensor networks. In: IEEE INFOCOM 2005, Miami, pp. 1976–1984 (2005)
3. Cardei, M., Wu, J., Lu, M., Pervaiz, M.O.: Maximum network lifetime in wireless sensor networks with adjustable sensing ranges. In: IEEE International Conference on Wireless and Mobile Computing, vol. 3, pp. 438–445 (2005)
4. Cardei, M., Du, D.-Z.: Improving wireless sensor network lifetime through power aware organization. Wirel. Netw. **11**(3), 333–340 (2005). Kluwer Academic Publishers
5. Tian, D., Georganas, N. D.: A coverage-preserving node scheduling scheme for large wireless sensor networks. In: ACM International Workshop on Wireless Sensor Networks and Applications, pp. 31–41 (2002)
6. Changlin, Y., Kwan-Wu, C.: A novel distributed algorithm for complete targets coverage in energy harvesting wireless sensor networks. In: IEEE International Conference on Communications (ICC), pp. 361–366 (2014)
7. Thippeswamy, B.M., et al.: EDOCR: energy density on-demand cluster routing in wireless sensor networks. Int. J. Comput. Netw. Commun. **6**(1), 223–240 (2014)
8. Heinzelman, W. R., Chandrakasan, A. P., Balakrishnan, H.: Energy-efficient communication protocol for wireless microsensor networks. In: Proceedings of the 33rd Annual Hawaii International Conference, Hawaii, vol. 2, pp. 10–17 (2000)
9. Open Networking Foundation: Software Defined Networking: the New Norm for Networks. Web White Paper. https://www.opennetworking.org/
10. Gante, A., Aslan, M., Matravy, A.: Smart wireless sensor network management based on software-defined networking. In: 2014 27th Biennial Symposium in Communications (QBSC), Ontario, pp. 71–74 (2014)
11. Tomovic, S., Radusinovic, I.: Performance analysis of a new SDN-based WSN architecture: In: Proceedings of 23rd Telecommunication Forum TELFOR 2015, Belgrade, Serbia, pp. 99–102 (2015)
12. OpenFlow Switch Specification v1.0.0. http://archive.openflow.org/documents/openflow-spec-v1.0.0.pdf
13. Costanzo, S.., Galluccio, L., Morabito, G., Palazzo, S.: Software defined wireless networks: unbridling SDNs. In: European Workshop on Software Defined Networking (EWSDN), Darmstadt, pp. 1–6 (2012)
14. Dijkstra, E.W.: A note on two problems in connection with graphs. Numerische Math **1**, 269–271 (1959)
15. Murata, T., Ishibuchi, H.: Performance evaluation of genetic algorithms for flowshop scheduling problems: In: IEEE Conference on Evolutionary Computation, vol. 2, pp. 812–817 (1994)
16. Heinzelman, W.B., Chandrakasan, A.P., Balakrishnan, H.: An application-specific protocol architecture for wireless microsensor networks. IEEE Trans. Wirel. Commun. **1**(4), 660–700 (2002)
17. Qun, Z., Gurusamy, M.: Maximizing network lifetime for connected target coverage in wireless sensor networks. In: IEEE International Conference on Wireless and Mobile Computing, Montreal, pp. 94–101 (2006)

SDN for 5G Mobile Networks: NORMA Perspective

Bessem Sayadi[1(✉)], Marco Gramaglia[2], Vasilis Friderikos[3], Dirk von Hugo[4],
Paul Arnold[4], Marie-Line Alberi-Morel[1], Miguel A. Puente[5], Vincenzo Sciancalepore[6],
Ignacio Digon[7], and Marcos Rates Crippa[8]

[1] Nokia Bell-Labs, Paris, France
bessem.sayadi@nokia.com, marie_line.alberi-morel@nokia.com
[2] IMDEA Networks Institute, Universisdad Carlos III de Madrid, Getafe, Spain
mgramagl@it.uc3m.es
[3] King's College of London, London, UK
vasilis.friderikos@kcl.ac.uk
[4] Telekom Innovation Laboratories, Berlin, Germany
Dirk.von-Hugo@telekom.de, Paul.Arnold@telekom.de
[5] ATOS Spain, Madrid, Spain
miguelangel.puente@atos.net
[6] NEC Laboratories, Princeton, USA
vincenzo.sciancalepore@neclab.eu
[7] Telefonica, Madrid, Spain
ignacio.digonescudero@telefonica.com
[8] TU Kaiserslautern, Kaiserslautern, Germany
crippa@eit.uni-kl.de

Abstract. To build a flexible and an adaptable architecture network supporting variety of services and their respective requirements, 5G NORMA introduced a network of functions based architecture breaking the major design principles followed in the current network of entities based architecture. This revolution exploits the advantages of the new technologies like Software-Defined Networking (SDN) and Network Function Virtualization.(NFV) in conjunction with the network slicing and multi-tenancy concepts. In this paper we focus on the concept of Software Defined for Mobile Network Control (SDM-C) network: its definition, its role in controlling the intra network slices resources, its specificity to be QoE aware thanks to the QoE/QoS monitoring and modeling component and its complementary with the orchestration component called SDM-O. To operate multiple network slices on the same infrastructure efficiently through controlling resources and network functions sharing among instantiated network slices, a common entity named SDM-X is introduced. The proposed design brings a set of new capabilities to make the network energy efficient, a feature that is discussed through some use cases.

Keywords: 5G architecture · SDN · NFV · QoE · Energy efficiency · Network slicing

© ICST Institute for Computer Sciences, Social Informatics and Telecommunications Engineering 2016
D. Noguet et al. (Eds.): CROWNCOM 2016, LNICST 172, pp. 741–753, 2016.
DOI: 10.1007/978-3-319-40352-6_61

1 Introduction

By just observing our daily routines we understand the impact that mobile connectivity has on our lives. This trend will be even clearer when new services that are envisioned in these days will come to market: self-driving vehicles or the Internet of Things (IoT) will further increase the need for expanded and faster network connectivity and increases the requirements on a mobile network. This goal cannot be achieved by just extending the current 4G architecture: a complete re-thinking of the mobile network system towards the 5G networking is needed.

The 5G NORMA[1] project aims at providing such a new architecture. Its main characteristics will be flexibility and adaptability required to match the available network resources with the fluctuating demand which is the utmost requirement to maintain the economic viability of future networks. Due to steadily rising energy costs a strongly related important key performance indicator addressed by the players in 5G is the energy efficiency as outlined in [1].

Driven by the rising of new technologies like Software-Defined Networking (SDN) and Network Function Virtualization (NFV), flexibility and adaptability can be achieved by using a virtualized infrastructure in which Network Function are allocated on demand by a centralized controller.

To provide the needed performance several levels of infrastructure may be deployed, ranging from the closest to the antenna (called *Edge Clouds*) to the most centralized (called *Central Cloud*). Network functions can be allocated at different levels according to the needed performance. This infrastructure is eventually shared among different *tenants* to provide different services, using *the network slicing* concept.

The network is sliced into many dedicated end-to-end virtual networks, each one handling a business case or a service while sharing the same network infrastructure. Based on customized SLAs, the owner of the network infrastructure should be able to allocate (or to sell) the required resources to each network slice. 5G NORMA leverages network softwarization and virtualization to implement a dynamic and a personalized network infrastructure resources allocation. In the 5G NORMA architecture the management of the network slices, running on the infrastructure, is in charge of a centralized controller. In this paper we describe how a network slice is controlled in order to achieve the target performance level for each business case or a service. In Sect. 2 we describe the 5G NORMA architecture, and then we focus on the Network Slice controller in Sect. 3. Section 4 describes the issues related to the coordination of different slices, while Sect. 5 shows the benefits of this architecture in terms of cost reduction and energy efficiency. Finally, Sect. 6 concludes the paper.

2 Slicing Enabled 5G NORMA Architecture

5G architecture should bring the required flexibility to support many services with different stringent requirements in terms of latency, throughput and availability. This

---

[1] http://5gnorma.5g-pppp.eu/.

requires rethinking the current mobile network architecture to move from current *network of entities* architecture to a *network of functions* architecture. 5G NORMA introduces an architecture leveraging the network slicing and multi-tenancy concepts allowing to deploy different network slices instances running on the same network infrastructure. Each one is tailored to the corresponding service and business needs.

2.1 Design Principles

The future mobile network architecture will be designed by the following the above principles which play an important role to fulfill the flexibility, scalability, context and security requirements:

- **Multi-tenancy** allows for several service providers operating on top of a shared infrastructure. The range of tenants is diverse, ranging from mobile network operators (MNOs) via over-the-top (OTT) service providers to companies from vertical industries. This also results in varying levels (depths) of service and resource control to be exposed to tenants.
- **A shared infrastructure** leverages the economies of scale to be expected when hosting multiple logical mobile networks. The infrastructure consists of heterogeneous hardware resources (general-purpose as well as dedicated, special-purpose hardware) and necessary software for hosting mobile network functions. The infrastructure as a whole is provided by several infrastructure providers, e.g., MNOs or 3rd party providers.
- **Efficient control frameworks** allow for a sufficient abstraction of controllable resources and functions and expose uniform control APIs on different abstraction and architectural levels. Thus, they allow for, e.g., cross-domain orchestration of network functions and services, flexibility in function decomposition and placement, and customized business service composition.
- **The fragmentation of administrative domains** increases complexity. Vertically, at least business service providers, network service providers and infrastructure providers have to be differentiated. Depending on the type of tenant, some or even all can collapse into one. Horizontally, multiple providers of each type co-exist.

2.2 Building Blocks Overview

The concept of network slicing is paramount within the 5G NORMA architecture. Therefore, the main 5G NORMA building blocks are related to this concept. More specifically, three families are envisioned (see Fig. 1): (i) related to the network slice life cycle and the interactions among different network slice instances, (ii) related to the management functions within a slice, and (iii) the mobility management module. In the sequel, each block will be introduced.

The Software Defined for Mobile Network Orchestrator (SDM-O) interfaces the network slices infrastructure to the business domain. The handling of slice creation requests (e.g., vehicular, IoT, possibly belonging to different tenants) is managed here. Requests come mapped to a set of KPIs according to the requested service: e.g., a vehicular

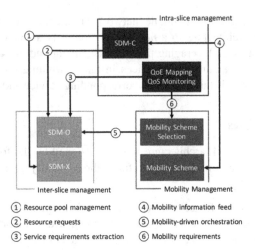

Fig. 1. The 5G NORMA connectivity and QoE/QoS management building blocks.

and a HD video slice requests will be associated to low latency and high bandwidth KPIs, respectively. These abstract requirements are translated to real network requirements that are used to build the chain of virtual Network Functions (vNF) and physical Network Functions (pNF) by using a template library. This process is similar to the one under investigation by the IETF Service Function Chaining (SFC) WG[2]. The needed vNFs are finally orchestrated by SDM-O, which has a complete view of the network: the optimal set of resources to be used and their location in the infrastructure is decided here.

On the other hand, the Software Defined for Mobile Network Control (SDM-C) directly manages the resources assigned to a network slice (there is a SDM-C in every network slice allocated by the SDM-O): it builds the forwarding path used to realize an SFC (Service Function Path, SFP, in the IETF SFC WG) while fulfilling the constraints and requirements defined by the SDM-O.

The information used to define those constraints is gathered from the QoE/QoS Mapping module, which also continuously analyzes the performance of a network slice and reports to the SDM-C. The configuration of a network slice may then be changed based on the QoS reporting. The re-configuration may happen at either vNF level or by reconfiguring forwarding paths by using an SDN interface. If the reconfiguration performed within a slice is not enough to fulfill the requirements assigned to a network slice, the SDM-C requests for more resources to the SDM-O.

Besides the Mapping & Monitoring module, also the Mobility Management module, in charge of selecting the right mobility scheme (e.g. Client based mobility, such as MIPv6, Network-based mobility, such as Proxy Mobile IPv6...) according to the service/user, reports to the SDM-C. The information exchanged through this interface is the one regarding the mobility of users, and it is used by SDM-C to deal with mobility-related aspects within the slice. The functional mobility of vNF between edge clouds is

---

[2] https://datatracker.ietf.org/wg/sfc/.

also managed by this module exploiting an interface with the SDM-O, which executes the migration.

The last major operation affecting a network slice is its reshaping (i.e., scale-in and scale-out operation). As mentioned above, these requests are managed by the SDM-O if they are related to computational resources. Otherwise, requests for shared resources (e.g., radio resources) are managed by the Software-Defined Mobile Network Coordinator named SDM-X.

2.3   5G NORMA Architecture

Figure 2 shows the architecture as specified so far in 5G NORMA [2]. The architecture differentiates between service layer functions which e.g. consider constraints and requirements related to demanded service and the policies applicable to the customer. The service layer itself is generally out of scope here. Major focus here is on the MANO (MANagement and Orchestration) layer mainly responsible for long-term allocation of resources for virtual network functions. The depicted architecture indicates at which functional entities the building blocks shown in Fig. 2 are logically located. Further details of the functional entities characteristics and features are described in [2]. The exact functionality is still for further specification within the 5G NORMA project. A typical example for newly defined entity in 5G NORMA is the SDM-C. The latter has two interfaces. The southbound interface (SBI) connects to all (dedicated) pNFs and vNFs. The northbound interface connects to the 5G NORMA-MANO components. For

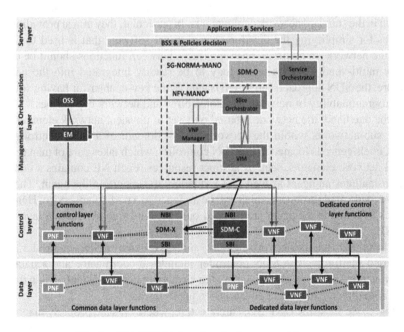

Fig. 2. 5G NORMA functional reference architecture

sharing pNFs and vNFs across multiple slices instances, a common entity named SDM-X is introduced as detailed in Sect. 4.

Another example for instantiation is the multipath feature of 5G which enables a device to be connected via multiple radio links and transport paths to enhance reliability, improve handover performance and save resources by efficient usage of different radio sites and/or technologies matching to the actual session demands. This functionality included in mobility management is realized at different vNFs in the Control Plane under control of one or multiple SDM-Cs resulting in new challenges for efficient operation.

3 Software Defined for Mobile Network Control Design for a Network Slice

Along the lines of SDN, 5G NORMA architecture incorporates the Software-Defined Mobile Network Control and Orchestration (SDM-C and SDM-O) concept which includes functions relevant for radio access and mobile core network. Starting by introducing the SDN concept, we describe in Sect. 3.2, our approach SDM-C highlighting the similarities and the new proposed features.

3.1 SDN Overview

The well-known approach of SDN has been recently introduced by researchers and data center architects in order to decouple the signaling layer (controller) from the physical resources. This directly implies a functional split between C-plane and U-plane. The Open Networking Foundation (ONF) focuses its activities on the adoption of the SDN paradigm in the standard network deployments. In particular, they aim at providing open interfaces for simplifying the development of novel software that is used to directly control the network. In addition, open interfaces between functions should be defined to allow multi-vendor access technology to be readily integrated into the network. Therefore, the SDN approach can be envisaged as the key-enabler for having software-based programmability of network functions (NFs) and network capabilities.

On the one hand, the network is based on a set of physical network elements (NE) forming sub-networks within the network control domain of a SDN controller. The network intelligence is located in the SDN controller, which takes care of managing the physical network elements, seen as abstract resources. Each NE contains a controller which instantiates an agent as well as a virtualizer to supporting the agent. The agent represents dedicated resources which will be assigned to a particular service. Hence, the resources assigned to the agent are mapped onto the hardware abstraction layer of the single NE.

On the other hand, the network elements are controlled by a SDN controller, placed on different platforms. The SDN controller includes a coordinator and a data plane control function (DPCF). Another agent with virtualizer may be instantiated in the SDN controller to directly support applications on upper layers.

In Fig. 3, we show the ONF-SDN architecture, where NFs are properly managed by the applications.

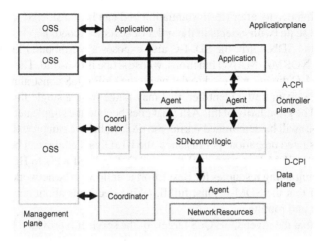

Fig. 3. The ONF-SDN architecture [4]

We can easily identify three different layers in the SDN architecture: (*i*) the data plane, wherein network entities expose their capabilities through D-CPI interface, (*ii*) the controller plane, the intermediate layer which maps the applications requirements onto physical resource requests while it feedbacks the infrastructure status to the application layer by means of A-CPI interfaces and, (*iii*) the application plane, where SDN-capable applications are developed.

It is clear that the current SDN architecture is well realized for fixed/transport network. For that, further modifications are needed to support RAN part of the network and forthcoming 5G features and to coordinate with legacy network controller to guarantee a smooth transition to the novel SDN paradigm.

3.2 Intra Network Slice Control

The inter-slice control comprises two main blocks. The first one is the SDM-C which is slightly different from SDN, introduced in Sect. 3.1, in case of mobile network through the co-existing of data and control plane. The second block is the QoE awareness which play an innovative role in 5G architecture design introduced by 5G NORMA.

3.2.1 Software Defined for Mobile Network Control
The Software Defined for Mobile Network Control (SDM-C) is a key function of the 5G NORMA architecture. It is assumed to have an SDM-C instance per network slice. It controls all of the network slice's dedicated pNFs, vNFs and associated resources (network, radio). The SDM-C allows for a fast reconfiguration (the right order will be assessed through the course of the project), to dynamically influence and optimize the performance of its network slice within the given amount of resources assigned to its network slice, i.e. at the time of the last (re-)orchestration made by SDM-O. The

(re-)configuration occurs after (re-)instantiation and can be considered to take place at a different time scale (with extents in the order of several seconds).

Following the SDN spirit, the SDM-C also exposes a Northbound Interface (NBI) towards the 5G NORMA-MANO functions, whose scope is two-fold. The 5G NORMA-MANO to SDM-C direction is used to define all the QoE /QoS constraints that have to be fulfilled for a given traffic identifier, that may range from a single flow to an entire network slice. The granularity of this API (that goes beyond the simple network function re-configuration) will be determined during the project, but we can provide some examples of its envisioned operation. For instance, the UL/DL scheduler can be dynamically configured by the SDM-C to provide the needed QoE-related KPIs to HD Video Users flows, while maintaining resources for Best Effort user flows. The network capacity may be another KPI that the SDM-C must fulfill, taking decisions about network function reconfiguration and routing.

In the case that the given QoE/QoS targets of the service(s) provided by its network slice cannot be met, the SDM-C may request re-orchestration. For that purpose, it uses the SDM-C to 5G NORMA-MANO interface to trigger a re-instantiation request (both of computational capabilities or shared resources such as frequencies or other shared network function).

3.2.2   QoE Awareness in 5G for Energy Efficiency

Quality of Experience (QoE), which is lately becoming the ultimate item to be delivered to end-users, is defined as "the degree of delight or annoyance of a person whose experiencing involves an application, service, or system" [3]. This contrasts with Quality of Service (QoS), which concerns objective and technical metrics at network (delay, jitter, packet loss, etc.) and application level (frame rate, resolution, etc.).

QoE-awareness refers to the ability of knowing about and react according to the user perceived quality (QoE) of a certain service. The 5G-NORMA network architecture will allow the development of QoE-aware mechanisms, being one of their main goals energy efficiency (see Sect. 5).

A key aspect to achieve QoE-awareness is the proper monitoring of the QoS factors. Conventionally, legacy and 4G mobile networks employ QoS measures based on objective system-centric metrics at network (such as delay, jitter or packet loss, etc.) and application level (frame rate, resolution, etc.). In 5G-NORMA the QoE will be then derived from these QoS metrics through a QoE/QoS mapping process using appropriate mathematical functions. Furthermore, the QoE model shall include the complete end-to-end objective system factors (network, application/service and terminal) and the subjective human influencing factors (expectations, likings, etc.) along with information about the user's context and business factors (Fig. 4).

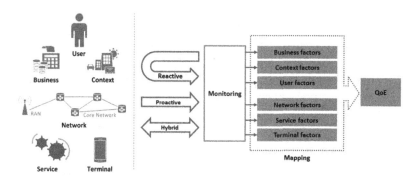

Fig. 4. Monitoring and mapping of the QoE influencing factors

QoE/QoS mapping and monitoring is strongly architecturally related to the SDM-C. The QoE engine described in Fig. 4 will be an application running in the Northbound of the SDM-C.

4 Software Defined for Mobile Network Control Design for a Set of Network Slices

To optimize the infrastructure usage and support different business cases or services, it is expected that many network slices will be instantiated on the same infrastructure.

According to NGMN, a set of network slices could share one or many network functions or network capabilities. One can cite, the scheduling function deployed to share the same spectrum among network slices. Thus, two network functions categories could be defined: dedicated or common ones, invoking different manner of control. For that purpose, our 5G NORMA design introduces a new component called SDM-X developed in the following section.

4.1 SDN Hierarchy Overview

In [4], the ONF provides its perspective on the relationship between SDN and NFV. According to the report, a SDN controller virtualizes resources under its control, and then orchestrates their shared use to satisfy the user requests. VNFs are among the resources available for a SDN controller, which brings together SDN and NFV. The goal is to structure the resources as an end-to-end service. This process is recursive, as SDN controllers at a higher hierarchical level can see lower level controllers as resources as well. One illustrative example is given in Fig. 5, showing a SDN hierarchy where the Green controller has a global perspective, while the Blue, Gold and Violet controllers have a local view of their sub-domains.

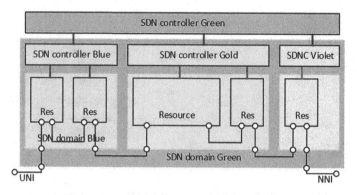

Fig. 5. SDN hierarchy example [4] (Color figure online)

The Green controllers orchestrates the resources offered by its subordinate SDN domains to offer a service for some given user-network interface (UNI) to some network-network interface (NNI), as well as coordinates the handoff between domains (alignment of physical port, protocol stack, security, etc.), which requires alignment between Green and the neighboring domains. Alignment may be achieved by provisioning, discovery, or negotiation.

Local SDN controllers will recursively operate in the same way with the resources in their domain (orchestrating, interconnecting and provisioning resource instances).

A similar dynamic occurs in 5G NORMA, where the Software-Defined for Mobile Network Coordination (SDM-X) manages network functions or resources shared among selected network slices. Those resources can be physical or virtual network functions, or wireless resources, like spectrum. The SDM-O decides which functions will be shared among network slices and provides the SDM-X with the relevant service policy. The SDM-X combines this policy with the received network MANO requirements and decides whether to modify or not a network slice capabilities, in order to fulfill the agreed Service Level Agreement (SLA) for the given tenant.

4.2   Software Defined for Mobile Network Coordination for Common Functions Control

5G NORMA defines Software Defined for Mobile Network Coordination (SDM-X) in order to control resources and vNFs sharing among instantiated network slices [5]. Based on service policies defined by the SDM-O (orchestrator) it can provide short term modifications to fulfill the SLA of a given tenant. Due to the decomposition and virtualization of core and RAN related NFs [2] the SDM-X controls and reacts on data flow requirements if the network slices share their allocated resources. The SDM-X coordinates the resources (e.g. computing power or radio resources), vNF and pNFs shared by network slices and decides, e.g. which network function, such as scheduling or ICIC schemes have to be used to avoid conflicts. It needs to tightly interact with all affected SDM-Cs which have access to the same PNFs and VNFs.

It is of paramount importance to have a common entity among network slices which provides scheduling decisions for efficient usage of wireless physical resources, such as frequency, time and transmission points. It has to coordinate and control the access of single network slice related SDM-Cs on common pNFs and vNFs to prevent conflicting decisions.

Based on [6], Fig. 6 shows an example, how the SDM-X needs to control inter-slice related radio resource allocation and RAN related network functions. Focus of the example is on RAN related aspects, while the core related functions are excluded for simplicity here. Figure 6 illustrates three instantiated slices A (green), B (magenta) and C (blue). Within each network slice A and C one data flow is given, while in network slice B, two data flows are concurrently served. Dependent on QoS requirements of the data flows and network slice related policies the SDM-X needs to decide which set of RAN related network functions the data flows need to have when network slices are allowed to use the same physical radio resources. Slice A and B use the same time, frequency grid while slice C dedicated resources are allocated based on the defined network slice C policy in time and frequency domain. Thus for these exclusively assigned resources the slice (C) specific SDM-C can decide autonomously on how to map radio resources to the flows and packets. However, slice C might use the same Tx points as slice A and B. Therefore the SDM-X needs to resolve upcoming conflicting requests of the slice individual SDM-Cs.

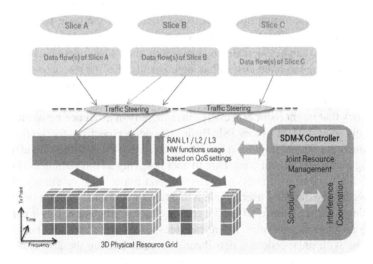

Fig. 6. SDM-X control of radio resources and RAN network functions (Color figure online)

5   Energy Efficiency Impact

The proposed 5G NORMA architecture leveraging NFV and SDN approaches can provide energy proportionality by gracefully adjusting the network resource to the current demand, to improve inter alia energy consumption.

SDM-O can allocate network functions at selected geographical locations exploiting the energy saving benefits of centralized scenarios, e.g. in terms of the power amplifier, air-condition, etc., and distributed environments, e.g. allocating network functions and content at the edge of the network, closer to the end user, can save energy for the mobile backhaul and core network. In addition, SDN enables a split of the control and data plane simplifying energy saving management of distributed functions and heterogeneous radio environments, while at the same time creates flexible service chains among virtualized network functions in where energy can be an additional optimization parameter.

To this end, the way that network and node resources are allocated by the SDM-O and controlled via SDM-C/X in a specified network slice has a clear impact on the overall energy consumption. The number of nodes, routing paths and associated virtual machines (VM) that will be allocated, to be active for a specific set of users request will define the energy footprint of the resource allocation. The power consumption of a node can be deemed as load depended and more specifically scales linearly with respect to the CPU usage [7]. To this end, the power consumption can be modeled as shown below based on the CPU utilization U_{cpu}.

$$P = \begin{cases} P_{idle} + (P_{max} - P_{idle}) \times U_{cpu} & , ifU_{cpu} > 0 \\ 0 & , otherwise \end{cases}$$

Where P_{idle} and P_{max} denote the power consumption of the node in the idle state and the maximum power consumption of the node (i.e., when the node is fully utilized) respectively. Since the utilization is in essence a function of time, i.e., $U_{cpu}(t)$, the total energy consumption can be calculated by integrating the above function over a required time interval. Noting the fixed cost of having elements idle, the resource orchestration should take this into account in order to minimize the number of low utilized nodes. At the same time, care should be taken in high node utilization regimes due to the potential QoS penalties that might incur especially since there is interference between the operations of different VM instances [8]. Besides node perspective, the ability to shift network functions via VM migration between centralized and distribute cloud platforms considering also the fronthaul contribute to energy saving as it is demonstrated in [9].

In addition to, QoE-awareness may have a great impact also on energy efficiency. QoE allows having satisfied users while allocating the minimal amount of resources, thus reducing costs, avoiding churn and increasing energy efficiency. This is particularly interesting in wireless networks, where radio resources are scarce. 5G networks will have to cope with unprecedented densification levels, causing the access network to account for the major energy consumption share [10]. In this scenario a moderate reduction in the data rates can lead to large energy (and therefore costs) savings as it is shown in [11].

6   Conclusions

In this paper we present the 5G NORMA perspective on controlling the network slices. We introduce the concept of Software Defined for Mobile Network Control (SDM-C) putting forward its role in controlling the intra network slices resources and its complementarity with the Orchestration component called SDM-O. We demonstrated the need for another controller named SDM-X in charge of configuring and controlling the common network functions (physical or virtual) between a set of network slices instantiated on the same infrastructure. The energy impact of 5G NORMA design is also discussed showing the direct benefit through the use of our concept SDM-C/X/O.

Acknowledgments. This work has been performed in the framework of the H2020-ICT-2014-2 project 5G NORMA. The authors would like to acknowledge the contributions of their colleagues. This information reflects the consortium's view, but the consortium is not liable for any use that may be made of any of the information contained therein.

References

1. NGMN Alliance, NGMN 5G white paper. https://www.ngmn.org/uploads/media/NGMN_5G_White_Paper_V1_0.pdf. Accessed Dec 2015
2. 5G NORMA: D3.1: functional network architecture and security requirements. https://5gnorma.5g-ppp.eu/dissemination/public-deliverables
3. Le Callet, P., Möller, S., Perkis, A.: Qualinet white paper on definitions of quality of experience. European network on QoE in multimedia systems and services, version 1.2, March 2013
4. Open Networking Foundation: Relationship of SDN and NFV, issue 1, Technical report 518. https://www.opennetworking.org/images/stories/downloads/sdn-resources/technical-reports/onf2015.310_Architectural_comparison.08-2.pdf. Accessed Oct 2015
5. Gramaglia, M., Digon, I., Friderikos, V., von Hugo, D., Mannweiler, C., Puente, M.A., Samdanis, K., Sayadi, B.: Flexible connectivity and QoE/QoS management for 5G networks: the 5G NORMA, Submitted to ICC (2016)
6. Deutsche Telekom AG, Deutsche Telekom's view on 5G, September 2015. ftp://ftp.3gpp.org/workshop/2015-09-17_18_RAN_5G/Docs/RWS-150033.zip
7. Beloglazov, J.A., Buyya, R.: Energy-aware resource allocation heuristics for efficient management of data centers for cloud computing. Future Gener. Comput. Syst. **28**(5), 755–768 (2012)
8. Pu, X., Liu, L., Mei, Y., Sivathanu, S., Koh, Y., Pu, C.: Understanding performance interference of I/O workload in virtualized cloud environments. In: Proceedings of the 2010 IEEE 3rd International Conference on Cloud Computing, pp. 51–58 (2010)
9. Sabella, D., et al.: Energy efficiency benefits of ran-as-a-service concept for a cloud-based 5G mobile network infrastructure. IEEE Access **2**, 1586–1597 (2014)
10. Auer, G., Giannini, V., Desset, C., Godor, I., Skillermark, P., Olsson, M., Imran, M.A., Sabella, D., Gonzalez, M.J., Blume, O., Fehske, A.: How much energy is needed to run a wireless network? IEEE Wirel. Commun. **18**(5), 40–49 (2011)
11. Andrews, J., Buzzi, S., Choi, W., Hanly, S.V., Lozano, A., Soong, A., Zhang, J.: What will 5G be? IEEE J. Sel. Areas Commun. **32**(6), 1065–1082 (2014)

Statistically Sound Experiments with OpenAirInterface Cloud-RAN Prototypes

CLEEN 2016

Niccolò Iardella[1(✉)], Giovanni Stea[1], Antonio Virdis[1], Dario Sabella[2], and Antonio Frangioni[1]

[1] University of Pisa, Pisa, Italy
{niccolo.iardella,giovanni.stea,a.virdis,frangio}@di.unipi.it
[2] Telecom Italia Lab, Turin, Italy
dario.sabella@telecomitalia.it

Abstract. Research on 4G/5G cellular networks is progressively shifting to paradigms that involve virtualization and cloud computing. Within this context, *prototyping* assumes a growing importance as a performance evaluation method, besides large-scale simulations, as it allows one to evaluate the computational requirements of the system. Both approaches share the need for a *structured* and statistically sound experiment management, with the goal of reducing errors in both planning and measurement collection. In this paper, we describe how we solve the problem with OpenAirInterface (OAI), an open-source system for prototyping 4/5G cellular networks. We show how to integrate a sound, validated software, namely ns2-measure, with OAI, so as to enable harvesting samples of arbitrary metrics in a structured way, and we describe scripts that allow structured experiment management, such as launching a parametric simulation campaign and harvesting its results in a plot-ready format. We complete the paper by demonstrating some advantages brought about by our modifications.

Keywords: LTE-A · Cloud-RAN · OpenAirInterface · Performance evaluation · Experimentation · ns2-measure

1 Introduction

Future 5G cellular networks will employ *virtualization* and *cloudification* of the Radio Access Network (RAN) [12], whereby the baseband processing is done on *virtual baseband units* (BBU) running on commodity hardware, leaving only antennas on site. On the other hand, software products, both commercial and open-source, are already available that emulate a software BBU compliant with the 3GPP standards. One such product OpenAirInterface (OAI), which runs an LTE protocol stack entirely implemented in software [2]. OAI also allows one to carry out experiments using hardware equipment and commercial terminals. The above two fact motivate a shift in the research paradigm, which is progressively based on *prototypes* of cellular networks. In fact, OAI has been and is being widely used in EU-funded and academic projects in the field of cellular

© ICST Institute for Computer Sciences, Social Informatics and Telecommunications Engineering 2016
D. Noguet et al. (Eds.): CROWNCOM 2016, LNICST 172, pp. 754–766, 2016.
DOI: 10.1007/978-3-319-40352-6_62

networks. The Flex5GWare EU project [6], where the authors of this paper are involved, aims at building cost-effective hardware/software platforms for 5G so as to increase the hardware versatility and reconfigurability, increase capacity and decrease the overall energy consumption. Within it, one of the proof of concepts will consist in evaluating resource allocation algorithms in a Cloud-RAN environment, which will be realized running a customized version of the OpenAirInterface software on virtual machines.

This implies the need to get credible *performance metrics* out of the OAI software, for both the cell and the user, and at several levels: what is the cell MAC-level throughput, how user application-level throughput varies with the number of users or interfering eNBs, how much energy is consumed, etc. It goes without saying that the above activity must be done with a long-term perspective, so as to keep the software maintainable, and ensuring that rigorous, unbiased and statistically sound results are obtained. In this respect, it has already been observed in [7, 8] that an *unstructured* approach to experiment management is often a major source of bugs, and ultimately affects the credibility of the results.

Unfortunately, OAI offers little in the way of a structured experiment management, leaving the task almost entirely to the user. First of all, emulation scenarios are defined in non-parametric XML files. This requires a user to manually change the XML file so as to modify the parameters (e.g., in order to vary the number of users), possibly in several parts simultaneously, which is error-prone. For instance, even generating a new replica of the same emulation scenario with a different random seed becomes non trivial. As far as measure gathering is concerned, OAI offers two basic ways: one is system logging printouts, which can be redirected to file and parsed (using standard tools such as `grep`). The other is a built-in dashboard, which shows the instantaneous situation at the physical level in terms of channel response and signal power. These tools, which were probably meant for different purposes – namely, logging/debugging for the first one, and debugging and providing a quick visual feedback regarding physical-layer parameters for the second, are not suited for a systematic performance evaluation. For instance, the throughput is computed having the *simulation duration* at the denominator, regardless of when the generator is actually started. This implies that – if generators are started at different times in the simulation – the throughput results are incorrect. More-over, there is no way to define a *warm-up* phase, where samples are not collected. Finally, the overhead of writing on file the entire system log (of which just a minor portion may be of interest) is non negligible.

In this paper we describe how to automate experiment management with OAI so as to make it faster, structured and less error prone. First of all, we show how to integrate an existing software, namely *ns2-measure* [7], into OAI. *ns2-measure* was originally developed for the ns2 simulator, and offers to researchers a framework for data collection and creation of statistically sound results. We describe the steps to compiling the two software together (something made slightly tricky by the fact that OAI is written in ANSI C, whereas ns2-measure is in C++), and the few, localized modifications required to OAI. This enables a user to gather a wide range of measures of interest in a seamless way, adding a negligible overhead to the OAI running time and memory consumption. Moreover, we describe intuitive, yet general scripts that can be used to generate para-metric emulation scenarios and aggregate performance metrics across a set of parametric scenarios to facilitate producing output graphs and tables. As for parametric scenario generation, our script describe the set of parameters that should vary across the scenarios (therein including the initial seed for the random generators when independent replicas are required) at a high level, and the script generates the XML scenario files to run the OAI emulation and manages their execution. As for aggregation of performance metrics, we show scripts that allow to compute means and related confidence intervals, taking measures from *ns2-measure* outputs or OAI built-in logging facilities.

The rest of the paper is organized as follows: Sect. 2 reports background information on OAI. Section 3 describes the ns2-measure software. In Sect. 4 we describe our tools and explain how to integrate ns2-measure with OAI. We report some example evaluation results in Sect. 5, and we conclude the paper in Sect. 6.

2   OpenAirInterface

OpenAirInterface (OAI) is an open-source platform for wireless communication systems, developed at Eurecom's Mobile Communications Department. It allows one to prototype and experiment with LTE and LTE-Advanced (Rel-10) systems, so as to perform evaluation, validation and pre-deployment tests of protocol and algorithmic solutions. OAI allows one to experiment with link-level simulation, system emulation and real-time radio frequency experimentation. As such, it is widely used to setup Cloud-RAN and Virtual-RAN prototypes. It includes a 3GPP-compliant LTE protocol stack, namely the entire access stratum for both eNB and UE and a subset of the 3GPP LTE Evolved Packet Core protocols [2].

OAI can be used in two modes: the first one is a real-time mode, where it provides an open implementation of a 4G system interoperable with commercial terminals, so as to allow experimentation. This requires using a software-defined radio frontend (e.g. the Ettus USRP210 external boards [3]) for airtime transmission.

The second mode is an emulation mode, where software modules emulating eNBs and UEs communicate through an emulated physical channel. In the emulation mode, scenarios are completely repeatable since channel emulation is based on pseudo-random number generation. In emulation mode, OAI can emulate a complete LTE network [1], using the oaisim package. Several eNBs and UEs can be virtualized on the same machine or in different machines communicating over an Ethernet-based LAN. The PHY and the radio channels are either fully emulated (which is time-consuming) or approximated in a PHY abstraction mode, which is considerably faster. In both cases, emulation mode runs the entire protocol stack, using the same MAC code as the real-time mode. This way, the oaisim package can be used to alpha-test and validate new implementations or sample scenarios, dispensing with all the problems that airtime transmission on a SDR frontend may bring about. Since the same code is used in the emulation and the real-time mode, a developer can then switch seamlessly to the real-time environment.

OAI includes the *OAI Traffic Generator* (OTG), which can be mounted on top of the LTE stack and used to run an emulation with different loads [4]. The generator includes predefined traffic profiles, such as device-to-device, gaming, video streaming and full buffer, and can be customized using OAI scenario descriptors (OSDs).

OAI's structure reflects the one of the LTE protocol stack: every layer of the stack is composed of one or more modules, implemented by one or more C libraries. Every layer or module uses calls to interface functions of other modules to retrieve status information and to encapsule/decapsule data. For example, the MAC scheduling module gets called by the main MAC module at every subframe and is implemented in *eNB_scheduler.c* and *pre_processor.c* files [9]. Every application in the OAI suite (such as oaisim and the real time eNB) instantiates and initializes the stack layers and the other modules it needs: oaisim, for example, makes use of other modules for the emulation capabilities, the most notable being the OTG, the OAI Channel Generator (OCG) which emulates the radio channel and the OAI Mobility Generator (OMG) that emulates the movements of the nodes. After the *init phase*, the application enters in a *loop phase* where modules and layers execute their functions on a per-subframe basis; lastly, before exiting, the main process deallocates the layers and possibly executes termination operations (e.g., output of performance stats).

OAI software uses three methods for the output of performance metrics:

- a graphical dashboard that can be optionally shown while the system emulation or the eNB implementation runs, which shows received/sent signal power, channel impulse and frequency responses, constellation diagram and PUSCH/PDSCH throughput (see Fig. 1).

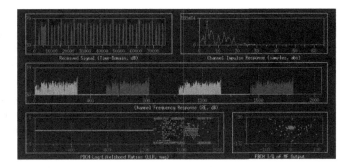

Fig. 1. Detail of the graphical dashboard showing in real-time the physical-level stats of a node in OAI system emulation.

- A series of prints in the standard output logging of the system emulation, which appear when the traffic generator is enabled, and show traffic-related metrics (sent and received bytes, application level throughput, one-way delay and so on).
- One or more files with PHY level stats on HARQ processes and DLSCH/PDSCH throughput.

All these methods are useful to get a rough idea of how the system behaves, but none of them, taken alone, is sufficient to profile it completely and effectively: the graphical dashboard is shown in real-time, it leaves no logs and it is destroyed once OAI terminates. The traffic generator stats have the disadvantage of being written on standard output together with the entire OAI log, so they must be collected using `grep` or other text search tools, which can be more and more impractical as the number of simulation runs and input parameters increases. The same can be said of the output files for the PHY stats. Another limitation of the traffic generator stats is that throughput is calculated on the *entire* emulated time, taking no account of the initial warmup time in which the system is running but the generator is not.

In general, OAI is missing a structured, flexible and extendable system for the collecting and managing performance measures. Different metrics get collected and shown in different ways and the only way for the user to keep track of experiment results is to tailor custom scripts to launch OAI and extract the desired data from the existing outputs. The time interval of samples collection cannot be selected and this makes the analysis of dynamic scenarios difficult when not impossible and leads to warped measurements when warmup time is a critical factor.

Moreover, built-in metrics might not be sufficient for research purposes. For example, the performance evaluation of MAC scheduling algorithms would need to keep track of resource block allocation, which is not among the built-in metrics. In this case, the lack of an efficient and robust metrics collection framework makes custom metrics hard to implement and collect, and even when they are implemented they are bound to the same limits of the built-in ones.

3   Ns2-Measure

The *ns2-measure* package [7] is a C++-based framework for collection of statistics. It was originally developed for the ns2 network simulator [10], offering an interface to TCL, its main configuration language. Its C++ API however, can be used for integration into any C/C++ code. The main goal of *ns2-measure* is to provide researchers a structured and ready-to-use tool for collection of statistically sound measurements. More in detail it can be used for both collecting samples of user-defined metrics during the simulation, and to estimate the average values or the probability density function (PDF) of the above samples. Metrics can be of three *types*, depending on how their samples are collected, as listed below:

- RATE, which are time-related and time-averaged, e.g. the throughput;
- CONTINUOUS, describing a continuous-time stochastic process (either discrete- or continuous-state), i.e. one whose trajectories are continuous in time. An example is the number of packets in the queue during the simulation, which is a discrete quantity (hence discrete-state) that varies at any time (hence continuous-time);
- DISCRETE, describing a discrete-time stochastic process, i.e. one whose trajectories are impulses. An example is the end-to-end delay of a flow, a continuous quantity measured at successive packet departure instants (hence discrete-time).

The framework also offers the user support for independent replicas, which are used to obtain statistically sound results (e.g., with associated confidence intervals).

Each metric (of any type) can be defined for more than one entity at a time within the system. This allows a user to obtain both system-wide and per-entity statistics. For example, when simulating an LTE network one might be interested in both a cell-level and a per-UE throughput, and both can be defined and sampled simultaneously. Data collection can be enabled and disabled *dynamically* at runtime by flipping the *collect* variable. This allows the user some (very much needed) freedom: for instance, she can define a warm-up time wherein statistics are not collected, or she can measure the throughput of intermittent applications in a meaningful way, by turning on throughput sample collection only when a burst of activity occurs.

The core element of *ns2-measure* is the *Stat* C++ class. It is responsible for creating data structures for each metric at the beginning, collecting samples while the emulation runs and producing output at the end. It also keeps a reference to the elapsed emulated time, to tag time-related metrics, such as the RATE ones. These operations are made available via three main C++ functions. The *Stat::command* instantiates the data structures for the user-defined metrics, activates and deactivates the collection and manages the output file. The available metrics are configured via file and are included into the system during compilation. The above data structures are implemented in the *Sample* class, which stores the measured samples for each entity, keeping track of their total, maximum and minimum values.

The *Stat::put* function is used to insert data collection probes within the code. This function takes as a parameter the name of the considered metric, the ID of the entity for which the sample is collected and the measured value, which will be stored in the appropriate instance of the *Sample* class, possibly updating the max and min values.

The *Stat::print* function is used to finalize an experiment. More in detail, it computes and stores to a file the estimated mean value of each metric. Files will also contain the run-id of the experiment, which can be used when multiple replications of the same scenario are run, e.g. to aggregate metrics across the various replications.

The output of experiments performed on complex and possibly large system can grow quite big in some cases. For this reason results are stored into *binary* files, thus reducing the occupancy with respect to text files. If needed, the results can be converted to human-readable text format using external tools that come together with the *ns2-measure* framework.

4   Contributions and Integration

In this section we first explain how to integrate *ns2-measure* with OAI, and then show scripts that automate experiment management, so as to facilitate running entire simulation campaigns.

4.1   Integrating ns2-Measure

As outlined before, the core of *ns2-measure* is the *Stat* class, which collects the raw samples from user-defined probes. It is a static C++ class which uses the method *Stat::command* to implement the TCL interface and interpret the commands specified in Table 1. The other main method is *Stat::put*, which implements the collecting probe and accepts as input parameters the name and type of the metric and the value of the sample. The metrics' names are defined in two headers, *metrics.h* (which contains macros with names to use when calling the *put* method) and *metric_names.h* (which contains human-readable names to be used when saving the output).

Table 1. Main commands of ns2-measure TCL interface.

`$ stat file< filename>`	Specify the output file
`$ stat on`	Enable sample collection
`$ stat off`	Disable sample collection
`$ stat print`	Print stats on output file

Since the *Stat* class has been developed with ns2 in mind, a certain effort of adaptation must be spent to port it on other simulators. In particular we need to work: (a) on code interoperability so that the *Stat* methods can be used in the new environment; (b) on the method that the *Stat* collecting probe must use to read the simulated time when acquiring samples, and (c) since build automation tools (such as CMake) are likely to be used, we need to make them aware of the new code. The following passages describe the specific interventions on OAI. However, they are general enough to apply to other C++-based simulation softwares.

Code Wrapping. OAI is mainly implemented in C and makes no use of the TCL language, so the static class must be modified so as to allow its methods to be called from the C code. To achieve this, we implemented a wrapping library which contains one C function per TCL command: for example *stat_cmd_add()* calls *Stat::command*, thus emulating the "add" TCL command and so on. Similarly, the *stat_put()* function wraps the *Stat::put* method (see Fig. 2).

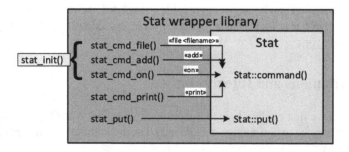

Fig. 2. The wrapping library contains one wrapper function per TCL command and a wrapper function for the probe. An initialization procedure is defined for the sake of convenience.

Simulated-Time Reading. On collecting a sample, the *Stat* class calls an ns2 method to read the current value of the simulated time. In OAI these calls must be replaced by a read to the *time_ms* variable, which is updated at every new subframe.

Build Tools. Since OAI uses CMake [11] as an automation tool for building, we added as a libraries the *Stat* class and the wrapping library, and we added these libraries to the OAI System Emulation target. To speed up testing, we added the ability to activate or deactivate the *ns2-measure* functionality at run time via configuration file, without having to recompile the target.

In order to use the new code, the OAI code must be modified in at least two points, namely the initialization phase and the termination phase, plus all the points where we want to put a probe in the loop phase. In the initialization phase we need to call the `file` command, specifying the name of the output file (again, the file name can be specified through a configuration file), and the `add` command, once per metric. In this phase one may want to activate measure collection using the `on` command (alternatively, this can be deferred to the time when the traffic is actually started). For the sake of convenience, all these operations are gathered in a *stat_init()* function. In the termination phase we need to call the `print` command so that the output metrics are calculated and the output is printed of the specified file. In the loop phase, probes are added where required. The procedure to add a new metric is thus quite straightforward, and consists of the following steps:

- define its name in the *ns2-measure* files *metrics.h* and *metrics_names.h*,
- add a call to *stat_add()* inside the *stat_init()* function (*stat_init.c*),
- add the probe using *stat_put()* wherever samples are to be collected, including the *stat.h* header. For example, if we need the number of resource blocks allocated by the scheduler, we need to insert a *stat_put()* call inside the *pre_processor.c* file.

Note that the target code must be recompiled *only* when new probes and/or new metrics are inserted, while *ns2-measure* can be (de)activated via configuration file.

4.2 Experiment Management Automation

The method used by OAI to define scenarios is XML files, the so-called *OAI Scenario Descriptors* (OSDs). These allow a very fine-grained customization of the emulation scenario, editing parameters such as the transmission power of the eNB antennas, the mobility model for the nodes and the profile of the traffic flows. However, OSDs do not support *variable parameters*, so when running an emulation campaign a different OSD must be prepared for each combination of parameter values.

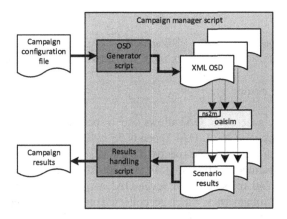

Fig. 3. Automated campaign management using handling scripts.

To fill this gap we implemented an automatization script package: the main script takes as input a configuration file where parameter values or ranges of values are specified; then, for every combination of parameter values it calls another script which generates a specific XML descriptor, and launches the OAI system emulation using that descriptor; lastly, another script parses the results from different runs and gathers them in a CSV file. For example, if we want to try the same scenario with 1, 2 or 3 UEs, we specify the parameter *numUEs* as {1, 2, 3} and the script will generate three different XML descriptors, launch OAI three times, and merge the three sets of results in a CSV file. This process is shown in Fig. 3.

5   Experimental Results

The purpose of this section is twofold. On one hand, we show that the integration of ns2-measure framework has a negligible impact on OAI performance. On the other hand, we exemplify the benefits that our framework brings to the user by showing that different (and unbiased) throughput results are obtained by allowing sample collection to start with the traffic generation (instead of at time zero), and by showing that comparing different metrics allows a user to get an immediate insight on the behavior of the system.

To assess the impact of the new code on performance, we evaluate the execution time and memory occupancy as a function of the traffic rate, both with and without *ns2-measure* samples collection activated.

We run OAI System Emulation (oaisim) on a machine with an AMD FX 8350 4 GHz CPU, 8 GB RAM, running Xubuntu 14.04.2, emulating an eNB sending downlink traffic to a UE, using increasing traffic rates. The main emulation parameters are summarized in Table 2.

Table 2. Main parameters for the performance evaluation campaign.

Parameter	Value
Emulated time	20000 TTIs (20 s)
# eNBs	1
# UEs	1
Mobility and position	Static - eNB and UE are 200 m apart
Traffic type	CBR: 800, 1600, 2400, 3200 kbits/s *at the application level*
# ns2-measure metrics	12

A note on traffic generation: OAI allows one to specify the size of generated traffic packets at the *application* level. OAI appends 55 bytes of TCP/IP headers and OTG metadata [9] to each packet. If we specify a packet size of 100 B and an inter-packet time of 1 ms, we obtain a data rate of $100 \times 8 = 800$ kbits/s at the application level, or $(100 + 55) \times 8 = 1240$ kbits/s at the IP level. OAI statistics refer to *IP-level* traffic.

We added 12 custom metrics, which get collected on a per-subframe basis. This adds 12 function calls to the *init* phase and $12 \times 1000 = 12000$ function calls per second to the *loop* phase. Each configuration/scenario is run three times with three different seeds, for a total of nine runs per configuration. To evaluate the execution time and the memory occupancy (more specifically the *maximum resident set size*, i.e., the maximum amount of memory the process allocates during its execution) we use the /usr/bin/time command [5].

Figure 4 shows the results for the execution time: the introduction of *ns2-measure* samples collection introduces minimal to null overhead. Also memory occupancy is unchanged, being about 800000 kB for every run.

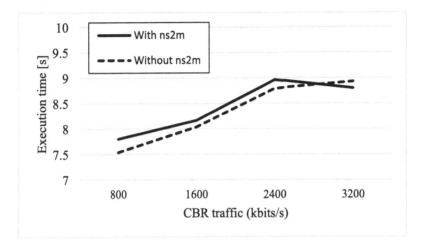

Fig. 4. Execution time of OAI system emulation, determined with /usr/bin/time, emulating 20 s of DL CBR traffic from an eNB to a UE.

We now show that our solution eliminates biases in throughput measurement. Since the traffic generator needs the underlying protocol stack and an active radio bearer to work, it needs to wait for the initialization of the stack and the establishment of the RRC connection. The OAI in-code documentation fixes the minimum starting time at 310 ms [14], and we chose a starting time of 500 ms in our experiments.

The native stat collection uses the entire emulation time to calculate traffic throughput and other rates, without considering the traffic starting time. Conversely, in our experiments, *ns2-measure* started collecting samples when the traffic started. In Fig. 5 we show the IP-level throughput of the UE, as measured by the native traffic generator and by *ns2-measure*. As expected, the values reported by *ns2-measure* are slightly higher, since the measurement interval is 500 ms shorter (as it should be).

This very experiment can also be used to show another benefit of using a flexible metric collection: a sub-linear behavior can be observed in the throughput curve, which suggests that the network approaches saturation as the offered load increases. This claim can be easily verified by collecting the number of resource blocks (RBs) allocated to the UE by the eNB scheduler (25 being the maximum number of RBs for the specific configuration). The number of RBs is shown on the right vertical axis, and clearly shows that the knee in the throughput is due to resource depletion. The same metric can also be used to infer the energy consumed by the eNB, according to well-established models

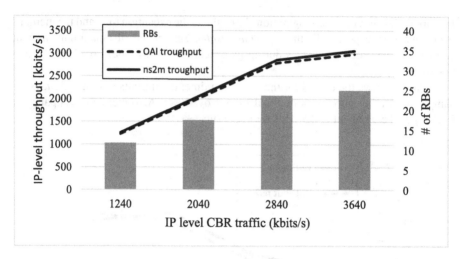

Fig. 5. IP-level throughput between an eNB and a UE as measured by OAI traffic generator and *ns2-measure* (left vertical axis); number of RBs allocated to the UE (right vertical axis).

of energy consumption [13], e.g. to evaluate the energy efficiency of the scheduling algorithm in use.

Moreover, while the statistics offered by OTG are calculated above the LTE protocol stack (i.e., at the IP level), with ns2-measure we can probe all the layers, e.g. to assess the overhead introduced by each of them. Figure 6 shows the throughput measured at different layers. As we expect, the closer we get to the physical layer the higher the throughput is, as more headers are added to the application payload.

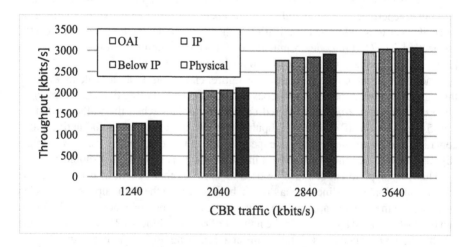

Fig. 6. Data throughput between an eNB and a UE, using different profiles of DL CBR traffic, as measured by ns2-measure at different layers of the LTE protocol stack.

6 Conclusions

This paper presented a set of tools to automate experiment management with a C-RAN prototype realized through OpenAirInterface. These tools allow a user to create a whole simulation campaign, i.e., to launch (possibly several replicas of) scenario where parameters vary, and to harvest the results obtained in the above campaign in a plot-friendly way. Having these tools spares a user time-consuming and error-prone tasks, which can be automated, thus enhancing the credibility of her simulations and increasing her productivity.

As a companion and complementary contribution, we integrated a structured and validated measuring framework, namely *ns2-measure*, into OAI. This allows one to define metrics in an easy way, and enable/disable measure gathering dynamically. On one hand, this speeds up debugging, since it allows a user to analyze the reasons of unexpected behaviors in the system by cross-checking different related metrics. On the other hand, this presents the user with a simple unified approach to harvesting measures, thus facilitating experimenting in the large (e.g., in teamwork).

Acknowledgements. The subject matter of this paper includes description of results of a joint research project carried out by Telecom Italia and the University of Pisa. Telecom Italia reserves all proprietary rights in any process, procedure, algorithm, article of manufacture, or other result of said project herein described.

This work was partially supported by the European Commission in the framework of the H2020-ICT-2014-2 project Flex5Gware (Grant agreement no. 671563).

References

1. Wang, R., et al.: OpenAirInterface - an effective emulation platform for LTE and LTE-Advanced. In: Proceedings of ICUFN 2014, pp. 127–132. IEEE, Shanghai (2014)
2. OpenAirInterface website. Url: http://www.openairinterface.org. Accessed Jan 2016
3. Ettus Research USRP B200/B210 Bus Series. Url: http://www.ettus.com/content/files/b200-b210_spec_sheet.pdf. Accessed Jan 2016
4. Hafsaoui, A., Nikaein, N., Lusheng, W.: OpenAirInterface Traffic Generator (OTG): a realistic traffic generation tool for emerging application scenarios. In: Proceedings of MASCOTS 2012, pp. 492–494, 7–9 August 2012
5. Kerrisk. M.: time(1) - Linux manual page. url: http://man7.org/linux/man-pages/man1/time.1.html. Accessed Jan 2016
6. Flex5Gware website: http://www.flex5gware.eu. Accessed Jan 2016
7. Cicconetti, C., Mingozzi, E., Stea, G.: An integrated framework for enabling effective data collection and statistical analysis with ns-2. In: Proceedings of WNS2 2006, Pisa, Italy, 10 October 2006
8. Perrone, L.F., Cicconetti, C., Stea, G., Ward, B.: On the automation of computer network simulators. In: Proceedings of SIMUTOOLS 2009, Rome, 3–5 March 2009
9. Virdis, A., Iardella, N., Stea, G., Sabella, D.: Performance analysis of OpenAirInterface system emulation. In: Proceedings of PMECT 2015, Rome, Italy, 26 August 2015
10. The Network Simulator - ns-2. Url: http://www.isi.edu/nsnam/ns/. Accessed Jan 2016
11. CMake. Url: https://cmake.org/. Accessed Jan 2016

12. C-RAN: The road toward green RAN. Technical report, China Mobile Research Institute (2011), Beijing, China, October 2011
13. Migliorini, D., Stea, G., Caretti, M., Sabella, D.: Power-aware allocation of MBSFN subframes using Discontinuous Cell Transmission in LTE systems. CLEEN 2013, Las Vegas, USA, 2 September 2013
14. Gitlab OpenAirInterface repository. Url: https://gitlab.eurecom.fr/oai/openairinterface5g/ blob/master/targets/SIMU/EXAMPLES/OSD/WEBXML/template_0.xml. Accessed Jan 2016

Erratum to: Utilization of Licensed Shared Access Resources in Indoor Small Cells Scenarios

Eva Perez[1(✉)], Karl-Josef Friederichs[1], Andreas Lobinger[1],
Bernhard Wegmann[1], and Ingo Viering[2]

[1] Bell Labs Research, Nokia, Munich, Germany
{eva.perez,karl-josef.friederichs,andreas.lobinger,
bernhard.wegmann}@nokia.com
[2] Nomor Research, Munich, Germany
viering@nomor.de

Erratum to:
Chapter 38: D. Noguet et al. (Eds.)
Cognitive Radio Oriented Wireless Networks
DOI: 10.1007/978-3-319-40352-6_38

The title of the paper starting on page 462 was incomplete and should read:
"Utilization of Licensed Shared Access Resources in Indoor Small Cells Scenarios"
This has been corrected.

The updated original online version for this Chapter can be found at 10.1007/978-3-319-40352-6_38

© ICST Institute for Computer Sciences, Social Informatics and Telecommunications Engineering 2016
D. Noguet et al. (Eds.): CROWNCOM 2016, LNICST 172, p. E1, 2016.
DOI: 10.1007/978-3-319-40352-6_63

Author Index

Printed in the United States
By Bookmasters